W9-CLG-170

COLLECTED WORKS OF ERASMUS

VOLUME 84

Erasmus, after Hans Holbein the Younger, c 1530
By permission of the Parma Gallery / Anderson-Giraudon

COLLECTED WORKS OF
ERASMUS

CONTROVERSIES

RESPONSIO AD EPISTOLAM PARAENETICAM
ALBERTI PII

APOLOGIA ADVERSUS RHAPSODIAS
ALBERTI PII

BREVISSIMA SCHOLIA

edited by Nelson H. Minnich

translated by Daniel Sheerin

annotated by Nelson H. Minnich and Daniel Sheerin

University of Toronto Press

Toronto / Buffalo / London

The research and publication costs of the
Collected Works of Erasmus are supported by
University of Toronto Press.

© University of Toronto Press Incorporated 2005
Toronto / Buffalo / London
Printed in Canada

ISBN 0-8020-4397-6

Printed on acid-free paper

Canadian Cataloguing in Publication Data

Erasmus, Desiderius, d. 1536
[Works]
Collected works of Erasmus

Includes bibliographical references and index.
Contents: v.84. Controversy with Alberto Pio / edited by Nelson H. Minnich;
translated by Daniel Sheerin;
annotated by Nelson H. Minnich and Daniel Sheerin.
ISBN 0-8020-4397-6 (v.84)

1. Erasmus, Desiderius, d. 1536 – Collected works. I. Title

PA8500 1974   199'.492   C74-006326-x rev

University of Toronto Press acknowledges the financial assistance to its
publishing program of the Canada Council and the Ontario Arts Council.

University of Toronto Press acknowledges the financial support
for its publishing activities of the Government of Canada
through the Book Publishing Industry Development Program (BPIDP).

Collected Works of Erasmus

The aim of the Collected Works of Erasmus
is to make available an accurate, readable English text
of Erasmus' correspondence and his
other principal writings. The edition is planned
and directed by an Editorial Board, an Executive Committee,
and an Advisory Committee.

Patribus magistrisque nostris
in Societate Iesu
sacrum

Contents

Illustrations

Acknowledgments

Many debts of gratitude have been incurred over the long years this volume has been in preparation. The editor and translator wish to acknowledge in particular the constant support they have received from those closely associated with the Collected Works of Erasmus project, especially James Kelsey McConica, csb, Chairman of the Editorial Board, and Ron Schoeffel, Chairman of the Executive Committee. In spite of his pressing responsibilities as President of the Pontifical Institute of Mediaeval Studies, Toronto, Professor McConica generously reviewed the manuscript in detail before it went into production. He was ably assisted in this task by Milton Kooistra, who double-checked references and provided updated cross-references to the cwe edition of Erasmus' works in English. The University of Toronto Press supplied copies of cwe publications and photocopies of sixteenth-century materials. John H. Munro, University of Toronto, provided the informative notes on monetary matters in this volume. As the project neared completion, valuable assistance was provided by Penny Cole, University of Toronto, who obtained various illustrations and helped with other tasks; by Joan Bulger, whose patience, good cheer, and careful eye guided the manuscript through the copyediting stages; and by Lynn Burdon and Philippa Matheson, who typeset this complex volume.

In support of their work Nelson H. Minnich and Daniel Sheerin received a Grant for Translation from the National Endowment for the Humanities in 1986. Minnich received various Faculty Research Grants from The Catholic University of America from 1993 to 1995, in 1997, and in 2003. With the support of a fellowship from the Lilly Endowment at the National Humanities Center in 2004 Minnich was able to complete his work on this volume.

Among those who helped with the typing of the volume were Helen Santini, Evelyn Smith, Gloria Wilkinson, Joseph C. Linck, Juliann Kreiger Heller, Sue Marie Korlan, and Sarah L. Lynch. Allison Kitchner provided invaluable editorial assistance. Christian Washburn and Gregory Finley did preliminary work on the index.

Many individuals helped to secure copies of rare materials. The editors are very grateful to the late Leonard E. Boyle, OP, Prefect of the Biblioteca Apostolica Vaticana; to Lena Cowen Orlin of the Folger Institute in Washington, DC; to Amélie Lefébure, Conservateur des collections, Musée Condé, Chantilly, France; to M.A. Hubrecht and Christine Abelé of Photographie Giraudon in Paris; to Brian Van Hove, SJ, currently of the Seminary of St Joseph and St Peter in Mission, Texas; to James K. Farge, CBS, of the Pontifical Institute of Mediaeval Studies, Toronto; to Jacques M. Gres-Gayer of The Catholic University of America; to Florence Clavaud, Conservateur of the Archives nationales in Paris; to Gianfranco Ravasi, Prefect of the Biblioteca-Pinacoteca Ambrosiana in Milan; to Leslie A. Morris, Curator of Manuscripts in the Harvard College Library, and William Stoneman and Roger Stoddard of the Houghton Library, Harvard University; and to Sister Janice Farnham, RSJM, of Weston Jesuit School of Theology in Cambridge, Massachusetts.

A number of scholars working in the field have provided encouragement, helpful advice, and gifts of their publications. They include Henk J. de Jonge of Leiden University in the Netherlands; Richard L. De Molen, founder of the Erasmus of Rotterdam Society; James K. Farge; Paul F. Grendler, *emeritus*, University of Toronto; Brian Krostenko, University of Notre Dame, and Erika Rummel, University of Toronto. In addition to sending offprints of some of his earlier studies of Juan Ginés de Sepúlveda, Julián Solana Pujalte of the Universidad de Córdoba provided a copy of his critical edition of Sepúlveda's *Antapologia* but unfortunately it arrived too late in the production process to be properly consulted. Fabio Forner, Università Cattolica del Sacro Cuore in Milan, sent an offprint of his important study 'Genesi ed elaborazione della "Responsio" di Alberto Pio a Erasmo,' but it too arrived late and only limited references could be made to it. Similarly his critical edition with a translation into Italian of Pio's *Responsio paraenetica* arrived when our volume was already in production and thus could not be utilized in the notes to that text.

Introduction

Few of Erasmus' Catholic critics were attacked with greater virulence than the Italian prince and diplomat Alberto Pio (1475–1531). Erasmus claimed that one should ignore the criticisms of one's opponents unless they unjustly accuse one of impiety, but he believed that this was precisely what Pio had done in denouncing him for having prepared the way for the Reformation and for sharing the same religious opinions as Luther. These charges evoked from Erasmus what has been described by Myron P. Gilmore as 'perhaps his most comprehensive and detailed justification against his critics' and, with some hyperbole, as 'one of the most savage compositions Erasmus ever wrote.'[1] Even had he wished to ignore Pio's criticisms, the stature of his opponent required a response. Most of Erasmus' other accusers were clerics, notably professors of theology and members of religious orders (the two groups Erasmus had most bitterly attacked over the years), but Pio was a layman with an international reputation as a skilled diplomat, a man of learning, a patron of humanists, a pious man who favoured moderate church reforms, and a relative by marriage and confidant of popes and cardinals.[2] When

* * * * *

1 For a survey of Erasmus' Catholic critics, see Rummel *Catholic Critics* especially II 115–26 (on Pio and his defender Sepúlveda). For Erasmus' need to rebut charges of impiety, see eg Allen Epp 778:237–8, 980:46–9 / CWE 778:264–5, 980:53–6, Allen Epp 2136:150–84, 2443:78–119. For these evaluations of Erasmus' writings against Pio, see Myron P. Gilmore 'Erasmus and Alberto Pio, Prince of Carpi' in *Action and Conviction in Early Modern Europe: Essays in Memory of E.H. Harbison* ed Theodore K. Rabb and Jerrold E. Seigel (Princeton 1969) 299–318, especially 313 and 317. See n3 below for why Gilmore may have overestimated the length and thoroughness of Pio's critique.
2 The principal scholarship on Pio's life includes: Guaitoli 'Sulla vita' 133–313; Semper; Vasoli *Pio*; Avesani et al, especially Cesare Vasoli 'Alberto Pio e la cultura del suo tempo' vol 46 3–42, and Albano Biondi 'Alberto Pio nella pubblicistica del suo tempo' vol 46 95–132; and Sabattini. Erasmus claimed that his protection from enemies

Pio spoke, important men listened, whether at the papal and imperial courts or later at Paris, where the celebrated Faculty of Theology acted as the guardian of Catholic orthodoxy. Pio's comments could not be treated lightly.

Erasmus' responses to Pio are notable not only for their virulence but also for their scope, length, and detail. Except for the criticisms of Noël Béda, the articles of the Valladolid Conference, and the censures of the Paris Faculty of Theology, Erasmus' critics usually addressed themselves to a single question or a few topics and works and hence Erasmus' replies to them were similarly limited. But in the Erasmus-Pio debate over twenty major topics were systematically treated and at least fifteen different works written by Erasmus were cited.[3]

Erasmus was eager to give the impression that he had produced a thorough answer to Pio's criticisms. After preliminary comments he began the core of his *Apologia*, written after Pio's death in 1531, with detailed responses even to the often brief marginal notes that Pio had appended to Erasmus' *Responsio*. To some criticisms Erasmus gave elaborate replies and to others answers of varying length, while still others he dismissed as irrelevant, repetitious, or already answered elsewhere.[4] His intent was to show that he was unscathed by any of his opponent's charges. So deep was Erasmus' animosity towards Pio that he could not resist lampooning his opponent's political misfortunes and his funeral, questioning Pio's abilities, character, and sanity, and even resorting to vulgar and puerile outbursts.[5] Pio's criticism elicited from Erasmus one of his most comprehensive defences. It was also his last major work against an individual opponent. He must have felt that, even in death, Pio posed a threat: his words had to be refuted, his reputation destroyed.

* * * * *

depended on the support of such good and influential people as Pope Clement VII; see *Spongia* LB X 1657A–B / ASD IX-1 178:309–19; Klawiter 207–8. Pio was a particularly dangerous enemy because he was an intimate friend and relative by marriage of this pope.

3 Erasmus' two principal replies to Pio fill 102 columns in LB IX. By way of comparison his replies to Pierre Cousturier fill 65 columns, to Diego López Zúñiga 73 columns, to Edward Lee 77 columns, to certain Spanish monks 79 columns, to the censures of the Faculty of Theology of Paris 139 columns, to Noël Béda 187 columns, and to Martin Luther 287 columns. See n85 below for the works cited.

4 For Erasmus' selecting out items for comment, see eg *Responsio* 8, 12; for his claims that Pio's statements do not apply to him, see *Apologia* 225, 231, 254, 298; for his holding that he has already answered this complaint elsewhere, see *Apologia* 265, 270, 273, 299; and for his replies to marginal notes, see *Apologia* 119–53.

5 Minnich 'Underlying Factors' 13–21

ALBERTO PIO

PIO'S EARLY YEARS, EDUCATION, AND RULE OVER CARPI

Alberto III Pio de Carpi was born on 23 July 1475, the eldest son of Lionello I, lord of Carpi in Emilia-Romagna, and Caterina Pico, who was the sister of Galeotto I Pico, lord of the principate of Mirandola that bordered Carpi on the northeast, and Giovanni Pico, the famous humanist and philosopher. Alberto's brothers were Leonello, born in 1477, who later managed Alberto's affairs for him in Carpi, and Teodoro (his natural brother), who entered the Franciscan order, became guardian of its house in Carpi, and then bishop of Monopoli in 1513. Their sister Caterina also entered the Franciscan order. In 1477 Alberto's father died and his cousin Marco, who shared the title of lord of Carpi, became his legal guardian. Perhaps in an effort to distance his ward from administrative interests, Marco agreed to Alberto's training in letters.[6]

Alberto was well educated. Apparently on the advice of her brother Giovanni, then studying at the University of Ferrara, Caterina Pico Pio in 1480 hired as tutor for Alberto and Leonello the humanist Aldo Manuzio [Aldus], who was probably then studying and teaching at Ferrara. Aldus came to Carpi to carry out his charge and laid out for the boys a programme of studies that emphasized the classical languages and literature, especially Greek studies, for Aldus was convinced that the best had been written by the Greeks and then borrowed by the Latins. In 1481, on the urging of Giovanni Pico, Aldus and his two pupils went to Ferrara with its rich cultural life, and when the Venetian army drew near the city in 1482, Aldus and the Pio boys took refuge with their uncle at Mirandola. Perhaps from Giovanni Pico Alberto derived his later interest in philosophy. However, Pico soon moved on to Pavia (1482–3), and then Florence (1484) and Paris (1485), where he continued his philosophical and theological studies and

* * * * *

6 Sammarini I 334; the genealogical tree compiled by Francesco Tarquinio Superbi (1738–77) is reproduced in Contini *Alberto III Pio* [7]; the Pio di Carpi family, traced from 1096 and still not extinct in 1821, is described in fascicle 16 of Litta (Alberto and his immediate family are on table III); Eubel III 248; on Pio's father's death and his uncle's guardianship and intentions, see Contini *Alberto III Pio* [6–7, 9]. On Pio's mother, see Alfonso Morselli 'Caterina Pico della Mirandola' in *Studi e documenti della Deputazione di storia patria per l'Emilia e Romagna* III-2 (Reggio Emilia 1939); and Morselli 'Il corredo nuziale di Caterina Pico (1494)' in *Atti e memorie della Deputazione di storia patria per le antiche provincie Modenesi* 8th series vol VIII (Modena 1956) 26, 42–3 (among the items Caterina brought with her to Carpi were books by Cicero, Virgil, Aesop, Justinian, and others).

compiled his famous syncretistic *Conclusiones sive Theses* DCCCC (Rome 1486), which he offered to defend in Rome, but six of the propositions were declared dubious and seven condemned outright by a papal commission in 1487. Pico fled to France and eventually settled in Florence under Medici protection. It is unlikely that Alberto followed his uncle's physical peregrinations, although he did come to share with him an intellectual appreciation of scholasticism and cabbalism. It is more probable that Alberto returned to Ferrara once peace with Venice was restored in 1484. There Alberto continued in his enthusiasm for letters; among his fellow students and friends were Jacopo Sadoleto and Ludovico Ariosto. Since the age for enrolling in the University of Ferrara was between sixteen and eighteen, and since Alberto returned to Carpi in 1494 when he was sixteen or seventeen, it is unlikely that he was formally a university student at Ferrara. Like many sons of the nobility, he probably studied with a private tutor or independent teacher, or else lived as a boarding student in the home of a professor. That Aldus remained Alberto's tutor for the whole period until Aldus departed for Venice in 1488 is also unlikely, since he claimed to have been Alberto's teacher for only six years. Aldus, however, remained devoted to his former student and informed the literary world of his special relationship with Alberto by dedicating to him such works as his Latin grammar (1493), the Aldine Greek edition of Aristotle (1495–7), the Latin text of Lucretius (1500), and two philosophical works (1514–15).[7]

* * * * *

7 On Giovanni Pico's concern for his nephew's education and on Aldus as Pio's tutor, see Sepúlveda *Antapologia* ciii^r; Florido *Apologia* 116:44; and Lowry *World of Aldus* 52–3, 56–8. On Giovanni Pico, see CEBR III 81–4, especially 82–3; and Vasoli *Pio* 18–23. On Pio's defences of scholasticism and cabbalism, see Pio XXIII *libri* 172^v–176^v; and Pio *Responsio paraenetica* 28^v. On his friendships at Ferrara, see the following CEBR entries: on Alberto Pio III 86–8, especially 87; on Sadoleto III 183–7, especially 183; and on Ariosto I 71. On Aldus as Alberto's tutor, see the entry on Manuzio in CEBR II 376–80, especially 377, and Lowry *World of Aldus* 52; Lowry puts Alberto on his own in Ferrara in 1484. In the dedicatory letter to Alberto that prefaces his *Institutiones grammaticae* (9 March 1493) Aldus states that his students Alberto and Leonello *sex annos et plus eo summa fide curaque docuimus*; see *Aldo Manuzio Editore* I 165. On the age and education of pre-university students in Italy, see Paul F. Grendler *Schooling in Renaissance Italy: Literacy and Learning, 1300–1600* (Baltimore 1989) 23–41; Sabattini 10 has Aldus returning to Venice in 1488. On Pio's exile from Carpi until 1494, see Vasoli *Pio* 22–3. On Aldus' dedications to Pio, see Lowry *World of Aldus* 75, 111; Luigi Balsamo 'Alberto Pio e Aldo Manuzio: Editoria a Venezia e Carpi fra '400 e '500' in Avesani et al, vol 46 133–66, especially 134, 137, 139, 147, 153; and Mario Emilio Cosenza *Biographical and Bibliographical Dictionary of the Italian Humanists and of the World of Classical Scholarship in Italy, 1300–1800* 6 vols (Boston 1962–7) v 2826, where these and numerous other dedications by humanists are listed. On the dedication to Pio of book 2 of Ludovico Bigi Pittori of Ferrara's *Opusculorum Christianorum libri tres* (Strassburg: Schürer 1508–9), see *Briefwech-*

Alberto's title to Carpi was often contested. Following the death of
Marco I Pio in 1418, the title of lord was shared by his heirs. Alberto's
father, Lionello, ruled with his cousin Marco II. For a brief period during Alberto III's minority (1477–80), the child shared the title of lord with
Marco II, but then the boy's claims were no longer recognized and in 1478
his mother arranged a marriage for him with Eleonora, the eldest child
of Niccolò de Correggio, a nephew of Duke Ercole d'Este of Ferrara, who
was also a condottiere, literary figure, and ruler of the territory bordering Carpi on the west. Nothing came of this political match. In 1485 Caterina, Alberto's widowed mother, married Rodolfo Gonzaga, uncle of the
youthful Marquis Gianfrancesco Gonzaga of Mantua, and moved to Luzzara in Mantua, where she eventually bore him a daughter, Paola (who
married Giovanni Niccolò Trivulzio of Milan but was widowed by 1513),
and a son, Luigi Alessandro, whose grandson was the Jesuit St Aloysius
Gonzaga. Alberto's stepfather was a ruthless man who had earlier tried to
poison his own brother and nephew, who had his former wife, Antonia
Malatesta, beheaded for infidelity, and who himself died by a French dag-
ger in the Battle of Taro (Fornovo) in 1495. Caterina was poisoned to death
by her servant in 1501. Apparently with the aid of the Gonzagas in Mantua
and the Picos in Mirandola, Alberto secured in 1490 from Emperor Frederick III investiture with half of the domain of Carpi. Marco and his sons,
however, forced Alberto to live in exile, at least part of the time with his
uncle Giovanni Pico in Florence. On the death of Marco II on 22 March 1494
Alberto returned to Carpi and for a while shared the rule with Marco's
son Giberto III. But civil warfare between the factions favouring Giberto or
Alberto broke out in May and June 1496. On 8 October Giberto met Emperor Maximilian I in Genoa and secured from him investiture with all of
Carpi. Alberto, forced into exile in Ferrara from 1497 to 1500, protested
that he had done nothing to deserve the loss of his rights, whereupon the
emperor appointed Ercole d'Este as arbiter. The duke tried to get Giberto
to restore to Alberto his half of the territory but met with armed resistance and Giberto's insistence that the cousins could not cooperate in the
government of Carpi. Ercole took advantage of this situation and in 1499
persuaded Giberto to cede to him his rights to half of Carpi in exchange
for lands and a title in the territory of Modena. This contract was ratified
by the emperor in 1500. In 1509 Alberto was able to convince the emperor

* * * * *

sel des Beatus Rhenanus ed Adalbert Horawitz and Karl Hartfelder (Leipzig 1886; repr
Nieuwkoop 1966) 15, 19.

to declare this contract null and to give him a new and full investiture of Carpi with the title of count. After prolonged negotiations the new duke of Ferrara, Alfonso I, agreed in April 1511 to sell to Alberto his half of Carpi for twenty-six thousand ducats. When Alberto failed to pay promptly, Alfonso retook by force his portion of the territory in August 1511. A decision of the emperor on 14 June 1512, however, required Alfonso to restore to Alberto the whole of Carpi and to repay him for the damages and expenses he had suffered. There was deep animosity between the two men. By 1513 Alberto was denouncing Alfonso to the emperor's closest adviser, Matthäus Lang, as 'that sordid and wicked man, the foulest enemy of all nobility and everything that is good,' a man unworthy of remaining duke of Ferrara. Alfonso reciprocated the animosity but agreed in 1514–15 to let Pope Leo x arbitrate his differences with Alberto over Carpi. This arbitration failed, and eventually Alfonso, through an imperial investiture, succeeded in dispossessing Alberto of all of Carpi just months before Pio died.[8]

Alberto tried to make his court at Carpi a centre of learning. He spent large sums of money on the purchase of scholarly books and manuscripts, buying, for example, the library of Giorgio Valla in 1500 for the large sum of eight hundred gold crowns. With the offer of estates, civil authority over the town of Novi, an ecclesiastical fiefdom located in the northern part of his principality, and the prospect of founding an academy, Alberto tried in vain to induce his former tutor to settle in Carpi, but Aldus chose to remain instead in Venice, working on his printing ventures. Eventually

* * * * *

8 Adriano Cappelli *Cronologia, Cronografia e Calendario perpetuo* 3rd rev ed (Milan 1969) 378; Contini *Alberto III Pio* [7, 9]; Litta 'Pio di Carpi' table III (the earlier name of the family was 'Figli di Manfredi,' which was changed to 'Pio' after a pilgrimage to the Holy Land by one of Alberto's ancestors – table I; the family added the title 'di Savoia' in 1450 when it was aggregated to the ducal family of Savoy as a result of Alberto II's services to that family – table II) 'Da Correggio' table III, 'Gonzaga di Mantova' table XVI; Sabattini 10–20, 35–6; Werner L. Gundersheimer *Ferrara: The Style of a Renaissance Despotism* (Princeton 1973) 260–1; Kate Simon *A Renaissance Tapestry: The Gonzagas of Mantua* (New York 1988) viii–ix, 65–6, 113; Sammarini 382–6; Vasoli *Pio* 22; Guaitoli 'Sulla vita' 133–62, 168, 187, 201–2, 303–4. For Pio's description of Alfonso I, see his letter to Matthäus Lang, from Rome, 16 August 1513, in the University of Pennsylvania Library, Rare Book Collection, Lea Collection, MS 414 Letters of Pio sec 4 (correspondence with the imperial court) letter 11 folio a. The twenty-six thousand ducats that Pio was required to pay to obtain control of Alfonso's half of Carpi were presumably Venetian ducats (3.559 g fine gold), which would have been worth about £5,850 sterling (at 54d each) or about 53,950 *livres tournois* (at 2.075 *livre* or 41s 6d *tournois* each); see CWE 1 314, CWE 12 650. For the arbitration effort, see Nelson H. Minnich 'The Healing of the Pisan Schism (1511–13)' *Annuarium Historiae Conciliorum* 16 (1984) 59–192 especially 180, 185, 192.

Alberto secured the services of one of Aldus' associates, the Cretan scholar Marcus Musurus, who from 1499 to 1503 served as Alberto's tutor, companion, and librarian, providing him with daily lessons in Greek literature. Alberto's next Greek tutor was probably Triphonio Bizanti from Cattaro in Dalmatia.[9]

Unlike his fellow princes, Alberto did not engage in the youthful aristocratic diversions of hunts, games, and love affairs, but devoted himself to transforming the fortress town of Carpi into an architectural and cultural centre. His initial efforts concentrated on systematizing the city's central piazza; along the east side he had the medieval palace reconstructed in a Renaissance style; on the north he had constructed a Bramantesque church that would eventually become the seat of a bishop; and along the western side he had built a sweeping loggia that joined the centres of civil and religious authority to the residential quarters of the city. In addition, the Franciscan church of San Nicolò was rebuilt and the city was encircled by a new wall.[10] Having failed to bring Aldus back to the territory of Carpi, Alberto invited the printer Benedetto Dolcibello da Manzi to set up shop in Carpi, and from his press came a number of scholarly works.[11]

* * * * *

9 The eight hundred gold crowns that Pio paid for Valla's library may refer to the French *écu à la couronne*, but it should be noted that this gold coin was last struck in 1474, and was superseded, in November 1475, by the very similar and slightly heavier *écu au soleil* (both having a fineness of 22.125 carats); in the early sixteenth century the *écu au soleil* was valued at 1.813 *livre tournois* (36s 3d *tournois*) and thus this sum was worth 1,450 *livres tournois* or about £170 sterling; see CWE 1 315, CWE 12 648, 650; note that the English gold crown was not introduced until 1526; see CWE 12 574. Lowry *World of Aldus* 59; Sepúlveda *Antapologia* ciii[r] / Solana 124; Deno John Geanakopolos *Byzantium and the Renaissance: Greek Scholars in Venice, Studies in the Dissemination of Greek Learning from Byzantium to Western Europe* (Cambridge, Mass. 1962) 125–7; on Marcus Musurus see CEBR II 472–3, especially 472; and Julia Haig Gaisser *Pierio Valeriano on the Ill Fortune of Learned Men: A Renaissance Humanist and His World* (Ann Arbor 1999) 310–11; on Bisanti, see Gaisser 268–9. Pio was so adept in Greek that he was admitted to Aldus' New Academy in Venice, whose members were required to speak Greek; see Geanakopolos *Byzantium and the Renaissance* 128–30. In his writings against Erasmus Pio cited Greek authors in their own language; see eg Pio *XXIII libri* 68[r]c (Hesiod), 200[v]G (Homer), 164[r]s and 222[r]P (Luke).

10 Erasmus *Responsio* 9 below; Sepúlveda *Antapologia* ciii[v] / Solana 126; Adriana Galli 'Alberto III Pio: L'impianto urbanistico rinascimentale' in *Materiali per la storia urbana di Carpi* ed Alfonso Garuti et al (Carpi 1977) 55–8, with the accompanying 'Sezione c di catalogo [della mostra nel Castello dei Pio giugno–ottobre 1977]' 60–9; and Galli 'Presenza religiosa: Chiese e conventi come elementi di organizzazione urbana fino al XVII secolo' ibidem 70–6, especially 72–5, with the accompanying 'Sezione D di catalogo' 82–8, especially 82–6

11 Vasoli 'Alberto Pio e la cultura del suo tempo' (cited in n2 above) 20–1; and Balsamo 'Alberto Pio e Aldo Manuzio' (cited in n7 above) 155–65.

View of the city of Carpi by Luca Nasi, reproduced in an
engraving of 1677 by the printer Paolo Abbate of Carpi, and copied
with some variations by the Modenese printer Bartolomeo Soliani
By permission of the Museo civico di Carpi

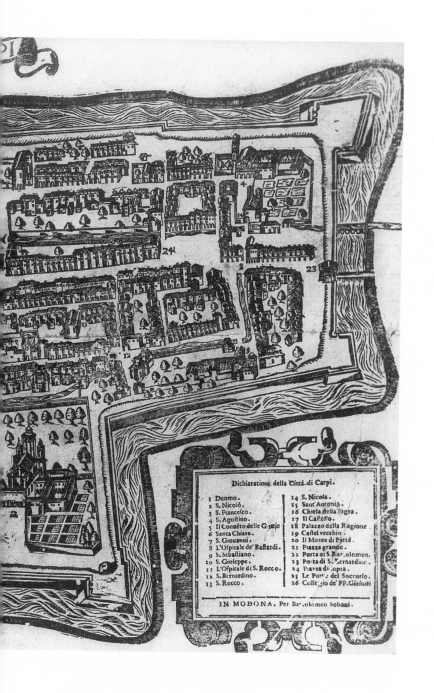

Dichiaratione della Città di Carpi.

1 Duomo.	14 S. Nicola.
2 S. Nicolò.	15 Sant'Antonia.
3 S. Francesco.	16 Chiesa della Sagra.
4 S. Agostino.	17 Il Castello.
5 Il Conuěto delle Grazie	18 Palazzo della Ragione.
6 Santa Chiara.	19 Castel vecchio.
7 S. Giouanni.	20 Il Monte di Pietà.
8 L'Ospitale de' Bastardi.	21 Piazza grande.
9 S. Sebastiano.	22 Porta di S. Bartolomeo.
10 S. Gioseppe.	23 Porta di S. Bernardino.
11 L'Ospitale di S. Rocco.	24 Piazza di sopra.
12 S. Bernardino.	25 Le Porte del Soccorso.
13 S. Rocco.	26 Collegio de' PP. Giesuiti

IN MODONA, Per Bartolomeo Soliani.

Whenever he was free of administrative duties, Alberto dedicated himself to the study of letters and philosophical disputations. Among the scholars who came to his court were Pietro Pomponazzi, who left his chair at Padua in the fall of 1496 to tutor Alberto both at Carpi and during his exile in Ferrara. Until his return to Padua in the fall of 1499, Pomponazzi taught Alberto and his humanist friend Celio Calcagnini a course in logic, training them in the methods of the late-fourteenth-century Oxford terminist logicians Richard Swineshead and William Heytesbury. Alberto, however, did not linger with Ockhamist logic but took up the study of such great scholastic theologians as Thomas Aquinas and John Duns Scotus. He was trained in Scotus' teachings by the Franciscan scholar Graziano da Brescia (d c 1505), who prior to 1495 had held the chair of Scotist theology at Padua and had come to Carpi by 1503 to tutor Alberto, and by the Spanish Augustinian friar Juan Montesdoch, who came from Bologna to Carpi in 1505 to lecture to the Franciscans there. In 1506 Dolcibello's press published both Graziano's commentary on the second book of the *Sentences* of Scotus and Montesdoch's preface to Friar Paulus Scriptoris' commentary on the first book of Scotus' *Sentences*. Among Alberto's other teachers were the philosopher Andrea Barrus and a hermit named Valerio who trained him in theology and sacred letters. Alberto was determined to master scholastic theology and became so skilled in debating both privately and publicly some of the most difficult questions that he gained the admiration of all. But as he grew older his tastes changed, and he devoted himself more to the study of Scripture and the writings of the church Fathers, especially Augustine and Jerome. By his frequent reading of the Bible Alberto gained a knowledge of the sacred texts that was unrivalled in his household. His personal piety is evident in his participation in the famous discussions of the Murano circle regarding the best form of Christian life.[12] Because of

* * * * *

12 Sepúlveda *Antapologia* ciii^r–civ^r / Solana 125–7. On Pomponazzi, see Bruno Nardi *Studi su Pietro Pomponazzi* (Florence 1965) 21, 55; and the letter of Celio Calcagnini to Erasmus, Ferrara, 6 July 1525, Allen Ep 1587:231–3 / CWE 1587:245–6. On Graziano da Brescia and Juan de Montesdoch (Montes de Oca), see H. Hurter *Nomenclator literarius theologiae catholicae* II: *1109–1563* 2nd ed rev (Innsbruck 1906) cols 994 and 1102 n1; and Solana 88–9 nn23 and 25. On the publication of their works in Carpi, see Balsamo 'Alberto Pio e Aldo Manuzio' (cited in n7 above) 161–3; and Vasoli *Pio* 33–4 nn41–2. Pio's teacher, 'Andreas Barrus magnus item philosophus' (Sepúlveda *Antapologia* ciii^v / Solana 125–6 n32), is likely Fra Andrea Baura (also Burius and Bauria) of Ferrara, an Augustinian hermit who was publicly denouncing, in print in Ferrara in 1513 and in sermons in Venice in 1517–19, the sins of the hierarchy and prophesying divine punishment; in 1521 he published a *Defensorium apostolicae potestatis contra Martinum Lutherum*, and in 1522 he was accused by Pio and others of having

Alberto's reputation as a man of learning who generously supported the scholarly labours of others, a number of humanists and philosophers dedicated their publications to him.[13] In many respects Alberto Pio was, as his contemporaries acknowledged, the Christian embodiment of a classical ideal, the 'philosopher-king,' and, despite his lay status and lack of university training, he was also hailed during his lifetime as *sommo thologo* [sic].[14]

PIO'S DIPLOMATIC CAREER

When Pio looked for some sphere of activity outside the confines of Carpi in which to exercise his remarkable talents and garner glory, patrons, and revenues, he settled on a diplomatic career. His mother's family and her second marriage had provided him with political and military support from Mirandola and Mantua against the encroachments of Ferrara. Pio strengthened his ties to Mantua by promising in 1494 to marry Margherita, the young natural daughter of the marquis, Gianfrancesco (1466–1519), whom he served as a diplomat in (1506–7). Perhaps because his stepsister, Paola, had married into the pro-French Trivulzio family of Milan, Pio was entrusted with missions to the court of King Louis XII, in whose army Gianfrancesco Gonzaga has served in 1503 as a general. While in the service of Mantua, Pio

* * * * *

preached Lutheran heresy in Ferrara the year before. On this fascinating figure, see 'Baura, Andrea' in *DBDI* VII 296–7; Hurter *Nomenclator* II col 1271 no 596; Friedrich Lauchert *Die italienschen literarischen Gegner Luthers* (Freiburg im Breisgau 1912) 238–9; Ottavia Niccoli *Prophecy and People in Renaissance Italy* trans Lydia G. Cochrane (Princeton 1990) 89–91, 105–8; and Alberto Pio 'Risposta della lettera o vero Invettiva di Alfonso secondo Duca di Ferrara contro Papa Leone decimo fatta da Alberto Pio Signor di Carpi, nella quale si vede quanto sia posseduto ingiustamente dalla casa d'Este sino all' anno 1522 [Rome, 6 January 1522]' in *Memorie storiche e documenti sulla città e sull' antico principato di Carpi* vol IV (Carpi 1886) 271–326, especially 310. On Pio's participation in the discussions of the Murano Circle, see Hubert Jedin 'Contarini und Camaldoli' *Archivio italiano per la storia della Pietà* 2 (1959) 51–117, especially 72, 75, 82, 86.

13 On some of the works dedicated to Pio, see Cosenza *Biographical and Bibliographical Dictionary* (cited in n7 above) IV 2825–6; and Vasoli *Pio* 28 n28, 31 n34.

14 Among those who referred to Pio as a philosopher-prince were Aldus and Girolamo Aleandro. See Aldus' dedicatory letters to Pio in *Aldo Manuzio Editore* I 17–18, 63, 127; and Aleandro's letter to Pio of c December 1517, where he calls Pio *et philosophus simul et princeps* in *Lettres familières de Jérôme Aléandre (1510–1540)* ed Jules Paquier (Paris 1909) 35 letter XIV. Erasmus was eager to deprive Pio of such a title by calling him *philosophaster* and thus disparaging his philosophical abilities; see Erasmus *Apologia* 169 below. The legal adviser Battista Papazzoni, writing a chronicle of Mirandola sometime prior to Pio's death in Paris, praised Pio's many accomplishments, adding that he was '*non solo philosopho eminentissimo, ma anche sommo thologo* [sic]' and had composed many works on both topics that had not yet been published. See Biondi 130.

secured funding for his first cousin Gianfrancesco Pico. In gratitude for
Pio's loyalty and support during his expedition against Bologna in 1506,
Pope Julius II on 26 February 1507 took Pio and his lands in Carpi un-
der the immediate protection of the Holy See and thus provided him with
one more safeguard against threats from Ferrara, which was a papal vassal.
Pio's diplomatic skills so impressed the French king, who was in northern
Italy tightening his control over Milan, that Louis XII decided to send him,
a person much in the pope's favour, to Rome in August 1507, together with
Guillaume Briçonnet, on a special embassy to assure Julius II of the king's
good will. After he returned to France in 1508 to report on his mission, Pio
and Etienne Poncher were entrusted with an important mission to Cambrai
to secure the friendship of Emperor Maximilian. So successful were they
that by the end of the year a formal treaty was signed uniting France, the
Holy Roman Empire, and other states into a league against Venice. In the
spring of 1509 Pio helped with the negotiations that allowed other states,
such as the papacy, to enter the league. He accompanied Louis XII on his
expedition to Italy and shared in the great victory at Agnadello on 14 June
1509. On learning that the pope might withdraw from the league, Louis
XII dispatched Pio in haste to Rome in February 1510, but, despite Pio's
efforts, Julius II made his peace with Venice and turned against France.
Pio returned to northern Italy where he served with the French army and
at times acted as its negotiator with the papal forces, represented by his
cousin Gianfrancesco Pico. Reports reached Louis XII that Pio was secretly
working for the enemy and had promised to hand over to the papal forces,
for a large sum of money, the French-controlled fortress of Mirandola. In
early June 1511 Pio was placed under arrest, but then released. Louis XII,
however, publicly denounced his treachery and dismissed him from his service. Within months Pio entered the employ of his feudal overlord, Emperor
Maximilian.[15]

* * * * *

15 On the support of Mantua and Mirandola, see Guaitoli 'Sulla vita' 134, 136. On the ter-
mination of Pio's betrothal to Eleonora in 1490 and his promised marriage to Margherita
Gonzaga in 1494, see Sabattini 10–11, 47; and Sammarini 334–6, 352 (on the promised
marriage), 394–5 (on the legitimization of Margherita by Emperor Maximilian in 1516
and his exhortation that the marquis provide her with a dowry befitting her status and
satisfactory to Pio); Alessandro Luzio *Isabella d'Este di fronte a Giulio II negli ultimi tre
anni del suo pontificato* (Milan 1912) 227–8 (Pio tells Ludovico Canossa in 1513 that he no
longer considers Margherita as his betrothed because of the changed political situation,
his absence from the area, the negotiations of the marquis for her to marry Agostino
Chigi, the Sienese banker in Rome, and Gianfrancesco's statement that he is grateful to
God that Pio is no longer to be his son-in-law); and Felix Gilbert *The Pope, His Banker,
and Venice* (Cambridge, Mass. 1980) 97 (for negotiations in 1511 between the marquis

For the next fifteen years (1512–27) Pio spent most of his time as an ambassador in Rome. After a mission to Venice during the winter of 1511–12 Pio was sent by Maximilian to the court of Julius II. With the assistance of Matthäus Lang he helped to negotiate the reconciliation of the emperor with the pope and a truce between the empire and Venice. Maximilian showed his gratitude by requiring Alfonso d'Este on 29 May 1512 to restore to Alberto the half of Carpi he had seized, by investing Alberto on 25 July 1512 with San Felice and other Modenese territories, and by attempting to install him in 1514 as the successor to Antonio Alberico II Malaspina, the marquis of Massa (1481–1519), who was without a male heir. King Charles I of Spain offered Pio in 1517 the governorship of Sicily, which Alberto declined in order to continue serving Maximilian in Rome. When Maximilian died in 1519 and Charles succeeded him, he did not renew Pio's commission as imperial ambassador.

During the next four years Pio moved from position to position and unintentionally succeeded only in jeopardizing his control of Carpi. On the urgings of Pope Leo X Pio reluctantly acted as the informal agent in Rome of Francis I of France – a service that earned him the enmity of the French king's rival and Pio's overlord, Emperor Charles V. During the *sede vacante*

* * * * *

and Chigi over his possible marriage to Margherita). On Paola Gonzaga's marriage to Giovanni Nicolò Trivulzio (d 1512), the eldest son of the French marshall Gian Giacomo Trivulzio (1441–1518), see Sammarini 382; and Guaitoli 'Sulla vita' 185, 207–8. On Pio's securing seven hundred *scudi* for Pico from the marquis, see the letter of Pio to Gianfrancesco, from Milan, 12 August 1506, BAV Vat Lat 8213 95^r. The Italian term *scudo* has the same meaning as the French term *écu*, that is, a coin with the emblem of a shield. What particular *scudi* these were cannot be ascertained. The first Italian gold *scudo* to be issued was struck at Genoa in 1508, when it was under French occupation, and thus two years after the transaction recorded in Pio's letter. With a fineness of 22.5 carats, this *scudo* contained 3.216 g fine gold, compared to the 3.369 g in the current French *écu au soleil* (then worth 1,269 *livres tournois*). Milan did not strike a gold *scudo* until 1520; Venice, not until 1528; and Florence, not until 1530; see Carlo M. Cipolla *La moneta a Firenze nel cinquecento* (Bologna 1987) reissued in English translation as *Money in Sixteenth-Century Florence* (Berkeley 1989) 12–15, 61–76. On Pio's service as ambassador of Mantua (1506–7), see Sabattini 29–33, and as ambassador of France (1507–11), Sabattini 33–40. On the protection given to Pio and Carpi by Julius II, see Guaitoli 'Sulla vita' 163. On the 1507 Roman embassy, see Guaitoli 'Sulla vita' 164; and Michel Veissière *L'evêque Guillaume Briçonnet (1470–1534): Contribution à la connaissance de la Réforme catholique à la veille du Concile de Trente* (Provins 1986) 64–6. On Pio's negotiations on behalf of the League of Cambrai, his 1510 mission to Rome, and his service with the French in northern Italy, see Guaitoli 'Sulla vita' 165–74, 178; Guicciardini *Storia d'Italia* II 333; and Cortesi 74^v (Pio travelled from French territory to Rome on horseback in only four days in February 1510). On Pio's alleged duplicitous dealings and his dismissal by the French king, see Guaitoli 'Sulla vita' 183, 186–90; Minnich 'Protestatio' 277–8, especially n39; and Sabattini 40.

of 1521 and the early pontificate of Adrian VI Pio served as papal governor of Reggio, Rubbiera, and Modena until the pope restored them on 6 November 1522 to Alfonso d'Este. Despite a formal imperial reinvestiture of Alberto with Carpi at the Diet of Worms on 15 May 1521, the emperor accused Pio of having surrendered his city to the French, and on 3 January 1523 forcibly confiscated all his lands and titles. Charles initially conferred these on his general Prospero Colonna (d 1523), whose rights passed to his son Vespasiano. Even the intervention of Pope Adrian VI on Pio's behalf proved of no avail. Taking advantage of a temporary weakness in the imperial garrison of Carpi, Pio's forces seized the city and reinstalled his regime (1 September 1523–9 March 1525).

Despairing of a reconciliation with the emperor, Pio returned to Rome in 1523 and formally took up the duties of French ambassador to the papacy. In 1524 he convinced Pope Clement VII to take an officially neutral position that favoured France; and when Francis I was captured at Pavia and all of Italy seemed doomed to Hapsburg domination, Pio persuaded the pope to become a signatory to the anti-imperial Treaty of Cognac (1526). When Francis I failed to supply the money and troops that were promised through his agent Pio, the war became a disaster for the Italians, especially the papacy. In May 1527 the city of Rome was sacked. Clement VII, a dozen or so cardinals, and some ambassadors such as Pio, together with their retainers, fled to Castel S. Angelo – about three thousand in all. By a capitulation agreement signed on 6 June Pio and some others were granted free passage from the fortress. Clement VII sent him at the end of June to the French court to handle the interests of the Holy See there. Pio set out for Lyon, where he temporarily resided, and by September he was at the royal court. He was honourably received and maintained at the expense of Francis I, who conferred on him naturalized citizenship, lands, and titles in France. He resided in Paris in the parish of Saint-Pol on the rue Saint-Antoine, close to the royal Hôtel des Tournelles, probably in one of the residences belonging to the king's palace complex. His efforts to regain Carpi with the help of Clement VII failed. As a reward for military services rendered, Charles V conferred it on Alfonso d'Este in 1527 and granted him formal investiture on 8 April 1530. Four months later, when drawing up his final will, Pio lamented the unreality of his assigning to heirs estates and properties he had lost because of his loyal service to Francis I. With deep sorrow that the destruction of his political and economic fortunes would leave his wife and daughters destitute, he commended them to the mercy and justice of Francis who had two years earlier agreed, despite Pio's protests, to Alfonso d'Este's possession of Carpi as part of the negotiations for the marriage of Renée Valois and Ercole d'Este. Pio's nemesis in Ferrara had triumphed and Pio's quarter-century career as one of the most distinguished diplomats of his age ended in humiliation, exile, and disgrace.[16]

* * * * *

16 This period of Pio's diplomatic career is traced by Guaitoli 'Sulla vita' 198–305; and Semper 8–18. On the offer of the governorship of Sicily, see the letter of Pio to Maximilian, from Rome, 19 March 1517, in the University of Pennsylvania Library, Rare Book Collection, Lea Collection, MS 414 Letters of Pio sec 4 (correspondence with the imperial court) letter 34. According to Cardinal Bernardo Dovizi da Bibbiena, in 1520 Pio represented at Rome the interests of Francis I 'con gran fede et amore' and 'con diligentia et con fede'; see Moncallero II 212–13, 215. In his letter to Gian Matteo Giberti from Carpi, dated 25 April 1522, Pio protested that he had agreed to serve Francis I in Rome only after repeated pressure from Pope Leo X, Dovizi, and the king, who offered him money, membership in the Ordre de Saint-Michel, and a personal armed guard, all of which emoluments Pio declined. His service to the French was not contrary to the interests of his overlord, Charles V, Pio protests, and immediately upon hearing of the enmity between Francis I and Charles V, who was allied with Leo X, Pio resigned his duties and left Rome. The intercepted letters that show that he was a traitor to the emperor were not written by him. If it were not for his illness, he would go in person to the emperor to argue his innocence. See Girolamo Ruscelli, ed *Delle lettere di principi, le quale o si scrivono da principi, o a principi, o ragionano di principi* rev ed 3 vols (Venice: Appresso Francesco Ziletti 1581) I 98ᵛ–100ʳ (letter to Giberti); and *Actes de François Ier* I 186 n1043 (on the offer to Pio of an annual pension of ten thousand *livres tournois* [a sum expressed in the French silver-based money of account, the *livre tournois*, with 20 *sols* or *sous* to the *livre*, and 12 currently circulating silver *deniers* to the *sou* (see CWE 1 328, 331–2); in 1519 this pension would have been worth about 4,820 gold ducats (or florins) or £1,084 sterling, a sum equivalent to 43,373.5 days' wages for an Oxford or Cambridge master carpenter, then earning 6d per day] and a commission as captain of one hundred lancers, dated 1 June 1519). Pio was temporarily reconciled with Charles V; see Sanuto XXXIII 24. In agreements reached on 30 September 1526 in Grenada and on 30 December in Gaeta between representatives of Charles V and Alfonso I d'Este, Ercole II d'Este was promised investiture, for himself and his descendants, of Carpi and Novi as dowries should he marry the emperor's natural daughter Margherita of Austria when she came of legal age. Vespasiano Colonna and his cousin Alfonso d'Avalos, the marchese del Vasto, are said to have agreed to this arrangement, renouncing their rights to the territories. See Sabattini 77; and ASV Fondo Pio 53 306ʳ⁻ᵛ. On the Sack of Rome and Pio's relocation to Paris, see Guaitoli 'Sulla vita' 290–2; Guicciardini *Storia d'Italia* V 148; Pio Pecchiai *Roma nel Cinquecento* Storia di Roma 13 (Bologna 1948) 433; Allen VI 200; and the letter of Giovanni Salviati to the papal nuncio with Charles V, no place given, 18 September 1527, in ASV SS Francia 1 40ʳ. On Pio's representing papal interests at the French court, see Pierre Hurtubise *Une famille-témoin: les Salviati* (Città del Vaticano 1985) 180 n84, 186. On 25 January 1527/8 Francis I bestowed on Pio the vicecomtés, lands, and seigniories of Conches and Bretueil, towns in Normandy located ten and seventeen miles respectively southwest of Evreux. The king then raised these feudal territories to the status of comtes and attached to them total revenues amounting to one thousand *livres tournois* (see above) that were drawn from these lands and from the *greniers* on salt for Evreux and Conches. See *Actes de François Ier* I 541–2 n2851. That Pio experienced great difficulty in securing local recognition of these titles and rights is evident from the repeated efforts of the king from 1528 to 1530 to override rival claims to these territories; see

Yet in his own lifetime and after Pio enjoyed a reputation as one of the leading diplomats of Europe: he had helped to negotiate the League of Cambrai, the Venetian-imperial truce, and the League of Cognac. Already in 1510 Paolo Cortesi listed him together with Janus Lascaris, Roberto Acciaiuoli, and Vicenzo Querini as outstanding diplomats, known for their learning, virtue, and understanding of rulers, who should be invited by

* * * * *

Actes de François Ier I 575 n3024, 589 n3093, 690 n3613, VII 464 n25745 (on the opposition of the lord of Rouville and of Jean Masseline who is called *vicomte de Conches et Breteuil*). By 18 July 1528 Francis I had conferred on Pio naturalized citizenship and an accompanying gift of two hundred *livres tournois* (see above) (*Actes de François Ier* I 583 n3066 and VII 464 n25742). That the king's efforts to provide financial assistance to Pio were not always effective is evident in Pio's letters to Guillaume (or Anne?) Montmorency, Paris, 23 May 1529 and 18 January, no year given, in Chantilly, Musée Condé, L series, vol 2 folio 187^{r-v} and L 3 229^r) in which he complained about his poverty and the lack of execution of the king's orders, and expressed a desire to be listed on the roll of those who received annually a regular pension. Guillaume's son Anne (1492–1567) had been *grand maître de France* since 1526. As Grand Master he supervised the officials of the royal household, ensured that ordinances were carried out, directed the police in the royal entourage, had charge of the keys, guaranteed security and fortifications, introduced ambassadors, and supervised buildings and artistic works in the royal palace; he was on intimate terms with the king; see H. Lemonnier *Les guerres d'Italie – La France sous Charles VIII, Louis XII et François Ier (1492–1547)* Histoire de France depuis les origines jusqu'à la Révolution, ed Ernest Lavisse, vol v–1 (Paris 1911) 206; and R. Doucet *Les institutions de la France au XVIe siècle* (Paris 1948) I 122–3. Pio continued to serve the king by passing on to him through Montmorency information of a political nature and the names of members of the Italian nobility who wished to enter the royal service; see Musée Condé, L series vol 10 folio 188^r, Paris, 20 January, no year given and L 12 56^r, Paris, 29 June, no year given (providing news); L 12 124^{r-v}, Paris, 16 October, no year given (recommending C. Colonna); and L 14 205^r, Paris, 5 April, no year given (recommending the nephew of a Neapolitan baron). On his residence in Paris, see Beaumont-Maillet 281 n209; David Thomson *Renaissance Paris: Architecture and Growth 1475–1600* (Berkeley 1984) 31–2, 196 n12; and Tuetey 102. Both Clement VII and Francis I negotiated over the lordship of Carpi. In 1530 Clement secured a promise from Charles V at Bologna to restore Pio, but the emperor later made this conditional on Pio's leaving the service of Francis I and working for him instead, something Pio rejected; see Guaitoli 'Sulla vita' 303–4. Francis I promised Alfonso d'Este, despite the vigorous complaints of Pio, that he could keep Carpi as part of the marriage contract between the king's sister-in-law Renée and the duke's son Ercole. To compensate Pio, the duke offered to pay him annually three thousand *scudi* (see n15 above) and agreed that after Pio's death his daughter would be heiress of all. Clement was said to agree to this arrangement; see ASV SS Francia I (Register of Cardinal Giovanni Salviati), letter to the nuncio at the imperial court, 6 November 1527 50^r–51^r and the spring 1528 report 127^r. See Pio *Will* 395 below. Alberto's nephew Rodolfo writing to Anne de Montmorency on 27 January 1531 lamented that his uncle had died leaving his house in total ruin; see Pier Giovanni Baroni ed *La nunziatura in Francia di Rodolfo Pio (1535–1537)* Memorie storiche e documenti sulla città e sull' antico principato di Carpi 13 (Bologna 1962) xxiv–xxv n14.

name to advise a future church council. In his *Dialogus Leo* (1513) Raffaele Brandolini has Cardinal Alessandro Farnese praise Pio for having earlier carried out his duties as French ambassador with prudence, integrity, and wisdom, then for representing imperial interests prudently and firmly, and for having always preserved the honour of the Holy See. Girolamo Aleandro in 1516 compared Pio to Plutarch, a man who excelled in both learning and political affairs: Pio busies himself with the affairs of all Europe, he has found favour not only with the emperor and the greatest kings but also with popes, and the greatest princes seek his opinion. Even Francesco Guicciardini, who was hostile to Pio, acknowledged his great talent and skill in diplomacy. The imperial representative in Rome in the early 1520s, Lope Hurtado de Mendoza, gave this striking assessment of Pio's diplomatic skills: 'Carpi is a devil, he knows everything and is mixed up in everything; the Emperor must either win him over or destroy him.' Even when Pio was in exile in Paris in 1529, the former ruler of Siena, Francesco Petrucci, still considered him to be 'a person of authority and full of every virtue and prudence, an Italian, and universal father and benefactor.' In his *Elogia virorum doctorum* Paolo Giovio attributed to Pio great eloquence in meetings where the most difficult matters were discussed and an outstanding and incomparable talent for negotiating the greatest affairs.[17]

Pio's remarkable success as a diplomat can be attributed to a number of factors. He was conscious of his own status as a prince and had a comfortable relationship with the rulers he served. The gentle gravity of his bearing led others to take him seriously but also to be attracted to him; even the unlettered comic Virgilio Nobili considered him a friend. Men valued what he said: his own bon mot comparing the avaricious to the crocodile who continues to grow as long as he lives was quoted by Cortesi. When entrusted with a mission, Pio devoted his full energies to accomplishing it, as, for example, when he went on horseback from the French court to Rome in a mere four days. He was diligent in gathering information, retained it with a tenacious memory, and could write detailed reports. He understood well

* * * * *

17 Cortesi 221[r]; Raphaelis Brandolini Lippi jr *Dialogus Leo nuncupatus* ... ed Francesco Fogliazzi (Venice: apud Simonem Occhi 1753) 79, 92, 97–9; letter of Girolamo Aleandro to Alberto Pio, no place given, c late 1516, in BAV Vat lat 8075 13[v]–14[r]; Guicciardini *Storia d'Italia* II 333; on Guicciardini's hostility towards Pio, see Guaitoli 'Sulla vita' 228–9, 248; Mendoza is quoted in Ludwig Pastor *The History of the Popes from the Close of the Middle Ages* trans Frederick Ignatius Antrobus et al, 40 vols (St Louis 1923–53) IX 269; the letter of Francesco Petrucci to Pio of 1 March 1529 is quoted in Guaitoli 'Sulla vita' 299; the quotation from Giovio's *Elogia* is cited by Semper 18 n10.

what would motivate people and was able to present his master's position in the most favourable light.[18]

Pio had the ability to win others' confidence. Louis XII sent him to Rome in 1510 with full powers to negotiate on his behalf. Maximilian entrusted to his sole care the affairs of the empire in Rome and removed as his other representative Julius II's nephew, Antonio Giuppo della Rovere, when Pio complained of his rival's ineptitude. Julius II, especially towards the end of his pontificate, placed the greatest trust in Pio. Pio also won the confidence of Leo X, into whose family he married. With Leo's cousin, Clement VII, Pio was on intimate and very affectionate terms.[19]

* * * * *

18 Cortesi 60[r] (for 'Albertum Pium hominem, doctrina et placida gravitate dulcem' and friendship with Nobili), 125[r] (for his crocodile comparison), 74[v] (for his remarkable trip on horseback); Giovio Elogia in Semper 18 n10.
19 On Louis XII's confidence in Pio, see Guicciardini Storia d'Italia II 333; for an example of Maximilian's exhortation that Leo place in Pio 'indubitatem ... fidem,' see ASV Archivio di S. Angelo Arm I–XVIII no 249, dated 23 April 1517 from Antwerp. For the controversy over Antonio Giuppo della Rovere as Pio's colleague, see the letters of Lang to Maximilian, from Viterbo, 16 November 1513, and from Rome, 1 December 1513, in Haus- Hof- und Staatsarchiv Vienna, Maximiliana 30 II 64[r], 118[r–v]; and the letter of Pio to Maximilian from Rome, 3 January 1514, ibidem 31 I 9[r]–10[r]. For Pio's protestations of openness and his ability to win others' confidence, see his letter to Lang, from Rome, 16 August 1513, University of Pennsylvania Library, Rare Book Collection, Lea Collection, MS 414 Letters of Pio sec 4 (correspondence with the imperial court) letter 11 folio b: 'Regarding Ferrara, about which he [Maximilian or his adviser Lang] abjures me to be willing to indicate to him freely the will of the supreme pontiff, I say that I have never concealed by silence nor have I dissembled on any matter, but I have always been open with him about all and whatever I knew or understood, openly and candidly, with the greatest sincerity, on whatever topics, just as is fitting, since I have constituted him as my singular patron and lord in whom I have uniquely placed all my hopes. Just as I have adopted and follow a certain state of mind, I not only cherish but ardently love him as a brother and father, and thus in this matter [of Ferrara] about which he begs so much to be informed, I am not changing my customary behaviour [of openness]. In truth the soul of Leo is not equally known to me in all matters as was that of Julius, whether because he is not by nature so open or because up to now I have not been joined to him by such familiarity as I was to Julius. Wherefore, I do not think his Holiness altogether confides in me as completely as would Julius, who in the last days of his life inwardly placed his complete confidence in me. Nevertheless, I am able to surmise (even if his Holiness has not opened up to me on this matter): I think he [Leo] wants to recover Ferrara and grant it as a fief to his brother [Giuliano].' Pio soon won over Leo's confidence, as Sanuto XXX 289 reports: 'The pope does nothing without his counsel,' cited by Odoardo Rombaldi 'Profilo biografico di Alberto III, conte di Carpi' in Alberto Pio III, Signore di Carpi (1475–1975) (Modena 1977) 33. While Pio may have been open, candid, and sincere with Emperor Maximilian and Pope Julius II, a number of his contemporaries saw him as treacherous and deceitful. In 1511 the French king, Louis XII, dismissed Pio from his services on charges that he had conspired to betray the fortress of Mirandola to papal forces; see Minnich 'Protestatio' 277. Emperor Charles V declared Pio to be a rebel for having served his rival Francis I; see Guaitoli 'Sulla vita' 237, 243, 246, 304, but 248–9 for a defence of Pio. Pio spent much of his life plot-

Pio's relations with the Medici became very close over time. His uncle Gianludovico Pio (d 1469) had married Aurante Orsini, the maternal aunt of Leo x. Alberto probably met members of the Medici family when, as a young man, he visited his distinguished maternal uncle, Giovanni Pico, in Florence. In October of 1494 he hurried to Modena to welcome the exiled ruler of Florence and his brother, who were reported to be lodging in disguise in the Hospitale della Misericordia over which Pio held patronage rights. Pio's noble blood, refined manners, cultural interests, and position as imperial ambassador helped him to gain entrance into the inner circle of Medici friends in Rome. But what brought him even closer to the Medici family was his marriage in February 1518 to Cecilia Orsini (1493–1575), the daughter of Franciotto (1473–1534), Leo x's first cousin and childhood friend.[20]

This marriage clearly met with the pope's approval. On 16 February Leo provided a banquet for the bride and groom at the Vatican and

* * * * *

ting against Alfonso d'Este, his great rival, who ultimately succeeded in dispossessing Pio of his hereditary lands and title; see Paolo Giovio *Le vite di dicenove huomini illustri* ... (Venice: Giovan Maria Bonelli 1561) 351^r (but misprinted 361^r). The posthumous verdict on Pio given by Jakob Spiegel, a former imperial bureaucrat, on 23 November 1531 is harsh: '[Pio] of Carpi was a man of restless temper and of a cunning and treacherous disposition, as I readily learned when I worked in the chancery. God spare him! But if He had taken him from among the living twenty years earlier, I am sure there would have been harmony among the monarchs of Christendom' (Allen Ep 2572:15–19). The knowledgeable historian Francesco Guicciardini, then in the papal service, considered Pio to be insincere and self-serving; *Storia d'Italia* II 333, III 24, 65. The d'Este ducal librarian, Ludovico Antonio Muratori (1672–1750), denounced Pio as a faithless servant of various princes who mocked and betrayed them all; *Delle antichità Estensi continuazione, o sia Parte Seconda* (Modena 1740) 336. A certain amount of irony and sarcasm may be suspected in Erasmus' comments on Pio's guileless and honest character (eg *Responsio* 5, 10 below). While Erasmus held that candour was a characteristic that befitted a prince, his contemporary Niccolò Machiavelli advocated the opposite. That Pio advised Leo on matters other than diplomacy is suggested by the pope's acknowledging in the bull *Salvator noster Jesus* of 19 May 1520, which set up in Rome a convent for penitent prostitutes, that Pio had informed him of such institutions in France; see Leo x *Salvator noster Jesus Christus Dei filius qui assumpta humanitate* (Rome 1520) Aii^v in Archivio di Stato di Roma Bandi/Buste 293 n34. Cardinal Dovizi claimed in 1518 that Pio was so loved and esteemed by Leo x that he obtained from the pope whatever favours were desired by the prince he represented; see Moncallero II 165. On Pio's relations with Clement VII, see the report of Stephan Rosin, from Rome, 17 July 1524, in *Acta Reformationis Catholicae* ed Georg Pfeilschifter I: *1520 bis 1532* (Regensburg 1959) 245 n13; and the letter of Giovanni Salviati to his father Jacopo, dated 1 January 1528, ASV SS Francia I 83^v: '*Nostro Signore el quale ha causa di amarlo [Pio] per le virtu sue, ma molto piu per essergli tanto affectionato che non si puo dire piu.*'
20 Vasoli *Pio* 23; Sammarini 357–8; Sabattini 11; Emmanuel P. Rodocanachi *La première renaissance: Rome au temps de Jules II et de Léon X* (Paris 1912) 13 n3; Litta 'Orsini di Roma' table IX.

personally graced their table. Franciotto, recently named a cardinal, provided as dowry the castles of Vaccone, Poggio, and Sommavilla in Sabina and three thousand ducats, and his papal first cousin added an additional nine thousand ducats. In March 1518 the pope confirmed the emperor's donation to Pio in 1512 of the Modenese towns of San Felice, Marano, and Fanano. Leo also conferred on Pio the lordship of Sarsina in the papal Romagna, and on 1 November 1519 he invested him with other former Malatesta lands in the border area between the Republic of Florence and the Duchy of Urbino, including Meldola, Polenta, Caminate, Collinella, Cuglianello, Ranchio, Ravicchio, and nine other towns. Pio was granted the right to use the papal keys in his coat of arms and his lands were taken under papal protection. Not only was Pio now a relative of the pope, but through his father-in-law he was related to four other cardinals. As part of the wedding celebrations Pio had a comedy staged in his house before honoured guests who included the pope, and he hired the Sienese entertainer Strascino to recite one of his farces.[21]

Pio's marriage to Cecilia was apparently a happy one, despite their age difference of eighteen years. She bore him two daughters, the elder, Caterina (b c 1519, and named perhaps after Pio's mother), and the younger, Margherita (b c 1527). Their son, Francesco, born in 1524, unfortunately died at a young age. Pio's wife, a strong Roman noblewoman, followed him to Carpi after the death of Leo x and threatened to resist force with force when in 1523 the imperial agents, having despoiled Pio of Carpi, now tried to infringe on his papal fiefdom of Novi to which the contessa had retreated. She was later reunited with Pio in Rome, took refuge with him in Castel S. Angelo during the Sack of Rome in 1527, and accompanied him into exile in France. That Pio's marriage was not merely a political alliance is suggested by his decree of 1522 forbidding the practice in Carpi of dressing brides in

* * * * *

21 On Pio's marriage, see Carlo Falcone *Leone X: Giovanni de' Medici* (Milan 1987) 86–7, 457–8; and Giovanni Alfredo Cesareo *Pasquino e pasquinate nella Roma di Leone X* (Rome 1938) 73, 239. On Cecilia's feudal dowry, see Litta 'Orsini di Roma' table IX. The three thousand ducats provided by Cecilia's father as part of her dowry were presumably Venetian ducats, and this sum would then have been worth £675 sterling or 6,225 *livres tournois*; see n8 above, CWE 1 314, and CWE 12 650. The nine thousand ducats provided by Leo were presumably also Venetian ducats, and would have been worth £2,025 sterling or 18,675 *livres tournois*. On Leo's gifts to Pio, see Guaitoli 'Sulla vita' 204, 215–16; Biondi 119–20; ASV Reg Vat 1200 315^r–318^r; Garuti 22, 33, 154–8; Philip J. Jones *The Malatesta of Rimini and the Papal State: A Political History* (Cambridge 1974) 242; P. Mastri *I Veneti e i principi Pio di Carpi in Meldola dal 1503 al 1531* (Bologna 1940); and G. Zaccaria *Storia di Meldola e del suo territorio, dal '500 ai primi del '600* (Meldola 1980) II. On Cecilia Orsini's father and relatives, see Hallman 158.

black since this implied that marriage was burdensome rather than a cause for joy. In his discussion of marriage in the *XXIII libri* Pio insists that it must be entered upon freely (not under duress from parents) for it brings one into a lifetime of intimacy, a union that can be sweet and joyous and happy, that promotes the common good and the procreation of offspring.[22] Given what is known about his own marriage and his attitudes towards this institution, one wonders what credence is to be placed in the slyly disparaging comments of some contemporaries, Erasmus included, that suggest that Pio was effeminate and homosexual.[23]

Pio's success as an ambassador may also be attributed at least in part to his connections to influential people at the papal court. Leo x's private secretaries, Pietro Bembo and Jacopo Sadoleto, were Pio's personal friends from their student days in Ferrara. Through his friendship with members of the Murano Circle Pio knew Querini, who became one of Leo's advisers.

* * * * *

22 On Pio's son, Francesco, his daughter Margherita (who married Girolamo Acquaviva), and his daughter Caterina (whose husband was Bonifacio Caetano di Sermmetta), see Hallman 158; Pio *Will* 390–2 below; and Contini *Alberto III Pio* [7, 36], where his son's name is given as Giulio. On Cecilia in Carpi and Novi, see Guaitoli 'Sulla vita' 241, 249; on her escape to Castel S. Angelo, see Pecchiai (cited in n16 above) 433. On the promotion of Giovanni Vincenzo Carafa(?) in 1527(?) to the cardinalate, Pio wrote in his own name and that of his wife to congratulate him; see BAV Ferrajoli 539 41ᵛ. On the wedding-dress decree, see Guaitoli 'Sulla vita' 238. On Pio's views on marriage, see Pio *XXIII libri* 226ᵛE.

23 In his study *Pasquino e pasquinate* (cited in n21 above) Cesareo has found many references to a man at the Roman court identified as *Savoia* who is of rustic origins but is now occupied with great affairs of state, a man who is very soft and refined, who uses perfume, ointments, powder of Cyprus, and stylish garb, who is the 'prince of all Ganymedes,' and who is no longer in Rome after the Sack. Given Alberto's family name of Pio di Savoia and the consistency of other clues, Cesareo concludes that Pio and Savoia are one and the same (236–7, 256–7). In the portrait of Pio by Bernardino Loschi, dated 1512, Pio is attired in a costly robe with nine golden cords and fur edging, his blond hair is shoulder length, and he has a modest beard and moustache; see Contini *Alberto III Pio* [24–5]. Erasmus, who as a young man shared a room and bed with Aleandro in Venice in 1508 (Allen Ep 2443:286) and was known to have been very close to him (Allen Ep 2570:73), cited reports that Pio preferred Aleandro to Erasmus because, among other things, the Italian was handsomer (*formosior*): Allen Ep 1479:133 / CWE 1479:146–7 and Allen Ep 1482:58 / CWE 1482:66. Erasmus went on to claim that Aleandro and Pio were in an intimate relationship: '*nec me fugit Alberto cum Aleandro necessitudinem esse intimam*' (Allen Ep 2077:51–2); '*iunctissimus*' (Allen Ep 2629:50); '*Aleandro mire addictum*' (Allen Ep 2042:17); and '*una anima*' (Allen Ep 2565:18–19). He twice cited in reference to them the statement by Satyricus in Juvenal 2.47 that there is a great concord among the soft/effeminate (Allen Epp 2371:37 and 2375:79–80). His ironic question in the *Apologia* (288 below) about whether Jerome, a man rendered effeminate by delicacies, had a pretty boy to fetch him a chamber pot, may also have been a snide allusion to Pio.

Pio helped Aleandro to secure his position as the personal secretary of Giulio dei Medici, the pope's cousin and closest adviser, who later became Pope Clement VII. Gian Matteo Giberti, the secretary of Giulio dei Medici and then of Leo X, and finally Datary (a major curial official) and adviser to Clement VII, was also a close friend of Pio.[24] These men were either literary figures in their own right or respected patrons of leading humanists.

PIO AS A SOCIAL AND CULTURAL FIGURE

Pio was an important figure on the Roman scene. He was an ambassador to the papal court and, through his marriage into one of the major noble families of Rome, the Orsini, related to the Medici popes, but he had other ties to the papal bureaucracy as well. He was named a member of the papal family of Julius II, an abbreviator of apostolic letters *de parco minore*, and a knight of St Peter.[25] He had a country villa at the foot of Monte Mario, and according to the census of 1526–7 he occupied a large building on the Via dell'Orso next to the church of Santa Maria in Posterola, near the Tiber River in the fashionable Regio de Ponte, not far from his wife's family palace of Monte Giordano, and with numerous cardinals, bishops, and bankers as neighbours. Pio's household contained sixty-four persons, among whom were Juan Ginés de Sepúlveda, a young Spanish philosopher who was re-translating Aristotle for Clement VII, and Francesco Florido from Sabina,

* * * * *

24 Pio claimed that he had worked to see that Aleandro entered the service of Giulio dei Medici, who is '*veluti alter pontifex*'; see his letter to Erard de la Marck, c 1517–18, in BAV Vat lat 8075 12ᵛ–13ʳ; and the French version with the alternate Latin wording of '*qui est alter papa et qui solus gubernat omnia*,' which has been assigned the date 20 January 1518 and is printed in *Documents concernant la Principaute de Liège (1250–1532)* ... ed Alfred Cauchie and Alphonse van Hove (Brussels 1920) II 63:7–8. Pio named Giberti as one of the five executors of his will and bequeathed to him 'in my memory and as a token of love that silver globe on which is depicted the representation of the whole world or, as it is called, a *mappamondi* and in which there is a clock' (Pio *Will* 402–3 below).

25 For Pio's status as a papal familiar and his appointment as an abbreviator on 19 February 1513, see ASV Reg Vat 990 204ᵛ–205ʳ and 1211 2ʳ; I am grateful to Dr Elliott Kai-Kee for providing the date of this appointment. Abbreviators *de parco minore* assisted the vice-chancellor in the sending out of apostolic letters by helping to compose shortened versions of supplications and of the pope's decision and also helped to assess the chancery fees. The term *de parco minore* refers to 'the placement of their desks in the chancery'; see D'Amico *Renaissance Humanism* 26–9. Pio was also among the original members of the college of the Knights of St Peter, founded on 20 July 1520; see Thomas Frenz *Die Kanzlei der Päpste der Hochrenaissance (1471–1527)* (Tübingen 1986) 272 no 49, where Pio's family name is mistakenly given as 'Cuperi.'

a young man who had studied Latin in Rome and was now employed as
Pio's amanuensis and servant.[26] Pio decorated his home in Rome with some
of the paintings he had commissioned over the years from Giovanbattista
Cima da Conegliano, Bartolomeo Veneto, and Bernardino Loschi.[27] Because
of his social standing, cultural interests and attainments, and reputation for
learning, Pio was welcomed into refined society and attended banquets and
gatherings of prelates and literary figures. He was apparently a member of
the academies sponsored by Johannes Goritz, Angelo Colocci, and others.
In *Coryciana*, a famous collection of poems gathered by Biagio Pallai in 1524
from a decade of celebrations of the feast day of St Anne, there are three
hymns in the form of prayers of thanksgiving for Pio's recovery from illness written by the humanists Giano Francesco Vitale of Palermo and Gaius
Silvanus Germanicus.[28]

* * * * *

26 For Pio's residence below Monte Mario, see Allen VI 200; and Pio *Will* 402 below (Pio
    left to his nephew Rodolfo Pio 'my vineyard with the house I have on the outskirts of
    the city of Rome at the foot of Monte Mario'). For the 1526–7 census, see *'Descriptio ur-
    bis': The Roman Census of 1527* ed Egmont Lee (Rome 1985) 67 no 3295, where Pio is mis-
    takenly listed as 'Roberto da Carpi.' For some of Pio's notable neighbours, see ibidem
    59–69 nos 2447–9, 2574, 2627, 2632, 2636, 2698, 2716, 2785, 3019, 3051, 3094–7, 3141, 3152,
    3160, 3286, 3321, 3353, 3473, 3481, 3518, 3548. On Sepúlveda's relationships with Pio at
    this time, see Sepúlveda *Antapologia* Aii^r, Bi^r, Diii^{r-v} / Solana 113, 116, 130–1; and Losada
    57–61. Kristeller *Iter italicum* I 380 notes a manuscript reference (Modena Est lat 370)
    that documents Sepúlveda's presence in Pio's household: 'Johan. Spagnolo philosopho …
    in Roma in casa del ill. sig. Alberto de Carpi.' On Pio's secretary Florido, see Sepúlveda *An-
    tapologia* Dii^v; Florido *Apologia* 116:11–22, 45–8; and Florido *Libri tres* liber III cap 4 264:36.
27 Contini *Alberto III Pio* [16–20, 24–6, 30–6]. Pio's will (which appears in translation 387–
    404 below) gives testimony to his interest in art. He bequeathed to 'the painter Master
    Bernard[in]o Loschi of Parma … food and clothing for the length of his life' (401).
    Further, he provided that the chapel of Santa Maria Magdalena in the church of San
    Agostino in Carpi be decorated with a *pictura decente* 'decent picture' (400), and that
    there be painted for the high altar of the church of San Nicolò, which Pio had had
    built in Carpi, an image of Christ seated on a throne, with to his right the Virgin Mary
    kneeling in adoration and behind her in the same pose San Nicolò, the church's patron,
    and to Christ's left images of St Francis and St Anthony of Padua, together with an
    image of Pio kneeling at the feet of St Francis who extends his arm either to Pio's head
    or his shoulders as if commending him to Christ (398). Pio is shown in a similar pose
    in two paintings, one attributed to Vincenzo Catena (see illustration 399) and the other
    to Mazzola Pietro Ilario (father of Parmigianino); see Contini [10–11].
28 In his letter of 1529 recalling the days of Leo X with their gatherings of learned men
    in the villas and gardens of Rome, Sadoleto lists among the luminaries of that time *il
    Savoja*. Pio is also cited in a popular sonnet, composed around 1516, celebrating the lead-
    ing cultural figures of the day; see Cesareo *Pasquino e pasquinate* (cited in n21 above)
    247, 249. For the three hymns (one by Giano Francesco Vitale of Palermo and two by
    Gaius Silvanus Germanicus), see *Coryciana* ed Biagio Pallai (Rome: apud Ludovicum
    Vicentinum et Lautitium Perusinum 1524) DDiii^r, EEi^v, EEiii^r. Although one was pub-
    lished in 1524, Vitale's poems (also called *voti* or prayer offerings) were composed by

Pio's circle of friends was not, however, limited to aristocrats, diplomats, prelates, painters, and poets. He also took part in gatherings where theological issues were discussed. One such gathering place was probably the palace of the learned Spanish cardinal Bernardino López de Carvajal (1455–1522). Pio and Carvajal shared similar intellectual interests, such as their concern for the Maronites of Lebanon, and Pio was involved in efforts to restore to the penitent cardinal the benefices he had lost for backing the Pisan Council. The Spanish scripture scholar Diego López Zúñiga is known to have taken up residence in Carvajal's household in 1521 and to have denounced Erasmus during after-dinner drinks at the cardinal's home. Zúñiga is also reported to have read sections from his *Erasmi Roterodami blasphemiae et impietates* at gatherings of the learned in Rome on several occasions in 1521. Given Pio's growing disenchantment with Erasmus, he would have found himself in sympathy with Zúñiga's views. Carvajal may also have been the person who introduced Pio to Juan Ginés de Sepúlveda, the brilliant young humanist, philosopher, and theologian who came to reside in Pio's palace, helped him with his first work against Erasmus, and defended his patron's reputation from Erasmus' attacks.[29]

* * * * *

11 December 1514, for in a letter of Michael Hummelberger to Beatus Rhenanus, from Rome, the *voti* written for 'Alberto, prince of the people of Carpi, recently seized by an over-heating slight fever,' are singled out for particular praise; see *Briefwechsel des Beatus Rhenanus* (cited in n7) 68, no 42. Vitale later wrote a poem in praise of Erasmus which acknowledged his noble genius and called him 'the Varro, of our age' (Varro was a Roman satirist and poet, 82–36 BC), while pointing out Erasmus' slippery and thorny mind; see LB I (19) (= sig **[6ʳ]). On Vitale, see Girolamo Tumminello 'Giano Vitale: umanista del secolo XVI' *Archivio storico siciliano* NS 8 (1883) 1–94; on Silvanus, see John D'Amico 'Humanism in Rome' in *Renaissance Humanism: Foundations, Forms, and Legacy* ed Albert Rabil, 3 vols (Philadelphia 1988) I 284; on Blosio, see idem *Renaissance Humanism* 134–7. On the Roman academies of this period, see 107–12, and Rodocanachi (cited in n20 above) 143–53. See also *Responsio* nn 126 and 431 below.

29 On Pio and Bernardino López de Carvajal's shared interest in the Maronites of Lebanon, see Teseo Ambrogio degli Albonesi *Introductio in Chaldaicam linguam, Syriacam, atque Armenicam et decem alias linguas* (Pavia: Iohannes Maria Simoneta 1539) 14ʳ (Maronite legates were housed in the residence of Carvajal, who commissioned Teseo to translate into Latin their Syrian liturgy); Jean Gribomont 'Gilles de Viterbo, le moine Elie, et l'influence de la littérature Maronite sur la Rome erudite de 1515' *Oriens Christianus* 54 (1970) 125–9, especially 127. Their residence is identified as the monastery of Santa Maria del Pace by Celestino Cavedoni 'Notizia letteraria di alcuni codici orientali e greci della R. Biblioteca Estense che già furono di Alberto Pio Principe di Carpi' in *Memorie di Religione, di Morale e di Letteratura* ser 3 vol 17 (Modena 1843) 1–19, especially 2, 5–6, 14. The Maronite monk Elia filius Abrahae stayed with Pio in 1515; he taught Pio Syriac and copied a number of Syriac manuscripts for him; see Alistair Hamilton 'Eastern Churches and Western Scholarship' in *Rome Reborn: The Vatican Library and Renaissance Culture* ed Anthony Grafton (New Haven 1993) 239. Cavedoni 'Notizia' 2 reports the

Alberto Pio, attributed to Bernardino Loschi, c 1512
This portrait depicts Pio as a young nobleman-scholar holding a book
opened, according to Gilmore, to Virgil's *Aeneid* 6.724–47, lines that speak of
the human condition and the passage of time. Behind the subject is
an imaginary Arcadian landscape with Muses and temples of Dionysus
and Phoebus Apollo; in the right corner a soldier listens to a woman,
perhaps Clio, the Muse of history.
By permission of the National Gallery, London

THE CONTROVERSY BETWEEN PIO AND ERASMUS

Pio's hostility to Erasmus developed over time. His first contact with Erasmus was probably in 1508 when he heard Aldo Manuzio praise him and apparently saw him together with Thomas Linacre at Manuzio's home in Venice. As Erasmus' literary fame increased, so too did Pio's esteem for him. He was especially impressed by Erasmus' collection of *Adages*, which he first read in 1516. Shortly thereafter he began avidly to read the *Praise of Folly* and was initially struck by its elegance and learning. But the more he read, the more disgusted he became with Erasmus' bitter denunciations of church practices and officials. Pio also sampled Erasmus' *New Testament* and some of the *Paraphrases*. He claimed to have read a few things from Diego López Zúñiga's writings against Erasmus; the rest he could have learned from any of Zúñiga's public recitations. Pio at first ignored Erasmus' responses to his critics, but because Zúñiga was in Rome, Pio did read

* * * * *

remarkable evaluation of Pio given by Elia at the end of one of his works: 'You [who read this manuscript] pray for the world, for the elect, and for the innocent legate of Emperor Maximilian, my lord Albert, of the holy city [Rome], a lover of Christ, whose name is of Carpi, and petition for him mercy from the Lord of Mercy that he gather him [Pio] with the just in his kingdom, for I have never seen a man like him, just, a lover of Christ, and a lover of guests, or one who is more merciful towards the poor. Lord, preserve his life in this world and in the next by the prayer of Mary the Mother of God and of all the saints. Amen. Amen.' Cavedoni 'Notizia' 12 has proposed that Pio considered compiling his own polyglot New Testament based on Syriac, Arabic, and Ethiopian texts (similar to the Complutensian Polyglot Bible [1514–17] sponsored by Francisco Jiménez de Cisneros, which used Latin, Greek, Hebrew, and Syriac texts; or the *Psalterium octuplex* [1516] of Agostino Giustiniani, which used the Massoretic Hebrew, Septuagint Greek, Arabic, Aramaic Targum, and three Latin versions, plus scholia) to criticize Erasmus' *Novum instrumentum* (1516), which was based on Greek texts and included a new Latin translation and annotations; Cavedoni is seconded by Giorgio Levi della Vida *Ricerche sulla formazione del più antico fondo dei manoscritti orientali della Biblioteca Vaticana* Studi e Testi 92 (Città del Vaticano 1939) 108. Their view is supported by Pio's commissioning c 1520 Giovanni Leone Africano (al-Ḥasan ibn Muḥammad ibn Aḥmad al-Wazzān al-Fāsī) from Fez, a Muslim convert to Christianity, to teach him Arabic and make a copy of the Pauline Epistles in Arabic; by his having Elia copy the four Gospels in Syriac in 1519; and also by his personal interest in the liturgical book of the Ethiopians sent to Leo x through the Portuguese ambassador Miguel de Silva c 1516; see Levi della Vida 105–7, 134–6; and Pio *XXIII libri* 96^rG. On the Maronites' appeal to Pio to intercede for them with Emperor Maximilian, see Ibrahim Harfus 'Al-Kunuz al-Makhfiyyah' in *Al-Manarat* 4 (1933) 591–8, especially 593–4. On Pio's efforts to have Carvajal's benefices restored to him, see Minnich 'Healing of the Pisan Schism' (cited in n8 above) 109–10. On Zúñiga residing with Carvajal, see Allen Ep 1330:42–7 / CWE 1330:48–51; on Zúñiga's public readings of his anti-Erasmian work in 1521, see the Introduction by Henk Jan de Jonge in ASD IX-2 23. On Carvajal's role in commissioning Sepúlveda to write the history of Cardinal Gil de Albornoz, see Losada *Sepúlveda* 319.

Erasmus' attacks on him. Two works that Pio read carefully and in their entirety were Erasmus' *Exomologesis* (1524) on confession and his *Diatribe* (1524) on free will. Of his other writings Pio read little prior to 1529. The opinion he formed was that Erasmus was the greatest humanist of his age, but that he was not a skilful philosopher or theologian. He also found the similarities between Erasmus' views and those of Luther to be so striking that he repeated Zúñiga's words that 'either Erasmus lutherizes or Luther erasmusizes.'[30]

In the period around 1520 Zúñiga was not the only one in Rome to hold hostile views on Erasmus, close to those Pio was forming. On the basis of his reading of the *Adages*, the *Praise of Folly*, and probably also the *Novum instrumentum*, the Roman humanist Battista Casali was simultaneously coming to a somewhat similar view of Erasmus as a careless scholar of Greek, a pseudo-Christian, and a heretic. In 1518–19 he penned an *Invective* which he never sent to Erasmus or published. To what extent, if at all, Casali influenced Alberto Pio is unclear – the Pio whom he lists among the Roman literary figures he knows has been identified as Giovanni Battista Pio, a professor of rhetoric at the University of Rome. Girolamo Aleandro, a noted humanist and secretary to the vice-chancellor of the church, whose career in the curia Pio had actively promoted in 1517–18, expressed in his 1521 nunciature reports from Germany to Giulio dei Medici positions remarkably similar to those Pio would express in writing five years later. In

* * * * *

30 On Pio's first contact with Erasmus in Venice, see Pio *Responsio paraenetica* 2ᵛB. On his reading of Erasmus' writings and his claim to have read only a few of Zúñiga's criticisms, see Pio *XXIII libri* 68ᵛ–69ʳF; on the similarities between Pio's and Zúñiga's views of Erasmus, see especially Henk Jan de Jonge 'Four Unpublished Letters on Erasmus from J.L. Stunica to Pope Leo x (1520)' in *Colloque Erasmien de Liège* Bibliothèque de la Faculté de Philosophie et Lettres de l'Université de Liège, fascicle 247 (Paris 1987) 147–160, especially 149 and 153; Zúñiga *Libellus trium illorum voluminum praecursor, quibus Erasmicas impietates ac blasphemias redarguit* (Rome: Antonius Bladus 1522) Gvʳ (bon mot), cited in Richard H. Graham 'Erasmus and Stunica: A Chapter in the History of New Testament Scholarship' *Erasmus of Rotterdam Society Yearbook* 10 (1990) 9–60, especially 15 n36. For Pio's use of Zúñiga's bon mot, see Pio *XXIII libri* 5ᵛ; and for Erasmus' response to such usage, see his *Supputatio* LB IX 519F–520A; *Responsio ad notulas Bedaicas* LB IX 708A; and *Responsio* 21 below. Even Luther in his 3 October 1519 letter to Johannes Staupitz used the term 'Erasmizant' to describe the sense and style of two Bohemian priests who had written to him; see WA *Briefwechsel* (1930) I 514. Clarence H. Miller has suggested that the edition of *Moriae encomium* that Pio used in composing his *Responsio paraenetica* in Rome was an Italian edition based on the Strassburg edition of 1512 that lacked the later additions and revisions (eg the edition of Giovanni Tacuino de Tridino published in Venice in April 1515 and the edition of Aldus and Andrea d'Asola published in Venice in August 1515, but not the Florence edition of 1518, which included additions from the Froben editions of 1516 and 1517); see ASD IV-3 27–8 nn 79 and 81, 43–4, 47.

these reports Aleandro claims that he has no personal animosity towards Erasmus and has not engaged in any earlier controversy with him. Rather he has always praised him and loved him sincerely because of his learning. But in Erasmus' writings he has found many things that are more contrary to the faith and more venomous than Luther's statements. Indeed, Erasmus is the 'great foundation of this heresy' and 'the tinder [fomes] of all evils.' Devotion to the Catholic faith and the church has impelled Aleandro to oppose Erasmus. However, those who denounce Erasmus to high church officials in Rome work under a disadvantage, he concludes, for whatever is said there is reported back to him. That Aleandro was the instigator behind Pio's later attacks on Erasmus is doubtful. What seems more likely is that, from the earliest years of the Reformation controversy, both men shared the view that Erasmus' writings were heterodox and pernicious.[31]

Erasmus received reports of Pio's hostile attitudes towards him. The first informant was none other than Aleandro, who was both Erasmus' friend from their days together at Manuzio's press in Venice and a protégé of Pio, from whom Aleandro had successfully sought assistance to advance his career at Rome. In his *Responsio* of 1529 Erasmus claimed that seven years earlier, presumably during their accidental encounter at the Wild Man Inn of Louvain in September 1521,[32] Aleandro (identified as the Bullbearer)

* * * * *

31 On Casali, see John Monfasani 'Erasmus, the Roman Academy, and Ciceronianism: Battista Casali's Invective' *Erasmus of Rotterdam Society Yearbook* 17 (1997) 19–54, especially 22, 29, 44–8, 52 n41. See the letters of Aleandro to Giulio dei Medici from Worms, 12 and 28 February and 1 March 1521, in Balan 55, 80–1, 101–2. In a letter of 19 January 1521 (Balan 40) Aleandro complains that the contents of his letters to the vice-chancellor are communicated to Erasmus' friends, whom he does not identify. The likely intermediaries of this information were Lorenzo Pucci and his secretary Paolo Bombace, both devoted friends of Erasmus. Pucci was an intimate adviser of the pope and one of the three cardinals (Pucci, Dovizi, and Giulio dei Medici) who lived in the papal palace. Bombace reported to Erasmus on his critics in Rome (eg Allen Ep 1213:30–5 / CWE 1213:33–41). On Pucci and Bombace, see their entries in CEBR I 164 and III 123. Another important informant on the activities of Erasmus' opponents in Rome from 1521 to 1525 was Jakob Ziegler; see CEBR III 474–6. On Erasmus' assertion that Aleandro was the instigator of Pio's attacks, see Heesakkers 'Erasmus' Suspicions.' On the probability that Aleandro was responsible for turning Theodoricus Hezius against Erasmus, see Henry de Vocht, *Texts and Studies about Louvain Humanists in the First Half of the XVIth Century: Erasmus – Vives – Dorpius – Clenardus – Goes – Moringus* Monumenta Humanistica Lovaniensia (Louvain 1934) 513–14. On Erasmus' early hostility towards Aleandro, see Klawiter 87–9 nos 119–24. For evidence that Aleandro was indeed deceitful in his conversation with Erasmus, see Aleandro's confession that he '*sempre dissimulai dextramente et trovai alcune bugiette officiose*' in his report of 1 March 1521 to Giulio dei Medici (Balan 101).

32 While Aleandro as papal nuncio in the Low Countries and Rhineland was attempting to secure implementation of the bull *Exsurge Domino* against Luther and his supporters,

let slip in the course of a lengthy conversation that there was in Rome a certain imperial ambassador who was hostilely disposed towards Erasmus. The name of this ambassador was not mentioned and only later did Erasmus come to the conclusion that this critic was Pio.[33] After a visit to Rome in 1522 Erasmus' correspondent Haio Hermann reported that there was a man there who preferred Aleandro to Erasmus in all things, apparently a reference to the close friendship between Pio and Aleandro. The two are later identified as the leaders of a pagan sodality of scholars who grumbled about Erasmus' failure to write against the Lutherans. In the summer of 1524 Erasmus noted that among his Italian critics who complained about his arrogance and shoddy scholarship was an old man who preferred Aleandro to Erasmus in everything – perhaps a reference to Pio, who ironically was probably eight years younger than Erasmus. By the spring of 1525 Erasmus had received reports that Pio was denouncing him at all the

* * * * *

he seems to have met Erasmus on three different occasions: in Cologne on 8 November 1520, when they discussed rumours of Erasmus' support for Luther (see *Spongia* ASD IX-1 150–1, and the summary of their conversation provided in Aleandro's report to Giulio dei Medici from Worms on 1 March 1521 in Balan 101–2); in Brussels, apparently at the beginning of July 1521, when Aleandro confronted Erasmus for spreading stories of Aleandro's Jewish origins (Balan 101–2); and in Louvain during September 1521, when Aleandro expressed his disapproval of those who openly criticized Erasmus, such as Zúñiga (Heesakkers 'Erasmus' Suspicions' 378–9). This last encounter would seem the most likely occasion for Aleandro to have mentioned in passing the criticism of the imperial ambassador in Rome. Because Louvain is about fifteen miles east of Brussels, it is possible that what he described as the Brussels meeting really took place at the Wild Man's Inn in Louvain. Allen x 53 suggests four encounters (Cologne, Bruges, Brussels, and Louvain). On Aleandro's asking for Pio's assistance to advance his career in 1517–18, see *Lettres familières de Aléandre* ed Paquier (cited in n14 above) 30–7 and the entry from Aleandro's journal, written in Greek: '1517, Dec.2 – At an hour a little before the first hour of the evening, I kissed the hand of cardinal Medici in the presence of the count of Carpi' in 'Journal autobiographique du cardinal Jérôme Aléandre (1480–1530). Notices des manuscrits de Paris et Udine' ed H. Omont, in *Notices et extraits des manuscrits de la Bibliothèque nationale et autres bibliothèques* XXXV-1 (Paris 1896) 1–116, especially 17.

33 Erasmus *Responsio* 16 below. For Aleandro as the Bullbearer, see *Responsio* n206 and Allen Ep 2565:13–5. Pio was the ambassador of Emperor Maximilian I (d 12 January 1519). His commission was not renewed by Maximilian's successor Charles V, who was represented in Rome as king of Spain by Pedro Urrea (d 1518). At the time of his election as emperor Charles was represented by Carroz, and thereafter the imperial ambassador was Don Juan Manuel, lord of Belmonte (near Cuenca), a skilful diplomat and an enthusiastic supporter of Erasmus, who held the post from 11 April 1520 to 13 October 1521. By October 1522 the emperor's representative was Luis de Cordova, duke of Sessa. See Guaitoli 'Sulla vita' 227; Pastor *History of the Popes* (cited in n17 above) VIII 8–9, 285, 288, IX 4, 164–5; Marcel Bataillon *Erasme et l'Espagne: Recherches sur l'histoire spirituelle du XVIe siècle* (Paris 1937) 257–8; and the entry on Juan Manuel in CEBR II 247–8.

social gatherings and dinner parties of Rome as a writer who was neither a philosopher nor theologian, who lacked solid doctrine, and who refused to come to a defence of the church beset by Lutheran attacks.[34]

Erasmus sought advice on how to deal with Pio. Complaining about the prince's reported hostility, he wrote on 13 May 1525 to his old friend Celio Calcagnini of Ferrara, who was devoted to the service of the d'Este family, the enemies of Pio. Calcagnini's reply must have come as a surprise. The Ferrarese humanist stated that he knew Pio from the time they had studied together under Pomponazzi and that there was no one more humane or more modest than this prince who was not known to disparage the well deserving. He advised Erasmus not to put credence in the reports of detractors who, even when they relay something true, proceed to embellish and exaggerate things and are wont in their jealousy to rail against the reputation of good men.[35] Calcagnini did not offer to intervene with Pio on Erasmus' behalf.

In a carefully written letter of 10 October 1525 Erasmus urged Pio to desist from attacking him.

TO HIS ILLUSTRIOUS HIGHNESS ALBERTO PIO, PRINCE OF CARPI, CORDIAL GREETINGS

Men of high birth should show a generous heart, and those who have been blessed by fortune have no cause to envy anyone. Friends of long standing who know you well are full of praise for the exceptional kindliness of your nature, which extends even to those who have done little to deserve your support. So I cannot understand what can have entered your head to make you turn against Erasmus. I hope you will pardon the frankness of my remarks. Possibly there is some confusion over names: I know how dangerous it is to trust rumours of this kind. So, if anything like that has happened, please consider this letter as not meant for you.

We hear the same story from many visitors from your part of the world, that there is a certain prince of Carpi at Rome, a good scholar and a man of influence among the cardinalate, who takes every opportunity to denounce Erasmus openly as no philosopher and no theologian, and as someone who lacks

* * * * *

34 Allen Ep 1479:130–9 / CWE 1479:146–56, Allen Ep 1482:21–2, 30, 46–7, 56–9 / CWE 1482: 25–6, 35, 52–3, 63–7, Allen Ep 1576:34–48 / CWE 1576:39–58, Allen Ep 1634:39–42 / CWE 1634:44–7; and later accounts Allen Epp 1717:34–6, 1744:129–32, 1804:249–56, and 1987: 1–6

35 Allen Ep 1587:229–48 / CWE 1587:243–63; and the entry on Celio Calcagnini in CEBR I 242–3

any solid learning. I get the same message from many of my correspondents, whose letters all point in the same direction.

I am not troubled, and never have been, when someone says things about me which I frankly admit, both in conversation and in print, to be true; on the other hand nobody upsets me so much as those who heap false and invidious praises on my head. If I ever put myself forward as a great theologian or philosopher, I would deserve to have my lion's skin stripped from me. But I make no great claims for myself and promise only what I can deliver, without setting myself up above others or undermining anyone else's reputation; why then, I wonder, should someone of good family and uncommon learning, that is, someone so blessed by fortune that he might feel contempt, rather than hate, for these apelike nonentities from the world of learning, stoop to belittling Erasmus; for I have always worshipped and venerated men like you, and, far from envying my equals, I delight in praising my inferiors in every way I can. In the subjects which I have tackled, I think I have acquitted myself tolerably well, especially if you bear in mind that I am a barbarian writing for barbarians. I have not so far dealt with any subject which requires a deep knowledge of philosophy, except the problem of free will. This I took up reluctantly and in response to many requests. I treated it as simply as I could, and if my work was deficient in learning, I do not think it lacked respect for the faith. So what is to be gained, I ask you, if a great man like yourself persuades us that Erasmus is a mere beginner in these noble fields of learning? There are many people in this category who nevertheless claim the rank of scholar. You would have been better employed unmasking those (and they are to be found everywhere in our schools) who teach young people nothing and make them believe they know everything.

When someone makes a personal judgment about my abilities, my writings, or my style, I accept the criticism without protest. But the story which this prince likes to repeat every time the cardinals give a dinner-party or scholars have a meeting – this I find more disturbing, for he tells them that all our present troubles began with Erasmus. This story is entirely without foundation, but could anyone have thought of a more damning indictment if I had poisoned both his parents? When the opening scene of the Lutheran tragedy began to unfold, to the applause of nearly the whole world, I was the first to urge my friends not to get involved, for I could see that it would end in bloodshed. And since certain people were putting it about, presumably with the idea of attracting others to their cause, that I was sympathetic to Luther, I published several books in which I made it clear that I never had, and never would have, any truck with Luther. I advised Luther himself to be careful how he treated the gospel message, so that no one could accuse him of yielding to ambition or spite, and to see that the whole thing did not end in civil strife. I even used threats to discourage Froben from publishing anything of Luther's. He did not publish his work, though that meant a considerable financial loss, for he preferred to take my advice rather than serve his own interests. In Germany and Switzerland all lovers of the humanities were enthusiastic supporters of Luther at the beginning. With a few exceptions, I turned all of them from my devoted friends into the bitterest of foes. On the one side I had the theologians, who, because of their hatred of the humanities, were doing everything possible to push me into a sect which they themselves believed should be condemned out of hand; on the other side I had the Lutherans, who were working in the same direction through wheedling, trickery, threats, and abuse, though their ultimate aim was different from that of the theologians. Yet in spite of this no one has yet been able to move me a finger's breadth from membership in the church of Rome. You would not be so totally unsympathetic to my attitude if you understood what so many of the regions of this country feel and what the princes are planning to do, or if you realized what trouble I could have caused, had I wished to put myself at the head of this movement. But I chose instead to expose myself, naked and unarmed, to attack from both sides rather than lift a finger in support of a sect which the Roman church does not recognize.

But 'Luther drew his inspiration from my books.' Luther himself stoutly denies it. In what he says about me and in his writings he takes the same position as the prince of Carpi, namely that I know no theology, apparently because I agree at no point with his own teaching. But let us suppose that he did derive something from my writings. Should I be held accountable for that, any more than Paul and Augustine, whose works he is fond of quoting in support of his point of view? Certainly when I was writing those works of mine, I had no idea that such a commotion would arise.

Some people say: 'Why did you not oppose this evil the moment it first appeared?' Because, like many others, I thought Luther was a good man, sent from God to reform the evil ways of men – though even then some of his ideas offended me and I spoke to him about them. Moreover, the world was full of universities, and I could see that Luther's fantasies were gaining a surprising amount of support, not only among ordinary people, but among princes, bishops, and even a number of the cardinals; so would it not have been sheer effrontery for an ordinary citizen like myself to fly in the face of such general support?

If I may speak freely, I shall tell you what was the original source of this disaster, at least as I understand it. It was the blatantly godless lives of some of the clergy, the arrogance of certain theologians, and the tyrannical behaviour of some monks – these were the things which brought such a storm of troubles upon us. When I say this, I should not like my strictures on the wicked to reflect upon the reputation of the good; nor should my general remarks be construed as an insult to any particular order. The first battles were against the study of languages and of Latin literature. I supported such studies in so far as their inclusion might further and support traditional learning, not, as the saying goes, 'to push the old professors off the bridge.' I thought they might serve Christ's glory; I did not want to bring the old paganism back to the modern world. Then, when the outcome of the battle was still in doubt, some monks turned the issue into a question of faith. This was a convenient target, which suited their purpose. I am sure you have heard of the sorry business concerning Reuchlin. From that time on, the resentment which scholars felt against the monks was more bitter. Soon, when the war between the devotees and enemies of the Muses was still raging at its fiercest, Luther came on the scene. Immediately the monks began to connect the supporters of the humanities with the Lutheran affair, hoping in this way to destroy both at the same time. So between the obstinacy of the one faction, with its steady decline from bad to worse, and the prejudices and untimely clamour of some of those on the other side, things have gradually sunk to their present sorry state.

That is the whole story, just as it is, without any embroidery. If the rumours I have heard are true, I must ask you to revise your view of the case and stop saying things like this which are dangerous to me. If on the other hand the rumours are false, please pardon my intrusion upon your Excellency's time with this mournful litany. I wish you every success and happiness.

Basel, 10 October 1525

Erasmus of Rotterdam in haste, with his own hand[36]

PIO'S RESPONSIO PARAENETICA (1526/9)

Pio did not respond immediately to Erasmus' letter. He received it in Rome on 14 November 1525, at a time when, as the ambassador of Francis I to Clement VII, he was desperately trying to win papal support for the French cause, weakened as it was by the recent Spanish coup in Milan and by the continued imprisonment of Francis I in Madrid after his capture at the battle of Pavia in February. As mentioned earlier, Pio was one of the principal secret negotiators of a new anti-imperial alliance that was eventually formalized in the Holy League of Cognac of 22 May 1526 and that included among its major adherents the vacillating Clement VII, who became an official

* * * * *

36 Allen Ep 1634:1–111 / CWE 1634:1–126 and nn

signatory on 4 June 1526. Helping to negotiate this alliance was Pio's principal preoccupation, and he claimed that his work on the league kept him from responding promptly. Also preventing him from devoting attention to Erasmus' letter was his poor health – Pio's attacks of gout were so severe that he could be confined to bed for weeks on end. Another factor contributing to the delay was Pio's concern that he lacked the literary skills needed to respond properly to so eloquent a humanist. Fearing that his silence could be interpreted as prideful disdain, Pio felt obligated to overcome his reluctance and devote what time and energy he could spare to writing a response. The surviving draft of his *Responsio paraenetica* gives ample evidence of the care that Pio put into composing an adequate reply. The folios are full of cancellations, substitutions of more elegant words and phrases, further elaborations of ideas and arguments, and even insertions of new sections. But there is no evidence that Pio was here recycling material he had earlier written against Luther, as Erasmus claimed. The only scholar to assist Pio in its composition was, according to his own testimony, the Spanish philosopher resident in Pio's household, Juan Ginés de Sepúlveda (1491–1572), who criticized various versions of the *Responsio paraenetica* just as Pio did for Sepúlveda's *De fato et libero arbitrio dogma Lutheri confutatur* that was published in Rome by Antonio Blado in June 1526. The only other person allowed to see the *Responsio paraenetica* during its composition was, also according to his own and Sepúlveda's testimony, Pio's youthful amanuensis Francesco Florido de Sabina. Work on Pio's response to Erasmus was eventually completed on 15 May 1526, one week before the conclusion of the League of Cognac.[37]

Having explained the reasons that led to his delayed response, Pio devoted the first third of his lengthy epistle (that in all filled ninety-nine folios in its first printing) to a detailed refutation of statements made by Erasmus in his letter of 10 October 1525. He begins by denying that he had a low opinion of Erasmus. From the first, when he saw him in Venice and heard Aldo Manuzio extol his literary talents, Pio formed a favourable opinion

* * * * *

37 Pio *Responsio paraenetica* 1^{r-v}A and 99^{v}N; on Pio's diplomatic activities, see above xxviii; Roger Mols 'Clement VII' in *Dictionnaire d'histoire et de géographie ecclésiastique* 12 (Paris 1953) cols 1175–1244, especially 1184–7. This draft of the *Responsio paraenetica* is preserved today in the Biblioteca Ambrosiana of Milan, Archivio Falcò Pio di Savoia, Prima Sezione MS 282 entry 6; and in Harvard University's Houghton Library, MS Latin 299 58^{r-v}. For examples of the addition of new sections through the insertion of new pages, see folios [47a] and [51a]. For a description of this manuscript, see Fiorina 70; and Sepúlveda *Antapologia* ci^v, Di^r–Ei^r / Solana 121–2, 127–33; and Forner 207–9.

of Erasmus. Over the years, as he became more familiar with his writings, Pio came so to esteem Erasmus that he considered him as superior to many contemporary Italian and Greek scholars for his eloquence and erudition, and as the leading humanist of the century. Since it is his custom to revere and support men of learning, Pio writes, he gladly praises Erasmus' talents in his conversations with scholars and poets and cardinals.[38]

If Pio lavishes praise on Erasmus for his skills as a humanist, he feels that his talents in the area of Aristotelianism are as yet modest. But if Erasmus devotes himself to disputation as he has to grammar, he can achieve fame also as a scholastic philosopher and theologian.[39]

Pio rejects Erasmus' claims that he was not responsible for the Lutheran revolt, but that the real blame belongs to the priests, friars, and theologians who attacked Latin letters. Pio grants that, because Erasmus never intended the tumult, he was not its cause, but some of his incautious writings became its great occasion insofar as others used them as fonts for watering Luther's poisonous weeds. Pio reports that many in Rome are persuaded that the opinions of Luther and Erasmus are so similar that either Luther erasmusizes or Erasmus lutherizes. The one member of the Roman circle of Erasmus' critics whom Pio cites approvingly by name is Diego López de Zúñiga. None of these critics, however, claims that Erasmus has deliberately incited mobs against the church.[40]

Erasmus' disclaimer that he is not nor ever has been a supporter of Luther's condemned teachings is challenged by Pio. Like many others in Rome, Pio feels that Erasmus has given grounds for the suspicion that he is a participant in, if not the author of, the religious disorders in Germany. Erasmus is eager for novelties. He improperly calls into doubt the church's defined teachings and common assumptions about the practices of the ancient church regarding the sacraments of penance, marriage, and the Eucharist, various religious practices, monastic life, the priesthood, the episcopacy, the cardinalate, and the papal office. Thus, for example, he ridicules religion in his *Praise of Folly*, undermines auricular confession in his *Exomologesis*, and tries to substitute for the inspired (if at times inelegant) words of the Scriptures his own humanistic rendering in the *Paraphrases*. When commenting on Luther, Erasmus either praises him or speaks of him

* * * * *

38 Pio *Responsio paraenetica* 3^v–5^rC
39 Pio *Responsio paraenetica* 4^r–vC
40 Pio *Responsio paraenetica* 2^r–vA, 5^r–6^vD. Pio is here echoing Zúñiga's clever depictions of the relationship between Luther and Erasmus; see n30 above.

in such a way that one can reasonably conclude that he often agrees with him, even if he disapproves of those elements in his message that can lead to sedition and tumult.[41]

By his failure to oppose Luther early on, Pio continues, Erasmus seems to be supporting him. Even if at first Erasmus saw Luther as a good man divinely sent to correct evils, why did he not take up his pen to oppose him when it became clear that Luther's message was incendiary and quickly spreading? Erasmus tries to excuse himself on the grounds that local civil and ecclesiastical authorities favour Luther, and that there are plenty of university theologians around whose responsibility it is to oppose him. The ancient Greek and Latin Fathers whom Erasmus praises, however, acted otherwise when confronted with the heretical teachings of the Arians that enjoyed the support not only of the people but of kings and emperors. At great personal danger to themselves, these Fathers openly opposed this heresy. And in our own day a number of professors of theology and even laymen such as Henry VIII of England have joined the battle line with their writings against Luther. Because of his skill as a prolific writer and his great influence among Germans, Erasmus could have been expected to contribute something significant to quelling the spread of heresy. But for fear of antagonizing Luther's supporters he has deserted his duty and kept quiet.[42]

Pio objects to Erasmus' contention that Luther derived his teachings from Paul and Augustine. While Luther does quote from the Gospels and Pauline Epistles, he uses these texts as honey to mask his poison. His teachings do not come from the Scriptures but are the resurrected heresies of men like Simon Magus, Montanus, Eunomius, Jovinian, Berengar, Wycliffe, and Hus, or they have been thought up by himself. Luther claims that his interpretation of the Scriptures is correct and that the church has been in error for the past thirteen hundred years, but, on the contrary, it is the divinely instituted church guided by the Holy Spirit and not a proud and inconsistent friar who has the authority to interpret the Scriptures. Luther also asserts that nothing should be accepted as solidly Christian unless it is expressly stated in the Scriptures. Christ himself neither left writings behind nor commanded others to write down his words. Only two of the apostles wrote a gospel and not everything Christ said was recorded with ink. The church in council determined which writings constitute the canonical Scriptures. Why not also accept its authority to

* * * * *

41 Pio *Responsio paraenetica* 6$^\text{v}$F–13$^\text{r}$I
42 Pio *Responsio paraenetica* 6$^\text{r}$E, 13$^\text{r}$K–15$^\text{r}$K

interpret their meaning and to promulgate or abrogate the laws contained in them? Erasmus' claim that Luther's teachings are the same as those of Paul and Augustine not only gives aid to Luther but perpetuates a falsehood since many have shown that Luther distorts the writings of these saints.[43]

Erasmus' suggestion that Luther was heaven-sent is, Pio claims, without a solid basis. Christ's Spirit rests on the meek and humble and is sure and constant. But Luther has shown himself to be proud and disrespectful and belligerent, and to change his own message. He has contradicted the consistent teachings of the ancient church Fathers and of later celebrated theologians and the beliefs of the whole Catholic community for the past thirteen hundred years. Without a sign or miracle to confirm his radical doctrines, he accuses the whole church of error. If by their fruits you will know them, what is one to conclude about his teachings that have led to the profanation of churches and monasteries, to the breaking of vows, to the violation of sacred rites, to the plundering of church goods, to the contempt for laws and civil authority, and to the terribly destructive peasants' wars that have recently raged in Germany? It is highly improbable that Luther was heaven-sent.[44]

From Pio's perspective Erasmus has yet to write effectively against Luther. Pio asserts that it is not enough that Erasmus has not openly joined the Lutheran camp. A loyal son of the Catholic church should defend her. While Pio seems not to know Erasmus' two-volume *Hyperaspistes* (1526–7), he acknowledges the existence of his *De libero arbitrio diatribe sive collatio* (1524), but criticizes it as too timid and gentle, as inept and lacking in doctrinal clarity. Indeed, the presentation is so confusing, he says, that some even feel that Erasmus is often agreeing with Luther's heretical views.[45]

Pio adamantly rejects Erasmus' attempt to shift the blame for the Reformation away from Luther and onto his opponents: theologians, members of religious orders, and priests. Over 40 per cent of Pio's *Responsio paraenetica* is devoted to this topic, and this section functions as a bridge between the first part that refutes each of the major assertions in Erasmus' letter and the latter part that attacks the principal premises of Luther's theology.[46] If the first part targets Erasmus, this middle section addresses at various times Erasmus, 'you Germans,' and Luther, and the final section, while attacking

* * * * *

43 Pio *Responsio paraenetica* 15[r]L–22[v]O
44 Pio *Responsio paraenetica* 22[v]P–26[v]Q
45 Pio *Responsio paraenetica* 27[r]–28[r]R
46 This middle section is Pio *Responsio paraenetica* 28[v]S–71[r]R.

Luther, is aimed at Erasmus. The effect of these various addresses is to mix together the cases of Erasmus and the Lutherans.[47]

Pio examines carefully the claim that the hostility of theologians towards humanistic learning contributed significantly to the outbreak of the Reformation. The case of Johann Reuchlin was cited by Erasmus as an example of the conflict between humanists and theologians. But Pio objects that this division into camps is not as clear as Erasmus claims. Pio, himself an ardent supporter of scholastic theology, personally intervened on Reuchlin's behalf with Leo x and found the *Letters of Obscure Men* somewhat entertaining, but he did not give them his approval. Some of the humanists who supported Reuchlin quickly lent their protection also to Luther, who thereby grew bolder and became their leader in attacking theologians. This is but one example of the generally negative effect that humanistic learning has had on Germany. Prior to the arrival of 'good letters' this northern land was known for its tranquillity, modesty, serious study of the best academic disciplines (theology, philosophy, mathematics, etc), and religious piety. But now civil and domestic conflicts rage and immorality and heresy are rampant. This is not the fault of all humanists, but only of those who failed to imbibe Italian gravity, good judgment, and wisdom at the same time they adopted humanistic learning from Italy. Their new skills in grammar and rhetoric have led some to become arrogant and bold, to show contempt for those less eloquent, and to apply their philological tools to sacred texts that had earlier been the preserve of theologians. They claim that theologians

* * * * *

47 Although the whole work is addressed to Erasmus, Pio occasionally employs the vocative case as an apostrophe – that is, he merely feigns an address to others. For examples of those whom Pio addresses in the three sections of the *Responsio paraenetica*, see section I 2^{V}B, 3^{r}B, 4^{r}C, 4^{V}C, etc (Erasmus); section II 47^{V}H (echoing Paul in Gal 3:1), 49^{r}H, 63^{V}P, ('you Germans'), 66^{r}Q (Erasmus), 59^{V}M, 67^{V}R, 68^{r}R, 68^{V}R, etc (Luther); section III 85^{V}D, 96^{V}L, etc (Erasmus). While the letter format may account for Pio's use of the second person in addressing Erasmus, his similar addressing of Luther and the Germans in this work written in 1526 indicates that he was slower than his contemporary Catholic controversialists in switching to the third person in their compositions. David V.N. Bagchi dates the switch to 1522, after Luther was officially condemned as a heretic by the papacy (*Decet Romanum Pontificem* 3 January 1521), the Faculty of Theology of Paris (15 April 1521), and the German Diet (12 May 1521). Given the prohibitions of canon law against disputing with heretics, the controversialists now referred to Luther in the third person and provided arguments in support of the official condemnations. See Bagchi *Luther's Earliest Opponents: Catholic Controversialists, 1518–1525* (Minneapolis 1989), 211–14, 251–2. Hubert Jedin dates the switch to 1525 and the publication of Johannes Eck's *Enchiridion locorum communium*, which treated numerous disputed topics in a dispassionate, impersonal, and academic style; see Jedin 'Die geschichtliche Bedeutung der katholische Kontroversliteratur im Zeitalter der Glaubensspaltung' *Historisches Jahrbuch* 53 (1933) 70–97, especially 78–80.

who lack their skills in classical languages have not interpreted Scripture properly. The techniques these humanists use to explain the fables of Aesop and the plays of Aristophanes are now applied to biblical passages. Pio concedes that this approach can be helpful in elucidating Christ's parables, and that the eloquence of the humanists can also be used to move the emotions of the laity, to inspire them and stir up their piety. But in the final analysis theirs is a 'rude' type of theology, unable to resolve difficult questions. For this task one needs scholastic theology, which was divinely invented and uses the tool of philosophy, the teacher of wisdom. By following its rules for correct thinking, using its principles, and making important distinctions, the theology of the Schools helps one to uncover truths, to remove errors, to come to some understanding of the church's teachings on the Trinity and other mysteries, and to make sense of abstruse scriptural passages. Everyone is not capable of doing scholastic theology but that is not a reason for disparaging it. Christ's message was not understood even by his disciples. Some of Paul's texts are too difficult for simple people. The help of professional theologians is needed to comprehend the Bible properly. While supercilious and immoral theologians are deserving of censure, some German humanists oppose all theologians, even the good ones, and throw their support behind Luther and Zwingli in an effort to appear consistent in their opposition to priests, religious, and the pope. But Luther does not want only to reform abuses but to overthrow the Catholic church and its religion. Some humanists understand his positions and agree with him. Erasmus' former friends in poking fun at the church are now Luther's allies and protectors. Blame for their going over to the Lutheran cause does not lie with scholastic theologians.[48]

Pio also rejects Erasmus' assertion that the intolerable tyranny of certain evil members of religious orders, notably those who turned the Reuchlin controversy into a question of orthodoxy, was in part responsible for the Lutheran tumult. This accusation, according to Pio, is motivated by Erasmus' unfounded contempt for monasticism. Pio cites scriptural, patristic, and even non-Christian classical sources to demonstrate that divine teachings and human reason justify the practices of celibacy, poverty, obedience, a common life, withdrawal from the world, and ascetical disciplines such as fasts and abstinences that are often part of the monastic way of life. Monks do not constitute an anti-Christian sect as claimed by this new bilingual school of theologians, trained in Latin and Greek. Christ did not condemn

* * * * *

48 Pio *Responsio paraenetica* 28^{v}s–37^{r}c, 67^{v}R, 68^{v}R. For an overview of the conflict between humanists and theologians, see Rummel *Humanist-Scholastic Debate*.

a separate, more demanding style of life. He praised the ascetical John the Baptist and chose some of his apostles from among the strict Pharisees. Good monks do not deserve to be attacked in blanket criticisms of monasticism. In concluding this section Pio points out to Erasmus, the exclaustrated Augustinian canon, that no one is a more bitter enemy of monasticism than he who has defected from it, and that those who cannot understand or attain its ideals and are found unworthy of its heavenly life are known to turn around and condemn it as an insane way of life so as to justify their own failures.[49]

While Erasmus has asserted that the other factors contributing to the beginning and progress of the Lutheran tragedy are the openly impious lives of certain priests, plus their hatreds and unrestrained complaints against the German friar, Pio launches into a defence of the priesthood itself against Luther's assertions that priests are no different from laity and have no special sacramental powers conferred on them by ordination. Pio claims that what lies behind the hostility towards priests is often jealousy of their wealth and power; he finds the opposition of German humanists to priests otherwise difficult to explain. Even the pagans honoured their priests. These humanists, however, will exempt from criticism all members of lay society and attack only priests. Complaints about the wealth of the clergy are unfounded. Secular clergy who do not have vows of poverty are allowed to have possessions of their own. For their service to the Christian people they deserve recompense. Material goods also help to maintain the majesty and propriety of the priesthood. To carry out its various missions the church needs possessions, which priests administer.[50]

Pio next turns to a defence of the church's high priest, the Roman pontiff. He claims that Erasmus has written that the prologue to the Lutheran tragedy is also to be found in the numerous priests and bishops who do not wish to be subject to the pope and pay him tribute.[51] But this statement is not contained in Erasmus' letter to him. From this point on, Pio is no longer responding directly to Erasmus' letter, but he addresses now Luther, now the German humanists, and most often Erasmus. Often, as in the issue of papal supremacy, Pio provides a defence of traditional Catholic teach-

* * * * *

49 Allen Ep 1634:86–98 / CWE 1634:97–112; Pio *Responsio paraenetica* 37ᵛD–42ᵛE. While Erasmus did live outside of an Augustinian monastery, he had a papal dispensation to do so and never left the order. See Richard L. De Molen, *The Spirituality of Erasmus of Rotterdam* Bibliotheca Humanistica at Reformatorica XL (Nieuwkoop 1987) 191–7.

50 Allen Ep 1634:86–7, 101–4 / CWE 1634:97–8, 115–18; Pio *Responsio paraenetica* 42ᵛF–52ᵛI

51 Pio *Responsio paraenetica* 66ʳQ

ings based on Scripture, the church Fathers, the beliefs and practices of the church over time, and common sense.[52]

Having disposed of Erasmus' claim that the Reformation was occasioned by the misdeeds of theologians, friars, and priests, Pio proposes three explanations of his own for why the German humanists oppose the papacy and support Luther. The first is jealousy of the power, wealth, and privileges of the popes. This is unjustified, for the Roman pontiffs have always been the principal supporters of men of learning. The popes employ with handsome stipends more men of letters in their bureaucracy than any other ruler of Christendom. Their universities at Rome, Bologna, and Perugia have on their faculties many famous and well-paid humanists. The Roman court is open to all men of talent. Cardinals are chosen from all the nations of Christendom. Even the papacy itself can be held by a German, as evidenced by Hadrian of Utrecht's election four years earlier. The Germans, however, never elect as emperor an Italian. Such jealousy is groundless.[53] The second explanation is German opposition to Roman taxes. But these taxes are reasonable and justified. Annates are paid by bishops but once in a lifetime, and Rome exacts nothing from the laity. Just as the laity pay their taxes to support civil government, so too do the clergy pay to support the church's hierarchy. Taxes paid to Rome sustain not only the pope but also the cardinals and central bureaucracy of the church.[54] Pio's third explanation is that many Germans have rallied around Luther's teachings on 'Christian liberty' and the equality of all believers because they find the church's laws too burdensome. Among the onerous laws they wish to abolish are those on fasting, tithes, annual confession, clerical celibacy, and monasticism. Luther's teachings on justification by faith lead many to conclude that they no longer need to perspire over works. These teachings are based on a flawed theological method. The peasants have been stirred up by Luther's message of Christian liberty and equality to rebel against the nobles, only to be slaughtered by them, also at Luther's urging. This most perfidious fellow seems to take delight in the death of over 120,000 victims so far. And like the heretics of old, Luther has founded a sect whose followers bear his name and he has thus achieved a perverse fame and glory as a leading heretic. His teachings on Christian liberty and equality are contrary to those of the church Fathers and councils

* * * * *

52 For his defence of the papacy, see Pio *Responsio paraenetica* 52[v]K–63[v]O.
53 Pio *Responsio paraenetica* 49[v]H, 52[v]K–53[r]K, 63[v]P–65[r]P
54 Pio *Responsio paraenetica* 65[r]Q–66[r]Q

and are not inspired by the Holy Spirit. Such are the teachings of a false prophet.[55]

In the last major section of his *Responsio*, some twenty-five folios that are addressed to Erasmus, Pio refutes Luther's teachings on a number of topics. Pio's primary purpose, however, is to defend traditional Catholic beliefs and practices. He shows how these are in conformity with Scripture and human reason and have been formulated and developed over the centuries with the assistance of the Holy Spirit. Among the topics Pio treats are church laws and traditions, sacred ceremonies, the divine office, religious music and images, the cult of saints, the canon of the Mass with its references to sacrifice, vows of all kinds and notably of celibacy, good works, the role of grace, and the insufficiency of faith by itself in an adult. At the end of this section Pio praises Erasmus for having argued so prudently and learnedly against Luther's denial of free will – thereby contradicting his earlier criticism that Erasmus had written so confusingly and ineptly on the topic that he appeared to be in agreement with Luther's views.[56]

Pio concludes his *Responsio paraenetica* with exhortations that others oppose Luther. He praises the universities, the pope and cardinals, and Charles v at the Diet of Worms for having publicly condemned Lutheran teachings. He urges the emperor and his vicar (his brother Archduke Ferdinand of Austria) to bring an end to the Lutheran heresy and singles out by name or title other German officials who have assisted and will assist them. And Pio renews his request that Erasmus write clearly and forcefully against Luther as the church Fathers did against the heretics of their day. Should Erasmus fail to act, his reputation will be forever stained. In concluding, Pio apologizes if he has said anything that might offend Erasmus, whose learning and virtue cause him to love him, and expresses the hope that this epistle that has grown into a volume may sanction a perpetual treaty of friendship between them.[57]

Pio did what he could to preserve this treaty of friendship by keeping his criticisms of Erasmus from being made public. Although his colleague Sepúlveda quickly published his work against Luther, Pio took careful measures to prevent the publication of his lengthy letter to Erasmus. Besides the rough draft, he kept only one good copy of it, and he made sure that no one could make their own copy of it or publish it without his knowledge.[58]

* * * * *

55 Pio *Responsio paraenetica* 66^{V}R–71^{r}R
56 Pio *Responsio paraenetica* 71^{r}S–96^{r}L
57 Pio *Responsio paraenetica* 96^{V}M–99^{V}N
58 Pio *XXIII libri* 48^{r}I, 69^V–70^{r}I

ERASMUS' INITIAL REACTIONS AND THE PUBLICATION OF PIO'S RESPONSIO PARAENETICA

Erasmus had mixed feelings about Pio's *Responsio paraenetica*. He received a copy of it by the beginning of September 1526. He claimed that he was initially much taken by Pio's courtesy in sending him a response that was so ample, thoughtful, humane, and flattering. He even thought of publishing it himself, but refrained because of what he interpreted to be hostile statements in it that portrayed him as involved in the Luther affair and that criticized his various writings, especially the *Moriae encomium* and *Paraphrases*. Erasmus began work on a reply and had completed 'a couple of pages' when he learned of the Sack of Rome (May 1527), after which he was no longer sure of Pio's whereabouts. How much progress Erasmus made on the draft of his reply is unclear. By October 1528 Erasmus proposed to send Pio a copy of his response, thus implying that it was at least in a close to presentable format.[59] He seems also to have made an effort to reciprocate Pio's expressions of friendship by stating in print: 'In my opinion Alberto Pio, Prince of Carpi, comes closer to Cicero's style of expression than Aleandro does. As yet he hasn't published anything.' This passing statement in the first edition of his *Ciceronianus* (March 1528), however, had the opposite effect by putting into motion a series of events that turned Erasmus and Pio's efforts at maintaining friendship into open animosity.[60]

Erasmus' brief comment in the *Ciceronianus* put pressure on Pio to publish his *Responsio paraenetica*. Having been alerted by Erasmus to the existence of a work by Pio written in a distinguished Ciceronian style, many scholars tried to purchase a copy from their local book vendors, only to be told that, as far as they knew, it was not in print. Pio himself now became the target of their importuning. The one good copy of his letter that he had carefully corrected was lost, together with his library and home furnishings during the Sack of Rome.[61] Due to the diligence of a member of his

* * * * *

59 Allen Epp 1744:130, 1804:249–53, 2066:60–5, 2080:1–2, 24–5; Erasmus *Apologia* 153 below
60 Erasmus *Ciceronianus* ASD I-2 670; CWE XXVII 420, 585–6 nn662–4
61 Pio *XXIII libri* 70ʳ1. Hermann von der Hardt claimed to have found during a visit to Rome a copy of the original unaltered version of the letter that Pio sent to Erasmus in 1526, which in 1577 was in the possession of Cardinal Guglielmo Sirleto. He printed this in his *Historia literaria Reformationis in honorem jubilaei anno MDCCXVII* (Frankfurt/Leipzig: Officina Rengeriana 1717) 112 (claim of discovery of the original version of the manuscript), 114–63 (edition of Pio's letter). A close comparison of this 1717 edition with the 1529 edition of Bade does not sustain von der Hardt's contention. Erasmus pointed out that scriptural proof-texts defending auricular confession that were not in the version sent to him were added to the Bade edition (*Responsio* 74 below). While the Milan manuscript

household, the original rough draft of the letter was rescued, together with his account books and diplomatic papers. When Pio moved to Paris, he took the rough draft with him. Because it was so flawed, disorganized, and confusing, however, with numerous editorial markings, corrections, deletions, and insertions, Pio felt justified in telling those who wanted to read his *Responsio paraenetica* that he no longer had a copy of it. They refused to accept this and continued all the more to pester him. In the end, worn down by their persistence, he admitted that he still had the rough draft. He refused, however, to let them read it in its confused condition. At first his illness prevented him from revising it, but eventually with great labour he took the task in hand and even improved on the text by inserting two or three scriptural proof-texts. When scholars offered to help him, Pio politely declined their assistance. In principle he was opposed to publishing private communications, but Erasmus' statements in the *Ciceronianus* had made the existence of the letter public, and Pio also wondered if Erasmus might not already have published it in Germany. Pio turned over the revised copy of the letter to his friend Josse Bade, the Parisian printer, who prefaced the work with both Bade's letter to the reader of 5 January 1529 and Erasmus' letter to Pio of 10 October 1525. The whole work was published at Bade's press by 7 January 1529 and presented to Pio in its completed form.[62]

* * * * *

of the rough draft (folio 7^r) does not contain these scriptural passages, both the Bade edition (Pio *Responsio paraenetica* 9$^{r–v}$H) and the von der Hardt edition (118) do. Similarly Erasmus noted the Greek word ἐπαμφοτερίζειν 'to play on both sides' in the text sent to him (*Responsio* 187 below), while both the Bade edition (Pio *Responsio paraenetica* 6^{v}E) and the von der Hardt (117) have the word ἀμφισβητεῖν 'to stand aloof and waver'; see *Responsio* nn 151 and 157 below. Erasmus also complained about the addition of the proof-text from Deut 4:2 (Moses' prohibition against adding anything to what he said) (eg *Responsio* 76, 140). But while the Milan manuscript (folio 8^r) does not contain this reference, neither does the van der Hardt edition (119). His version also has other minor differences from the Bade edition; for example, compare the wording of the concluding section in Bade (Pio *Responsio paraenetica* 99$^{r–v}$N) and von der Hardt (163). It would thus seem that the von der Hardt edition is not the version sent to Erasmus but a corruption of the Bade edition. On Sirleto's private library and its fate, see Georg Denzler *Kardinal Guglielmo Sirleto (1514–1585), Leben und Werk: ein Beitrag zur nachtridentinischen Reform* Münchener Theologische Studien I, Historische Abteilung 17. Band (Munich 1964) 70–2; and Giovanni Mercati *Codici Latini Pico Grimani Pio e di altra biblioteca ignota del secolo XVI esistenti nell' Ottoboniana e i codici Greci Pio di Modena con una digressione per la storia dei codici di S. Pietro in Vaticano* Studi e Testi 75 (Città del Vaticano 1938) especially 1–23, 44–52, 79–91, 106–42, etc.

62 Pio *XXIII libri* 70^{r}I–70^{v}K and Pio *Responsio paraenetica* Aii^r–Aiv^v. Pio had reason to wonder whether Erasmus might have already published his *Responsio* on his own, for he had reprinted Sancho Carranza's *Opusculum* when responding to it in 1522; see Allen Ep 1277:22n / CWE 1277:27 n8.

Frontispiece of an illuminated manuscript of Pio's *Response*, a French translation of his *Responsio paraenetica* dedicated to Baron Guillaume de Montmorency, c 1529–30. Chantilly, Musée Condé, ms 187, folio 1[r]

By permission of the Musée Condé / Art Resource, New York

At some point during the next two years Pio also issued for circulation in manuscript form a French translation of his *Responsio paraenetica*. A preface discusses the translation. It says that not all but the greater part of the work has been translated. Indeed, a comparison of the original Latin of the *Responsio paraenetica* with the translation reveals that the French version is more succinct, dropping repetitions, elaborations, expansions of honorific titles, whole clauses, and even sections of material. On occasion it even adds new arguments. Because these revisions were either made by Pio himself or approved by him, this French translation should be considered an authorized text. The preface goes on to state that, in order to make the *Responsio paraenetica* easier to follow, it has been divided up into chapters, each with its own title and a brief description of its contents. An examination of the manuscript shows that it has been divided into fifty-three chapters (often coinciding with the lettered subdivisions in the Bade 1529 edition). In addition to a title and short summary, each chapter is furnished with the Latin *incipit* of Pio's original text. The preface also affirms that the purpose of the translation is to ensure that those who do not understand Latin can see, judge, and come to know the great error, weakness, and deviance of the immoral Luther, the enemy of the Catholic faith and of the holy decrees of the church. At least one of the persons for whom the translation was made is identified in the dedicatory epistle.[63]

That Pio addressed the dedicatory epistle to Guillaume de Montmorency (1452–1531) is not surprising. They were close friends, both having served Louis XII during his Italian campaigns and both being knights of Francis I's elite Ordre de Saint-Michel. They were also mutual friends of Louise de Savoie, the queen mother: Montmorency was one of her *chevaliers d'honneur*, while Pio was considered her cousin. Montmorency was a highly respected baron, serving early on as *général des finances* and a member of the royal *conseil* of Francis I and also exercising influence through his son, Anne (1492–1567), the Grand Master of France since 1526, and consid-

* * * * *

63 This manuscript is preserved in Chantilly, France, Musée Condé MS 187 (709, XIX C16) and is described below (Introduction cxxxiii–cxxxiv entry 2e). Its composition is assigned to 1529–31 because the translation is clearly based on the Bade text published in 1529 and was completed during Pio's lifetime, given the existence of his dedicatory letter. Pio died on 8 January 1531. For examples of the differences between the Latin original and the French translation, see Minnich 'Debate on Images' 383–4 n5. The French translation adds another proof-text in the section on saints as intercessors (chapter 42) by citing Luke 5:8; see Pio *Response* – Chantilly MS 164[v] and compare with Bade 82[r] B[BB]. The preface has been transcribed in Pio *Response* – Chantilly Plon 169. The division into titled chapters is not found in the Bade edition, despite claims to the contrary in the *Response* – Chantilly Plon 170.

ered the king's 'most powerful royal adviser' until his falling out with him in 1541. The queen mother and the Montmorencys were leading defenders of the Catholic church in high government circles. In order to be better informed about Lutheran heresies and how to refute them, Guillaume asked Pio to provide him with a French translation of the *Responsio paraenetica*. This request for a translation was not unusual; Guillaume Montmorency also had a translation of the history of the Roman wars by Appian (fl AD 160) made for him in 1515 by Claude de Seyssel, and his son Anne was also known to have commissioned French translations of Latin works and to have had them decorated with miniatures and illuminations. As Grand Master of France Anne was in charge of artistic works at court.[64]

With appeals to antiquity Pio in his dedicatory letter urges Guillaume to defend the Roman Catholic church. He begins by citing a passage from the eighth satire of Juvenal (c 60–c 140), *Stemmata quid faciunt?*, which attacks pride in one's ancestry that is not accompanied by imitation of the qualities which made one's forefathers famous. Some honour their ancestors with depictions of their great deeds on tapestries and stained glass, but they do not adorn their own lives with the virtues of their forefathers. Pio notes that the Montmorencys are one of the oldest and most distinguished families of France. He mentions in particular their ancestor who, according to an ancient chronicle, on hearing the preaching of Saint Denys of Paris, converted to Christianity, bringing with him into the church his family, subjects, and many others. Over the centuries the Montmorencys have ensured that the faith not only survived but flourished in France. To this end they have fought for the church and have founded and endowed abbeys, collegiate churches, and prebends. Guillaume himself has rebuilt and re-endowed the church in Montmorency. He has always loved the church and is concerned that the barque of Peter is battered by tempests and troubles. When persecuted, however, the church becomes more flourishing, triumphant, and exalted, like gold tested in the fire, as noted by the ancient Fathers Origen

* * * * *

64 On Guillaume de Montmorency, see the entry in *Nouvelle biographie générale* ed M. Le d'Hoefer (Paris 1861) xxxv col 341; on his request for the translation, see Pio *Response* – Chantilly Plon 169. On the friendship of the Montmorencys with Pio, see the letter of Rodolfo Pio to Nicola Raince, Rouen, 4 May 1535, in Baroni ed *La nunziatura in Francia* (cited in n16 above) 154; and note the use of the familiar form *tu* in Pio's dedicatory letter (Pio *Response* – Chantilly Plon 170). On Seyssel's translation, see Brigitte Bedos Rezak *Anne de Montmorency Seigneur de la Renaissance: La France au fil des siècles* (Paris 1990) 22. On Anne Montmorency, see the entry in OER III 85; and Lépold Delisle 'Traduction d'auteurs grecs et latins offertes à François I et à Anne de Montmorency par Étienne Le Blanc et Antoine Macault' *Journal des Savants* (1900) 476–92, 520–34, especially 534. On the Grand Master of France see n16 above.

and Hilary. Today the church suffers persecution at the hands of the condemned Luther and his conspirators, who seriously wound the Christian religion. Guillaume has asked Pio to have a French translation made of his Latin *Responsio* so that the baron can learn how frivolous, baseless, and destructive are the Lutheran teachings and learn how to refute and confound these heresies with the help of Sacred Scripture. Pio hopes that this task, assisted by the translation of his *Response*, will reinvigorate the baron in his senior years. The church will weather the present persecution because it is sustained by the unshakeable promise of Christ to St Peter: 'I have prayed that your faith may not fail' (Luke 22:32). May the baron live to see the church repaired and rendered more beautiful than ever. May it please him to accept the gift of this book from a man who is elderly like him. And may they both rejoice to see the Catholic faith flower and bear fruit and heresies be confounded, especially in this most noble and Christian kingdom of France. May God always preserve it in good peace, fidelity, and justice.[65]

Pio's gift was furnished with an elaborate frontispiece depicting the Montmorency coat of arms encircled by the chain worn by knights of the Ordre de Saint-Michel, framed by nine smaller panels, four containing flowers,

* * * * *

65 Pio's dedicatory letter has been transcribed in Pio *Response* – Chantilly Plon 168–9. The transcription modernizes to some extent the capitalization of letters, punctuation, division into separate words, and use of accents. The transcription is reliable, except for the reading of *diziné* (168) instead of *dirivé* (MS 2ᵛ); in the margin at the beginning of his letter Pio cites in Latin the opening words of the eighth satire of Juvenal: '*Stemmeta quid ... Aemilianos corvini*'; for a slightly different Latin text and English translation, see *Juvenal and Persius* trans G.G. Ramsay (Loeb rev 1940, repr 1969) 58–9. Pio also cited Juvenal's eighth satire when treating sacred images in his *XXIII libri* 145ʳx. The dedicatory letter makes allusions to information about Guillaume de Montmorency's life that is known from other sources. The baron devoted much of his energies and resources to rebuilding churches. He demolished the chapel at the Chantilly chateau and rebuilt it, dedicating it to St James and St Christopher and decorating it with two stained-glass windows, one depicting his wife, Anne Pot, with their three daughters and the other himself with his five sons. He also began in about the year 1515 the reconstruction of the collegial church of St Martin in Montmorency, leaving the task for his son Anne to complete. This church became the burial site of the Montmorency family, Guillaume himself being interred there following his death on 24 May 1531; see Bedos Rezak (cited in n64 above) 21–2. According to the legendary life of St Denys, the nobleman he converted when he was preaching in Paris was named Lisbius, who also suffered martyrdom when he refused to renounce his new faith. This account was incorporated into subsequent chronicles. See Charlotte Lacaze *The 'Vie de St. Denis' Manuscript (Paris, Bibliothèque Nationale, Ms fr. 2090–2092)* (New York 1979) 17–18. A copy of the chronicle *Les Grandes Chroniques de France dites la Chronique de Saint Denis depuis les origines fabuleuses jusqu'àu couronnement de Charles VI* was kept in the Condé collection; see Frédéric Vergne *A Princely Library: The Manuscript Collection of the Duke d'Aumale, Chateau de Chantilly*, trans Nina McPherson (Paris 1995) 348.

berries, and birds, and five incorporating grotesques of griffins, dolphins, birds, horns, and weapons. The dedicatory letter was also beautifully illuminated. The book was thus properly attired to travel in aristocratic circles, but whether it ever did is not clear. Erasmus and most Pio scholars ever since have been unaware of its existence.[66]

ERASMUS' RESPONSES TO PIO: LETTER TO PIO (1528) AND RESPONSIO (1529)

During the more than two-year period between receiving Pio's letter and learning of its impending publication Erasmus was formulating in his private correspondence with friends and associates the various responses he would later make to Pio. His Roman informants confirmed that while Pio was still at the papal court he had accused Erasmus of secretly favouring Luther and of providing the occasion for the Lutheran tumult. Erasmus also learned that Pio's animosity towards him was such that he forced the young German humanist Georg Sauermann to delete from his publication any praise of Erasmus. Erasmus' Paris informant, Louis Berquin, described Pio as being similar to other Italians who hoped to win glory by having found some fault in Erasmus, as being likely to react very negatively to any criticism, and as perversely misinterpreting many of Luther's teachings, especially those on faith and good works. From such reports, and from his own reading of the *Responsio paraenetica*, Erasmus concluded that Pio was jealous of him, that he had written against him at the instigation of others such as Girolamo Aleandro, and that he was ready to resort to extreme measures to oppose him. Erasmus questioned Pio's abilities. He noted that, although the Italian was only a layman, he prided himself on being an Aristotelian, tried to pass himself off as a theologian, and derived his materials from discussions with trained theologians.[67]

When Erasmus learned from Berquin that Pio's admonitory letter to him was circulating among scholars in Paris and that there was talk of publishing it, Erasmus tried to prevent its publication, at least in the form in which he had received it. Berquin reported in mid-October 1528 that he had passed on a message through a member of Pio's household that publication should be delayed until Pio heard from Erasmus. Berquin warned Erasmus in Basel that, should he decide to send his response to Pio or have

* * * * *

66 Pio *Response* – Chantilly MS 1[r] (frontispiece), 1[v] (illumination for dedicatory epistle)
67 Allen Epp 1717:4–6, 1719:34–6, 1744:129–34, 1804:248–56, 1840:78–83, 1987:1–10, 2042: 16–20, 2066:60–79

it printed, he should be careful neither to offend nor to flatter excessively this prince who was very close to Clement VII and held in high esteem by the pope's nephew, Cardinal Giovanni Salviati, who functioned in Paris as a papal legate. Writing to Berquin on 23 December, Erasmus informed his friend that he had decided to respond to Pio with only a letter, and the letter he addressed to Pio (Ep 2080), written that same day, was entrusted to Berquin for delivery. This letter begged Pio not to publish his admonitory letter.

ERASMUS OF ROTTERDAM TO ALBERTO, THE MOST ILLUSTRIOUS PRINCE OF THE PEOPLE OF CARPI, GREETINGS

When your small book was delivered to me, most illustrious Prince Alberto, I immediately began σχεδιάζεσθαι 'to draft' some sort of reply. But in the meantime the crash of Rome's fall dismayed the souls of all. Various reports about you were in circulation. Even if the place where you were residing could have been determined with absolute certainty, there was no passable route for a letter.

Recently I learned from the letters of friends that you are carrying out a legation with the most Christian king, and that the small book is making its rounds through the hands of many, and that in a very short time it will be published by the printers. In the hope that the prestige of your name, or the balance of your argument, or the charm of your style could enchant some, perhaps, towards sounder thinking, I would myself long ago have seen to its publication, were it not for the fact that your attack on me is so savage and so relentless. In fact, since the charge you allege is a capital one, the degree of civility of the language does not matter very much. It is as if a prince should condemn someone to death by an honorific address.

How I wish that you had challenged me in an area where I could have been permitted to contend with you in a contest only of cleverness or eloquence! I would have considered it great enough glory to have competed with Alberto, even if I were to come out of the contest defeated. As it is, you are pressing against me a case that is such that, if I keep silence, I admit a capital crime, but to which I could not reply without, first, causing offence to you and, secondly, without moving a Camarina which it would be better not to touch, given that, all around, circumstances are otherwise quite sufficiently irritated. Beyond this, I strongly advise you not to plunge ahead with the publication of the small book, or, if you do not wish to have wasted this effort, soften that section in which you deal with me.

If this γραμματοφόρος 'letter-carrier' had informed me [of his departure] even three days earlier, I would have sent my *Responsio* after rereading your small book. Nevertheless, I will do this soon through another, if anyone reliable becomes available. After you have read it [the *Responsio*], you will determine according to your prudence what you will judge to be in the interest of both parties.

It is scarcely safe to commit anything to letters. But if an opportunity were given to speak face to face or if you were to be resident in this region, either my judgment of you is altogether mistaken, or you would consider that Erasmus should be raised up by your kind regard rather than weighed down by your repeating the calumnies of others. Farewell.

Given at Basel, 23 December 1528

Given Erasmus' obvious attempt to prevent publication of Pio's admonitory letter, it is interesting to note that his later recollection of this letter mentions only his appeal to Pio to soften harsh statements about him, denies outright that he ever asked Pio to refrain from publication, and adds the curious claim that he had urged Pio to use stronger weapons against the doctrines of Luther. The history of this letter took an even stranger turn when it was deliberately not given to Pio, because Berquin feared that Pio might find in it some new opportunity for venting his spleen.[68]

When Erasmus received from a friend a copy of Pio's printed *Responsio paraenetica*, he rushed to reply. The book arrived on 9 February, and, as Erasmus complained, if he was to have his reply available for sale at the forthcoming Frankfurt Book Fair, he needed to have it written and printed by 22 February. Other commitments prevented him from giving full attention to the task. Indeed, at about this time Erasmus was also seeing through

* * * * *

68 The person who sent Erasmus a copy of the Bade edition of the *Responsio paraenetica* was probably Louis Berquin, but it could also have been Ludwig Baer of Basel or Gervasius Wain of Memmingen; see Peter G. Bietenholz *Basel and France in the Sixteenth Century: The Basel Humanists and Printers in Their Contacts with Francophone Culture* (Geneva / Toronto 1971) 169–70, 178. About a month after the completion of his own *Responsio* Erasmus wrote to Antonio Salamanca that he had learned of Pio's presence in France 'ex amicorum litteris' (Allen Ep 2118:19–20). Erasmus' claim (Erasmus *Responsio* 8 below) that he had learned of Pio's whereabouts only three days earlier (ie on 9 February 1529) does not seem to be borne out by a letter sent to him by Berquin (Allen Ep 2066:60–2, dated by Allen c 13 October 1528). Berquin, in response to Erasmus' news that he had composed a reply to Pio, reports that Pio is living in Paris: 'You wrote that you have composed a reply to the letter of Alberto Pio, prince of Carpi, who is living here, but that you have not sent it, because you did not know where he was living.' See also Allen Epp 2066:60–79, 2077:46, 2080:1–31 (the letter of 23 December 1528 here translated), 2291:25–9; and Erasmus *Apologia* 153–4 below. For Erasmus' earlier successful efforts to suppress material he considered not helpful to his cause, see Allen Epp 732:53n and 776:19–20 / CWE 776:23–4 (regarding Dorp's letter) and *Responsio* n68 below (for the prohibitions on Zúñiga's publications).

the Froben presses at least eight other works: his editions of Seneca's *Opera*, Chrysostom's *Opuscula*, Lactantius' *Liber de opificio Dei*, Augustine's *Opera* (volume VIII), and his own *Loca quaedam in aliquot Erasmi lucubrationibus per ipsum emendata*, appended to his second edition of the *Apologia ad monachos Hispanos*, plus his treatises *Vidua christiana*, *Responsio adversus febricitantis libellum*, and *Responsio ad notulas Bedaicas*. Unable to give his undivided attention to Pio's work, Erasmus claims to have written extemporaneously, completing the composition of his *Responsio* in five or six days on 13 February 1529, the date he placed at the end of the *Responsio ad epistolam paraeneticam clarissimi doctissimique viri Alberti Pii Carporum principis* 'The Reply to the Hortatory Letter of the Most Illustrious and Most Learned Alberto Pio, Prince of Carpi.'[69]

While his claim to have written hurriedly seems true, given the ramblings and repetitions of material, his other assertions are beset with difficulties. He seems to suggest that he started writing on 12 February, dates the completion of his work 13 February, and yet states that he devoted six (but elsewhere five) days to its composition. This account does not add up. What is more likely is that, upon a careful reading of the printed version of Pio's *Responsio paraenetica* soon after its arrival, Erasmus noted that not only

* * * * *

69 Erasmus *Responsio* 8 below (Erasmus' receipt of *Responsio paraenetica*, and his deadline and commitments), 103 (scarcely six days given to this task plus others, extemporaneous work, completion date); Allen Epp 2108:15 (six days), 2118:24–5 (five days given to the tasks of rereading the book and preparing a response). Six of the eight books on which Erasmus was working bear the publication date of March 1529, the same date assigned to his *Responsio*; see vander Haeghen I 175, 178, 180; II 11, 38, 50; Allen VIII 25, 41, 47, 55, 61, 69, 146; and Erasmus *Apologia* 154 below. Erasmus had clearly finished work on his *Responsio* by 25 February, as is evident from Allen Ep 2108:15. Pio openly questioned Erasmus' statement that he had composed the work in six days, suspecting that his response had been prepared long in advance; see Pio XXIII *libri* 71^{r-v}M: 'I too can form a view [reacting to *Responsio* 3 below], one far more reasonable and likely, that you had been working long before on the reply you made to me (the one you boast you completed in the space of six days!). This is more likely to be true, since I had written to you almost three years earlier, and you took so much offence from my letter as perhaps never before, as your letter attests. Who, really, are you going to persuade that you were able to check your headstrong disposition so as to postpone writing until now, especially as your skin is so delicate that not even a fly can touch it without drawing blood? What is far more likely, Erasmus most learned, is what a number of your friends who live here have repeatedly said you had done, that is, you wrote a reply much earlier, though you did not publish it before learning that our reply had been published ...' The fact that Erasmus on occasion cited an earlier version of the *Responsio paraenetica* instead of the version printed by Bade provides supporting evidence for Pio's conjecture; see Forner 222–3. Was Erasmus' claim of composing the *Responsio* in six days of labour merely a literary allusion to God's creation of the world in six days (Gen 1:31)?

had Pio not mitigated any of his criticisms of him, but had strengthened his arguments by the insertion of additional materials – Erasmus identified at least four such instances. Given the demands on Erasmus' time and energies to see through the press the works already scheduled for publication, he was in no position to write a fresh rejoinder. Instead, the response that Erasmus had earlier prepared, which was probably in the form of a private, conciliatory communication, he now reshaped so as to hold Pio up to public ridicule by a potent mix of seeming flattery and biting sarcasm. His rush to complete work on his *Responsio* so that it could be printed in time for the Frankfurt Book Fair resulted in his failure to correct some internal inconsistencies. Thus he claims towards the beginning of the *Responsio* that he had only three days earlier (ie, on 9 February 1529) learned of Pio's whereabouts, but then asserts towards the end that, as soon as he had heard of Pio's residency in France, he had sent a letter to him (dated 23 December 1528) in time to halt his publication of the *Responsio paraenetica*. He also claimed to have already answered complaints about his position on auricular confession and then later devoted a section to defending the *Exomologesis*. Erasmus thus seems not to have submitted the *Responsio* as a whole to a final revision. Was he so distracted by his other projects then being printed that he could not find time for a final revision of his hurriedly revised response to Pio, a polishing task he usually found burdensome and tended to neglect? Was he in such a rush to meet the impending deadline that, as soon as he completed a section of the *Responsio*, he handed it over to the pressmen to set in type and print, and thus was unable to revise earlier sections already printed? But then in how much of a hurry was he if he found time not only to compose and print his *Responsio* to Pio, but to have printed with it his *Responsio ad notulas Bedaicas* that constituted pages 81 to 127, or signatures m to r, assigning to both works the same publication date of March 1529? If Erasmus succeeded in completing the task of revising and printing his *Responsio* in six days, Pio was nonetheless correct in challenging the impression Erasmus had tried to convey that he had begun work on the *Responsio* only on 12 February, rather than merely revising a response that he had worked on over the course of many months.[70]

* * * * *

70 Erasmus *Responsio* 27, 87, and 93 (examples of repetition); *Responsio* 8 (three days ago on 9 February); *Responsio* 103 (13 February for date of work); *Responsio* 33, 51, 74, 76, and 91 (examples of differences; see also Allen Ep 2118:26–8); *Responsio* 6–7, 101, and *Apologia* 154 (inconsistency about when he learned of Pio's whereabouts); *Responsio* 50, 66–72, and 74 (he has already responded yet makes a new defence of his position on auricular confession). Pio was well justified in doubting that Erasmus had penned his response in a mere six days: Pio *XXIII libri* 71[r]M. Although Erasmus claimed that his

While the revised version retained passages that were flattering to Pio, Erasmus probably reworked his draft portrait of him so that the Italian prince now appeared gullible, vain, stupid, unfair, lazy, deceitful, and insane. According to Erasmus, Pio was pressured by others to be their spokesman. He agreed in order to please them and display his Ciceronian eloquence. But unlike Cicero, who would begin temperately, argue robustly, and conclude forcefully, Pio opened with savage accusations, argued weakly, and summed up unconvincingly. Erasmus mocked Pio's arguments: do you write this in jest or in earnest? who is not awed by such powerful reasoning? how distressed I am that you do not use more effective arguments to support papal power! Pio's behaviour is not that of a learned man: he repeats silly jingles, and, although endowed with so abstruse a knowledge of philosophy and theology, he does not write against Luther whose teachings he finds most objectionable. Contrary to his reputation as a man of singular fairness, Pio brings charges against Erasmus that are false, attributing to him statements that are not to be found in his writings, distorting his views by a clever choice of words, and repeating the criticisms of others without checking out their veracity. Indeed, Pio's criticisms are but repetitions of what he has heard bandied about during drinking bouts by ignorant pseudo-monks and irresponsible theologians. He seems also to have borrowed heavily from the catalogue of complaints compiled by Zúñiga, but Pio did not bother to read Erasmus' refutation of them. The reason Pio is so ready to use others' materials is because he has grown accustomed to the easy life and lacks the strong stomach needed to read on his own and with care Erasmus' writings. In numerous passages of pointed sarcasm Erasmus speaks of the candour, sincerity, guilelessness, and straightforwardness of Pio's character. The malicious deceit of Pio is most evident in his use of flattery, which like honey smeared on the rim of a cup of hemlock, disguises the deadly potion. His treatment of Erasmus, for whom he protests affection and esteem, is so hostile that one must

* * * * *

Responsio had to be published by 22 February, the *editio princeps* bears the publication date of March 1529. The *Responsio* was surely printed by 7 March since Felix Rex took a copy of it with him when he left Basel for Speyer. There he presented it, together with copies of the colloquies augmented by the *Ciceronianus* and of the *Epistolae selectae*, to Johann von Vlatten; see Allen VIII 99 and Ep 2130:28–36. Froben's press had the capacity to print Erasmus' work relatively quickly, since its six presses could print up to forty-eight pages per day; see Allen Ep 1683:16–17 / CWE 1683:20n. Erasmus' *Responsio* consisted of eleven quires (sigs a-l) with four folios in each, measuring 18.5 x 13.5 cm. Thus, it was because of the eight other works then being printed that Erasmus found himself under the extraordinary time pressures he spoke of; the *Responsio* by itself could have been more easily managed.

wonder about his sanity.[71] The six days that Erasmus supposedly gave to writing his *Responsio* probably consisted for the most part in reshaping, disfiguring, and adding these dark colours to his earlier portrait of Pio. This material he scattered throughout his rejoinder.

Although hurriedly revised and at times rambling and repetitious, Erasmus' *Responsio* nonetheless retains an overall coherent structure. Its major divisions are an introductory section, a denial of the truthfulness of Pio's charges, a review of Erasmus' relations with Luther, detailed rejoinders to specific accusations, responses to minor questions left unanswered, and a conclusion.

In the introductory section Erasmus treats a number of topics, most of which relate to the printing of Pio's *Responsio paraenetica*. He claims that it was Pio's intent all along to publish it and that it consists for the most part of recycled materials: complaints Pio had heard from others about Erasmus and a treatise he had earlier composed against Luther. Erasmus' letter to Pio, which prefaces the work, is printed inaccurately. Bade's letter to the reader falsely claims that Erasmus has a bad reputation among the cardinals of Rome. The reason Erasmus has not responded earlier to Pio's letter was that he did not know the prince's whereabouts following the Sack of Rome. Because of the pressure of deadlines Erasmus has been unable to give the *Responsio paraenetica* a careful reading, but he has nonetheless rushed to give some response, treating only those sections that touch directly on himself.[72]

Erasmus denies the validity of the charges Pio has made against him. He claims that Pio has unjustly mixed him up with Luther and unfairly accused him of paving the way for Luther's attacks on the church. Pio's charges are not based on a careful reading of Erasmus' works, but are derived from conversations he had with those who wanted him to be their spokesman. From reports Erasmus has received over the years, he has been able to piece together that Pio, whose hostility towards him goes back at least to 1521, is a member of a circle of Erasmus' Roman critics that includes Aleandro and Zúñiga. Much of what Pio says is but a rehash of Zúñiga's

* * * * *

71 For examples of the various ways in which Erasmus portrayed Pio in his *Responsio*, see the following: with flattery (3, 6, 8, 10); his motives in writing (59, 65, 84); his weak arguments (44, 81, 84); his failure to act like a learned man (21–2, 28); his unfairness (59, 63, 66, 101, 102); his laziness and negligence (11, 18, 51–2, 74); his malicious deceit (10–12, 29); his insane behaviour (102). Pio summarizes Erasmus' portrayal of him in *XXIII libri* 67[v]c.

72 Erasmus *Responsio* 3–8. Pio explicitly denied that he had written the material against Luther earlier and then inserted it into his letter to Erasmus (Pio *XXIII libri* 47[r]a and 70[v]L). The surviving draft of the letter supports this contention, for it provides no evidence of such insertions; see Introduction xlviii above.

complaints, which Erasmus has already answered. Thus Pio's charges are not only unfair, but also tardy and unproductive.[73]

As a bridge to the next section, Erasmus openly admits that he has made mistakes over the years. He has rushed into print without proper revision and has yielded to others' importunity in allowing such writings as his *Moriae encomium* to be published. He has also taken on tasks for which he lacked the requisite abilities. Had he known what the future would bring, he would never have written or done certain things. When rereading his works he finds things that displease him, but he has yet to come upon anything that agrees with one of the condemned positions of Luther and there is much that opposes Luther's teachings.[74]

Erasmus argues that he is not a supporter of Luther. It is a lie that he differs from him only in style and not in substance. So serious are his disagreements with Luther's censured teachings that Luther himself has violently attacked him. Erasmus rehearses the history of his relations with Luther: how he hoped that Luther might bring about the desired reform of the church; how he counselled his friends not to get involved and avoided involvement himself; how he delayed writing against him because it was not his responsibility as a grammarian, he lacked the proper training, and he could not clearly discern whether Luther was guided by the Spirit, what his intentions were, and if his teachings were in error. Erasmus has tried to avoid partisan behaviour, while those who wrote against Luther often did so stupidly or ineffectively, and even the University of Paris delayed rendering its judgment against Luther. Meanwhile, ignorant friars, who were hoping thereby to destroy Erasmus, were denouncing him for failing to take up his pen against Luther. When at last Erasmus did write his *De libero arbitrio diatribe* (1524), it became obvious how little influence he had with a man who had already rejected the authority of the pope, the emperor, and leading universities. Nonetheless, Erasmus is still criticized as someone who is at odds with the church's teachings. Erasmus insists that he has always remained in communion with the church and that the only things he has criticized are abuses, excesses, and superstitions, but never the office or the institution itself. Regarding church teachings, Erasmus has questioned only those things as yet undefined and not yet adequately explained. He has called upon the learned to provide better arguments for controverted Catholic teachings. What he has written for tranquil study has unfortunately become ammunition for stirring up the masses against him.

* * * * *

73 Erasmus *Responsio* 9–18
74 Erasmus *Responsio* 18–22

His critics hurl at him the proverb that a pig has disputed. All the good things Erasmus has written are ignored; what is collected is whatever can be twisted for slander. Pio's charges are empty, glittering falsehoods.[75]

At the core of Erasmus' *Responsio* is a detailed refutation of the individual charges Pio had made. Erasmus has not diminished papal authority by pointing out that it was neither acknowledged nor exercised in the ancient church. It is also true that in antiquity the title of cardinal did not exist, priests and bishops were considered equals, emperors convoked church councils, and blessed rather than consecrated bread was used in liturgies. Erasmus insists that he considers matrimony a sacrament and does not esteem marriage over continence and celibacy. While he has criticized certain abuses, he does not condemn or abhor ecclesiastical laws, the institution of monasticism, the canonical hours, and chanting itself. He has replied elsewhere and often to his earlier observations on auricular confession. These charges are, according to Erasmus, the recycled complaints of Zúñiga and he has already given them an adequate response. If he has let slip out things better left unsaid, he has often removed them from later editions of his works. It is better to put these things to rest and not force Erasmus to defend them. Erasmus then argues for his use of the ancient church as the norm rather than Pio's position that the doctrine and practice of the church developed under the guidance of the Holy Spirit. Erasmus finally provides detailed defences of his *Novum Testamentum* (his use of the terms *instrumentum* and *fabula*), the *Moriae encomium*, the *Exomologesis*, and his *Paraphrases*. He concludes this section by summarizing the evidence that Pio has produced for his claim that Erasmus and Luther are often in agreement. So weak is Pio's argument made to appear that Erasmus holds it up to ridicule.[76]

The last major section of Erasmus' *Responsio* is a hodgepodge of rambling and often repetitious materials. He rehearses his role in the Luther affair and asks why Pio did not write against Luther nine years earlier from the safety of Rome. He reiterates his claim that Pio exaggerates Erasmus' abilities and influence, and repeats his excuses for being tardy in writing against Luther. Given that Pio revised his *Responsio paraenetica* of 1526 before publishing it in 1529, Erasmus criticizes him for not having updated it to take into consideration the two-volume *Hyperaspistes* (published in Basel probably during March 1526 and August 1527). He suggests that Augustine's writings on free will provide some basis for Luther's teachings. He

* * * * *

75 Erasmus *Responsio* 22–38
76 Erasmus *Responsio* 38–85

also argues for a knowledge of ancient languages by theologians and for the Scriptures to be made available to ordinary Christian people. He defends Germans against the accusation that they are more interested in empty eloquence than in the pursuit of wisdom. He complains that Pio keeps urging him to write against Luther when he has already done so at length and has proven that he does not agree with Luther. Erasmus states that he has written to Pio to urge him not to publish his *Responsio paraenetica* and is amazed that the letter was not delivered to him. If Pio has good will towards Erasmus, why does he try to make him appear to be a Lutheran? Nonetheless, Erasmus accepts Pio's invitation to friendship and apologizes for his hastily written response to Pio's exquisitely wrought book.[77]

Erasmus' *Responsio* is the work of a master rhetorician who uses anonymous reports, partial truths, and artful dodges to portray himself as the innocent victim of Pio's unjustified attacks. Without Pio's *Responsio paraenetica* open to the passage in question, the reader of Erasmus' rebuttal does not know if he is accurately reporting Pio's charge and responding directly to it. Thus, without Pio's text as a corrective, the reader may be led to think that Pio espouses the obscurantist position that a theologian does not need a knowledge of ancient languages. Pio's position, however, is that when dealing with such difficult topics as the Trinity the theologian needs to complement his linguistic skills with those of dialectical reasoning so that he can argue correctly and avoid heresy. Erasmus also tries to make Pio seem unreasonable and uninformed for urging him to write more against Luther when Erasmus has already written an additional two-volume work, the *Hyperaspistes*, that Pio ignores. But citing this work on free will does not answer Pio's real charge that Erasmus has thus far restricted his dispute with Luther to this one issue. Again, because Erasmus, like many of his contemporaries, often holds up the ancient church as the model to be followed, he is less than straightforward when he claims that his pointing to ancient teachings and practices that are at variance with those of the contemporary church does not, as claimed by Pio, diminish the validity and authority of the modern church's doctrines and practices. Pio likewise accuses Erasmus of acting irresponsibly when he points out in print weaknesses in the arguments that support Catholic teachings at the very time when Catholic theologians are engaged in a desperate combat with Protestant critics. Pio claims that Erasmus did not communicate this information privately to the Catholic controversalists, nor did he provide them with stronger arguments.

* * * * *

77 Erasmus *Responsio* 85–103; for the dates of publication of the two-volume *Hyperaspistes*, see Allen VI 262, VII 116.

Nonetheless, Pio contends, Erasmus acts as if he were doing the church a service, while he is really helping its opponents. Erasmus' account in the *Responsio* of his relations with Luther over the years has ignored or minimized the support he gave to him in the crucial early phase. And instead of making his *Responsio* strictly an answer to Pio's charges, Erasmus prefers at times to attack his critic's motivations and intellectual abilities, claiming that these characterizations of his opponent come from the reports of unnamed informants in Rome and Paris.[78] Given Erasmus' artful way of dealing with Pio, the reader needs to be wary and not accept all of his statements at face value.

Events in Basel that provided the immediate, local context for the composition and publication of Erasmus' reply to Pio may help to explain why he was at such pains to deny that he held positions similar to those of the Protestants. On the very day (9 February) that Erasmus received a printed copy of Pio's *Responsio paraenetica*, in which he is accused, among other things, of supporting Luther, whose teachings have led to massive civil unrest in Germany, the city of Basel was convulsed by iconoclastic riots. While Erasmus was working on his reply, the city government not only abolished the Catholic practices of the veneration of images and the celebration of the Mass, but also ordered all citizens to attend Protestant church services. Since these measures in effect proscribed a public expression of the Catholic faith and suggested that the city's residents were practising Protestants, Erasmus resolved by 2 March to take up residence elsewhere, but because of illness he waited until 13 April to take a boat down the Rhine for Catholic Freiburg. He had protested to Pio that he was a loyal son of the church, and now at considerable personal inconvenience he gave symbolic and concrete expression to this claim by distancing himself from an officially Protestant milieu.[79]

In his correspondence with friends Erasmus commented on his recent exchange with Pio. He criticizes Pio's *Responsio paraenetica* both for having

* * * * *

78 See *Responsio* 96–7 (languages and philosophy), 95 (topics other than free will), 38–42, 45–6, 48 (ancient church differs from contemporary one), 69 (Erasmus points out weaknesses), 88 (early support for Luther), and 4, 13–16 (unnamed sources of Erasmus' charges).

79 Allen VIII 22, 137; since these measures put an end to religious dissension by enforcing a uniform Protestant practice, Erasmus' decision to leave would seem to be based not primarily on a dislike of turmoil (as suggested by Allen), but rather on an aversion to intolerance and a desire to continue practising his Catholic faith. On the events in Basel, see Carlos M.N. Eire *War Against the Idols: The Reformation of Worship from Erasmus to Calvin* (New York 1986) 53, 117–19; and Lee Palmer Wandel *Voracious Idols and Violent Hands: Iconoclasm in Reformation Zurich, Strasbourg, and Basel* (Cambridge 1995), 149–89.

tried to prove that Erasmus was the author of the Lutheran tumult and for refuting Luther's teachings with weak arguments so unworthy of Pio's talents and learning that one is led to conclude that he never gave serious thought to the issues. Indeed, Erasmus suspects that Pio was not the only author of the work, but had been assisted by the Parisian theologians who provided scriptural proof-texts, and that he composed his *Responsio paraenetica* with the help of hired hands. The Catholic faith, Erasmus claims, is not well supported by such a work and Pio has hurt his own reputation by publishing it. Nonetheless, Erasmus concedes, it has succeeded in alienating to some extent Clement vii's regard for Erasmus. Perhaps in response to Pio's urgings that he write more against the Protestants, Erasmus published in the autumn of 1529 his *Epistola contra pseudevangelicos*. In this work Erasmus developed with good effect Pio's analogy between the stages of human growth and the progress of the church over time. But Pio seems to have been ignorant of this work and Erasmus soon dropped this line of argument. While Erasmus claimed that his initial printed response to Pio had been carried out in a civil and honorific manner, and that he had written nothing that could provoke Pio's anger, Erasmus also acknowledged that their exchange had ended in rage.[80]

PIO'S XXIII LIBRI

Upon receiving a copy of Erasmus' *Responsio* Pio hesitated about responding. After many days of reflection he concluded that piety demanded that he give answer, for Erasmus had not only defended his old errors, but had added new ones, and Pio's failure to refute these could lead others to accept Erasmus' views on the grounds that Pio's silence seemed to confirm their veracity. His failure to respond could also have been interpreted as an admission by Pio that he had misrepresented Erasmus' positions. Despite his physical weakness as the result of illness, Pio took up the taxing burden of responding. To minimize the labour he resorted to brief rejoinders placed

* * * * *

80 Allen Epp 2108:13–16, 2110:31–6, 2112:45–50, 2118:28–31, 2378:42, 2443:329–36. On the *Epistola contra pseudevangelicos*, see Minnich 'Underlying Factors' 40–1. Erasmus' correspondent Alfonso Valdes wrote to complain that not only have princes disturbed the political world with their endless warfare, but now in the person of Alberto Pio, prince of Carpi, they even invade the world of letters and attack Erasmus out of jealousy because he benefits the Christian republic; Allen Ep 2163:132–7. The Italian professor of Roman law at Bourges, Andrea Alciati, sympathized with Erasmus' complaints and explanations and claimed that Pio would achieve little among those who also read Erasmus' *Responsio*; Allen Ep 2394:46–53.

in the margins of Erasmus' work, but this format not only limited the fullness of his response but also seemed to diminish its seriousness and authority. In addition, the space restrictions of the narrow margins prevented him from quoting Erasmus' earlier statements back to him when Erasmus denied that he had ever held certain positions and thus implied that Pio was a liar. Reluctantly Pio decided to give a fuller and for him more burdensome response. In the preface of the *xxiii libri* he rehearsed the factors which had forced him to respond, assigning some of the blame to Erasmus' injudicious statement in the *Ciceronianus*. He then began the task of refuting Erasmus' views by criticizing in detail one work, the *Moriae encomium*, which was the first of Erasmus' offensive works. When this proved a tedious chore and required over ten large folio pages to answer only this one of Erasmus' numerous writings, Pio apparently abandoned the work-by-work refutation for a thematic approach.[81]

Pio organized the remainder of his response around some twenty controversial topics. These were grouped under such headings as fasts, monasticism, church ceremonies and decoration, the cult of and images of the saints, especially the virgin Mary, the differences between humanistic and scholastic theology, the authority of Scripture, a defence of Trinitarian teachings from Arian attacks, church offices and regulations, vows and celibacy, matrimony and confession, faith and good work, and the morality of warfare, swearing, and lying. Each of these twenty topics was treated singly in its own chapter or book, the lengthiest being that devoted to the veneration of saints and relics (52 pages) and the briefest dealing with confession (6 pages).[82]

* * * * *

81 Pio *xxiii libri* ãii[r], 69[v]1–70[v]K, 71[v]N–72[r]O, 73[v]C. On Pio's chronic illnesses, see Pio *Responsio paraenetica* 1[v]. During his stay in Paris Pio continued to suffer from chronic gout; see Guaitoli 'Sulla vita' 230, 295; and the letter of Giovanni Salviati to Uberto da Gambara, no place given, 29 January 1528, in ASV SS Francia I 92[r]. His illness when he was serving at the papal court was the occasion of a poetic prayer by Giano Vitale and a prayer and *eucharisticon* by 'C. Silvanus Germanicus' published by Biagio Pallai in his collection *Coryciana* (Rome: apud Ludovicum Vicentinum et Lautitium Perusinum 1526) DDiii[r]–EEiii[r]. For examples of his earlier illnesses (eg fevers), see the letters of Marino Caraccioli to Duke Massimiliano Sforza, Rome, 27 July 1514, and of Pio to Emperor Maximilian, Spoleto, 16 April 1515, in Haus- Hof- und Staatsarchiv – Wien, Maximiliana 32 1 108[r] and 33 III 101[r]. From his fortieth year (c 1515) until his death Pio was so incapacitated by gout that he had to retire to bed for months on end, so ill that he could neither read nor write; see Sepúlveda *Antapologia* Di[r] / Solana 127–8; and Moncallero II 212, 215 (reports of 1520).

82 Some of the following sections are reprinted in whole or in part from Minnich 'Underlying Factors,' courtesy of the *Erasmus of Rotterdam Society Yearbook*. For a list of the chapter/book titles, see Pio *xxiii libri* ãii[r]. Book 9 on veneration of the saints is on 147[r]–

This approach had its advantages. By beginning each chapter with a series of incriminating quotations excerpted from a variety of Erasmus' writings, Pio could quickly demonstrate that he was not a liar nor had he engaged in deceptive selectivity by attributing to Erasmus objectionable positions. Erasmus' own words condemned him. But because Erasmus' statements could at times disguise their destructive message behind wit and rhetorical flourishes, Pio tried to expose the dangerous suppositions underlying the quotations and answer the likely arguments Erasmus would use in claiming that Pio had misunderstood him. To provide the needed antidote to Erasmus' subversive teachings, Pio marshalled evidence drawn from common sense and careful natural reasoning, from Scripture, the church Fathers, and ecclesiastical authorities, and from the *sensus fidelium* or practices of Christians from ancient times until his own that had been approved by miracles and the decrees of popes and councils – all in an effort to confirm or defend the church's orthodox teachings and traditional practices. Pio's method of refutation was borrowed in part from the scholastics, who would also state the opinions and arguments of their adversaries, clarify and analyse them, present and argue their own position, and then refute their adversaries' arguments one by one and answer their possible objections.[83] Because Pio was fearful that yielding to Erasmus on one issue of Catholic orthodoxy would open the door to questioning all of the church's

* * * * *

172^v, and book 19 on confession is on 227^v–230^r. Pio claimed that these twenty chapters would demonstrate that the views Erasmus expressed in his *Moriae encomium* were also present in his other writings (84^{r}F). The titles of the chapters do not always adequately convey the range of topics treated under each heading. Thus, for example, book 6 'On a Defence of Ceremonies Adopted by the Church' includes treatments of festivals, singing, and candles, while book 7 'On the Decoration of a Church' treats such themes as funerals and bells. In his *Erasmi Roterodami blasphemiae et impietates ...* (Rome: Antonius Bladus de Asula 1522) Zúñiga also organizes his criticisms of Erasmus into thematic chapters, eg 40 in book I, 47 in book II, and 36 in book III. While their organization of the topics differs, all of the themes Pio treats were written on earlier by Zúñiga, who also wrote on topics Pio ignored, eg the Minims and Minorites (Aiii^r), extreme unction (Biii^r), and indulgence preachers (Eiv^r, Fiii^r). Pio, however, highlighted the questions of virginity and celibacy by giving them separate treatment (book 17) and not relegating them to a subsection of monasticism. To what extent Pio may have borrowed from Zúñiga's earlier work deserves a separate study. Pio's criticisms are also similar to those registered at the Valladolid Conference of 1527 (LB IX 1022E–1091D passim).

83 Pio *XXIII libri* 68^{r}D, 69^{r}F, 72^{r}O, 73^{r-v}AB, 84^v–85^{r}B, 99^{r}M, 107^{v}I, 137^{r}G–138^{r}K, 195^{r}B, 223^{r}V; for a description of his methodology, which is similar to that of the scholastics, see especially 84^{r-v}A, which is translated below in *Apologia* n444; see also Translator's Note on the Texts cxlvi below. For an example of the wide range of Pio's argumentation, see Minnich 'Debate on Images' 388–91, 402–12.

teachings, he assumed at times very defensive and rigid positions, out of character with his generally balanced presentations.[84]

In *XXIII libri* Pio also tried to answer the charge that he had not read Erasmus' writings. Both in the preface and at the end of book 3 Pio traced his growing familiarity with these publications. The first of Erasmus' works to command his approving attention was the *Adagia*, and that work he perused around the year 1516. Shortly thereafter he eagerly read the *Moriae encomium*. While he enjoyed its wit, he was offended by many passages in it. He also mentioned reading the book on confession (*Exomologesis*), the *De libero arbitrio*, the translation of the New Testament, and the *Paraphrases*. Of Erasmus' apologetical works he read nothing except for Erasmus' response to the criticisms of Zúñiga. Of Erasmus' other writings, Pio admitted that he was ignorant of their titles and content and did not feel himself obligated to read them – their number seemed infinite and was ever expanding. But once he received Erasmus' *Responsio* to his own work, Pio did read his *Responsio ad notulas Bedaicas*, which was published together with the rejoinder to Pio. He obviously went on to study Erasmus' major work against Noël Béda, the *Supputatio*, given the numerous citations he made from it. At the end of book 3 Pio lists ten works or categories of Erasmus' writings that he finds particularly offensive in addition to the ones cited in the preface: the *Colloquia, Enchiridion, Ratio, Encomium matrimonii, Institutio principis christiani*, scholia, annotations, prefaces, epistles, and apologies. The excerpts from Erasmus' writings that he quotes in books 4–23 were taken almost exclusively from the works cited in the preface and book 3.[85]

In addition to Erasmus' writings, Pio made references in the *XXIII libri* to the works of contemporary Catholic controversialists. He cited the anti-Lutheran writings of Henry VIII and Noël Béda and the criticisms of Erasmus by Maarten van Dorp, Diego López de Zúñiga, Jacques Masson, Pierre

* * * * *

84 Eg Pio *XXIII libri* 178[r]c

85 Pio *XXIII libri* 68[v]–69[r]F, 84[r]F. My rough estimates suggest that Pio cited these works in the following order of frequency: *Annotationes in Novum Testamentum* (over 60 times); the life of Jerome and scholia on the epistles of Jerome (34 times); *Colloquia* (30 times), with the most frequently cited colloquies being Ἰχθυοφαγία (8), *Naufragium* (7), and *Peregrinatio* (4); *Ratio* (24); *Enchiridion* with the preface to Volz (22); *Supputatio* (16); *Encomium matrimonii* and *Institutio christiani matrimonii* (11); *De esu carnium* (8); *Institutio principis christiani* (5); preface to St Hilary (= Ep 1334) (3); *Ciceronianus* (2); *Exomologesis* (1); and *Responsio adversus febricitantis libellum* (1). All of book 3 is dedicated to a detailed criticism of the *Moriae encomium*. For the works of Erasmus' adversaries that Pio read, see Minnich 'Debate on Images' 385–91. For Pio's statement on not attempting to read the numerous apologetical works of Erasmus, see Erasmus, *Apologia* n549 below.

Cousturier, and Frans Titelmans in such a way as to indicate that he had read them.[86] His laudatory statements about the writings of Johannes Heigerlin (Faber), Johannes Maier (Eck), and John Fisher show that he knew about their Reformation writings but it is unclear whether he had actually read them.[87] Unmentioned contemporaries who may have been a source for some of Pio's ideas include: Juan Ginés de Sepúlveda, who critiqued early drafts of Pio's *Responsio paraenetica* and whose *De fato et libero arbitrio* against Luther received similar attention from Pio;[88] Josse Clichtove, who visited Paris at this time, was an admirer of Pio's uncle Giovanni Pico della Mirandola, and also wrote against Luther, Oecolampadius, and Erasmus;[89] and conservative members of the Faculty of Theology of the University of Paris who investigated Erasmus' writings, such as Jacques Berthélemy, Jacques Godequin, and Valentin Lievin, OP. Notably critical of Erasmus was Pierre de Cornes, OFM, faculty member and guardian of the Franciscan monastery in Paris, who preached at Pio's funeral.[90] Although not a contemporary, Thomas Netter, OCARM (d 1450), had written a huge work against the followers of Wycliffe and Hus, the *Doctrinale*, that was published in Paris by Pio's friend Josse Bade from 1521 to 1532. When refuting Erasmus' positions, Pio would make occasional references to similar views held by Wycliffe and Hus.[91]

While Pio's *XXIII libri* clearly builds on his earlier *Responsio paraenetica*, it is not always an improvement. Except for the attack on Erasmus' *Paraphrases*, the same topics on church teachings and practices that were

* * * * *

86 Pio *XXIII libri* 58^{V}b, 63^{V}f (Henry VIII); 69^{V}H, 230^{r}A, and 237^{V}I (Béda); 56^{r}l (Dorp); 48^{V}p, 51^{r}o, 54^{r}n, 69^{r}F (Zúñiga); 58^{r-V}s,b (Masson); 59^{r-V}o,s (Cousturier); 60^{V}i (Titelmans). For a fuller treatment of Pio's contemporary sources, see Minnich 'Debate on Images' 388–90.
87 Pio *XXIII libri* 50^{V}n (Faber); 50^{V}n, 52^{V}a (Eck); 63^{V}f (Fisher)
88 Sepúlveda *Antapologia* DIV^r, EI^r / Solana 131, 135
89 Entry on Josse Clichtove CEBR I 317–20, especially 319; Farge *Biographical Register* 92–3; Jean-Pierre Massaut *Josse Clichtove, l'humanisme et la réforme du clergé* 2 vols, Bibliothèque de la Faculté de Philosophie et Lettres de l'Université de Liège 183 (Paris 1968) II 44 (visits to Paris) and 405 (influence of Pico); Pio *XXIII libri* 54^{r}l (reference to Oecolampadius' eucharistic views, perhaps from Clichtove)
90 Farge *Registre* 97 no 95A, 111 no 159A, 127 no 141C, 135 no 149E, 141 no 159A, 158 no 179C; Farge *Biographical Register* 42–3 (Berthélemy), 198–9 (Godequin), 281–2 (Lievin), 110–12 (de Cornes); entry on Petrus de Cornibus CEBR I 341–2
91 For examples of Pio's references to Wycliffe and Hus, see Pio *XXIII libri* 52^{V}d, 53^{r}e, 62^{r}g, 64^{V}f, etc; for the Paris printing of Netter's *Doctrinale*, see Philippe Renouard *Bibliographie des impressions et des œuvres de Josse Badius Ascensius, imprimeur et humaniste 1462–1535* 3 vols (Paris 1908; repr New York 1967) III 387–90; on Netter's importance, see Margaret Harvey 'The Diffusion of the "Doctrinale" of Thomas Netter in the Fifteenth and Sixteenth Centuries' in *Intellectual Life in the Middle Ages: Essays Presented to Margaret Gibson* ed L. Smith and B. Ward (London 1992) 281–94.

discussed in the *Responsio paraenetica* are treated again in the later work, and a number of new issues, such as the Trinity, and the morality of swearing and lying, are added. *XXIII libri* avoids the *Responsio paraenetica*'s extended digressions on the teachings of Luther and his fellow Protestants and focuses instead on refuting Erasmus' opinions and on supporting Catholicism with carefully argued theological essays that reflect a deep affection for and appreciation of the church's traditional beliefs and practices. While this later work is generally better organized, there are clear signs that large sections of it never benefited from a careful revision. Thus, for example, in book 20 on faith and works, Pio not only quotes the same text from the *Enchiridion* in two different versions, but confuses a citation from the *Supputatio* with one from the *Annotationes* and does not follow, as he claims he does, the order of the citations when commenting on them individually. Book 22, on swearing, is poorly organized and very repetitious. Pio seems to have rushed to complete his first draft by the eve of Easter 1530. He intended to revise the whole work, but was able to accomplish this only as far as the middle of book 9 when death overtook him on 8 January 1531.[92]

The *Responsio paraenetica* and the *XXIII libri* also differ somewhat in style. While Pio continues to alternate between the first-person singular and plural when addressing Erasmus, *XXIII libri* does not have sections directly addressed to Luther. The real audience is not Erasmus, but educated Catholics whose faith Pio wishes to confirm. It is doubtful that Pio thought he could win over Erasmus to his views. Erasmus' insulting rejoinders to his *Responsio paraenetica* have put Pio on his guard; a wary coolness pervades his comments when he is addressing his opponent. Gone is the literary flair of those earlier passages in which Pio had hoped to win over Erasmus to an ardent defence of the church. To prove that he is not a liar in having accused Erasmus of holding dangerous views Pio resorted to a tedious listing of quotations and comments on them. To avoid prolixity he did not attempt to list all of Erasmus' offensive statements; by refuting a few representative passages, he hopes to provide a response to the others. While *XXIII libri*

* * * * *

92 On the manner in which *XXIII libri* builds on the *Responsio paraenetica*, see Pio *XXIII libri* 196^{r}D, 230^{v}B; on the *Paraphrases*, see Pio *Responsio paraenetica* 10^r–13^{r}I; on the Trinity, swearing, and lying, see Pio *XXIII libri* 181^{v}P–186^{r}Y, 244^{v}A–251^{r}V; on the avoidance of the earlier digressions, see Pio *XXIII libri* 134^{r}A; on Pio's desire not just to refute Erasmus but to confirm Catholic teaching, see Pio *XXIII libri* 195^{r}B; for the sections of book 20 that need revisions, see Pio *XXIII libri* 230^{r}A and 232^{r}G; for book 20 see Pio *XXIII libri* 244^{v}A–248^{r}K; for the date of the completion of the draft, see Pio *XXIII libri* 252^{r}Z; for the comment on how far the revisions had progressed, see Pio *XXIII libri* 171^{r}H.

impresses the reader with its theological erudition, it often lacks the literary verve of the earlier work.[93]

PIO'S THEOLOGICAL PERSPECTIVE

Permeating and shaping the theological argumentation of both the *Responsio paraenetica* and the *XXIII libri* is Pio's conviction that the church has been guided effectively over the centuries by the Holy Spirit. He cites as justification for this belief a series of scriptural proof-texts: 'I have prayed for you, Peter, that your faith may not fail' (Luke 22:32); 'I am with you always until the end of time' (Matt 28:20); 'I will not leave you orphans' (John 14:18); 'When the Spirit of truth comes, he will teach you all things' (John 16:13); and 'Holy Father, keep them whom you have given to me. But I pray not only for them but also for those who will believe in me through their words' (John 17:11, 20). Pio considers it wicked to claim that the Spirit of Christ has deserted the church for thirteen hundred years, or that the Spirit has spoken lies, or lain hidden in a jug until Luther drank from it. Luther's message is not confirmed by the signs and miracles that Christ promised would accompany the preaching of the Gospel (Mark 16:60), but his teachings have led instead to the breaking of vows, the profanation of churches, the violation of civil and ecclesiastical laws, and the slaughter of numerous peasants proclaiming Luther's false doctrine of 'Christian liberty' – such are the fruits of a bad tree (Matt 7:17–20). God in his goodness would never allow all Christian people, the whole church, to be deceived and wander in spiritual blindness for so long a time regarding so many things that pertain to the integrity of faith and religion.[94]

* * * * *

93 For examples of Pio alternating within the same sentence between the first person singular and plural, see Pio *XXIII libri* 221^{V}M, 227^V–228^{r}H; on addressing only Erasmus, see Pio *XXIII libri* 134^{r}A; on commenting on each quotation in turn, see Pio *XXIII libri* 230^{V}B; on answering representative statements, see Pio *XXIII libri* 134^{r}A. Pio's indignation does on occasion lead him to engage his rhetorical skills: eg Pio *XXIII libri* 204^{r-V}v. Not until 1552 was there a comprehensive and systematic listing of the offensive passages in Erasmus' writings. It was compiled by the Dominican John Henten (1500–66) of Louvain on the orders of the Faculty of Theology, to be sent to the Council of Trent to guide it in its 'most arduous business'; see Bruce Mansfield *Phoenix of His Age: Interpretations of Erasmus: c 1550–1750* Erasmus Studies 4 (Toronto 1979) 41. Henten's work became the basis for the *Index expurgatorius* published in Antwerp in 1571; see Jésus Martinez de Bujanda, ed *Index d'Anvers 1569, 1570, 1571* Index des livres interdits 7 (Sherbrooke, Québec 1988) 808–30. The lists of works and passages to be expurgated that were published by various authorities are reprinted in LB X 1781A–1844C (Spanish and Roman), 1844C–D (Alexander VII), and 1844D–E (Council of Trent).

94 Pio *Responsio paraenetica* 24^{V}P–25^{V}Q

Pio identifies particular groups that have been guided by the Spirit over the centuries. Christ revealed his message only slowly to his followers and did not explain every truth, but left many things to be established by the power of the Holy Spirit. At the beginning, the Apostles, St Paul, and the elders of Jerusalem were given the special assistance of the Spirit for determining church practices and beliefs. This Spirit has continued over the centuries to guide and govern the church, distributing his diverse gifts when and where he pleases. He has not revealed all the mysteries at once, nor expressed everything in the same way. Truths have unfolded over time and special persons have helped in the process. Just as in the Old Testament God sent to his people in their times of need leaders and prophets full of the Spirit, so too in New Testament times Christ has always provided his church in times of distress with holy and learned men guided by his Spirit. He has inspired not only the church Fathers of the early centuries but holy and learned men in various lands and times. To help theologians discern the truths of the Gospel in an exact and more fitting way, the scholastic method was 'divinely invented.' Dialectical reasoning has been applied with good results to such difficult theological questions as providence, the Trinity, the nature of Christ, and his real presence in the Eucharist. Natural reason and the action of the Holy Spirit have provided for the needs of the church. Christ's Spirit is notably present in councils that define church teachings and his guidance extends to the whole Christian people, whose public consent to conciliar definitions is important, if not essential, for determining if something is divinely instituted.[95]

The Holy Spirit's special assistance to the popes is not a theme developed by Pio. He asserts that Peter had a power superior to that of others for defining church teachings and he insists that the Apostolic See has always and most solidly sustained the purity of the faith. Indeed the integrity of the faith stands by reason of the solidity and wisdom of the apostolic See. But Pio does not claim for the popes a special inspiration from the Holy Spirit. Indeed he leaves open the possibility that some decisions may be in error because of a lack of proper information. But such decisions are not on major issues, and the popes have been guided over the centuries by the opinions of orthodox Fathers and saints. Although Pio cites the scriptural text on Peter's receiving a special revelation from God to recognize Christ's status as the Son of God, he does not develop the theme of a special illumination for Peter's successors. Rather, he is very reluctant to concede that a

* * * * *

95 Pio *Responsio paraenetica* 19[r]N, 20[r]N, 22[v]O, 30[r]X–Y, 32[r]Z, 53[v]K, 60[r]M, 68[v]R, 72[r]T, 73[r]T, 96[r]K; *XXIII libri* 102[r–v]T, 178[v]E, 204[v]V, 228[v]X

doctrine could be based merely on a pope's claiming a private inspiration that is not recognized publicly as certain and manifest. The ultimate test for validity is the public acceptance of a decree by the whole Christian people over time. Pio mentions Leo x's condemnation of Luther's teachings and notes that these heretical beliefs have not been accepted outside Germany, nor by the highest authorities within the Empire.[96]

Because of his conviction that the Holy Spirit has effectively guided the church over the centuries, Pio readily acknowledges that it has changed over time, indeed properly so, and hence Erasmus' criticism of current beliefs and practices based on an appeal to antiquity is rendered almost irrelevant. Against various forms of the principle of *sola scriptura*, Pio argues for a divinely inspired tradition that is open to change.[97]

In Pio's system all church practices and even our understanding of church doctrines are susceptible to change. He notes that many things are merely human traditions. When the situation that gave rise to them changes, they too should adapt to the new circumstances. With the help of the Holy Spirit the church eliminates whatever human traditions are found to be superstitious, empty of meaning, or chilling to piety. In addition to the mutable practices of human invention, there are those of divine institution, expressly prescribed in the Bible; yet one must not assume that they are forever immutable. Thus the church no longer keeps holy the Sabbath to commemorate the seventh day of creation (Exod 20:8–10), but rather Sunday to recall the day of Christ's resurrection. Today's preachers no longer go forth without purse or sword (Luke 22:35–6), but are appropriately supplied with material things. The prohibition of the Council of Jerusalem against consuming blood or what has been strangled (Acts 15:29) is no longer observed. Such prescriptions were intended to last only for a time and have been properly abrogated under the guidance of the Holy Spirit. Things considered of divine institution and defined as such by the church come either from sacred Scripture, or from an oral tradition originating with Christ and the apostles and successively passed on and received over the centuries, or from a fresh outpouring of the Holy Spirit that has been confirmed by the church. But even the things defined by the church as being of divine institution can be changed or abrogated if the church is instructed to do so by a new divine revelation that is open, certain, and manifested repeatedly in various ways. When responding to such a revelation, the church is not so much changing the old or instituting something new on its own, but rather

* * * * *

96 Pio *Responsio paraenetica* 22^{v}O–23^{r}P, 61^{r}N, 65^{r}P, 97^{r}M; Pio *XXIII libri* 196^{r}E, 228^{r-v}I–K
97 Pio *Responsio paraenetica* 6^v–7^{v}F, 18^{v}N

promulgating a divine sentence. Pio provides no example of such a new revelation.[98]

The changes that Pio describes fit into a pattern of progress from what is unclear, implicit, and poorly understood at the beginning towards what is more clearly expressed, perfected, and defined in his own days. Pio advances what we would call today a developmental theory of Christian practice and doctrine and he explains his views with a biological analogy. The church is like a human being in having stages of development. At the time of its birth and infancy its features are not yet formed and its body is feeble. This babbling babe in swaddling clothes is not an adequate model in all things for later generations. Over time the church grew and entered its childhood and adolescence. What had been instituted by Christ and the apostles became better understood, more clearly defined, and more fully perceived and put into practice. Teachings not explicitly found in the Bible were formulated and approved under the guidance of the revealing Spirit. Thus Christ's descent into hell became part of the Apostles' Creed, his relation to the Father was defined as *homoousion* (of one substance) by the Council of Nicaea, and the Spirit was declared to proceed from the Father and the Son (*filioque*). As the church grew in strength, it gradually admitted, approved, and promulgated practices that would have been difficult for it to put into effect at its origins, such as requiring auricular confession and giving effect to the doctrine of papal primacy. As the church entered its adulthood and grew robust, it perfected and defined more of its beliefs and practices. Pio does not continue the analogy into old age and death. While pointing out to Erasmus, and other proponents of antiquity as the norm for the contemporary church, how inappropriate it would be for an adult to dress in child's clothing, Pio is careful to concede that the early church was under the influence of the spiritual ardour of Christ and his Apostles, but for how long he does not know. Of greater importance than the antiquity of a practice for discerning whether something is of divine institution and promotes religion is the opinion of learned and holy men who are guided by the Spirit.[99]

* * * * *

98 Pio *Responsio paraenetica* 51^{r}I; Pio *XXIII libri* 90^{r}M, 106^{v}I, 225^{v}C, 228^{v}I–K. Prior to the decree of the fourth session of the Council of Trent (8 April 1546) declaring that revelation ended with the death of the apostles, theologians contemporary with Pio, such as John Fisher, Bartholomaeus Latomus [Steinmetz], and Johannes Mensing taught similar ideas about post-apostolic revelations; see George H. Tavard *Holy Writ or Holy Church: The Crisis of the Protestant Reformation* (New York 1959) 159–62.

99 Pio *Responsio paraenetica* 7^{r-v}F, 18^v–19^{r}N, 20^{r-v}N; Pio *XXIII libri* 95^{r}E, 190^{r}G, 204^{v}V, 228^{r}H. Although not alone in advancing a theory of the development of church doctrine and

Pio's skills as a theologian have impressed scholars. In recent times Hubert Jedin (1900–80), a noted historian of the Council of Trent, has commented on Pio's position on the question of sacred images: 'Up to the Council no one had better confronted the problem in its philosophical-religious aspect with the profundity and care of the count of Carpi.'[100] In Pio's own times his adversary Erasmus was so taken by the wealth of the scriptural and patristic sources cited and by the subtlety of Pio's syllogistic reasoning that he doubted that a mere layman who was not university trained could have composed this work.[101] Indeed, impunging Pio's authorship of these anti-Erasmus publications became one of Erasmus' principal ways of counter-attacking. Erasmus' accusations will be examined in some detail, but before ending our treatment of *xxiii libri*, the question of its authorship needs to be addressed.

THE AUTHORSHIP OF XXIII LIBRI AND ITS FIRST PRINTING

Erasmus raised a number of questions about the authorship of *xxiii libri*. That Pio was capable of composing this work is supported by what is known about his talent, training, and theological interests.[102] That Juan Gínes de Sepúlveda helped him with it, notably by polishing its style, is highly unlikely, not only because of Sepúlveda's explicit denials, but also because he remained in Rome while Pio wrote in Paris.[103] That the busy papal nuncio in Venice, Girolamo Aleandro, spent his time collecting material from Erasmus' writings to send to Pio in Paris is both improbable and disavowed with an oath as completely false and indeed silly in the nuncio's letter to

* * * * *

practice, Pio seems to have been unusual in employing an analogy based on the stages of biological growth. For a brief overview of earlier and contemporary exponents of doctrinal development, see John W. O'Malley 'Giles of Viterbo: A Sixteenth Century Text on Doctrinal Development' *Traditio* 22 (1966) 445–50, reprinted in his *Rome and the Renaissance: Studies in Culture and Religion* (Aldershot, Hampshire 1981) entry III; Tavard *Holy Writ or Holy Church* (cited in n98 above) 151–71 for Pio's contemporaries who held that the Holy Spirit continues to inspire the church over time; and John F. D'Amico 'A Humanist Response to Martin Luther: Raffaele Maffei's *Apologeticus*' *Sixteenth Century Journal* 6 (1975) 37–56, especially 53–4 and reprinted in his *Roman and German Humanism, 1450–1550* (Aldershot, Hampshire 1993) entry IX.

100 Hubert Jedin 'Entstehung und Tragweite des Trienten Dekrets über die Bilderverehrung (1935/63)' in his *Kirche des Glaubens, Kirche der Geschichte: Ausgewählte Aufsätze und Voträge* 2 vols (Freiburg 1966) II 460–98, especially 466

101 Allen Ep 1804:254–6; Erasmus *Apologia* 156–7, 181, 260, 284 below

102 For a treatment of this issue, see Minnich 'Debate on Images' 380–5.

103 Minnich 'Debate on Images' 386–7

Erasmus.[104] That Pio was assisted by members of the Faculty of Theology at the University of Paris was dismissed as 'far from the truth' by Pio's personal secretary Francesco Florido,[105]although his statement does not preclude the possibility that Pio was helped indirectly by the theologians. To facilitate the deliberations that led to their 1526 and 1527 censures of some of Erasmus' works, each master of theology was furnished with a set of notes, written on small sheets of paper, containing controversial statements excerpted from Erasmus' writings by specially deputized members of the faculty. Given the numerous sets of these Erasmian quotations in circulation in Paris during the time Pio was in that city, and the concerns he shared with members of the faculty, whose writings critical of Erasmus he praised, it would not be surprising if Pio were helped indirectly by having access to their notes.[106] That Pio hired research assistants to comb Erasmus' works for controversial passages was dismissed as patently false by Florido, who insisted that he was the only person in Pio's domestic employ at Paris who acted as his amanuensis, reader, researcher, and editor.[107] When Pio died, leaving the *XXIII libri* only partially revised, it fell to Florido and the publisher Josse Bade to bring the draft manuscript to a publishable state with

* * * * *

104 Allen Epp 2329:105–7, 2371:34–6, 2375:78–9, 2379:110–13, 2411:49–50, 2414:14–16, 2466: 98–9, 2572:11–15, 2638:1–25, 2639:7–48, 2679:21–33; Hugo Laemmer, ed *Monumenta Vaticana historiam ecclesiasticam saeculi XVI illustrantia ex tabulariis Sanctae Sedis Apostolicae Secretis una cum fragmentis Neapolitanis ac Florentinis* (Freiburg 1861) 99 no 74; on Aleandro's Venetian nunciature, see Franco Gaeta *Un nunzio pontificio a Venezia nel Cinquecento (Girolamo Aleandro)* Civiltà Venezia, Saggi 9 (Venice/Rome 1960)

105 Florido *Libri tres* 264:28–41: 'At id tam sinistre accepit Erasmus, ut immerentem heroëm, preter decorum calumniari non sit veritus, sed scommatibus paroemiisque tanti viri maiestatem devitans, atque sapissime repetens, Parisienses Theologos alios ei phrasim, alios loca, alios alia suggessisse: quod tamen a vero plurimi abesse cum omnes qui Lutetiae Parissorum cum paulo familiarius noverant, tum ipsa constantius quam ceteri omnes asserere queo: quando per ea tempore, quibus is princeps illud opus scribebat neminem qui literis operam dedisset in domestica consuetudine habebat, preter me unum iuvenem et admodum, quem Rome ... et Ioannes Sepulveda Cordubensis ... pro illo scripsit defensione, non dissimulat.' Bade, however, refers to more than one *familiarus* in his note on Pio *XXIII libri* 171ʳH.

106 For a fuller treatment of this, see Minnich 'Underlying Factors' 18.

107 See above n105. Erasmus, however, claimed to have first-hand testimony of this assistance (Allen Ep 2810:104–6), and secondary evidence in letters from friends telling him that Pio was assisted by mercenaries who looked for passages in his works that seemed to agree with Luther's views (eg Allen Ep 2328:42–50); also a report from Gerard Morrhy to Erasmus mentioned various helpers, and in particular a certain promising youthful Frisian, Gerard by name, who for some months was employed by Pio to comb Erasmus' annotations on the New Testament for offensive passages (see Allen Ep 2311:20–35).

the greatest accuracy and fidelity possible.[108] That Pio's manuscript was subjected at this stage to major revisions by the theologians and friars of Paris
is not borne out by any notable difference in the styles of the sections Pio
personally revised and those he did not, or by the obvious need of editorial work in some of the later unrevised chapters.[109] The reasons advanced
by Erasmus for questioning Pio's authorship of the XXIII *libri* are without
substance.

The printing of Pio's XXIII *libri* seems to have been done in stages, and
Erasmus was given not always reliable reports by his informants on how
the volume was progressing. By June 1530 he knew that Pio had put together a huge tome. At the end of November Erasmus mistakenly claimed
that the book had already been printed and could not be recalled, yet he
seemed to be urging Lorenzo Campeggi to stop its distribution. In his later
correspondence Erasmus repeated the story that the book was being printed
while Pio still lived. He suggested that the Franciscans under the leadership of the guardian of their Paris monastery, Pierre de Cornes, were chiefly
responsible for seeing that the whole work was printed. In his letter to Jacopo Sadoleto of 7 March 1531 Erasmus reported that people who had seen
the published tome said it was huge. In a letter of 16 April he claimed that
the Paris theologians were also involved and suspected that Noël Béda had
a notable part. When Erasmus received a copy of the printed work is not
clear, but on 20 August 1531 he reported that he had responded in a few
words (*perpaucis*). The copy sent to him did not include the index. When it
arrived, he wrote a separate rejoinder to it.[110]

* * * * *

108 Florido *Libri tres* 264:26–8: '*Quod sane opus, cum autor morte praeventus, summam ei manum
imponere non potuisset, utcunque partim eo, partim a nobis recognitum in lucem exiit.*' In his
letter to the reader of the XXIII *libri*, Bade describes his role as: '*quanta maxima potui,
et fide et accuratione coimpressi: strenuam admodum operam in eis ex archetypo transcribendis
et recognoscendis, mihi navante ornatissimo cum literis cultis, tum moribus iuvene, Francesco
Florido natione Sabino et Comitis ipsius, dum inter mortaleis agebat, per multos annos contu-
bernali et amanuensi*' (Pio XXIII *libri* sig ãi[v]). But at 171[r]H Bade states that Florido was not
alone: '*omnia tamen scripsit [Pius], et ab eius familiaribus fideli solertia ex Archetypo exscripta
nobis sic tradita sunt.*'

109 Erasmus *Apologia* 237, 254–5 below; Allen Ep 2486:36–41; even when the editor of the
French translation of Pio's *Responsio paraenetica* altered the text by dropping and adding
material, he apparently did not make any substantive changes. See the example of the
minor alterations to book 8 on sacred images noted in Minnich 'Debate on Images' 383–4
n5.

110 Allen Epp 2329:96–8, 2411:55–6, 2441:66–70, 2443:336–42, 2466:99–101, 2486:36–40;
2522:79; Erasmus *Scholia* m2[r]; letter of Campeggi to Jacopo Salviati (secretary of Clement
VII) 20 December 1530, from Cologne, '*Forsi non seria male che l'opera che'l dice del s. Alberto
non si publicasse*' in *Nuntiaturberichte aus Deutschland nebst ergänzenden Aktenstücken, erste
Abteilung 1533–1559, I. Ergänzungband 1530–1531: Legation Lorenzo Campeggios 1530–1531
und Nuntiatur Girolamo Aleandros 1531* ed Gerhard Müller (Tübingen 1963) 201

An examination of the Bade edition suggests that Pio's work went through three principal stages of production. The first section contained Erasmus' letter to Pio of 10 October 1525 and Pio's *Responsio paraenetica* of 15 May 1526. These were printed on the first forty-six folios, comprising octavo quires a to e and sexto quire f. The decision to use the smaller quire probably indicates a desire to bring this phase of the printing to a clean conclusion with no blank pages. No indication is given as to the title of the whole work and Pio's *Responsio paraenetica* has not yet been designated as book 1. The numbering of all subsequent books is clearly indicated both at the beginning of each book after its title and in the running heading on the front of each folio. Could this initial exchange between Erasmus and Pio have been printed earlier as an independent publication – as a second edition of the *Responsio paraenetica* – but then made part of a larger work? The second stage of production probably began while Pio was still alive, given the epigram at the end of the preface addressed to him as though he were still living, and the insertion into the text on folio 171^r, about three-quarters of the way into the volume, of a statement by Bade that Pio had reread and approved everything up to this point, but that because of his death Bade had to depend on a copy of Pio's original draft faithfully made by members of his household. The 'Conclusion of the Work and End of the Disputation' is numbered as book 24. As yet no indication is given as to the title of the whole work. The numbering of the folios and octavo quires is continuous with that of the first section and ends with folio 252^r (misprinted as CCXLII) and with the only other sexto quire lettered I. The colophon, dated 9 March 1531, notes that Pio had died in January and that a favour and privilege were printed on the back of the first page. The third and final section to be printed contains the introductory material, printed on an octavo quire lettered ã. That this was printed some time later is suggested by a number of factors. Whoever composed this material did not take care to harmonize it with all that was printed earlier. The title that is finally given to the whole work, *XXIII libri*, while reflecting the decision to call the chapters *libri* and Pio's own references to them as *libelli,* ignores the numbering of the conclusion as book 24. The royal privilege granting Bade exclusive rights in the French realms to be the sole printer and vendor of the work for four years is dated 15 March 1531, and thus was not issued until six days after the date given in the colophon for the printing. The table of contents and index are so extensive, consisting of twenty-five columns of entries with their folio and letter subsection designations, that they must have required a good deal of time to compile. That Erasmus wrote his response without having this introductory quire is evident in his apparent ignorance of the work's title; in his anger at not knowing who had assisted Pio whereas Bade in his letter to the reader clearly identifies Florido as Pio's long-time attendant (*contubernalis*) and secretary (*amanuensis*) who helped edit the work; in his treating the conclusion as book 24; and in his statement that the index was sent to him after he had written his response.[111] While others may have given to the XXIII *libri* over time its title, index, format, and a readable presentation of the material that Pio did not live to revise, this evolving volume nevertheless deserves to be considered the principal literary legacy of Alberto Pio, and not the work of a team of unnamed researchers and writers, as claimed by Erasmus.

PIO'S DEATH AND FUNERAL AND ERASMUS' MOCKERY OF THEM

The death of Pio was not unexpected. As his health continued to fail, he not only worked to put his manuscript into publishable form, but also attempted to provide for the welfare of his family and of his own soul. The earlier negotiations to have Alfonso d'Este grant Pio an annual subsidy of three thousand *scudi* and to recognize Pio's daughter as heir to Carpi after his death never secured Pio's consent, although d'Este and Clement VII were agreeable to these terms. Because Pio lost Carpi as a result of the services he rendered to Francis I, the French king tried to compensate him in various ways. While Pio's French fiefdoms of Conches and Bretueil could not pass to his daughters, the king promised him a large sum of money which could be willed to them. At the time of Pio's death the king still owed him 98,864 *livres*. As readily available dowries, Pio left for each of his daughters in the bank of San Giorgio in Genoa twenty thousand ducats.

The principal way in which Pio provided for his approaching death was by drawing up a will. On 21 July 1530 he dictated his will at his home in the presence of eight witnesses: Bishop Lorenzo Toscani, his father-confessor Friar Martin Rogerii OFM, another Franciscan friar, two secular clerics, and three laymen. He set up as its executors important persons who were personal friends: Anne Montmorency (Grand Master of France and son of his good friend Baron Guillaume), Antonio Pucci (former papal nuncio to France and bishop of Pistoja), Gian Matteo Giberti (fellow promoter of the papal-French alliance and bishop of Verona), Toscani (for-

* * * * *

111 Pio XXIII *libri* 72ᵛ (epigram), 251ʳv (*pro libro* XXIIII), 252ʳ (colophon), ãiʳ (title), 165ʳ[v], 213ᵛʀ (*libellis/libello*), ãiʳ (royal privilege), ãiiʳ–viiiʳ (table of contents), ãiᵛ (description of Florido); Erasmus *Scholia* m2ʳ (Index sent to Erasmus after he wrote the *Apologia*); in Erasmus' *Apologia*, Froben edition of 1531, page 5 (aa3ʳ): *D. Erasmi Roterodami Apologia brevis ad vigintiquatuor libros Alberti Pii quondam Carporum comitis*, Erasmus adds seemingly as part of the title what is in fact the *incipit* of the *Apologia*: *Nimirum Albertus Pius rem perficit.*

mer papal nuncio to France, fellow northern Italian residing in Paris, and bishop of Lodève), and his first cousin Gianfrancesco Pico (count of Mirandola). The lengthy will made detailed provisions for the disposition of his property, the relocation of family tombs, gifts of money and personal items, petitions to the king and the pope for compassion on his heirs, and prescriptions for his last rites, funeral, burial, and prayers for the repose of his soul.

Pio divided his feudal estates and property among his close family members and other relatives. He made his brother Leonello heir of the imperial fiefdom of Carpi, with one of Leonello's sons (first Rodolfo and then Alberto) as his brother's single designated heir; his daughters Caterina and Margherita were given the ecclesiastical territories of Novi and Meldola and various properties in Carpi and Mantua, with the elder daughter, Caterina, receiving a greater share to encourage her marriage prospects; his wife, Cecilia, was to be provided income so that she could live in an honest and decent state of widowhood and her dowries were to be restored to her should she remarry; and his half-brother Teodoro was to receive the enjoyment while he lived of other named properties in Carpi. Pio begged his heirs to maintain peace and concord among themselves and threatened them with the loss of specific properties should they sue each other over the inheritance. In the event that these close relatives did not leave legitimate male heirs, the estates were to be divided equally between one of the sons of his first cousin Gianfrancesco Pico and one of the sons of his half-brothers (that is, a son of Gianfrancesco Gonzaga, lord of Luzzara, or of Luigi Alessandro Gonzaga, lord of Castelgoffredo, Solferino, and Castiglione) – provided the heirs changed their names and coats of arms exclusively to those of Pio di Carpi. Should there be no heir who accepted such terms, Pio willed his estates to the Roman church provided it established a bishopric in Carpi and named its bishop the count of Carpi. If the bishop failed to reside in Carpi, his annual revenues beyond five hundred ducats were to go to establishing and maintaining an observant Benedictine monastery in Carpi. He required of his heirs, his brother Leonello and his daughters, Caterina and Margherita, that they pay off any and all debts which he may have incurred.

Pio made provisions for his own tomb in Paris and for those of his relatives already interred in Carpi. He asked to be buried with his own tombstone among the Friars Minor in Paris. In the place in the church of San Francesco in Carpi that had been designated for his own burial, Pio ordered that the marble tomb of his grandfather Alberto II be located. In the church of San Nicolò in Carpi where his father Lionello is buried, he asked that an epitaph in his own memory be set up, that the office of the dead be said for him, and that the bones of his infant son, Francesco, be transferred there.

On the assumption that Carpi would be recovered, Pio made generous bequests. He left a thousand ducats, plus an annual payment of one hundred ducats, to complete construction of the major church of Carpi and required his heirs to complete the construction of the monastery of Augustinian canons attached to that church, a project to which his father Lionello had committed him by his will and to which Alberto now committed his heirs. To this monastery he gave the Church of Our Saviour next to his palace in Carpi and half the furnishings from his own private chapel in Paris, ordering his daughters to complete work on the church. To the various monasteries, churches, and confraternities of Carpi he bequeathed specific sums of money for construction, repair, decoration (including the commissioning of religious paintings), and the catch-all category of fabric. He also provided for the establishment of a college with an honest stipend for a teacher to instruct twelve poor but talented boys from Carpi, plus two or three boys from elsewhere, in grammar and the arts.

Even if Carpi were not recovered, Pio ordered his heirs to provide for his faithful servants and their families. Among those singled out for specific bequests were the family of his former chancellor, the master of his house, three familiars and their heirs, his cook, and a painter. He urged his heirs to favour those who had been his supporters in Carpi and had therefore suffered reprisals. It is interesting to note that Francesco Florido, his faithful secretary, amanuensis, researcher, and editor of the *XXIII libri*, is not mentioned in the will.

Personal gifts were given to several persons as tokens of his affection: to his half-brother Teodoro a silver cross, to his sister a book of hours and a rosary, to his niece a rosary and a *pax*, to his friend Giberti a silver globe or *mappamondi*, to his nephew Rodolfo his home and vineyard at the foot of Monte Mario in Rome together with his books, art works, and antique collection. Duplicate copies of his books and codices he gave to the library of the monastery of San Nicolò in Carpi.

Pio begged King Francis and the king's mother, Louise, and also his friend Pope Clement VII, to take pity on his wife and daughters whom he felt he was leaving in dire straits.

Pio also made detailed provisions for his death and funeral. Having commended his soul to God and specific intercessor angels and saints and having protested his orthodoxy and desire to receive the church's last rites, he asked that Franciscan friars console him on his deathbed, clothe him in the garb of their order in which robe he asked to be buried according to their rituals, and serve as the porters of his bier. He requested that the

Franciscans grant him a final resting place among them. The various funeral rites and specific place of burial he left to his executors to decide. He also provided for the saying of prayers for the repose of his soul, both in Paris and in Carpi, at the time of the funeral and afterwards.[112]

* * * * *

112 On Pio's loud protests (*gridi al cielo*) over Francis I's agreement to Alfonso d'Este's possession not only of Carpi but also of Novi di Modena as part of the conditions for the marriage of Renée Valois to Ercole d'Este, see ASV SS Francia I 50^r and 51^r (letter of Giovanni Salviati to the nuncio at the imperial court, 6 November 1527). On d'Este's offer of an annual subsidy of three thousand *scudi* (see n116 above) and recognition of Pio's daughter as heir to Carpi, see ASV SS Francia I 127^r (spring 1528). On the dowry money in Genoa, see Biondi 126; in the early sixteenth-century the sum of twenty thousand ducats, presumably Venetian ducats, would have been worth £4,500 sterling or 41,500 *livres tournois*; from 1517 the name of the Venetian ducat was changed to *zecchino* [*sequin d'or*], and by the date of Pio's last will, in 1530, its value had risen to 56d sterling and 2.275 *livres tournois* [45s 6d *tournois*], so that this sum would then have been worth £4,667 13s 4d sterling or 45,500 *livres tournois* (see CWE 12 650). On Francis I's financial debt to Pio, see Rémy Scheurer ed *Correspondance du Cardinal Jean du Bellay* I: *1529–1535* (Paris 1969) 231 n2; and *Actes de François Ier* VII 612 n27603 (the king orders the treasurer and receiver-general of Languedoc to pay henceforth each year to the widow of Alberto Pio ten thousand *livres tournois* from the tax income on salt, to commence with the next quarter period beginning in July, until the full sum of 98,864 *livres*, 16 *sous*, and 6 *deniers tournois* has been paid [these sums are expressed in the French silver-based money of account, the *livre tournois*, with 20 sols or *sous* to the *livre*, and 12 currently circulating silver *deniers* to the *sou* (see CWE 1 328, 331–2); the full sum (98,864.825 *livres*) would have been worth (from 1526) 43,457 ducats or £10,140 sterling, the equivalent of 405,600 days' wages for a master carpenter at Oxford or Cambridge, still earning only 6d per day (see CWE 12 694–9); the annual pension of ten thousand *livres tournois* to Cecilia would have been worth about 4,396 gold ducats (or florins) or £1,026 sterling, the equivalent of 41,040 days' wages for an Oxford or Cambridge master carpenter]; II 392 n5717 (a similar order, dated 22 April 1533, to pay to Alberto's and Cecilia's daughters, Caterina and Margherita, the sum of twenty-five thousand *livres tournois* from the same amount owed [this sum would have been worth 10,989 ducats or about £2,564 sterling, the equivalent of 102,564 days' wages for a master mason at Oxford or Cambridge]); VII 538 n26615 (a similar letter with the amount of ten thousand *livres* per annum [see above] to be paid to the daughters). For Pio's other provisions for the welfare of his wife, daughters, and future grandsons regarding their rights to feudal properties and lands around Carpi, Mantua, and Meldola, see also the full translation of his will 391–6 below. He also made provision for the support of a number of religious foundations, for example, one thousand ducats (worth £233 6s 8d sterling or 2,275 *livres tournois*) to the major church in Carpi to complete its construction, plus an annual payment of one hundred ducats (worth £23 6s 8d sterling or 227 *livres* 10 *sous tournois*). After Pio's death his widow and daughters returned to Rome. His older daughter, Caterina, married Bonifacio Caetani di Sermonetta, inherited her father's palace in Campo dei Fiori, and was a generous patroness of the early Jesuits in Rome; his younger daughter, Margherita, married in 1544 Giangerolamo Acquaviva and their son, known as Blessed Rodolfo Acquaviva, was a Jesuit martyr in India. See Litta 'Pio di Carpi' table III and 'Acquaviva di Napoli' table V; and Pietro Tacchi-Venturi *Storia della Compagnia di Gesù in Italia* (Rome 1910) I 661 no 160 n8. The early Jesuits who studied in Paris had a high opinion of Pierre de Cornes, OFM, who preached at Pio's funeral.

Pio enjoyed a special relationship with the Franciscans in life and death. His paternal great-aunt Camilla, his sister Caterina (Sister Angela Gabriella), his niece (Sister Caterina Angelica), and his half-brother Teodoro all belonged to that order, as did one of his early teachers of Scotism, Graziano da Brescia. As lord of Carpi Pio had sumptuously renovated the Franciscan church of San Nicolò and then invited the order to hold its general chapter there in 1521, paying personally for all of the expenses of the meeting. In gratitude the Fransicans granted a *fratellanza* between the order and the Pio family. As death drew near, Pio consummated his alliance with the Seraphic Order by adopting the garb of a Franciscan friar. Three days later, on 8 January 1531, he died, the cause of death being variously given as complications related to gout (*articulari morbo*), the plague (*peste*), and congestion (*sanguinolenta*).[113]

Both Francis I and the Franciscans of Paris were determined that Pio be honoured in death. On learning of his demise the king, then at the royal Château des Saint-Germain-en-Laye five leagues from the capital, wrote on 10 January to the officials and people of Paris urging that the funeral for his much beloved cousin and devoted servant be conducted 'with every honour possible.' In a similar letter to the court of the Parlement de Paris the king explained more fully why honour should be shown to Pio: 'the former count of Carpi was a stranger, had performed many services for the king, had left behind his lands and feudal rights, was a personage of letters and of great esteem, [and a member] of his order.' Francis I also sent his royal councillor and master of the palace, Robert de la Martonnye, who met with the city officials on the following day to arrange the ceremonies according to the king's wishes.[114] But because Pio had donned the Franciscan robe just prior

* * * * *

113 On Pio's family members who were Franciscans, see Litta 'Pio di Carpi' table III. On Pio's renovation of San Nicolò, see Adriana Galli 'Alberto III Pio: L'impianto urbanistico rinascimentale' in *Materiali per la storia urbana di Carpi* ed Alfonso Garuti et al (Carpi 1977) 55–64, especially 56, 67 nc57; and Galli 'Presenza religiosa' 71–80, especially 74. On the special brotherhood between the Pio family and the Franciscan order, see Guaitoli 'Sulla vita' 232. On the granting to a *confrater* the right to be buried in the habit of a Friar Minor, see John Moorman *A History of the Franciscan Order from Its Origins to the Year 1517* (Oxford 1968) 355. On Pio's adopting the Franciscan habit three days prior to his death, see Allen Ep 2441:72–3. On the claim that Pio also promised to become a Franciscan, see Erasmus *Exequiae seraphicae* ASD I-3 687:44–7 / CWE 40 1000:40–1001:3. On the cause of his death, see ASD I-3 687:30–4 / CWE 40 1000:26–30; Sepúlveda *Antapologia* Di^r / Solana 127 (gout); Tuetey 102 n1 (plague); Pio *XXIII libri* āviii^v poem of Florido (congestion). Because Florido lived in Pio's household and was his personal assistant and secretary, more credibility should be given to his account. On 8 January 1531 as the date of Pio's death, see Guaitoli 'Sulla vita' 307; and Biondi 126.

114 The king's letter is transcribed in Tuetey 101–2. His reference to Alberto III as his cousin may perhaps be explained by the facts that the king's mother was Louise de Savoie and

to his death and had asked that the burial rituals of that order be followed, his wishes as well as the decisions of his will's executors would have to be respected.

The funeral was held on Monday, 16 January, with elaborate ceremony. Having assembled at the Hôtel de Ville at eight o'clock in the morning, the mourners made their way north to Pio's residence on the rue Saint-Antoine near the royal palace of les Tournelles. There at nine o'clock the procession formed to accompany Pio's corpse across town to the great Franciscan church of Sainte-Marie-Madeleine, according to the royal wishes. Pio's body was dressed in a Franciscan habit with the face and hands uncovered.[115]

The composition of Pio's funeral cortège reflected traditional practices, the instructions in his will, the decisions of his executors, and the particular wishes of the king. Leading the procession were the customary representatives of the four mendicant orders: first the Franciscans, and then the Dominicans, Augustinians, and Carmelites.[116] Next came delegations from five religious establishments in town: the Frères de la Charité Notre-Dame, known as Billettes from the location of their hospital on the rue des Billettes, whose chapel commemorating a miraculous host became a very popular pilgrimage shrine;[117] the Hermits of Saint Guillaume, popularly called the Blancs-Manteaux because they took over from 1297 to 1618 the church

* * * * *

that Pio's grandfather Alberto II and his family were aggregated to the family of Savoia in 1450 and thereafter bore the full title of Pio di Savoia; see Fiorina 66, MS 262 n7; and Litta 'Pio di Carpi' table II. Pio may also be considered a cousin of the king because Francis I's maternal (Louise de Savoie) aunt's (Filiberta de Savoie) husband (Giuliano dei Medici) was Alberto Pio's wife's (Cecilia Orsini) father's (Franchiotto Orsini) first cousin; see Litta 'Duchi di Savoja' table XII. For Francis I's instructions to the court of the Parlement of Paris, see Archives nationales (Paris), Registres du Parlement de Paris, Conseil, coté X/1a/1534 72ʳ. I am grateful to James K. Farge of the Pontifical Institute of Mediaeval Studies (Toronto) for having called this reference to my attention, to Jacques M. Gres-Gayer of the Catholic University of America for helping me to contact the Archives nationales, and to Dr Florence Clavaud of the Centre d'accueil et de recherche des Archives nationales for having graciously provided me with a photocopy of the register entry.

115 Tuetey 102–3. For similar funeral cortèges, but with apparently fewer members of religious orders (especially Franciscans), see the descriptions of the funerals of Louise de Savoie on 17 October 1531 (Tuetey 129–32), and of Jean de la Barre on 5 March 1534 (Tuetey 176–7). For a brief description of the cortège of Francis I on 21 May 1547, see Robert J. Knecht *Francis I* (Cambridge 1982) 420–1. Pierre de Cornes gave a speech at the welcoming ceremonies for the cardinal legate on 20 December 1530 (Tuetey 94).

116 Tuetey 102

117 On the Billettes, see de Sivry and Champagnac I cols 178–9; for a fuller account of the Miracle des Billettes in its various renditions and popular reception, see Marilyn Aronberg Lavin 'The Altar of Corpus Domini in Urbino: Paolo Uccello, Joos Van Ghent, Piero della Francesca' *Art Bulletin* 49 (1967) 1–24, especially 3–10.

of Notre-Dame formerly belonging to the white-mantled Servites;[118] the Augustinians of the cloister of Sainte-Catherine-de-la-Couture (also called du Val des Écoliers because of their school), which was near Pio's residence;[119] a delegation from the foundling Hôpital du Saint-Esprit on the rue de Martroi;[120] and finally a delegation from the hospital l'Hôtel-Dieu with its Augustinian brothers and Franciscan sisters.[121] Following these members of religious orders came the mourners *de corps et de vins*, and after them up to one hundred of the city's poor in mourning robes and carrying torches.[122] Next, accompanied by torch bearers, marched a delegation from the city's militia, composed of four archers, four cross-bowmen, and four halberd carriers. The next group was made up of priests from the churches of Saint-Paul (Pol) and Saint-Gervais, parishes near Pio's residence whose pastors were doctors of theology at the University of Paris, Etienne Le Roux of Saint-Gervais being known as a stalwart opponent of heresy.[123]

The rest of those marching in the procession, except for the Franciscan friars actually carrying the bier, were laity. First, in the section of mourners around the bier were the officials and servants of Pio's household, all dressed in mourning clothes. Among these was probably his faithful secretary, Francesco Florido da Sabina.[124] The next group consisted of nine gentlemen dressed in mourning garb, eight of them mounted on large horses each draped with a long black saddle-cloth decorated with a cross of white taffeta, and each gentleman carrying an item that highlighted Pio's status as a member of the old aristocracy – a noble of the sword. The items were displayed in the following order: a square banner (bannière carrée), Pio's coat-of-arms, his battle colours (guidon), his shield and horn, his standard, his spurs, his glove and sword, his helmet and hat, and finally a sym-

* * * * *

118 Sivry and Champagnac II col 179
119 L.H. Cottineau *Répertoire Topo-Bibliographique des Abbayes et Prieures* 3 vols (Macon 1935–70) II col 2204
120 Sivry and Champagnac II col 332
121 Augustin Renaudet *Préréforme et Humanisme à Paris pendant les premières guerres d'Italie (1494–1517)* 2nd ed rev (Paris 1953; repr Geneva 1981) 15, 440–2
122 On the traditional role of professional mourners and of the poor in a funeral procession, see Philippe Ariès *The Hour of Our Death* trans Helen Weaver (New York 1981) 165–8; on the noisy manifestation of sorrow then considered appropriate, see Johan Huizinga *The Waning of the Middle Ages* trans F. Hopman (1924; repr Garden City, NY 1954) 52. The mourners were called *de corps et de vins* apparently because they were rewarded for their services with a ration of meat and wine.
123 The churches of Saint-Gervais and Saint-Pol were near Pio's residence, both in the ninth arrondissement; see Sivry and Champagnac II col 161; on their curés, see Farge *Biographical Register* 191 (Adrien Geneau of Saint-Pol), 270 (Etienne La Roux of Saint-Gervais).
124 Tuetey 102

bol of his membership in the king's Ordre de Saint-Michel.[125] After these mounted gentlemen Pio's corpse, elevated on high on a bier, was carried by eight Franciscans with another friar at each of the four corners. Carrying the four corners of the pall cloth that was draped over the bier were four distinguished noblemen: Renzo da Ceri (Lorenzo Orsini dell'Anguillara, lord of Ceri), a famous Italian general and a relative of Pio's wife, Cecilia Orsini Pio (he was the son of Cecilia's great-grand-uncle Lorenzo Orsini's daughter Giovanna, who married Giovanni di Ceri); François II de la Tour, vicomte de Turenne and former French ambassador to Rome; Alfonso de San Severino, duke of Somma, and father of Violante, who married Giulio, a soldier in the service of Francis I and the great-grandson of Lorenzo Orsini); and Luigi Carafa, prince of Stigliano. Two of these Italian noblemen had, like Pio, found refuge in France after their military support for the disastrous League of Cognac negotiated in part by Pio.[126] Following the bier were three gentlemen dressed in mourning clothes who were likely members of Pio's extended family and are identified only by their first names in French: Octovein, Cord, and Canilley. The first, accompanied by Sebastiano Giustiniani, the Venetian ambassador to France, was undoubtedly Ottavio Orsini, lord of Monterotondo and the brother of Pio's wife, Cecilia. The second, who remains unidentified beyond the name Cord,

* * * * *

125 Tuetey 102. It is not clear what is meant by 'l'ordre du Roy': while it may have designated the king's letter to the Paris officials ordering them to honour Pio's corpse, it was more likely a symbol of Pio's membership in the king's exclusive Ordre de Saint-Michel, which was limited to thirty-six very high-ranking nobles and kings. The chevaliers of the order wore a special collar of gold scallop shells from which hung an oval medallion depicting the archangel Michael thrusting Satan into hell. On this order and its garb, see Desmond Seward *Prince of the Renaissance: The Golden Life of François I* (New York 1973) 106. On Pio as a knight of Saint-Michel, see Guaitoli 'Sulla vita' 231 (Pio turned down the offer in 1520) and 246 (he accepted in 1523); and Tuetey 102 n1, quoting Francis I who called Pio 'chevalier de son ordre.'

126 That *signeur Rance* is Lorenzo Orsini is likely given his nickname in French of Raynce and his presence at the French court; see Sanuto (Venice 1899) LIV cols 121, 391; and Émile Picot *Les italiens en France au XVIe siècle (Extraits du Bulletin Italien de 1901, 1902, 1903, 1904, 1917, et 1918)* (Bordeaux 1901–18) 29–30. The *seigneur Rance* who welcomed the cardinal legate into Paris on 20 December 1530 is identified by Tuetey 94 n9 as Renzo da Ceri. Orsini, who tried to defend Rome in 1527, shared Pio's fate both in having to take refuge in Castel Sant' Angelo and in having his release negotiated; see Guicciardini *Storia d'Italia* v 148. On de la Tour, see Tuetey 94 n8; and Émile Picot *Les français italianisants au XVIe siècle* (Paris 1906; repr New York 1968) I 83. On Orsini and Carafa fighting together in Puglia for the League, see Guicciardini *Storia d'Italia* v 244. On Luigi Carafa as prince of Stigliano, see Litta 'Carafa di Napoli' table XLI; on Alfonso de San Severino as duke of Somma, see *Actes de François Ier* II 400–1 n5756; and Picot *Les italiens en France au XVIe siècle* 14. Sabattini 81 identifies the four bier escorts as the prince of Melfi, Giancorrado Orsini, Renzo da Ceri, and Claudio Rangone.

was assisted by the prince of Nice, perhaps a member of the Grimaldi de Beuil family; and the third was probably the Camillo who later appears as closely associated with Pio's widow, who was accompanied in the procession by the marquis de Moncharche, that is, Ferrente Caraffa, count of Montesarchio.

After these noblemen came officials of the royal and city governments. On the right marched the royal agents, some deputies of the Parlement and of the Chambre des comtes; on the left were the officials of the city of Paris: the provost of the merchants, the alderman, the registrar, the procurator, councillors, quarteniers, and citizens. Following them were many gentlemen, all dressed in their best outfits.[127]

The procession wound its way from the northwestern section of Paris by the Bastille, through the city centre, crossed the Ile de la Cité, and ended at the Franciscan monastery located by the city wall in the southeastern part of Paris. Many of the citizens of Paris could have witnessed this funeral cortège for it followed the principal streets of the city: down the rue Saint-Antoine, across the pont Notre-Dame, turning onto the rue Saint-Pierre-des-Arcis, crossing the inner courtyard of the palace and exiting by another gateway of the palace, proceeding along the rue de la Harpe, and turning right onto the Grand rue that goes directly to the Franciscan monastery.

At the monastery's church of Sainte-Marie-Madeleine there was a solemn funeral service. Music was sung and a requiem high Mass was celebrated by Gui de Montmirail, Benedictine abbot of Saint-Magloire, titular bishop of Megara in Greece, and auxiliary bishop of Meaux. During the Mass the Guardian of the Franciscan monastery, Pierre des Cornes, a

* * * * *

127 On Ottavio Orsini as the brother of Pio's wife, Cecilia, and present in Paris in 1531, see Litta 'Orsini di Rome' table IX; *Correspondance du Cardinal Jean du Bellay* ed Scheurer (cited in n112) I 232 nn2–3; and *Actes de François Ier* VII 658 n28062. On Grimaldi, see Sanuto LIV cols 391, 702; and Pio *Will* 396 below. For references to a Camillo who seems to have accompanied Cecilia Orsini in Rome, see Baroni ed *La nunziatura in Francia* (cited in n16 above) 175, 259–60. Cecilia had an uncle named Camillo, of illegitimate birth, who was given an estate by his uncle Rinaldo; see Litta 'Orsini di Roma' table IX. If 'Cord' was also a relative of Cecilia, perhaps he is Giancorrado Orsini, lord of Bommarzo, a noted *condotierre*, and Cecilia's brother-in-law; see Litta 'Orsini di Roma' table IX. On Giustiniani, see the entry on Sebastiano Giustiniani in CEBR II 103; he was an early friend of Erasmus and More. On Ferrente Caraffa as count of Montesarchio, see *Actes de François Ier* II 401 n5756. The deputies named to represent the court of the Parlement of Paris at the funeral and burial of Pio were Denys Poillet (president), Nicole Hennequin, Jaques Le Roux, Martin Sumet, Michel Gilbert, Claude Dezasses, Francoys de Lage, Robert Berseau, Jaques LeClerc (called *doctrer*), Jaques Spifame, Léon Lescot, Jehan de Loungueil, and Maurice Buliond *conseilleurs*; see Archives nationales (Paris), Registres du Parlement de Paris, Conseil, coté x/1a/1534 77ʳ.

doctor of theology at the University of Paris and an ardent enemy of heresy, preached a solemn sermon in praise of Pio.[128] After the funeral service Pio's body was buried in the choir area of the church, where were also buried numerous royal princes, descendants of kings and queens, noblemen, and magistrates. Francis I contributed the services of his favourite artist, Gian Battista di Jacopo (il Rosso Fiorentino), who designed a marble sarcophagus with a bronze statue of Pio, in military garb, reclining on a couch with his head propped up by his bent right arm; he is holding in his left hand an opened book, and at his feet lie some volumes carelessly placed. This monument was erected in 1535 and gilded seven years later. It is considered one of the artistic treasures of Paris and is today at the Louvre.[129]

A funeral such as this, provided for one of his major opponents, was sure to be reported to Erasmus, and it became grist for his literary mills. In his letters he commented on how Pio had put on the Franciscan garb three days before his death and, so clothed, his corpse had been carried through the streets of Paris with much religious pomp to the Franciscan monastery where he was given the full burial rites of a friar. Erasmus went on to criticize not only Italian religiosity but also Franciscan practices that were so tainted with worldliness and superstition that they provoked the hostility of many. Pio's burial in the habit of a friar inspired Erasmus to extend his description of Pio's fallen fortunes from former lord of Carpi, to exile, then 'sycophant' (here meaning 'denouncer'), and finally Franciscan. In the opening paragraphs of the *Apologia*, Erasmus' response to the XXIII *libri* which he completed by late August 1531, he repeated some of these statements, but he reserved for his colloquy *Exequiae seraphicae*, published in September 1531, the full force of his satirical attacks.[130]

In this colloquy Erasmus mocked Pio's funeral. As noted by Bierlaire, it was perhaps Johann Koler in his letter to Erasmus of 26 June 1531 who had suggested the idea of writing a colloquy about Pio's funeral, since it seemed so similar to what Erasmus had described in his earlier colloquy 'The Funeral' (1526). In his letter to Julius Pflug of 20 August Erasmus com-

* * * * *

128 Tuetey 103; Eubel III 240; Farge *Biographical Register* 110–12
129 Tuetey 103; Beaumont-Maillet 358, 361, 420 n794. The marble sarcophagus was decorated with flowers in relief and rested on two stone blocks whose base was decorated by the bronze heads of lions. There was also a large stone pedestal. On the sarcophagus was the following inscription: 'To Alberto Pio de Savoia prince of Carpi, follower of the fortune of King Francis, whom prudence rendered most famous, learning made him immortal, and true piety brought him to rest in heaven. Scarcely fifty-five years old. His most sorrowful heirs erected this monument in 1535' (Beaumont-Maillet 282 n216).
130 Allen Epp 2441:66–77, 2443:338–40, 2466:107–14, 2522:66–7; Erasmus *Apologia* 112, 117 below; ASD I-3 14, with the colloquy itself on 686–99 / CWE 40 996–1032

plained that Pio's alliance with the Franciscans put a burden on him and that he had responded but briefly in his *Apologia*. Pio's death prevented him from responding more copiously and sharply. The colloquy format with its veiled references to persons and events, however, relieved Erasmus of these difficulties and thus he penned his satirical *Exequiae seraphicae*. Speaking through the persona of Philecous, Erasmus holds up to ridicule the piety of Pio (here Eusebius) as manifested in the funeral. Far from imitating the humility of Saint Francis, Pio showed pride (*elatus*) in planning such a funeral, which is called a theatrical production (*spectaculum*) that makes one sad. Perhaps for dramatic effect, as well as to provide a springboard for his later attacks on Franciscan practices, Erasmus adds details not found in other accounts. Thus Pio is not only clothed in a Franciscan habit, but is given a tonsure and is said to have promised to enter the order if his life is spared. The friars prepare his corpse for burial by bathing it, crossing its hands, baring its feet, tilting its head towards the shoulder, and annointing its face with oil. By describing bruises on the hands and feet and a tear in the left side of his robe Erasmus suggests that Pio was made to appear to have the stigmata. The friars are the only ones mentioned as marching in Pio's funeral cortège and they are said to have made mournful noises as they marched. The whole rationale for being buried in Franciscan garb is then subjected to hostile criticism. While Erasmus does not state here that the burial in Franciscan garb marked the final stage of Pio's degradation, he does expand on the stages of his decline – from ruler to private citizen to exile to virtual beggar to sycophant.[131] By concentrating on the Franciscan aspects of the funeral Erasmus gives a distorted account of what happened, but also avoids criticizing Francis I and the powerful executors of Pio's will, who helped design the funeral, and the many noblemen and government officials who marched in the procession. This colloquy, of course, was not Erasmus' principal response to Pio – that is found in his *Apologia*.

* * * * *

131 Franz Bierlaire *Erasme et ses Colloques: Le livre d'une vie* Travaux d'humanisme et Renaissance 159 (Geneva 1977) 108–10, especially 109 nn287–8; Allen Epp 2505:55–60, 2522:79–80; Erasmus *Exequiae seraphica* ASD I-3 686–9:7, 13, 21–2, 41–5, 51–2, 62–9, 97, 104 / CWE 40 1000:2, 8, 16–18, 36–41, 1001:7–9, 18–29, 1003:19, 34. Erasmus responded separately to Pio's comments on Aldus as Erasmus' teacher, and his charge that Erasmus lived in Venice as a parasite at Aldus' expense, in a companion colloquy, *Opulentia sordida*, which implicitly criticized Aldus for failing to protect Erasmus, whom he had invited to live in his household, from the stinginess of Aldus' father-in-law, Andrea Torresani d'Asola; see Bierlaire 111–12; the entry on Andrea Torresani in CEBR III 333; and ASD I-3 676–85 / CWE 40 982–91.

ERASMUS' APOLOGIA

From the very first pages of his *Apologia* Erasmus signalled the tack he would take in responding to Pio's XXIII *libri*. In the edition published by Froben in 1531 the *Apologia* is given two titles. The title-page reads: *Desiderii Erasmi Roterodami Apologia adversus rhapsodias calumniosarum querimoniarum Alberti Pii quondam Carporum principis quem et senem et moribundum et ad quiduis potius accommodum homines quidam male auspicati ad hanc illiberalem fabulam agendam subornarunt* 'The Apology of Desiderius Erasmus of Rotterdam against the Patchworks of Calumnious Complaints by Alberto Pio, Former Prince of Carpi, Whom, Although Elderly and Terminally Ill and Better Suited for Any Other Undertaking, Certain Ill-Starred Men Have Clandestinely Incited to Enact This Farce.' On page 5 of the Froben edition the work was given a shortened title by which it is more commonly known: *Apologia brevis ad vigintiquatuor libros Alberti Pii quondam Carporum comitis* 'A Brief Apology to the Twenty-Four Books of Alberto Pio, Formerly Count of Carpi.' Both titles reflect themes also found in Erasmus' earlier correspondence: Erasmus claims that Pio was duped by others into writing these falsehoods about him, and he mocks his opponent for having been deposed as prince of Carpi and for being ill suited by old age and illness to engage Erasmus in combat. In the index to the *Apologia*, on pages 3 and 4 of the Froben edition, Erasmus cites passages (also identified by marginal notes) to support his accusing Pio of slandering, lying about, misquoting, truncating, distorting, misunderstanding, and raving madly against excerpts from his writings.[132]

    In his prefatory remarks Erasmus tries to explain why he has found it necessary to respond to a critic who has died. He does not, however, provide justification for the repeated and at times scurrilous attacks he makes throughout this work against an opponent who can no longer defend himself. He reiterates and amplifies his earlier charge that Pio was not the true author of his books, claiming that others have done the research, provided the arguments, and even polished the literary style. In addition, Erasmus says that things that Pio had never even seen were undoubtedly inserted into his writings. He asserts that Pio and his helpers have often employed a flawed methodology and he claims to have exposed their clear slandering, bare-faced lies, distortions of correct statements, falsification and cunning

* * * * *

132 Erasmus *Apologia* (Froben 1531) sig aa[r] or page 1 and sig aa3[r] or page 5. The index is on sig aa2[r]–aa2[v] or pages 3 and 4 of the Froben 1531 edition; they are reproduced in a revised format by Le Clerc in LB IX 1195E, F–1196E, F.

abbreviation of quotations, and excessive tirades and witticisms based on their misunderstandings of his writings.[133]

Having made these introductory observations, Erasmus proceeds to comment selectively on each section of Pio's *XXIII libri*. Responding to book 1, a reprinting of Pio's 1526/9 epistle to him, Erasmus expresses surprise at Pio's negative reactions, for he feels that he had used great courtesy and tact in writing his rejoinder. He repeats his charges that Pio had been urged to write this work by others, had agreed to do so out of a desire for literary fame, and had been helped extensively in the production of this letter.[134] When treating book 2, Pio's marginal notes on Erasmus' *Responsio* (1529), Erasmus claims that he will 'extract a few items' for comment and then proceeds to write rebuttals to over sixty notes.[135] In the section answering Pio's preface Erasmus tries to defend himself and his earlier criticisms of Pio. He disparages his opponent's qualifications by describing him as a mere layman, untrained in canon law, unacquainted with theological literature, and weakened by a life of luxury at courts. He also contends that he has tried to protect Pio's reputation by blaming his helpers for whatever is hateful in his letter and attributing to Pio whatever seems commendable.[136] The numerous examples of character defamation, documented above, clearly show that Erasmus does not take this approach consistently either in his *Responsio* (1529) or in his *Apologia* (1531).

There are discernible patterns in the ways Erasmus comments on books 3–23. In his rejoinders to the earlier books he quotes from Pio (but not always accurately, and at times he falsely states that Pio did or did not say certain things)[137] and then provides ample rebuttals. In doing so he has composed coherent essays that could stand on their own. As the task becomes more tedious, Erasmus resorts more to paraphrasing Pio's words and ideas, and it is increasingly difficult for the reader to follow Erasmus' arguments unless he has before him a copy of the *XXIII libri* opened to the section on which Erasmus is commenting. Erasmus admits that he is treating

* * * * *

133 Erasmus *Apologia* 107–14 below
134 Erasmus *Apologia* 114–19 below
135 Erasmus *Apologia* 119–53 below
136 Erasmus *Apologia* 153–8 below and again on 181
137 For examples of inaccurate attributions and misquotations, see Erasmus *Apologia* 317 below (where Erasmus falsely claims that Pio cited certain scriptural texts) and compare Erasmus *Apologia* 318 with Pio *XXIII libri* 213ᵛs (where Erasmus falsely claims that Pio never considered what makes a vow solemn), Erasmus 333 with Pio 223ᵛx (rewording Pio's quotation and failing to acknowledge Pio's allowance of moderate sexual activity by married Eastern-rite priests), and Erasmus 333 with Pio 221ᵛɴ (on the Holy Spirit inspiring the church to adopt a better course).

Pio's text in a selective and summary way, avoiding particulars.[138] He dismisses his opponent's arguments as wordy, impudent, and ignorant trifles. Claiming that Pio has misunderstood his point or that he has already answered a criticism elsewhere, Erasmus skips over whole sections of Pio's writings.[139] However, another reason for his ignoring long sections of Pio's work is his recourse to judicious silence when he has no good response to offer.[140] When Pio has exposed a scholarly error, Erasmus becomes desperate, lying to hide his mistake and accusing Pio of slander.[141] For example, when Pio quoted one of his statements questioning the scriptural bases for a bishop's exercise of secular jurisdiction, Erasmus refuses to admit the potentially subversive nature of his own words and instead attacks the supposed notecard on which the excerpt was written, denouncing it as 'shitty.'[142] His repertoire of responses even includes swearing an oath that would condemn him to hell if his protestation of innocence were not true.[143] As noted by Rummel in her brilliant analysis of Erasmus' debating strategies, he had so mastered the classical techniques of argumentation that they became second nature to him and he fell under the spell of his own rhetoric.[144]

ERASMUS' THEOLOGICAL PERSPECTIVE

Erasmus' response to Pio's *XXIII libri* was not just an exercise in the rhetorical techniques of an apologia. When he was defending his criticism of church practices and beliefs, Erasmus was guided by two at times antithetical theological considerations: 1 / that, like all institutions, the church and its piety have degenerated over time and need to be restored to the original evangelical model; and 2 / that to promote the salvation of the Christian people,

* * * * *

138 Eg Erasmus *Apologia* 158, 133 below. He is similarly selective in the *Responsio*; see 8, 12 below.

139 Erasmus *Apologia* 229, 254, 273, 299 below

140 For examples of significant sections of Pio's argument that Erasmus ignores, see Pio *XXIII libri* 187[v]D–194[r]T (compare Erasmus' excuse in *Apologia* 299 below), and Pio 221[v]O–222[v]Q (Erasmus *Apologia* 332–3). In the latter example, Erasmus is avoiding Pio's challenge to come up with more evidence for his claim that the apostles were married, beyond the references to Peter's mother-in-law and Philip's (the Apostle's?) daughters. See also Rummel *Catholic Critics* I 89.

141 Erasmus *Apologia* 324 below. But in Erasmus *Declarationes ad censuras Lutetiae vulgatas* LB IX 879D, Erasmus clearly identifies Clement as living at the time of the apostles.

142 Erasmus *Apologia* 293 below; Pio *XXIII libri* 186[v]A

143 Erasmus *Apologia* 300 below

144 Rummel *Catholic Critics* II 152

many things can and should be changed, but not necessarily always according to the model of the primitive church. One way of determining if something can and perhaps should be retained as true and valid is whether or not it has been approved and accepted by the consent of the faithful. A good part of Erasmus' scholarship, however, seems bent on demonstrating that this consensus either has not existed from the beginning or has not been unanimous over time. He thus throws into question the validity of numerous church practices and beliefs.

Building on his earlier historical and literary scholarship, Erasmus documents changes in the church's doctrinal teachings, liturgical practices, and disciplinary rules. He claims that the church's authority is so great that it can interpret, relax, and dispense with even biblically based prescriptions out of consideration for the faithful's spiritual needs. Some prescriptions, divinely inspired, were intended to last only for a specific time. Thus the apostolic church's prohibitions on eating the meat of strangled animals and on consuming products made of blood (Acts 15:28–9) were meant to last only until the opposition of Jewish converts was no longer a factor.[145] Some of the requirements laid down by St Paul (Titus 1:7) for promotion to the episcopal office have also been altered over time.[146] On a whole series of doctrinal and disciplinary questions, which Erasmus notes, the church was in doubt for a long time. After lengthy consideration it finally deter-

* * * * *

145 Erasmus to Jacob van Hoogstraten, 11 August 1519, from Antwerp, in Allen Ep 1006: 198–201, 239–43 / CWE 1006:208–12, 253–9; Erasmus *Annotationes in Novum Testamentum* LB VI 696A–C

146 Erasmus *Annotationes in Novum Testamentum* LB VI 696C points out how the church no longer observes Paul's admonition against consecrating recent converts or violent men as bishops. The recently baptized Christian raised to episcopal office may be a reference to Henrique, who, although not a convert, was new in the faith, being the grandson of Nzinga, the Congolese king baptized as John in 1491. In 1518 Leo X allowed Henrique to be consecrated as coadjutor to the bishop of Funchal; see François Bontinck 'Ndoadidiki Ne-Kinu a Mumemba, premier évêque Kongo (c 1495–c 1531)' *Revue africaine de théologie* 3 (1979) 149–69, especially 160 (Leo X saw the new bishop as someone who could more easily convert his fellow Congolese, but he wanted him accompanied by persons learned in theology and canon law to strengthen him in the teachings of the Lord). On the Spanish *conversos* of Jewish origins raised to the episcopacy in the fifteenth century, see Tarsicio de Azcona *La elección y reforma del episcopado español en tiempo de los reyes católicos* (Madrid 1960) 218–24; and for the former chief rabbi of Burgos, Solomon ha-Levi (1352–1435), also known by his Christian name of Pablo de Burgos, bishop of Cartagena (1403) and then of Burgos (1415), see B. Netanyahu, *The Origins of the Inquisition in Fifteenth Century Spain* (New York 1995) 168–206. Erasmus' citing of the pirate who became a bishop could be a reference to Baldassare Cossa, who was said to have engaged in piracy before becoming a cleric and who eventually was elected as the Pisan pope, John XXIII (elected 1410, deposed 1415). See G. Mollat 'Jean XXIII' *Dictionnaire de Théologie Catholique* VIII-1 (Paris 1924) col 642.

mined a number of teachings: on the relationship of *homoousion* between the Father and Son; on the double procession of the Holy Spirit; on the baptism of infants and non-repetition of this sacrament; on the requirement of auricular confession; on considering matrimony a sacrament; on transubstantiation; on purgatory; and, some say, on the immaculate conception of Mary, which was proclaimed at the recent Council of Basel.[147] The statements of popes give further evidence of change since some even contradict each other and the common teaching of the church. Thus, the date for celebrating Easter was computed in various ways by the ancient popes;[148] medieval popes gave conflicting rulings on marriage;[149] John xxii in 1323 reversed the teachings of Nicholas iii in 1279 on the poverty of Christ and his apostles;[150] and John xxii was attacked by the University of Paris in 1333 for his denial of the immediate particular judgment of the deceased.[151] Erasmus also notes that, even if the Roman See has changed its positions over

* * * * *

147 Erasmus *Annotationes in Novum Testamentum* LB VI 696C-D, 699E–F; *Hyperaspistes I* LB X 1258D–E, 1262D, 1279B–C / CWE 76 119, 128, 169–70. The rump thirty-sixth session of the Council of Basel on 17 September 1439 (at the thirty-fifth session on 7 May 1437 the council had been transferred to Ferrara) ordered the acceptance of the Immaculate Conception of Mary as a pious doctrine consonant with the cult and faith of the church, with right reason, and with Sacred Scripture, and it authorized the liturgical celebration of its feast. See Hefele-Leclercq VII-2 1071.

148 Erasmus *Annotationes in Novum Testamentum* LB VI 696E; for a brief overview of this issue, see Karl Baus *From the Apostolic Community to Constantine* vol 1 of *History of the Church* ed Hubert Jedin and John Dolan (1962; English trans, New York 1965) 268–72.

149 Erasmus *Annotationes in Novum Testamentum* LB VI 696F–697B; for a study of medieval canonical legislation on marriage, see James A. Brundage *Law, Sex, and Christian Society in Medieval Europe* (Chicago 1987) 176–416.

150 Erasmus *Annotationes in Novum Testamentum* LB VI 696E–F; for a discussion of the contradictory papal teachings on the poverty of Christ and his apostles, see John Moorman *A History of the Franciscan Order from Its Origins to the Year 1517* (Oxford 1968) 313–17; and Brian Tierney *Origins of Papal Infallibility 1150–1350: A Study on the Concepts of Infallibility, Sovereignty and Tradition in the Middle Ages* Studies in the History of Christian Thought 6 (Leiden 1972).

151 In a series of sermons beginning on 1 November 1331 Pope John xxii (1316–34) advanced the unusual theory that the souls of the just, before the resurrection of the body, do not enjoy the beatific vision. He invited theologians to debate this teaching, and scholars from the University of Paris formally protested against it on 19 December 1333. On 3 December 1334 John xxii in the presence of his cardinals recanted his views and died the next day. In 1336 Benedict xii (1334–42) by the constitution *Benedictus Deus* affirmed the traditional teaching that the souls of the just do enjoy immediately after death the beatific vision. See Guillaume Mollat *The Popes at Avignon, 1305–1378* trans Janet Love (London 1962) 21–3, 28; and John xxii *Les Sermons sur la Vision Béatifique* ed Marc Dykmans, Miscellanae Historiae Pontificiae 34 (Rome 1973). Erasmus claimed that John xxii taught his strange views, not as something he personally and privately held, but in published writings and by decree; see *Annotationes in Novum Testamentum* LB VI 697C.

the years, the church has not erred on the most important questions of faith and religion.[152]

Erasmus believes that the Holy Spirit is with the church. The Spirit binds together the faithful in charity. He did not reveal everything at once, but illuminates the meaning of obscure scriptural passages whenever it is necessary for our salvation. Some things are left abstruse for now and may not be properly understood for many centuries.[153] The Spirit is also selective regarding those to whom he reveals his truths. Some people think they are interpreting the Scripture correctly, but in fact are twisting its meaning out of human passions. Those who clearly had the Spirit as their guide were the prophets, John the Baptist, and the apostles. In post-apostolic times, Erasmus holds, it is more probable that the Holy Spirit is given to the holy, those who are ordained to church office, and gatherings of such persons in councils. All of the members of the hierarchy should share in formulating the final doctrinal statements that encapsulate the consensus of the church. The Spirit has assisted over the centuries those holy and learned Fathers who taught the Christian people the meaning of the Scriptures and thus drove away heresy. When the church is called upon to decide the meaning of a scriptural passage, it seeks the advice of the learned who by their expertise are able to discern the truth. The church has a special regard for the interpretations of those holy Fathers whom it has esteemed over the centuries for their faith, charity, and wisdom.[154] It is likely that they were guided by the Spirit, but Erasmus also finds numerous occasions where they have contradicted each other. Thus, as James McConica has observed, 'the ascertainable essentials of doctrine were few indeed.'[155]

Where Erasmus finds the surest evidence of the guidance of the Holy Spirit is in the consensus of the whole Christian people. While the prescrip-

* * * * *

152 Erasmus *Annotationes in Novum Testamentum* LB VI 696D–E
153 Erasmus to Jacob van Hoogstraten, 11 August 1519, from Antwerp, Allen Ep 1006:192–3 / CWE 1006:201–2; Erasmus *Explanatio symboli* ASD V-1 216:335; *Enarratio in Psalmum XXII* LB V 315E–F, *Hyperaspistes* I LB X 1310C–D / CWE 76 240; and James K. McConica 'Erasmus and the Grammar of Consent' in *Scrinium Erasmianum* ed Joseph Coppens (Leiden 1969) II 81, 84–5
154 Erasmus *Apologiae contra Stunicam* (2) LB IX 365C; *De libero arbitrio* LB IX 1219D / CWE 76 17–18; *Hyperaspistes* I LB X 1301C–E, 1303D–E, 1304D–F, 1310C–D / CWE 76 221–2, 224–5, 227–8, 242–3; McConica 'Grammar of Consent' (cited in n153 above) 85–9
155 Erasmus *Hyperaspistes* I LB X 1303E / CWE 76 224–5. On the disagreements and mistakes of the holy and learned Fathers, see, for example, Erasmus to Jean II de Carondelet, 3 January 1523, Basel, Allen Ep 1334:84–9, 794–9 / CWE 1334:91–6, 833–8; and Erasmus *Annotationes in Novum Testamentum* LB VI 583F–584D, 793E–F, 794C–E; McConica 'Grammar of Consent' (cited in n153 above) 97.

tions of popes and the rulings of councils carry great weight with Erasmus, they become most binding when they are accepted over time by the Christian people. He does not, however, feel that one can merely look at what is the current practice of the church and assume that it enjoys divine sanction. The 'merry consensus' of the masses at a particular point in time can be contrary to the teachings of Scripture. But traditional beliefs and practices that have withstood the test of time are to be accepted, for Erasmus is convinced that the Holy Spirit would not allow the whole church to be in significant error for a long period of time.[156]

Erasmus repeatedly affirms his adherence to the Catholic church. He says he obeys especially those decrees that were issued by general councils and approved by the consent of the Christian people. He embraces these as oracles from God, even in those cases where he cannot follow the reasoning behind them. And he pledges that he will not violate a constitution of the church unless necessity has relaxed that law.[157] When he points out problems with the argumentation used to justify a particular doctrine or practice, he claims that he is not trying to define a new position, but rather to open up an issue for debate, and that he defers to the judgment of the church.[158]

Given these theological convictions, at times very similar to Pio's and always within the wide boundaries of pre-Tridentine Catholic orthodoxy, Erasmus nonetheless found himself at variance with his Italian opponent on many of the twenty major topics he debated in his *Apologia*.

COMPARISON OF PIO'S AND ERASMUS' THEOLOGICAL
PERSPECTIVES

Both Pio and Erasmus believed that the Holy Spirit has guided the Catholic church so effectively that it has not erred on the central teachings of the faith. But where Pio felt that the church has defined or approved with the assistance of the Holy Spirit numerous beliefs and practices, Erasmus held that relatively few of them enjoyed such guarantees of divine approbation. Both men held that over the centuries the Holy Spirit has revealed the full message of the Gospel only gradually and has done so through the teachings of holy and learned men. But where Pio found agreement among these

* * * * *

156 Erasmus *Responsio* 35 below (*concordes hilaresque*); *De libero arbitrio* LB IX 1219D–E, 1220C–D/ CWE 76 17–20; *Hyperaspistes* I LB X 1262A–B, 1279B / CWE 76 127, 169
157 Erasmus *Hyperaspistes* I LB X 1258D–E, 1262A–B / CWE 76 118–19, 127
158 Allen Ep 1006:265–6 / CWE 1006:283–5

teachers stretching from his own day back to the church Fathers, Erasmus was quick to point out a variety of opinions in patristic sources. Pio's very positive view of the scholastic method as a divine institution that helps the church clarify its doctrines was not shared by Erasmus, who faulted scholasticism for promoting discord and allowing theologians to waste their time and talents debating useless topics. While Pio saw the Holy Spirit as guiding the apostolic see so effectively that it has always and solidly sustained the purity of the faith, Erasmus questioned such guidance, for the Roman see has on a number of occasions changed its positions. Whereas both agreed that the surest indication of the validity of a church teaching is its acceptance by the Christian people over time, Erasmus was reluctant to conclude that, because something is currently accepted by the people, it must be true, for he feared what he called a 'merry consensus' of the masses that was not based on the Scriptures and would pass with time. Pio seems to have felt that a widespread consensus could not have developed unless the Spirit were behind it. Both men recognized that changes in church teachings and practice, even those originally based on scriptural injunctions, could be valid if inspired by the Spirit. Erasmus, however, tended to see change more typically as degeneration from the supposed purity and freedom of the primitive church, while Pio saw things becoming clearer, better understood and implemented, and improved over time. Erasmus did not respond directly to Pio's theory of new revelations that could abrogate or alter even divinely constituted things defined by the church. He preferred to base his positions on what Scripture and the church Fathers have taught. Pio's views on the development of church doctrine and practice based on his biological analogy remained consistent throughout his controversy with Erasmus. Erasmus seems to have toyed with these ideas, temporarily adopting them in his 1527 letter to Nikolaus von Diesbach and in his 1529 letter to the Strassburger Pseudo-Gospellers, but in the end he abandoned these ideas for they contradicted his calls to return to the ancient church as the true model.[159]

The differences in Erasmus and Pio's views on how the passage of time has affected the church's beliefs, institutions, practices, and piety reflect larger cultural and religious thought patterns. Erasmus was thoroughly imbued with Renaissance concepts about the superiority of antiquity. He yearned to return to a supposedly purer form of Christianity. In this regard he was to some extent a prisoner both of the secular metaphysical model that required one to return to the source to be restored and of the

* * * * *

159 For a fuller treatment of this topic, see Minnich 'Underlying Factors' 40–2.

related humanistic myth of a restoration of the golden age – the classical cyclical view of history.[160] Pio, while acknowledging the need to reform corrupted morals and abuses, was confident that the Holy Spirit had effectively guided the church over the centuries, and he produced a variety of scriptural texts to justify this conviction. Given his biological model of growth, change is unidirectional. The church cannot and should not try to return to the forms of belief and practice of its youth. In holding this position Pio was espousing the biblical view of history, which is linear.[161]

Pio and Erasmus also had differing views on the task of the theologian and on how the theological enterprise should be conducted, especially at a time when the church's teachings were subject to serious attack. Pio felt it was incumbent on a Catholic theologian to mount a full defence of church doctrine.[162] Erasmus maintained that he could better serve the church if he retained his objectivity and pointed out the weaknesses in the traditional arguments used to support these teachings. But Pio wondered at how well Erasmus loved the traditional church and practised scholarly objectivity when he typically undercut the rationale for the church's beliefs without providing new and better arguments, and did so, not in private communications with Catholic theologians in which he warned them of problems, but in publications that he knew the church's opponents would carefully examine for fresh ammunition for their heretical views. Whereas Pio had a deep respect for tradition, Erasmus suspected that tradition recorded a deterioration from a primitive purer form of Christianity. Because of his conviction that the Holy Spirit had effectively guided the church over the centuries, Pio held that one can legitimately argue that, when the church has defined a doctrine, there must be either evidence supporting this teaching in Scripture and the church Fathers or at least nothing in these sources that prevents the church from coming to such a definition.[163] Given his

* * * * *

160 On the metaphysical principle of returning to the source to be restored, see John O'Malley *Giles of Viterbo on Church and Reform: A Study in Renaissance Thought* Studies in Medieval and Reformation Thought 5 (Leiden 1968) 105–6 n5. On the Renaissance's adoption of the cyclical view of history, see Peter Burke *The Renaissance Sense of the Past* (New York 1969) 87–9.

161 On the linear biblical view of history, see, for example, John L. McKenzie 'Aspects of Old Testament Thought' in *The Jerome Biblical Commentary* ed Raymond E. Brown, Joseph A. Fitzmyer, and Roland E. Murphy, 2 vols (London 1968) II 736–67, especially 755–6 nos 11–17; David M. Stanley and Raymond E. Brown 'Aspects of New Testament Thought' in ibidem II 768–99, especially 777–82 nos 64–92; and Hans Urs von Balthasar *A Theology of History* (1959; English trans New York 1963) especially 123–48.

162 Pio *Responsio paraenetica* 71^{v}s

163 Pio *XXIII libri* 178^{v}E, 228^{r}H–I

belief that all earthly things deteriorate over time, Erasmus was inclined to find corruption in the current church and to look to the ancient church as the norm for a purer faith. The differences between Erasmus and Pio also extended into another area. Pio found little problem with the externals of religion, for he recognized that human nature needs a material expression of interior dispositions, and at the core of Christianity is the doctrine of the incarnation.[164] Erasmus espoused a Platonist perspective that held that the higher form of spirituality was that which was freer from material factors.

Differences in personality affected Erasmus' and Pio's theology. Pio seems to have been overly serious by disposition: Cortesi commented on his 'placid gravity.' His sense of propriety was particularly offended when Erasmus selected in his colloquy Ἰχθυοφαγία a fishmonger as the representative of the Catholic position on abstinence. Pio repeatedly faulted Erasmus for treating religious topics in a less than serious manner. Erasmus, however, saw humour as a most effective tool for overcoming people's resistance to examining certain religious topics from a fresh perspective; it was the honey that helped the medicine go down.[165] Their attitudes towards changes in the church may also have been influenced by other psychological and even sociological factors beyond their theological differences. As a member of the aristocracy and an official spokesman for some of the most important rulers of Christendom, Pio was inclined to defend the status quo. As the illegitimate son of a priest Erasmus may well have favoured changes in a system that he viewed as unfair and oppressive.

ERASMUS' BREVISSIMA SCHOLIA

Erasmus wrote against Pio's position once more in his *In elenchum Alberti Pii brevissima scholia*, published by Froben in early 1532. In the preface to this work Erasmus recounts how he had already written his *Apologia* to Pio's XXIII *libri* when the title-page and index to the book arrived. When he saw that he was being portrayed in these folios as a Lutheran, he realized that he had to respond, for most people read only the title and index of a book and base their judgments on them. Although Erasmus also received the letter of Josse Bade addressed to the reader, which was printed on the back of the title-page, he acts as if he had not read it. Bade clearly

* * * * *

164 Pio *Responsio paraenetica* 75[r-v]v; XXIII *libri* 123[r]D–E
165 Cortesi 60[r]; Pio XXIII *libri* 209[r]G (fishmonger), 209[v]G and 216[r]z (Erasmus lacks gravity); Erasmus *Scholia* entry 103

states that Francesco Florido had helped him to edit the work, yet Erasmus protests that he has no idea who composed the index. Erasmus states that he has selected from the index 'a few items' for comment. Each of these few items, numbering 122, is restated and then followed by a brief, often one-sentence rejoinder. In his responses Erasmus employed the same strategies he had used when debating Pio himself. He accuses the author of the index of lying about his views. What is being attributed to Erasmus, he charges, is not his own position; rather, he was reporting the views of others or merely raising topics for discussion. He does not criticize church practices themselves but only abuses related to them. He expresses indignation at how his own views are being misunderstood and urges his critics to vent their spleen instead at Folly. Although he knows that Pio was not the author of the index, he nonetheless accuses him of dying in his slander.[166]

Erasmus' *Brevissima scholia* attracted little attention from the scholarly world. The work said almost nothing that was new and was occasionally flippant and sarcastic in tone. Even Jean Le Clerc failed to include it in his edition of the *Omnia opera*.[167] Yet the fact that Erasmus felt himself compelled to respond to an index to Pio's work, even though it was not written by Pio himself, suggests how great a threat to his reputation Erasmus feared the criticisms that emerged in the course of his debate with Pio could be.

LATER EDITIONS OF PIO'S WORKS AGAINST ERASMUS

Erasmus suggested that Pio had received assistance in the production of his XXIII *libri* from Venice. He claimed that, while Pio still lived, his former client Girolamo Aleandro, now the archbishop of Brindisi, had been

* * * * *

166 That Erasmus' *Scholia* was printed by early March 1532 is evident from a letter of Bonifacius Amerbach to Jacopo Sadoleto, bishop of Carpentras near Avignon, in which Amerbach lists the *In elenchum eiusdem (Alb. Pii) scholia* among the works (*nuper editis*) which he is sending to Sadoleto by way of a book dealer in Lyon; see *Amerbachkorrespondenz* III 108–9; Ep 1610:35–44; Erasmus *Scholia* 363 below (preface); entries 21, 29, 36, 56 (lies); entries 6, 7, 61, 81 (reporting others' positions); entry 83 (raising a topic for discussion); entries 10, 58 (criticizing only abuses); entry 25 (indignation); entries 3, 5, 17, 24, 40, 41, 88, 89, 105 (let Folly take the blame); entry 69 (Pio raving); and entry 47 (Pio dying in his slander); Pio XXIII *libri* ai^v (Bade's letter).

167 Erasmus *Scholia* entries 37, 43, 53, 85, 105 (flippant); entries 28, 47, 49, 57 (sarcasm); see LB 1195–6. Vander Haeghen does not even list the *Scholia*. It was, however, considered important enough for Amerbach to send it along with other works by Erasmus to Sadoleto (see n166).

active in Venice collecting material from Erasmus' writings for Pio's use. According to Erasmus, not only had Aleandro given excerpts to Pio and encouraged his animosity to Erasmus, but Aleandro's role in the composition of the work was so great that he should be considered its principal author. While Erasmus' accusations may seem preposterous, and were flatly denied by Aleandro, Erasmus did receive reports from Etienne Dolet, who in 1531 served as secretary to the French ambassador in Venice. According to Erasmus' summary of these no longer extant reports, Aleandro was carefully scrutinizing his writings in order to attack them himself or provide ammunition for Pio's refutation, but perhaps Erasmus drew this conclusion from less precise comments in Dolet's letters. Dolet may have reported only that the papal nuncio in Venice, unnamed, was engaged in activities critical of Erasmus' writings, and Erasmus may have surmised that the nuncio was Aleandro, and that his activities consisted in combing Erasmus' publications for excerpts to send to Pio – a charge Erasmus had made earlier against Aleandro. The first problem with Erasmus' accusation relates to chronology. If Dolet was reporting events contemporaneous with his 1531 sojourn in Venice, it would seem that any material Aleandro sent to Pio, who died on 8 January 1531, would have arrived too late for incorporation into the Paris volume. The second problem relates to the identity of the papal nuncio in Venice. While the papal nuncio was indeed a great admirer of Pio, he was Altobello Averoldi, bishop of Pola, and Aleandro's predecessor as resident nuncio to Venice. The activity that Averoldi, who died on 1 November 1531, engaged in that was critical of Erasmus was not collecting unfavourable excerpts from his writings but seeing to a reprint in Venice of Pio's *xxiii libri*. The work was published on 5 September 1531 by Luca Antonio Giunta of Florence. In his prefatory letter to Averoldi, dated 2 September, Giunta recounts how the nuncio preferred Pio above all other princes who have written against the Lutheran heresy and saw Pio as his champion in his own efforts to oppose the Protestants.

What other factors led Averoldi to promote a Venetian edition of Pio's *xxiii libri* are not clear. He does not seem to have been a personal friend of Pio. Perhaps Aleandro during the years that he was in Venice (1529–31), brought Pio's Paris publications to the nuncio's attention. Averoldi may well have been looking for a way not only to combat heresy but also to restore his own reputation as a defender of orthodoxy. Giampietro Carafa had been openly critical of Averoldi's failure to secure the secular government's co-operation in the punishment of heretics, for Averoldi was unwilling to put pressure on a reluctant Venetian government that was already cool towards Rome. Praising Pio as the great champion of the Catholic faith and encouraging the republication in Venice of his anti-Erasmus and anti-Luther writings may have seemed to Averoldi to be a way of opposing heresy without offending the officials of the Venetian republic and of proving that he had not neglected his own duty to protect the Catholic religion.[168]

Within months of Pio's death book 1 of the *XXIII libri*, the *Responsio paraenetica*, was also reissued, in a translation into French. This time the intended audience was one person, King Francis I of France, and the work was not printed but handwritten on vellum and beautifully illuminated. The frontispiece depicts Guillaume de Montmorency presenting Alberto Pio, in a teaching pose, to the French king, seated on the throne. The manuscript is also decorated with colourful illuminated letters incorporating flowers, leaves, berries, birds, dolphins, putti, and the like, and with drawings of branches and carved staffs that fill up the space to the end of a line or help set off a Latin quotation, either a text from the Bible or an *incipit* from a section of the *Responsio paraenetica*. Not only does this manuscript resemble in physical appearance (for example, the presence of the elaborate frontispiece honouring the intended recipient and similar motifs in the illuminations, although there is a much more lavish use of decoration) the French translation of the *Responsio paraenetica* presented by Pio to Guillaume de Montmorency, but a close comparison of the two texts reveals that the work presented to the king is indeed a copy of that translation. Although the handwriting used in the king's copy is similar to that found in the manuscript presented to Montmorency, a number of the initial letters (for example, D, E, H, I/J, N, and S) are written in a different style and point to another copyist being employed. This scribe has on occasion altered the spelling of words (for example, dropping the *l* in *angles* and substituting *s* for *z*

* * * * *

168 Allen Epp 2466:99 (Aleandro *praecipuus* among Pio's helpers), 2639:7–14, 40–2. By putting credence in these and other reports about Aleandro and Pio, Erasmus was failing to follow his own advice; see Erasmus *Responsio* 13–14 below. Aleandro seems to be denying that he helped Pio when he was in Venice, where he stopped to negotiate a peace before continuing on to Brussels to take up his post as nuncio to the imperial court, arguing that he was too ill and too busy during his Venetian stay to do anything literary (Allen Ep 2639:16–21). He does not refer explicitly to his activities in his earlier stay in Venice in 1529–30. On his predecessor, see Gaeta *Un nunzio pontificio a Venezia* (cited in n104 above) 13–15; and 'Averoldi, Altobello' DBDI 4 (Rome 1962) 667–8; for Giunta's letter to Averoldi, see Pio *XXIII libri* (Venice: Luca Antonio Giunta 1531) *iv. On Carafa's complaints of 4 October 1532 and Averoldi's efforts, see the memorial to Pope Clement VII, translated by Elisabeth G. Gleason in her *Reform Thought in Sixteenth-Century Italy* American Academy of Religion, Texts and Translations Series 4 (Missoula 1981) 55–80, especially 58–60. See also the entry on Aleandro in CEBR I 28–32, especially 30–1. On Dolet as Jean de Langeac's secretary from the beginning of his ambassadorship to Venice, see Richard Copley Christie *Etienne Dolet: The Martyr of the Renaissance, a Biography* (London 1880) 36.

Frontispiece of a richly illuminated manuscript of Pio's *Response*, 1531,
a French translation of Pio's *Responsio paraenetica* dedicated to King Francis I.
Paris, Bibliothèque nationale, Fonds français, Ancien fonds, ms 462, folio 1ᵛ
The miniature depicts Baron Guillaume de Montmorency, on the left,
presenting Pio in a teaching pose to the enthroned king. The Latin poem reads:
'Renowned France does homage to its star-spangled king / while the rest of the world
trembles 'neath the soldiery of France. / A variety of emblems bloom on
the rulers of diverse lands; / 'twas heaven that sent gleaming lilies to the
French. / May God Almighty grant long life, I pray, to the French king, /
and at his life's close bestow on him a place among the stars.'
By permission of the Bibliothèque nationale de Paris

and *d* for *t*), reversed word order, used singular instead of plural pronouns and verbs, added words for clarity's sake, and changed prepositions (for example, *en/es* to *au/aux* and *sur* to *contre*). Although the scribe may have improved the style of the translation, he did not produce a better text for he is known to have inadvertently dropped a line or a word when copying.[169] The overwhelming similarities between the two translations should not come as a surprise because the man who presented Francis I with this translation of Pio's *Responsio* was none other than the elderly baron Guillaume de Montmorency, although he did not identify himself by name in his presentation letter.

In his dedicatory letter the senior Montmorency urges the king to utilize better the baron's talents and experience and to read, understand, and have published this French translation of Pio's *Responsio*. Montmorency describes himself as an elderly and knowledgeable royal servant whose talents are not being employed by Francis I, as they were by his three predecessors, especially Louis XII. He wants to be of service to the king and thus offers him the gift of this book. After handling the fatiguing business of deciding the affairs of the realm, the king customarily looks at or has read to him chronicles and secular histories that recount the deeds of noble and virtuous men of both ancient and modern times. These men provide models on how to live virtuously and avoid prudently various inconveniences, perils, and dangers. An uneducated person from the countryside might overlook or despise things he should not. The book he offers the king is, Montmorency claims, significant. It was recently composed by one of the king's subjects, familiars, and friends, the noble count of Carpi, whom God wishes to have in paradise. Pio's book deals with the controversy over the improper opinions of Erasmus and the heresies of Luther. Echoing Pio's statements in *his* dedicatory letter to Montmorency,

* * * * *

169 Bibliothèque Nationale de Paris, Fonds français, Ancien fonds, MS 462, described in *Catalogues de la Bibliothèque Impériale, Departement des Manuscrits, Catalogue des manuscrits français* 5 vols (Paris 1868–1902) 1: *Ancien Fonds* (Paris 1868) 46 no 462. The folios measure 32.8 x 21.3 cm, the frontispiece 15.0 x 17.5 cm. This manuscript, cited as Pio *Response* – Paris MS, should be dated to the early months of 1531, given the death of Pio on 8 January and the reference to it in the dedicatory letter, and also the death on 24 May of Guillaume de Montmorency, who presented the manuscript to the king. On the death date of Montmorency, see *Correspondence du Cardinal Jean du Bellay* ed Scheurer (cited in n112 above) 1 208 n 2. The conclusions about the differences in the handwriting and literary style of the copyists are based on a comparison of their transcriptions of chapter 42 on saints as intercessors; see Pio *Response* – Chantilly MS 160^v–166^r and Pio *Response* – Paris MS 182^r–187^r. The line *'comme nous indignes de venir envers luy'* of Chantilly MS 164^v is missing from Paris MS 185^r.

the baron claims that by means of biblical testimonies Pio refutes most of these heresies and shows their weak foundation. Heretics establish their teachings, as it were, on the water's edge. Their bases are vanity, vainglory, arrogance, luxury, and wickedness. Those who truly know these heretics know the evil reputation of their damnable lives which beats against, inundates, and brings to ruin their foundation. The only true, firm, and beautiful rock is Christ without whose help all human efforts are without effect, profit, or long duration. May the king accept this small book of Pio's and consider not the gift but the good will of its giver. Montmorency wishes the king to have always a true zeal and love for the happiness, holiness, and permanence of his realm. May he disregard the roughness of the book's translation and consider instead the true sense and intention of its words, which promote the honour of God and the salvation of true Christians, and may he from his bounty wish the book to be communicated to all. May God, the shaper and doer of all works, grant the king his grace always to govern his kingdom well, and after this mortal life to see in paradise his royal predecessors reigning in the glory of God almighty.[170]

The French translation of Pio's *Responsio paraenetica* seems to have had little effect on the king, who did not see to its publication, despite his protestations of admiration and affection for its author. The king also continued to vacillate in his attitudes towards the suppression of heresy, despite pressure from his staunchly orthodox ministers Anne Montmorency and Antoine Duprat and from the papal nuncio Cesare Trivulzio. Not until the public challenges of Nicolas Cop's inaugural speech as rector of the University

* * * * *

170 Pio *Response* – Paris MS 2^r–4^r. The reference to heretics establishing their teachings on the water's edge is probably meant to suggest that they lack a firm foundation given the erosion caused by waves, currents, and floods. The reference to ignorant folk from the countryside may be an allusion to the reputation of the French aristocracy for loving warfare and detesting learning, considering it an insult to be called scholarly. Although raised in the countryside at Cognac, Chinon, and Amboise, Francis I's military training was supplemented by lessons in letters. Under his mother's supervision the youthful Francis had chivalric and moralistic works read to him. This practice continued when he became king. It was Francis I's custom to rise at ten o'clock, attend mass, and dine at about eleven o'clock, during which time he conversed with lords and scholars. A lengthy reading usually followed the meal – often chivalric romances or accounts of the fall of Troy, the revolt of the Maccabees, and the invasion of Italy by Charles VIII. He also listened to humanist works in French translation because he was a poor Latinist; see Desmond Steward *Prince of the Renaissance: The Golden Life of François I* (New York 1973) 104–6; Dorothy Moulton Mayer, *The Great Regent Louise of Savoy, 1476–1531* (New York 1966) 21–36; and Robert J. Knecht *Francis I* (Cambridge 1982) 4–5.

of Paris in 1533 and the Placard Affair in 1534 did the king clearly identify himself with the cause of eliminating open heresy.[171]

The hope that a Castilian translation of Pio's *xxiii libri* would help to stop the beginnings of heterodoxy in Spain led to the publication on 1 January 1536 of a work entitled *Libro del muy Illustre y doctissimo Señor Alberto Pio Conde de Carpi: que trata de muchos costumbres y estatutos de la Iglesia, y de nuestra religion Christiana mostrando su autoridad y antiguedad: contra las [malditas] blasphemias de Lutero, y algunos dichos de Erasmo Rotherodamo. Traduzido de Latin en Castellano, para vtilidad de muchos hombres sabios: que por carecer de lengua Latina son priuados de doctrina tan fiel, y prouechosa* 'The book of the very illustrious and most learned lord Alberto Pio, count of Carpi, which treats many customs and statutes of the church and of our Christian religion, arguing from authority and antiquity, against the (cursed) blasphemies of Luther and some statements of Erasmus of Rotterdam. Translated from Latin into Castilian for the utility of many wise men who for the lack of the Latin tongue are deprived of doctrine so faithful and advantageous.' The publisher was Miguel de Eguía, printer to the University of Alcalá de Henares.[172] The purpose behind the translation was set out both in the dedicatory letter and in the prologue.

The dedicatory letter, dated 1 November 1535, was addressed to Don Juan Téllez Girón, the second count of Urueña and lord of Peñafiel and Osuna. This grandee's brother-in-law was the constable of Castile and his eldest son briefly led the revolt in 1520 of the Santa Junta of the Comunero. In the dedicatory letter Juan is thanked for having defrayed the costs of printing the volume. He is said to have done this as a Christian lord for the honour of God and the conservation of the authority of the holy church. The author of the letter is not identified.[173]

The prologue addressed to the reader by the Dominican translator of the work explains the significance of Pio's *xxiii libri*. The devil has been

* * * * *

171 Knecht *Francis I* 244–52; James Farge *Orthodoxy and Reform in Early Reformation France: The Faculty of Theology of Paris, 1500–1543* Studies in Medieval and Reformation Thought 32 (Leiden 1985) 204–7; Lucien Febvre and Henri-Jean Martin *The Coming of the Book: The Impact of Printing 1450–1800* trans David Gerard, ed Geoffrey Nowell-Smith and David Wootton (London 1976) 272.

172 The title is given twice in this work, on the title-page (sig 1^r) and just before the prologue (sig 2^v). The word *malditas* is added to the title on sig 2^v. The colophon on folio 166^v states: 'Acabole de imprimir en castellano en la Villa de Alcala de Henares en casa de Miguel de Eguya primero de Enero de Mil y Quinientos y Treynta y Seys Años.'

173 Pio *Libro* sig 2^r; on Juan Télley Girón and his family, see Henry Latimer Seaver *The Great Revolt in Castile: A Study of the Comunero Movement of 1520–1* (Boston 1928) 26, 63–4, 169 n1.

(1536)

Title-page of *Libro* (Alcalá de Henares: Miguel de Eguía 1536),
a Spanish translation of Pio's *XXIII libri*
By permission of the Folger Shakespeare Library, Washington, DC

busy throughout history, the translator says, trying to lead men away from God, and among his captains and soldiers are heretics: the Ebionites, whom St John the Evangelist refuted; the Arians, Sabellians, Manichaeans, and Pelagians, whom church Fathers such as Augustine, Jerome, and Ambrose confounded. Currently heretics like Luther rebel against the rule and government of the Catholic church. They cleverly conceal their malice, act as friends, and use pleasant words, taking as their instruments of persuasion the tongues of eloquent, gracious, and prestigious men. Those who are truly wise do not give authority to these mouthpieces. Among such spokesmen for Luther is Erasmus of Rotterdam. Who can deny that he promotes and defends manifest and pertinacious heresy, boldly following his own judgment whether in human or in divine and ecclesiastical matters, without regard for the divine majesty or reverence for the church's teachings? He claims that he has good intentions and seeks to serve God, but his words disprove this and are aimed at engendering doubt and inducing tepidity. He tries to undermine the authority of the church and religious customs. He does not wage open warfare, but resorts to skirmishes and undermining the walls, as do those who attack a city. He uses ambiguous words and claims that he is attacking superstitions and only seeking to entertain. The devil astutely uses his writings. Erasmus often resorts to the excuses that he teaches what was held by the early church and that he submits his writings to the judgment of the current church. But rejoice, oh reader, for there are learned and holy Catholic writers who publish books that expose the deceits of the devil, arouse flaccid consciences, and restrain the pursuit of novelties. Among such authors is Alberto Pio di Savoia, count of Carpi.[174]

The translator continues: Pio has provided an antidote to the deceits of heretics. An Italian of distinguished lineage who lived out his Catholic faith, he devoted himself to sound and salutary study. In the midst of much adversity and in poor health, he toiled for the honour of God and the love of his church by writing this *grande volumen* which he could not complete without much labour. It is both eloquent and very learned, citing the Bible, the teachings of the saints, the whole history of the church, philosophy, and

* * * * *

174 Pio *Libro* sig 2ᵛ–3ᵛ. The identification of the translator as a member of the Order of Preachers comes from the records of the Inquisition, which eventually banned the book; see Marcel Bataillon *Erasme et l'Espagne: Recherches sur l'histoire spirituelle du* XVIᵉ *siècle* (Paris 1937) 456–7 n3; and Eugenio Asensio 'El Erasmismo y las corrientes espirituales afines: Conversos, Franciscanos, Italianizantes' *Rivista de filologia española* 36 (1952) 31–99, especially 79. See also Andrés II 570 n206.

other human sciences. He not only refutes Luther and Erasmus but provides much doctrine that is useful to a Christian of any state or condition.[175]

The author of the prologue ends by explaining how he translated the volume. Because Pio included Erasmus' *Responsio* and wrote marginal comments in answer, he has also translated this work of Erasmus, but he has not translated Erasmus' *Apologia*, not out of any scorn for him, but because his intention is to provide doctrine that is solid and useful for souls. He laments that much of the literature that is in Castilian deals with scandalous romances. By translating Pio's volume into the vernacular he will make a useful work more widely available. As a translator he has been careful to respect the special qualities of the Castilian tongue and has therefore avoided a literal translation. He has tried, nonetheless, with God's help, to be faithful to the sense of the original.[176]

Perhaps in order to render Pio's work more readily accessible, the translator made two slight changes in the organization of its material. He assigned book numbers to the preface and conclusion and divided the now twenty-five books into two sections: part I contains books 1–4 (the original first three books plus the preface) and part II contains the newly numbered books 1–21; but the chapter on faith and works has been moved to the penultimate position, thus retaining its original numbering as 20, and the conclusion is listed as book 21.[177]

That the noted Erasmian Miguel de Eguía became the publisher of this extensive attack on Erasmus demands an explanation. By 1524 Eguía had become the leading printer in Alcalá and from his presses came Latin editions of Erasmus' *Enchiridion* and *Paraphrases* on the four Gospels. He was also an active member of the illuminist circle around Fadrique Enriquez, the admiral of Castile, who sought to evangelize his estates. Eguía's activities eventually came to the attention of the inquisition. He was brought in for questioning in November 1531 and apparently quickly confessed to his errors. He was not formally convicted of heresy and was released around June 1533. Rummel suggests that he published Pio's attack on Erasmus as a way of exonerating himself from the suspicion of heresy.[178]

It is ironic, therefore, that the Supreme Council of the Inquisition on 6 September 1536 ordered the suppression of this 'dangerous' book. The

* * * * *

175 Pio *Libro* sig 3[v]
176 Pio *Libro* sigs 3[v]–4[r]
177 Pio *Libro* sig 1[v]
178 Asensio 'El Erasmismo' (cited in n174 above) 78–81; John E. Longhurst *Erasmus and the Spanish Inquisition: The Case of Juan de Valdes* University of New Mexico Publications in History 1 (Albuquerque 1950) 30–1; Rummel *Catholic Critics* II 105

reason given was the poor quality of the translation. But after a careful comparison of the original and its translation Eugenio Asensio has concluded that the Castilian version was faithful to the Latin and the claim of a faulty translation was merely a pretext. One can only speculate as to the real reason for its suppression. Perhaps the inquisitors did not wish to have Erasmus' *Responsio* available in a Castilian translation, even a version heavily annotated with hostile comments by Pio. Or perhaps political considerations entered into their decision. Pio was repeatedly praised in the book's title, prologue, and colophon, and referred to as count of Carpi. The king of Spain in his capacity as Holy Roman Emperor had, however, declared Pio a rebel and had stripped him of this title. Because Pio's volume had been printed under a royal licence signed with the *real nombre* and sealed by the lords of his royal council, some pretext was needed for suppressing it. In the prologue the translator had unwittingly furnished this by stating that he had not provided a word-for-word translation.[179] Erasmus seems to have been unaware of the Spanish edition of Pio's volume.

LATER PHASES OF THE CONTROVERSY: STEUCO, SEPÚLVEDA, LUTHER, LANDO, AND FLORIDO

The death of Pio did not end the controversy. Erasmus complained that his adversary had inflicted his stinger and fled.[180] Smarting from Pio's final assault, Erasmus retaliated by writing three responses (*Apologia, Exequiae seraphicae*, and *Brevissima scholia*) full of insults and counter-charges to which the deceased prince could not respond. Just as friends of the former count of Carpi had reprinted his *XXIII libri* in Latin and produced another copy of the French translation of the *Responsio paraenetica* as well as a Castilian version of *XXIII libri*, so too there were others who took up their pens in his defence. One of the first to do so was Guido Steuco (1497/8–1548), who took the name Agostino when he joined the Augustinian canons at Gubbio in Umbria, Italy. Steuco was a scholar known for his love of books – he was librarian (1525–9) of the collection of Domenico Grimani in Venice and succeeded Girolamo Aleandro in 1538 as papal librarian – and an admirer of Pio. He probably used the library that Pio had left in Rome and eventually

* * * * *

179 Bataillon *Erasme et l'Espagne* (cited in n174 above) (1937) 456 n3; Asensio 'El Erasmismo' (cited in n174 above) 79–81; Pio *Libro* sigs 1ʳ, 2ᵛ, 3ᵛ, x5ᵛ [= folio 166ᵛ] (title of count of Carpi), sig 1ᵛ ('La quale fue impressa con licencia de su Magestad por cedula firmada du su Real nombre y señalada por los Señores de su Real consejo')
180 Allen Epp 2441:71, 2443:346, 2466:102; *Apologia* 112 below

he received a portion of it from Pio's nephew and heir, Rodolfo.[181] Writing to Erasmus on 25 July 1531 from Reggio in Emilia, where he was prior of the Augustinian canons, Steuco scolded him for responding compulsively to any criticism and thus wasting his time and energies in writing *apologiae* instead of more serious works. These defences of his reputation, Steuco wrote, often rehash what is scandalous and descend to exchanges of abuse. Such behaviour is unbecoming a Christian, much less a priest, and even less a very learned man who should exhibit modesty. This reprehensible conduct is most evident in his hate-filled ravings against Pio for his 'crime' of dying. Erasmus ridicules and gloats over Pio's loss of Carpi, Steuco charges, and then mockingly traces his fall from prince to exile, sycophant, and Franciscan. Pio was a sycophant only in the technical sense of that Greek word because he summoned Erasmus to stand before the court of justice. Erasmus urges others to be modest, yet he immediately resorts to publishing an apologia if someone slights him or fails to lavish praises on him. Instead of extending sympathy to Pio because of his misfortunes, Erasmus seems to be urging, even pushing, him on towards greater tragedies. Instead of explaining to others how Pio's faith and piety manifested themselves at the end, Erasmus mocks his Franciscan garb. Such scoffing hardly befits a Christian. Given the date of the letter, and the details here cited, Steuco was probably responding, not to Erasmus' attacks on Pio in the *Apologia* or *Exequiae seraphicae*, but to the fuller jibes found in his 21 August 1531 letter to Julius Pflug published a month later. Erasmus did not reply to Steuco's letter, either when it was sent to him privately or when it was published two years later.[182]

* * * * *

181 On Steuco, see Allen IX 204–5; the entry on Agostino Steuco in CEBR III 285–6; Ronald K. Delph 'From Venetian Visitor to Curial Humanist: The Development of Agostino Steuco's "Counter"-Reformation Thought' *Renaissance Quarterly* 47 (1994) 102–39; and Rummel *Catholic Critics* II 135–40, especially 135, where she states that Steuco was given a portion of Pio's library by Alberto's nephew, and not by Alberto himself, as claimed by Allen (VI 201). In his will (Pio *Will* 402 below) Pio states: 'I leave to my above-mentioned nephew Rodolfo Pio ... all my books and codices of whatever kind and also all the statues and paintings and monuments of antiquity that were mine and belonged to me in whatever place they may have been and I urge him, the town of Carpi having been recovered, to set up a library with the above-mentioned books for the adornment and benefit of the above-mentioned town just as I had planned to do, handing over to the library of San Nicolò of Carpi a share of the books of which he has duplicates.' For a study of the fate of Pio's library, see Mercati *Codici Latini Pico Grimani Pio* (cited in n61 above).

182 Allen Ep 2513:659–722. Allen IX 289 and Ep 2513:690n states that Steuco's letter was first published in 1533, probably in a revised form that took cognizance of Erasmus' insulting comments about Pio in a letter to Julius Pflug of 20 August 1531 that was published in his *Epistolae floridae* of September 1531. It is in that letter, and not in the

The scholar who provided the fullest defence of Pio was Juan Ginés de Sepúlveda (c 1490–1573). This distinguished Spanish humanist and philosopher had studied theology at the Spanish College of San Clemente in Bologna (1515–23), became a member of the household of Alberto Pio in Rome in 1523, but remained in Italy after the Sack of 1527 when Pio went to Paris. Sepúlveda was accused by Erasmus of having helped Pio write the XXIII *libri* by polishing up its Latin style. In January 1532 Sepúlveda's *Antapologia pro Alberto Pio Comite Carpensi in Erasmum Roterodamum* was published in Rome by Antonio Blado, and the work was reissued on 22 March 1532 by Nicolas Prévost at the press of Antoine Augereau in Paris. Erasmus had already received a copy of the Paris edition when he received a personal letter dated 1 April 1532 from Sepúlveda in Rome, together with an emended copy of the Rome edition. Sepúlveda explained that, after his *Antapologia* was printed in Rome and before it was openly distributed, some impartial scholars had examined the book and recommended that certain words which could needlessly irritate Erasmus and Italian scholars be tempered or expunged. Sepúlveda agreed, and the text was accordingly altered by erasures. Similar changes were not made in the Paris edition. Sepúlveda wanted Erasmus to know the reason for the strange erasures, lest he wonder at the book's unusual appearance.[183]

* * * * *

Apologia or the colloquy *Exequiae seraphicae*, that Erasmus claims that becoming a Franciscan was the final stage of Pio's degradation; see Allen Ep 2522:66–80, especially 67; Erasmus *Apologia* 112–13, 117 below; *Exequiae seraphicae* ASD I-3 687:21–2, 40–2 / CWE 40 1000:16–18, 36–7. On the lack of a response to Steuco's letter, see Allen IX 289.

183 Solana lxxv–lxxvii has identified copies of the Roman edition with various words (eg *caeci furoris, temere, Italorum, stultissime,* etc) cancelled out in ink (just as Sepúlveda reported) and he has also found in the colophon of a copy preserved in the Archivio storico comunale of Carpi (lamina v) evidence of a 1532 Roman printing 'apud Antonium Bladum' subsequent to that of 'Mense. Ianuario.'; Losada 37–69, 305–7; the entry on Juan Ginés de Sepúlveda in CEBR III 240–2; for the accusation that Sepúlveda had polished up Pio's work, see Allen Epp 2324:99–102, 2375:76–7; Erasmus *Apologia* 119, 359 below; on Erasmus' earlier high hopes for Sepúlveda as a humanist writer, see his *Ciceronianus* LB 1015E / ASD I-2 691:2 / CWE 28 429; on the Paris edition, see Losada 361–3; and Jeanne Veyrin-Forrer 'Autour d'une édition clandestine de "Colloques d'Erasme" (1532)' in her *La lettre et le texte: Trente années de recherches sur l'histoire du livres* Collection de l'École normale supérieure de jeunes filles 34 (Paris 1987) 51–62, especially 59; Allen Epp 2637:1–21; 2652:1–2. Erasmus' critic Theodoricus (Thierry/Dirk) Adriani (Adriaans) Hezius (of Heeze), in the hope of winning over Gerard Morinck to an anti-Erasmus stand, sent him in 1537 a copy of Sepúlveda's *Anatapologia*. In response Morinck wrote: 'I am very pleased with the modesty of the man [Sepúlveda]. Whether any one of those who have drawn a pen against Erasmus has a similar modesty, I am not yet aware; although the most learned and at the same time most noble man Alberto Pio

Both in the cover letter and in the *Antapologia* Sepúlveda explains the reasons that led him to write. He insists that he is motivated neither by a hatred or malevolence towards Erasmus nor by a desire for fame. Indeed, as someone who has been repeatedly praised in Erasmus' writings, Sepúlveda feels only benevolence and affection for him. Men of letters should be reconciled to one another. Sepúlveda is certainly not out to damage Erasmus' reputation. Rather, it is his intention to restore another's reputation, that of his former patron and friend Alberto Pio, whom Erasmus has falsely and bitterly denounced and cursed, even though he is dead and can no longer defend himself. Sepúlveda is *de jure* bound to come to the aid of Pio, who called him into his princely service while he was still a student at Bologna, a patron who always treated him in a humane and loving way. Sepúlveda remembers Pio as a man of great wisdom, learning, gravity, piety, humanity, generosity, virtue, and skill in managing affairs. Piety demands that Sepúlveda honour that memory. He also has a duty to set the record straight since Erasmus has named him as the one who had polished the Latinity of Pio's *xxiii libri*.[184] Sepúlveda will comment on the many, often false, accusations made by Erasmus.

Sepúlveda writes that Erasmus has unfairly accused Pio of lacking learning and talent in the areas of letters, philosophy, and theology. But this charge flies in the face of Pio's international reputation as a man of outstanding eloquence, erudition, and wisdom. Sepúlveda, who lived for years in his household, claims that Pio's learning was well known to him. He traces Pio's education under such personal tutors as Aldus, Pietro Pomponazzi, Juan Montesdoch, Marcus Musurus, and other distinguished scholars who taught him letters, philosophy, and theology (scriptural, patristic, and scholastic). Pio's dedication to serious study continued over a lifetime, and popes and cardinals can testify to his exceptional erudition and eloquence. It is thus a great lie that Pio lacked learning.[185]

Erasmus also claims that Pio was accustomed to criticizing him in the presence of others as someone who lacks solid doctrine and is neither a philosopher nor a theologian. Sepúlveda admits that Pio, like many other Italians, wondered about Erasmus. When trying to describe what kind of

* * * * *

for the same reason became somewhat known to me earlier. In this regard he is not inferior [to Sepúlveda]. Would that all the others, whether enemies or friends of Erasmus, would use the same sobriety'; see de Vocht *Louvain Humanists* (cited in n31 above) 517–18 Ep 2:8–14.

184 Allen Ep 2637:4–5, 12–13; Sepúlveda *Antapologia* Aii^r–v, Av^v, Bi^r–Bii^r, Mii^v / Solana 113–18, 162–3

185 Sepúlveda *Antapologia* Aiv^v, Bi^v, cii^v, ciii^r–civ^v / Solana 116–17, 122–7

philosopher Erasmus might be, they tended to see him as a follower less of Aristotle and more of Lucian, who exhibited little knowledge of or concern for philosophy, morality, theology, and religion. While Italians are generally jealous of Erasmus' remarkable skills as a Latin stylist, Pio was not. He admired Erasmus for his eminent erudition and eloquence. On the question of style Erasmus has justly criticized the Italians for their excessive Ciceronianism.[186]

Erasmus makes the ridiculous charge that Pio wrote against him in the hope of gaining thereby fame and glory. What vanity on Eramus' part! If Pio's opinion of Erasmus' abilities was as low as Erasmus claimed, what possible glory could Pio gain from attacking so weak an opponent? It would not bring glory to a prince, already well regarded for his talents and learning, to argue publicly with Erasmus. Pio did not debate him out of any thirst for glory![187]

The claim that Pio entered the controversy not on his own initiative but on the urgings of others is a fantasy concocted by Erasmus with no basis in reality. It implies that Pio, a distinguished diplomat, was weak-minded and easily manipulated by others. A man who was so overwhelmed by the burdens of his ambassadorial duties and poor health that he could rarely enjoy a moment of quiet would be the one least likely to be asked and most easily able to resist such urgings.[188]

Why did Pio write? According to Sepúlveda, he completed the *Responsio paraenetica* out of his sense of personal dignity and good manners, lest he be considered unlearned, being unable to respond, or haughty, deeming Erasmus unworthy of his reponse. His intention was not to hurt Erasmus but to defend truth and the Christian religion. He undertook to write the XXIII *libri* because of the provocative nature of Erasmus' *Responsio*. He felt that he had been attacked by Erasmus. In fact, Erasmus, in keeping with his typical way of responding immediately to anyone who criticized him, denounced Pio as an impudent ignoramus, a trifler, liar, and calumniator. No wonder Pio thought that he needed to protect his reputation. He also noted that Erasmus had failed to correct many of his positions in the *Responsio*. Topics that Pio had treated succinctly in his earlier work he could now develop more fully, and he was especially concerned to warn simple Christians about suspect passages in Erasmus' writings.[189]

* * * * *

186 Sepúlveda *Antapologia* Biii^v–Biv^v, ci^v, Liv^v / Solana 119–20, 161
187 Sepúlveda *Antapologia* Biii^{r-v}, Liv^v / Solana 118–19, 160–1
188 Sepúlveda *Antapologia* Biii^r, civ^r / Solana 119, 127
189 Sepúlveda *Antapologia* Bii^v, ci^v–cii^r, Fi^v, Liv^v / Solana 118, 121–2, 137, 160–1

An accusation, repeatedly made by Erasmus, was that Pio's role in the composition of the *xxiii libri* was so minimal that he contributed only his name to the work. Pio did not even read the writings he criticized, Erasmus charged, but relied instead on excerpts taken from them by the monks. The scriptural proof-texts used in his arguments were similarly supplied by others. On what grounds does Erasmus make these charges? Hearsay from unnamed sources. But Sepúlveda provides counter-testimony from an identified and very knowledgable person with whom he recently spoke in Rome, that is, the young Italian who was Pio's personal secretary (Francesco Florido da Sabina). Florido reported that he was the only one to give assistance to Pio and that his role consisted of reading aloud and copying out whatever passages Pio indicated. It is a common practice of princes to employ such secretaries, and Pio's use of these services should not be ascribed to any lack of industry on his part, but was necessitated by the severe pain in his joints caused by gout which prevented him from holding even a book or pen. Sepúlveda puts greater credence in the secretary's testimony than in the hearsay Erasmus has used to attack Pio. Besides, the premise underlying these charges supposes that Pio was dull and sluggish of mind, but anyone acquainted with Pio knows that he was intelligent enough to read by himself, to recognize a controversial passage, and to refute its errors. He was capable of examining on his own something pointed out to him by another.[190]

Sepúlveda reviews with particular interest the charge that others provided Pio with Latin words and polished up his style. According to the reports repeated by Erasmus, Pio hired a Spaniard to bring literary harmony to the hodge-podge of materials contributed by others. It is in this context that Erasmus stops making only vague accusations that numerous monks and theologians supposedly helped Pio and finally identifies someone by name, 'Sepulvelus, a learned man and a good Latinist.' While grateful for what seems at first a compliment, Sepúlveda notes that, not only has Erasmus misspelled his name, but that later he implicitly retracts the praise when he asserts that the variety of styles still evident in the *xxiii libri* proves that Pio was not its author. 'Sepulvelus,' hired for providing a literary polish to the work, must have been singularly inept in his labours if he failed to remove such dissonance. But this great disparity of styles claimed by Erasmus does not in fact exist, although there are indeed a few theological terms left in a scholastic formulation, instead of being trans-

* * * * *

190 Sepúlveda *Antapologia* cii[r], di[r], dii[v], diii[v], eii[r], eiii[r] / Solana 122, 127–8, 130–1, 134–5

lated into Ciceronian expressions. Pio was concerned with both content and style. He tried to put the ideas of Aquinas, Scotus, and Alexander into clear Latin words while avoiding the obscurity of a pure Ciceronianism. Unfortunately, death overtook him before he could complete his task. It should also be noted that Pio's Latin style is different from Sepúlveda's.[191]

What role, if any, did Sepúlveda have in the writing of Pio's two works against Erasmus? Sepúlveda admits that he had helped his patron by criticizing a draft of the *Responsio paraenetica*, and Pio reciprocated by suggesting revisions Sepúlveda might make in the *De fato et libero arbitrio Dogma Lutheri confutatur*, which he was then writing. Having a learned friend provide this valuable service before a work is published is not unusual. Among the ancient writers, Cicero and Pliny followed this practice. The unfinished draft of Pio's book did not, however, make the rounds of Rome. Sepúlveda swears that in the Eternal City only he and Pio's personal secretary were allowed to read the draft. A contribution on his part to the composition of Pio's *XXIII libri* was rendered impossible by geography. All the while that Pio resided in Paris and worked on that book, Sepúlveda remained in Italy, mostly in Rome. His presence in Italy was surely known to Erasmus, for it had been reported to him in a letter from Alfonso Valdes. Neither Sepúlveda nor any other Spaniard helped Pio to write his *XXIII libri*.[192]

Having reviewed all this evidence, Sepúlveda concludes that it is beyond all belief and contrary to the truth to claim that Pio's writings were composed not by his own learning but by the talent and labour of others.[193] From defending the reputation of his patron and friend, Sepúlveda now turns to admonishing Erasmus. He criticizes him for rushing into print. Erasmus would do well to spend more time on composition, especially when treating serious topics involving philosophy and theology. Nature teaches that what takes longer to produce generally lasts longer. Writers are not praised for the quantity but for the quality of their writings. Some of the greatest writers of antiquity wrote little and revised much.

* * * * *

191 Sepúlveda *Antapologia* cii[r], Di[v], Diii[r], Ei[v]–Eii[r] / Solana 122, 128, 130, 133–4. By putting scholastic terminology into classical Ciceronian Latin, Pio was parting ways with his uncle Giovanni Pico and borrowing the practice of such adopted 'Roman' humanists as Ermolao Barbaro and Paolo Cortesi. See Quirinus Breen *Christianity and Humanism: Studies in the History of Ideas* ed Nelson Peter Ross (Grand Rapids, Mich. 1968) 8, 15–25; and D'Amico *Renaissance Humanism* 155–60.

192 Sepúlveda *Antapologia* Di[r]–Ei[r] / Solana 128–33. For examples of Erasmus' charge that Sepúlveda had polished up Pio's literary style, see Allen Epp 2261:69–71, 2328:42–5, 2329:99–102, 2375:76–7; and Erasmus *Apologia* 119 below. On Sepúlveda's journey to Genoa, c July 1529, see Allen IV 622.

193 Sepúlveda *Antapologia* Eiii[r] / Solana 134–5

Erasmus himself has often revised his edition of the New Testament (four times so far), yet ironically whenever someone offers suggestions on how it can be improved, Erasmus resents the advice. He is also too quick to repeat improbable and false charges made against his opponents. He not only fails to check their accuracy but invents his own accusations not based on fact.[194]

Erasmus also writes in a manner unbecoming a Christian. Although ever ready to cite Christ's teachings, he does not imitate him who bore abuse patiently and did not respond in kind. Instead, Erasmus takes childish delight in attacking others. He writes as might a comedian cursing a slave or someone from the dregs of society. Through the mouth of Folly he mocks Christian practices and officials, claiming that she speaks wisely. When criticized for this, he reverses his position and insists that Folly speaks stupidly and ineptly. The Christian religion should not be the target of jokes, especially by a Christian. Yet Erasmus speaks more hostilely about Christianity than would conceivably Lucian or Averroes. This is evident in his treatment of monasticism, the cult of the saints, celibacy, and church ceremonies involving feast days, music, candles, and decorations. In his *Moriae encomium* and *Colloquia* he has dealt with these topics with mocking frivolity and disrespect.[195]

Erasmus' writings on religion have had deleterious effects. He claims that his intention is the correction of superstitious practices and moral abuses. His writings, however, lack the reflection and gravity required of such works and often contain singular teachings that have furnished Luther with the occasion for disturbing the church and have led to a wider revolution.[196]

Given the fact that for many years now upright and learned men, after careful examination of Erasmus' writings and mature reflection, have pointed out in a friendly way numerous passages that are suspect, doubtful, dangerous, and lacking in gravity, it is amazing that Erasmus ignores their admonitions and instead accuses them all of ambition, anger, and hatred. Pio is not alone in being so attacked. Why is Erasmus not solicitous that his writings be acceptable to all Christians both learned and simple? St Paul warns us not to scandalize our brethern. Why does Erasmus not put aside strife, remove from his writings what offends, and revise them

* * * * *

194 Sepúlveda *Antapologia* Bi[v], Biii[r], Eiii[v]–Fii[v] / Solana 117–19, 135–9
195 Sepúlveda *Antapologia* Aiv[r–v], Bii[v], Eiii[r], Fii[v]–Fiv[r], and Giv–Liv[r] / Solana 115, 118, 134–5, 138–40, 142–60 (for his analyses of how Erasmus treated specific topics)
196 Sepúlveda *Antapologia* Liv[r–v] / Solana 160

in such a way that no one can draw an evil or impious interpretation from them? Instead, he wastes his time writing apologiae that put forward ingenious interpretations of what to others would be offensive statements and then insists that he has written everything with a good intention. Why does he not follow the example of Augustine, who, having dispensed with defending his reputation, was more concerned with Christian piety and truth, and wrote retractions? May Erasmus heed the criticisms of men like Pio and acknowledge his errors.[197]

Erasmus' response to Sepúlveda's admonitions was predictable and yet also surprising. Rather than write retractions, he wrote to his friends denouncing Sepúlveda as the 'vainglorious' one, the successor of Zúñiga, and now the haughtiest of all the Spaniards. Erasmus claims that Sepúlveda has read none of his publications, but has repeated and exaggerated what he heard others say. His *Antapologia* Erasmus dismisses as truly dull, full of mournful ditties and hateful statements written to favour Pio, a work that is exceptionally stupid and insulting. But instead of dashing off a bitter rejoinder, he claims that the book is unworthy of a response and he has more urgent things to do. In a brief but delayed letter to Sepúlveda of 16 August 1532 Erasmus begins by saying that he admires the learning, talent, and eloquence of someone like Sepúlveda, who is devoted to the muses and to Christ. He laments, however, that Sepúlveda's talents are devoted to the themes treated in his book. Should Erasmus respond to this work, he does not see what can come of it other than discords of which the world is more than full. He has thus thought it wiser not to respond.[198]

Despite his three written responses to Pio's *xxiii libri* (the *Apologia, Exequiae seraphicae,* and *Brevissima scholia*), Erasmus continued to feel burdened by that 'egregiously stolid volume' and by the fame of its author, who had traduced him in Rome and tarnished his reputation in France. He would not back down from his insistence that Pio had been helped in writing that work, claiming that some of Pio's assistants had either written or come in person to Freiburg to confess their involvement. His attention, however, shifted away from their theological disagreements and focused instead on literary issues. In letters to friends he points out that Pio had misunderstood Erasmus' use of the word *anathema,* [literally 'something set up'], thinking that he had meant papal excommunications instead of *ex-voto* or votive offerings (often figurines presented as testimonies to favours received), such as those set up in Italian churches. Erasmus resented Pio's assertion that

* * * * *

197 Sepúlveda *Antapologia* Liv^v–Mii^r / Solana 160–1
198 Allen Epp 2652:1–4, 2701:1–8, 2810:90–3, 2906:16–19

Aldus had been Erasmus' teacher in Venice and adamantly denied that he had taught him either Latin or Greek. He was also upset that Pio had stirred up among the Italians a hatred against him because he had noted in his *Ciceronianus* the impiety of some of their scholars.[199]

Pio's criticisms of Erasmus were not only defended by his friends but even praised by someone Pio had insulted as 'most foul' and 'inspired by the spirit of Lucifer,' namely Martin Luther. Writing to Nikolaus von Amsdorf around 11 March 1534, Luther complimented the prince of Carpi for correctly denouncing Erasmus as a supporter of Arianism because of his statement in the preface to his edition of Hilary: 'We dare to call the Holy Spirit God, something the ancients did not dare.' Luther accuses Erasmus of implying that the doctrine of the Trinity and the Christian religion itself are based on human authority and that all religions are nothing but fables. Erasmus is guilty of unheard-of pride and is the devil incarnate whose statements, devious lies, cannot be trusted. Even though Luther is aware that Pio is often hostile to him, nonetheless in their evaluations of Erasmus they found some agreement. Because multiple editions of this letter of Luther were published in 1534, Erasmus felt compelled to respond with his *Purgatio adversus epistolam Lutheri* (1536), in which he noted the irony that Luther's hatred of Erasmus was so great that he praised Pio. Erasmus accused Luther of committing the same mistake as Pio, namely, of failing to read the writings he attacked.[200]

Even the deaths of these adversaries, Pio in 1531 and Erasmus in 1536, did not end the controversy. In 1540 two Italians briefly revived the dispute by publishing in Basel, where Erasmus had died four years earlier, contrasting descriptions of Pio. The former Augustinian friar and wandering Milanese humanist Ortensio Lando (c 1512–c 1554) described

* * * * *

199 Allen Epp 2613:31–6, 2615:66–8, 2721:39, 2810:98–108, 3002:670–8 (Erasmus' Protestant-leaning correspondent Giovanni Angelo Odoni agreed with his views on Pio), 3032:233–5, 260–2, 528–35, and Allen XI xxiv (Erasmus claims that Pieter de Corte got from Pio his mistaken views on Aldus as Erasmus' teacher).

200 WA *Briefwechsel* VII 34–5 no 2093:224–60. Erasmus' statement is in Allen Ep 1334:444–6 / CWE 1334:475–7: 'We dare to call the Holy Spirit true God, proceeding from the Father and the Son, which the ancients did not dare.' Pio *XXIII libri* 182^VR later cites Erasmus' text as: 'We dare to call the Holy Spirit God, something the ancients did not dare.' Pio goes on to note that if Erasmus is thereby defending Arian teachings, 'he should be considered by far more impious than Luther, indeed, and the greatest blasphemer.' For Pio's vilification of Luther, see Pio *Responsio paraenetica* 23^rP: 'For no one recently is more foul,' and 24^VP: 'Or who (unless not only inspired, but rather seized, by the spirit of Lucifer) would dare to say that ... the whole church has departed from true piety thirteen hundred years ago?' For Erasmus' response to Luther, see his *Purgatio adversus epistolam Lutheri* LB X 1545C–D / ASD IX-1 418:437–9.

in his *In Desiderii Erasmi funus dialogus lepidissimus* a dream reported by the German interlocutor Arnoldus in which Erasmus was surrounded by angels in heaven while his various critics suffered in hell. Notable among the critics was an Italian nobleman who was well known to the cardinals and died as an ambassador in France.[201] In 1540 there appeared a collection of writings by Florido, the former personal secretary of Pio, who had lived in his household both in Rome and in Paris. When listing those who had written against Erasmus, Florido includes Pio, that 'most excellent prince,' who was tutored in his youth by Aldus, excelled in philosophy and theology, and was skilled in the diplomatic affairs of popes and kings. Although provoked by Erasmus, he had responded with a work of which 'nothing more modest, nothing more sensible could be expected from so learned a prince.' When Erasmus continued the controversy, this 'greatest hero' laboured despite physical pain and approaching death to write a learned theological response. Florido also testified that Pio wrote his work without the assistance of others, whatever Erasmus claimed. Florido alone assisted him with the services one might expect from a secretary.[202]

The controversy took a final ironic twist when church authorities seemed to question the orthodoxy of Alberto Pio. The condemnation of the Castilian version of his *XXIII libri* by the Spanish Inquisition in 1536 was incorporated into a number of the indexes of forbidden books, each apparently repeating what had been condemned in an earlier index. Thus Pio's work was proscribed in a 1551 Spanish index and in the 1559 index of the Spanish inquisitor-general, Fernando Valdes, in the 1570 index of Antwerp, in the 1547, 1551 and 1581 Portuguese indexes, and in the 1583 index of the Spanish inquisitor-general, Gasparo Quiroga. Beyond Spanish Hapsburg

* * * * *

201 On Lando, see Myron Gilmore 'Anti-Erasmianism in Italy: The Dialogue of Ortensio Lando on Erasmus' Funeral' *The Journal of Medieval and Renaissance Studies* 4 (1976) 1–14; Silvana Seidel Menchi 'Sulla fortuna di Erasmo in Italia: Ortensio Lando e altri eterodossi della prima metà del Cinquecento' *Schweizerische Zeitschrift für Geschichte* 24 (1974) 537–634; Seidel Menchi 'Chi fu Ortensio Lando?' *Rivista storica italiana* [Naples] 106 (1994) 501–64; the entry on Ortensio Lando in CEBR II 286–7; Rummel *Catholic Critics* II 145; and Bruce Mansfield *Phoenix of His Age: Interpretations of Erasmus: c 1550–1750* Erasmus Studies 4 (Toronto 1979) 103–8, especially 104 (where the monks do for Erasmus' corpse the opposite of what they did for Pio's: the funeral procession is turned into a riot, the funeral eulogy is contradicted, the corpse is disinterred and mutilated, and the monument is thrown down), and 108 (Pio appeals from hell to Erasmus in heaven for succour).

202 On Florido, see Remigio Sabbadini 'Vita e opere di Francesco Florido Sabino' *Giornale storico della letteratura italiana* 8 (1886) 333–63; Florido *Apologia* 116; and Florido *Libri tres* 127–8, 262–5, especially 264:20–39.

lands, however, Pio's *xxiii libri* found a more favourable reception among Catholic authorities. In 1566 Duke Albrecht of Bavaria ordered the monasteries in his territory to have in their libraries the works of sincere and wholesome Catholic writers, among whom he listed Albertus Pius.[203]

A CHRONOLOGICAL SURVEY OF
THE PRINCIPAL TEXTS

1 Erasmus' letter to Alberto Pio of 10 October 1525. This letter was first published with a number of typesetting errors at the beginning of Pio's *Responsio* (items 2b, 2c, 2d below). It was published in a corrected version in Erasmus' *Responsio* (items 3a, 3b, 3c, 3d below), and in a revised form in Pio's *xxiii libri* (items 4a, 4b, 4c below). For a critical edition of this letter, see Allen Ep 1634, and especially his introduction, where he traces the history and peculiarities of the various editions.

2 *Alberti Pii Carporum comitis illustrissimi, ad Erasmi Roterodami expostulationem Responsio accurata et paraenetica, Martini Lutheri et asseclarum eius haeresim vesanam magnis argumentis, et iustis rationibus confutans.* 1526/9

* * * * *

203 Franz Heinrich Reusch *Die Indices Librorum Prohibitorum des Sechzehnten Jahrhunderts, gesammelt und herausgegeben* (Tübingen 1886; repr Nieuwkoop 1961) 231, 318, 332, 357, 432. For an updated version of these indexes, see the series of Jésus Martinez de Bujanda *Index des livres interdits* (Sherbrooke, Québec / Geneva) eg vol v: *Index de l'Inquisition espagnole 1551, 1554, 1559* (1984), for 1551, see pages 248–9 no 61 and for 1559, see page 451 no 432; vol vi: *Index de l'Inquisition espagnole 1583, 1584* (1993), for 1583, see pages 567–8 no 1711, for Portuguese indexes of 1547, see page 567 no 127, of 1551, see page 567 no 37, and of 1581, see page 567 no 95; vol vii: *Index d'Anvers 1569, 1570, 1571* (1988), for 1570, see pages 407, 695, nos 668 and E432. Among those who defended or praised Pio's writings against Erasmus were the following: the future cardinal Bartolomeo Guidiccioni writing in Rome c 1538, whose treatise *De ecclesia et emendatione ministrorum eorumque abusuum per generale concilium facienda*, dedicated to Pope Paul III, repeatedly cited Pio's *xxiii libri* as the work of a major Catholic writer who effectively refuted Lutheran errors on a host of topics (*Concilium Tridentinum xi: Tractatuum pars prima*, ed Vincentius Schweitzer [Freiburg im Breisgau 1966], 234:2, 7, 9–10, 12, 17, 19, 22, 31, 35, 235:3, 5, 238:4, 241:2); the Italian Dominican Ambrogio Catarino dei Politi in his *De certa gloria invocatione ac veneratione sanctorum disputationes atque assertiones catholicae adversos impios* (Lyon: apud Mathiam Bonhomme 1542) 8 (Pio broke the arrogance of Erasmus by his pious and grave censure), 26 (Pio wrote against Luther and Erasmus with praise), 9, 56, 60 (Erasmus wrote against Pio); the French bishop Robert Ceneau in his *Pro tuendo sacro coelibatu axioma catholicum* (Paris: apud Joannen Roigny 1545) 209 (defends Pio's claim that a monastic profession prevents a subsequent marriage); and the Spanish Franciscan Antonio Rubio in his *Assertiones catholicae adversus Erasmi Roterodami pestilentissimos errores* (Salamanca 1567) in Mansfield *Phoenix of His Age* (cited in n201 above) 32 (Pio had the wolf Erasmus by the ear) and Andrés II 599 and 630 n2.

(a) The surviving rough draft of part of this work with the basic text in the hand of a scribe and with numerous corrections and insertions often in the autograph hand of Alberto Pio is preserved in the Biblioteca Ambrosiana of Milan, Archivo Falcò Pio di Savoia, I Sezione, Parte I (Famiglie e Persone), Scatola 282, Documento 6 [vecchia segnatura 254, 6]. Fiorina 70 describes this manuscript on paper as consisting of fifty-seven folios in all, of which the final two are blank. According to the ancient numeration, the codex is missing folios 1, 2, 4, 5, 11[actually 12]–16, 58–63; folios 8 and 9 are both numbered 8; an unnumbered folio with writing only on the recto side is inserted before folio 47^r and after 51^v; and an unnumbered folio with writing on both sides is inserted between folios 48 and 49. The final numbered folio is 67. For the reconstructed provenance of this manuscript, see Forner 207–9. Folio 58^{r-v}, which is missing from this Milan manuscript, is preserved in the Houghton Library of Harvard University in Cambridge, Massachusetts as MS Latin 299. It is of irregular shape with a width varying from 21.5 to 22.0 cm and a length of 29.0 to 29.3 cm. Two-thirds of the paper is water-stained. In the upper left corner of the recto side is a note in a later hand: 'Fragmento della lettera ad Erasmo di Roterdamo e Luthero scritta nel 1478 [sic] e stampata in Parigi nel 1531'; and in the upper right corner: '1478. Correzioni e postille del celebre letterato filosofo ed autore della lettera Alberto Pio.' At the bottom of the recto folio in the middle is '.M.'. The Curator of Manuscripts in the Harvard College Library, Dr Leslie A. Morris, in a letter of 23 June 1997, reported that it had been 'purchased from Goodspeed's Bookshop in Boston on 18 December 1964, with funds given by Curt H. Reisinger.' A clue to its earlier provenance is what appears to be a catalogue entry glued to a piece of paper kept with the folio in the MS Latin 299 folder. It reads: '2675. Pio (Albert), prince de Corps, Modenais, théologien, connu par ses discussions avec Erasme. Minute avec corr. autogr. 1531. 2 p. in-fol. Fragment de sa lettre addressée à Luther et à Erasme; il reproche aux protestants de fermer leur temples en semaine, de ne pas les ouvrirs à ceux qui veulent y prier; ils ne permettent aux fideles que l'oraison secrète.' Goodspeed's Bookshop is no longer in business; its uncatalogued records, consisting of twenty-seven cartons and twenty-two card files in eleven units (in all thirty-five linear feet), were donated by George T. Goodspeed to Harvard University and are housed at Houghton Library. Carton 27, entitled Ledgers of Goodspeed's Bookshop, contains a record of Harvard University's purchase of this folio on card 15, with the following entry: '[Date:] Dec. 12 64 ... [Ledger No:] 2263 ... [Reference:] Houghton Lib. ... [Debit:] 165.00 ...' Roger

Stoddard, the Curator of Rare Books at the Houghton Library, who formerly worked at Goodspeed's, could not provide any information about the folio's provenance, but doubts that it was purchased directly from a French catalogue for resale in America. The Biblioteca Ambrosiana manuscript without Pio's autograph corrections and additions has been transcribed as an appendix in Alberto Pio da Carpi *Ad Erasmi Roterodami expostulationem responsio accurata et paraenetica* ed Fabio Forner, 2 vols, Biblioteca della Rivista di storia et letteratura religiosa. Testi e documenti 17 (Florence 2002), II, 543–611.

(b) A manuscript copy of Erasmus' 1525 letter to Pio and of Pio's *Responsio paraenetica*, written in a gothic cursive hand, with various corrections in a humanistic hand identified as probably that of Pio by Forner (221), and penned on paper with French watermarks dating from the early sixteenth century, is preserved in the Herzog-August-Bibliothek in Wolfenbüttel under the collocation number Guelf. 85.4.2 Aug. 2°. Forner speculates that this manuscript is evidence of the existence of a now lost manuscript (beta) that Pio brought with him to Paris that is different from the corrected Ambrosian version in at least seventy instances in that Pio adds biblical, patristic, and contemporary references (Forner 218 n86). On the basis of this manuscript a clean copy of Pio's *Responsio paraenetica* was made and circulated in Paris among his friends. The manuscript copy of the *Responsio paraenetica* that Pio sent to Erasmus (alpha) is very similar to the Wolfenbüttel manuscript. See Forner 209, 217–24. Pio, however, refers to the manuscript rescued by his servant during the Sack of Rome and brought to Paris as 'that first archetype [ie original]' manuscript and states that the clean copy '*illud castigatum*' was lost together with his library and home furnishings. In Paris he revised and corrected the archetype and added two or three scriptural proof-texts, that is, he 'corrected errors, removed superfluous material, and added in places what was opportune' (Pio *XXIII libri* 70^{r}1). Thus, the Wolfenbüttel manuscript is the revised copy of the Ambrosian manuscript (no need for a hypothesized beta manuscript) Pio made in Paris, with scriptural citations later inserted into the Bade printed version.

(c) The first printed edition, bearing the above title, was published in Paris by Josse Bade with the colophon: 'Finis Sub prelo Ascensiano ad Septimum Idus mensis Ianuarii. Anno. M.D.XXIX. calculo Romano.' It contains: title-page (sig Ai^r), letter of Bade to the reader dated 5 January 1529 (sig Aii^{r-v}), Erasmus' letter to Pio of 10 October 1525 (sigs Aiii^r–iv^v), Pio's *Responsio paraenetica* (sigs ai^r–niii^v or folios I^r–XCIX^v), and a final folio of *erratula* (recto only). It consists of quarto quire A, sexto quire n, and octavo quires a, b, c, d, e, f, g, h, i, k, l, m, and an unnumbered folio of *erratula* at the end. The folios measure c 14.5 x 18.5 cm. This Bade edition is catalogued in Renouard *Imprimeurs* II 251 (no 625). There is a copy in the Biblioteca Apostolica Vaticana. For the locations in libraries in Europe and America of this and other editions of the works here cited, see Forner 211–13, especially nn 69 and 70.

(d) An apparently pirated edition, printed in Paris by Pierre Vidoue and bearing the same title, has on folio 144^r the colophon: 'Sub Prelo Vidouaeo ad Septimum Idus mensis Maii. Anno M.D.XXIX.' This work consists of octavo quires: a, a[bis], b, c, d, e, f, g, h, i, k, l, m, n, o, p, q, r, s; folio numbers begin with sig a[bis]i^r (= folio 1^r) and continue to sig sviii^r (= folio 144^r). The book contains: title-page (sig ai^r), letter of Bade to the reader (sigs ai^v–aiii^r), an index of the principal arguments arranged according to the progressive order in which they appear in the book together with their folio numbers (sigs aiii^v–avi^r), the letter of Erasmus to Pio of 10 October 1525 (sigs avi^v–aviii^v), Pio's *Responsio paraenetica* (sigs a[bis]i^r / folio 1^r–sig sviii^r / folio 144^r), blank (sig sviii^v / folio 144^v). The folios measure c 15.8 x 10.3 cm. Vidoue's press in Paris had republished in July 1522 Diego López Zúñiga's *Annotationes contra Erasmum* together with Erasmus' *Apologiae contra Stunicam* (1), and in March 1523 reprinted large sections from Zúñiga's *Annotationes contra Erasmum* together with Erasmus' *Apologia contra Santium Carranzam*. Another publisher, F. Peypus in Nuremberg, reprinted Zúñiga's *Conclusiones . . . in libris Erasmi Roterodami* together with Erasmus' *Apologiae contra Stunicam* (4) against it in 1524/5. Although Erasmus had in 1520 published at the Hillen press in Antwerp Edward Lee's *Annotationes* together with his *Apologia qua respondet invectivis Lei*, he came to the conclusion that he should not provide his opponents with another edition of their criticisms and henceforth tended to resist this practice. Thus Vidoue's 1529 publications of the works of Pio and Erasmus were probably unauthorized; see ASD IX-2 35–6, 51; Allen IV 624 (Appendix XV Ep 1:71–4); and CWE 71 xxvii. The Vidoue edition of Pio's *Responsio paraenetica* is often found bound together with the Vidoue edition of Erasmus' *Responsio* that was published a month later; for example, see the copies in Paris, Bibliothèque Nationale, inventaire D 27886; and in Washington, DC, Folger Library, PA8518 P55 1529 cage.

(e) A French translation of the *Responsio paraenetica* was produced on Pio's orders and presented to Baron Guillaume de Montmorency (1452–1531), who had requested it (Pio *Response* – Chantilly Plon 169: 'A ceste cause, acceptant ton bon vouloir et humble commandement, ay voulu te faire ce livre translater de latin en françois'). Because the translation was based on the Latin edition published by Bade (167, citing folio 203ᵛ: 'Imprimé à Paris le 4ᵉ[7] jour de janvier 1528 [1529 n.s.]'), it was finished some time between 7 January 1529 (date of the Bade edition) and 8 January 1531 (date of Pio's death). The codex is of vellum, consisting of 203 folios that measure c 31.3 x c 22.8 cm with 24 to 27 lines per page. The translator is anonymous, but his handwriting has been identified by Charles Samaran as the same as that used in a document dated from after 1509 and preserved in Paris, Arsenal 2037, folio 21. The codex contains an elaborate frontispiece (folio 1ʳ), a dedicatory letter to Montmorency (folios 1ᵛ–4ʳ, transcribed in Pio *Response* – Chantilly Plon 168–9), a title and brief preface (ibid 169–70), and the translation itself, divided into chapters and terminating in folio 203ᵛ. This manuscript is preserved in the Musée Condé at Chantilly as MS 187 (formerly 709, XIX C16). For further descriptions of this manuscript see Institut de France, Musée Condé *Chantilly. Le Cabinet des livres. Manuscrits.* I *Theologie – jurisprudence – sciences et arts* (Paris 1900) 167–70; Charles Samaran and Robert Marichal, *Catalogue des manuscrits en écriture latine portant des indications de date, de lieu ou de copiste* 7+ vols (Paris 1959–[85]) I: *Musée Condé et bibliothèques parisiennes* (1959), 19 (verbal description of codex) and I: *Planches* (plate CLVI for a photographic example of the handwriting of the Arsenal document); and *Catalogue général des manuscrits des bibliothèques publiques de France. Paris. Bibliothèques de l'Institut: Musée Condé à Chantilly. Bibliothèque Thiers, Musées Jacquemart-André à Paris et à Chaalis* (Paris 1928) 44.

(f) Another French translation of Pio's *Responsio paraenetica*, dating from early 1531, which was based on item 2e above and was presented by Guillaume de Montmorency to King Francis I, exists in a manuscript on vellum with a miniature on the frontispiece and illuminated letters. The folios measure 32.8 x 21.3 cm. The codex consists of 221 folios: frontispiece with miniature (folio 1ᵛ), dedicatory letter of Montmorency to Francis I (folios 2ʳ–4ʳ), blank folios (folios 4ᵛ–5ʳ), translator's preface (folio 5ᵛ), *Responsio paraenetica* (folios 6ʳ–221ᵛ). The *Responsio paraenetica* has been divided into 53 numbered chapters (but chapter 30 is listed as 31, and the starting point of chapter 36 is not identified). This manuscript is in the Bibliothèque nationale de Paris, Manuscrits, Fonds Français, Ancien, MS 462. It is briefly described in *Catalogues de la Bibliothèque impériale, Departement des manuscrits, Catalogue des manuscrits français* I: *Ancien Fonds* (Paris 1868) 46 no 462.

(g) The *Responsio paraenetica* was reprinted as book 1 of Pio's *XXIII libri* (see items 4a, 4b, 4c below). In the Bade edition of 9 March 1531 it can be found on folios i^r–xLvi^v. In Luca Antonio Giunta's edition of the *XXIII libri*, published on 5 September 1531, book 1 occupies folios 2^r–35^r.

(h) Pio's *Responsio paraenetica* translated into Castilian was published in the 1536 Spanish edition of his *XXIII libri*, known as *Libro* (see item 4c below). Together with Erasmus' letter to Pio of 1525, it constitutes *Libero primero* of *Parte primera*, folios i^r–xlv^r.

(i) Apparently mistakenly thinking that he had found the original manuscript version of Pio's *Responsio paraenetica* as it existed before the Paris revisions, Hermann von der Hardt published in his collection *Historia literaria Reformationis in honorem jubilaei anno MDCCXVII* (Frankfurt/Leipzig: in officina Rengeriana 1717) 114–63 a transcription of the manuscript he had found in Rome. This manuscript, which had in 1577 belonged to Cardinal Guglielmo Sirleto, represents, according to Forner (cited in item 2j below) xlvii–xlix, a subsequent revision of item 2b that circulated in Paris prior to the Bade publication (item 2c) in 1529.

(j) A critical edition with a translation into Italian of the *Responsio paraenetica* has been published by Fabio Forner as Alberto Pio da Carpi *Ad Erasmi Roterodami expostulationem responsio accurata et paraenetica* 2 vols, Biblioteca della Rivista di storia et letteratura religiosa. Testi e documenti 17 (Florence 2002) I, 1–327 (Latin text and Italian translation), II, 331–541 (notes)

3 *Desiderii Erasmi roterodami Responsio ad epistolam paraeneticam clarissimi doctissimique viri Alberti Pii Carporum principis.* 1529

(a) The *editio princeps* of this work was printed together with *Eiusdem notatiunculae quaedam extemporales ad naenias Bédaicas* in Basel by Froben and Nicolaus Episcopius during March 1529. Added at the end of the full title are these promotional words: 'Nihil horum non novum est.' The *Responsio* is found on quarto quires b–l. The book consists of quarto quires a, b, c, d, e, f, g, h, i, k, l, m, n, o, p, q, and r; the pagination in arabic numbers begins on sig bi^r with Erasmus' *Responsio* to Pio. This work lacks a colophon and its dimensions are c 18.5 x 13.5 cm. The book contains: title-page (sig a1^r); blank (sig a1^v); letter of Erasmus to Pio of 1525 (sigs a2^r–a4^v), with an explanation on sig a4^v that it was deemed good to reprint this letter because the version Pio had earlier printed was 'depravatae'; Erasmus' *Responsio* to Pio (sig b1^r/page 1–sig l4^v/page 80); Erasmus' *Responsio ad notulas Bédaicas* (sig m1^r/page 81–sig r4^r/page 127); Froben's printer's mark (sig r4^v/page [128]). There is a copy in Cambridge, Massachusetts, at Harvard University's Houghton Library; for other copies, see Solana xc.

(b) In his work *Catalogue of Books Printed on the Continent of Europe, 1501–1600, in Cambridge Libraries* (Cambridge 1967) I 405 H.M. Adams lists as no 828 an anonymous edition with the truncated title *Responsio ad epistolam paraeneticam Alberti Pii. Eiusdem notatiunculae quaedam extemporales ad Naenias Bédaicas*. It consists of fifty-six folios, sigs a–g8, in octavo format, bearing the publication date of April 1529, and is preserved in the university library. Given the foliation and signatures, this is most likely an earlier edition of the identifiable Pierre Vidoue edition of June 1529; see item 3c below. For other copies, see Solana xc.

(c) Bearing the full title is an apparently pirated edition of this work published by Pierre Vidoue in Paris with the date on the title-page of 'Anno M.D.XXIX. Mense Iuni.' The colophon states: 'M.D.XXIX. Sub praelo Vidouaeo.' The folios measure c 15.8 x 10.3 cm. The book is composed of fifty-six numbered folios, that is, octavo quires a, b, c, d, e, f, g (plus perhaps two blank folios, hi–hii). It contains: title-page (sig ai^r), blank (sig ai^v), the letter of Erasmus to Pio of 1525 (sig aii^r/folio 2^r– sig av^r/folio 5^r), Erasmus' *Responsio* (sig av^v/folio 5^v–sig gviii^r/folio 56^r), blank (sig gviii^v/folio 56^v). There is a copy in Washington, DC, at the Folger Library; for other copies, see Solana xc.

(d) This work was included in *Desiderii Erasmi Roterodami Operum Nonus Tomus Complectens Ipsius Apologias Adversus Eos Qui Illum Locis Aliquot, In suis libris, aut non satis circumspecte, aut malitiose sunt calumniati, quarum nomenclaturam versa pagina indicabit*, published in Basel by Froben in 1540. The colophon states: 'Basiliae in officina Frobeniana per Hieronymum Frobenium et Nicolaum Episcopium anno M.D.XL.' The folios measure c 34 x 23 cm. This volume reprints the letter of Erasmus to Pio of 1525 on pages 892–3/sigs Fff2^v–3^r and the *Responsio* on pages 894–916/sigs Fff3^v–Hhh2^v. There is a copy in Washington, DC, at the Folger Library; for other copies, see Solana xc.

(e) The work was also reprinted in 1706 by Jean Le Clerc in volume IX of the Leiden edition of the *Opera omnia* published by Peter vander Aa (= LB IX 1095A–1122E).

4 *Alberti Pii Carporum Comitis illustrissimi et viri longe doctissimi, praeter praefationem et operis conclusionem, tres et viginti libri in locos lucubrationum variarum Desiderii Erasmi Roterodami, quos censet ab eo recognoscendos et retractandos. 1531*

(a) The *editio princeps* of this work was published in Paris by Josse Bade of Assche with this statement on the colophon: 'Imprimebat autem haec Iodocus Badius Ascensius in clarissima Parrhisiorum academia, cum gratia et privilegio a tergo primae chartae expressis: septimo Idus Martias, sub pascha M.D.XXI.' Since Bade had received a royal monopoly for four years for the publication of this work in French realms, it was not followed immediately by a pirated Vidoue edition. This volume consists of octavo quires: ã, a, b, c, d, e, g, h, i, k, l, m, n, o, p, q, r, s, t, v, x, y, z, A, B, C, D, E, F, G, H, and of sexto quires f and I; foliation in Roman numerals begins on sig ai^r (folio I^r) and continues to sig IVi^r (folio CCLII, misprinted as CCXLII). The folios measure c 32.5 x 21 cm. The volume contains: title-page (sig ãi^r), letter of Bade to the reader and grant of a royal monopoly (sig ãi^v), table of contents and index of *argumenta* arranged alphabetically (sig ãii^r–ãviii^r), poem of F. Florido in the form of a dialogue between this volume and its reader (ãviii^v), letter of Erasmus to Pio of 1525 (sig ai$^{r–v}$/folio I$^{r–v}$), Pio's *Responsio paraenetica* [= *liber* I] (sig aii^r/folio II^r–sig fvi^v/folio XLVI^v), Erasmus' *Responsio* with marginal notes by Pio [= *liber* II] (sig gi^r/folio XLVII^r–sig iiiii^r/folio LXVI^r); *Praefatio, libri* III–XXIII, and *Conclusio* (sig iiiii^v/folio LXVI^v–sig IVi^r/folio CCLII^r [mistakenly written CCXLII], blank (sig IVi^v/folio CCLII^v). This Bade edition is catalogued in Renouard *Imprimeurs* II 265 (no 669). There is a copy in Washington, DC at the Folger Library.

(b) A Venetian edition was printed by Luca Antonio Giunta of Florence. It bears the same title, except for adding the letter *h* in *Rhoterodami*, and it inserts on the title-page under the woodcut of Giunta's fleur-de-lis printer's mark these promotional words: 'Cunta haec candide lector diligenter considera, nam universum ferme Lutheri dogma in his confutatum invenies. Eme et fruere bonis avibus.' Its colophon states: 'Venetijs in Aedibus Lucae antonij Iuntae Florentini Anno Domini. M.D.XXXI. Nonis Septembris.' The volume consists of sexto quires A, B, C, D, E, F, G, H, I, K, L, M, N, O, P, Q, R, S, T, V, X, Y, Z, AA, BB, CC, DD, EE, FF, GG, HH, and of octavo quires * and II. Foliation begins with sig Ai^r/folio 2^r and ends with sig IIvii^r/folio 193^r. The folios measure c 31.5 x 22 cm. The volume contains: title-page (sig *i^r), letter of Giunta to Averoldi, who is called 'apostolicae sedis in toto Venetorum dominio Legato' (sig *i^v), table of contents and index of *argumenta* arranged alphabetically (sig *ii^r–sig *vii^r), poem of F. Florido (sig *vii^v), letter of Erasmus to Pio of 1525 (sig *viii$^{r–v}$), Pio's *Responsio paraenetica* [= *liber* I] (sig Ai^r/folio 2^r–sig FV^r/folio 35^r), Erasmus' *Responsio* with marginal notes by Pio [= *liber* II] (sig FV^v/folio 35^v–sig LV^v/folio 51^v), *Praefatio, libri* III–XXIII, *Conclusio* (sig LVi^r/folio 52^r–sig IIvii^v/folio 193^v), blank (sig IIviii$^{r–v}$/folio 194$^{r–v}$). This edition is listed in Paolo Camerini *Annali dei Giunti* I: *Venezia* part 1 (Florence 1962) 244–5 no 345. I am grateful to Paul F. Grendler for this reference. There is a mutilated copy in Washington, DC, at the Catholic University of America's Mullen Library.

(c) A Castilian translation of Pio's *XXIII libri* bears the title: *Libro del muy Il-lustre y doctissimo Señor Alberto Pio Conde de Carpi: que trata de muchas cos-tumbres y estatutos de la Iglesia, y de nuestra religion Christiana, mostrando su autoridad y antiguedad: contra las [malditas] blasphemias de Lutero, y algunos dichos de Erasmo Rotherodamo*. Following the title-page are the mostly promotional words: 'Traduzido de Latin en Castellano, para utilidad de muchos hombres sabios: que por carecer de lengua Latin son priuados de doctrina tan fiel, y prouechosa. Con priuilegio.' The word *malditas* is added to the title repeated on sig 2ᵛ. Part of the colophon states: 'Acabole de imprimer en castelleno en la Villa de Al-cala de Henares en casa de Miguel de Eguya primero de Enerode Mil y Quinientos y Treynta y Ses Años.' This volume consists of octavo quires *, A, B, C, D, E, F, G, H, I, K, a, b, c, d, e, f, g, h, i, k, l, m, n, o, p, q, r, s, t, v, of sexto quire x, of a quarto quire of preliminary material with signatures 1–4, and of the two-folio quire L. The folios measure c 29.0 x 20.5 cm. This volume contains: title-page (sig 1ʳ), table of contents (sig 1ᵛ), letter to Don Juan Téllez Girón dated 1 November 1535 (sig 2ʳ), pro-logue of the translator addressed to the reader (sig 2ᵛ–4ʳ), corrigenda (sig 4ᵛ), table of *sentencias* and *historias notables* (sig *iʳ–viiiᵛ), letter of Erasmus to Pio of 1525 (sig Aiʳ/folio iʳ–sig Aiiʳ/folio iiʳ), and Pio's *Re-sponsio paraenetica* (sig Aiiʳ/folio iiʳ–sig Fvʳ/folio xlvʳ) [the letter and re-sponse = *Libero primiero* of *Parte primera*], Erasmus' *Responsio* with Pio's marginal notes [= *Libero segundo*] (sig Fvᵛ/folio xlvᵛ–sig Iiiʳ/folio lxviʳ), the other books [part I books 3–4, part II books 1–21] (sig Iiiᵛ/folio lxviᵛ–sig xvᵛ/folio clxviᵛ), blank (sig xviʳ⁻ᵛ/folio clxviiʳ⁻ᵛ). There is a copy of this volume in Washington, DC, at the Folger Library.

5 Desiderii Erasmi Roterodami *Apologia adversus rhapsodias calumniosarum querimoniarum Alberti Pij quondam Carporum principis, quem et senem et mori-bundum et ad quiduis potius accommodum homines quidam male auspicati, ad hanc illiberalem fabulam agendam subornarunt.* 1531

(a) The *editio princeps* published by Froben in 1531 bears this title on its title-page (sig aa1ʳ/page 1), but an alternate title at the beginning of the *Apologia* (sig aa3ʳ/page 5): *Desiderii Erasmi Roterodami Apologia bre-vis ad vigintiquatuor libros Alberti Pii quondam Carporum Comitis.* The colophon on page 285 reads: 'Basileae, in officina Frobeniana, per Hi-eronymum Frobenium et Nicolaum Episcopium. Anno M.D.XXI.' This book consists of octavo quires aa, bb, cc, dd, ee, ff, gg, hh, ii, kk, ll, mm, nn, oo, pp, qq, rr, ss; pagination begins on sig aa3ʳ with page 5 and continues to sig ss7^r/page 285. The pages measure c 16.0 x 10.3 cm. The book contains: title-page (sig aa1^r), blank (sig aa1^v), index with references to page numbers (sig aa2^{r-v}), the *Apologia* (sig aa3^r/page 5– sig ss7^r/page 285), errata (sig ss7^v), blank (sig ss8^r), Froben printer's mark (sig ss8^v). There is a copy of the book in Washington, DC, at the Folger Library; for other copies, see Solana xci.

(b) Vander Haeghen 10 lists an edition of this work published in Antwerp by Michael Hillen in 1531. The British Library has a copy of this edition, which is 160 pages long in octavo format; see British Museum *General Catalogue of Printed Books: Photolithographic Edition to 1955* vol 67 (London 1960) col 929, shelf no 1009.a.22; see Solana xci.

We have been unable to locate copies of a second edition of the *Apologia* published by Froben and Episcopius in 1532, or of a third edition published together with Erasmus' *Dilutio eorum quae I. Clithoveus scripsit* also by Froben in 1532, according to vander Haeghen 10. In his letter to Jacopo Sadoleto at the beginning of March 1532 Bonifacius Amerbach mentions an early 1532 Froben edition that contained three works of Erasmus: '*dilutionem eorum quae Jud. Chlictoveus scripsit, adversus rhapsodias querimoniarum Alb. Pii et in elenchum eiusdem scholia*'; see *Amerbachkorrespondenz* III 108–9 Ep 1610:38–40.

(c) Paolo Camerini *Annali dei Giunti* I: *Venezia* Parte prima (Florence 1962) 252 lists as no 357 a Venetian Giunta edition with two titles: on the title-page *Des. Erasmi Roterodami ultima apologia adversus rapsodias calumniosarum querimoniarum Alberti Pii quondam Carporum comitis illustrissimi nuper diligentissimi me [L.A.G.] excusa*; and on folio 2^r *D. Erasmi Roterodami apologia brevis ad vigintiquatvor libros Alberti Pii quondam Carporum comitis*. The colophon on folio 34^r states: 'Venetijs in Aedibus Lucaeantonij Iuntae Florentini Anno Domini M.D.XXXII. Die quinto Ianuarij.' Camerini describes the format as '2.0 cc. 34 rom. a–e6 f4,' and the year as 1533 according to the modern system. Angelo Scarpellini, in his study 'Erasmo e Alberto Pio principe di Carpi' in *Ravennatensia* IV: *Atti del Convegno di Ferrara (1971)* Centro Studi e Ricerche sulla Antica Provincia Ecclesiastica Ravennate (Cesena 1974) 659–92, especially 674 n7, gives alternate spellings of some words (eg *Roterodani, adversum, calunniosarum, Corporum,* and *diligentissime*) and describes the format as 'piu di 60 fitte pagine in 4° (674). In their study 'Erasmus Holdings of Roman and Vatican Libraries' in *Erasmus in English* 13 (1984) 2–29, especially 10, Marcella and Paul Grendler note that there are two copies of this edition in the Biblioteca Apostolica Vaticana with the collocation numbers v.R.G. ss. Padri II. 101 int.2 and Barber. v.VI.1. For other copies, see Solana xci.

(d) The *Apologia* was reprinted in vol IX of the Froben 1540 *Opera omnia* (see 3d above): its index is on sig ннh2ᵛ/page 916, the *Apologia* itself on sig ннh3ʳ/page 917–sig nnn5ʳ/page 981.

(e) Vander Haeghen 10 states that the *Apologia* was also reprinted in an octavo format in 1551 *in officina Frobeniana*. We have been unable to locate this edition.

(f) The *Apologia* was reprinted by Jean Le Clerc in 1706 in the Leiden edition published by Peter vander Aa (LB IX 1123A–1196[F]). The index to the *Apologia* is at the bottom of cols 1195–1196.

6 *In Elenchum Alberti Pij brevissima scholia per eundem Erasmum Roterodamum.* 1532

(a) The *editio princeps* (and the only known edition) of this work was published by Froben in 1532 as the third and last item in a collection of Erasmus' smaller works: 1 *Dilutio Eorum quae Iodocus Clithoveus scripsit adversus Declamationem suasoriam matrimonij,* 2 *Epistola Eiusdem De delectu ciborum, cum scholijs per ipsum autorem recens additis,* 3 *In Elenchum* ... The colophon of the volume states: 'In officina Frob. M.D.XXXII.' This collection is composed of octavo quires a, b, c, d, e, f, g, h, i, k, l, m, n; pagination begins with sig a2ʳ (= page 3) and ends with sig m1ᵛ (= page 178); there is no pagination for the *Brevissima scholia.* The volume contains: title-page (sig 1ʳ), blank (sig 1ᵛ), *Dilutio* (sig a2ʳ/page 3–sig d6ᵛ/page 60, *Epistola de delectu* [of 1522] (sig d7ʳ/page 61–sig h8ᵛ/page 128), *Scholia defensoria* (sig i1ʳ/page 129– sig m1ᵛ/page 178), *Brevissima scholia* (sig m2ʳ–sig n8ʳ), corrigenda for the *Dilutio* on the bottom of sig n8ʳ, Froben printer's mark (sig n8ᵛ). The folios measure c 16.8 x 11.7 cm. This collection is bound together with the *Declarationes Desiderii Erasmi Roterodami, ad Censuras Lutetiae vulgatas sub nomine Facultatis Theologiae Parisiensis* with its colophon: 'Basileae in officina Frobeniana apud Hieronymum Frobenium et Nicolaum [sic] Episcopium. M.D.XXII.' in the copies in Washington, DC, at the Folger Library; in Cambridge, Massachusetts, at Harvard University's Houghton Library; and in London, at the British Library. For other copies, see Solana xcii.

Vander Haeghen does not give a separate listing for this work. The statement of Bonifacius Amerbach (see item 5b above) suggests that there may also have been a second edition published in early 1532 together with Erasmus' *Dilutio* and *Apologia adversus rhapsodias Alberti Pii,* but we have been unable to locate such an edition. The *Brevissima scholia* is not included in the 1540 Froben *Opera omnia.* The *Brevissima scholia* is scheduled to be reprinted in 2004 together with an unedited French translation in the volume *Circuler et voyager ou les index à l'epoque humaniste* ed Alexandre Vanautgaerden and Jean-François Gilmont in the series Nugae humanisticae sub signo Erasmi (Turnhout: Brepols).

Translator's Note on the Texts:
How to Read Them

THE TEXTS

RESPONSIO

The translation of the *Responsio* is based on a collation of the original print-ing by Froben in 1529, the only source, so far as we know, with indepen-dent authority, and the text printed in LB IX. Variations between the two texts are indicated in the footnotes, with the preferred reading given first. Where significant differences occur (there are nineteen in all), the read-ings and punctuation from the 1529 printing are usually superior and have been adopted here. The superior readings provided in LB (three instances) seem to be based on later common-sense corrections of typographical errors rather than on any authoritative witness.

APOLOGIA

The translation of the *Apologia* is based on a collation of the original print-ing by Froben in 1531, again, apparently, the only source with independent authority, with the text in LB IX. Here, too, variations between the two texts are indicated in the footnotes, with the preferred reading given first. Again, when significant differences between 1531 and LB occur (thirty-three in all), the usually superior readings and punctuation of 1531 have been adopted. Superior readings provided by LB (ten instances) seem, again, to be derived from common-sense correction of typographical errors (or, in one case, from checking the Erasmian work being quoted) and not from another authori-tative witness to the text.

SCHOLIA

The translation of the *Scholia* presented here is based on the unique version of the text available in the original printing by Froben in 1532.

CONVENTIONS

Square brackets are used to enclose editorial insertions in the translation. These include translations of Greek and Latin words left in the original languages; the Greek or Latin words that have been translated by the preceding English word or phrase; supplementary words inserted by the translator to clarify meaning; and insertions of marginal materials into the text. [...] indicates an ellipsis in a quotation not indicated by the one quoting it. Angle brackets <> are used within quotations to enclose editorial insertions of needful words from sources that were carelessly quoted.

SOME SUGGESTIONS FOR READING THE TEXTS

RESPONSIO

This is the easiest of the three works to read and follow. Like Pio's *Responsio paraenetica* which occasioned it, Erasmus' reply takes the form of an extended epistle. In it Erasmus has observed epistolary conventions in form and tone, and this long letter shows a greater degree of literary polish and care in composition than the longer *Apologia*. The *Responsio* can be read with a degree of understanding without close consultation of Pio's *Responsio paraenetica*, for Erasmus makes many direct references to Pio's work and quotes from it often. But a good number of his references to Pio's volume are vague, and in a number of instances he provides only a partial representation, or even a misrepresentation, of what Pio had written. The notes provide clearer indications of the passages to which Erasmus is referring or alluding, and they contain many translated passages of Pio's work in order to give the reader a more informed and balanced view of the exchange. In addition, notes have been provided, where it seemed useful, with translations of Pio's later reactions (see below) to the passage at hand. Much more is to be gained from a reading that involves regular consultation of the notes.

APOLOGIA

Since the *Apologia* is a response to Pio's XXIII *libri*, its structure is largely determined by the long and complex work to which it replies. The 1531 edition of the *Apologia* helps the reader keep track of the section of XXIII *libri* to which Erasmus is replying at any particular point by providing a number of reader's aids including running titles at the tops of pages, marginal rubrics that identify the book to which a particular section of Erasmus' work corresponds, and headings inserted into the text. Our format does not allow us to reproduce most of these features, but we have taken the liberty of inserting the running titles and marginal rubrics into the text, so that the reader will have a better idea of the subject of a particular section of the work.

Special problems

Erasmus' reply to Pio's book 2 presents the reader with special problems. Book 2 contains Pio's marginal comments (*scholia*) on Erasmus' *Responsio*, the first phase of his second effort against Erasmus (see Pio's remarks on *XXIII libri* 72ʳN and compare the publication of Prierias' *Epitoma* with marginal notes by Luther and Melanchthon [CC 41 147–8] or Erasmus' re-publication of *Epistola de esu carnium* with his own '*scholia defensoria*' [ASD IX-1 65–89]). Pio's literary executors published a reprint of Erasmus' *Responsio* that reproduced Pio's marginal comments; each comment is keyed by a successive letter of the alphabet to the corresponding letter inserted into the text of Erasmus' *Responsio*. One alphabetical series follows another, for there are some 253 comments, varying in length from a few words to an entire paragraph.

Although Erasmus replied to only a fraction of the comments, he devoted more space in his *Apologia* (12.7 per cent) to answering them than to anything else. Erasmus explained at the conclusion of the Index he prefixed to his *Apologia* his use of marginal citations to identify the comment of Pio that he is rebutting (see *Apologia* 109 and n8 below). Since we could not reproduce these citations in the margins of the translation, we provide the citations in the notes, and indicate in parentheses the passage in Erasmus' *Responsio* on which Pio had commented. As it is clear that Erasmus wanted the reader to examine the comments to which he was replying, we have included translations of substantial portions of them in the notes. Useful reading of this section of the *Apologia* requires constant consultation of these footnotes and of Erasmus' *Responsio*.

Pio's *XXIII libri* is, according to its title, directed 'against passages of diverse works by Desiderius Erasmus of Rotterdam which he [Pio] thinks he should review and revise.' Pio provides, however, a great deal more than a list of scandalous passages. He tells Erasmus in the Preface how he came to realize that in order to prove that Erasmus had written the scandalous assertions he claimed, he would have to quote the passages which he had earlier only summarized, and how he also came to realize 'that these [passages] could not be brought forward without providing an antidote, by which to analyse and refute the arguments with which you attack the doctrines and approved practices of the orthodox' (72ʳo). Thus, Pio's criticisms of Erasmus are everywhere accompanied by demonstrations of the truth of what Pio takes to be the orthodox position.

Now in book 3, his critique of the *Praise of Folly*, Pio takes up Erasmus' offensive statements in *Moria* as they occur in that work, *seriatim*, providing criticism and refutation as he goes along. And in his reply to book 3 Erasmus follows Pio's structure, refuting his criticisms as they occur, and so this section makes for fairly straightforward reading.

But at the end of book 3 Pio signals his intention to proceed thereafter topic by topic, 'each section [ie book] containing its own topic [derived] from the passages quoted' (84^{r}F), and by way of preface to book 4 Pio describes to his adversary what will be his method throughout the rest of *xxiii libri*:

> The reader must forgive me if some repetitiveness gives offence in the course of his reading, particularly in my reporting your statements. Necessity has driven me to adopt what others commonly do in a disputation with an opponent: first they quote the adversary's statements and the arguments with which he supports his views; then they repeat these same statements one by one as each is answered in turn. I must do the same in my discussion with you. I review all your passages that have to do with the same topic, so that your views may be more evident and more readily grasped, for these passages clarify, support, and explain one another when any one of them is at all unclear. Next, since I mean to reply to each of these, it is surely necessary to repeat at least a summary of each, and sometimes the entire statement, to provide a clearer understanding of my replies which are intended to refute them. Nor should the reader be critical when he hears me discussing a great many topics that I touched upon in my earlier letter [*Responsio paraenetica*], since the present work was undertaken both to prove on the basis of your own words that I wrote nothing deceitful or slanderous, as you charge, in my summary of your opinions that I criticized there, and also to refute the arguments with which you have maintained and supported your views, something I forbore to do in my letter, since that was not its intent and the passages themselves had not been quoted in it. (84^{r}A–84^v)

In fact, Pio's procedure follows the pattern of scholastic disputation more closely than he suggests, for the structure of the succeeding books replicates the scholastic *quaestio* or *articulus*. Pio first quotes multiple passages from Erasmus' writings in order to present his opponent's views on the topic at hand. This corresponds to the '*Quod non' probatio* (also called *obiecta, argumenta*) that opens the *articulus*. Next, Pio provides, with elaborate argumentation and extensive scriptural and patristic testimonia, a presentation of what is, in his view, the orthodox position. This corresponds to the *Respondeo* section, the 'body' of an *articulus*. Finally, Pio reviews some or all of the Erasmian passages quoted earlier, refuting them one by one. This corresponds to the *Solutio obiectorum* or *responsiones*.

But just as the scholastic authors could over-extend, obscure, and clutter this relatively straightforward structure, so too Pio often vitiates the structure of his work by his prolixity and manic thoroughness, and by indulging his penchant for stylistic flourishes, displays of erudition, and excursive invective or encomium.

But if Pio's work is a challenge to the reader's patience, Erasmus' reply to it is a real challenge to the reader's understanding. Erasmus replies to Pio *carptim* [selectively], that is, to only a limited number of points, features, and details of Pio's vast work, and he is often less than careful to inform the reader as he moves from one target to another.

Erasmus and Froben provide some help to the reader in the headings and marginal notes mentioned above that inform the reader of the section of Pio's work being answered. Erasmus also employs various reader's aids to signal his focus on Pio's handling of specific passages quoted from his works. In two instances (his replies to books 5 and 6) he inserts the heading LOCA [Passages] into the text. In these sections and in his replies to many other books (4, 7, 10, 13, 15, 16, 17, 20) he numbers the passages with numerals placed in the margins. In some cases he numbers all the passages, but in others he numbers only the first few. In replies to other books the passages are not numbered, but are indicated, and only sporadically, by the use of quotation marks in the margins.

But the use of mechanical aids like these to keep readers informed of the progress of Erasmus' reply to Pio's XXIII *libri* is inconsistent and intermittent, and so also are Erasmus' verbal indications of just what portion of Pio's work he is addressing. Sometimes he keeps his readers well informed, and his text can be read and understood without reference to Pio's volume. The fact that in replying to Pio's book 2 he provides marginal citations of Pio's comments suggests that he intended or preferred readers to have Pio's volume open before them. In other instances it is difficult to know what Erasmus may have intended, that is, whether he meant his readers to be looking back and forth from Pio's XXIII *libri* to *Apologia* or not, but it is clear that good sense cannot be made of his words without careful consultation of Pio's book. For this reason pains have been taken to provide in the notes constant citation, summary, and, when needed, quotations from Pio's work, in order to orient readers and to equip them to form some judgment of the accuracy of Erasmus' representation of his opponent's statements and intent. While the translator must apologize for thus cluttering the notes, it seemed better to disfigure the volume and, perhaps, try the readers' patience, than to leave them sometimes in the dark and always dependent on Erasmus for a view of his remarkable adversary.

SCHOLIA

The Index to Pio's *XXIII libri*, which Erasmus refers to as an *Elenchus*, actually bore the more pedestrian title of *tabella*, a standard epithet for an index in early printings. There are two *tabellae*, the first, only one column in length, is a table of contents (ãii^{r1}). The second, which covers twenty-five columns and fills up the quire, is entitled: 'An excellent index [*tabella*] of the notable statements and of the whole mélange [*farrago*, from Juvenal 1.86] in the following work, carefully collected in alphabetical order.'

Erasmus explains in a prologue that he is troubling to reply to the *Index* because even lazy readers will look at indexes and title-pages, and so they are effective and notorious vehicles for creating prejudice. He may also have felt challenged to compose these *scholia* because of the close similarity of the form of the summaries of Erasmus' supposed views in entries in this *index sententiarum* to the summaries of his views formulated by Zúñiga, Béda, the Theology Faculty of Paris, et al in their condemnations of Erasmus. In the *Scholia* Erasmus responded very selectively, for from this vast index of statements and topics he chose only 122 entries for a reply. Some of these do not mention him by name, although most do. Erasmus exercised an uncharacteristic restraint in ignoring the very many additional entries in the Index which criticize him by name. Erasmus' arrangement of the *sententiae* follows the alphabetical arrangement of the Index to Pio's *XXIII libri*, but he also numbered each item and assigned the same number to his comment which follows it.

In this translation each Index entry is followed by the folio number and section letter as originally given in the Index of Pio's *XXIII libri*, but here enclosed in brackets. Often an entry is accompanied by a note that refers the reader to the passage(s) either in Pio's *XXIII libri* translated in a note or in Erasmus' *Responsio* or *Apologia*, citing the page numbers of this volume.

THE REPLY
TO THE HORTATORY LETTER OF
THE MOST ILLUSTRIOUS AND MOST LEARNED
COUNT ALBERTO PIO, PRINCE OF CARPI

Responsio ad epistolam paraeneticam
clarissimi doctissimique viri
Alberti Pii Carporum principis

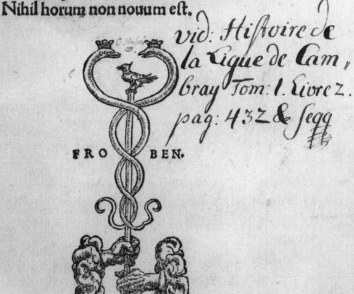

DES · ERASMI

ROTERODAMI RESPONSIO AD
epistolam paræneticam clarissimi doctissimiʠ ui
ri ALBERTI PII Carporum principis.

EIVSDEM NOTATIVNCVLAE QVA=
dam extemporales ad Nænias Bedaicas.
Nihil horum non nouum est.

*vid: Histoire de
la Ligue de Cam,
bray Tom: 1. Livre 2.
pag: 432 & seqq*

FRO   BEN.

BASILEAE AN. M. D. XXIX.
MENSE MARTIO.

Title-page of Erasmus *Responsio ad epistolam Alberti Pii* (Basel: Froben 1529)
By permission of the Houghton Library, Harvard University, Cambridge, Mass

THE REPLY
OF DESIDERIUS ERASMUS OF ROTTERDAM
TO THE HORTATORY LETTER OF
THE MOST ILLUSTRIOUS AND MOST LEARNED
ALBERTO PIO, PRINCE OF CARPI

How I wish that fortune had granted me, Alberto, most erudite of princes,[1] some more agreeable reason for writing to you! But, in fact, this less than fortunate situation has not been, in the end, altogether unfortunate. For it has provided in exchange for a letter,[2] one not very long and quite rough and unpolished, a full-size book[3] that has been brought to a high gloss with very great learning and equal wit. For all that, I am also not sure that I ought to congratulate myself that a personal letter, one addressed to an individual, has been repaid by a book, which appears to have been written for the public, and not for me alone. I would rather imagine that you did this to seize an opportunity: that is, your argument against Luther's sect has been ready for some time now,[4] but it was upon being challenged by my letter that you combined the complaints of certain parties against me with your refutation of Lutheran doctrines,[5] and, as the saying goes, '[white-washed] two walls from the same bucket.'[6] For I do not want to suspect so renowned, learned, and humane a man, indeed to suspect Alberto Pio, of intentionally confusing my case with Luther's, making use of the same ploy

* * * * *

1 Pio's education, studies, and standing in the world of learning are discussed in the Introduction xvii–xviii, xx–xxv, and xxxvi–xxxviii.
2 That is, Ep 1634, which is translated in the Introduction xliv–xlvii
3 Pio *Responsio paraenetica* 2ʳA defends himself for sending 'almost a full-size book under the title of a letter'; in his preface Bade describes Pio's work as 'a very careful and virtually book-length reply' (Aiiʳ).
4 On the question of whether or not Pio had an already prepared attack on Luther and his followers that he inserted into his reply to Erasmus, see Introduction xlviii.
5 For the proportions of Pio's work devoted to Erasmus and Luther, see Introduction xlviii–lvi.
6 See *Adagia* I vii 3: *Duos parietes de eadem dealbare fidelia* 'To whitewash two walls out of the same bucket' LB II 263B–C / CWE 32 68: this means, says Erasmus, 'to earn thanks twice over for the same thing, and lay two people equally under an obligation to you by a single act ... A similar adage in Greek is ... applied to those who in party strife make overtures to both sides.'

as some pseudotheologians[7] who routinely join my name to his in their sermons and conversations, thus exploiting, obviously, the naivety of the ignorant masses.[8] It is, of course, due to the contrivances of these men[9] that the situation has advanced even to this gale of controversy, which will soon spread more widely (may I prove a false prophet!) unless it is checked by sounder counsel.

But although, in keeping with the agility of your wit, you pile up and pack in abundant considerations to make it seem that you were forced to act as you did, in fact, as the case itself makes plain, you could have answered my letter with very few words. It required nothing of you but that you set aside the opinion you have of me and curb those remarks which are as groundless as they are uncomplimentary to me – if, that is, the rumour which had flown hither[10] was true. If not, you were to regard my letter to you as not having been written.

Someone else might, perhaps, complain strenuously that you have published[11] for the world my carelessly composed personal letter, and that you have not only failed to curb the remarks I was asking you hereafter to curb, but have behaved even more savagely by publishing them through the printers,[12] at one stroke both making a display of my ignorance and barbarism and hailing me up on a capital charge despite my innocence. Certainly I spared you from having to make any defence[13] of all the criticisms

* * * * *

7 *Pseudotheologi* is a term of opprobrium Erasmus used to describe those teachers of scholastic theology who were more concerned with dialectics and disputations than with a spiritual knowledge that transforms one's life. For descriptions by Erasmus of true and false theologians, see *Ecclesiastes* LB V 940B and the annotation on Matt 11:30 Reeve 54 for the 1522 addition.

8 An old complaint; see Ep 1144:24–7 (13 September 1520): 'When the bull had appeared that commands them to preach against Luther, two or three of these barefoot bullies passed a decree in their cups to this effect, that they would couple me with Luther and defame me too in public'; the same claim is made in Ep 1192:36–7.

9 Compare Erasmus' earlier denunciation of the *monachi* 'monks' and *theologi monachales* 'theologians who think like monks' in Ep 1313:37–40.

10 See Introduction xliv–xlv.

11 Erasmus' Ep 1634 was printed virtually as a preface in Pio *Responsio paraenetica* Aiii^r–iv^v.

12 Both of Pio's works against Erasmus were published by Josse Bade of Aasche; see n15 below and Introduction cxxxii item 1c, cxxxvi item 4a.

13 At both the beginning and the conclusion of his letter to Pio (Ep 1634:9–11, 119–23) Erasmus offers Pio what would now be called 'plausible deniability' by suggesting the possiblity that the rumour which has attributed criticisms of Erasmus to Pio were false; throughout Erasmus assumes a posture of humility before the alleged criticisms of his ability and achievements, undertaking to defend himself at length (43–95) only against the charge of his responsibility for the Lutheran disturbance.

that you were reported to be making publicly against my ability, judgment, learning, and eloquence.

You ought at least to have taken care to have my letter printed exactly as it was sent to you. For there I read *penitus* instead of *penitius*, *nostrarum* instead of *nostratium*, *nosceres* instead of *nosses*; likewise, *in nobis revocarent* for *nobis revocarent*, *indignatus* for *indignans*; again, *hic coepit* for *hinc coepit*; and next, *negotio se involvere* for *negotio involvere*.[14] This last might seem to have been corrupted by someone purposely, for the addition of the pronoun wholly alters the meaning. But I realize that these are to be blamed on the printers, and not on you, as also is the fact that, although you are refuting all the chief tenets of Luther's sect, the whole book has on its title-page *Reply to the Letter of Erasmus*, as if whatever is taken up in the book is aimed at me.

Now as to our friend Bade's attaching a short preface, it is his privilege and his practice to do so, and, unless I am mistaken, that has nothing to do with you. But there are also some remarks in it that are hardly in keeping with my generous treatment of him.[15] Among these is his writing that I have had a bad reputation among the purple-clad fathers,[16] whereas I have never heard that any of the cardinals has spoken of me less than cordially. In fact, I have letters from many of them, written in a most complimentary way, and I have never had occasion to complain on this score. In addition, he interprets the praises that your good nature[17] has chosen to lavish on me in excess of any merit of mine as if you generously meant in this way

* * * * *

14 Erasmus had printed a copy of his entire letter to Pio immediately preceding his *Responsio* A2ʳ–4ᵛ. Allen based his edition of Ep 1634 upon this version, and all of the errors Erasmus mentions here (with the exception of *penitus* for *penitius*) are reported in Allen's *apparatus criticus* for lines 64 (twice), 95, 18, 98, 101, in that order; *penitius* is a problem (see Allen Ep 1634 intro). The word *penitus* occurs in Allen Ep 1634:3 in both Pio's and Erasmus' version. Perhaps the error *penitus* for *penitius* also crept into Erasmus' authorized version of the letter.

15 Bade's preface is in Pio *Responsio paraenetica* Aiiʳ⁻ᵛ; on Bade's career and his troubled relations with Erasmus, see CEBR I 79–81; and Renouard *Imprimeurs* II 6–24.

16 Not an exact representation of Bade's words (*ad expostulationem Erasmi Roterodami se Romae apud patres purpuratos male audisse*... 'to the expostulation of Erasmus of Rotterdam complaining that he [Erasmus] has had a bad reputation among the purple-clad fathers'). 'Purple-clad fathers' is a phrase current in the Erasmus-Pio exchange; see Ep 1634:13–14, where Erasmus refers to Pio as *doctum magnaeque apud purpuratos patres autoritatis*, and Pio *Responsio paraenetica* 3ᵛC, where Pio addresses Erasmus' report that *ad te delatum esse, me apud purpuratos patres nomino tuo detraxisse*. For further information on Erasmus' more important supporters among the cardinals, see CEBR on Albrecht von Brandenburg I 184–7; Lorenzo Campeggi I 253–5; Erard de la Marck II 382–5; Domenico Grimani II 132–4; Antonio Pucci III 122–3; Lorenzo Pucci III 123–4; and Thomas Wolsey III 460–2.

17 See Introduction xxxiii n19.

to soften what was in fact a scolding. [18] And yes, it is quite true that Paul challenged Peter to his face, as he [Bade] writes;[19] but nowhere did Paul thrust a slanderous accusation into Peter's face, though the Greeks interpret this passage somewhat more agreeably, and in my opinion more correctly, than St Augustine, who imputed [to Peter] a perverse zeal for bringing the gentiles into servitude to the Law.[20] Next he also renders the verdict that you are quite right to summon me to sing a palinode.[21] But you will deny responsibility for any of these matters, and I do not intend to press charges against a most humane prince because of trifles like these.

In fact, what I fear is that it is you who may seem to have the more just grounds for complaint, in that it has been a year,[22] unless I am mistaken, since you sent me your book, and yet I have not had the courtesy to reply. But at that time no letters could get through from here to Italy, and not long after I had received your book the crash of the fall of Rome overwhelmed the spirits of all here, at least those of the pious.[23] Vague rumours were circulating about you: some reported that you were sharing the peril of Clement VII; others, again, said that you had made provision for

* * * * *

18 Bade praises Pio's restraint (*modestia*) in his gentle reply to Erasmus, and remarks: 'This most wise prince gives abundant indication that the glory of Erasmus' renown and the salvation of his soul are both quite important to him, as he honours him with very great praise and urges him with so very apt an exhortation to take a more public stand with the Catholic church' (Pio *Responsio paraenetica* Aii[v]).

19 Bade had cited, in connection with his praise of Pio's restraint, that of St Peter: 'though Paul had resisted him to his face and declared him blameworthy [alluding to Gal 2:11], nonetheless he wrote of him "as our most beloved Paul has written to you according to the wisdom given him" [2 Pet 3:15]' (Pio *Responsio paraenetica* Aii[r]).

20 In his extensive annotation on Gal 2:11 LB VI 807D–808B / Reeve-Screech 571–4 Erasmus provides a history of the exegesis of this verse and rejects Augustine's explanation; the interpretation of Gal 2:11 was one of the points of disagreement in the epistolary controversy between Jerome and Augustine (see n21).

21 Erasmus alludes to Bade's words 'he [Pio] advises him, properly and in a kindly way, that he should either revise these things or explain their meaning' (Pio *Responsio paraenetica* Aii[v]). See *Adagia* I ix 59: *Palinodiam canere* 'To sing a palinode' LB II 356A–D / CWE 32 214–15: 'To sing a palinode is to say the opposite of what you said before and adopt the contrary opinion'; the background for the use of this term in an ancient controversy very familiar to Erasmus is provided by Kathleen Jamieson 'Jerome, Augustine and the Stesichorean Palinode' *Rhetorica* 5 (1969) 353–67.

22 Erasmus gives the same inaccurate chronology in a letter to Antonio Hoyos de Salamanca written 10 March 1529 (Allen Ep 2118:16–17): 'Alberto Pio had sent me a year ago the book he has now published in Paris.' Erasmus had received Pio's *Responsio* by early September 1526; see Allen Ep 1744:130; and Introduction lvii.

23 On Erasmus' cool reaction to the Sack of Rome, which began on 6 May 1527, see André Chastel *The Sack of Rome* trans Beth Archer, Bollingen Series xxxv 26 (Princeton 1983) 129–36, 268–71 nn49–75; and Allen Ep 1938:1–2.

ALBERTI PII

Carporum Comitis Illuſtriſſimi, ad Eraſmi Rotero⸗
dami expoſtulatiōe reſpōſio accurata & parænetica,
Martini Lutheri & aſſeclarum eius hæreſim veſanam
magnis argumentis,& iuſtis rationibus confutans.

dabo electis

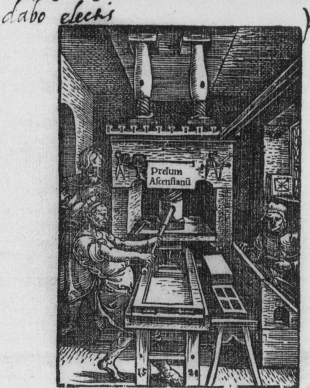

Vęnundatur Io.Ioco Badio Aſcenſio.

Title-page of Pio *Responsio paraenetica* (Paris: Bade 1529)
By permission of the Biblioteca Apostolica Vaticana

your interests and safety by timely flight.[24] But not even a hint of where you were living got through to us. So, since I was overloaded with too many other labours, it seemed best to postpone this task until I had learned where you were for certain.

I finally found this out three days ago, from the letter of a friend who sent along with it the printed version of your book.[25] This occurred on 9 February, as the printers were getting ready for their departure for the Frankfurt Fair[26] on 22 February. Thus ten days were left for the composition of a reply, and I was busier than almost ever before. At that time the print shop was in amazing commotion, dripping perspiration from six presses, and I the sole supplier of material.[27] I was in distress because of my irresolution. I was afraid that, if I were to put off the courtesy of a reply, I should seem to be doing so either because I had taken offence or out of contempt for you. However, I was discouraged by the extraordinary size of your book and by the limited time available, which seemed hardly sufficient for a careful rereading of your work. But your singular kindness encouraged me to reply to your letter if only extempore,[28] but to do so selectively, only in connection with those elements that touch upon me in particular. You, in keeping with your accustomed affability, will take this in good part.

First, as to your excuse of pressing activities,[29] I feel that we share it, to this extent: you are engaged in the more important and glamorous

* * * * *

24 See Introduction xxviii.

25 See Introduction lxv.

26 In the early sixteenth century the annual book fair at Frankfurt am Main was the principal meeting place of those engaged in book production and sales. See Friedrich Kapp *Geschichte des deutschen Buchhandels bis in das siebzehnte Jahrhundert* (Leipzig 1886) I 448–521 especially 455–8; and Lucien Febvre and Henri-Jean Martin *The Coming of the Book: The Impact of Printing 1450–1800* trans David Gerard, ed Geoffrey Nowell-Smith and David Wootton (London 1976) 228–33. For a 1574 description of this book fair, see *The Frankfort Book Fair: The Francofordiense Emporium of Henri Estienne* ed James Westfall Thompson (Chicago 1911) 3–123 especially 32–3 (historical account), 154–76 (Estienne's description of the book fair), and 168–79 (description of the scholars and poets who attended it).

27 He means, of course, the publishing house of Hieronymus Froben (1501–63), who succeeded his father Johann (c 1460–1527). On Erasmus' relations with the Frobens and for further bibliographical references, see their entries in CEBR I 58–63. At this same time Erasmus was busy with at least eight other works being printed on Froben's press; see Introduction lxvi.

28 In Allen Ep 2118:32 Erasmus refers to this work as an *extemporalis Responsio*; the extempore character of Erasmus' work generally is discussed by D.F.S. Thomson 'The Latinity of Erasmus' ed T.A. Dorey *Erasmus* (Albuquerque, N Mex 1970) 119–21.

29 Alleged by Pio *Responsio paraenetica* 1ᵛ–2ʳA

activities,[30] but I in more prolonged and laborious ones. Of course, it saddens me especially that we share the excuse of poor health,[31] for your character, Alberto, is such that you deserve continually flourishing health, inasmuch as you have rejected all the pleasures to which renown of ancestry and abundance of wealth usually afford temptation,[32] and have joined outstanding eloquence to an erudition complete in every respect. I rather suspect that the very illness of which you complain is, to a great extent, the result of your scholarly activities. Certainly I am myself so overwhelmed by learned labours that I may neither be ill nor die, nor enjoy the services of physicians; my health is so ruined that I must wrestle from time to time with death.[33] Thus for some months I have scarcely had sufficient free time to read your book, though I was quite eager to do so. As it is, I have rushed through it instead of giving it a thorough reading. Please believe me in this above all, excellent sir, for I am quite trustworthy, and I would not hesitate to furnish an oath, however solemn, should you require it, that I refrain from furnishing proofs, lest, while attempting at some length to demonstrate that I have no leisure, I show by that very fact that I enjoy an abundance of it.

Now you will not, I think, expect me to return like for like,[34] to repay a book with a book. But even if the leisure were available in abundance, very many of the charges that you heap up are such that either they do not pertain to me at all, or they constitute a tardy and thus unprofitable critique, or they are, finally, of the kind that I may neither admit without disgracing myself,

* * * * *

30 A note of sarcasm may be suspected in this statement because Pio at this time was living as an exile in Paris, forced to stand on the sidelines while his patrons, Francis I and Clement VII, were entering into peace treaties with Duke Alfonso d'Este and Emperor Charles V that would sacrifice Pio's rights over Carpi and his other territories. See eg the letter written by Giovanni Salviati to the papal nuncio to the emperor on 6 November 1527, in ASV, SS Francia I 50^r (Alfonso d'Este allowed by François I to keep Carpi), 51^r (Pio angered at the French-Ferrarese agreement); and Guaitoli 'Sulla vita' 295.

31 Alleged by Pio *Responsio paraenetica* 1^{v}A: 'not to mention poor health, which has troubled me often and painfully for very many years'; on Pio's illness, see Introduction lxxv n81.

32 Sepúlveda *Antapologia* ciii^v remarks on Pio's rejection of the pleasures of youth and wealth in favour of scholarly pursuits.

33 Erasmus suffered as a youth from quartan fever, in middle age from kidney stones, and as an old man from gout and arthritis. In the winter of 1529 he suffered an abdominal abcess or ulceration of the bladder brought on by a stone. See H.N. Cole 'Erasmus and His Diseases' *Journal of the American Medical Association* 148 (1952) 529–31; Hyacinthe Brabant 'Erasme, ses maladies et ses médecins' in *Colloquia Erasmiana Turonensia* ed Jean-Claude Margolin 2 vols (Toronto 1972) 1 539–68 especially 548; and Margolin *Erasme: Humaniste dolent* (Québec 1971).

34 See *Adagia* I i 35: *Par pari referre* 'To return like for like' LB II 41A–E / ASD II-1 150–1 / CWE 31 83–5.

nor refute without giving scandal to many and without, more grievous still, endangering that cause which we both support.[35] For in these times everything is so inflamed that, wherever you turn, you realize that you are walking upon fires hidden under deceptive ash, whatever you touch, you find abscess and canker, and beneath each stone a sleeping scorpion.[36]

But there are a number of things scattered through your book, items that were read earlier in Rome by many[37] and are now being read everywhere by all, that are of the sort that, should they come into the hands of those who have studied my works and are familiar with the whole tenor of my life, are likely to do considerable damage to that reputation for honesty and fairness[38] that makes you dear to all good men, just as the brilliance of your good fortune and your eminence in learning make you remarkable. For my part, I think the love of many is preferable to their admiration. I would otherwise have wanted this book of yours that is now in print to be widely received, in the hope that perhaps at least your prestige, or your learning, or your eloquence, or your generosity and fairness, the sort exhibited by no one so far, might recall some to more pious resolves. For we have witnessed[39] how much some have aggravated this illness by their maladroit attempts to apply a cure.

But I really wish that in your writing you had been restrained enough not to have mixed up my case with the affair of Luther, and that you had won substantial acclaim for yourself through your refutation of impiety without spattering me unfairly with any opprobrium. For although I find it annoying to be praised in Luther's company, it is more distressing by far to be accused along with him, the accusation being, in effect, that I paved the way[40] for the attack he launched on the church.

But I know that you did this neither by design nor with malice; rather, as I suspect, your good nature was persuaded that this is the case by those

* * * * *

35 Perhaps the defence of the church or 'good letters' (but see Allen Ep 1804:252–3).
36 *Adagia* I iv 34: *Sub omni lapide scorpius dormit* 'Under every stone sleeps a scorpion' LB II 163E–164B / ASD II-1 434 / CWE 31 344: 'The adage warns us that a man should beware of speaking heedlessly in the presence of fault-finders and slanderers.'
37 Pio refutes this remark in his later note on this passage: 'You are making a false supposition, for in Rome no one was given access to it [the book]' (Pio *XXIII libri* 48ʳi).
38 Given Pio's reputation for having betrayed his former French employer and his imperial sovereign and his present ruined condition as a deposed and dispossessed exile in Paris, a certain amount of sarcasm must be suspected here; see Introduction xxxiii n19.
39 *Vidimus* 1529 *Videmus* LB
40 *ut ego viam struxerim*: see Allen Ep 2892:44 where Erasmus gives, as the summation of the accusations accumulated against him by that time (December 1533), 'he [Erasmus] paved the way for Luther' (*strauit viam Luthero*).

with whom you associate.[41] For those of guileless and honest character (the sort suitable for princes) study to deceive no one,[42] and thus are not quick to suspect anyone. It is for this reason that they are easily deceived by others through the pretence of virtue. Nor can I bring myself to believe that your activities allow you so much leisure[43] (and that you are possessed of so strong a stomach, accustomed as you are to what is most refined!) that you had time even to glance at, let alone examine, my trifles, even if you had wished to, or that you wished to do so even if you had had time. I feel that I have much reliable evidence for concluding that you have drawn many of the things that you hurl at me not from my works but from the conversations of men who are far removed from the sincerity of your disposition, men whose particular aim is to stage their own play through hired actors.[44] And, as I was just now saying, none are more vulnerable to the deceits of the wicked than those who, because of the sincerity of their own characters, do not know how to weave a snare for anyone. This is no surmise on my part. Rather, I have proof from the experience of my own misfortunes, for I have often been pitched into very serious controversies because I estimate the characters of others on the basis of my own.

Moreover, with your customary generosity you award me in this case such an abundance of praise – not at all what I could accept, but more than has been awarded me by any of the others who have extolled my mediocre

* * * * *

41 Erasmus repeatedly claims that the principal instigator of Pio's attacks on him was Girolamo Aleandro. See eg Allen Epp 1719:34–6, 1840:81–2, 2077:51–2, 2329:105–6, 2375:78–9, 2379:110–14, 2411:49–50, 2443:336–7; Heesakkers 'Erasmus' Suspicions' 371–84; and André Godin 'Erasme, Aléandre: une étrange familiarité' *Actes du Colloque International Erasme (Tours, 1986). Etudes Réunies* ed Jacques Chomarat et al (Geneva 1990) 249–74. Erasmus also accused certain theologians in Rome of being behind Pio's attacks; those theologians in Paris (Béda, people like Béda, and certain Franciscans) and Sepúlveda, whom Erasmus accused of assisting Pio in the production of his second reply, may also have been involved in the revisions of his first reply. See eg Allen Epp 1804:253–4, 2118:26–7, 2329:96, 2375:76–7, 80–1, 2443:337–8, 2486:39–40, and Introduction xxxviii–xlii, cxxiii.

42 See Introduction xxxiii n19.

43 A sarcastic dig at Pio's current condition as an underemployed, dispossessed exile in Paris.

44 *subornatas personas*: see Allen Ep 1744:120–4 (dated by Allen [c 2 September] 1526): 'Paris has sent us this year four books which are simply libellous and obviously insane. The effort is being managed, as people suspect, with the agreement of the theological fraternity. They procure surrogates for the staging of this farce, some vainglorious men who are clearly lacking in common sense.' Erasmus made a similar accusation in the case of Zúñiga in Ep 1416:21–3 (8 February 1524): 'It is stupid monks who put forward this mountebank, in pursuit of their own interests, not the Holy Father's'; cf his remark in his *argumentum* to Jerome's Ep 57: *Ruffinus subornaverat Palladium quendam Galatam, aduersus Hieronymum qui illius interpretationem calumniaretur*' (HO 3 166ᵛ).

ability (if, indeed, it is a mediocre ability) in a way that leads to the resentment of others and my own disgust – that, far from easing my burden in this exchange, you have increased it terribly. For the reader will think thus to himself: 'So great a gentleman, one so well disposed towards everyone else, would not be writing such things about Erasmus unless he knew more than he writes.'

So that I may not seem to write this altogether without justification, I shall offer, briefly, a few selected illustrations. I shall pass over, for the moment, your abundant compliments. If you are offering them to me sincerely, my debt to you for this kindness is all the greater, as I own that the praise is undeserved. But if, as I rather suspect, you seek to temper the wormwood of a rebuke, as if with honey smeared on the rim,[45] then I embrace your prudent compassion. But I care nothing for such palliatives, nor am I a child, nor wholly unacquainted with the affairs of men or, if you like, uninstructed in wicked artifices of this sort. In fact, it is not wormwood you are offering but a deadly poison, served up in this enticing guise of eloquence, as if in a gem-encrusted goblet.[46] The gist of your honeyed words is really this: to relieve Luther of animosity and convince the world that this whole deplorable calamity should be blamed on Erasmus alone. Since this is more savage than if you were to charge me with poisoning, what difference does it make to me how charmingly you accomplish it, save that the more gently you proceed, the more lethal a poison you administer, as if you were mixing hemlock in wine?

Thus I am sure that while pleading so very unpopular a case I shall obtain from you in return that indulgence for plain speaking[47] which you implored from me in your cordial, private remonstrance. But what good is it to have obtained what one may not employ? In this respect I have not yet determined, distinguished sir, whether I ought to congratulate you on your good luck or commend your foresight. For you have undertaken to present a drama, in itself full of appeal, in such a way that, even should you produce it badly, the theatre will applaud you nonetheless. For earlier, at Rome,

* * * * *

45 See Lucretius *De rerum natura* 1.936–42; see also *Adagia* I viii 57: *Melle litus gladius* 'A sword smeared with honey' LB II 320D–321A / CWE 32 157: 'A sword smeared with honey is applied to flattery that can do real harm.'

46 Gregory the Great compares heretics to those who serve up poison in a goblet with honey on its rim (*Moralia in Iob* 5.11.28 CCSL 143 237). See also *Adagia* I viii 58: *Letale mulsum* 'A deadly honey-brew' LB II 321A–B / CWE 32 158.

47 Requested by Pio in *Responsio paraenetica* passim, eg 5ᵛC, 6ʳE, and especially at the very conclusion of his work 99ᵛN.

in what is, of course, your home stage, you polished and virtually published this book.[48] You are now performing the same play in a new production at Paris,[49] where you know there are a number of men like Béda,[50] men like Cousturier,[51] and pseudomonks,[52] who will favour with both thumbs, as the saying goes,[53] whatever you write, so long as it is against Erasmus. Moreover, you are engaged in praising the sort of men whom it is dangerous to criticize fairly, and a source of favour to praise falsely.

I will not seek additional time,[54] even though I must plead my case extempore. I shall not, therefore, be asking exactly what were the criticisms of me made by certain learned men there. I am alone in giving no credence to what a certain party announced here to everyone, with no distinction between friend and foe. I regard as fantasies[55] the matters indicated in letters[56] and reported in the conversations of the learned: what this or that

* * * * *

48 Erasmus seems to be saying that, because a draft of Pio's *Responsio* was circulated in Rome, it was 'virtually published,' even if it was not until 1529 that it was printed and sold in Paris. Pio XXIII *libri* 69^v–70^{r}1 insisted that he had never intended to publish his 'inept' private letter and had taken care that no one else could publish it without his knowledge. Sepúlveda *Antapologia* EI^r swore that only he and Pio's personal secretary, Francesco Florido, read the *Responsio* in Rome before it was finished. How widely it circulated once it was finished is unclear.

49 On the publication of Pio's slightly revised *Responsio* in Paris, see Introduction lviii; for the circulation of Pio's draft in Paris, see Allen Ep 2066:62–5.

50 *Beddaicos*: Noël Béda (Bédier) d 1536/7, the leading conservative Catholic theologian at the Sorbonne and opponent of both Erasmus and Luther; see CEBR I 116–18; on Béda and his Sorbonne circle, see Rummel *Catholic Critics* II 29–59.

51 *Sutorios*: Pierre Cousturier (Le Cousturier, Sutor), d 1537, a celebrated theologian at the Sorbonne until he entered the Carthusian order in 1509. He subsequently wrote against Lefèvre d'Etaples, Erasmus, and Luther; see his entry in CEBR I 352–3; and Rummel *Catholic Critics* II 61–79 (on Cousturier and Clichtove).

52 *pseudomonachi*: a word from Jerome (Ep 57.2 CSEL 54 505:12–13) that is Erasmus' favourite epithet for corrupt members of religious orders, not limited to members of the monastic communities, and more often referring to the mendicants.

53 See *Adagia* I viii 46: *Premere pollicem. Convertere pollicem* 'Thumbs down. Thumbs up' LB II 315–16 / CWE 32 149, where Erasmus cites Horace Ep 1.18.66: *Fautor utroque tuum laudabit pollice ludum* / 'Horace uses "both thumbs" to express whole-hearted support.' To favour with both thumbs is to signify complete support by turning both thumbs downwards.

54 Literally, 'nor shall I ask that the *clepsydrae* be increased for me,' alluding to Pio's reference (*Responsio paraenetica* 2^{r}A), in defence of the length of his *Responsio*, to the ancient practice whereby a greater number of *clepsydrae*, vessels of water for timing presentations in court, were allotted to the defendant than to the plaintiff.

55 *pro somniis* 1529 *pro summis* LB

56 An example of such a letter is Allen Ep 1791 of 13 March <1527> from Pedro Juan Olivar; lines 36–64 report remarks critical of Erasmus attributed to Baldesar Castiglione,

one said, with names, places, and dates; how one disparaged my eloquence and another, though allowing that my style is somehow tolerable, denied my general good taste; others say that I am not a Ciceronian,[57] with them a charge more dreadful than murder;[58] quite a few who seek a reputation for wit keep calling me 'Porrophagus'[59] and 'Errasmus'[60] instead of Erasmus; some would also remove me from the ranks of the orthodox.

I have pamphlets worthy of Orestes[61] that were circulating in Rome[62] among the papal secretaries,[63] and I have no need of a title-page to identify the author – his very face would disclose the man to me less than his style.[64]

* * * * *

Andrea Navagero, Alessandro d'Andrea, and Benedetto Tagliacarne. For further information on these persons, see CEBR: Olivar III 31–2; Castiglione I 279–80; d'Andrea I 376; Navagero III 8–9; and Tagliacarne III 305–6.

57 For the background of the Ciceronian controversy and Erasmus' colloquy of 1528 attacking the Ciceronians, see the Introduction by Betty I. Knott in CWE 27 324–36; Emile V. Telle 'Erasmus' *Ciceronianus*: A Comical Colloquy' ed Richard L. De Molen *Essays on the Works of Erasmus* (New Haven, Conn 1978) 211–20; Jean-Claude Margolin, 'Alberto Pio et les Cicéroniens italiens' *Medioevo et umanesimo* 46 (1981) 225–59; and d'Ascia *Erasmo e l'umanesimo romano.*

58 In Allen Ep 1875:156–7 Erasmus says of the 'pagans' in Rome, 'and they say that not to be called a Ciceronian is far more shameful than to be called a heretic.'

59 An allusion, arising in Roman literary circles, to Erasmus' alleged abuse of the particle *porro*; see CWE 1479:139–40: 'There has been great merriment over the expression "*porro* fanatic," as though I use it frequently or in such a clumsy manner'; Allen Ep 1479:125–6n explains 'Ie, *porrophagus*, literally "voracious for *porro*." This is mockery of Erasmus for his tendency to start paragraphs with *porro*, "furthermore."'

60 See Ep 1482:52–5: 'In Rome they call me Errasmus, as though authors in your part of the world ... never made a mistake'; see Ep 1482:52n for Zúñiga's use of this and another perversion of Erasmus' name, *Arasmus*. Erasmus provides a list of his unflattering nicknames in Allen Ep 2468:91–3.

61 Orestes pursued by the Furies was a commonplace for frenzied insanity; see Erasmus' note on the quotation from Persius 3.118 in *Adversus Iovinianum* 1.1: 'For it is in mythology that Orestes was pursued by the avenging Furies on account of the murder of his mother' (HO 3 24ᵛ). Compare Erasmus' remark in Ep 1744:131–4: 'For it has not yet been published, but it is flying through the hands of the papal palace. Not even Orestes driven by the Furies could write anything more insane. Nor is the author's identity unknown to me. The style suits his character, for I know him from personal association.'

62 For the circulation of this document, the *Racha*, in Rome, see Allen Ep 1804:248–9: 'You will receive Aleandro's pamphlet. It is flying anonymously through the hands of the papal secretaries'; and see also Erasmus' letter to Clement VII, Allen Ep 1987:6: 'the second book is current among your staff.'

63 For a list of the apostolic secretaries and of the pope's domestic secretaries at this time, see Hofmann *Forschungen* II 121–4.

64 This unnamed opponent of Erasmus, whose style clearly identifies him, was, in Erasmus' opinion, Girolamo Aleandro (1480–1542), former friend and roommate of Erasmus at the Aldine Press (see especially Allen Ep 2443:285–97), humanist turned papal diplomat, archbishop of Brindisi, and suspected author of the *Racha*; see CEBR I

I do not mind Zúñiga's[65] two pamphlets,[66] as unlearned as they are impudent, which were published there against me and sold,[67] contrary to the prohibition of popes and cardinals.[68] I am not at all disturbed that that accomplished young man Georg Sauermann[69] was compelled to delete whatever contained a complimentary mention of Erasmus from the book that he was preparing for publication there, although you were supreme in literary

* * * * *

28–32; DBDI II 128–35; and Eugenio Massa 'Intorno ad Erasmo: una polemica che si credeva perduta' in *Classical, Medieval and Renaissance Studies in Honor of Berthold Louis Ullman* ed Charles Henderson JR (Rome 1964) II 435–54. Massa, who once held that the author of the *Racha* was Egidio Antonini de Viterbo, has now conclusively shown it to be a work of Aleandro, as Erasmus himself was convinced; see Allen Ep 1987:6–8; and Gilmore 'Italian Reactions' 65–70. Godin in his study 'Erasme, Aléandre: une étrange familiarité' (cited in n41 above) 266 n77 argues for Zúñiga as the author of the *Racha*.

65 Diego López Zúñiga (Stunica), d 1531, 'the scourge of Erasmus' (LB IX 358F), a Spanish cleric who had worked on the Complutensian Polyglot Bible, criticized Erasmus' biblical erudition and other opinions in six works published between 1520 and 1524; see CEBR II 348–9; H.J. de Jonge ASD IX-2 3–43; Bentley *Humanists and Holy Writ* 198–211; and Rummel *Catholic Critics* I 145–77.

66 Three pamphlets appeared in Rome in 1522–3: the *Erasmi Roterodami blasphemiae et impietates* (April–May 1522); *Libellus . . . praecursor* (March–July 1522); and *Conclusiones principaliter suspecte et scandalose . . .* (September–November 1523); see de Jonge ASD IX-2 42. In 1524 two other pamphlets were published in Rome: *Assertio ecclesiasticae translationis novi testamenti a soloecismis quos illi Erasmus Roterodamus impegerat*; and *Loca quae ex Stunicae annotationibus, illius suppresso nomine, in tertia editione novi testamenti Erasmus emendavit*; see Bentley *Humanists and Holy Writ* 224.

67 See the account of the sale of Zúñiga's works in Rome by street vendors (translated by de Jonge ASD IX-2 27) from Erasmus' *Apologiae contra Stunicam* (4) LB IX 385A–B.

68 Church law forbade the publication of books in Rome unless permission were first obtained there *per vicarium nostrum et sacri palatii magistrum*, or in the dioceses from someone deputized by the local bishop and from the local inquisition (see the decree *Inter solicitudines* of 4 May 1515 of the Fifth Lateran Council, Mansi 32.913A–B). Here Erasmus was probably referring to the more specific prohibitions against the publication of Zúñiga's criticisms imposed by Cardinal Francisco Jiménez de Cisneros in Spain, who urged him instead to send on his comments in a private letter to Erasmus; and in Rome by Leo X, by the cardinals during the *sede vacante* of 1521–2, by Adrian VI, and by Clement VII; see Epp 1416:16–18, 1418:26–8; Allen Epp 2443:351–5; LB IX 384E–385B; and de Jonge ASD IX-2, 18, 23, 26, 28.

69 Georg Sauermann (c 1492–1527), a cleric from Breslau and imperial proctor at the court of Rome, who died from injuries suffered during the Sack of Rome in 1527. His friendship with Erasmus apparently dates from his days at Louvain in 1520. Sauermann was a member of the literary circle around Goritz (see n431 below). The publication in which statements favourable to Erasmus were alleged to have been edited out on the urging of Pio was perhaps Sauermann's oration dedicated to Adrian VI according to Allen Ep 1479:13on, or, more likely, his *Expositio in epistolam Pauli ad Romanos*, according to Bauch 174; Michael Erbe in CEBR III 198 suggests that the work was 'in connection with the Ciceronian discussions in Rome.'

circles at that time, and so, if what people write is true, it was in effect
your pen that deleted[70] those passages.[71] I am not complaining that I have
been cheated of praise, for praise does nothing but burden one with envy[72]
– deleting those passages was a service to me – but I am reckoning the bias
of those who cannot abide my being praised. But I readily forgive them this
as well, although they would hardly be behaving thus with impunity were
it not for the fact that my whole attention, now that I am near death, is devoted to making my conscience acceptable to Christ, disregarding all else.
Many things must be overlooked by one who is hastening elsewhere.

I am quite aware that it is characteristic of human nature that people
are more ready to communicate bad news than good,[73] and no one attends
to this sort of gossip with less relish than I. But among such people there is a
certain young man[74] who is quite dear to you and an unequalled admirer of
your learning. In addition, another stoutly sworn comrade of yours, when
he was presenting the pope's case to my countrymen seven years ago,[75]
blabbed the following secret to me (this is his particular excellence, that he
can repress nothing): that there is in Rome a certain imperial ambassador
who is hostile to me. Now at that time, indeed, I knew neither that you
were in Rome nor that you were serving as imperial ambassador. At any
rate, he did not mention a name, though he would have had I pressed him.
But from the indications he blurted out, and from details later discovered,

* * * * *

70 *iugularat*; for *iugulare* = delete, see *Adagia* I v 57: *Stellis signare. Obelo notare* 'To mark
   with stars. To brand with an obelus' LB II 204A–205B / ASD II-1 530–4 / CWE 31 435–6.
71 See Allen Ep 1840:77–81 (22 June 1527): 'Georg Sauermann, a young man of extraordinary talent, was getting ready to print some writing or other. Alberto of Carpi – for he
   is the leader of that group – deleted whatever had been added in praise of Erasmus. I
   do not blame this on Alberto, but upon the one [Aleandro] who can endure that neither God nor any human save himself be praised. For Alberto, as I hear, holds him in
   the highest regard.' Pio later said (*XXIII libri* 48ᵛp) that what has been reported to Erasmus in this connection is a lie: 'I have never laid eyes on Sauermann, so far as I know,
   nor did I see a word of his book.' Also see *XXIII libri* n63 on the *Apologia* 119–21 below.
72 Compare Ep 1634:20–1.
73 Erasmus is here reacting to Pio's ornate development of this commonplace in *Responsio
   paraenetica* 3ᵛB.
74 Probably a reference to Juan Ginés de Sepúlveda (c 1490–1573); see CEBR III 240–2;
   Erasmus first mentions Sepúlveda, misspelling his name as 'Sepulvelam Hispanum,' in
   Allen Ep 2261:69 (31 January 1530); see Introduction xxxvi, xlviii.
75 Aleandro was charged on 22 July 1520 with the publication and implementation of the
   bull *Exsurge Domine*. He worked primarily in the Low Countries and at the court of
   Charles v; see J.K. Sowards in CWE 71 xl–xliii; and Paquier *Jérôme Aléandre* 142–50. Aleandro's mission ended with the death of Leo x on 1 December 1521. Erasmus' 'seven
   years ago' suggests, perhaps, some work on his reply before the publication of Pio's
   work, indeed, shortly after his receipt of the manuscript copy of it in 1526.

I have drawn the sure conclusion that you are the one of whom he spoke.[76] But in this sort of situation it is my practice gladly to trick myself, so that I do not know what I know, so much do I dislike a quarrel with learned men.

I approve your generosity in so lovingly heralding your man Zúñiga,[77] although I do not think that even the Spaniards[78] will read your words 'a good-natured fellow' and 'a man not at all envious' without laughter – not if they have read his books packed with vanity, vainglory,[79] and obvious slander.[80] So far as religion is concerned, he cited only a few passages in support of his slander,[81] but, as the case itself shows, he did so in neither a religious nor a scholarly way.

* * * * *

76 Pio ceased to be imperial ambassador in Rome on the death of Maximilian I on 12 January 1519. Maximilian's grandson, Charles V, was elected emperor on 28 June 1519 and did not reappoint Pio as his ambassador to the papal court; see Guaitoli 'Sulla vita' 227. For more background on these rumours and Erasmus' conclusion that Pio was the critic mentioned, see Ep 1479:57n; Ep 1576:46–9 (to Celio Calcagnini, 13 May 1525) re Pio's criticisms of Erasmus and Ep 1587:243–63 (Calcagnini's reply in praise and defence of Pio); and Ep 1634:12–17 (Erasmus' letter to Pio of 10 October 1525).

77 Pio's praise of Zúñiga and defence of his behaviour (*Responsio paraenetica* 6ᵛᴇ) include the expressions *homini candidi ingenii* and *homo minime invidiosus*, quoted here; see Jacob Ziegler's description of Zúñiga to Erasmus (Ep 1260:258–9): 'Hatred and spite [*invidia*] for you and self-love for himself – these are the mainsprings of his whole mind.'

78 See the account of Zúñiga given by Juan Luis Vives in a letter to Erasmus of 1 April 1522 (Ep 1271:89–98): 'I have never yet spoken of Zúñiga to any Spaniard without his saying how much he dislikes him – arrogant, boastful, slanderous, malicious beyond anything one could explain or believe ... He sounds like one of the Furies, not a human being.'

79 See the words of Zúñiga's preface to his *Blasphemiae et impietates* (Rome: Antonius Bladius de Asula 1522) Aiiʳ⁻ᵛ: 'I consider it enough, and more than enough, of an accomplishment that I was the first to thrust out the serpents from the works of Erasmus as if from their dens, that I was the first also, to the extent of my ability, to crush their heads, and that I have shown clearly to the reader that the man is not only a Lutheran but the standard-bearer and leader of the Lutherans'; see Erasmus' comment on these words (LB IX 373C–D, 380F–381A). Zúñiga was similarly boastful in a letter in Vergara (printed in Allen IV 625–8); Vives was shown this letter, and remarked of it (Ep 1271:88–9): 'Such a letter neither Thraso nor Pyrgopolynices [outrageous braggarts from Terence's *Eunuchus* and Plautus' *Miles gloriosus*] would have written.'

80 In the letter to Leo X prefacing the second book of the original, much fuller, and never published version of the *Blasphemiae et impietates* (1520), Zúñiga claimed: 'In fact, whatever disorder we observe in Germany due to the most impious Luther and his most pitiable followers should be considered without a doubt to have been received from Erasmus. That man, most holy Father, is the source and cause of these woes [*Caput ille horum est et causa malorum, Pater Beatissime*]. From that man, as from a fountain, have flowed forth all the Lutheran impieties.' See Henk J. de Jonge 'Four Unpublished Letters on Erasmus from J.L. Stunica to Pope Leo X (1520)' in J.-P. Massaut ed *Colloque Erasmien de Liège* (Paris 1987) 147–60 especially 153.

81 *in calumniam* 1529 *ad calumniam* LB; Pio says Zúñiga acted 'stirred by love of religion' *Responsio paraenetica* 6E.

But I shall subtract nothing from Zúñiga's praise, nor will I be responsible for moving that Camarina[82] again. No, so far as I am concerned, I am ready with a heartfelt amnesty to strike a compact of Christian good will with the man, with the elimination of every trace of earlier controversies. I will say only this: if he had touched upon only such matters as were truly deserving of criticism, or if he had kept himself within the bounds of the civility worthy both of liberal studies and of a Christian, then he would have won more authentic praise from all the right-minded and respectable, even as he has now attained universal notoriety.[83]

But at this point, by taking care not to scratch anywhere the sore of former controversy, I am knowingly and willingly surrendering a large part of my case. For although you are bringing up against me things that he has already hurled at me, you are nevertheless ignoring my replies to him,[84] either because you considered them insignificant, or, as I rather think, because you have not read them. I will have to put up with this disadvantage throughout my whole argument. As it is, although there is nothing among the particulars of your indictment that I have not rebutted over and over again in published books, you present them just as if no response at all had been made. But it was your duty in fairness either to accept my explanation if it seemed right, or to refute it if you found it in any respect unacceptable. If I go on to repeat here the statements that I have made so often in various places, what limit will there be to the book? Yet if I keep silent, I shall appear to admit whatever you claim, especially with those who have no contact with my works, and, of course, none have less contact with them than those who rage most boisterously against my writings.

* * * * *

82 Camarina is said to be a swamp near a town of the same name in Sicily; see *Adagia* I i 64 *Movere Camarinam* 'To move Camarina' LB II 51 / ASD II-1 174 / CWE 31 107–8; Erasmus explains the background of this arcane expression: '*Movere Camarinam* is to bring evil upon oneself.' He alludes to the verse quoted and translated from Stephanus in the adage ('Move not Camarina, for it is best not moved') in his letter to Pio of 23 December 1528 (Allen Ep 2080:18–20): 'moreover, I would again be moving Camarina, which it were better not to touch, since matters are otherwise more than enough exacerbated'; this is apropos of Erasmus' reluctance to reply to Pio. Note Erasmus' use of this favoured expression in a variety of contexts: in connection with the harmful reform measures of the Lutherans in his letter to Melanchthon from Basel on 10 December 1524 (Ep 1523:88–9); with Aleandro's supposed troublemaking in Paris in a letter of 7 November 1531 (Allen Ep 2565:5–6); and with Erasmus' controversies with the Lutherans in a letter of 2 December 1531 (Allen Ep 2579:23).

83 See Ep 1294:7–9; and Erasmus' observation in his *Apologiae contra Stunicam* (2) LB IX 357A: 'he has come into the worst possible favour with all the learned and sound.'

84 For Erasmus' works in reply to Zúñiga, see de Jonge in ASD IX-2 41–3.

But before I reply to the allegations of certain parties, I will follow the Wise Man's dictum: 'The just man, at the beginning of his discourse, is his own accuser.'[85] I admit that in publishing my efforts I have gone wrong in several ways.[86]

The first fault is clearly mine: whatever I publish is better termed a miscarriage than a birth. This vice is deeply ingrained in me. I cannot endure the tedium of revision, delays in printing, or diligence in polishing up what is misshapen. And I myself have very often paid the penalty for this fault, no slight penalty, as I lick my work into shape[87] a second time more wearisomely than if I had exercised due care at first.

The second fault I share with my friends, who have often abused my readiness to oblige by pressing me to work on material that I am not suited to handle, or material that it would be better not to handle at all. This occurred first in the case of the *Praise of Folly*,[88] and then in translating the New Testament,[89] although all the best people vote approval of this later work, for they realize that this effort was required for understanding the Greek commentaries, since the text of all of them disagrees with the text of ours that is in general circulation. The Spaniards did the same thing to the Old Testament,[90] but caused less resentment

* * * * *

85 The words of Solomon, in Prov 18:17, quoted here in the Old Latin version
86 Erasmus' admission here of carelessness and haste in the preparation of works for publication, and of embarking upon projects for which he was unsuited at the urging of his friends, is anticipated in his *Catalogus* of 30 January 1523 (Ep 1341A:47–64); for Erasmus' frequent acknowledgment of his being overhasty in publication, see Allen Ep 1352:92n.
87 *relambens ... refingens*: probably an allusion to the remark attributed to Virgil in the so-called Donatus Life, para 22 (*Vitae Vergilianae antiquae*, ed C.G. Hardie 2nd ed [Oxford 1966] 11:83–4), where Virgil's slow rate of work on the *Georgics* is described: 'saying ... that like a female bear he brought forth a poem and then licked it into shape'; this lore is also transmitted by Jerome (CCSL 76A 848; PL 26 400B); see an earlier use of this conceit by Erasmus in Ep 1236:125–6.
88 For a study of Erasmus' revisions of the *Praise of Folly*, see Michael A. Screech *Ecstasy and the Praise of Folly* (London 1980) 1–11, 241–9; and ASD IV-3, 29–64. The final of seven revisions by Erasmus appeared in the 1532 Froben edition.
89 Erasmus' annotated edition of the New Testament was initially entitled *Novum instrumentum* (see n266 below); it went through various editions in his lifetime: Basel: Froben 1516 (see facsimile edition *Erasmus von Rotterdam Novum Instrumentum, Basel 1516: Faksimile-Neudruck* ed Heinz Holeczek [Stuttgart–Bad Cannstatt 1986]), 1522, 1527, and 1535. There was also a Louvain edition, published by Theodoric Martens in 1519. See *Catalogus* 14:8–21 (Ep 1341A:482–95) for Erasmus' account of the meddling of his friends in his work on the New Testament, and his revised editions of it.
90 For two studies of the Complutensian Polyglot Bible project, see Melquíades Andrés Martín *La teología española en el siglo XVI* (Madrid 1977) 63–71; and Bentley *Humanists and Holy Writ* 70–111 (on the New Testament). Although the first volume of the

and, a fact over which I rejoice, greater profit to scholarship. For these reasons it has come about that my writing has been more prolific than successful.

In addition, while I confess[91] that I have done and written certain things that I would not have been about to do or write had I been able to divine that this epoch was going to dawn, I also affirm that my soul has always recoiled from a zeal for revolution and from inciting to insurrection, as it recoils even now, and I hope never to repent of this attitude.

With so much said by way of preface, I shall come to the complaints people make. You are conciliatory in reporting them, but you do so in such a way that you appear to be of the same opinion. First you declare that I provided the opportunity for this commotion, although you acquit me of being its cause.[92] And yet, according to the Greek proverb, 'for the wicked, nothing is lacking save an opportunity.'[93] If they seize their opportunity from Sacred Scripture, or from the books of ancient and orthodox writers, it is no wonder that they have seized it from my writings, and twist what was said rightly for the correction of morals to the advancement of their own revolt.

With this as your point of departure, you go on to say that, however much they may differ about the rest, all men agree on this: the Lutherans have drawn upon my springs to irrigate their sprouting poisons, or they

* * * * *

Polyglot was printed in January 1514, it was not published until after Leo x's approbation had been secured on 22 March 1520.

91 The following is a restatement of Ep 1634:84–6.

92 The second part of Pio's reported indictment was, as Erasmus tells it (Allen Ep 1634: 41–2 / CWE Ep 1634:46–7), *quicquid est huius tumultus, ex Erasmo natum esse* 'All our present troubles began with Erasmus.' Pio had corrected this (*Responsio paraenetica* 5ʳD): 'I admit that among learned friends in social gatherings, something was said similar to that statement, but not so savage, namely that certain of your books and annotations provided a grand opportunity for this conflagration.' This may, indeed, have been what Pio had been saying, or else he took his cue from Erasmus' own words, for in his letter to Pio (Allen Ep 1634:69 / CWE Ep 1634:78), Erasmus provides the following by way of an *anticipatio*: *Sed 'occasionem Lutherus hausit e libris meis'* 'But "Luther drew his inspiration from my books."' Pio, for his part, goes on (*Responsio paraenetica* 5ʳD) to lecture Erasmus on the difference between *causa* and *occasio*, hence Erasmus' words here. Pio had written (*Responsio paraenetica* 5ʳD): 'No one in our company has said that you were the cause (*te causam fuisse*),' but shortly afterward he wrote (*Responsio paraenetica* 6ʳE): 'But if you permit me to speak frankly, I will say that you were at fault [*in causa fuisse*], and that your blame was great, with the result that many came to suspect that you were a participant in, not to say the author of, this discord.'

93 A variant of *Adagia* II i 68: *Occasione duntaxat opus improbitati* 'Rascals have all they need save opportunity' LB II 432 / CWE 33 54.

have gathered their deadly seeds from my gardens.[94] 'There are so many points of agreement,' you observe, 'in the positions of each, that either Luther appears to erasmusize, or Erasmus to lutherize.'[95] I am, perhaps, a blind judge of my own works, nor is it, according to Socrates, the work of the same man both to give birth and to evaluate what has been born.[96] But even so, as I reread my works, though I come upon some offensive items, still I find no teaching that agrees with the condemned positions of Luther, but instead countless teachings that explicitly disagree with his. This is so manifest that it may truly, as the saying goes, be touched with the fingers and felt with the feet.[97] Furthermore, what is more foolish than this proverb, born amid the drinking bouts of the pseudomonks, 'either Erasmus lutherizes, or Luther erasmusizes.'?[98] A jingle that deserves to be

* * * * *

94 Paraphrasing Pio's words, *Responsio paraenetica* 5ᵛD
95 A verbatim quote from Pio *Responsio paraenetica* 5ᵛD. Zúñiga coined this slogan in his *Libellus trium illorum praecursor* (Gvʳ) where, announcing his proposed publication *Erasmi ac Luterii parallela*, he cites as the general opinion: 'there is so much agreement with Luther on the part of Erasmus (despite his notable efforts to suggest otherwise) that one could fittingly say about the two of them, as antiquity said of Plato and Philo, ἢ ἔρασμος λουτερίζει, ἢ λουτέριος ἐρασμίζει, that is, either Erasmus lutherizes or Luther erasmusizes.' As Zúñiga indicates, he modelled his slogan on that current in antiquity about Philo of Alexandria and Plato, best known from Jerome's *De viris illustribus* 11.7: 'Of him it is commonly said among the Greeks ῾Η Πλάτων φιλωνίζει ἢ Φίλων πλατωνίζει, that is, "either Plato follows Philo or Philo Plato" – so great is their similarity in thought and style' (Ceresa-Gastaldo 98). This bon mot was common coin; see eg Aldus' use of it in his dedicatory letter to Pio of Aristotle's and Theophrastus' *Physica*, published in 1497 (*Aldo Manuzio Editore* I 16); and Erasmus had incorporated it into his *Adagia* II vii 71 LB II 631C–E / ASD II-4 132–3 / CWE 34 35–6. Erasmus cites the slogan as Zúñiga's in his *Responsio ad notulas Bedaicas* (printed along with his *Responsio* to Pio in 1529, LB IX 708A): 'illud Stunicae proverbium, "aut Erasmus Lutherissat, aut Erasmusissat Lutherus."'
96 See Plato *Theaetetus* 150A–151D.
97 Perhaps reacting to Pio's use of a like expression (*Responsio paraenetica* 5ᵛD): *in hoc tamen consentiunt omnes, quod tangi velut digito potest* ...
98 Erasmus quotes the slogan in Greek: ἢ ᾽Ερασμὸς λουθερίζει, ἢ Λουθῆρος ἐρασμίζει. See Erasmus' account of a similar slogan in his *Apologia* against Zúñiga's *Blasphemiae et impietates* (LB IX 373F): 'This expression was babbled in his cups by some lone camel [Egmondanus], but quite a few bawlers of the same quality had long since scattered it around Louvain. "There is nothing," they say, "in the books of Luther which is not the same in those of Erasmus."' Erasmus later (1534) calls such slogans *proverbia* ἀδελφικά 'friars' proverbs' when he describes the collection of them provided by Nikolaus Ferber of Herborn OFM, in Allen Ep 2956:39–41: 'Erasmus is the father of Luther; Erasmus laid the eggs, Luther hatched the chicks; Luther, Zwingli, Oecolampadius, and Erasmus are the soldiers of Pilate who crucified Jesus.' Pierre des Cornes, the Parisian Franciscan who preached at Pio's funeral (see CEBR I 341–2), produced a witticism that identified the lion and dragon of Ps 90:13 with Luther and Erasmus (Allen Ep 2205:214–17).

echoed by you, a man of renown, and learning too! Erasmus lutherizes as much as a nightingale sings 'cuckoo.'

But come, granting that I may be blind in the case of my own writings,[99] how is it that, although there have been ever so many from all sides who strive most passionately to show that I am in agreement with Luther, no one has yet been able to prove that even in a single doctrinal teaching – I mean the condemned ones – I am in agreement with him? If this were the case, do you think the Lutherans would have kept quiet about it, or even Luther himself, whose *Bondage of the Will* shows, as does nothing else, how violently he attacks me?[100]

And who are the ones who pronounce that we agree on so many positions? Of so many slanderers there has not yet been one who could prove agreement on even a single doctrinal teaching. Of course, there sallied forth recently, along with some of his co-conspirators,[101] Béda, the Parisian theologian, a man so bereft of judgment, as the situation reveals, that he lacks

* * * * *

99 Perhaps a reaction to Pio's invitation (*Responsio paraenetica* 6^{V}E): 'read your own works, but as if they were another's. You will realize that men of the Christian faith, men desirous of peace, are deeply disturbed, with good cause.'

100 Among the many studies of the Erasmus/Luther debate, see the following: from an ecumenical-theological perspective, McSorley; from a historical-cultural perspective, John O'Malley 'Erasmus and Luther: Continuity and Discontinuity as Key to their Conflict' *Sixteenth Century Journal* 5 (1974) 47–65; from the perspective of the French Catholic school of *Nouvelle théolgie*, Georges Chantraine *Erasme et Luther: Libre et serf arbitre, Etude historique et théologique* (Paris: 1981); and a literary analysis, Marjorie O'Rourke Boyle *Rhetoric and Reform: Erasmus' Civil Dispute with Luther* (Cambridge, Mass 1983).

101 For an overview of Erasmus' Parisian critics, see Rummel, *Catholic Critics* II 29–79. Although Noël Béda was frequently in the forefront of efforts to have the theological faculty of the University of Paris censure Erasmus' writings, he was not alone. Among the faculty members who were active in formal investigations critical of Erasmus' works were the following: Jacques Berthélemy (see James K. Farge, *Biographical Register of Paris Doctors of Theology 1500–1536* Subsidia Mediaevalia 10 [Toronto 1980] 42–3, and Farge *Registre* 97, 127, 141, 177–8); Guillaume Duchesne (CEBR I 410; but despite Erasmus' claims that Duchesne was Béda's principal collaborator, he is seldom named as such in the *registres*); Jacques Godequin (Farge *Biographical Register* 198–9, and *Registre* 6, 14, 141); Jean Gillain (Farge, *Biographical Register* 194–6); Etienne Loret (Farge *Biographical Register* 286–9); and Tristan Ravault (Farge *Biographical Register* 388–9, and *Registre* 135, 137). Among the Paris doctors who were not active in the faculty's anti-Erasmian deliberations but who published works critical of Erasmus' writings were Josse Clichtove (CEBR I 317–20) and Pierre Cousturier (CEBR I 352–3). Béda's principal published works against Erasmus were *Annotationum in Jacobum Fabrum Stapulensem libri duo: et in Desiderium Erasmum liber unus* ... (Paris: Badius 1526), and *Apologia adversus clandestinos Lutheranos* ... (Paris: Badius 1529). Josse Bade also printed in 1531 and again in 1534 the *Determinatio Facultatis Theologiae in schola Parisiensi super quam plurimis assertionibus D. Erasmi Roterodami*, for which Béda worked very hard.

even common sense.[102] What has he accomplished by his detailed list of so many errors but to make a laughing stock of his peculiar combination of stupidity with a deranged zeal for slander? If people are allowed to twist sound statements so that they have an unorthodox meaning, and drag every similarity of opinion into suspicion,[103] then there is a greater number of such statements in the writings of Jerome, Augustine, and especially Paul than in mine.[104] So they are absolutely shameless liars who noise it about that what Luther and I write differs not in substance but only in style.[105]

But, you say, I have touched upon certain matters as if recalling into doubt things about which doubt is forbidden.[106] I did touch upon them, not so as to weaken them, but to the end that those more learned than I might establish with more solid arguments matters that seemed not yet fully demonstrated. Moreover, those works of mine, such as they are, were read in tranquil study, with no commotion, until some stupid preachers[107]

* * * * *

102 Alluding to Horace *Satires* 1.3.66. Support in the theological faculty of Paris for Béda's censures of Erasmus was significant. On 16 May 1526 the faculty voted unanimously to prevent the publication of Erasmus' *Colloquies* and to ask the Parlement to see to their suppression (Farge *Registre* 137). No conclusion about a lack of interest should be drawn from the low attendance at the meeting of 12 October 1527. It was a preliminary inquest, and such work was typically done in small groups (Farge *Registre* 180 n44, correcting Allen VII 233). After meeting three times a week for two-hour discussions each time, the faculty finally condemned a list of errors in Erasmus' writings on 16 December 1527 – despite Erasmus' epistolary campaign to prevent this (Duplessy *Collectio judiciorum* II-1 53–77; and Allen VII 233–4). The Sorbonne's censures of 1526 and 1527 were not published until July 1531 (Farge *Registre* 177 n36). Erasmus thus seriously misread the situation in Paris: opposition to him in the Sorbonne was substantial and resulted from a detailed examination and lengthy discussion of his views. Béda's 'co-conspirators' were the vast majority of the faculty.
103 This is a procedure of which Erasmus had complained much earlier: see Ep 1202:237–58.
104 A restatement of the defence offered in Ep 1634:82–4
105 Perhaps a reaction to Pio's report (*Responsio paraenetica* 5ᵛD) of the account by Erasmus' critics of the difference between Erasmus and Luther: 'But this is the difference: what you remark upon by way of admonition or expression of a doubt, he defines; what you touch upon in a modest way, he handles quite arrogantly; where you introduce a scruple, he passes judgment'; see the similar accusation of Nikolaus Ferber of Herborn OFM, who joined the name of Erasmus to the reformers, and, as Erasmus reports, said of their writings: 'there is no difference, except that I [Erasmus] crack jokes, they proceed seriously' (Allen Ep 2906:66–7).
106 Pio (*Responsio paraenetica* 6ʳE) speaks of critics of Erasmus who read in his writings many items which call into doubt matters long ago defined; see n134 below.
107 For 'preachers' Erasmus uses in a deprecatory way the Greek term κραγέται; for examples of mendicant preachers who denounced Erasmus from the pulpit, see Rummel *Catholic Critics* I 121–44. If Erasmus is referring specifically to members of the Order of Preachers, then he probably had in mind men like Vincentius Theoderici, Eustachius de Rivieren (van Zichem), and Laurens Laurensen (see ibidem 131–4). For his comments on other prominent Dominicans, see *Responsio* nn123, 176, 201, 203, and *Apologia* n288 below.

began to blather them about before the ignorant masses. If I should go on here to describe to you their outcries, at once factious and insane, you would scarcely believe that such monstrosities are concealed among those who keep on advertising themselves to the people under such splendid and grand labels. But they have accomplished nothing save that they have rendered themselves all the more hateful to sensible and learned men, and have won over to my side some unlettered persons to whom I have before been either unknown or not particularly dear. These reckon that the person against whom they tend to babble so stupidly, so compulsively, so insanely has to be a good man.

But 'at the outset,' you say, 'Luther's recklessness was not unwelcome.'[108] Even before his books appeared, I energetically dissuaded my friends from getting involved in this business, and I succeeded with many.[109] Not content with that, I at once asserted in published letters that I had nothing to do with Luther.[110] All the Germans who approached me while I was living at Louvain, men whom I now realize were even then pursuing their present course,[111] I sent packing in such a way that they could easily understand that my mind completely recoiled from this business. I was the first one of all to sniff out the fact that the fellow's spirit had been corrupted by ambition. Since I thought he could still be cured, I warned him about this, but even then only when challenged by his letters.[112]

These, of course, are proofs that I liked the fellow's recklessness! And yet, at the start, what you term recklessness was being called piety by the majority of people. Even if I had been of this opinion, why should I be singled out for accusation? Leo x forgave those who had at first been drawn into Luther's party.[113]As it is, I never approved of Luther's performance.

* * * * *

108 Pio *Responsio paraenetica* 5ᵛ D: 'nor at the outset was the fellow's recklessness unwelcome to you'; Pio reports this as the statement of unnamed informants; see also Ep 1127a:51–6, 66–104.

109 See the more elaborate version of this defence in Ep 1634:51, 58–61.

110 A repetition of Ep 1634:53–5; Erasmus is here trying to hide his initial cautious support for Luther. For a survey of the Erasmus-Luther relationship, see Cornelis Augustijn *Erasmus: His Life, Works, and Influence* trans J.C. Grayson (Toronto 1991) especially 119–45.

111 For Erasmus' early relations with the Lutherans, see Wallace K. Ferguson's introduction to his edition of the *Acta Academiae Lovaniensis contra Lutherum* in *Erasmi opuscula* 304–11.

112 For example, Epp 933, 980, 1127a, 1443, 1445, Allen Ep 1688

113 In the version of the bull *Exsurge Domine* published by Johann Eck, Leo x threatened with excommunication seven persons by name if they did not abjure within sixty days from the publication of the bull the forty-one errors listed therein. Of the seven indicted, Bernhard Adelmann (humanist and canon of the cathedral of Augsburg), Willibald Pirkheimer (humanist and city councillor of Nürnberg), and Lazarus Spengler (lawyer and city secretary for Nürnberg) eventually requested and received

The play that I thought was being staged was not altogether displeasing, but I shared this mistaken judgment with countless men and with many leaders of church and state.[114] The world had long been sighing over the collapsed morals of the church, nor was there anywhere a devout man who did not desire the restoration of Christian piety.[115]

'But,' you say, 'you were not willing to unsheathe your pen against him.'[116] The whole world would have changed its mind, of course, at the bidding of my pen! And at this point I am imagined to be possessed of unbelievable influence and prestige with the Germans because of my extraordinary eloquence, surpassing wit, and wondrous learning![117] Obviously, I am praised only when resentment is heaped upon me.[118]

First I ask you, O man most just, what great hardihood would it have been, when the world contained so many universities, so many outstanding theologians, so many learned bishops, if I, a private person,[119] and one untrained in the entire scope of theology – for so they declare me to be – had leapt forth like a second David to challenge Goliath to single combat,[120] and brought down upon my head a roar of disapproval from the entire world? But fancy that I would have dared this deed, what wisdom

* * * * *

absolution. See Bäumer 'Lutherprozess' especially 45–6; Harold Grimm *Lazarus Spengler: A Lay Leader of the Reformation* (Columbus, Ohio 1978) 38–49; and Lewis Spitz *The Religious Renaissance of the German Humanists* (Cambridge, Mass 1963) 177–9. See n207 below.

114 A repetition of the argument presented in Ep 1634:90–5; Pio had already rejected this defence (*Responsio paraenetica* 13ᵛκ).

115 See also Ep 1033:151–94.

116 Pio nowhere says this in quite these words, but see *Responsio paraenetica* 6ʳᴇ: 'nor did you oppose your pen to the Lutheran madness'; and 13ʳκ: 'because you did not oppose yourself at the beginning to the Lutheran madness.'

117 Erasmus is here sarcastically throwing back to Pio his praise of Erasmus' learning and ability, and his reference to Erasmus' effectiveness 'because of the favour, because of the remarkable influence, that you enjoy among the Germans' (*Responsio paraenetica* 13ʳκ); see n430 below.

118 See Ep 1576:39–45 where, just before complaining of Pio's remarks about him, Erasmus says of the Catholic theologians hostile to him: 'And yet these men know well enough what a defeat I could have inflicted on them, had I been willing to lift even a finger on behalf of the Lutherans. Now, when the merits of theologians are being discussed, Erasmus is dismissed as a dullard; but when someone wants to damage his reputation, then it is he and he alone who could have put out this whole conflagration if he had chosen to lift his pen'; the same point is made in Ep 1578:34–6.

119 A point made earlier in Allen Ep 1634:84 / cwe Ep 1634:94: *me privatum homuncionem* 'an ordinary citizen like myself.' Pio had already rejected this defence (*Responsio paraenetica* 13ᵛκ), and shortly observed that both universities and men moved by a zeal for piety, *etiam privato officio*, had condemned Luther in learned books (14ʳκ).

120 2 Sam 17:45–7

would it have been to draw my sword against a man about whom I was really in doubt as to what he was up to, or by what spirit he was being led? Again, suppose that I had taken a stand, what was I likely to have achieved? Surely nothing but what I brought about with the publication of my *Diatribe*:[121] [I spurred] 'the horse,' as the saying goes, 'into the level ground.'[122] I aroused his fierce temper to harden his doctrinal positions, and I gave his adherents grounds for another victory celebration.

What success had your countryman Prierias?[123] Only, of course, that he was silenced at Rome! The University of Paris was keeping quiet,[124] and Erasmus was to have pronounced against Luther? I was at that time, I admit it, to some degree in good standing with the learned. But what real influence could I have had, as not only a private person, but one lowly in every respect? Imagine, however, that I had considerable influence. Would those men have yielded to it who yielded not a bit to the influence of universities, of popes, of the emperor?

* * * * *

121 *De libero arbitrio diatribe* (Basel: Froben 1524). Because of Luther's violent attack on this work, Erasmus wrote a two-volume reply, *Hyperaspistes diatribae adversus 'Servum Arbitrium' Martini Lutheri* I (Basel: Froben 1526), II (Basel: Froben 1527); see n460 below.

122 *Adagia* I viii 82: *In planiciem equum* ... 'The horse to the plain ...' LB II 327–8 / CWE 32 169: 'Used whenever a man is encouraged to do the thing he is best at and most enjoys doing'; see also *Apologia* n151 below. Compare Erasmus' use of this proverb in his own defence for not writing against Luther in Ep 1236:92–4.

123 Silvestro Mazzolini OP (1465–1527), known as Prierias because of his birthplace, Priero in the Piedmont. He was a professor of Thomistic theology at the Sapienza University in Rome, and from 1515 Master of the Sacred Palace, in which capacity he led the effort to have Luther condemned. Prierias was silenced as part of an agreement worked out by Karl von Miltitz whereby both sides in the controversy would refrain from public debate. Among the most recent studies on Prierias, see Michael Tavuzzi *Prierias: The Life and Works of Silvestro Mazzolini da Prierio, 1456–1527* (Durham, NC 1997) 104–15, 119–22, 178–9; CEBR III 120–1; and Peter Fabisch 'Silvester Prieras (1456–1523)' *Katholische Theologen der Reformationszeit* I ed Erwin Iserloh, Katholisches Leben und Kirchenreform im Zeitalter der Glaubensspaltung 44 (Münster 1984) 26–36.

124 The Sorbonne first rendered its judgment against Luther on 15 April 1521, two years after the Leipzig debate; see Duplessy *Collectio judiciorum* I-2 365–74. For studies on the Sorbonne's stance against Luther, see Nathanaël Weiss 'Martin Luther, Jean Eck, et l'université de Paris d'après une lettre inédite, 11 septembre 1519' *Bulletin de la société de l'histoire du protestantisme français* 66 (1917) 35–50, 348–9; Léon Cristiani 'Luther et la faculté de théologie de Paris' *Revue de l'histoire de l'église de France* 32 (1946) 53–83; David Hempsall 'Martin Luther and the Sorbonne, 1519–1521' *Bulletin of the Institute of Historical Research* 46 (1973) 28–40; and James K. Farge *Orthodoxy and Reform in Early Reformation France: The Faculty of Theology of Paris, 1500–1543* Studies in Medieval and Reformation Thought 23 (Leiden 1985) 125–30, 165–9. The earlier condemnations of the University of Cologne on 30 August 1519 and of the University of Louvain on 7 November 1519 and the later articles of the Sorbonne of 4 November 1523 are printed in Duplessy *Collectio judiciorum* I-2 358–61, 374–9.

Finally, for this sort of work eloquence, if I have any at all, would not have been as useful as a knowledge of theology.[125] If you think my minimizing my knowledge of theology is a ruse, I shall cite against you that very weighty author Zúñiga. You cannot escape his authority after you have honoured the man with such a splendid testimonial. I can also cite against you many from the Roman sodality[126] who barely credit me with human intelligence. But now, so that I can be loaded down with even more

* * * * *

125 While Erasmus had studied scholastic theology off and on since 1496 at the Sorbonne, he received his doctorate in divinity not from there, but from the University of Torino on 4 September 1506; see the degree-granting document in Grendler 'How to Get a Degree' 65–9; and the comments of John Monfasani 'Aristotelians, Platonists, and the Missing Ockamists: Philosophical Liberty in Pre-Reformation Italy' *Renaissance Quarterly* 46 (1993) 233 n30. Erasmus only rarely mentions his doctoral degree, eg in his letter to Servatius Roger of 4 November 1506, Ep 200:9. Despite his years of study and academic degrees Erasmus tended to play down his own skills in academic theology (see *Apologia ad Fabrum* LB IX 66B [misprinted as 50B]), but on occasion he could even highlight his status as 'sacrae theologiae professor' (see the salutation of Ep 396) and defend his use of grammar in treating theological questions (see Ep 456:142–58). On his playing down his qualifications, see Rummel *Erasmus' Annotations* 28–9. But see also *Apologia* 260 below where Erasmus attacks Pio, who is neither a theologian nor a priest, for criticizing Erasmus, who is both.

126 'sodalitas Romana': It is not clear to which group Erasmus is referring here and on 87 below (*litteraria sodalitas*). For a brief overview of the evolution of the Roman academies, see John F. D'Amico 'Humanism in Rome' in *Renaissance Humanism: Foundations, Forms and Legacy* ed Albert Rabil JR 3 vols (Philadelphia 1988) I 264–95 especially 274–9. The Roman Academy had been constituted a formal, religious sodality at its refounding in 1478 (see D'Amico *Renaissance Humanism* 92, 96), and is referred to by one of its members, Pietro Marsi, as *sodalitas nostra litteraria* (quoted by D'Amico *Renaissance Humanism* 278 n66), but Erasmus may be using *sodalitas* in the non-technical sense of 'association.' By the 1520s the Roman academies had become informal gatherings of humanists, often under the leadership of a particular humanist who opened his house or garden as a centre for the discussion of literary or historical questions. Prior to the Sack of Rome the two most prominent groups were under the patronage of Johann Goritz (Johannes Corycius c 1455–1527, CEBR I 348) and Angelo Colocci (1467–1549; CEBR I 330–1). While Erasmus lists Goritz among his supporters in Rome (Ep 1342:362–3), he reports that Colocci was critical of him (Epp 1479:32–4 and 1482:34–5). Regarding the hostility in Roman humanist circles towards Erasmus on literary and religious questions, see Silvana Seidel Menchi 'Alcuni atteggiamenti della cultura italiana di fronte a Erasmo' in *Eresia e Riforma nell'Italia del Cinquecento* (De Kalb, Ill / Chicago 1974) 71–133 especially 89–116; and D'Amico *Renaissance Humanism* 138–42. Among those known to have written against Erasmus are Aleandro (*Racha*), Pio (*Responsio paraenetica* and XXIII *libri*), and Battista Casali (*Invectiva in Erasmum Roterodamum*). On Casali, see CEBR I 276–7; and John Monfasani 'Erasmus, the Roman Academy, and Ciceronianism: Battista Casali's *Invective*' *Erasmus of Rotterdam Society Yearbook* 17 (1997) 19–54, which provides the background and context (19–30), plus a transcription (33–40) and translation (41–9) of the *Invective*, together with historical-literary notes (49–54); on Pio's role in the Roman academies, see Introduction xxxvii and n431 below.

ill will, O immortal God, into how big an elephant I am suddenly changed from such a tiny fly![127]

Was there, then, no one in this vast commotion of the world who could have been exposed to the hazard of battle except Erasmus? With what greater success might you have advanced, Alberto, for, in addition to the ornaments of good fortune, in addition to prestige and a glory unstained by any ill will, you are possessed of so abstruse a knowledge of philosophy and theology, such great eloquence, and restraint. Surely in this way you would have become known to the world[128] more auspiciously than by repeating the complaints of others about Erasmus.

But where, I ask, are these men you mention, these keen critics of my works, who know how to compare statement to statement, premise to premise, word to word, and syllable to syllable?[129] I shall say nothing of Lee[130] and Zúñiga with whom I desire peace, if they allow it, or, if that is too much, at least a truce. Just read the pamphlet *De Racha*[131] (I am sure you have read it!) and you will see the glorious example of one such critic. Read Béda's wild censures, and you will see what these critics are like. But if it disgusts you to read these works, do not, O best of men, provoke hostility towards me by recalling, not to speak of approving, the suspicions of such people. If I were here to recall for you what those censors have babbled in

* * * * *

127 *Adagia* I ix 69: *Elephantum ex musca facis* 'You make of a fly an elephant' LB II 359 / CWE 32 219: Erasmus is here turning back on him Pio's own words (*Responsio paraenetica* 14[r]K): 'But now you shrink yourself into a gnat, now enlarge yourself into an elephant'; see n453 below. Flies and elephants have a history in the Erasmus-Luther debate; see Marjorie O'Rourke Boyle *Rhetoric and Reform: Erasmus' Civil Dispute with Luther* (Cambridge, Mass 1983) 1–4.

128 See Louis de Berquin's comment on Pio (Allen Ep 2066:66–7): 'And, as far as you are concerned, the man, vainglorious after the manner of the Italians, wanted to appear to have been able to find some fault with Erasmus.' Erasmus was quite ready to see a quest for glory as a principal motive of his critics; see eg his remark about Cousturier in Allen Ep 1687:59–61: 'Now he has the one thing he was seeking. The name of Pierre Cousturier flies through the world; it had been an obscure one before, even in Paris.' He had similar comments about Edward Lee (see n130 below), sometimes put into the mouths of Erasmus' defenders whose letters Erasmus edited and published; see Rummel *Catholic Critics* I 108, 110–13.

129 A sarcastic, in part verbatim, reference to Pio's description (*Responsio paraenetica* 6[r]E) of learned and pious men who read Erasmus carefully, 'comparing statements to statements, premises to conclusions, words to words.'

130 Edward Lee (c 1482–1544), English cleric trained at Oxford (BA) and Cambridge (MA, BD), who criticized Erasmus' *Novum instrumentum.* He later went on to become a diplomat and archbishop of York. See CEBR II 311–14. The dispute between Erasmus and Lee has been analysed by Robert Coogan in his *Erasmus, Lee and the Correction of the Vulgate: The Shaking of the Foundations* (Geneva 1992).

131 On the *Racha* and its author, see n64 above.

sermons, in public lectures, at banquets, in meetings, in the classroom, in private conversations, you would certainly repeat the proverb 'a pig passed judgment.'[132] And yet these are the men who are most offended by my writings! Since, as I believe, you are continually in the company of such men because of the similarity of your studies, it would perhaps be inopportune for me to quote from the comedy: 'evil communications corrupt good manners';[133] but, even so, malevolent conversations becloud not a little the celebrated magnanimity of your nature.

You go on to give an inventory of my sins. You say[134] there are many passages in my works that 'are scandalous; that call into doubt matters long ago defined; that deprive sacraments of their importance, and the ordinances of the fathers of their authority; that clamour against the office of the pope and the honour of the priests; that show insufficient reverence for the sacred rituals; that speak ill of the monks, poke fun at their vows, and condemn their excellent observances.' I am amazed that Alberto can even repeat such obviously empty charges, for I am convinced that these things were written by your pen, but with the bias of others.[135] I could rebut all of these with a single reply, were I to say what I could say quite truthfully: all of these are glittering falsehoods.

But to reply concisely to each charge: If I have called anything into doubt again, either it was something that seemed not yet to have been adequately considered, or about which there existed no definition of the church. I openly affirm that, whenever I do this, I have no desire to anticipate prejudicially the determinations of the church. I have not deprived the sacraments of the church of the least bit of their importance; rather, to the extent I could, I enhanced it. As to the ordinances of the Fathers, I do not know which ones you are referring to.[136] I assign an appropriate authority to the regulations of the church, unless, perhaps, one assigns too little authority

* * * * *

132 Erasmus quotes an adage: ὖς ἔκρινεν. This is perhaps a variation on *Adagia* I i 41: *Sus cum Minerva certamen suscepit* 'A sow competed with Minerva' LB II 43 / ASD II-1 156 / CWE 31 90–1.

133 See *Adagia* I x 74: *Corrumpunt mores bonos colloquia prava* 'Evil communications corrupt good manners' LB II 388–9 / CWE 32 267–8.

134 Erasmus quotes the following verbatim from Pio (*Responsio paraenetica* 6[r]E); however Pio gives this, not as his own opinion, but as that of the piously disposed who read Erasmus' works with care.

135 Pio in fact does describe the indictment above as the opinion of others; see preceding note.

136 For a fuller description of the charge, see book 15, 'De constitutionibus ecclesiasticis et legibus ac traditionibus humanis,' of Pio's *XXIII libri* (206[r]A–210[r]N), and the 1526 Determination of the Sorbonne on the *Colloquia* in LB IX 952B–C.

to them if he does not consider them equal to divine commandments. I shall speak of the pope's dignity shortly. If one who reminds priests of their duty as opportunity presents itself has clamoured against their honour, then I confess that I clamour frequently, but I share this clamour with Jerome,[137] with Cyprian,[138] with Chrysostom,[139] with Bernard.[140] But why adopt a defence based on the ancients,[141] since you, in your own book, included many frank criticisms of priests, theologians, and monks,[142] for all that you praise their calling extravagantly.[143] Why, indeed, are the divine Scriptures pondered except that they rail against the evil ways of all men? Even you would not be afraid to criticize certain ceremonial elements received by the church, such as the *cantus praefractus*, as you call it, and the roar of organs.[144] I have taken nothing from the dignity of the priests, but

* * * * *

137 Jerome (d 420) was Erasmus' hero; see Eugene F. Rice JR *St. Jerome in the Renaissance* (Baltimore 1985) chapter 5 '*Hieronymus Redivivus*: Erasmus and St Jerome' 116–36. Erasmus had produced a monumental edition of the works of Jerome (Basel: Froben 1516); see translations of a selection of Erasmus' introductions and notes to this edition and a description of the contents in CWE 61. Jerome had a critical view of the higher clergy in the church of Rome, having spent the years 382–5 in that city in the service of *his* hero, Pope Damasus (366–84).

138 In the course of his episcopate Cyprian (d 258) engaged in controversy with two popes, Cornelius (d 253) and Stephen (d 257); Erasmus had produced an edition of Cyprian's *Opera* (Basel: Froben, 1520), which he dedicated to Cardinal Lorenzo Pucci (Ep 1000).

139 John Chrysostom (d 407), whom Erasmus called 'that most honey-tongued preacher and indefatigable herald of Christ' (Allen Ep 2359:8). Erasmus' involvement with Chrysostom's works is described by W. Lackner 'Erasmus von Rotterdam als Editor und Übersetzer des Johannes Chrysostomos' *Jahrbuch der österreichischen Byzantinistik* 37 (1987) 293–311.

140 Erasmus had mixed feelings towards Bernard of Clairvaux (d 1153). Allen Ep 1142:45n lists references to Bernard in Erasmus' letters; see also nn 199 and 260 below.

141 See the similar line of argument with similar constellations of the Fathers in Ep 1202:16–21 (Jerome, Bernard); Ep 1313:97–100 (Jerome, Chrysostom, Bernard); Ep 1469:33–6 (Jerome, Bernard, Cyprian); Ep 1479:178–9 (Jerome, Cyprian).

142 Reacting to Erasmus' words in Ep 1634:97–9, Pio *Responsio paraenetica* 35^v–36^{r}B admits the existence of stubborn and arrogant theologians, corrupt priests, and perverse and proud monks.

143 Pio's defence of priests is to be found in *Responsio paraenetica* 42^{v}F–52^{v}I, of theology and philosophy on 30^{r}X–33^{v}Z, and of religious orders on 37^{v}D–42^{v}E. Pio was a firm supporter of religious life. His natural brother Teodoro was a member of the Observant branch of the Franciscans; Alberto esteemed him and obtained for him from Leo X the bishopric of Monopoli in Puglia. In 1518 Pio convinced the Observants to hold their next general chapter in Carpi. It met there in the church of S. Nicolo on 18 May 1521, and Pio personally paid the Franciscans' expenses. See Guaitoli 'Sulla vita' 225–6, 232, 307.

144 Pio *Responsio paraenetica* 78^{r}z: 'Wherefore, to say what I think, I do not consider really worthy of approval the very frequent use of the *cantus perfractus* in which a variety of voices are heard, but the words cannot be made out, nor yet so great a roar of organs.' In reporting the word as *praefractus*, Erasmus had perhaps misread Pio, but

your people, perhaps, measure the dignity of priests with a gauge different from mine. I have always revered the rituals of the church, but occasionally I censure those which are extravagant, superstitious, and silly. I do censure those people, of whom the world is full, who neglect true devotion and place their trust in rituals, sacred rituals to be sure, but devised to the end that from them we should advance to better things.

Show me one monk whom I have attacked as a monk. There are none whom I revere more, nor would I more gladly live among any than among those who are truly dead to the world and live according to the Gospel's rule, were it not for the fact that my physical frailty renders me unsuited for any common life.[145] If I sometimes warn young men not to enter precipitously into an unknown manner of life, while they are virtually unknown to themselves,[146] if I censure monks who traverse sea

* * * * *

this is impossible to determine because the manuscript copy of the *Responsio paraenetica* sent to Erasmus has not survived, while Pio's own draft in Ambrosiana MS Falcò Pio 1-i-282/6 lacks the relevant folio 59^r (= Pio *Responsio paraenetica* 78^{r}z); see Introduction cxxxi. In the index to the *XXIII libri* (ãii^{v}b), the editor explains the term *cantus perfractus* with the clause 'which they call *discantum.*' That Erasmus interpreted Pio's *cantus perfractus* to mean polyphony is clear from his use of the term *musicam polyphonon* (219) later when he was referring to Pio's criticism of a particular type of contemporary church music. *Cantus firmus* was the term used for traditional plainchant. *Perfractus* suggests that other voices break through or violate the single melodic line. *Praefractus* would seem to imply that these voices abruptly break in at the beginning or end of the melodic line but do not infringe upon it. Pio's criticism of multiple voices singing simultaneously so that the words become unintelligible was also shared by Erasmus, by the Sorbonne (*Declarationes ad censuras Lutetiae vulgatas,* LB IX 898C–D and 899B), and by many humanists. This complaint helped to shape the Tridentine reform of liturgical music; see Karl Gustav Fellerer 'Church Music and the Council of Trent' *The Musical Quarterly* 39 (1953) 576–94. Erasmus claimed that, while he approved of polyphony, provided modesty and moderation were used, Pio condemned it outright (LB IX 899C).

145 Erasmus felt himself both physically and mentally unsuited for the common life of a monastery; see Epp 296:24–34 and 447:499–501. The physical strains of the monastic life that Erasmus found so difficult would seem to have been primarily fasting, abstinence, night watchings, and physical labours; see Allen Ep 2771:59–62. For those suited to this life style he later praised the benefits of monasticism and claimed that he would prefer to live in a monastery if his health so permitted; see Allen Ep 1887:47–9.

146 This same argument was used by Erasmus as grounds for a dispensation from his religious rule; see Allen *Compendium vitae Erasmi* (1 49:68–70). To this and the points that follow, compare Erasmus' earlier statement on the subject in the prefatory letter to Paul Volz in the 1518 edition of the *Enchiridion* (Ep 858:521–9): 'Not but what these men will find it more desirable that recruits to the religious life should be honourable and genuine rather than numerous. And would that it had been provided by law that no one under the age of thirty should put his head into that kind of noose, before he has learnt to know himself and has discovered the force of true religion. In any case those who take the Pharisees as the model in their business, and course over land and sea that

and land[147] to lure the naivety of youth cunningly into their snare,[148] if I point out in what true religious life consists, am I then attacking monks? I nowhere attack either the monastic order or calling. I have always been on friendly terms with communities in which regular discipline flourishes; and every really observant monk has always been most dear to me. I quite frequently feel compassion for those religious who bring themselves into disrepute more and more daily. But if the world's esteem for certain of them has grown cold, they should blame this on their own deeds and not on my books. What are the excellent observances that I condemn? None, unless they mean by 'excellent observance' whatever has been generally adopted. Although I have made replies about all these matters to a great many others, and to Zúñiga as well,[149] you still recount them as if no reply had been made at all, and croak to me the old complaint.

Of course there are introduced as evidence letters[150] in which I either praise Luther or handle him so gently that I seem to many to play on both sides [ἐπαμφοτερίζειν], even though I am perceived as condemning the revolt.[151] Is it any wonder that my words were tentative when the matter was still in doubt? Or is someone necessarily in collusion[152] with an Arius if he esteems the man's ability and knowledge of the Scriptures?

* * * * *

they may make one proselyte, will never be short of inexperienced young men whom they can get into their net and try to persuade.'

147 See Matt 23:15.
148 For similar statements, see Ep 447:737–9, Allen Ep 2136:198–200, *Declarationes ad censuras Lutetiae vulgatas* LB IX 941B–E, and *De contemptu mundi* 204 ASD V-1 84 / CWE 66 174.
149 *Apologiae contra Stunicam* (2) LB IX 360C–F, 364E–365B, 366E–367C.
150 Pio *Responsio paraenetica* 5ᵛD describes Erasmus' critics as basing their critique of his handling of Luther on Erasmus' own letters, *epistolas quasdam tuas allegantes*. It should be noted that some of the letters that caused Erasmus the most trouble were published without his permission; eg Ep 939 did not appear in Erasmus' collections of letters but was published anonymously (Allen III 527), and Ep 980 had an unauthorized publication in July 1519 in Leipzig and was reprinted in an expurgated version by Erasmus in October of that year (see Allen III 606 n36).
151 Pio *Responsio paraenetica* 6ʳ–ᵛE: 'this increases their suspicion: whenever you happen to mention Luther, you either praise him, or speak of him so uncertainly, so obscurely, so gently, so carelessly, that people are compelled, for the time being, to think that you are to a great extent in collusion with Luther, or, surely, to put it precisely, ἀμφισβητεῖν [you stand aloof and waver], though you are perceived as condemning the insurrection and rebellion.' For the variants ἀμφισβητεῖν and ἐπαμφοτερίζειν, see n157 below.
152 Reacting to the accusation (*Responsio paraenetica* 6ᵛE) that people are compelled, on the basis of Erasmus' treatment of Luther in certain letters, to believe that Erasmus *magna ex parte cum Luthero colludere*.

I have never expressed approval of Luther's headstrong opinions, neither in jest nor in earnest. Perhaps it has been your good luck, or rather my bad luck, that only those letters which are supportive of this false charge have come into your hands, but different ones are being conveyed[153] to the Lutherans![154]

Thus it seems that you have experienced in the case of my writings what you gravely and prudently describe[155] as what usually occurs: what is worse is reported to us, what is better is passed over in silence. Although in countless passages I give due deference to the authority of the church, to the sacraments, to the ordinances of the Fathers, to devout priests, to holy rituals, to monks worthy of the name, to excellent observances, and I do this so frequently that, should I wish to collect it all together, a huge book would be produced, for all that, none of these is reported to you or has come to your attention, but only those passages which can be twisted for the purposes of slander.

Now imagine that I am writing privately to a prince whom I know to be wholly in favour of Luther.[156] Should I not temper my words to accomplish my purpose, so long as my purpose pertains to the Christian religion? If, in your opinion, a man seems 'to play on both sides' (or, an expression which later had more appeal for you, 'to stand apart and waver' [ἀμφισβητεῖν])[157] when, never removing his eyes from the goal of evangelical

* * * * *

153 *perferantur* 1529 *perferuntur* LB
154 Eg Erasmus' letters to Melanchthon: Ep 1523:27–38 (Basel, 10 December 1524) and Allen Ep 1944:3–8 (Basel, 5 February 1528); to Zwingli: Ep 1384:3–23 (Basel, 4 September 1523); and to Bucer: Allen Ep 1901:36–64, 81–102 (Basel, 11 November 1527)
155 See n73 above.
156 Erasmus did in fact write to Luther's own prince and principal supporter, Frederick III (the Wise), elector of Saxony, and to his brother and successor John I (the Steadfast); see eg Epp 939 and 1670. During a personal interview with Frederick III at Cologne on 5 December 1520, Erasmus wittily described Luther's only error to have been touching the pope's crown and the monks' bellies. This comment was very significant for solidifying the elector's support for Luther; see Spalatin's account in Mangan *Life* II 158–9, 180. The *Axiomata* which Erasmus wrote for Frederick in November 1520 were clearly favourable to Luther; see *Erasmi opuscula* 332–3, 336–7 / CWE 71 106–7.
157 Erasmus earlier and here quotes Pio as using the term ἐπαμφοτερίζειν; Bade's print of Pio's work reads ἀμφισβητεῖν; this probably indicates a change from the manuscript version originally sent to Erasmus; the Ambrosian MS draft of Pio's work does not include this section. The meaning assigned to ἀμφισβητεῖν is based on one given in the *Suda*, the *editio princeps* of which (Milan 1499) was, incidentally, dedicated to Alberto Pio. The *Suda* (*Suidae Lexicon* ed A. Adler [Stuttgart 1971] I 157) has '1763. Ἀμφισβητεῖν: not to agree, but, for example, to go apart and be in doubt,' as opposed to the more usual meaning of 'to dispute' (see Liddell and Scott s.v., though note there too the literal meaning 'go asunder, stand apart'). ἀμφισβητεῖν may recall the legendary serpent,

piety, he approves, on each side, what should be approved and condemns what should be condemned, then I admit that 'to play on both sides' is what I do even now. And yet I have not departed even a finger's breadth from the communion of the church, nor have I ever given myself to any faction, although I have been pushed this way by countless machinations.[158]

So the charge that I condemn nothing in Luther except his rebellion[159] is plainly false.[160] But, even granting that it is true, does one who hates rebellion seem to you to hate a slight evil? How much of the spirit of the world has long ago crept into the church? How many things are there over which men to whom Christ's glory is dear have long been sighing? How close to extinction is the spark of evangelical energy? But the man who would see these things corrected, and that by the authority of princes and popes lest tumult break out and the situation grow worse, is he impious?[161] It is not he who never departs from the right that is a Metius,[162] but rather

* * * * *

the ἀμφίσβαινα; see Erasmus' discussion of the *amphisbena* in ecclesiastical politics in *Parabolae* LB I 607D / ASD I-5 264 / CWE 23 241:11–16.

158 For Erasmus' own inner forces that could have led him to join the Protestants, see Allen Ep 2136:147–84. For Erasmus' later description of his having been attacked by conservative Catholics while he opposed leading Protestants, yet of his remaining loyal to the Roman church, see Allen Ep 2443:94–119. Erasmus claimed that many important persons had tried to persuade him to join the Lutheran movement, but that he in turn tried to reconcile them to the pope; see eg such a letter from Hutten, Ep 1161:1–44, and Erasmus' letters, Epp 1195:138–44, 1337a:31–3, and 1418:13–26. At Basel he was threatened by the Protestants; see Ep 1585:32–5. He claimed earlier in a letter to Pio that Lutherans had tried to win him over by flattery, threats, arguments, and strategems; see Ep 1634:67–70. For its part, the Catholic party urged Erasmus to take an open stand against Luther. Already in 1520 Leo X was promising him a rich bishopric if he would write against Luther; see Epp 1141:36–7 and n, 1180:11–14. Adrian VI also attempted to get Erasmus to take up the defence of the church (see Ep 1324:32–41, 90–104), as did his successor, Clement VII (see Ep 1443B:25–9). Others also urged Erasmus to write against Luther, eg Marino Caraccioli, Girolamo Aleandro, Jean Glapion, Duke George of Saxony, and King Henry VIII; see *Catalogus* (Ep 1341A:1362–74), and Ep 1415:62–4.

159 Erasmus was concerned not only with the civil disturbances that arose in part from Luther's teachings, but also with the content of some of those teachings. He accused Luther and his German followers of fostering sedition years before the Peasants' War; see Ep 1195:37–47, 113–17. In Allen Ep 1688:28–32 (11 April 1526) he accused Luther directly of having a rebellious temperament: 'with your temperament that is so arrogant, headstrong, and rebellious you are upsetting the whole world with fatal dissension ... you are equipping the unprincipled and radical for rebellion.'

160 We prefer the punctuation of 1529, a full stop, over the interrogation mark of LB.

161 We prefer the punctuation of 1529, an interrogation mark, over the full stop of LB.

162 Metius Fufetius, dictator of the Albans, a stock figure of disloyalty and vacillation; see Livy 1.27.5–7; Valerius Maximus 7.4.1: 'the flank of the Roman army having been abandoned, [Metius with his army] took up a position on the neighbouring hill, becoming an on-looker instead of an ally in the battle, so that he might pounce on the defeated or

he who wheedles Christ with titles and rituals but in reality serves the world.

But if we observe that the nature of human affairs is such that they always slip towards the worse, while the anus, as the saying goes, is prevailing,[163] and by a certain course and momentum of their own they drag even the best thought out institutions towards the very worst practices, I ask you what use are the Scriptures if we are not to be allowed through them to recall an almost wholly degenerate piety back to its original model? What will become of the light if it is put under a bushel?[164] What will become of the salt if it has no sharpness?[165] Let us get rid of the Scriptures and carouse in merry consensus; let us abolish Christ and enjoy this world while we can! Not all understand what this means. For not only there, but also in all regions of the world, there are people who measure Christianity by a few rituals, who have regard for nothing but the world, who never give a serious thought to the Philosophy of Christ.[166] Nor is this fact now kept hidden. Impiety has made such progress[167] that it displays itself openly, bringing back here plain Judaism,[168] there unconcealed paganism.[169] A great many

* * * * *

attack the weary victors' (*Factorum et dictorum memorabilium libri novem* ed Karl Kempf [Stuttgart 1966] 345).
163 See *Adagia* I x 90: *Podex lotionem vincit* 'The arse beats all efforts to wash it' LB II 394B / CWE 32 276: Erasmus quotes the adage here in Greek: τοῦ πρώκτου περιγενομένου.
164 Matt 5:14–15, and parallels
165 Matt 5:13, and parallels
166 The expression 'philosophy of Christ' is defined and explained in Erasmus' *Paraclesis*, and has been the subject of scholarly study; see LB V 137D–144D especially 139B–144D / Holborn 139–49; and Margaret Mann Phillips *Erasmus and the Northern Renaissance* (London 1949) 40–85.
167 *profecit* 1529 *proficit* LB
168 Erasmus viewed Judaism as a religion of laws and external observances. See Shimon Markish *Erasmus and the Jews* trans Anthony Olcott (Chicago 1986) 7–26; Markish 'Erasmus and the Jews: A Look Backwards' *Erasmus of Rotterdam Society Yearbook* 22 (2002) 1–9, especially 3–4; Guido Kisch *Erasmus' Stellung zu Juden und Judentum* Philosophie und Geschichte 83–4 (Tübingen 1969); and Hilmar Pabel 'Erasmus of Rotterdam and Judaism: A Reexamination in the Light of New Evidence' *Archiv für Reformationsgeschichte* 87 (1996) 9–37.
169 Erasmus accused humanists in Rome, such as Tommaso Inghirami (1470–1516), of being so enamoured of Ciceronic Latinity that they preached sermons full of pagan mythology and lacking clearly Christian content; see the *Ciceronianus* LB I 994F–999E, 1020C–1021A, and 1025C–1026B / ASD I-2 640–7, 701–2, 709 / CWE 28 387–96, 438–9, 447–8; see also Erasmus' allusions to the alleged paganism of the Roman humanists in Epp 1488:13–14, 1489:23–5, 1496:202–4, Allen Epp 1717:4, 1719:34–5, 1875:155. In Ep 1581:124–33 Erasmus explained to Béda why paganism has returned among the Italians. See also Kenneth Gouwens 'Ciceronianism and Collective Identity: Defining the Boundaries of the Roman Academy, 1525' *Journal of Medieval and Renaissance Studies* 23 (1993) 173–95 for

have grown so unaccustomed to a zeal for piety that they have cast away even the pretence of piety.

I think that you, a learned, good, and prudent man, understand what I am saying here, and what I am leaving unsaid, as I serve the cause of public tranquillity at the expense of my own case. Not all who go to Rome are doing the business of religion, nor are all doing the business of Christ who call Luther a pestilential beast.[170] And how do we know whether God sent this beast to reform our morals?[171] I can say this with less risk of suspicion, because he has played the beast against no one more ferociously than against me.[172] And yet I still seem to be acting in collusion with him! And while I am fighting it out with this beast, the Cousturiers and Bédas attack me from my flank, prompted, of course, by a zeal for religion.[173]

You say: 'You leapt to the fray too late.'[174] I wish that charge could really stick to me. What I fear is that I was premature, unless we are to do Christ's business with resources different from those with which I see it being done so far. I am keeping silent about many things lest I reopen any sore, given that matters have deteriorated to this extent. Call to trial on this charge the Parisian theologians who were late in publishing their *Articles*, and published nothing but articles.[175]

Is anyone who has not unsheathed his pen against Luther a Lutheran? I have myself refrained from partisan behaviour, and I have deterred others. What need was there for me to write against Luther, when the first of all to write against him was Prierias, the Master of the Sacred Palace,[176] and you

* * * * *

an example of pagan imagery used in a sermon in the papal chapel given in 1525 by Pietro Alcionio.

170 Pio, perhaps echoing the opening words of *Exsurge Domine*, terms Luther a *bestia* (*Responsio paraenetica* 14ᶠκ).

171 For similar statements, see Epp 1495:8–12, 1522:13–17, 1526:138–42, and Allen Ep 1672:95–9.

172 Erasmus complained to Luther himself against the harsh treatment he had received from Luther's pen; see Allen Ep 1688:6–24. He also wrote to Emperor Charles v about the wrath of Luther and his followers; see Allen Ep 1873:6–12.

173 Erasmus complained of the attacks of the Spanish, which came as soon as he had taken on Luther; see Allen Ep 1873:16–19. For Erasmus' complaint that Cousturier and Béda attacked him while he was fending off Luther, see also Allen Ep 1687:39–130. For a list of Erasmus' Catholic opponents who attacked him from the rear, see Allen Ep 2443:94–102.

174 These precise words do not occur in Pio's work; Erasmus may intend them as a more lively paraphrase of Pio's second reason why it is Erasmus' own fault that he is suspected of contributing to the Lutheran revolt (*Responsio paraenetica* 6ᶠE): 'secondly, because you kept silent, and did not oppose your pen to the Lutheran madness.'

175 See n124 above.

176 The writings of Silvestro Mazzolini (Prierias) against Luther were: ... *in presumptuosas*

know very well how successful he was?[177] None aid Luther's cause more than those who have shouted stupidly or written unconvincingly against Luther.

But who imposed upon me the duty of joining battle with Luther? The only one was Nicolaus Egmondanus,[178] when, in the presence of the rector of Louvain, after he had babbled as much abuse[179] as he pleased against me, he summoned me to this enterprise with the words: 'You have written[180] for Luther, now write against Luther.' A pretty stipulation, that I should admit having done what it had never occurred to me to do, and then pay the penalty for it by writing! And already the shouts of the monks were flying about: 'Let us pit Erasmus against Luther, and then we will destroy them both.'[181] Who would not be surprised that Erasmus could not be enticed by such blandishments to write against Luther? Then the one who delivered the document of censure[182] began immediately to scatter among his people the remark: 'We will accomplish nothing unless

* * * * *

Martini Luther conclusiones de potestate pape dialogus ... (Rome: E. Silber 1518); *Replica* ... *ad Fratrem Martinum Luther* ... (Rome 1518); *Errata et argumenta Martini Luteris recitata, detecta, repulsa et copiosissime trita* (Rome 1519; reprinted Rome: Antonius Bladius de Asula 1520); the third book of the *Errata* was later published separately as ... *Epithoma responsionis ad Martinum Luther* (Perugia: F.G. de Perusia 1519); and *Modus solennis* ... *ad* ...*convincendum Luteranos* ... *anno 1519 compositus* ... *anno 1553 revisus* ... (Rome: Jordanus 1553); see Klaiber 239–40. Prierias' criticism of Luther's teachings focused on the question of obedience to the authority of papal teaching, even though there was no consensus at that time on the extent of that authority. Luther's responses were effective on this point. See Horst 127–62; and Jared Wicks 'Roman Reactions to Luther: The First Year (1518)' *The Catholic Historical Review* 69 (1983) 521–62 especially 529–31. In addition, see n123 above.

177 For Erasmus' earliest negative views of Prierias' writings against Luther, see Ep 872:18–24, 1167:456–8; Allen Epp 1875:141, 1909:118–19, and 2445:46–7; and for his evasive statement to Prierias himself, Ep 1337A:60–4. See also n123 above.

178 Nicolaas Baechem of Egmond (d 1526) was a Carmelite friar and early opponent of Erasmus at the University of Louvain; see CEBR I 81–3. For a detailed account of his accusations and Erasmus' responses, which were presented in person before the rector of the university, Godschalk Rosemondt, in October 1520, see Erasmus' letter to More, Ep 1162, especially lines 230–1, where he recounts the same statement: '"Right," he said, "you have written in support of Luther, now write and attack him."'

179 *conuiciorum* 1529 *vitiorum* LB

180 *scripsisti* 1529 *Conscripsisti* LB

181 That conservative Catholics wanted Erasmus to write against Luther is indisputable; that their intention was thereby to destroy both is an assertion which is difficult to prove from independent sources.

182 Aleandro was the apostolic nuncio entrusted with publishing the bull *Exsurge Domine* in the Low Countries. See Erasmus' vituperative account of Aleandro and of his behaviour at the University of Louvain on this occasion in *Acta Academiae Lovaniensis contra Lutherum* in *Erasmi Opuscula* 317–28 / CWE 71 101–5.

first we do away with Erasmus.'[183] He continually avoided meeting with me,[184] though he suborned people to find out what I thought about him or about the Luther business. Indeed, he left nothing untried to destroy me, although he had quite different orders from Leo x.[185]

But in good order you finally proceed to name certain things whereby I have done a lot of damage to the piety of the weak.[186] I do not know whom you are calling weak, unless, perhaps, you count Zúñiga and Béda, or surely Luther, among the weak, for I have scandalized these men most particularly. 'What good was it,' you say,[187] 'to point out that in the early period the power of the pope was not acknowledged or exercised,[188] that

* * * * *

183 Independent confirmatory evidence for this statement is difficult to find. Erasmus claimed that Aleandro, when he came to Germany to promulgate the bull against Luther, came with the intent to eliminate Erasmus first and then move against Luther; see Allen Ep 2565:9–11. From other statements in Erasmus' correspondence it is clear that he also felt that he was viewed by his opponents as the principal obstacle to eliminating the Lutheran problem (Ep 1157:9–12) and as more harmful than Luther, who drew his poison from Erasmus' bountiful udders (Ep 1196:558–60). The dispatches from Aleandro to Rome show that the nuncio did indeed see Erasmus as the instigator of Luther's attacks and as the source of heretical ideas; see Seidel Menchi 61 and documents cited there.

184 According to Erasmus' account, written to von Hutten in 1523, while Aleandro was the papal nuncio in Brabant during September 1520, he intentionally avoided meeting Erasmus. Discussions earlier in Rome, the Brabant, and Cologne had apparently turned Aleandro against Erasmus. When Aleandro returned to Cologne in October and November to meet Charles v, he repeatedly extended an invitation to Erasmus, who was also in town, to dine with him. Erasmus declined the invitation for fear that Aleandro might poison him. After one of these dinners, however, they did speak together for several hours. Later, in September 1521 in Brussels, they met again and spoke for five hours. See *Spongia* ASD IX-1 148–51 especially 150:696 / LB X 1645A–E / Klawiter 177–9); Ep 1188:40–4, and Paquier *Jérôme Aléandre* 165–71, 280–2.

185 For a summary of Aleandro's instructions from Leo x, see Paquier *Jérôme Aléandre* 146–9. The papal commissioning briefs are reprinted in *Causa Lutheri* II 438–45. Erasmus' name does not appear in these documents.

186 Pio *Responsio paraenetica* 6^{v}F: 'For just as you have one way or another helped those of more rugged temperament and more mature judgment towards advancement of knowledge by the many comments you have published as the products of your flourishing ability, abundant knowledge of a variety of fields, and wide reading, so also have you done harm, quite a lot of it, to the piety of those who are weaker.'

187 Pio *Responsio paraenetica* 6^{v}F

188 *agnitam aut exertam*: see the letter of 19 January 1524 to Lorenzo Campeggio, discussing Zúñiga's accusations (Ep 1410:23–6): 'As to the sovereignty of the pope I have never doubted; but whether that sovereignty was recognized in Jerome's day or exercised [*agnita ... aut exerta*] is a doubt I do raise somewhere when prompted by the context, I think in my published notes on Jerome'; Ep 1410:8n identifies relevant notes in Erasmus' edition of Jerome. In his *Apologiae contra Stunicam* (4) LB IX 386C Erasmus had written of the *praeeminentia* of the pope: 'it was, however, neither

Pope Damasus was called *summus sacerdos* not *summus episcopus*,[189] that Augustine, and Jerome too, wrote to the bishop of Rome as to the equal of the rest of the bishops,[190] while ignoring many things which were written by the same authors to the opposite effect?' and so on.

Here again, most illustrious sir, I beg you to listen patiently to the truth. I am forced either to find you wanting in honesty or to suspect that you have not read my works. For although I have made what I consider to be a satisfactory reply in respect to your accusation, both to Zúñiga[191] and to other denouncers, you repeat the accusation just as if no reply had been made.

As the occasion arose, I made annotations in support not of one position only but of the contrary position as well. But you excerpt the first for the purpose of slander, and you deny that I have done the second. I did this out of no hostility towards the Roman pontiff, but to provide those who argue about this matter with something to think about, and to provide

* * * * *

acknowledged by all the churches, nor always exercised.' A little later he declares: 'But pretend that I had doubts [about papal power], the doubt was about whether the power was acknowledged and exercised, not about where it had been granted' (LB IX 387C).

189 Damasus, pope of Rome, 366–84, Jerome's great patron; in a note on Jerome's letter *Novum opus* Erasmus had remarked, 'it should be noted that he calls Damasus *summum sacerdotem*, not *summum pontificem*' (HO 4 13ʳ), and in another note he observed, 'you see that among the ancients he is not called *summum*, but *Romanum*' (HO 1 106ᵛ). Zúñiga noted this in *Blasphemiae et impietates* (Eiiᵛ), under the rubric 'De summo pontifice'; see Erasmus' reply to Zúñiga on this very point (*Apologiae contra Stunicam* (4) LB IX 386A–C, where Erasmus cites Ep 843:564–7, an indirect reply to Edward Lee, as the passage in which he wrote that the title *summus pontifex* did not exist in the time of Damasus. In his reply to Zúñiga Erasmus wrote: 'He [Damasus] was not called *summus pontifex*, but bishop of the city of Rome. Or am I denying that he was *summus* because he was not then called *summus*?' (LB IX 386B–C); see n489 below.

190 Erasmus had written in an *antidotus* in his edition of Jerome: 'Moreover he seems to make all bishops equal among themselves as being all of them alike successors of the apostles, for he makes the bishop of Gubbio equal to the bishop of Rome. Nor does he think any bishop less than another, except to the degree that one excels in humility' (HO 3 150ᵛ). Zúñiga *Blasphemiae et impietates* (Eiiʳ) quotes this under the rubric 'De aequalitate Episcoporum.' Under a rubric 'De autoritate sedis Apostolicae' (Fivʳ) Zúñiga cites Ep 843:477–81: 'Not that I am against Augustine, but because it is clear that he did not recognize the authority of the Roman See which we now accord it; all the more so as, when writing to Innocent, he addresses him as brother and gives no hint of his supreme eminence, but treats him as a colleague'; Allen Ep 843:434n cites Augustine Ep 175.7 as the passage to which Erasmus refers.

191 See Erasmus' reply to Zúñiga (*Apologiae contra Stunicam* (4) LB IX 385F–389A; for his response to certain Spanish monks, see *Apologia adversus monachos* LB IX 1087B–F.

an opportunity to temper what was offensive. In fact, there are more passages[192] in my works that support the authority of the pope than passages that call any aspect of it into doubt. And in the second edition, if there was anything of this sort to scandalize a weak person, or, to put it better, an overly strong person, I either removed it or tempered it.[193]

I see, however, that only those things which are conducive to slander were reported to you. But Johann Eck saw different passages, for he cites my testimony from the same annotations in support of the office of the pope.[194] But, as I said, I did not do this in only one passage. Elsewhere as well, in a great many writings, I assign to the Roman pontiff primacy over the entire church, and I do so to the accompanying roars of those who are hostile towards the Roman pontiff.[195]

What contradictory views! One group says that I am undermining the authority of the pope, the other that I am fawning on it. Have many not disputed about the power of the pope in the past, does this dispute not continue daily? On this topic do not theologians north of the Alps disagree with their colleagues to the south in certain respects?[196] Did Ockham not sometimes pick away at the authority of the Roman pontiff?[197] Did John Gerson not do

* * * * *

192 Eg Epp 1039:104–24, 1217:49–58, 162–3, Allen Ep 2136:205–7; LB IX 255C–D, 269A–D, 362A–F, 388A–F, 1056E–F, and 1067A–D

193 In his 1516 *Annotation on the New Testament*, Erasmus expressed surprise that Matt 16:18 was applied to the Roman pontiff, but added mitigating remarks in the 1519 edition (given here in brackets { }): 'Accordingly, I am amazed that there are people who twist this passage to make it refer to the Roman pontiff. {Surely it does apply most particularly to him as the leading figure in the Christian faith, but it applies, not to him only, but to all Christians, as Origen neatly points out . . .}'; see LB VI 88C–F / Reeve 71.

194 See Eck's *De primatu Petri adversus Ludderum . . . Libri tres* (Paris: Petrus Vidovaeus September 1521) 3.5.6ʳ: 'Erasmus of Rotterdam also points this out in his notes, observing with learning and prudence that "Jerome seems entirely to feel that all churches ought to be subject to the See of Rome"'; Eck also cites approvingly the scathing *Dialogus Iulii et Genii*, that is, Erasmus' *Julius exclusus* (1.31.49ᵛ). That Eck was unaware of its Erasmian authorship is suggested by his failure to identify the writer of the *Julius exclusus* and also to list Erasmus among those who had written against Julius II (1.44.70ʳ); see also Erasmus' *Apologiae contra Stunicam* (4) LB IX 388A. Erasmus' authorship of the *Julius exclusus* is a scholarly issue on which there is still controversy; see Michael J. Heath's summary of the arguments in CWE 27 156–60.

195 So too in his *Apologiae contra Stunicam* (4) LB IX 388A–B: 'Why does he not report the many passages in which I call the Roman pontiff truly *summus*, in all ways *maximus* . . . I have a worse reputation among the Lutherans for no other reason than that I am rather lavish in praise of the Roman pontiff.'

196 For a study of the differences between the Gallican ecclesiology of the north and the Dominican papalism of the south, see de la Brosse. Erasmus pointed to these differences in Ep 1337A:57–9 and *Apologiae contra Stunicam* (2) LB IX 370F.

197 William of Ockham OFM (d 1347), especially his *Dialogus de imperio et pontificia potes-*

the same from time to time,[198] not to mention Bernard?[199] Has a certain theologian[200] of the Sorbonne not already replied to Cardinal Cajetan's book?[201]

And yet the dignity of the Roman pontiff continued unshaken for so many centuries, but only ten words of Erasmus – not an assertion, but a passing remark on a passage in the sources which needs scrutiny – had the power to cause the whole world to rise up against the pope? And so many passages from the opposite point of view, passages where I resolutely and reverently defend the Roman pontiff's dignity, had no force at all? Your people pretend that these statements do not exist, but they do not escape the notice of the Lutherans, and I have a terrible reputation among

* * * * *

tate and *Octo quaestiones de potestate papae.* See Gordon Leff *William of Ockham: The Metamorphosis of Scholastic Discourse* (Manchester 1975) 614–43; and Charles Zuckerman 'The Relationship of Theories of Universals to Theories of Church Government in the Middle Ages: A Critique of Previous Views' *Journal of the History of Ideas* 36 (1975) 579–94, especially 585–6.

198 Jean le Charlier de Gerson (d 1429); for his ecclesiological views, see de la Brosse 107–45. Erasmus refers indirectly to Gerson's critical stance in *Apologiae contra Stunicam* (4) LB IX 388B, and openly in Allen Ep 1902:251–4.

199 That is, Bernard of Clairvaux, who combined support for the papacy with regular admonition; see n140 above and n260 below.

200 Erasmus here refers to Jacques Almain who published in 1512 his *Libellus de Auctoritate Ecclesiae seu sacrorum Conciliorum eam representantium editus a magistro Jacobo Almain contra Thomam de Vio* (Paris: Jean Granjon 1512). This work, commissioned by Louis XII and the Sorbonne, was an answer to Cajetan's . . . *de auctoritate papae et concilii utraque inuicem comparata* (Rome: Marcellus Silber 1511). These works are translated in *Conciliarism and Papalism* ed J.H. Burns and Thomas M. Izbicki (Cambridge 1997) 1–200. See Aleandro's letter to Erasmus, Ep 256:38–60; de la Brosse 70–8, 185–310; Francis A. Oakley 'Almain and Major: Conciliar Theory on the Eve of the Reformation' *American Historical Review* 70 (1969) 673–90, and Oakley 'Conciliarism in the Sixteenth Century: Jacques Almain Again' *Archiv für Reformationsgeschichte* 68 (1977) 11–32; and Harry J. McSorley 'Erasmus and the Primacy of the Roman Pontiff: Between Conciliarism and Papalism' *Archiv für Reformationsgeschichte* 65 (1974) 37–54 especially 52–3.

201 Tommaso de Vio OP (1469–1534), better known as Cajetan (Cajetanus), eminent Thomist, master general of the Dominican order (1508–18), cardinal (1517), bishop of Gaëta (1518), and legate entrusted with negotiating Luther's recantation. His principal works on papal authority were: *Tractatus de comparatione auctoritatis papae et concilii* [a variant on the title cited in n200 above] (Rome 1511) and *Apologia tractatus de comparata auctoritate papae et concilii* (Rome 1512), both written against the Pisan Council, and *De divina institutione pontificatus* . . . [= *De primatu Ecclesiae Romanae ad Leonem Decimum*] (Rome: Marcellus Silber 1521), written against Luther. While Erasmus criticized the anti-Pisan works as too immoderate (Ep 1033:157–61), he praised the anti-Lutheran work as a model of objectivity, brevity, and erudition (Ep 1225:215–20). For Cajetan's ecclesiological views, see Horst 27–54. On Cajetan and Erasmus, see CEBR I 239–42; and Christian Dolfen *Die Stellung des Erasmus von Rotterdam zur scholastischen Methode: Inaugural-Dissertation* (Osnabrück 1936) 92–3. According to McSorley 'Erasmus and the Primacy' (cited in n200 above) 51–3, Erasmus did not understand Cajetan's views and may not have read his reply to Almain's *Libellus de auctoritate.*

them for this very reason; I am stoned with abuse as a flatterer of the Enemy of the church, for so do they call him whom we call the Prince of the
church.

Furthermore, since it is established from the records of the ancients
that the Roman pontiff did not employ that power which he employs now,
is someone who distinguishes between power granted and power exercised
weakening the authority of the pontiff or strengthening it instead? He is
strengthening it, of course, for he shows the way in which those [contrary
arguments] can be refuted.

But if it is the expressions 'acknowledged' and 'exercised'[202] that offend your ears, Cardinal Cajetan used the same expressions in his latest
book[203] in which he defends the pope's majesty. So why am I blamed for
what is praised in him? I have a bad name among the Lutherans[204] for no
reason more than that I want the authority of the pope to be everywhere in
good repair. I wish that all Roman pontiffs valued their office, in those respects in which it is appropriate, as much as I desire that nothing be taken
away from papal authority! But they should watch out that those who have
so much regard for the pope's office and protect it with the weapons of
this world do not serve it far worse than those who attack it openly. I have
learned from my own experience that no enemy causes more trouble than
a friend who loves unwisely and maladroitly. Cajetan's overzealous book
stirred up more hostility towards the papal name than Luther's attacks.[205]

And little enough to counteract Luther's view was done by the Bullbearer [Διπλωματοφόρος];[206] he behaved quite arrogantly, made threats
against the learned and great, and filled everything with his pretensions.

* * * * *

202 See n188 above.
203 *De divina institutione pontificatus Romani Pontificis super totam ecclesiam a Christo in Petro*
(Rome 22 March 1521), critical edition by Friedrich Lauchert in *Corpus Catholicorum* x
(Münster 1925). For Cajetan's distinction between the power given and the power exercised, see ibidem 86–7. Cajetan admitted that not all church Fathers acknowledged the
Roman pontiff as Peter's successor and as pastor of the whole church, but so many did
that this teaching is sufficiently clear; see ibidem 89.
204 See n195 above.
205 It is difficult to determine which reactions Erasmus is here considering. Luther gave
no notice to this treatise. Peter Martyr Vermigli dismissed it lightly. On the Catholic
side, Cochlaeus praised it for its argumentation and scriptural citations. Later Catholic
theologians have also had high praise for it; see Lauchert's introduction (n203 above)
xvi–xvii nn1–3; and Horst 168–9. Yet Erasmus claims in his *Declarationes ad censuras
Lutetiae vulgatas* that papal theologians wrote against this book, eg LB IX 920D.
206 Διπλωματοφόρος is Erasmus' epithet for Aleandro throughout this work; later, writing to
John Choler, 7 November 1531 (Allen Ep 2565:13–15), Erasmus remarks about Aleandro:
'Now he is even angrier because in my reply to Alberto Pio he is criticized here and
there under the name *diplomatophorus.*'

In my hearing he said: 'The Roman pontiff has repeatedly humbled countless dukes and counts; he will easily humble a few lousy grammarians.'[207]
Another time he said: 'The pope can say to the Emperor Charles: "You are
a day-labourer."' Is this defending the dignity of the papal name, or making it hateful by undiplomatic pronouncements? His associate[208] said in my
hearing: 'We will find out all about the Duke Frederick,'[209] and with exactly the facial expression with which stern schoolmasters threaten boys
with switches.

Due regard must be given to the papacy, who is denying it? But no
less regard must be given to the flock for which Christ shed his blood. Everything is full of authority, dignity, dominion, principality; but meanwhile
the world needs shepherds, it languishes from hunger for the word of God.
Men wander like lost sheep because they have no one to light the way with
examples of a devout life, no one to teach, admonish, rebuke, exhort, and
console from the divine Scriptures.

Do you really believe we do not know what the Italian people are
like? I am speaking of the common people. Italy would have quite a devout people if the priests would attend to their duty. We know only too
well the situation of our own people. What is the cause? Surely it is that
the priests do anything but what they should. The churches are silent, and

* * * * *

207 Erasmus uses the term *grammatista* as the pejorative variant of *grammaticus*, probably
on the basis of Suetonius *De grammaticis et rhetoribus* 4:3: 'There are those who distinguish the *litteratus* from the *litterator* – as the Greeks distinguish the *grammaticus* from
the *grammatista* – and express thereby the judgement that the former is completely educated, the other only moderately so' (ed and trans Robert A. Kaster [Oxford 1995] 9,
and discussion on 86–93). The bull *Exsurge Domine*, promulgated by Aleandro and Eck,
mentioned by name Martin Luther, Andreas Bodenstein of Karlstadt (Wittenberg professor of theology and canon law), Johannes Dolz of Feldkirchen (later a preacher at
Zwickau), Johannes Wildenauer of Cheb (a disciple of Erasmus and Luther, who wrote
against the likelihood of a later marriage of St Anne to Cleopas), Bernhard Adelmann,
Willibald Pirkheimer, and Lazarus Spengler (see n113 above). From Rome's perspective
the three most prominent academics/humanists of this group were probably Luther,
Bodenstein, and Pirkheimer. The three 'lousy grammarians,' however, were probably
Adelmann, Wildenauer, and Spengler.
208 Apparently Marino Caracciolo, whom Leo x sent to Charles v as nuncio to deal with
political matters while Aleandro handled the Luther question; see CEBR I 264–5; Allen
I 35:11–15; and Paquier *Jérôme Aléandre* 148–9 n1.
209 For the efforts of the papal delegates to obtain his cooperation or to depose this Saxon
elector, see Paquier *Jérôme Aléandre* 158–62; and Smith *Erasmus* 235–6. It seems unlikely that Erasmus could recall nine years later the exact words of Aleandro and his
colleagues. If this conversation ever did take place, it would probably have been in
Cologne on 9? November 1520, when they conversed for several hours and parted in
apparent friendship (LB x 1645C–D / ASD IX-1 150:701–10). In none of the other likely accounts of this conversation are such statements about Charles v and Frederick of Saxony
found; see Epp 1141:36–7, 1188:40–4; and *Responsio* 16, 37–8 above, 87, 92 below.

masses infrequent, to say nothing of the quality of the clergy's lives. The Lord's flock has those who will skin and slaughter it, but none to feed and cherish it, or at least very few. These are not matters of particular concern to those whiners. Instead, one computes how much his revenue is from tithes or indulgences, another how much is subtracted from his emoluments because of dispensations, annates, confirmations, privileges, exemptions,[210] suits, duties, feudal dues, and the like, the common sources of increase of the wealth and prestige of bishops.

Even so, I have never begrudged the priests these things. They deserve honour, and an income sufficient to their dignity, but not one that is abundant to the point of luxury. But I should wish that they would be as mindful of what they owe the people as they are of what the people owe them. If they would do this, they would find, unless I am mistaken, that they would regain their popularity. I do not know how long-lasting and stable an authority can be that is protected by no other resources than threats and penalties. Here you are being less than a gentleman by drawing me into a situation in which silence puts me at a disadvantage and a reply is unsafe.

But I ask you, are these not excellent reasons which your people allege to prove that I was the cause of this entire uprising? I mentioned that a power was at one time not yet exercised. I made a note to the effect that the Roman pontiffs were once called by titles different from those by which they are called now. But there is nothing in these which is diametrically opposed to the points that you review in support of the supreme pontiff.[211] All the same, so terrible a charge should have been proved by other arguments. For my part, I am so enthusiastic about the proof-texts[212] that you quote in support of the papacy, with such learning, and elegance too, that I regret that your argumentation was not more effective.

Of course, for us who acknowledge the primacy of the bishop of Rome, especially in spiritual matters, with our whole hearts, these are quite enough, or, to be more precise, they are superfluous. Your reply should have been directed to the attacks made against papal primacy by others in their statements and writing. There is no need for me to remind you what these attacks are like; sometimes we barely have anything we can say in reply to them.

* * * * *

210 *exemptionibus* 1529 *exceptionibus* LB
211 Pio provides an extended defence of papal primacy against the Lutherans in *Responsio paraenetica* 52^{v}K–66^{r}Q.
212 Pio *Responsio paraenetica* 7^{r}F and 52^{v}K–59^{v}L

But if you think that a man who speaks out against evil pontiffs for the purpose of admonition[213] is injuring the majesty of the pontiff, then you ought to complain against Cyprian, against Jerome, indeed against yourself, for in your own book you say many things against impious priests, non-observant monks, and captious theologians.[214]

You say: 'It is not the place of grammarians to reprove the way of life of pontiffs.'[215] But you are saying this against others, not against me. And I do not disapprove of what you say, or even of what Cardinal Cajetan wrote, that it is lawful only for monarchs to admonish the pontiff, and gently at that, if he is plainly exercising a tyranny over the church.[216] But how will you shut the mouths of the people? By threats alone?

A similar charge is that somewhere I stated briefly that the title 'cardinal' was unknown at the time of Jerome.[217] So what then? Does the entire cardinalate collapse forthwith? Granted that the name existed, surely the current type of cardinal did not.[218] The titles 'subdeacon' and 'archdeacon'

* * * * *

213 *monendi* 1529 *medendi* LB
214 See n142 above.
215 Not to be found verbatim in Pio; Pio delivers an invective against the arrogance of *litteratores* and *grammatistae* in *Responsio paraenetica* 34^v–35^{r}A; he reproaches the *litteratores* for their ingratitude and implied hostility towards the Holy See on 52^v–53^{r}K.
216 Cajetan held that secular princes and prelates of the church are allowed to resist and impede the abuse of power by a pope. They should openly accuse him of crimes, argue with him, refuse to obey his wicked commands, and call upon illustrious persons to rebuke him. In the case of a certain and indubitably heretical pope to be deposed, the emperor, cardinals, and prelates can assemble a council for this sole purpose. See his *De comparatione auctoritatis papae et concilii cum apologia eiusdem tractatus* ed Vincent M. Jacques Pollet Scripta Theologica I (Rome 1936) 108 no 230, 178–80 nos 411–12; see also McSorley 'Erasmus and the Primacy' (cited in n200) 49; and *Apologia adversus monachos* LB IX 1087C.
217 The following pages contain replies, point by point, to a long indictment given by Pio in *Responsio paraenetica* 7^{r}F; we will quote the entire indictment here, and refer back to this note as each topic comes up: 'What [point is there] in saying [1] that cardinals did not exist in the time of Jerome; [2] that bishops and priests were once equals; [3] that the church has decreased because of wealth [a compression of the formula 'the church has increased in wealth but decreased in virtue']; [4] that councils were once usually convoked by emperors? What point [in saying] [5] that matrimony was not counted among the sacraments of the Church which are properly called sacraments; [6] in finding fault with the prayers of the canonical hours; [in saying] [7] that auricular confession was not the ancient practice; [8] that the Eucharist was customarily received by each in his own home? How erroneous also [9] to extol marriage with such praise that you prefer it to celibacy and continence? What point [10] in censuring ecclesiastical chants and rituals; [11] in slandering monks, and, what is more grave, their calling; and [12] in saying that their institutions are the product of weakness? What point [13] in expressing a loathing for ecclesiastical decrees and human ordinances?'
218 See *Vita Hieronymi Stridonensis* 587–90: 'Not only had that splendour and dignity of the cardinals which we see today not yet come to be, but in those days, I believe, the name

do not exist in the apostolic writings, just as that of 'archbishop' does not. Is their class for this reason condemned? The category of 'monk' began to be known in Rome under Anastasius.[219] The monastic state is not for that reason necessarily condemned, is it? Did no one before me investigate when and from what source the order of cardinals arose? Has no one complained of the great number and the pomp of cardinals?

The books of the popes[220] show that they were established when the world recognized the Roman church as the leader of the whole Christian commonwealth, and one had recourse to Rome if there was anything that

* * * * *

of cardinal had not even existed' (*Erasmi opuscula* 155 / CWE 61 36); and when defending his rejection of a letter as inauthentic, Erasmus observed: 'Didn't he [the forger] even take care not to call one who was not a bishop a *sacerdos*, since in that age only bishops were called *sacerdotes*? I cannot imagine why he did not call him a cardinal too, since in those times this name was still unheard!' (HO 2 195ʳD).

219 *Anastasius* is perhaps a typographical error for *Athanasius*. This argument seems to be connected to Erasmus' account of the monastic vocation of Jerome's friend Marcella, 'who was the first woman in Rome to have the courage to profess the monastic life when that very word was exceedingly offensive to Roman ears. This burning desire took possession of her above all because of the influence of Athanasius and other Egyptian priests' (*Vita Hieronymi Stridonensis* in *Erasmi opuscula* 155:599–602 / CWE 61 36); this is based on Jerome Ep 127.5 CSEL 56 149:5–10. These developments, however, antedate the pontificate of Anastasius I (399–401), of whom Erasmus reports only his regulation of the reception of foreign clergy at Rome (HO 1 28ʳ).

220 *Pontificum libelli*: a vague reference, perhaps alluding to the famous *Liber pontificalis* supposedly compiled by Pope Damasus at the urging of Jerome, but in reality a sixth-century collection of biographies of popes, beginning with the pontificate of St Peter. It was later, and at various times, continued, down to the death of Pius II (1464). See Louis Marie Olivier Duchesne ed *Le 'Liber Pontificalis': Texte, introduction et commentaire* 3 vols, rev ed Cyrille Vogel (Paris 1955–7). Bartolomeo Sacchi of Piadena [= Platina], a Latin stylist and historian whom Erasmus respected (see CEBR III 100–1), included in his *Liber de vita Christi et omnium pontificum* (Venice: J. de Colonia and J. Manthen 1479) references to what seems to be the origins of the cardinalate, ie Pope Evaristus (112–21) dividing the parishes (*tituli*) in Rome among the priests and setting the number of deacons at seven, and Pope Marcellus I (307–9) establishing twenty-five parishes (*tituli*) in Rome; see his *The Lives of the Popes from the Time of Our Saviour Jesus Christ* trans W. Benham 2 vols (London nd) I, 20, 64. Thomas Netter (c 1377–1430) claimed to have found in the *Liber pontificalis* these same two references (*Liber pontificalis* ed Duchesne and rev Vogel I 126 (6:3) and I 166 (31.3). He discusses the possible etymology of the name cardinal, but is more concerned with the antiquity of its institution: 'Et quomodocumque sit de nomine, tamen quod tantum est curandum, res ab antiquo profecit'; see his *Doctrinale antiquitatum fidei catholicae ecclesiae* 3 vols, ed Bonaventura Blanciotti (Venice: Typis Antonii Bassanesii ad S. Cantianum 1757–9; repr Farnborough, Hants 1967) I cols 526–527B. This work, in an edition by Carmelites in Paris, was first printed by Bade in Paris from 1521 to 1532; see Renouard *Imprimeurs* II nos 491, 526, 696. Erasmus here seems to be arguing for a much later establishment of the cardinalate. The first appearance of the word *cardinalis* in the *Liber pontificalis* is not until the pontificate of Stephen III (768–72); see Duchesne/Vogel ed I 476.

required the advice of the learned. For this work twelve cardinals were chosen, men equipped with learning, good morals, and a knowledge of languages, those who are now called penitentiaries.[221] These still have the work, but others have the name and the eminence, and they are now equated with the majesty of kings. There is no point in reporting how great is the worldwide complaint about these matters. I only questioned the apostolic origin of the dignity of cardinals, and I am the cause of the whole revolt, because, when the occasion arose, I made a brief note to the effect that the title seemed to be rather recent!

Yet there are countless places in my works in which I write in a most honorific way either to cardinals or about cardinals. Those who are hostile to the Roman pontiff rage at these passages, but your people pretend that they do not exist. A brief note has so much influence with them that they think this whole revolt should be blamed on me. I wish that the cardinals, however many they are, would shine with a true dignity, and thus free the world twice from dissensions. I wish that the supreme pontiff, who all alone surpasses the dignity of many kings in wealth and fanfare, would have the same degree of power based on spiritual authority, so that he could bring his children into harmony.

I will enter no farther into this cave, the sort into which you invite me all too often. For if I wanted to discuss here the bases for the real dignity of church leaders, and to compare to it the morals and doings of these times, you realize in your prudence, I think, how vast a field for discourse would lie open to me. But far be it from me to throw oil into the furnace.[222]

If you approve of Zúñiga because he raged at me, as if from a wagon,[223] on account of some few words which were rather critical of the

* * * * *

221 The notion that the establishment of the college of cardinals was related to the office of penitentiary seems to date from the Middle Ages. It is not based on the *Liber pontificalis;* the first mention of *penitentiarius* there (Duchesne/Vogel ed [see n220 above] II 454:2–3) is to Raymond of Peñafort (c 1175–1275). Modern scholarship dates the origins of the office of penitentiary to the late twelfth or early thirteenth century; see Charles H. Haskins 'The Sources for the History of the Papal Penitentiary' *The American Journal of Theology* 9 (1905) 421–50 especially 423; and Emil Göller *Die päpstliche Pönitentiarie von ihrem Ursprung bis zu ihrer Umgestaltung unter Pius V* Bibliothek des königlichen Preussischen Historischen Instituts in Rom 3–4, 7–8 (Rome 1907–11) especially I, part 1: *Darstellung* 85–6, 129–31; and *Apologia* n139 below. On the origins of the cardinalate, see Stephan Kuttner ' "Cardinalis": The History of a Canonical Concept' *Traditio* 3 (1945) 129–214; and Carl Gerold Fürst *Cardinalis: Prolegomena zu einer Rechtsgeschichte des römischen Kardinalskollegiums* (Munich 1967).

222 *Adagia* I ii 9: *Oleum camino addere* 'To pour oil on the fire' LB II 71E / ASD II-1 220–1 / CWE 31 151

223 *Adagia* I vii 73: *De plaustro loqui* 'Wagon-language' LB II 290F–291C / CWE 32 110–11, referring to the licence given to satirical criticism in ancient comedy, which was performed

Translator,[224] you will allow, in fairness, that those who love the spiritual beauty of God's house[225] with their whole hearts may sometimes give a brief indication of what they long for.

I said that once presbyters were equal to bishops.[226] Yes, I said that, but in dignity, not in function. In fact, it was not I who said this, but Jerome.[227] In his work you also read that bishops are fathers, not masters.[228] He is also the one who made the statement, which you charge me with,[229] that the church has increased in wealth and decreased in virtue.[230] (What would he say if he lived in these times?) And are not many things of this sort repeated in the decrees of the popes?

I wrote that once councils were customarily convoked by the emperors.[231] Is this not clear from the histories? In fact, Jerome requires among the proofs that a council has been lawfully assembled that the name of the emperor by whose command it was assembled be produced.[232] I am pointing to what was the usual practice of old; I am not condemning what is done now, nor would what I point out have been unknown to any learned person, even if I had kept silence. And what I am talking about is not so very ancient. The emperor presided at the councils of Constance and Florence.[233]

* * * * *

from a wagon that served as a movable stage. See also Pio XXIII *libri* n39 on *Apologia* 115 below.

224 Erasmus is probably referring to Zúñiga's *Assertio ecclesiasticae translationis*; see n66 above. On the identity of 'the Translator' of the Vulgate, see n374 below; for studies of some of Erasmus' criticisms, see Bentley *Humanists and Holy Writ* 162–73; Rummel *Erasmus as a Translator* 99–101; and Rummel *Erasmus' Annotations* especially 17, 30, and 135–6, where she points out that Erasmus denied that the Vulgate was the translation used by Cyprian, Hilary, Ambrose, Augustine, or Jerome, although he nonetheless accepted it, in spite of its errors, as authentic because it was the text received by the custom of almost the whole church.

225 Ps 25:8

226 See n217 above, item [2]. For Erasmus' statements, see HO 3 150ᵛC; *Institutio christiani matrimonii* LB V 652D; and *Annotationes* LB VI 863C–D / Reeve *Galatians to Apocalypse* 619.

227 *Comment. in Ep. ad Titum* 1.1.5 PL 26 562C–563A

228 Jerome Ep 82.5 CSEL 55 119, cited by Erasmus in *De esu carnium* LB IX 1208B / ASD IX-1 39:608–9

229 See n217 above, item [3].

230 Jerome *Vita Malchi monachi* 1 PL 23 53C

231 See n217 above, item [4]. In a comment on the passage from Jerome cited in the next note Erasmus wrote: 'Note, reader, that of old councils were customarily convoked by order of the emperors' (HO 3 104ᵛD); Erasmus alludes to this ancient practice in *Julius exclusus* 427–8 *Erasmi opuscula* 90.

232 *Apologia contra Rufinum* 2.19 CCSL 79 56:45–6

233 At the Council of Constance (1414–18), Sigismund (the king of the Romans 1410–33; emperor 1433–7), presided (1415). At the Council of Ferrara-Florence-Rome (1438–45), the Byzantine emperor, John VIII Palaeologus (1425–48), presided (1438–9). The Latin

Everywhere I count matrimony among the church's sacraments properly so called,[234] but Peter Lombard[235] does not, Durandus is unsure,[236] and this matter would not have disturbed anyone if it has not been made a cause

* * * * *

emperor only once presided at the Council of Constance, while the Byzantine emperor acted as leader of the Greek delegation. On the presidents of the Council of Constance, see Hefele-Leclercq VII-1 167, 190 (John XXIII presided at the first and second sessions); 202 (Cardinal Pierre d'Ailly presided at the third session once John XXIII left Constance); 207, 209 (Cardinal Giordano Orsini presided at the fourth and fifth sessions); 215 (Cardinal Jean de Brogny, cardinal-bishop of Ostia, presided at the sixth and all subsequent sessions until the election of Martin V); and 296–7 (at the fourteenth session the emperor-elect, Sigismund, initially presided while Gregory XII's procurators, Carlo Malatesta and Giovanni Dominici, reconvoked the council in the name of the Roman pope, after which Cardinal de Brogny reassumed its presidency). On Eugenius IV as the president of the Council of Ferrara-Florence, see Joseph Gill *The Council of Florence* (Cambridge 1959) 107 ('The thrones were arranged in such a way that the Pope's [on the north side of the church] was a little in advance of an empty one allotted to the Holy Roman Emperor, with the cardinals and the rest of the western prelates and members arrayed in order behind, while the throne of the Byzantine Emperor [on the south side of the church] corresponded exactly with that of the western Emperor, and those of the Patriarch and the rest of the Greeks were placed accordingly'); and Gill's *Personalities of the Council of Florence and Other Essays* (New York 1964) 111–13 (the Byzantine emperor insisted on equality, disputed over heights of thrones); 21–2, 119 (the Byzantine emperor convened the council and was head of the Greek delegation, but never intervened on theological questions). According to Pio *XXIII libri* 54[r]g: 'It is amazing how you assert things you know and things you do not with equal aplomb, but I would not expect you to lie about a matter of such common knowledge! Surely some rascal among your assistants reported this to you. If you read the *acta* of the councils you mention, you will realize that your statement is a plain lie. How could the emperor preside at the Council of Florence when he was not even there? Yes, the Constantinopolitan prince attended, only to request help against the Turks from Eusebius [ie Eugenius] IV. Who presided over the Council of Constance? Sigismund was there, indeed, for he had travelled the world to eliminate the schism, but it was John XXIII who presided, as long as he was present. After his departure one of the cardinals, usually the cardinal of Ostia, the dean of the Sacred College, presided. On one occasion, however, the fathers were assembled with no one presiding – when Gregory XI [ie XII] abdicated the papacy [on 4 July 1415], as he indicated through envoys – but with the elimination of the schism and the election of Martin as pope [on 11 November 1417], he always presided. But the emperor never presided, as the records of the council attest.'

234 See n217 above, item [5]; for Erasmus' statements on the subject, see, for example, *Declarationes ad censuras Lutetiae vulgatas* LB IX 843E–844A; *Apologia adversus monachos* LB 1068D–1070C / ASD IX-2 210–12, n784; Allen Ep 2136:212–13. The crucial expression here is 'properly so-called'; see Emile V. Telle *Erasme de Rotterdam et le septième sacrament* (Geneva 1954) 260–1, 267, 273, 290.

235 The Master of the Sentences (d 1160/4); this question arises from his treatment of matrimony in *Sententiae in IV libris distributae* 4 d2 c1, and d26 (*Spicilegium Bonaventurianum* IV, V) (Grottaferrata 1971, 1981) II 239–40, 416–21.

236 Durant of Saint-Pourcain OP (d 1334), in his *In Petri Lombardi Sententias Theologicas Commentariorum libri IIII*, 4 d26 q3 ('An matrimonium sit sacramentum') (Venice: Ex Typographia Guerraea 1571; repr Ridgewood, NJ 1964) 367–8

célèbre by melodramatic outcries. Though I have written replies about this slander to Zúñiga[237] and Sancho[238] and Lee,[239] nonetheless you report this among the proofs in your argument that I am depriving the sacraments of their importance, while you ignore my repeated denials of this false accusation.

You say: 'What point is there in finding fault with the prayer of the canonical hours?'[240] To this I answer nothing but that whoever reported this to you reported a plain lie. Not even in a dream[241] has it ever occurred to me to condemn the prayers of the hours, even if sometimes, as occasion arose, I censure certain people who are self-satisfied when they have mumbled mindlessly through those prayers.

About private confession[242] I have replied quite as often,[243] and yet the same ditty is warbled at me just as if I had made no reply.

And what danger is there if the Eucharist was once received differently[244] from the way it is received now? Was no one going to know this fact unless I pointed it out? Or is it not permitted to make notes based on the monuments of the ancients describing how the rites of the church came

* * * * *

237 See *Apologiae contra Stunicam* (2) LB IX 369C, (3) LB IX 381F–382A, (4) LB IX 389E–390C.

238 Sancho Carranza de Miranda (d 1531); see CEBR I 273–4. In 1522 he had published his *Opusculum in quasdam Erasmi Roterodami Annotationes* and had been answered by Erasmus in *Apologia ad Caranzam* LB IX 429A–432E especially 429A–430C.

239 See *Responsio ad annotatones Lei* LB IX 225A–228D

240 See n217 above, item [6]; some of Erasmus' criticisms of chanting the canonical hours are reviewed by Pio in book 6 of his *XXIII libri*; see Erasmus *Apologia* 217–21 below.

241 See *Adagia* I iii 62: *Ne per somnium quidem* 'Not even in a dream' LB II 135D–136A / ASD II-1 370–2 / CWE 31 286–7.

242 See n217 above, item [7]; Pio devoted book 19 of his *XXIII libri* to Erasmus' views on confession; see *Apologia* 337–41 below.

243 For Erasmus' replies, see n326 below.

244 See n217 above, item [8]; Pio criticized Erasmus for pointing out 'that the Eucharist was customarily received by each in his own home.' Erasmus had commented in 'A Defence of Jerome's Books against Jovinian' in his edition of Jerome: 'We conclude from this passage that of old anyone who wished customarily received the Eucharist at his home' (HO 3 51ʳB). Erasmus does not address this specific issue (Pio took it up again in book 6 of his *XXIII libri*, and Erasmus replied to it then; see *Apologia* 222–3 below and the accompanying notes), but he speaks here in vague terms, *aliter sumebatur quam nunc sumitur*, which might include this practice and other features of the reception of the Eucharist. Erasmus' most famous statement on the reception of the Eucharist under both species is in Ep 1039:125–35 (1519; published in *Epistulae ad diversos* in 1521), where he seems to favour such reception as the form instituted by Christ; this passage was cited by the Spanish monks (see *Apologia adversus monachos* LB IX 1066A–E). Erasmus' views on the Eucharist are presented in a well-documented account by John B. Payne *Erasmus: His Theology of the Sacraments* (Richmond, Va 1970) 126–54.

gradually into their present form? Among the ancient Christians the synaxis was celebrated more than once [a day],[245] and it is probable that it was celebrated without bread being consecrated into the Body of the Lord.[246] To know these matters and the like leads to an understanding of the ancient orthodox. Nowhere, not by one syllable, have I criticized the rites of the contemporary church.

Obviously at this point it became necessary to increase the number of slanders, for you made an addition to the earlier version of your book, the one you had sent me. It could be that someone like Béda interpolated this into your book without your knowledge, for you say: 'How erroneous also it is to praise marriage so much that you value it more highly than celibacy and continence!'[247] What impudent liar supplied this to your honest mind? I praise continence in countless passages, but short of abusing matrimony. But I am not surprised that you have encountered none of these passages, since you do not read my works.

If it seemed to you a good idea to recite to me a list of charges like this, it would have been simpler to transcribe here all the chapter headings of Zúñiga.[248] Surely if you had time to read his slanders, you should also

* * * * *

245 *Synaxis* is a generic term for a liturgical assembly. It can refer explicitly to the eucharistic assembly (see C. Miller's note to *Moria* ASD IV-3 146:398), and this is Erasmus' usual practice; see eg *Christiani hominis institutum* 44: 'Mysticus ille cibus (Graeci dixere *synaxin*)' (LB V 1358E / CWE 85 98–9), and his explanation of this usage in *Modus orandi* LB V 1102C–D. But *synaxis* in Latin and Greek is also used to refer to assemblies for the liturgy of the Hours; see A. Blaise *Dictionnaire latin-français des auteurs chrétiens* (Turnhout 1954) sv; and Geoffrey William Hugo Lampe *A Patristic Greek Lexicon* (Oxford 1961) sv E – hence, perhaps, 'more than once a day.' Erasmus seems to be using the term here to refer to agape meals; Pio later insisted upon a distinction between Eucharist and agape (see Erasmus *Apologia* 223 and n690 below, and also *Apologia* n159 below).

246 See Erasmus' annotation of 1527 on 1 Cor 10:16 LB VI 711E / Reeve-Screech 488: 'For the early Christians regarded all bread as sacred because of the memory of the Lord's Supper, and when they shared a cup of wine among themselves, they gave thanks to the Lord, even though the sacramental consecration did not always take place.' In the 1535 edition this comment was accompanied by the marginal note: 'Of old the synaxis was sometimes celebrated without the consecration of the Body and Blood.' Erasmus had written in a letter to Cuthbert Tunstall in 1530: 'Next it is clear that there existed in the time of the apostles a synaxis which lay people held among themselves, with prayer and blessing, and it is likely that they used to call that bread the body of the Lord, as often in Scripture also the same term is applied to the sign and the thing signified' (Allen Ep 2263:72–6).

247 See n217 above, item [9]; Pio devoted book 17 of his *XXIII libri* to Erasmus' views on virginity and celibacy; see Erasmus *Apologia* 319–33 below.

248 That is, in his *Erasmi Roterodami blasphemiae et impietates* (Rome: Antonius Bladius 1522); see the description of the arrangement of this work in Rummel *Catholic Critics* II 166–7.

have examined my reply. You admit that it had the salutary effect of making my words less damaging, but it also had the salutary effect of making it clear that those censures were sheer fabrications!

As to their slander against the *Declamation on Matrimony*,[249] I have already answered many, even to the point of nausea. But this is their practice: they ignore my words and just keep on repeating their own, and thus, of course, they will prevail. But I fear that by their outcries and malice they will bring it about that this general calamity will eventually travel there. Then they will stop their rampage. But I hope that the kindness of Christ will bring these disordered revolts to a happier outcome for us, as once Minerva is believed to have done for the Athenians.[250] Surely a temperate man like you ought either to refrain from repeating such terrible charges or to prove them with sounder arguments.

The report that I censure the chants and rituals of the church[251] is also false. Occasionally I do criticize the style of chants, or reliance on them, or some other abuse occurring in this area; but there is nothing wrong with this, since you do the same thing.[252]

It is an equally empty charge that I am critical of monks, and it is an even emptier charge that I criticize their calling;[253] unless, perhaps, you

* * * * *

In the prefatory letter to the work (Aii^{r–v}, reproduced by Erasmus in his reply, LB IX 372A–D), Zúñiga virtually invited such a use of his work: 'that if there shall be any who might want to refute his [Erasmus'] mad and rash assertions, the blasphemies and impieties with which he seems to be stuffed, they may have ready at hand what are particularly to be refuted in him, as a result of my printing them out.'

249 His *Encomium matrimonii* (Basel 1518). See Erasmus' reply to the criticisms of Jan Briart of Ath in his *Apologia pro declaratione de laude matrimonii* (Basel 1519) reprinted in LB IX 105E–112A; and the *Appendix de scriptis Clichtovei* (1526) in LB IX 811E–814D, which is Erasmus' reply to Clichtove's attack on *Encomium matrimonii* in the 1526 *Propugnaculum ecclesiae*; note Erasmus' later (1532) defence of the *Encomium matrimonii* against the attacks of Clichtove in *Dilutio eorum quae Iodocus Clithoveus scripsit adversus Declarationem Des. Erasmi Roterodami suasoriam matrimonii* ed Emile V. Telle, De Petrarque à Descartes 15 (Paris 1968).

250 See *Adagia* I viii 44: *Atheniensium inconsulta temeritas* 'Thoughtless and headstrong like Athenians' LB II 314B–315A / CWE 33 147–8; Erasmus says of this proverb that 'it is suitable for those whose designs are not well thought out, but turn out well.'

251 See n217 above, item [10]; book 6 of Pio's *XXIII libri* reviews Erasmus' statements about church music and ritual; see Erasmus *Apologia* 209–21 below; *Apologiae contra Stunicam* (4) LB IX 390C–E; and *Declarationes ad censuras Lutetiae vulgatas* LB IX 889A–892B, 897E–902C, 936B–C.

252 See n144 above.

253 See n217 above, item [11]; book 5 of Pio's *XXIII libri* reviews Erasmus' views on religious life; see *Apologia* 192–209 below.

mean that the monks are not the only ones being admonished. I think there is a big difference between a buffoon and a monk.

I am not sure where I attributed their calling to weakness,[254] and maybe this was suggested for argument's sake under another persona. The opinion is surely a likely one, for the one who cannot be corrupted by wealth seems stronger than the one who casts it away lest he be corrupted. The one who is locked up so as not to be corrupted by association with men seems weaker than the one who lives among good men and bad and still maintains his integrity, as we see the apostles did. But what sacrilege would there be if I had said that the rules of monks are due to weakness, since the same can be said without impiety about the Law of Moses, which Paul[255] calls a pedagogue? But a pedagogue is provided, not for adults, but for children and for the weak.[256]

Do you see, illustrious sir, how these very unconvincing arguments do not measure up to the terrible charges set out at the beginning of your case? And nevertheless, all the while you are winning favour with bishops, cardinals, priests, theologians, and monks!

With the same kind of truthfulness you write that I loathe ecclesiastical decrees and human ordinances,[257] whereas I frequently express my abhorrence for those who thoughtlessly disdain the customs and regulations of the ancients. I know that you are repeating the slanders of others, but they do not deserve to be reported by your pen, believe me. For there is a danger that some may perhaps come forward who will suspect that this heap of wanton lies has originated with the one by whom they are being repeated.

* * * * *

254 See n217 above, item [12]; Erasmus, in fact, reports this as Jerome's view in the *antidotus* to the *Adversus Vigilantium*: 'Moreover, as to his writing that the monastic life has to do with weakness, as if it did not have to do with perfection, he applied this to his own person for the sake of modesty. In fact, one who flees every occasion of sin, and does not reckon himself to be good unless he is not allowed to do evil, seems rather to be acknowledging his frailty, and to lack confidence in his own strength' (HO 3 58ʳ). This argument, which was also used in the fifteenth century by Erasmus' hero Lorenzo Valla, has medieval antecedents; see Mario Fois *Il pensiero cristiano di Lorenzo Valla nel quadro storico-culturale del suo ambiente* Analecta Gregoriana 174 (Rome 1969) 287–8.

255 Gal 3:24

256 For Erasmus' evasive response to the Sorbonne's complaint that he seems to be claiming that the laws and ceremonies of the church are pedagogic for beginners but that they do not apply to those strong in spirit, see *Declarationes ad censuras Lutetiae vulgatas* LB IX 952B–D.

257 See n217 above, item [13].

It is beneficial for the people to be admonished not to disdain the just, fair, and devout ordinances of princes and bishops; but it is necessary that they be admonished to pay heed first to God. Even if there are no unfair episcopal ordinances, no superfluous ones, none that look to revenue rather than to piety, even so the admonition that no such ordinances be enacted is a salutary one. As it is, the world is full of complaints along these lines. The authority of the bishops must be provided for, but the care of the flock must not be neglected in the meantime, for the people do not exist on account of the bishop, but the bishop on account of the people. But far from anyone having learned through me to oppose his bishop, I even teach that evil bishops should be obeyed, so long as their commands do not make people wicked.[258]

Are the dignity of bishops and the prestige of monks so wobbly that they are going to collapse if anyone reminds those men of their duty in a general way? But I, in fact, do not so much admonish them as point out what is in the monuments of the ancients. Are a large number of Jerome's books, then, to be burned, are a great many passages to be erased from Gratian's *Decretum*,[259] is Bernard's *De consideratione* to [Pope] Eugenius[260] to be done away with on this account?

In addition, you list these items not only starkly out of context, but also differently from the way they are found in my works, without even giving the citations of the passages whereby the slanders could be refuted. Oh, in your prudence you wanted to modify statements that could scandalize the weak! It is not the weak, glorious sir, who are scandalized by my writings, but rather certain wicked men who are partisans of the world, whose least concern is for true religion; rather, one is anxious for his authority, another for his renown, another for wealth, another for ease and luxury, another for all these at once.

But granting I had let slip certain things that it would have been better not to have said, what good does this complaint of yours do? That you

*****

258 This is the way Erasmus interpreted Matt 23:1; see his annotation (LB VI 117E / Reeve 91) and *Methodus* LB V 113E / Holborn 253. Erasmus wondered at Hutten's position that all bishops who failed to carry out their office properly were to be deposed; see *Spongia* LB X 1654F, 1657C–D, 1658B / ASD IX-1 174:196–7, 178:329–40, 180:375–8 / Klawiter 202, 208–9, 210.

259 Gratian (d c 1179) compiled a vast collection of patristic, conciliar, and papal statements which he arranged thematically and sought to harmonize in his commentary. This collection, the *Concordantia discordantium canonum*, usually known as the *Decretum*, was the basic source-text for canon law.

260 Written for his sometime disciple Pope Eugenius III (1145–53); Erasmus mentions this work in Ep 1202:18–19, and quotes from it in Ep 1609:102–5. See nn 140 and 199 above.

challenge me to defend them? I would prefer that my case there were less strong, if I am to be forced to give a defence here. No, the affairs of man have festered quite enough without a fingernail irritating this abscess.[261] What is less beneficial than to denounce these things that were better put to rest? I have removed quite a few things such as these from my books in later editions,[262] not because they were indefensible, but because the controversy was not worthwhile. And you act as if this has not taken place.

What foul favour Zúñiga obtained from the bishops of Germany by collecting and publishing those passages by means of which he was attempting to incite the whole human race against me! Why? Because those of Luther's party embraced his book, calling it a 'florilegium,' but sensible bishops would have preferred that those passages remain hidden in my writings rather than that they be excerpted and put on display for all. Accordingly, one and then another of the leading figures wrote to me from the Diet of Nürnberg asking that I make no reply at all to Zúñiga.[263] It seems to me to have happened by some evil fate that no enemy has harmed me more severely than certain friends who support me ineptly and achieve nothing but to increase the hostility of my enemies against me. But throughout my plea I am repeatedly compelled to weaken my own case, as I look out for the interests of the common peace.

Next, I do not see the point of your bringing up the business of the church's infancy and the stages by which it has grown up to this point,[264] for it was never my intention to recall the rites of the church to that infancy. It is always necessary to combat a decline in morals, so that they do not deteriorate too much from that original simplicity. And as the stages of a life span go, so, perhaps, there is a sort of old age of a doddering church. But there is nothing better for Christians than to be always in the state of becoming childlike again and always striving after their original sincerity.

* * * * *

261 *Adagia* I vi 79: *Tangere ulcus* 'To touch on a sore spot' LB II 252E–253A / CWE 32 53: 'To rouse or touch on a sore spot is to arouse pain, and to mention something which may hurt us very much.'

262 For Erasmus' revisions in his scriptural works, see Bentley *Humanists and Holy Writ* 203–5, and Erasmus' earlier statement at 39 above; see also Ep 1415:95–8.

263 We know that Zúñiga's *Blasphemiae et impietates* had reached Germany by June 1522 (see Epp 1289:6–16, 1290:1–13, 1291:24–43). Wolfgang Capito reports in a letter to Erasmus from Nürnberg, 5 June 1522 (Ep 1290:44–5): 'My prince's advice [that is, Archbishop Albert of Mainz] is that you should not reply to Zúñiga, and I urge the same thing'; see James M. Kittelson *Wolfgang Capito: From Humanist to Reformer* Studies in Medieval and Reformation Thought 17 (Leiden 1975) 86.

264 Pio *Responsio paraenetica* 7ᵛF: 'for our religion, like other things, had in many respects its infancy and childhood' etc. See also Introduction lxxxiii.

Who ever heard from me that the Roman pontiff should be put in his place, or that cardinals should be eliminated, or that bishops should be made equal, both among themselves, and with presbyters, or that the church should be despoiled of its wealth, or that matrimony should be removed from the number of the sacraments, or that confession should be abolished, or that the Eucharist should be received at home, or that chants and rituals should be eliminated from the church, or that monastic life for men and women should be overthrown, or that the regulations of the Catholic church should be rejected? But if I have neither said nor written any of these, why are you bringing such odious charges against me?

Even so, there are many things that have crept into human life in these areas, the correction of which would be in the interest of Christian piety. I yield that sphere of endeavour to the leaders of the church, but I think I might give an admonition here and there, especially when Sacred Scripture provides the occasion. Again and again observe, most illustrious prince, what a Camarina[265] you would be stirring here if you were dealing with a man to whom the peace of Christendom was not more dear than his own reputation. 'But,' you will say, 'Luther teaches these things.' So he teaches them! He gathered no seeds of this sort from my gardens, unless you are likewise going to put the Gospel on trial and the apostolic Epistles and the prophets, from which he has collected more than from my works.

I did not think it would be brought up as a charge against me[266] that once, I think, following Jerome,[267] I used the word *instrumentum* instead of *testamentum*,[268] whereas as a general rule, in countless passages, I use the term *testamentum*. If I had done this, as you say, for the sake of ostentation, what is ostentatious about the word *instrumentum*? Nor do I know for sure why Jerome sometimes preferred *instrumentum* to *testamentum*, unless because the term *instrumentum* more clearly refers to books, whereas *testamentum* has to do with wills and contracts. St Augustine[269] indicates this distinction clearly, and he is not at all averse to the term

* * * * *

265 See n82 above.
266 Pio *Responsio paraenetica* 7ᵛF: 'What point is there in altering approved and customary ecclesiastical terms except to show off, for example, to say *instrumentum* instead of a *testamentum* of Scripture, even if you have precedents for doing so?'
267 Allen (VII 140 n523) cites *In Ecclesiasten* 11.2 CSEL 72 345; see also Jerome Ep 55.1.2 CSEL 54 486 and Ep 108.26.2 CSEL 55 344.
268 Erasmus defended his use of *instrumentum* in Allen Ep 1858:519–36: (23 August 1527), where he says that the substitution occurs twice, at most, in his works. *Novum instrumentum* was, of course, the title under which Erasmus published the 1516 edition of the New Testament, changing *instrumentum* to *testamentum* in subsequent editions.
269 See *Contra duas epistulas Pelagianorum* 3.4.12 CSEL 55 498.

instrumentum, just as Jerome was not. Moreover, I think I altered this very expression. But if it had remained, just as it remains in the books of others, what crime was that? Or were revolts going to break out for this reason also?

But an even greater outrage is that I used the expression *evangelii fabula* [the gospel story].[270] A curse on these devils who bring such lies to your ears, which deserve to be occupied with better tales. Who ever read *evangelii fabula* in my writings?[271] What display of erudition is it, after all, if I used the expression *fabula* instead of 'the carrying on of an action'? The *emphasis*[272] of the word was my goal, not its novelty. If I had had available another word that would equally well have indicated what I had in mind, I would have avoided the noun *fabula*. The Gospel does not avoid the verb *fabulari* [to talk],[273] and yet the entire church totters because somewhere I use the noun *fabula*? Jerome is not afraid to call the history of Samson and Delilah a *fabula*.[274] In fact, the word *fabula* is far more flexible, as it corresponds to a Greek word, which is *drama* [δρᾶμα], that is, the carrying on of an action performed by different characters, each one playing his part, just

* * * * *

270 Pio *Responsio paraenetica* 7ᵛF: 'What is the point of employing in sacred matters words altogether inappropriate and wholly profane, as when you say *fabula evangelii*? What an inane, profitless, and harmful display of this kind of erudition!'

271 Erasmus had used the expression 'totius Christi fabulae' in *Methodus* LB V 91C / Holborn 209, but had been attacked by Béda for his use of it in his paraphrase on Mark 11:7, where he says of the details of the entry into Jerusalem on Palm Sunday: 'The apostles played a supporting role in this act [*fabula*], even though they did not understand its meaning' (LB VII 243B / CWE 49 136). Erasmus' use of *fabula* and his exchange with Béda are discussed by Peter G. Bietenholz *Historia and Fabula: Myths and Legends in Historical Thought from Antiquity to the Modern Age* (Leiden 1994) 147–8; Georges Chantraine *'Mystère' et 'Philosophie de Christ' selon Erasme* Bibliothèque de la Faculté de Philosophie et Lettres de Namur 49 (Gembloux 1971) 274–95 discusses Erasmus' use of the expression 'Christi fabula.' In 1512 a heretic named Herman Ruissvich was burned to death in The Hague after being found guilty by an inquisitor of returning to various erroneous opinions and scandalous actions he had abjured in 1499, among which was that 'he traduced the Christian Faith as a *fabula*, and similarly the books of the Old Testament and Gospel of the New Law'; see Duplessy *Collectio judiciorum* 1-2 342. By using the term *fabula* Erasmus was opening himself to misunderstanding.

272 *Emphasis* is a technical term frequently brought forward by Erasmus in his apologias to defend his choice of words. Erasmus explains the term in a note on Jerome HO 1 169ʳB: 'when there is in the words a certain unspoken meaning it is called "emphasis." As if, should one say "You believe a Cretan?" he does not merely mean a man born in Crete, but a liar, in keeping with the character of the people.' In *Methodus* he provides a series of examples of *emphasis* from Scripture (LB V 975C–E / Holborn 273.23–274.23).

273 Luke 24:15: *Et factum est dum fabularentur*; Erasmus cites this text in each of his explanations of *fabula* cited above.

274 *Commentarius in epistolam ad Philemonem* PL 26 609B; Erasmus quotes this passage from Jerome in his *Supputatio* against Béda LB IX 563E.

as when Philemon is said to have completed the *fabula* of his life.[275] But I have replied quite fully to the *fabula* slander, to Béda as well as to others. From this passage also, excellent sir, I conclude that you have obviously not read my works.

Next my *Folly* is cited for its share, because, as you declare, it has caused the 'perdition of many.'[276] I have never heard that a fly has perished because of my *Folly* – only that it has produced a lot of hatred for me, but that was among the theologians and pseudomonks. And yet no other book had been received with greater applause, I do not mean only from the young, but also from princes,[277] and from the leading figures of the church. Indeed, I learned from the letters of friends that Leo x read this work through and liked it.[278] At any rate, I did not write on this theme from my own inclination, nor did I think it should be published. It was a concession, such as it was, to the wishes of my friends. Moreover, as to your view that it is important to my reputation that this work be wholly erased from human memory,[279] we are surely on the same side in this. That had already almost taken place, for the silence about *Folly* was so great[280] that I myself forgot that it had been written. But your recalling it revives that memory and does not allow it to escape the minds of men. Its publication

* * * * *

275 See *Adagia* I ii 35: *Supremum fabulae actum addere* 'To add a last act to the play' LB II 83C–E / ASD II-1 249–50 / CWE 31 177, where Erasmus quotes from Apuleius' *Florida* 16.17 the anecdote of the death of the comic poet Philemon.

276 Pio's critique of the *Moria* is in *Responsio paraenetica* 7ᵛ–8ᵛG; Pio remarks parenthetically in 7ᵛG: '(I say this with sorrow because of the perdition of many.)'

277 See Erasmus' account of the *Moria* in the *Catalogus* Allen I 19:7–10: 'though I had such disdain for this work that I did not consider it worth publishing ... nonetheless scarcely any other was received with greater applause, especially among the nobility.'

278 Epp 673:7–9, 739:14–15, 749:16–21; but in a note which he later wrote about this claim (XXIII *libri* 55ᵛi) Pio observed: 'As to what you say about Leo, it was your friends who were applauding you. For I know that Leo approved neither your boldness nor the work itself, though he did not altogether disapprove of some of the witticisms and the keenness of your wit.'

279 Pio *Responsio paraenetica* 8ᵛG: 'none of your works has brought you so much praise as this *Moria* has taken away. Wherefore, it really should be your wish that all copies of it perish, and that the memory of it, if possible, be altogether effaced.'

280 From 1515 to 1519 Erasmus repeatedly responded to criticism of his *Moria*; see eg Epp 337:1–702 (the reply to Maarten van Dorp), 597:4–18, 622:23–33, 739:5–30, 749:4–36, 967:197–207. After a gap of three years he answered Zúñiga's attack on it (*Apologiae contra Stunicam* (2) LB IX 360C–371F; and in 1526 the attack of Cousturier (LB IX 805D–806E). For other references to criticism of the *Moria*, see Ep 1581:438–40, Allen Epp 1706:6, 2465:288–309, 2566:83–4. See also Myron P. Gilmore '*Apologiae*: Erasmus' Defenses of Folly' in *Essays on the Works of Erasmus* ed Richard L. DeMolen (New Haven, Conn 1978) 111–23; and H.J. de Jonge 'Four Unpublished Letters' (cited in n80 above) 156–60.

cannot be undone, so nothing remains but for it to be lulled to sleep by silence.

Here again I recognize my fate: *Folly* was conveyed into your hands, but its defence was not, the one in which I reply to Dorp,[281] who was the very first to complain about *Folly* and so became a celebrity.[282] But I nowhere defended it without, to tell the truth, preferring that it had not been published.[283] Yet so far there has been no more unfair critic of *Folly* than those whose judgments you repeat. Some have called it foolishness, others a dangerously outspoken book, but no one has imputed to it an impiety as great as we would scarcely expect, as you say, from Porphyry or Julian.[284]

But what are the 'baneful seeds,' what the 'lofty trees' laden with 'pestilential fruit,'[285] that you are telling me about, or what is the destruction of so many that you mention? I have heard of no one so hypercritical that he said that anyone was the least bit worse off because of *Folly*. The views of many have been set straight; the melancholy of a great many has been dispelled. Wherefore I am even more amazed that these things occurred to you, if, indeed, you are writing them on the basis of your own views, or, if you are repeating the opinions of others, that you judge them worthy to smear on your pages. For they are rife, most learned sir, with falsehoods both many and obvious. These falsehoods are not your own, to be sure, but they are published abroad by your pen to please those for whom, apparently, you are singing this tune. You are not singing to me, at any rate, either because you know that it is too late, or, as I

* * * * *

281 Ep 337 / CWE 71 7–30
282 Maarten van Dorp (1485–1525) was not a scholar who suddenly became famous for having criticized Erasmus' *Moria* in 1514. Prior to this event his lectures on Aristotle were published (1512); he also wrote introductory material for others' books (1512–14), published his own *Dialogus* and *Tomus Aululariae* (1513), proofread Erasmus' edition of Cato (summer of 1514), and was editing Agricola's *Dialectica* (published in January 1515, four months before Erasmus received a copy of his letter, Ep 304, complaining about the *Moria*). See CEBR I 398–404 especially 400, and Jozef Ijsewijn's splendid edition *Martinus Dorpius orationes IV* (Leipzig 1984).
283 Epp 337:29–32, 144–51, 739:5–9, 749:12–13, and more general statements in 1202:264–7 and Allen Ep 2136:275
284 Pio *Responsio paraenetica* 8ʳG: 'what more dreadful thing could Porphyry or Julian the Apostate have written.' Porphyry (d c 305), the Neoplatonist polymath, author of an attack *Against the Christians*; Julian the Apostate, emperor 361–3, author of a work *Against the Galileeans*. The two are linked as typical enemies of Christianity by Jerome in *In Matthaeum* 1 (re Matt 9.9; CCSL 77 55) and (with Celsus as a third) in *De viris illustribus* 7 (Ceresa-Gastaldo 58).
285 Pio *Responsio paraenetica* 7ᵛG: 'By means of this book you have sown such baneful seeds that from the field thus tainted have grown, spontaneously, lofty trees which have borne pestilential fruit ...'

think, because you are well aware that I have replied to complaints of this sort.

First, I admit that there are certain elements that were provided for no other purpose than to delight the reader with laughter. These passages are scandalous to none except certain people so grim, so difficult, and so estranged from the Graces[286] that they neither know how to joke themselves, nor tolerate others who do. Besides, no state of life is condemned there, no discipline is rejected;[287] rather, the abuses in each is made known, not by way of grim rebuke, but through inviting fun. But you seize on this charge in passing, as if it were an acknowledged fact and had no need of proof, although everyone knows that I censure neither disciplines, institutions, nor professions, but point out the abuse in each. For my main point in this discourse was to show that there is no class of men free of folly, and so it is the vices of humanity, not of individuals, that are being exposed through humour. Plato approved of this technique of correction, and for this reason sanctioned rather generous toasts at banquets.[288] And the Germans of today seem to have no other purpose than this for bringing to stately banquets some clever buffoon who is to reprimand individual guests with witty remarks.[289] All the while many hear of their own private misdeeds, but they cannot take offence, either because these are reported by a fool, or because it occurs at a feast, or because no one at all is spared. But *Folly* reprimands no one by name but me.[290]

Is someone who shows what should be avoided in any particular situation actually teaching that there is nothing serious, nothing dependable, nothing to be sought after in life?[291] On the contrary, he brings it about that the things which one should seek after are sought more successfully.

* * * * *

286 'a Gratiis alienos'; see *Adagia* II vi 18: Ἄμουσοι 'Strangers to the Muses' LB II 588D–589A / ASD II-4 30–2 / CWE 33 299–300.
287 Pio *Responsio paraenetica* 7^v–8^{r}G: 'if you had [only] censured all the arts, all human studies, all states of life, finally, all things human . . .'
288 Erasmus had used this argument in his letter to Dorp (Ep 337:101–5); ancient sources are cited in CWE 71 144n28.
289 Erasmus' statement would seem to be confirmed in the careers of the professional buffoons Claus Narr and Wigand von Theben of Kalenberg and of the mythical figure Till Eugenspiegel who were invited to banquets to provide merriment; see Enid Welsford *The Fool: His Social and Literary History* (London 1935) 42–3, 45, 143.
290 Folly mentions Erasmus by name passim, but see especially *Moria* chapter 58: 'The second place in this flock [of Greek pedants], if not the actual leadership, certainly belongs to my friend Erasmus, whom I mention by name from time to time by way of a compliment' (LB IV 491C / ASD IV-3 182:974–5 / CWE 27 144).
291 Pio *Responsio paraenetica* 8^{r}G: 'if you had [only] harmed the youth, especially, with the

Who ever gave so much weight to the jests of *Folly* that on the basis of it he convinced himself that human happiness does not exist at all, and was rendered more sluggish in the pursuit of virtue and in engaging in life's activities?[292] Either you give too much weight to trifles of this sort, or you have a less than generous opinion of human intelligence.

To what, then, illustrious sir, do these words of yours refer: 'There are many things in the life of men at which we may laugh, I admit; but he is an exceedingly unfair appraiser of human affairs who condemns everything indiscriminately'?[293] Does *Folly* criticize anything commendable? You philosophize correctly that human affairs are such a mixture of act and potency, that is, of the complete and the incomplete, that there can be nothing so pure that it does not have some admixture of corruption.[294] Since this is the case, we must struggle all the more with corruption.

You go on to say that I have spared neither the sacred rites, nor religion, nor even Christ himself.[295] Who dared to convey this heap of lies to a man as important as you? It is superstitious rituals I criticize there, not sacred ones, and I advance the cause of religion when I declare that feigned religion must be avoided. But let such hateful impiety be far from my soul, that I would utter any abuse against Christ, whose name I revere with my whole heart. I feel that I would rather meet death countless times than by any word slight the author of our salvation and his name, which is to us the source of salvation, to the angels the object of adoration, and to the demons the cause of fear. Whoever peruse my writings easily perceive, unless they are altogether blinded by hate, that my feelings towards Christ are anything but impious.

You continue: 'You would allow yourself to make sport of things divine.'[296] This is, of course, quite false, but it is on foundations of this sort that your discourse rests. This, I suppose, is what it is to speak in the Cicero-

* * * * *

opinion that there is nothing certain, nothing solid in life, nothing to be sought after for its own sake ...'

292 Pio *Responsio paraenetica* 8^{v}G: 'when we profess that nothing in life is deserving of praise, we preclude human happiness altogether, and weaken mankind's pursuit of the virtues to a great extent.'

293 Pio *Responsio paraenetica* 8^{r}G

294 Pio *Responsio paraenetica* 8^{v}G: 'and the philosophers teach that all things mortal consist of act and potency, that is, of the complete and the incomplete.'

295 Pio *Responsio paraenetica* 8^{r}G: 'But as it is, since you have spared not even the sacred rites, nor religion, nor even Christ Jesus himself ...'

296 Pio *Responsio paraenetica* 8^{r}G: 'To serious men it seems base if anyone jokes even about serious profane matters. You allow [*permittis*] yourself to make sport of things divine?'

nian manner![297] Prince Alberto, I regard you as a man possessed of Christian piety, and thus I may more frankly declare to you what I think. You will more readily find the sort who make sport of things divine in Italy, among persons of your own rank, and indeed in most esteemed Rome,[298] than among us. I tell you, I could not endure even to share a meal with such people.[299] (As always, I beg you to take no offence at my outspokenness.)

I am not particularly troubled by vulgar jeers, but forgive me if I cannot endure witticisms of this sort. What, indeed, do you mean by 'making sport of things divine'? Do they make sport of things divine who criticize the superstitious cult of the saints, who stigmatize people who, void of charity, assign the totality of religion to cult, dietary observances, and bits of prayer that they do not understand? Do you think someone who exposes the stupidity of those who translate Aristotle[300] incorrectly because they

* * * * *

297 This and the many other caustic references to Cicero which occur passim in the rest of the work can be viewed as having their remote occasion in Erasmus' antipathy towards the Italian Ciceronians (see n57 above), and their immediate occasion, perhaps, in Bade's remark (in Pio *Responsio paraenetica* Aii^r) on one of the splendours of Pio's work, 'manifold learning seasoned with Ciceronian majesty.' Florido (see Introduction xxxvi–xxxvii, xlviii, cxxiv, cxxix) tells us that it was Pio's practice to listen daily to the reading of *Ciceronianum aliquid* (Florido *Apologia* 116). Their proximate occasion may be Pio's attack on Erasmus (*Responsio paraenetica* 11^vı) for correcting by his *Paraphrases* the Holy Spirit who failed to speak eloquently enough: 'in your Paraphrases you take on the *persona* of the evangelist, the *persona* of Paul, or of the rest of the apostles who wrote anything, even, indeed, the *persona* of Christ preaching. You bring him on at times, contrary to all propriety, like Hortensius or Cicero pleading a case with studied speech and polished oratory.' Erasmus' retort is that not Erasmus but Italians typically sacrifice religion to classical rhetoric; indeed the Roman Ciceronians are cryptopagans. On the Roman humanists as cryptopagans, see, for example, Epp 1489:24–5, 1496:202–4, Allen Epp 1717:4, 1719:34–6 ('At Rome that pagan sodality of learned men for a long time now complains about me, its leaders, so they report, being Aleandro and Alberto the former prince of Carpi'). See also d'Ascia *Erasmo e l'umanesimo* 20–1; and n298 below.

298 See Erasmus' sustained attacks on the Ciceronian literary circle of Rome, eg in the following: LB I 995B–996A / ASD I-2 640–2 / CWE 28 388–9 (it confects literary alternatives for Christian terms); LB I 997E–998C / ASD I-2 644–5 / CWE 28 392–3 (it shows preference for classical wisdom over that derived from the Bible); LB I 999D–E / ASD I-2 647 / CWE 28 396 (it shows preference for the imagery of pagan mythology over biblical accounts); LB I 1020C–1021A / ASD I-2 700–2 / CWE 28 437–8 (Christian themes are treated in a pagan way); and LB I 1025C–1026B / ASD I-2 708–9 / CWE 28 447–8 (Ciceronianism is contaminated with paganism). In Allen Ep 2465:470–6 (27 March 1531) Erasmus gives a summary account of blasphemies heard by himself and others in Rome; see also n169 above.

299 Perhaps an allusion to 1 Cor 5:11

300 An *ad hominem* element, given Pio's attachment to Aristotle (see n427 below), and his hint (*Responsio paraenetica* 4^r–^vC) that Erasmus' reputation might have benefited if Erasmus had worked on Aristotle. For Erasmus' glowing praise of Aristotle's various writings, see his commendatory preface to the *Aristotelis opera* ed Simon Griener (Basel: Johann Bebel, May 1531), dated 27 February 1531 (Allen Ep 2432, especially lines 54–66).

misunderstand him is making fun of Aristotelian philosophy? You would not say so, unless I am mistaken.

You say: 'What more savage statement could be made with a view to the utter destruction of the Christian religion than that it is wholly based on folly?'[301] Perhaps there is something like that in *Folly*,[302] but it was expressed quite differently. This is something you do quite often – you make my statements more shocking by your choice of words. This is rhetoric, to be sure, but it is certainly not theology!

Paul writes: 'If anyone seems wise among you, let him become foolish that he may be wise.'[303] He also speaks of the folly of the cross,[304] and attributes folly to God,[305] and says that there are not many of the wise[306] among the disciples of Christ. Ergo, as many as are not wise are foolish. Who is so foolish that he does not understand that here the wisdom of the world is being distinguished from the wisdom of the Gospel? If this distinction is maintained, nothing more Christian could be said than this statement that you attack so ferociously.

Likewise, as to the statement that 'Christ always took delight in buffoons, old folks, and children, but rejected the grave and learned,'[307] either it is not in *Folly*, or it was expressed in different words.[308] If you forgive me the resentment these words cause, I ask you, by your good fortune, what more religious or holy statement could be made? Did not Christ reveal himself to shepherds?[309] Did he not cling to the embrace of Symeon?[310] Did he not embrace children[311] and urge the apostles to follow their example?[312] Did he not choose simple men as his apostles? Did he not, in fact, treat

* * * * *

It is worth noting that Pio's protégé Sepúlveda had been involved in translating Aristotle for some time. His translation of the *Parva naturalia*, completed in 1522, was dedicated to the future Clement VII, and the translation of the section *De incessu animalium* was dedicated 'Ad Albertum Pium, patronum meum'; see Angel Losada 'Juan Gines de Sepúlveda, traductor y commentarista de Aristoteles' *Revista de Filosophia* 7 (1948) 501–36; 8 (1949) 109–28.

301 Pio *Responsio paraenetica* 8ʳG
302 See *Moria* LB IV 499D / ASD IV-3 189:141–3 / CWE 27 149
303 1 Cor 3.18
304 1 Cor 1:18
305 1 Cor 1:21; see *Moria* LB IV 496B / ASD IV-3 186:73–5 / CWE 27 147.
306 1 Cor 1:26
307 Pio *Responsio paraenetica* 8ʳG
308 See *Moria* LB IV 497B / ASD IV-3 186:91–187:94 / CWE 27 148.
309 Luke 2:8–20
310 Luke 2:28
311 Matt 19:13–15, and parallels
312 Matt 18:2–4, and parallels

Herod with contempt?[313] Did he not inveigh with great severity against the way of life and arrogance of the scribes and Pharisees? With what high priest, with what scribe or Pharisee, did he have a close friendship, except when through the Gospel they became childlike again? Even today Christ does not accept everyone who is wise, powerful, or learned in the eyes of the world. I use the term 'world' just as Paul used it: 'God has chosen the foolish things of the world in order to confound the wise.'[314] The idea is a pious one, and since that is evident, it is unfair to hunt up a false charge based on the words.

But Folly abuses quotations from Sacred Scripture.[315] As if certain priests and monks do not appropriate almost all of their witticisms from the sacred books![316] As when they say: 'What is charity? It is the cloak of the monk. How so? Because it covers a multitude of sins!'[317] And when one was asked whether he liked wine, he said: 'It goes up higher.' Then the other said: 'It goes up higher? Then you will have glory.'[318] But Folly twists the Scriptures with far more restraint, and yet she does not do so without having first sought pardon[319] because her quotes will be malapropos [οὐδὲν πρὸς ἔπος],[320] and in the peroration of this passage she again begs pardon.[321] But I have already said that I do not wish to defend *Folly* in every respect. I shall allow *Folly* to be accused up to a point as a work of foolishness, thoughtlessness, or frivolity, but I shall defend it energetically against these charges you make.

Maybe among the Italians there are people so weak in faith that they abandon Christ because of a playful book. Among us even women are more steady than to be shaken by jests of this sort, and no one has yet come

* * * * *

313 Luke 23:9
314 1 Cor 1:27
315 This objection is not found explicitly in Pio.
316 Compare Erasmus' observation in the *Methodus* LB V 129A / Holborn 287: 'As to the contemporary practice of certain people who, anxious to appear droll, distort the mystic words to make buffoonish jokes, this is not only unlearned, but in fact impious, and deserves to be punished.'
317 See 1 Pet 4:8.
318 See Luke 14:10.
319 *nec tamen id facit nisi praefata veniam*; compare *Moria: principio veniam a theologis praefatae* 'Let me begin first by asking pardon from the theologians' LB IV 488E / ASD IV-3 178:904 / CWE 27 142.
320 See *Adagia* I v 45: *Nihil ad versum* 'Nothing to do with the verse' LB II 199C–F / ASD II-1 518–20 / CWE 31 422–4; quoted in *Moria* LB IV 488B / ASD IV-3 178:890 / CWE 27 142.
321 *Moria* LB IV 496A / ASD IV-3 186:64–6 / CWE 27 142

forward in these regions with the view that these jokes proceeded from impiety of soul, as you write.[322]

For many years *Folly* enjoyed a brisk and wide circulation. It was even translated into French.[323] It was a favourite in princely palaces. And no one was raging against it, with the exception of a few pseudomonks who were handled differently from the way they wished.[324] Even among the monks all the truly religious liked the very *Folly* to which you are so violently hostile. Yes, I have frequently been alerted by letters from Rome[325] that Prince Alberto is at his most eloquent when censuring *Folly*. If this is the result of some ardent zeal for religion, I can only approve of your strong feeling, though I think that your verdict will please no one save those who would dislike the Gospel if it were published under the name of Erasmus. In the case of people like this hatred and prejudice have wholly distorted every opinion.

Again and again I am astonished that you pile up against me so many and such hateful charges in so fierce and virtually partisan a manner, just as if I have not rebutted any of them in books I have already published. Another example of this is your amazing exaggeration of the complaint about confession,[326] whereas there is no point here that I have not blunted in published writing. In any event, if it was your desire to be considered articulate and forceful, you should first have refuted those works whereby I repelled

* * * * *

322 Pio *Responsio paraenetica* 8ᵛG: 'it is no wonder that you are accused of having acted thoughtlessly by the many who take this as having proceeded from the impiety of your soul.'

323 Joris van Halewijn (c 1470–1536/7) translated the *Moria* into French by 1517 and a version (that by Halewijn?) was published on 2 August 1520 in Paris by the press of Pierre Vidoué under the title *De la declamation de louenges de follie*; see vander Haeghen 123; Epp 598:16–18, 641:6–9, 739:5–10, 1013A 7–10; and the study of C. Matheeussen 'La traduction française de l'*Eloge de Folie* par Georges d'Halluin et la traduction anonyme parisienne de 1520' *Humanistica Lovaniensia* 28 (1979) 187–98.

324 See Erasmus' comment in *Catalogus* 19:10–11: 'Its frankness offended only a few monks, and they the worst sort, and some of the more picky theologians.'

325 The surviving letters to Erasmus from Rome do not mention Pio as a critic particularly of the *Moria*. Perhaps to protect the identify of his informants in Rome, Erasmus destroyed some of their letters. Among his important informants were Pierre Barbier for c 1522–4 (CEBR I 93–4), Paolo Bombace for c 1513–27 (CEBR I 163–5), Haio Herman for 1522 (CEBR II 157–8), and Jakob Ziegler for 1522–5 (CEBR III 474–6). Erasmus informed More in 1527 that Pio was especially hostile to the *Moria*; see Allen Ep 1804:252. By this date he had already received from Pio in Rome a copy of Pio's *Responsio paraenetica*; see Allen Ep 1744:130 (September 1526).

326 Pio's complaint about Erasmus' treatment of confession covers *Responsio paraenetica* 8ᵛ–10ʳH; book 19 of Pio's *XXIII libri* reviews Erasmus' statements about confession; see *Apologia* 337–41 below.

the slander of certain people, unless, perhaps, you do not think anyone will read this work of yours except those who either have not encountered mine, or who condemn things blessed and cursed alike with the same censure. After I have answered Lee at great length, and replied carefully to Zúñiga, and quite a few others about confession,[327] is it really fair, I ask, to sing again to me that old complaint from the carousals and harangues of the pseudomonks just as if I had never said anything about the matter?

But I find your fairness even more wanting when you throw up to me, not what I have written, but what is commonly noised about by certain troublemakers. You say I bring together the advantages and disadvantages of confession, and leave the reader virtually free to form his opinion on 'whether the practice of confession should be retained or abolished.'[328]

First, see how falsely this was reported to you. I assert that the disadvantages arise not from confession itself but from the corrupt practices of men. My words are as follows: 'That as much advantage as possible be derived from confession, and that as few abuses as possible be allowed, we observe that these abuses arise not so much from the thing itself as from its corruption by men, even as there is almost nothing in human affairs so devout, so, if I may say so, heavenly, which the debased morals of men do not turn to their own destruction';[329] and shortly after: 'We show briefly what great benefits confession provides, and next how great a bane to true piety can arise from it sometimes among men.'[330] You hear me plainly ascribing the advantages to confession and the disadvantages to the corrupt practices of those who employ a good thing badly. And right at the very beginning of the work I say the following: 'whether it be more expedient that it be left as it is because of the countless advantages that we see proceeding from it, or be eliminated because of the countless evils that we find in existence through the occasion provided by confession due to the fault of penitents and confessors.'[331] You hear 'occasion,' not 'cause,' you hear of the fault of men, not a charge against confession. You ignore the fact that these things were said repeatedly in the preface and quoted again

* * * * *

327 For Erasmus' replies to his critics, see the following: to Lee, LB IX 255A–262E; to Zúñiga, LB IX 389B–D; and to many others, for example, *Exomologesis* LB V 145A–170D, Epp 1153:78–86 (letter to Godschalk Rosemondt), 1299:66–74 (letter to Josse Laurens), 1300:15–31 (letter to Jerome vander Noot); Allen Ep 2136:214–20 (letter to Ludwig Baer); and CWE 71 118–24 (a previously unpublished reply to Taxander).

328 Pio *Responsio paraenetica* 9ʳH

329 From *Exomologesis* LB V 146B

330 From *Exomologesis* LB V 147B; *ostendimus* in 1529 and LB; *ostendemus* in text of *Exomologesis* in LB

331 From *Exomologesis* LB V 145B

in the apologies, and you describe me as reporting confession's advantages and disadvantages. You do not appear to have learned these things, Alberto, from those who, as you write, examine my writings minutely, comparing statement to statement, meaning to meaning, and syllable to syllable.[332]

Equally silly and slanderous is your writing that I leave the reader a free choice as to whether confession is to be retained or abolished. I ask you, in what words did I show that this was my idea? I am speaking there of the authority of the Roman pontiff and of the consensus of the Christian people,[333] that is, of the authority of the entire church. If confession has been enjoined by the church, can it not be abolished by the authority of the same church? Or do you think that any reader is the whole church?

Nor, for that matter, do I assert there that confession was established by men. Rather, I argue on the supposition that this has been granted, with the following said by way of preamble: 'Then, if anyone grant that it was instituted by men . . .'[334] Is the Dutchman not being quite clear enough here, even if not sufficiently eloquent?[335]

Nor am I disputing here whether confession, as we now practise it, is of human origin or was instituted by Christ,[336] or whether it should be retained or abolished. Rather, I point out that questions of this sort are being

* * * * *

332 An allusion to Pio's earlier description of Erasmus' critics; see n129 above.
333 This recalls *Exomologesis* LB V 145B: *auctoritas Romani Pontificis, et populi Christiani consensus*.
334 From *Exomologesis* LB V 145B
335 Erasmus is being sarcastic in his use of *Batavus* (= Dutchman) here and in 103 below (*Batava epistola*); see n504. He used the term in a posture of humility in a letter of 1519 to Cardinal Campeggio (Ep 996:45–7): 'About my gifts (but I am a true Dutch dumpling) and what I have written, you are wrong, of course, wrong in the kindness of your heart, and I am grateful.' But in Roman circles it was used as a term of abuse against him: by Zúñiga in 1520 (see Erasmus' comments on this in *Apologiae contra Stunicam* [1] LB IX 286C, 287A / ASD IX-2 66:121, 68:152–7); by Casali in 1522 (*Invectiva in Erasmum Roterodamum* ed Monfasani [cited in n126] 45: 'that Dutch ear of yours [*auris illa tua Bataua*] is exceedingly delicate'); by Christophe de Longueil in 1526 (Allen Ep 1706:5–7: 'you have invented a new style of speech in far Holland [*in extrema batavia*]); and by Benedetto Tagliacarne (Allen Ep 1791:57–8 [1527]). Erasmus is alluding here to the proverbial rustic speech of the Dutch, as he does also in comments on *Auris Batava, Adagia* IV vi 35 LB II 1083F–1084E, where he also notes that the Dutch prize integrity of life over brilliance of erudition. See Ari Wesseling 'Are the Dutch Uncivilized? Erasmus on the Batavians and His National Identity' *Erasmus of Rotterdam Society Yearbook* 13 (1993) 68–102, especially 73 (where this particular use of the term is treated) and 89–102 (where the adage *Auris Batava* is reproduced, translated, and annotated).
336 In 1526 Erasmus raised the question of by what authority confession was instituted, and said that he suspended his own judgment on the question, but viewed confession as instituted by Christ; see *Detectio prestigiarum* LB X 1570A–B / ASD IX-1 258:587–91.

raised in other circles. You will say: 'What good did it do to bring up this matter?' – as if anyone were in the dark as to what I am talking about! And yet it was important for me to warn the reader not to look for something beyond what I had undertaken.[337] I knew quite well that a great many would be looking for this. But I omitted it because this matter could not be handled except in a large book, and because it had been treated in detail by many, for scarcely any question is more noised about among the scholastics.[338] I discussed what would be supportive of each side, that is, both those who doubt and those who are firmly convinced.

You say: 'You ought, at least in passing, to have disclosed your own opinion.'[339] I candidly declare my conviction: I term it a very salutary and particularly devout practice. Even when disputing against Lee,[340] I declare that it is an excellent thing, which proceeded from the spirit of Christ and has been approved by the long-standing agreement of the whole world. I declare that it should be maintained with the same reverence as if it had been instituted by Christ himself, and I mean not any form of confession, but that which is now in use among Christians, with all of its accidentals. If I could prove that it had been instituted by Christ, how very gladly would I do it!

You will have me to cheer you on if you can make a definitive pronouncement on something that has been discussed by the theologians of old and recently by especially learned men, one of whom is Jacques Masson.[341] But none of these has succeeded as I would have wished. Indeed, Masson leaves us virtually a free choice to believe that confession either was instituted by Christ, or has been derived from compelling passages of Scripture, or has been established by a universal determination of the church. And you are demanding of me that I pronounce in favour of its having been instituted by Christ!

Imagine that I have made such a pronouncement: who would believe it unless I prove it? And I consider silence a safer course than to under-

* * * * *

337 *Exomologesis* LB V 145B: 'You will look for none of these, O reader, in this book.'
338 For an overview of the debates among scholastic theologians and reformers on these issues, see Thomas N. Tentler *Sin and Confession on the Eve of the Reformation* (Princeton 1977) 57–70, 352–63.
339 Pio *Responsio paraenetica* 9ʳH
340 LB IX 255C
341 The *De confessione secreta* of Masson (Jacobus Latomus) together with his treatises on the church, the obligation of human law, and other controverted questions were published as a single volume in Antwerp by Michaël Hillen in 1525. Although the treatise on private confession was ostensibly aimed at refuting the opinions of Johannes Oecolampadius, Erasmus' *Exomologesis* was also its target; see Rummel *Catholic Critics* II especially 10–12.

take this task with trifling sophistries.[342] One who keeps silence can appear capable of having prevailed had he wished to do so. He who attempts to establish a point, and does not succeed in the attempt, is more persuasive against the matter he wishes to prove than the one who openly argues against it. Note, too, in what era and in what region[343] I had to publish that book. Even so, orthodox men thanked me for it, such as it was.

You say: 'What is being very obstinately attacked by heretics cannot be left in doubt.'[344] So, what heretics maintain is always something impious! But the more it was being attacked, the more just was my plea for declining the task. That job demands a man of leisure, a stout, learned, famous theologian. I would have weakened the cause of confession.

Finally, how do you dare bid me to convince others when I am not myself sufficiently convinced of it, and am still waiting for someone to convince me? Does a man who wants proof that this type of confession was instituted by the Lord himself, but at the same time asserts that it should be devoutly maintained, seem to be in agreement with the heretics? What council has declared this? The celebrated king of England, who has handled this subject better than you seem to, declared no more than that it is probable that confession was instituted by Christ.[345]

* * * * *

342 *Exomologesis* LB V 147A–B: 'In this book I neither agree nor contend with those who maintain that it [confession] was instituted by mere men; I am, rather, more inclined towards that party which believes that it was instituted by Christ, and I will gladly fight for this position when I am fitted out with the appropriate arms of Scripture and argument, lest I make the case weaker if my effort is met with failure. "He sits still who fears that he may not succeed" [Horace *Epistles* 1.17.37]. For it is better to leave the case to others, untouched, than to bring it into greater peril by handling it badly. I leave to others what requires an extraordinary warrior, and have taken for myself what is suitable to someone from the ranks, namely to point out some matters which will bring it about that confession will be more beneficial.'

343 Erasmus resided in Basel and published through the Froben press located there. This city was then going through a turbulent religious revolution with a legalized violent suppression of Catholicism; see Rudolf Wackernagel *Geschichte der Stadt Basel* III (Basel 1924) 490–511.

344 Pio *Responsio paraenetica* 9ʳH: 'you leave this [that is, whether confession was instituted by Christ] in doubt and, accordingly, whether it should be numbered among the true sacraments. This is wholly inappropriate, particularly at this time when this is being very obstinately attacked by heretics.'

345 In his work on the sacraments, *Assertio Septem Sacramentorum or Defense of the Seven Sacraments* trans Thomas Webster, ed Louis O'Donovan (New York 1908) 332–3, Henry VIII treated the divine institution of the sacrament of reconciliation very briefly, stating only that: 'For my Part, let Luther say what he will, I will believe that Confession was instituted and is preserved by God himself; not by any Custom of the People, or Institution of the Fathers.' In the critical edition of the *Assertio* by Pierre Fraenkel, Corpus Catholicorum 42 (Münster 1992) this text is on p 179.

But if by 'heretics' you mean the Lutherans, you know how much I disagree with them. For the majority of them make the issues of whether to confess or not, and to whom to confess, matters of individual choice. A good number of them even abominate confession as something impious. But if people who are in some doubt as to how confession was introduced are heretics, then you must take issue with the king of England, whose learned piety you rightly praise,[346] and with Masson, to say nothing of the others.

Nor do I quite understand what you mean by 'indeed, accepted as instituted by Christ.'[347] If 'as' [ut] indicates analogy, you are saying the same thing as I, for the beginning of my book reads: 'Whether introduced by the ancients or not, it gradually gathered force, so that its authority is as great as if [quasi] instituted by Christ.'[348] But if 'as' indicates the reason [for accepting it], then you ought to have corroborated this with other proofs.

You continue: 'although you say that it cannot be denied that he who has made a good confession is more secure.'[349] I do not say even this in an unqualified way, but concessively. My words are as follows: 'For however much the opposition may argue with many great proofs that this form of confession was not instituted by the Lord Jesus himself, and that so great a burden cannot have been imposed on men by a mere man, it surely cannot be denied, etc.'[350]

Thus it was my desire that the book be read by two categories of people, those who are firmly convinced that confession must not be abolished, and those who are doubtful of the possibility or benefit of any innovation by the rulers of the church in confession (which you call a sacrament, while in the opinion of many it is a sacramental), to the end that, for the time being at least, they may derive greater benefit from confession,

* * * * *

346 Pio's praise of Henry VIII and his *Assertio* is in *Responsio paraenetica* 14^{V}K. This passage, *O Regem celsissimum ... propagando depugnasse*, was included in an edition of the *Assertio* published in Rome at the press of F. Priscianensis Florentinus in 1543; a manuscript copy is preserved in Biblioteca Apostolica Vaticana (BAV), Ottob. Lat. 415, fols 38^V[really 37^V]-38^r. Erasmus' praise of the *Assertio* taken from his letter to Duke George of Saxony (Ep 1313:74–92) is found on 37^r. On 37^{r-v} is a statement of John Fisher on the *Assertio*. These tributes to Henry VIII's work are missing from a printed edition in BAV, Stamp. Barb. G.III.37, but are included in a later reprinting at Lyon by the press of Guiliel[mus] Rouillius in 1561; see J. De Reuck *Bibliotheca Erasmiana Bruxellensis* ed G. Colin and R. Hoven (Brussels 1993) number 590.
347 Pio *Responsio paraenetica* 9^{r}H
348 From *Exomologesis* LB V 145A–B
349 Pio *Responsio paraenetica* 9^{r}H
350 From *Exomologesis* LB V 145C

and not rush into a decision until the church has made a declaration about this.[351] I have barred a third class of men from reading that book, namely those who insist that confession is superfluous, or even impious. I remand those stubborn men to other teachers; at that time I had neither the leisure nor the forces for joining battle with them.

Here I am passing over the arguments and evidence[352] by which you attempt to prove to me that one is obligated to believe absolutely that this type of confession was instituted by the Lord. I shall say nothing about these, for the time being, except that I am distressed that they are not more sound, especially on account of those who stubbornly condemn confession altogether.

But you go on to deliver a grand pronouncement which condemns my comparing the advantages and disadvantages [of confession]. You say: 'For even if these things appear to us to be altogether disadvantageous, nevertheless, because they have been established' thus, 'they must be maintained sincerely, and retained steadfastly.'[353] I beg you to hear me out patiently in turn.

What is the benefit of my pointing out the disadvantages,[354] disadvantages that arise not from the matter itself, as I said, but from human failings? Is it not so that the disadvantages may be avoided, and that the things which have been established may be better grounded in the meantime? If abuses must be tolerated in the case of confession, then why is there a daily outcry against people who are unfaithful in matrimony, who approach the Lord's Table unworthily, who abuse public office? For I am not abolishing the practice of confession, but pointing out what must be avoided in its practice. Doing this, of course, strengthens confession; it does not undermine it. You can see, I think, that your remarks were as beside the point as they were elegant.

After completing your discussion of these matters, you attach a conclusion. You say: 'Whereas we are surely claiming overmuch for our own judgment when we dare equally in every case to pass judgment on what is

* * * * *

351 At the eighth session of the Council of Florence (1439) the Armenian and Latin churches together declared penance to be one of the seven sacraments; see COD 541, 548. That Erasmus may have known of the existence of this decree of the Council of Florence which lists the seven sacraments is suggested in his *Apologiae contra Stunicam* (2) LB IX 369C and implied in *Apologiae contra Stunicam* (4) LB IX 390B–C, where he mentions a decree of the Council of Florence on matrimony as a sacrament.
352 Provided by Pio in *Responsio paraenetica* 9ʳ–ᵛH
353 Pio *Responsio paraenetica* 10ʳH
354 In *Exomologesis* Erasmus lists at least nine disadvantages; see LB V 153C–156A.

LB IX 1114A

beneficial and what is not beneficial, on what is upright and what otherwise, on what is correct and what is perverted, on what is divine and what is human, and do so both in matters defined and in those about which judgment is awaited, as also when we criticize practices, laws, customs, rites, and rituals that have long been accepted, because you do these very things quite often, as your books bear witness, it is only appropriate that many accuse you of arrogance and declare that you are a radical reformer.'[355] When you say 'many declare,' you confess candidly that you are saying these things on the basis of the opinions of others, not your own. What acceptable sense can these words of yours make otherwise?

If these are your views about the Lutherans or about some people or other in Italy,[356] then they have nothing to do with me. But if you are talking about me, I do not really appear to you to pass judgment alike about everything, do I? In this very book I avoided expressing an opinion, and entrusted final judgment to others; in fact, everywhere I defer to the judgment of the church. 'In matters defined,' you say, as if you have already proved that this has been defined! Show me where I have criticized anything that is correct as being corrupt, where I have rejected as merely human something that is divine, where I have criticized pious practices, laws, and so on, as you say I do ever so often, though you cite no passage.

I think that Zúñiga's Catalogue[357] served you as an oracle, not withstanding that those who know you, by reputation at least, were expecting a far more solid, fair, and vigourous disputation.[358] Indeed, so far you have nothing to offer but what is bandied about in their drinking bouts by ignorant pseudomonks and irresponsible theologians. These men do not blush at putting forward from time to time tales taller than mountains, and their sole purpose in doing so is to destroy my writings, so that their singular foolishness may not be found out. But from you[359] what should have been expected except what is true, considered, and fair?

* * * * *

355 Pio *Responsio paraenetica* 10ʳʜ; 'radical reformer' translates Erasmus' *rerum novarum studiosum* which weakens slightly Pio's *rerum novarum studiosiorem*.

356 It is unclear to whom Erasmus is here alluding. The call for personal repentance and an insititutional reform of church and society was widespread in Italy and influenced in part by Protestant ideas. For a brief overview of this reform movement, see Elisabeth G. Gleason 'Evangelism' OER II 82–3.

357 See n248 above.

358 To this and like criticisms of Pio's work at 83 and 102 below, compare Erasmus' judgment in Allen Ep 2118:28–31 (10 March 1529): 'Perhaps I shall appear a less than fair judge; nevertheless this book is a far cry from the intelligence and erudition that I was expecting from Alberto.'

359 *A te vero* 1529 *At vero* LB

A page from a 1526 draft version of Pio's *Responsio paraenetica*,
in the hand of his secretary but with Pio's autograph insertions.
Milan, Biblioteca Ambrosiana, Archivio Falcò Pio di Savoia, 1 sezione,
parte 1 [Famiglie e persone], scatola 282 6, folio 7[r]
A fair copy of this version was apparently sent to Erasmus, whose complaint
that the version published in Paris in 1529 had added new material is supported
by a comparison of the texts; for example, the Paris edition of 1529 has added
at this point proof-texts not present on this page of the draft.
By permission of the Veneranda Biblioteca Ambrosiana, Milan

Next you enter upon another topic.[360] Although I made more than adequate reply on this point to Pierre Cousturier,[361] not to mention the others, nonetheless you handle it as if it were a wide-open and untouched subject. You will offer the excuse that my book had not yet reached you. But I am sure that you saw it in Paris, where you added many scriptural proof-texts to your work.[362] What, then, shall I do? Shall I repeat here what I said there? What could be more bothersome? Shall I keep silent? The people who have not read my works will applaud you as victor. I shall keep to the middle way, and remark on certain matters briefly.

You say that many condemn my expounding Sacred Scripture more clearly and elegantly through paraphrase, because the Scriptures scarcely admit of commentaries, let alone paraphrase.[363] Now in this connection you are, I think, sophisticated enough to know that Sacred Scripture consists above all of its mystic meaning rather than of words which are likely to have been partly altered by Ezra,[364] and secondly, that it is situated[365] in the languages in which it was handed down by its original writers. I have changed nothing in these languages. In the third place, there are translations

* * * * *

360 That is, Erasmus' paraphrases, taken up by Pio in *Responsio paraenetica* 10^r–13^r1

361 *Adversus Petri Sutoris ... debacchationem apologia* (Basel: Froben 8 August 1525), reprinted in LB IX 739A–804E, with the appendix on 805A–812D. On Cousturier, see CEBR I 352–3.

362 In Allen Ep 2118:26–7 Erasmus claimed, on the basis of a collation of the manuscript version of Pio's work sent to him earlier with the printed version, that Pio had added these scriptural proof-texts with the help of Parisian theologians. See 76 and 91 below for examples cited by Erasmus. That Pio did, in fact, insert additional proof-texts is evident from a comparison of his treatment of confession in the printed *Responsio paraenetica* 9^r–^vH with the earlier version in the Milan MS, fol 7^r. See also n376 below and *Apologia* n16 below; and Forner 218.

363 Pio *Responsio paraenetica* 10^rI: 'I will also attest to the fact that a great many disapprove of your scheme of publishing paraphrases on Sacred Scripture, not because they deny the skill and elegance of your writing, but because they disapprove of your boldness, saying that the Scriptures do not allow paraphrase, since they scarcely admit of commentaries and retain their majesty, which is amazingly diminished when you alter anything to make the language more elegant or more obvious.' An interesting parallel is to be found in Erasmus' explanation of his failure to produce a paraphrase on the Apocalypse: 'I have completed all the paraphrases on the New Testament, except the Apocalypse, which entirely refuses to admit of paraphrase and almost of commentary' (Ep 1432:29–32). For an overview of the controversy regarding Erasmus' paraphrases, see Sem Dresden ' "Paraphrase" et "commentaire" d'après Erasme et Alberto Pio' *Medioevo e umanesimo* 46 (1981) 207–24.

364 Jerome ascribes to Ezra the form of Hebrew letters currently in use (*Prologus in libro Regum, Biblia sacra iuxta vulgatam versionem*, 3rd ed [Stuttgart 1983] 364), and suggests that one may call Moses the author of the Pentateuch or Ezra its redactor: *sive Moysen dicere volueris auctorem Pentateuchi, sive Ezram ejusdem instauratorem operis, non recuso* (*De perpetua virginitate B. Mariae adversus Helvidium* PL 23 190A).

365 *sitam esse* LB *situm esse* 1529

into another language. We know from the writings of the ancient theologians that these varied, both among the Syrians and among the Greeks and Latins. In fact, St Augustine,[366] far from condemning this variety, believes that it is quite helpful for the understanding of the Scriptures. And so, inasmuch as the New Testament was translated into the type of Latin that was the common language of the masses at that time, whereas now Latin is not in use except among the educated, was it a sacrilege if, given these altered circumstances, I translated the New Testament into better Latin? Surely Jerome dared to do this in the case of the Old Testament, and he won praise for it. In my case, I did not change the vulgate translation; instead I translated what is found in the Greek manuscripts,[367] particularly those employed by the ancients, Chrysostom,[368] Basil,[369] Gregory Nazianzen,[370] and their like.

You will say: 'It is beneficial that the universal church agree on a text.' You can see that I do not disturb that agreement. The church has its own translation. My only goal is that it have its translation in a more accurate form, and understand it more correctly.

But there are two types of translation,[371] one which transfers a book from one language into another, a second which explains the meaning of Scripture under consideration. If you decree that commentaries and all expositions must be banned, I will allow my *Paraphrases* also to be banned.

* * * * *

366 *De doctrina Christiana* 2.11.16, 2.12.17–18 CCSL 32 42–3; Erasmus alleges the same argument in the Apologia prefixed to his edition of the New Testament LB VI **2ᵛ / Holborn 167:22–5.

367 Compare Erasmus' reply to Zúñiga on this point (*Apologiae contra Stunicam* (1) LB IX 287C–D / ASD IX-2 7:175–99): 'But even more irresponsible is his writing that I condemn the ecclesiastical translation [ie the Vulgate], since everywhere I shout, as Plautus says, to the point of hoarseness, that I am translating what is contained in the Greek manuscripts.'

368 See n139 above.

369 Basil the Great (d 379); for Erasmus' long involvement with Basil's works, culminating in his editorship of the Froben *Basilii opera* of 1532; see Allen's prefatory note to Ep 2611.

370 Gregory Nazianzen (d c 390); Erasmus was to publish Willibald Pirckheimer's translation of Gregory's sermons in 1531; see Allen Ep 2493 and *Apologia* n214 below.

371 Erasmus is not contrasting here liberal and free translation (for Erasmus' views on this subject, see Rummel *Erasmus as a Translator* 27–8, 92, 122, 124–5), but translation (metaphrase, version) and interpretative paraphrase; note eg Erasmus' words in Ep 1274:37–43: 'In a version, the sense is rendered literally; in paraphrase, it is legitimate to add something of your own as well that may make the author's meaning clearer. The scope allowed to the writer of paraphrase can easily be seen by anyone who compares Themistius with Aristotle. For a paraphrase is not a translation but something looser, a kind of continuous commentary in which the writer and his author retain separate roles.' In Ep 1274:40–3 and Allen Ep 2871:22–3 Erasmus cites Themistius' paraphrase of Aristotle as a classic example of paraphrase; Themistius' work is described briefly by Roberts *Biblical Epic* 54–5.

But if you allow them, then a paraphrase is nothing other than a type of commentary,[372] one more suitable for the busy or the fastidious reader. And if you are going to condemn Basil, Nazianzen, Cyprian, Hilary, and Jerome because they expounded the Scriptures in a more elegant style than they had been transmitted, then I will allow myself to be condemned because I have explained the content in a somewhat more polished way[373] than the apostles have transmitted it, or than the Latin translator, whoever he or she was,[374] rendered it.

If you acknowledge the absolute truth of these points, then there is no room for the statements you declaim with a sort of Ciceronian amplification, namely that the Holy Spirit has no need of my polishing, that there are as many mysteries as there are words in Scripture, and that not a single letter is to be meddled with,[375] all as if I were a new translator of the Scriptures. Indeed, at this point, if it please heaven, you have even made an addition to your earlier book of the testimony of Moses from the Book of Deuteronomy: 'Thou shalt make no addition to the word which I am speaking to you.'[376] But the Translator did this in many passages, as when in the case

* * * * *

372 See the dedicatory letter to Charles v (January 1522) in the paraphrase on Matthew (Ep 1255:41–2): 'for a paraphrase is a kind of commentary.' The phrase is repeated in the dedicatory letter to Ferdinand in the paraphrase on John (Ep 1333:422), and in the dedicatory letter to Henry VIII in the paraphrase on Luke (Ep 1381:441–3). See also Epp 1274:41–3, 1342:1025–9; *Supputatio* LB IX 528B; and 81 below.

373 In his reply to Béda (*Supputatio*) Erasmus described his paraphrases as using the rhetorical flow of an oration (LB IX 540E), as following the practice of a teacher of grammar who uses plainer words to explicate a poetic text (LB IX 521A), and as avoiding any studiously affected eloquence unless the text calls for it (LB IX 530B). On the nature of the paraphrases, see the Introduction to CWE 42.

374 The identity of the translator of the Latin version of Scripture in general use was a matter of controversy from the thirteenth century. Jerome's authorship, particularly of the translation of the New Testament, was denied by many, including Erasmus, but defended by Zúñiga, Frans Titelmans, Jan Driedo, and Agostino Steuco; see Eugene F. Rice JR *Saint Jerome in the Renaissance* (Baltimore 1985) 173–99.

375 Pio *Responsio paraenetica* 10ʳI: 'for they say that it is outrageous for any human to wish to improve the work of the Holy Spirit, so that where he wished to be brief, the paraphraser would be prolix, where he wished to be copious and abundant, the paraphraser would be brief and compressed. They assert that the divine Spirit uttered the mysteries of his Wisdom with the words he approved, and in the order and style that were most suitable. In this idiom, as many words as you count, so many are the mysteries you reckon up. Wherefore, not even one letter is to be meddled with.' Pio may have deployed a cliché from Jerome Ep 53.9.6, where he says of Revelation: 'Apocalypsis Iohannis tot habet sacramenta, quot verba' CSEL 54 463:9–10. As a cabbalist Pio also found hidden meanings in the words, syllables, even accent marks of Scripture. Erasmus criticized such cabbalistic speculations; see eg *Annotationes* LB VI 928D / Reeve *Galatians to the Apocalypse* 664; and *Apologia* 263 below.

376 Apparently Pio had added Jer 23:29 and Deut 4:2, the texts cited here; both proof-texts

of *tabita cumi*[377] he added 'I say to you.' The Canon of the mass has added 'for' in the words of consecration,[378] something not found in the gospels. Thus, what Moses commanded about not making additions has to do not so much with letters, syllables, and tittles as with the meaning of a statement.

Do you not hear preachers far and wide explaining to the people a sentence read from the Scripture by means of a winding circuit of words? You will say that this is permissible, provided the sources are left intact. What then? Have I destroyed the sources of Scripture? This is what I am forced to listen to again and again to no good effect! Just provide us with uncorrupted sources, and I will give up changing a single word. But if it is established that the manuscripts are corrupt, certainly in some passages, you will not, will you, forbid my effort to restore them, so long as no harm is done to the vulgate edition? In fact, Leo x approved this effort of mine;[379] Adrian also gave his approval;[380] nor has it offended any intelligent theologians, and many truly learned and decent men have privately thanked me in their letters for my toils in this field.[381] At Louvain, when I was on the point of publishing that work,[382] and many were complaining considerably

* * * * *

are in *Responsio paraenetica* 10^{v}1, but not in the Milan ms fol 8^r. See similar additions identified in n362 above and *Apologia* n16 below.

377 Mark 5:41. On the variant *tabit[h]a* against *t[h]alit[h]a*, see CWE 69 196 n32; on the addition of the words *tibi dico*, see Erasmus' annotation on Mark 5:41 (LB VI 171C–D / Reeve 125). The argument used here against Pio is borrowed from Jerome (Ep 57.7 CSEL 54 512:9–12). See the discussion of Erasmus' views on the introduction of explanatory words and phrases into the translation of obscure scriptural texts in Rummel *Erasmus as a Translator* 97–8.

378 That is, *Hoc est enim corpus meum, Hic est enim calix sanguinis mei*; see A. Hänggi and I. Pahl *Prex Eucharistica* Spicilegium Friburgense 12 (Fribourg 1968) 433–4.

379 Ep 864, which Erasmus referred to in his letter to the reader in the fourth and fifth editions (LB VI ***1^v and ***3^v), and in his *Apologia de 'In principio erat sermo'* (LB IX 112E–F)

380 Of the extant letters of Adrian VI, the only one that supports this statement, and then only in a very vague manner, is Ep 1324:55–60, which praises Erasmus' many writings, but not the *Novum Testamentum* by name. In his own writings Erasmus claimed that as a cardinal in Spain Adrian had urged him through Pierre Barbier to do for the Old Testament what he had done for the New; see eg Epp 1571:50–2, 1581:325–9, 748–50; and *Apologia adversus Petrum Sutorem* (LB IX 753F). In his letter to the reader in the fourth and fifth editions of the *Novum Testamentum* Erasmus claimed that Adrian VI supported his scriptural work (LB VI ***3^v).

381 For examples of letters by 'truly learned and decent men' who thanked Erasmus for his scholarly labours on the paraphrases, see Ep 794:86–7 (everyone praises the paraphrases), Ep 1443B:3–8 (Clement VII thanks Erasmus for his paraphrase on Acts), Allen Ep 1906:78–9 (all good men), Allen Ep 2394:36–7 (Alciati), Allen Ep 3002:36–43 (Odonus), and Allen Ep 3072:78–9 (Pellican).

382 At Louvain from the press of Dirk Martens Erasmus published his paraphrases on Paul's epistles – Romans (1517), Galatians (1519), 1 and 2 Corinthians (1519), and Ephesians

about some of the rest of my works, no one addressed even a word of caution to me about the *Paraphrases*.

Where, moreover, did you light upon those babblers who say, who declare, who suspect, that my goal is for my *Paraphrases* to be read in place of divine Scripture in the churches?[383] Since this is so extraordinarily absurd, I will not refute it, even though many things are read in the churches that are less suited to the liturgy than my *Paraphrases*.

You say that no one has dared to do this to the Sacred Scriptures since religion came into existence.[384] If the expounding of Sacred Scripture is in itself forbidden, then you must condemn along with me those who first dared to produce commentaries on it, those who later on dared to discuss it with philosophical arguments, and in this connection you must condemn, along with Alexander of Hales,[385] Thomas,[386] and Scotus,[387] the whole multitude of the scholastics,[388] some because they were originators, some because they were imitators.

Of course, there are different types of commentary. What Erasmus dared to do, Juvencus dared to do before him in verse,[389] and he did not

* * * * *

(1520) – and on Peter and Jude (1520), the canonical epistles (1520), and Hebrews (1521); see vander Haeghen 143–4, 146; Allen IV 180, 283, 416, 436; and the Introduction to the paraphrases on Romans and Galatians CWE 42 xx–xxii.

383 Pio *Responsio paraenetica* 11^{V}I: 'there is no lack of those who suspect that it was your idea that eventually people, weary of reading that somewhat rustic language, but charmed by the polish and eloquence of yours, would reject that and adopt yours to take its place in public readings.' Erasmus explicitly denied that he ever intended his new translation (much less his paraphrases) of the New Testament to be read in church, replacing the traditional version; see *Apologia de 'In principio erat sermo'* LB IX 112E–F, 114B. He insisted that his paraphrases were not to replace the Gospels, but were mere commentaries on them; see eg the dedicatory letter to his paraphrase on Luke, Ep 1381:444–7.

384 Pio *Responsio paraenetica* 11^{V}I: '[they] assert that no one, Hebrew, Greek, or Latin, so far as anyone knows, since religion first came into existence has dared to do this with Sacred Scripture, except for you alone.'

385 Alexander of Hales OFM (d 1245), the *doctor irrefragabilis*

386 Thomas Aquinas OP (d 1274), the *doctor angelicus*

387 John Duns Scotus OFM (d 1308), the *doctor subtilis*

388 Pio had remarked (*Responsio paraenetica* 4^r–^{V}C) on the contempt of the *professores bonarum litterarum* for scholastic philosophy. For a study of this conflict, see Rummel *Humanist-Scholastic Debate*.

389 Juvencus (c 330) was the author of a paraphrase of a harmony of the gospels in Latin hexameters, the *Evangeliorum libri IIII* CSEL 24; the *editio princeps* was printed at Deventer by Richardus Pafraet c 1490; Allen II 323 reports that it was probably printed for use in the school there since three Deventer printings are known from the last decade of the fifteenth century. The works of Juvencus and other early Christian paraphrasers of Scripture in verse to whom Erasmus refers in his defence are described in Roberts *Biblical Epic*.

fail to win his share of praise. His poem is, in fact, a paraphrase. After Juvencus, Arator dared to do the same thing to the Acts of the Apostles,[390]
and not so long ago Aegidius Delphus, a theologian from the Sorbonne, did
this to the Psalms and a number of the other books of Sacred Scripture.[391]
Only a few months ago at Louvain the Franciscan Frans Titelmans,[392] still
a young man, did the same thing as I with the Epistles of St Paul, except
that instead of a 'paraphrase' he calls it an 'elucidation' (although the term
'paraphrase' is somewhat more modest), and he does this to the applause
of the theologians and the whole Franciscan order. I hope that he has engaged so successfully in this type of work that my *Paraphrases* will be cast
aside, and people will wear out his *Elucidations* with handling; only let true
piety grow to the extent that my reputation has waned! Now you would
do well to decide for yourself whether you wish to absolve or condemn
me along with all those who have dared to do or to approve what I have
done.

But you are scandalized by the fact that I present a *persona*[393] speaking
differently in paraphrase from the way he does in his own work. How silly,
as if the ancient commentators on the Scriptures, whenever they expound
the gist of a passage in their own words, are not doing the same as I, except
that a paraphraser does continually what they do at intervals! As if among

* * * * *

390 Arator (fl c 544), the author of a paraphrase on the Acts of the Apostles in Latin hexameters (CSEL 72); the version printed by Aldus in *Poetae Christiani veteres* II: *Sedulii, Iu-
venci, Aratoris opera* (1501) is usually given as the *editio princeps*, though Marcel Bataillon
(*Erasmo y Espana* 2nd Spanish ed [Mexico City 1966], 27) reports a Salamanca edition of
c 1500.

391 Gillis van Delft (d 1524); for his career see CEBR I 382–3. Erasmus here refers to his
Septem psalmi poenitentiales, noviter per E.D. metrice compilati (Antwerp 1501) and *Epistola
divi Pauli ad Romanos decantata ... Quinque psalmi Davidici decantati* (Paris: Josse Bade
1507 / Barbier and Maarnef Brothers c 1508). As early as 1519 Erasmus was citing the
precedent of Juvencus and Gillis van Delft in defence of his scriptural projects; Arator
was added to the 1527 edition to form the triad repeated here (Holborn 167:25-30).

392 Frans Titelmans (d 1537); for his career and his controversy with Erasmus, see CEBR
III 326–7. Erasmus refers here to his *Elucidatio in omnes epistolas apostolicas* (Antwerp:
M. Hillen 1528, 1529); for his later verse elucidations of Scripture, see Klaiber
284–7.

393 Pio *Responsio paraenetica* 11^{V}I: 'it makes indeed a very great difference whether one
interprets the intact text according to one's own understanding of it, as commentators do, or, assuming the *persona* of the author, one tempers the words of the Holy
Spirit according to one's pleasure, which is what happens in the case of paraphrases.
Since, then, you have assumed in your paraphrases the *persona* of the evangelist, of
Paul, or of the rest of the apostles who wrote anything ...' Erasmus defined paraphrase in Ep 1274:41-3: 'For a paraphrase is not a translation but something looser, a
kind of continuous commentary in which the writer and his author retain their separate
roles.'

the Italians, the French, the British, and the Germans, preachers do not do the very same thing, so that they can explain to the people a terse statement in the Scripture at greater length and in other words.

But what do you mean when you say 'as he pleases'?[394] Does one treat of Scripture 'as he pleases' when, in expounding it, he follows all the most approved Doctors of Mother Church, as I have done constantly?[395]

Not content with these words, you go on to say: 'In fact, you sometimes bring on the *persona* of Christ preaching in a very unsuitable style, as if he were Hortensius or Cicero pleading a case in studied language and artificial speech, and, what is even more revolting, saying a great many things not contained in Scripture, in words adapted to merely human wisdom.'[396] These are your words. To reply briefly: there are in my *Paraphrases* none of the trappings of declamation that you ascribe to them, with deliberate exaggeration, your purpose being, unless I am mistaken, to conclude your plea more handily. Nowhere do I assign to Christ speech that is studied and highly wrought, since there is no work of mine that I completed more extempore,[397] nor was this the appropriate venue for you to bring up Cicero or Hortensius.

If the meaning agrees with Christ's, then your complaint about the style is unfair, unless you are condemning all the writers I have listed. Christ never spoke in verse, and yet he speaks thus in Juvencus! At least the same indulgence should be afforded to me as is given to all commentators on Scripture. As it is, if I were to add nothing to what is found verbatim in Scripture, I would be neither a paraphraser nor an expositor. I am content if what I add does not disagree with the meaning of the Scriptures. Further, if you were going to be so outspoken, you should have cited the passages that you are condemning.

So chant Solomon's statement forbidding anything to be added to the divine utterance[398] either to those who teach contrary to Scripture, or else to those who are not afraid to assemble a great many resources for the

* * * * *

394 See n393.
395 Ep 1333:415: 'In this work I have followed the most approved Doctors of the church'; see also *Supputatio* LB IX 527D ('as best I could'), 540D ('imitated them').
396 Pio *Responsio paraenetica* 11ᵛI; the Hortensius mentioned by Pio and Erasmus is Q. Hortensius Hortalus (d 50 BC), Roman orator and Cicero's chief competitor in this field.
397 *Ex tempore* 1529 *in tempore* LB. On the extemporaneity of all of Erasmus' work, including that on the New Testament, see nn 28 and 86 above.
398 Pio *Responsio paraenetica* 12ʳI: 'regardless of the teaching of Wisdom in Proverbs: "Every word of God is fire-tried, a shield to those who hope in him. Do not add anything to his words, lest you be rebuked and found to be a liar [Prov 30:5–6]."'

explanation of the divine books from secular studies, particularly from the
sophisms in the works of Aristotle and Averroes.

But somewhere I admit that I am changing the style of the heavenly
Preacher.[399] Though[400] others make this the basis of slander, you, surely, no
less versed in rhetoric than in philosophy, ought to have recognized the
rhetorical device whereby I spoke somewhat facetiously indeed, but said
nothing save that I was expounding the thoughts of Paul in such a way that
they might be understood more readily and with less fatigue.

In your prudence you realize, I think, that the matters you add concerning the Holy Spirit (that he dictated the Scripture word for word, and
heaven forbid that any Christian be willing to oppose his authority)[401] do
not apply to me at all. Oh, you declaimed these points smartly and elegantly, but they are beside the point.

You also deal at some length with the point that there is a great difference between a commentator and a paraphraser,[402] because a commentator leaves the text intact, but the paraphraser alters it. Consult the church's
books, and see if I have altered a single one. Furthermore, if I did not use
different words from the sacred books, I would be neither paraphraser nor
commentator. Come, then, and answer me what difference it makes whether
I present Paul speaking in my way, or a commentator says 'this is what Paul
means,' and goes on to give his own explanation. And yet, as we have said,
they do this quite often, and in the *persona* of Paul, or of Christ, whenever
they explain the gist of a passage in a summary way.

But, as to your next point, that someone who has lapsed into error by following the authority of a commentator can readily be recalled
to the right path by the assistance of the Scriptures,[403] but that there is
no way to put right someone who has been misled by my paraphrase, I
pray you by your better self, illustrious sir, did you write this as a joke
or in earnest? Do heretics not inflict harm enough on the Christian religion? And they expound the Scriptures erroneously, not by paraphrase, but
in commentaries! 'What sort of utterance has escaped the barrier of your
teeth?'[404] And how is one who has fallen into error in these commentaries

* * * * *

399 See Ep 710:19–53.
400 *ut ut* 1529 *ut* LB
401 Pio *Responsio paraenetica* 12^{r}I: 'and will any paltry human strive to go against the chosen
plan of the divine Spirit who dictated all those things (as they say) word for word?'
402 Pio *Responsio paraenetica* 11^{v}I (briefly), 12^{r}I (at greater length)
403 Pio *Responsio paraenetica* 12^{r}I
404 ποῖόν σε ἔπος φύγεν ἕρκος ὀδόντων, a Homeric formula; see *Iliad* 4.350, 14.83; *Odyssey* 1.64,
5.22, 19.492, 23.70.

to be set right? Through the authentic text of Scripture. I agree. But if anyone has picked up an error from my paraphrase, can he in no way be set right? How can this be? Because an authentic text of Scripture nowhere exists? Indeed, the text still survives, even among the vast mob of commentators.

How can one refute the error of those who have written paraphrases of Aristotle?[405] Is this not done from Aristotle himself? For his work is extant just the same as if one had written a commentary instead of a paraphrase. What, then, prevents one who has gone wrong being corrected through the still extant Scripture? Indeed, it was for this purpose that I made note of the chapter numbers in the margins, so that the reader could more conveniently make comparisons.

I can hardly think that you were serious also when you wrote that there is a risk that those who have, as it were, sworn allegiance to me would follow my influence.[406] Where does this influence now dwell? I used to enjoy considerable favour among the learned Germans before this evangelical gale arose. Favour has quickly given way to enmity.[407] Nor have I noticed anywhere those whom you describe as having sworn allegiance. Over and over, again and again, I cry out in my introductions that a paraphrase is nothing but a type of commentary, and I have no wish that more store be set by a paraphraser than by any commentator.[408] If people detect any mistake, they should impute it to me, and not to the *persona* of the one speaking. Every danger, if any really threatens, is precluded by these prefaces.

You philosophize again about the syllables and letters of divine Scripture,[409] matters that do not apply to me at all. For if what you declare is

* * * * *

405 Pio himself had brought up the fact that commentators and paraphrasers of Aristotle have often erred (*Responsio paraenetica* 12ʳi). See n371 above.

406 After making the point that commentators and paraphrasers of Aristotle have often erred, Pio (*Responsio paraenetica* 12ʳ–ᵛi) goes on to ask: 'if this should have happened sometimes to you, and it could easily have happened, what of those who have encountered only so much of Sacred Scripture as is contained in your paraphrases and have fallen into error from reading your paraphrases, how could they be restored to soundness once they have been imbued with views which you had erroneously followed? Those who have, as it were, sworn allegiance to you, they could not, as is the case with commentaries, direct their attention to the author's own words before them.'

407 Compare Allen Ep 2134:166–8 (25 March 1529): 'and I who earlier was the Star of Germany and the Champion of Authentic Piety, now, the thumb, as they say, reversed, I am Balaam and Doctor Donkey.'

408 See n372 above.

409 Pio *Responsio paraenetica* 12ᵛi: 'For we know that the Hebrews are wont to approach Sacred Scriptures with such reverence, and hold that they contain so much power, that

true, then the Old Testament should not have been translated into Greek, nor the New Testament into Latin.[410] How many letters, how many mysteries are lost in this process! But these points are, as I said, irrelevant to me, for I leave the Scripture completely untouched.

But what is that hateful sort of composition by which all is altered and twisted?[411] Is this why you saw fit at the beginning to request indulgence for plain speaking,[412] so that you could say these things which, in consideration of your honour, I would much prefer had no place in this disputation, so smart and so Ciceronian?

The additional remark in your conclusion, that those whom only a pleasant style entices to the Scriptures[413] are unworthy to read them,[414] is, perhaps, true. But it is nonetheless a work of piety to entice the squeamish and the particular to the love of Scripture by some allure, a practice of which St Augustine also heartily approves.[415] But there is no good reason to argue with you at greater length about these matters now, particularly since I have answered these cavils so often.

* * * * *

they maintain that there are mysteries not only in the words and the word-order, but even in the syllables and in the very letters, so much so that they used to count the number of letters in books, when they obtained them from scribes, and if they found that there was even the least bit in excess or wanting, or any transposition, they used to reject the manuscript out of hand.' See n375 above and *Apologia* n933 below. Erasmus had contempt for the cabbalistic learning that was so popular in the learned circles of the papal court; see eg Ep 798:20–6 and his attack in the *Moria* LB IV 475B–477A / ASD IV-3 164:611–165:621 / CWE 27 133. See Gundersheimer 'Erasmus, Humanism, and the Christian Cabala' 38–52; and Jerome Friedman *The Most Ancient Testimony: Sixteenth-Century Christian-Hebraica in the Age of Renaissance Nostalgia* (Athens, Ohio 1983) passim. On cabbalistic learning at the papal court, where Pio was a prominent figure for many years, see John W. O'Malley *Giles of Viterbo on Church and Reform: A Study in Renaissance Thought* (Leiden 1968) 74–99; and Charles L. Stinger *The Renaissance in Rome* (Bloomington, Ind 1985) 306–8. Pio's uncle Giovanni Pico della Mirandola was a major exponent of the cabbala and Pio himself was depicted, according to Gilmore, as a kind of magus in the portrait probably by Loschi in the National Gallery in London (see illustration xxxix above).

410 Erasmus' argument here goes back to an addition (1527) to the *Apologia* prefixed to the *Novum instrumentum*: 'Nor am I rejecting the view expressed by some, that mysteries lie hidden in the letters and diacritical marks, so long as they acknowledge the hyperbole involved. Otherwise, one should not drink of the Scriptures except from its original sources, for one who translates them into another language is forced to deviate far and wide from the letters and diacriticals' (LB VI **2ᵛ / Holborn 169).

411 *immutantur* 1529 *mutantur* LB. Pio *Responsio paraenetica* 12ᵛI: 'It is for this reason that no one of the Christian Fathers has made bold to explain the divine Scriptures in the sort of composition in which all is altered [*immutantur*] and twisted.'

412 See n47 above.

413 *ad eas* LB *ad eos* 1529

414 Pio *Responsio paraenetica* 12ᵛ–13ʳI

415 Erasmus has in mind, perhaps, *De doctrina christiana* 4.13.29 CCSL 32 136.

It remains to examine your conclusion. Whereas I made a brief comment in my notes about papal power and its not having been employed, and noted that of old the pope was called *summus sacerdos* 'the highest priest,' not *summus pontifex* 'the highest pontiff' (something[416] I can scarcely believe I wrote); whereas I have expressed a doubt that the title 'cardinal' existed in the past; whereas in *Folly* I satirize the corrupt morals of mankind; whereas in the pamphlet on confession I do not pronounce that it was instituted by Christ (that I was not dealing with that topic then and do praise it devoutly elsewhere notwithstanding); and, finally, whereas I took liberties with the New Testament (but, however, with a view to writing a more handy sort of commentary), for these and for like reasons it is quite plain that I do not recoil completely from the programme of the Lutherans. Who would not be in awe of such a robust recapitulation?

Because I am convinced that you are not writing any of this from conviction, but have sought out a theme on which to demonstrate your great ability in Ciceronian eloquence, I take no offence, most renowned Alberto, but do I wish you had undertaken to do this on a more auspicious subject.

Whereas Cicero is somewhat temperate in opening, robust in argumentation, and forceful in summation, you open your case savagely, argue weakly, and sum up impassively. For you had promised that you would prove that I was the opportunity, the cause, the instigator, and the leader of the entire Lutheran tempest,[417] that my writings deprive the sacraments of their importance, that I place a higher value on matrimony than on celibacy, that I disparage the ordinances of the Fathers and the office of the pope, that I ridicule monasticism, that I ridicule the liturgy, that I criticize the prayers of the hours. Which of these have you proven, even with weak arguments? After so savage an opening statement, after so enervated a proof, how lame a conclusion!

'But so much,' you say, 'for the first reason, which has caused many

* * * * *

416 *quod* 1529 *quo* LB
417 Erasmus overstates his point. Pio (*Responsio paraenetica* 5ʳ–ᵛᴅ) indicates that no one in his circle said that Erasmus is the cause, but he does argue that Erasmus' works provided an opportunity for Luther. Pio also says (6ʳᴇ) that it was Erasmus' fault that the suspicion had arisen that he was a partner in, or the author of, the discord. He nowhere calls Erasmus the 'leader,' save by implication. Erasmus is, in fact, here recycling almost verbatim his summary description of Pio's endeavour, written 23 December 1528 to Berquin (Allen Ep 2077:45–6): 'trying hard to prove that I was the opportunity, the cause, the instigator, and leading figure of this whole business' (*me fuisse occasionem, causam, autorem et principem totius huius negocii*). In the *Responsio* the Latin text here reads *me fuisse occasionem, causam, auctorem ac ducem totius Lutheranae tempestatis.*

to suspect that you do not reject absolutely the Lutheran programme.'[418] It is a suspicion, not a judgment, and not of all, but of many, that I do not 'recoil absolutely.' How very little this conclusion corresponds to the opening of your case!

But what are you calling a programme, recalling people to the original sincerity of Christian devotion? For a long time now all the most esteemed people have been longing for this, not I alone. But I undertook nothing radical, unless it was radical to offer advice in passing about certain matters. But my longing for this was not such that I wanted remedies to be sought through insurrection, as was bound to take place if matters were managed not by the influence of the princes but by the recklessness of the people. But if by programme you mean Luther's teachings, then just as the highest virtue is very close to vice, or as the greatest generosity is close to prodigality, and the greatest sternness to cruelty, even so is the highest truth close to falsehood.

You move on to the charge of silence.[419] As is your practice, you deal with this aspect as if I have given no account at all of my silence, whereas I have done so many times. But it is a lie that I kept silent at the beginning.[420] In a document that I published immediately I declared that I had nothing to do with Luther.[421] I handled all those who approached me as if to sound me out in such a way that they realized that I did not approve of what Luther was trying to do.[422] I kept those friends with whom I had any influence

* * * * *

418 Pio *Responsio paraenetica* 13ʳκ
419 Pio brought up the issue of Erasmus' 'silence' in *Responsio paraenetica* 6ʳᴇ, and takes the matter up in detail on 13ʳ–15ʳκ.
420 Pio *Responsio paraenetica* 13ʳκ: 'by the voices of how many do you think you are being accused, Erasmus, because you did not oppose yourself to the Lutheran madness at the outset?'
421 Erasmus did not write openly and explicitly against any of Luther's doctrinal positions until 1524 when he published his *De libero arbitrio*. He claimed in Ep 1342:1022–56 that he had already published on this theme in his paraphrase on the Epistle to the Romans (1517) and that his statements there are known to differ from Luther's. The early writings against Luther here cited would seem to be published letters; see Ep 1352:72–82. For examples of statements by which Erasmus hoped to distance himself from Luther, see Epp 939:69–71, 126–38 (14 April 1519, to Elector Frederick of Saxony), 961:36–9 (1 May 1519, to L. Campeggio), 967:86–114 (18 May 1519, to Wolsey), and 1033:52–215 (19 October 1519, to Albrecht of Brandenburg). In his letter to Jean Glapion from Basel, c 21 April 1522 (Ep 1275, especially lines 23–5) Erasmus speaks of having begun to write a small book against the Lutheran affair, but of having discontinued work on it because of poor health. On this topic see Epp 1275:7n, 1341A: 1338–1416.
422 This section is a restatement of 24 above; see, for example, Erasmus' *Catalogus* 31:31, 35–8; Epp 1183:144–51, 1195:138–43; but also see Ep 1384:95–8, where Erasmus states that he seems to teach almost everything that Luther teaches.

because of popularity or prestige from getting involved in this business.[423]
I advised Luther himself to act on the purest of motives.[424] I admitted quite
frankly before Frederick, duke of Saxony, that I did not approve of Luther's
endeavours.[425] And I did all this immediately, at the very beginning.

Tell me, Alberto, by the shades of Aristotle, is this what you mean by
keeping silent? Why do you not, instead, sing this old song to the Sorbonne,
which was so late in publishing its articles against Luther?[426] Why do you
not sing it to yourself? For though you burn with so great a zeal for religion, though you have so much influence and charm, so much eloquence,
and are armed (this is what really counts!) with a knowledge of Aristotelian
philosophy[427] worked up with the sweat of so many years, only now, at last,
after the lapse of nine years,[428] are you leaping into the arena, despite the
fact that it would have been safe for you in Rome to hurl every sort of
missile at Luther as if from the citadel of religion. And you wanted me to
do in Germany, what you, grand figure that you are, did not dare to do in
Rome?[429]

* * * * *

423 For example, Ep 1143:10–32; *Spongia* LB X 1650C / ASD IX-1 162:961–3 and the passages
listed in 162n / Klawiter 190–1

424 Epp 980:53–8, 1033:55–8, 1167:161–71, 1445:15–16; *Spongia* LB X 1650F–1651A / ASD IX-1
163:986–93 / Klawiter 192

425 Ep 939:71–4, 135–8 (but the whole tone of the letter is an exhortation to Frederick to
protect Luther). For a very different account of this meeting, see Smith *Erasmus* 235–6;
and the *Axiomata* Erasmus wrote for Frederick (*Erasmi opuscula* 329–37 / CWE 71 xlii–iii,
106–7).

426 See XXIII *libri* 62ʳm and n124 above.

427 In his letter of 30 March 1529 to Thomas More (Allen Ep 1804:254–6) Erasmus remarks
about Pio: 'For he, a layman, has ambitions to be considered a theologian, and is wonderfully pleased with himself because he is an Aristotelian.' All of the volumes of the
famous Aldine Aristotle were dedicated to Pio (see Lowry *World of Aldus* 75), as were
the remaining items of philosophy on Aldus' list. See the beginning of Aldus' dedication of his edition of Lucretius to Pio (January 1519): 'Long ago I determined, O Alberto, glory of princes, glory of the learned of this age, to dedicate to you all the books
on philosophy, as many as should come forth from our house into the hands of scholars' (repr in *Aldo Manuzio, editore* I 152). See especially Charles B. Schmitt 'Alberto Pio
and the Aristotelian Studies of his Time' *Medioevo e umanesimo* 46 (1981) 43–64, repr as
item 6 in *The Aristotelian Tradition and Renaissance Universities* (London 1984).

428 Luther's views were formally condemned in *Exsurge Domine* (1520) and he was excom-
municated for refusing to recant by *Decet Romanum Pontificem* (1521). Pio's book, although circulating in manuscript form since 1526 (see Allen Ep 1744:129–30), was not
printed in Paris until 1529.

429 Erasmus repeats here what he wrote to Celio Calcagnini in 1525 (Ep 1576:46–53): 'And
now Pio of Carpi has launched a campaign of vilification. People write to tell me that
at every meeting and every dinner-party at Rome he belittles me, claiming that I am
no philosopher and no theologian and that I possess no genuine scholarship at all. Why

At this point you again exaggerate my learning, eloquence, and influence,[430] but I am never decorated with these lavish praises except when they serve to lay the groundwork for an accusation against me. But away with these tricks! The literary sodality whose *coryphaeus* you were in Rome[431] pronounced a far different verdict upon me;[432] far different too was the negative judgment pronounced by the Bullbearer, the better part of your soul, and, in my view, a large part of this disaster.[433]

* * * * *

does he "look on in silence with his arms folded while this wild boar is laying waste the vineyard of the Lord?" Why is there no one at Rome with the courage to do what I did in the very heart of Germany with no thought of reward except that of a clear conscience?'

430 Pio *Responsio paraenetica* 13ʳκ: 'inasmuch as you could have provided this [repression of the unheard-of arrogance of Luther] far more suitably than the rest, not only because of the eloquence, wit, facility in writing, learning in many languages, and varied and manifold scholarship with which you are endowed ... but because of the extraordinary influence that you enjoy among the Germans'; compare the words of Adrian VI in his letter to Erasmus of 1522, as he urges him to 'an attack on these new heresies': 'You have great intellectual powers, extensive learning, and a readiness in writing such as in living memory has fallen to the lot of few or none, and in addition the greatest influence and popularity among those nations whence this evil took its rise' (Ep 1324:29-32).

431 *Coryphaeus* (ie leader), a popular term with Erasmus, probably borrowed from Cicero *De natura deorum* 1.59. On the Roman academies, see n126 above. Alberto Pio does not appear among the members of a Roman academy listed by Casali (quoted in Silvana Seidel Menchi 'Alcuini atteggiamenti della cultura italiana di fronte a Erasmo [1520–1536]' in *Eresia e Riforma nell' Italia del Cinquecento: Miscellanea* I Biblioteca del 'Corpus Reformatorum Italicorum' ed Luigi Firpo et al [DeKalb, Ill 1974] 101 n148; the name Pio occurs last in the list, but Seidel Menchi identifies that Pio with Battista Pio of Bologna); nor is Alberto Pio's name found in the list given by Sadoleto in his letter of 1529 to Colocci (in Federico Ubaldini's *Vita di Mons. Angelo Colocci* ed Vittorio Fanelli, Studi e Testi 256 [Città del Vaticano 1969] 67–75), unless he is the 'Savoia noster' mentioned on 74; nor is he in the obituary list of Goritz's circle ('Corytianae Academiae Fato funci, qui sub Leone floruerunt,' printed by Fanelli 114–15), though 'Georgius Sauromannus Germanus' (see nn 69 and 71 above) does appear there. Pio was, however, the subject of three poems in *Coryciana*; see Introduction xxxvii. Erasmus may have no precise or formal group in mind, but only his Roman critics. He mentions their *coryphaeus* in Allen Ep 1794:46 (23 March 1527) (*coryphaeum et incitatorem*), but Allen (Ep 1794:46n) correctly identifies this figure with Aleandro; in Allen Ep 1719:35–6 Erasmus identifies Aleandro and 'Alberto quodam Principe Carpensi' as its leaders (*ducibus*); in Allen Ep 1840:80 (22 June 1527) Erasmus identifies Alberto Pio as the *princeps* of the sodality in question.

432 See 13–17, 27, and nn 60 and 126 above.

433 Aleandro, who saw to the promulgation of *Exsurge Domine* in the lower Rhineland, was considered by Erasmus to be an intimate friend of Pio (by innuendo a homosexual relationship is suggested; see eg Allen Ep 2077:51–2, and also Allen Epp 2371:37, 2375:80, where Erasmus quotes Juvenal 2.47, 'great is the harmony among the effeminate'), and also the principal instigator of the complaints against Erasmus (eg Allen Epp 1840:78–83, 1987:6–10, 2329:106). See also nn 41, 62, and 182–5 above; and Heesakkers 'Erasmus' Suspicions' 371–84.

What if I had leapt into the arena at once? By means of my antidote I would have routed the baneful serpent (your term for Luther, I think), and driven him to slink into his cave.[434] A man who did not yield to the censures of three universities,[435] who did not yield to a frightful papal bull,[436] who did not yield to the edict of the emperor[437] (a prince so very great that the least portion of his domain is an empire), a man who did not yield to the violent passions of men and to a world threatening its end, this man, struck at once with sudden terror by my writing, would have hidden himself in his cave? Surely a Ciceronian eloquence would be required to convince anyone of this! In fact, I know full well that you are not yet yourself convinced of what you are attempting to convince others.

Where did I write that I found the prologue of the tragedy of Luther not disagreeable?[438] I wrote that it was not altogether offensive,[439] but that it was inoffensive in such a way that I still complained; in fact, I wrote that

* * * * *

434 Pio *Responsio paraenetica* 13ʳ⁻ᵛκ: 'Why should you have passed by with closed eyes, and allowed that most poisonous serpent to pour out his venom and infect all of Germany, when, with your antidote, you could have put him to flight and compelled him to slink into his cave.'

435 Luther was censured by the universities of Cologne on 30 August 1519, Louvain on 7 November 1519, and Paris on 15 April 1521; see Bäumer 'Lutherprozess' 33–5. The three censures are printed in Duplessy *Collectio judiciorum* 1-2 358–9 (Cologne), 359–61 (Louvain), 365–74 (Paris). On Paris see n124 above.

436 Probably the bull *Exsurge Domine* of 15 June 1520, which threatened Luther with excommunication, rather than the actual bull of excommunication, *Decet Romanum Pontificem*, of 3 January 1521; for a critical edition of both, see *Causa Lutheri* II 364–411, 457–67. For examples of Erasmus' statements criticizing the severity of the papal bull, see Epp 1153:142–8, 1313:67–8, and *Erasmi opuscula* 336:11–12.

437 The Edict of Worms of 26 May 1521; for a critical edition of it see *Causa Lutheri* II 510–45.

438 Pio *Responsio paraenetica* 13ᵛκ: 'But if at first the prologue of this tragedy (as you yourself admit) was not disagreeable, because, with many others, you thought Luther to be a good man divinely sent to correct thoroughly corrupt morals . . .'; Pio is conflating (and misrepresenting) two statements in Ep 1634: 'When the opening scene of the Lutheran tragedy began to unfold, to the applause of nearly the whole world, I was the first to urge my friends not to get involved, for I could see that it would end in bloodshed' (Ep 1634:49–52); and 'Some people say: "Why did you not oppose this evil the moment it first appeared?" Because, like many others, I thought Luther was a good man, sent from God to reform the evil ways of men [see Allen VI 202 n78 for other statements of this view] – though even then some of his ideas offended me and I spoke to him about them' (Ep 1634:87–90).

439 For some of Erasmus' statements on Luther, see Epp 939:71–4, 967:93–6, and 980:23–7 (favourably reports others' praise of his innocence of life); 983:11–13 (hopes Luther's message will not lead to discord); 1033:52–80 (one should support Luther insofar as he is innocent); 1202:292–5, 1300:87–93, 1313:9–22, and 1526:12–14 (Luther started off with applause but failed to practice moderation as urged by Erasmus); 1358:5–15 and 1384:95–8 (Erasmus agrees with him on much); 1495:8–12, 1522:13–17, and 1526:200–3 (Luther as the bitter medicine to cure the body of the Christian people). See also n421 above.

I never approved of it. How, indeed, could it have been that I complained if I approved?

But I behaved as a spectator.[440] Yes, but in the company of so many thousands who watched in silence, whereas I complained immediately. It was not for me to drive the man from the stage, but for those who were watching from the orchestra. It is among these, my dear Alberto, that you were sitting. You had a much greater power to check the evil, for you could do in safety what had otherwise to be done at peril of one's life. You write, and you certainly write with devotion,[441] and for that reason you are finally publishing a pamphlet which you have laboured over for so many years, now, when it is almost too late. But you did not dare to do this except in Paris, where you are even safer than you were in Rome. Indeed, take stock of your own courage, as you charge me with timidity. You attack no one in your book except Erasmus who is overloaded with unpopularity, Luther who has been transfixed (your expression),[442] and Hutten who is dead,[443] now that the opposing party has virtually overwhelmed Luther.

* * * * *

440 Pio *Responsio paraenetica* 13^{v}k: '... but you behaved as a spectator when you ought to have been an assailant'; '... but even then you remained in between as a spectator [*medius et spectator perstitisti*].' Compare the views of anonymous critics of Erasmus reported by Celio Calcagnini in 1525 (quoted in Allen Ep 1576:2n from Calcagnini's *De libero animi abritrio*): 'But although neither I nor any scholar was ever able to believe this about that man [Erasmus], for he had provided countless proofs of his piety and had toiled so much on Scripture, even so, to bolster against us their suspicion they used to bring up the very stubborn and long silence of the man: "Would he watch," they used to say, "the wild boar devastating the Lord's vineyard and destroying the vessels of the Tabernacle [compare opening of *Exsurge Domine*], and acquiesce in this in silence with arms folded, unless he either finds pleasure in the spectacle or considers that it is none of his business? Especially since he has so many resources, has received from the Lord so many weapons whereby he could accomplish a noble work and come to the aid of truth as it struggles now with both horns." And so [Calcagnini continues] I rejoice for us, I rejoice for my Erasmus, that with this one book [the *De libero arbitrio*] he has shut the mouths of this rabble.' Erasmus quotes part of this indictment in Ep 1576:6–9, and later applies it to Pio (quoted in n429 above). Compare this with the letter to Floriano Montini, 16 May 1525, printed by Erasmus as a preface to Froben's edition of Calcagnini's work (Ep 1578:28–31): 'Some censure me for treading a middle course (*incedere medium*). To tread a middle course between Christ and Belial [2 Cor 6:15] is, I admit, a heinous sin, but to steer midway between Scylla and Charybdis is in my view simple prudence.'

441 Here we suggest a change from the punctuation of 1529 and LB: '... *cui tutum erat facere, quod alioqui cum vitae periculo faciendum scribis, et profecto pie scribis. Atque ideo libellum ...*' to '... *cum vitae periculo faciendum. Scribis, et profecto pie scribis, atque ideo ...*'

442 Pio *Responsio paraenetica* 14^{r}k describes Luther as transfixed (*confossus*) by the public determinations of the universities and the books of private persons.

443 Pio *Responsio paraenetica* 36^{v}b describes Hutten as 'a most fit comrade for Catiline [Luther],' and concludes: 'But now the Lord has taken that firebrand from our midst.'

You say: 'The more severe the disease, the more quickly it had to be checked.'[444] True enough. But at the beginning he seemed to the majority to be the antidote to diseases. Many were unsure how he would turn out. If you do not accept my excuse here, then you are accusing all the learned, the universities, and yourself most particularly, along with me.

You are correct in writing that the Arian heresy was by far the most dangerous.[445] But that vast conflagration arose from the malice of the clergy, for they drove a learned man from their company because of envy. In the same way, this Lutheran tempest has up until now been exacerbated by the factious outcries of certain people, and it will grow even worse unless better remedies are provided than the Cousturiers and Bédas have supplied.

But, Greeks and Latins alike, Hilary among them, had the courage to write against Arianism.[446] None of them, however, wrote so promptly against Arius as you wanted me to write against Luther, and Hilary unsheathed his pen against Arius a bit later[447] than I unsheathed mine, so many times now, against Luther.

Indeed, I am forced to complain of your lack of fairness here. You declaim quite violently about my silence, in spite of the fact that I published the *Diatribe* many, many years ago,[448] in spite of the fact that I combated

* * * * *

Ulrich von Hutten died at the end of August in 1523 on the island of Ufenau in Lake Zurich. See CEBR II 216–20, especially 219.

444 Pio *Responsio paraenetica* 13^{v}K: 'on the contrary, the more enthusiastically the poison was being drunk by people in increasing numbers and of greater renown, so much more carefully and quickly should it have been checked and resisted with a suitable antidote, as being even more baneful and contagious.'

445 Pio *Responsio paraenetica* 13^v–14^{r}K cites the Arian heresy and the vigorous reaction to it.

446 Pio *Responsio paraenetica* 14^{r}K: 'but Greeks, Italians, and Gauls alike met this obligation [of opposing Arianism] with the greatest care and constancy, not without considerable disadvantage to themselves, as you know very well.'

447 The reference is to Hilary, bishop of Poitiers (d 367), celebrated for his opposition to Arianism. Erasmus had edited Hilary's works (Basel: Froben 1523), and explained Hilary's 'silence' in the provocative dedicatory letter (Ep 1334:130–6): 'And indeed he himself testifies in several places that for a long time he was silent. He observed this silence either because the soul even of that great man felt some doubt amid such widespread discord in the world or because when there was no hope that the better side could win he thought it preferable to look to a calm silence rather than to exacerbate, not remove, the world's general depravity by untimely boldness.' Erasmus soon used Hilary's delayed reaction to Arianism to counter criticisms of his own delayed reaction to Luther, the same ploy we see here; see eg a letter of 21 September <1524> to Duke George of Saxony (Ep 1499:23–5: 'If I have hitherto not refuted any of Luther's doctrines in a book written for the purpose, Hilary had remained silent even longer while the Arians overran the world.' In a reply to Erasmus Duke George describes Hilary as: 'an impressive precedent for your silence' (Ep 1520:38–9).

448 *De libero arbitrio diatribe* (Basel: Froben 8 September 1524); see vander Haeghen 20.

Luther's *Bondage of the Will* with two full-sized books,[449] in spite of the fact that in so many passages scattered through my works I teach ideas different from his. And yet you disclose that you have read the *Diatribe* in the course of this very controversy.[450] You will say 'when I began to write this, I had not yet seen it.' Very well, but you should have been alert, and removed from your prosecution what was irrelevant. And indeed, if you did not find it too much trouble when in Paris to insert much additional material, especially scriptural quotations,[451] then you ought not to have been reluctant to remove those passages which increase my burden of enmity among my antagonists just as they arouse a suspicion of your lack of judgment among the learned and sensible. You had not yet seen the *Hyperaspistes*; but you certainly had seen it when you were producing this, or at least before you sent your book to me, and yet at the end of your book you urge me to a work of this very kind, as if I had provided nothing at all.[452]

I have called myself a gnat, not reflecting my own view, but on the basis of the opinion of others. Under different circumstances they usually credit me with nothing, neither talent, nor learning, nor judgment, but it is amazing how huge an elephant I suddenly become whenever they want to load me down with enmity. You, however, make me at one moment an elephant, at the next moment a gnat.[453]

But where do I assume to myself the censorship of the entire world?[454] Is this because *Folly* deals with human behaviour, but in so courteous a way that each can read about his own vices without pain? Your remarks are impressive and clever, but beside the point and untruthful as well.

You justify the universities by saying that they eventually did their

* * * * *

449 *Hyperaspistes diatribae adversus servum arbitrium Martini Lutheri* I (Basel: Froben, March 1526), II (Basel: Froben, September 1527); see Allen VI 262–3, VII 116. Pio denied having seen or read this work (*XXIII libri* 63ʳx). Erasmus' assertion that Pio must have seen the *Hyperaspistes* is a reasonable conjecture given the fame and availability of the work, but there is no documentary evidence for this assertion. For the various other contemporary editions (Antwerp, Krakow, Leipzig, and Paris), see vander Haeghen 109–10.

450 Pio praises and criticizes Erasmus' *De libero arbitrio* in *Responsio paraenetica* 27ʳR; see nn 470, 472, and 473 below.

451 See nn 362 and 376 above.

452 See n472 below.

453 Pio *Responsio paraenetica* 14ʳK: 'But at one moment you shrink yourself into a gnat, at another you raise yourself up to be an elephant. You are a paltry man [*homuncio*, Erasmus' own word; see Allen Ep 1634:84] when Luther is to be attacked, as if it were to be a fight with Hercules holding his club; but you play Atlas when you assume to yourself the censorship of the entire world'; and see n127 above.

454 See n453.

duty, and you say to me: 'But you have not yet brought up any siege engines to attack this beast.'[455] This is absolutely false, but even so you proceed in this entire section with amazing forcefulness, just as if it were an acknowledged fact. That, of course, was Cicero's practice. In fact, your friend the Bullbearer, in reply to my remark in conversation that some people were furious because I was not writing anything against Luther, said: 'Why should you have to write against Luther?'[456] Nevertheless, this is the main point of your entire indictment. Moreover, although at that time I was hoping for a happier outcome (for I am not a prophet), nonetheless, if any reputable person had urged this work on me diplomatically, I would not have shirked my duty.

Here is my entire plea in a nutshell. Unless you have proved that Erasmus is the one who by learning, eloquence, talent, judgment, and prestige would be more effective than countless theologians, scholars, and bishops, unless you have demonstrated that Luther can be smashed with a minimum of trouble, then no blame attaches to me that would not be shared by all the rest. And you bring this charge against me alone!

But to bring my own epilogue to its conclusion: I am charged with not having written against Luther. I was among the first to write. 'But you did not publish articles.' I did not, but then I am not a university. And if promptness is to be praised in this connection, the University of Paris was the last of all to do this, though it is the first in prestige. Furthermore, if silence is made a criminal charge, note how many universities you are condemning with a single judgment. Only three universities wrote articles,[457] the rest remain under indictment. 'You did not attack the man with a book.'[458] My *Diatribe* contradicts this; it was published long ago, and more than once. 'But not fiercely enough.'[459] Luther denies this, for he did not reply more bitterly to anyone than to me.[460] 'But then you fell

* * * * *

455 Pio *Responsio paraenetica* 14[r]K

456 When and if Aleandro ever made such a statement is not confirmed by other sources, eg *Spongia* LB X 1645A–1646C / ASD IX-1 150:693–151:763; Epp 1188:30–2, 37–42, 1195:56–71, Allen Ep 1238 introduction, Epp 1244:2–38 especially 33–4, and 1342:67–80, 115–25, 152–66. Indeed, Erasmus stated explicitly to Botzheim that as papal nuncio Aleandro had exhorted him to write against Luther; see *Catalogus* 35:11–15.

457 See n435 above.

458 Pio *Responsio paraenetica* 14[r]–[v]K mentions private persons, professors, religious leaders, and Henry VIII as having attacked Luther with books.

459 Pio *Responsio paraenetica* 27[r]R criticizes the *De libero arbitrio* on this score; see n470, but also n473 below.

460 Luther's reply, *De servo arbitrio*, published in November 1525, denounced Erasmus as ignorant about this theological question and as being an atheist, a vacillating sceptic,

silent.'[461] On the contrary, I replied with two volumes in which you would find neither boldness nor severity lacking. 'But Prierias wrote before you did.'[462] Yes, I admit it, but he wrote in such a way as to weaken the case for indulgences. 'Cardinal Cajetan wrote.'[463] He wrote more subtly, to be sure, but so far as the advancement of the cause goes, no more successfully.

It would have been the height of rashness for me, a private person, to unsheathe my pen against him before the censures by the universities, before the pope's lightning bolt, before the emperor's edict. After these it was redundant to write, for it was unlikely that the man who not only failed to yield to so many authorities but was even moved to greater boldness[464] would yield to me alone. If you say now that I would have had greater influence than the censures of the universities, than the bull of the pope, than the books of the theologians, than the edict of the emperor, then, forgive my boorishness, no one will believe you, nor, I think, do you yourself believe what you assert.

Finally, the outcome itself shows what success I might have had. Read Luther's *Bondage of the Will*, read my *Hyperaspistes*, and you will realize that what I am saying is the truth.[465] So much in passing for your allegation.

As it is, there are many matters that it is neither safe for me to commit to writing nor beneficial to the cause we both support. For the unhappy character of our times is such that it is more dangerous to oppose the interests of men than the glory of Jesus Christ. I am sure that you too, in your prudence, can imagine what I am saying by my silence. If I had seen any band of men who, setting aside their personal passions, were doing the work of Christ with single-minded devotion, believe me, I would have run to them, through rivers, through bogs, through rocks, through fire, and I would be running even today, if only to die among them. Those who contemplate the beauty of God's house with the eyes of the spirit and then turn these same eyes to the criteria whereby the dignity of the church is now reckoned up cannot but sigh from the bottom of their hearts. But I stop here.

* * * * *

and a satirical Epicurean; see WA 18:605, 609, 652–3; Erasmus' complaints about this in Allen Epp 1670:30–7, 1688:14–15, 1690:20–2; and *Purgatio adversus epistolam Lutheri* LB X 1558A / ASD IX-1 482:76–8.

461 Pio *Responsio paraenetica* 27ʳR warns Erasmus of the dangers of having disagreed openly with Luther in only one book, and on only one topic; see n472 below.

462 Silvestro Mazzolini wrote several works against Luther. For their titles and dates, see n176 above.

463 See nn 201 and 203 above.

464 *ad audenda majora* LB *ad audienda majora* 1529

465 For a critique of Erasmus' doctrinal positions on grace and free will and of Luther's reponses to them, see McSorley.

After deciding that you have made it quite clear that I lacked the will to write against Luther, you strain with rhetorical virtuosity to prove that I did not lack the ability, and in the process you provide a summary treatment of all of Luther's teachings.[466] If you are doing this to upbraid me for not writing, I have already written; but if you intend to equip me for writing,[467] it is already too late. This is so not only because you are providing the weapons for what has already been done, but also because the situation has long ago gone beyond human power, and the business is being done not with books but with weapons. If laws are void in the midst of arms,[468] this is even more the case with books of controversy. Nothing remains but for us to recognize the hand of God summoning us to repentance, and for all of us, whoever we are (for we all stand in need of the glory of God),[469] to seek refuge in his mercy.

If, however, it was your purpose to rout all the ranks of the heretics in a single battle, and to destroy them in a general slaughter, I would very much have preferred that this part of your book had been either equipped with arguments so solid that it would shut the mouths of all the impious, or else so successful that it would recall all men to sounder thinking. But you would have accomplished this more easily if you had addressed by name the enemy array, or, surely, its leader, instead of me.

For the rest, since there is nothing here that is found in my writings, I have nothing to say in reply, except to touch selectively upon certain matters that you chose to scatter through your discussion, though I do not quite see what you had in mind in doing so.

You criticize the *Diatribe* because in the beginning and at the end I am more temperate than is appropriate.[470] Luther did not think so, as is made

* * * * *

466 Pio *Responsio paraenetica* 15^{r}L–26^{v}Q

467 By admitting that Pio may have had such an intention when he wrote this and the later section (*Responsio paraenetica* 71^{r}S–96^{r}L) refuting Luther's teachings, Erasmus weakens his earlier argument (3–4) that Pio had already written this attack on Luther and merely appended it to his reply to Erasmus in order to embroil him in the Lutheran controversy. Or perhaps Erasmus is here referring only to the earlier section (Pio *Responsio paraenetica* 15^r–26^v).

468 Imitating Cicero *Pro Milone* 4.10: *Silent enim leges inter arma*

469 Alluding to Romans 3:23

470 Pio *Responsio paraenetica* 27^{r}R: 'In that book, though you argue a lot, you establish nothing definitively, and that is what the situation really required. You did destroy Luther, although you handle the madman more gently than the topic and the insanity of this most stubborn heretic demanded. Indeed, you are overindulgent to him, and seem to fawn on him. Forgive my frankness, but your temperateness seemed to me, and to many, inappropriate, naive, and timorous. Thus, the beginning and end of your book cannot be approved as is the rest.'

clear by the book with which he answered me.[471] But let us grant that I was too temperate, surely you cannot complain of this in the *Hyperaspistes*, yet you do not express your approval of it, ignoring the fact that it was published.

Unless, you say, I write against Luther again, I will have increased people's suspicion of me,[472] now that I have shown in the *Diatribe* what I could have done if I had wished.[473] What good does it do you, a man who, as you write, are extremely busy, to pour out in vain here such a flood of words? What you say all men desire, insist on, and demand of me[474] I have already provided in the two books. You act as if this did not happen, so as not to miss the opportunity for a long declamation in Ciceronian style.

You note in passing that I do not spare even the dead,[475] but go ahead and pass judgment on their works. If this is something I have in common with all writers, particularly with your Aristotle, for whom you have, quite properly, such high regard, why should I alone be blamed for it?

* * * * *

471 That is, the *De servo arbitrio*

472 After praise of the *De libero arbitrio*, quoted in part in the next note, Pio goes on to say (*Responsio paraenetica* 27[r]R): 'If you do the same in other doctrinal areas, there will no longer be any way that people can suspect that you are in agreement with Luther. But if you neglect to do this, then by publishing this book you have, instead, increased their suspicion. For many will think that if you disagreed equally on the other points, you would likewise refute the other positions; you seem to approve of them by your silence, since you condemn this one only.' Erasmus here tries to obscure this point about the need to refute Luther's views on topics other than free will by insisting that he has already written two works against Luther, the *Diatribe* and *Hyperaspistes*, but failing to mention that both works dealt with the topic of free will.

473 Pio *Responsio paraenetica* 27[r]R: 'For what you could do against him, if you wanted to, you have shown in your book *De libero arbitrio* in which you do not vex Luther, or only disquiet him, but you lay him low, you slaughter him.'

474 Pio *Responsio paraenetica* 27[r]–[v]R: 'Wherefore all who love you, all who contend on behalf of Catholic truth, desire this especially, ask and demand it of you.'

475 Pio *Responsio paraenetica* 28[r]R: 'You have not even spared the dead, but have gone ahead and passed judgment on their works. Now, however, will you, alive and present, keep silence and allow the living, present, blasphemous Martin to rage out of control, and not check him with your rod [Prov 22:15, 29:15]?' The identity of the dead criticized by Erasmus is not clear. Pio faulted him for rejecting the opinions of the ancient and holy Fathers (eg *XXIII libri* 88[r]G, 178[v]E, 182[v]R). Because Pio does not refer to Erasmus' work against Ulrich von Hutten, the *Spongia*, it is doubtful that he had in mind someone so recently dead who had been criticized by Erasmus. Hutten's *Expostulatio* was later (100–1) mentioned by Erasmus. Erasmus' reply to Hutten, the *Spongia*, went to press in mid-August of 1523, but the printing was not finished until 3 September. It was published in October even though Hutten had died at the end of August; see Ep 1389:5–8, 81–6. It should be noted that Erasmus would also publish an attack on Pio months after Pio had died.

Neither the apostles, you say, nor Augustine provided any opportunity.[476] I shall allow you to win this point if you can prove that Augustine provides no support for Luther on the question of the freedom of the will, not to mention the rest.[477] But one whose sound statements are violently seized upon as an opportunity does not furnish that opportunity. Then why should I be said to have afforded one, when no such thing can be shown in my books?

You seem to treat of humanistic studies as if it were my position that a simple teacher of rhetoric or of literature can dispute about matters of faith better than someone trained in the more weighty studies.[478] But I have advanced no such opinion. It is beyond controversy that a knowledge of languages is necessary for a theologian,[479] but whether

* * * * *

476 Pio *Responsio paraenetica* 28ʳ–ᵛR: 'But as to your defence [Ep 1634:81–4], that it should not be blamed on you if the Lutherans have taken their opportunity from your works, since the same blame could attach to Paul and Augustine, on whom Luther draws virtually to prove his doctrines, very many do not accept this defence. They say that they [Paul and Augustine] furnished no opportunity, but that Luther impudently snatched certain statements of theirs which he had, of course, twisted, as proofs of his ravings. They say that you, however, have furnished a great opportunity, since Luther takes most of his material from texts that you had noted and had raised questions about earlier. But a great many do believe you when you write that you never suspected, when you wrote these things, that a tumult of this sort would break out.'

477 For three studies of Luther's agreement or not with Augustine on questions of salvation, see Walther von Loewenich *Von Augustin zu Luther: Beiträge zur Kirchengeschichte* (Wittin / Ruhr 1959) 9–87, 161–90; McSorley 63–110, 217–73, 297–369; and Scott S. Ickert *Norms of Orthodoxy and Their Use in Jacob van Hoogstraten's Literary Attack against the Theology of Martin Luther (1521–1526)* (PHD dissertation, Catholic University of America 1985). Erasmus saw much agreement between Luther and Augustine; see also Epp 1033:92–4 and 1634:83–4.

478 Pio, in response to Ep 1634:102–18, takes up the question of the relationship of *bonae litterae* to theological studies in *Responsio paraenetica* 29ᵛv–31ʳz. Erasmus' complaint is based, perhaps, on 30ʳx: 'Therefore I cannot at all agree with those who maintain that this study of *bonae litterae* is more suitable to the acquisition of true theology than the study of more substantial disciplines. Let the most articulate rhetorician, the most precise grammarian, the most creative poet come forward, and let them strive with all the force of eloquence. Will they be able at all to explain the most profound mystery, the ὁμούσιον of the Trinity ...?'

479 Erasmus is reacting here, perhaps, to Pio's assertion in *Responsio paraenetica* 29ᵛY: 'But just as no one with an education would deny that all good *artes*, all areas of study, are the handmaids of theology, so it is certainly true that eloquence is of benefit to it. Yet eloquence is more useful for teaching and expressing theological ideas than for acquiring a knowledge of theology. But it is well established that philosophy is absolutely necessary, both for acquiring this knowledge and for expounding the ideas'; Pio mentions languages only in the course of observing (34ᵛA) that 'all of these [*litteratores* and *grammatistae*] ridicule and despise those who are less eloquent, even though they are the most precise philosophers and wise theologians, and they consider them unworthy to discuss Scripture because they are not trilinguals, or at least bilinguals.' Erasmus had presented his views on the necessity of knowing the biblical languages in his *Methodus*;

human philosophy[480] is required seems to be, in any event, a matter for dispute.

I have long ago given a clear indication of my opinion about making the Bible available to the common people,[481] so let the cabbage so often reheated[482] not make the reader sick. As to your statement that pearls should not be cast before swine, and that what is holy must not be given to dogs,[483] you have a low opinion of the Christian people if you consider they deserve to be compared to pigs and dogs. Surely Christ saw fit to reveal his mysteries without discrimination to the simple and uneducated, and yet lately we have begun to hear that the masses should be barred altogether from the reading of the Bible.[484]

I am afraid that it will seem to some an unfair distribution that you claim wisdom linked with eloquence for your Italians, while you leave nothing but an empty eloquence for my countrymen,[485] who, as you put it, prefer the pursuit of eloquence to the pursuit of wisdom.[486] And yet

* * * * *

see Margorie O'Rourke Boyle *Erasmus on Language and Method in Theology* (Toronto 1977); and Rummel *Humanist-Scholastic Debate* 112–23.

480 An allusion to Pio's encomium of philosophy in *Responsio paraenetica* 30^{V}Y–33^{V}A

481 Pio *Responsio paraenetica* 31^{r}z: 'Therefore I do not approve the position of those who think that Sacred Scripture should be made generally available to all indiscriminately, and should be divulged and expounded in the vernacular, barbaric language.' Erasmus proposed the view that all Christians and even Moslems should be able to read and discuss the Scriptures in his *Paraclesis* (LB VI *3^V / Holborn 142:15–28).

482 *Adagia* I v 38: *Crambe bis posita mors* 'Twice-served cabbage is death' LB II 196D–197B / ASD II-1 512–14 / CWE 31 417–18. Erasmus quotes the use of a related expression in Juvenal 7.154 and observes: 'He means by "cabbage served up again" an utterly boring speech which has to be listened to over and over again.'

483 Pio *Responsio paraenetica* 33^{r}z: 'But not even for that reason [widespread ignorance in matters of faith] does it seem good that every type of scriptural text be published for the masses, neither should pearls be cast before swine, nor what is holy (as the Lord says [Matt 7:6]) be given to dogs.'

484 Erasmus defended his advocacy of the common man having access to the Bible against the complaints of the Sorbonne; see *Declarationes ad censuras Lutetiae vulgatas* LB IX 872A–873D.

485 Pio *Responsio paraenetica* 35^{r}A: 'But they [the Germans] declare that *bonae litterae* have migrated from Italy to Germany. The Romans rejoice (the Germans deny this) that the influence of *bonae litterae* has been extended, and that the Germans are sharers in it. But they [the Romans] grieve that the Germans have not drunk of the Romans' gravity and prudence, and the Italians' wisdom, as much as they have their elegance of discourse.' For an overview of the differences between Italian and northern European culture and piety and of how Erasmus viewed and was viewed by Italians, see Gilmore 'Italian Reactions' 61–115. For some statements of Erasmus that are hostile to Italian claims to cultural superiority, see Epp 112:10–13, 321:9–15, 569:58–82, 880:5–7, 1110:71–3, 1165:50–3, 1187:6–11, etc.

486 Pio *Responsio paraenetica* 29^V–30^{r}v: 'It was determined that it [the pursuit of eloquence] produces more harm than good, unless it has been linked to wisdom. This has been proven in a modern instance by those Germans, articulate men rather than truly

nowhere are there more people who spend their entire lives doing nothing but polishing their style than in Italy. Among our countrymen the study of philosophy flourishes more than among you, as does the study of theology, which among you has been almost entirely relegated to a few monks.[487]

You call heretics those who, out of disrespect, refer to the pope as the bishop of Rome.[488] Whether I used this title somewhere I do not quite remember.[489] Surely, if I did, I did not do so as an insult. The ancient orthodox used to bestow this title as an honorific, and the popes used to designate themselves by this title.

There is, however, no reason, most eminent prince, for you to entreat me not to take offence because you appear to instruct me.[490] I wish that

* * * * *

eloquent, who have given themselves over to Luther. If they had cultivated wisdom as much as elegance in speech, hostility towards certain theologians would not have overcome their reverence for the true and the upright, and their desire for revenge would not have overcome the cause of religion and love of country.'

487 Erasmus' statement is basically true. Pre-Reformation Italian universities had no faculties of theology in either the medieval or modern sense of the term. The *collegia theologorum* present in some were not teaching faculties but promotion boards that set standards and conferred degrees. Those who taught theology in the Italian universities were often mendicants, did not provide a coherent body of courses, and were members of the arts faculties. The systematic teaching of theology was carried out in the *studia* of the various mendicant orders. See John Monfasani 'Aristotelians, Platonists, and the Missing Ockhamists: Philosophical Liberty in Pre-Reformation Italy' *Renaissance Quarterly* 46 (1993) 247–76 especially 253–6; Grendler *Universities* 351–92.

488 Pio presents a vigorous defence of the papacy in *Responsio paraenetica* 52ᵛκ–66ʳᴘ, twice bringing up the appellation 'bishop of Rome'; on 53ʳκ he says of the humanists among Luther's supporters: 'they have suffered no injury, no harm from the supreme pontiff, whom they call "the bishop of Rome" with, of course, a view to diminishing his authority'; and on 61ʳɴ he remarks: 'All these most prudent countries and peoples recognize a highest priest and supreme pontiff, and they acknowledge that he is the one whom the heretics call the bishop of Rome.'

489 Erasmus had pointed out ancient practice in a note on Jerome's Ep 1: '*Romanum episcopum*: You see that among the ancients he was not called *summus*, but *Romanus*' (HO 1 106ᵛc). See also Ep 843:564–9. Zúñiga repeats these words almost verbatim as the third of his *Conclusiones principaliter suspectae ... in libris Erasmi Roterdami* (text quoted in LB IX 381C). In his reply to this (LB IX 386C) Erasmus says that he does not know Zúñiga's source, but supposes it to be a letter in which he corrected the error of a critic who had termed Pope Damasus *summus pontifex*. Erasmus himself uses the expression *Romane urbis episcopo* in *Vita Hieronymi* (*Erasmi opuscula* 155:584) and in his note on Jerome's Ep 15: '*Hieronymus ... consulit Damasum Romanae urbis episcopum*' (HO 3 59ᵛc). And see nn 189 and 190 above.

490 At *Responsio paraenetica* 98ᵛɴ Pio shifts his focus back to Erasmus after his long attack on Luther, explaining: 'While I have presented a wide-ranging discussion of these issues, I have not done so because I have any suspicion that you are unaware of any of the matters I have brought forward (for I know that you can readily assemble both many

you had time[491] to teach me much more. In this respect, at least, I am most grateful. How grateful the Lutherans will be I do not know.

I do wonder why at the end of your book you sing again that tune which has so often been sung to me in vain, that is, that even after the *Diatribe* I should again unsheathe my pen against Luther, and do so fiercely.[492] Since you cannot be ignorant of the fact that I did this long ago, I do not quite see why you have chosen to pretend this did not happen. If you had written this before seeing the books, it would have been far better to alter this final flourish. This would have better served the reputations of each of us: mine, because I am being vilified without justification; yours, because you will appear either so lazy that you were unwilling to undertake so much work, or so dishonest that you pretend that something which is quite well known to everyone does not exist.

And meanwhile you threaten, please God, that, unless I do as you say, the consequence will be that a suspicion of Lutheranism will cling to me, that I will seem to be in agreement with Luther in ever so many respects, that reputable men will declare that I was the originator of these vast evils, and that, when I am dead, I will carry this reputation with me to the grave.[493] What if I do as you say? Will none of these happen? None, you say. But I have long ago done what you are urging. In fact, if what you have tried with such zeal to prove is true, that Luther gathered his deadly seeds from my gardens,[494] how will any writings of mine, however violent, remedy this evil? Can what has been done be undone? Yet, if I rage against Luther, I will have confirmed this suspicion, for I will be said to have atoned for my earlier failing by this service. But I have never written

* * * * *

more and better suited arguments against the outrage, if only you give it your attention), but because they came to my mind as I touched upon these points.'

491 A sarcastic comment, given Pio's forced retirement to Paris as an exile and underemployed former diplomat.

492 Pio *Responsio paraenetica* 98^{v}N: 'sharpen your pen against Luther, and write things worthy of Erasmus, not veiled, not convoluted, not ambiguous, not slippery but clear, not slack but violent, as the arrogance, impiety, and blasphemies of this fellow demand'; and 99^{r}N: 'Only gird yourself with your pen [Ps 44:4] as if with the sword of the Spirit [Eph 6:17], but not with a two-edged one [Heb 4:12]; victory is in swiftness.'

493 Pio *Responsio paraenetica* 99^r–^{v}N: 'If you fail to do this (pardon my bluntness), you will never prevent many people from suspecting you, many sound men from declaring that you were the original author of vast evils. You will not escape carrying the mark of heresy with you when you die, nor prevent succeeding generations from saying, when you are mentioned, that Erasmus would have been a great man had he not in a way supported Luther's defection. And thus by many will the glory of your name forever be diminished.'

494 Pio's words at *Responsio paraenetica* 5^{v}D, dealt with by Erasmus earlier, 21 above

nor shall I ever write a syllable against Luther on this account! Let these people lie, invent, fabricate whatever they like!

In addition, I am afraid that you may be found wanting in consistency, since at the beginning you professed that you have always thought and spoken of me in a complimentary way, and yet at the very beginning of your argument you hurl against me the following accusations: I diminish the dignity of priests and pope, I ridicule the monastic life, as well as the many other charges mentioned above. If you do not believe these things about me, what is the point in going to so much trouble to prove the calumnies of others? If you do believe them, how is it consistent that you have always thought and spoken in a complimentary way about one of whom you believe so many dreadful things?

In fact, a great many will feel a lack of fairness in your singing to the world this malevolent song at an unpropitious time. Eight years ago certain men with malicious tongues[495] were attaching to me a suspicion that I was not altogether uninvolved in Luther's business. Despite the fact that this was completely untrue, the slander was able to be viewed as in one way or another likely. At that time it would have appeared an act of kindness if you had challenged me to declare what I really thought. But now, after the fact that Luther and I agree on nothing has become better known than is good for me, after I have been pelted by so many books,[496] after I have aroused against myself the hatred of so widespread a faction, after I have battled away like a gladiator with Luther himself, how ill timed is this your complaint or, I should say, your exhortation?

You are complaining of things that cannot be changed. You exhort me to do things that I have already done. The first is superfluous, the second is ridiculous, and both are ill timed. Accordingly, just as the publication of your book will stir up a lot of hostility towards me, particularly among those who approve of anything that is an attack on Erasmus (people who took an extraordinary delight in Hutten's *Expostulation*,[497] even though they

* * * * *

495 Among the persons who early on accused Erasmus of being involved in the Luther affair were the following: certain wicked theologians Ep 993:52–6 (1519); certain very stupid men, Ep 1033:61–3 (1519); theologians at Louvain, Dominicans in Antwerp, a Franciscan in Bruges (Nicolas Bureau, the auxiliary bishop of Tournai), Epp 1144:31–53 (1520) and 1192:34–59 (1521); see also Rummel *Catholic Critics* I 133 (Vincentius Theoderici OP), 135, 137–8 (Nicolaus Baechem OCARM), 158–60 (Sancho Carranza de Miranda), and 164–5 (Diego López de Zúñiga).

496 For recent surveys of how Erasmus was attacked in the writings of his enemies, see Rummel *Catholic Critics*; and Cornelis Augustijn *Erasmus: His Life, Works, and Influence* trans J.C. Grayson (Toronto 1991) 135–60.

497 Erasmus claimed that Hutten's *Expostulatio* would be very gratifying to Erasmus' ene-

wanted to appear absolutely hostile to Luther), so also, I fear, among men of unimpaired, sound judgment your book will considerably diminish the good opinion of you which had obtained in the minds of many, that is, that Alberto is a gentleman blessed with unusual fairness and generosity. For you will appear to have mixed up my case with that of Luther with a cunning that is anything but gentlemanly, and, as it were, to have joined one thread onto another.[498] You heap up a lot of considerations so as to appear to have been provoked into doing this by my letter, but the facts tell another story.

They will note in addition that your charge is far more fierce than your proof, with the result that you appear to have written not for impartial judges but only for my enemies. No one, surely, will excuse your having published at an inopportune time, your not changing items which the facts told you should be changed, and your pretending that the two books of the *Hyperaspistes* were not published after the *Diatribe*.

Perhaps you will say that I am now committing the same fault that I reproach in you, since I am advising you that a book which has already been printed and is on sale ought not to have been published. I would have advised you in good time if I had found out where you were, and as soon as I heard that you were living in France, I wrote a letter to you about this matter. I am quite amazed that it was not delivered to you.[499] That letter, you see, was looking out for your good name no less than for mine. But

* * * * *

mies, such as van Hoogstraten, Baechem, and many others; see Ep 1356:27–33. Whether this was the case is not clear; on Protestant reaction, see Klawiter 19.

498 *Adagia* I viii 59: *Linum lino nectis* 'You join thread with thread' LB II 321B–E / CWE 32 158–9): Probably to be taken here in the sense of joining together things that are discordant: 'Proclus, in discussing the questions raised by statements in Plato, uses it in one passage in a way that makes it seem applicable to a man who couples together incompatible things.'

499 Allen Ep 2080 (Basel, 23 December 1528). Berquin had written to Erasmus from Paris on about 13 October 1528, telling him that Pio was active in Paris; see Allen Ep 2066:60–2. Erasmus later admitted that this letter was never delivered to Pio because its carrier feared that the letter might provide Pio with a new opportunity for venting his spleen against Erasmus; see *Apologia* 153 below. The carrier alluded to here is Berquin, as we learn from Allen Ep 2291:25–9 (27 March 1530), a letter written after Berquin's tragic death in 1529: 'The letter to Tussanus was, in fact, sent, but I suspect that it was suppressed by Berquin, something he was wont to do from time to time. I had also written to Alberto Pio; he [Berquin] admitted that the letter was not delivered through his fault, his reason being that he judged it inexpedient.' Pio *XXIII libri* 65ᵛK states: 'Certainly the [letter] was not delivered to me, about which failure I wonder, since you forwarded that to some trusted friend and nevertheless I was in Paris already many months beyond a year before the epistle was published.'

since these events cannot be recalled, we should try to make the best of the situation.

I do not really understand your asking me to forgive you[500] if you appear to have said anything too intemperate, violent, or bitter against me, or your telling me I should not imagine that any such thing was said against me, but should set it all down to the account of Martin, since 'no one, provided he is zealous for the Christian religion, can hear him mentioned without burning with anger and relaxing the reins of propriety,' because, you say, 'to temper one's language against so impious, so snarling a heretic seems to be contrary to all decency, and not just contrary to piety.'

You will forgive my amazement, most renowned Alberto. What sort of zeal is this, to be so angry at a heretic that you rage against an orthodox man? This is a mark of the insane, to strike a friend when they are angry with an enemy. But the ill also,[501] when they are in excruciating pain, attack their dearest ones with curses or even blows. If any such thing is going on here, then I regret that I have been so very close to you. But perhaps I am not following your meaning, and you meant to express something else with these words, for you are not, I think, really afraid that I would consider that anything said against Luther applies to me.

If your good will towards me were as great as you assert,[502] it would have been kinder to interpret my words in a favourable way instead of distorting sound statements so that they would appear to be in agreement with Luther's, not to mention the fact that you mention many scandalous things that you could not have found anywhere in my writings.

Furthermore, if you are as hostile towards Luther as you want to appear, I wish you had manifested this fierce energy in your proof rather than in your introduction. There it is a hand-to-hand fight, and the spectator waits to see who will stand and who will fall. But in the proof your speech is frequently so watered down that you seem not so much to prove your point as to want to obtain it by entreaty. Read your argument again, but as if it were someone else's, and you will find that this is the case. It is best either not to provoke Luther or to prosecute your case with the most convincing arguments. But my advice about this is now too late. In short, I wish that you revealed to us how very talented you are with some other

* * * * *

500 The following is a paraphrase and quotation of Pio *Responsio paraenetica* 99[V]N.
501 On Pio's chronic illness, see Introduction lxxv n81.
502 Pio *Responsio paraenetica* 99[V]N: 'So far as you are concerned, please be convinced that, because of your talent, learning, and singular virtue, I am moved to a love and respect for you that few will match.'

topic, or at least that you had handled the topic you chose in a different way.

In closing, I gladly embrace your invitation to a bond of friendship,[503] and I promise that I, though inferior in all other respects, will nowhere be your inferior in kindness and ready good will. But you must, in turn, forgive me, most generous sir, for in the course of thrusting savage calumnies away from myself I shall seem to have said certain things in a manner more outspoken than is in keeping with your dignity. I wish that this were the sort of controversy in which one could have taken at the same time both your dignity and my safety into account. In addition, I beg you to take it in good part that I am answering your exquisitely wrought book with an ignorant and obviously Dutch letter.[504] I was afraid that if I had kept silent someone might construe it as done out of contempt for you. But scarcely six days were spent on this work,[505] and a great part of that time had to be allocated to other activities. Whatever you see is extempore. But even if abundant time had been available, I am not so ignorant of my limitations that I should wish to vie either in learning or in eloquence with a man of the sort that I am always accustomed to revere rather than rival.

Farewell.

Basel, the ides of February 1529

* * * * *

503 Pio *Responsio paraenetica* 99ᵛN: 'and this letter will strike between us an everlasting bond of most holy friendship.'

504 A restatement of Erasmus' earlier apology (3, 5–6); here it corresponds to Pio's equally commonplace apology (*Responsio paraenetica* 98ᵛN): 'Therefore, so as not to dull your refined ears any longer with my coarse discourse, I will conclude.' For the 'Dutch' reference, see n335 above.

505 See Introduction lxvi–lxvii. Erasmus had similarly lamented the loss of six days of working time in reading and replying to Hutten's *Expostulatio* (LB X 1672F / ASD IX-1 210:141 / Klawiter 248). For a discussion of the significance of such a claim, see de Jonge in ASD IX-2 267 n636.

THE APOLOGY AGAINST THE PATCHWORKS OF CALUMNIOUS COMPLAINTS BY ALBERTO PIO, FORMER PRINCE OF CARPI

*Apologia adversus rhapsodias
calumniosarum querimoniarum
Alberti Pii quondam Carporum principis*

DES⯈ ERASMI ROTE⸗
RODAMI APOLOGIA ADVER⟋
fus rhapfodias calumniofarum querimo⟋
niarum Alberti Pij quondam Carporum
principis, quē & fenem & moribundum
& ad quiduis potius accommodum ho⟋
mines quidam male aufpicati, ad hanc il⟋
liberalem fabulam agendā fubornarunt.

FRO BEN

BASILEAE IN OFFICINA
FROBENIANA.

ANNO M. D. XXXI.

Title-page of Erasmus *Apologia adversus rhapsodias Alberti Pii* (Basel: Froben 1531)
By permission of the Folger Shakespeare Library, Washington, DC

THE APOLOGY
OF DESIDERIUS ERASMUS OF ROTTERDAM
AGAINST THE PATCHWORKS[1] OF
CALUMNIOUS COMPLAINTS BY
ALBERTO PIO, FORMER PRINCE OF CARPI,
WHOM, ALTHOUGH ELDERLY AND
TERMINALLY ILL AND BETTER SUITED
FOR ANY OTHER UNDERTAKING,
CERTAIN ILL-STARRED MEN
HAVE CLANDESTINELY INCITED
TO ENACT THIS FARCE

Of countless passages, this INDEX[2] will point out some manifest instances in order to make evident just how much credence should be given to Pio's impious slanders.

* * * * *

1 Erasmus chose as his pejorative epithet for Pio's XXIII libri the Greek loanword *rhapsodia*, the primitive or etymological meaning of which is a poem composed of fragments sewn together. Although Erasmus does twice describe Pio's work as a mosaic (see 116, 198 below), the language he more commonly uses to describe the XXIII libri is derived from textile: *assuo* 'sew on,' *attexo* 'weave/sew on,' and *contexo* 'weave/sew together,' and he describes Pio's work implicitly as a *cento*, a usual Latin term for a patchwork (281 below, and see *Adagia* II iv 58). Erasmus calls Pio's alleged helpers *rhapsodi*, not meaning performers of segments of the Homeric poems, a classical use of this term (see OCD sv 'rhapsodes'), but the scroungers and sewers who gathered scraps for the XXIII libri and stitched them together: *Post haec nescio quis ex Pii rhapsodis attexuit fragmentum* ... 'Next some one of Pio's "stitchers" sewed on a fragment ...' (222 below). Finally, Erasmus actually offers an example of what he means by a *rhapsodia* when he calls Gratian's *Decretum* a *rhapsodia* (338 below). Thus, 'patchwork' seems a more appropriate translation for *rhapsodia* in this case than other pejorative meanings (eg nonsense, blather, rigmarole) that *rhapsodia* might have in later Greek usage.

2 This Index occupies sig aa2^{r–v} of the *editio princeps*; in LB it is printed at the conclusion of the work at the foot of cols 1195–6, though Erasmus refers to it in the *Apologia* 113–14 below as a 'prefatory index' [*index quem praefixi*]. The page numbers refer to pages in the *Apologia* where he also provided marginal notes indicating the passages referred to here in the Index. The page numbers printed in italics and enclosed in brackets refer to the pages in this volume where the passages may be found.

A lie: page 18 [*page 123 in this volume*], likewise 22 [*126*], also 23 [*127*], and from there onwards 24 [*128*], 25 [*129*], 26 [*130*], 27 [*132*]. Likewise on page 35 [*139*] two lies. Again 37 [*141*], 38 [*143*], 39 [*144*]. Next 87 [*186*]. Next 92 [*190*]. Next 101 [*198*]. Next 116 [*209*]. Next 127 [*217*]. Next 147 [*233*]. Next 190 [*273*]. Next 191 [*274*]. Next 263 [actually 268] [*345*]. Again page 264 [*341*]. Also 273 [274] [*350*].

Also in the Index under the name of Erasmus[3] he has placed a huge heap of remarkable lies.

Toadying and silly, page 28 [*131*]

Hateful criticism of things he does not understand, page 29 [*134*]. Again page 39 [*145*]. Next 102 [*199*]. Again page 103 [*199*]. Also 148 [*233*], 150 [*236*], 250 [*326*]

Fails to understand his own terms, page 31 [*135*]

He crashes on a rock already pointed out, page 252 [253] [*328*].

Irrelevancies ['Aπροσδιόνυσα],[4] page 32 [*136*]

Repeats things I have obviously refuted, page 35 [*140*]; he does this frequently elsewhere as well.

Absolutely silly hyperboles, page 53 [*157*]

He hisses bitterly without justification, 36 [*140*].

He is fierce in promising things he never delivers, page 79 [*180*].

He gives a garbled quotation, page 85 [*185*]. Also page 250 [230] [*308*][5]

A malicious distortion, page 163 [*246*][6]

Stupid abuse, page 95 [*192*]

He falsifies a quotation, page 107 [*202*]. Again page 128 [*218*]. Shortly after on page 129 [*220*]. Likewise page 136 [*225*]

* * * * *

3 This index (termed *tabella* and *index* in XXIII *libri*, but hereafter referred to by Erasmus' name for it, *Elenchus*) is the 'remarkable index [*tabella*], arranged in alphabetic order, of notable opinions and of the whole mélange in the ensuing work' prepared to accompany the publication of Pio's XXIII *libri* (see Introduction lxxxvii and Translator's Note cxlviii). The sequence of events must have been that Erasmus received a copy of XXIII *libri* without the title-page or the Elenchus and proceeded to compose his *Apologia*. Once his *Apologia* was set in type, he was able to prepare his own Index to it (see preceding note). He must have received the voluminous Elenchus at some point after his *Apologia* was completed, probably after it had gone to press, and before his own Index was set in type. He contented himself for the moment with this glancing attack on the Elenchus, either postponing or, perhaps, deciding later to reply to it with his own *In elenchum Alberti Pii brevissima scholia* 361–85 below (see Introduction cviii).

4 See *Adagia* II iv 57: *Nihil ad Bacchum* 'What has this to do with Bacchus?' LB II 541D–542D / CWE 33 219–21: 'when someone brings forward trifles which have nothing to do with the matter in hand.'

5 Erasmus makes this same accusation in a marginal note on 311 n1193 below, but fails to include it here in his Index.

6 This entry is missing in LB IX 1195–6.

He falsifies quotations he does not understand, page 176 [259]. Likewise page 194 [278]
He gives an abridged quotation, page 134 [223].
He raves for no reason, page 166 [248], also 213 [293].

In addition, the very inconsistency in the use of words proves the charge that the work was put together and amassed by the toil of many,[7] as when one reads there *ita quod* instead of *ita ut*, *pluries* instead of *frequenter*, *indecere* instead of *indecens esse*, *morosus esse* instead of *immorari*, *successive* instead of *paulatim*, *virtualiter* instead of *pro re ipsa*, *plurificare* instead of *multiplicare*, and many, many other cases of this sort.

In the replies to Pio's notes, the first number indicates the page, the second the side of the page; the added letter indicates the passage itself.[8]

* * * * *

7 Erasmus cites this collection of non-Ciceronian, indeed post-classical, usages, many of them redolent of scholastic Latin, in support of his allegation of multiple authorship of Pio's *XXIII libri*. He makes a similar point when he complains of the usage *conatur suadere* (see n675 below), and he comments on the change of style at 192 below. Juan Ginés de Sepúlveda criticized Erasmus for blaming the presence of 'some few words ... which exhibit a theological simplicity of style rather than Ciceronian refinement' on the work of multiple authors rather than on Pio's failure to revise his work before his death. Sepúlveda noted that when Pio was dealing with a theological issue, he preferred speaking more clearly (*clarius*), if in less idiomatic Latin (*minus latine*), with words borrowed from Thomas Aquinas, Duns Scotus, or Alexander of Hales (*Antapologia* Ei^v–ii^r / Solana 133). Indeed, Erasmus himself, in the *Ciceronianus* has Nosoponus agree that Pio approaches Ciceronian style 'in so far' as anyone can who has involved himself since youth with theology and philosophy' (ASD I-2 671 / CWE 28 420). On humanist attitudes towards scholastic terminology, see Rummel *Humanist-Scholastic Debate* 35–40, 177–83.
8 Erasmus is referring here to the notes Pio had written in the margins of Erasmus' *Responsio*, which was reprinted, along with Pio's notes, on folios 47^r–66^r of Pio's *XXIII libri* (see illustration 120 below). These notes had been assigned consecutive letters of the alphabet, so when Erasmus responds to a note, he identifies it by a marginal note that indicates the number and side of the folio and the letter of the note: eg 51.fa.2.p. = folio 51, facies 2 [or verso], note p. When Pio's notes are cited or quoted here, we give Erasmus' citation of their locations in *XXIII libri* and indicate in parentheses the passage in Erasmus' *Responsio* on which Pio is commenting. N61 below is an example of this kind of citation.

[PROOEMIUM][9]

Alberto Pio has, of course, accomplished something absolutely clever by imitating the Parthians and taking flight after shooting his arrow,[10] so that I am prevented from striking back. That is, he died before the work was printed,[11] indeed, in my opinion, before it was even finished. I have ascertained[12] that it was his practice to employ a variety of employees[13] for what he wrote against me, some to excerpt passages from my works,[14] others to polish up his style,[15] some to supply proof-texts from Scripture (with

* * * * *

9 1531 UPPER MARGINS: 'Prooemium.' on pages [5]–13 [= (roughly) 110–19 in this volume]

10 See *Adagia* 1 i 5: *Infixo aculeo fugere* 'To flee after planting the dart' LB II 27B–F / ASD II-1 117–18 / CWE 31 53–4, where Erasmus mentions the Parthians who 'cast their javelins against the enemy and then wheel their horses and take flight'; see n21 below.

11 Pio died on 8 January 1531; his *XXIII libri* was published on 9 March 1531.

12 Erasmus learned of this assistance given to Pio from a letter of Gerard Morrhy, dated 16 April 1530, which mentioned a letter sent to Morrhy from a certain Frisian named Gerard who claimed to have been hired by Pio to examine Erasmus' writings for errors, especially in the area of New Testament scholarship; see Allen Epp 1744:130, 2118:16–31, 2311:19–35. Erasmus also claimed that some of those Pio hired to help him with this project later either wrote to Erasmus or confessed to him face to face their role, but Erasmus provided no names; see Allen Ep 2810:104–5.

13 In his *Antapologia pro Alberto Pio* Sepúlveda claimed that Pio, like most princes, had in his employ a secretary, elsewhere identified as Francesco Florido Sabino. See Introduction cxxiv and Florido *Libri tres* 116:21. On Florido, see Remigio Sabbadini 'Vita e opere di Francesco Florido Sabino' *Giornale storico della letteratura italiana* 8 (1886) 333–63. In this passage Erasmus repeats many of the charges made earlier elsewhere, but he also adds the new charges that Pio's assistants polished his style and that Pio was unfamiliar with Scripture. Erasmus had praised Pio's literary style and learning: eg *Responsio* 3, 10 above. In the March 1528 and March 1529 editions of his *Ciceronianus* Erasmus praised Pio's Ciceronian style, but in the October 1529 and March 1530 editions he added 'but it is said by some to be a known fact that the work was shaped by another's hand' (ASD I-2 598:1n / CWE 28 420, 586 n664). There is a certain irony in Erasmus' charge that Pio had helpers in writing his works. Many of Erasmus's friends over the years provided him with materials for such famous works as his *Adagia, Novum instrumentum,* and edition of Jerome. See eg LB II 405B–F; Epp 334:129–37, 456:300–1; Allen Ep 864 introduction; and 163 below.'

14 Sepúlveda objected that Erasmus was here implying that Pio never read Erasmus' works but only the passages excerpted by the friars for Pio to criticize. Sepúlveda refuted this contention by noting that no evidence was produced to support it and that this conjecture was at variance with what was known of Pio's character: he was not so sluggish and dull as to need others to tell him what he should condemn; he was trained in theology and intelligent enough to recognize a controversial statement and to know how to refute it. In direct contradiction to Erasmus' feeble conjecture is the testimony of Pio's secretary Florido, who denies that Pio worked from notes supplied by others. See Sepúlveda *Antapologia* Eii^r–Eiii^r / Solana 134–5 and Florido *Libri tres* 264. Pio himself also denied this charge (*XXIII libri* 70^vK). Erasmus repeated the charge in his *Purgatio adversus epistolam Lutheri* LB X 1545C / ASD IX-1 458:437–9.

15 Erasmus later (118–19 below) cited only one person by name as someone who had polished up Pio's prose, namely, Sepúlveda. Sepúlveda claimed that he was personally

which he was, of course, unfamiliar),[16] some to provide subject matter and arguments, others to bring the views of the theologians to his attention.[17] For this reason I have come to suspect that a great number of things were patched on under the name of the deceased which the fellow himself never saw.[18] These contrivers seized upon the smokescreen of a grand reputation[19] as a vantage point from which to thrust their absurdities upon the

* * * * *

flattered by having so famous a person as Erasmus thus testify to his literary skills in Latin, but he denied having any such role in Pio's work. He pointed out that he was in Italy during the whole time that Pio was writing *XXIII libri* – a fact that should have been known to Erasmus, for Alfonso Valdes had written to him telling him that Sepúlveda and Francisco de Zúñiga had visited Piacenza in the entourage of Cardinal Quiñones. Apart from this trip, Sepúlveda had remained in Rome during the whole period in question. See his *Antapologia* Di[v]–Dii[v] / Solana 128–30. Sepúlveda also noted an inconsistency in Erasmus' position: how could Pio have hired someone to polish his prose and yet a variety of styles still remain? Erasmus' examples of diverse styles focus on a few terms of scholastic theology that had not yet been transposed into the proper Ciceronian expression. The reason for their survival, according to Sepúlveda, was that death overtook Pio before he could rework these scholastic terms. Pio's own ability to write polished prose was evident in his eloquent speeches as ambassador of the emperor and of the kings of France. See *Antapologia* Civ[r], Ei[r]–Eii[r] / Solana 127, 133–4.

16 See Introduction xxii, xxiv–xxv for a detailed account of Pio's unusual abilities in Scripture and theology that refutes Erasmus' insinuation. Erasmus claimed that Pio inserted some scriptural material into his first *Responsio* later, in Paris, when he was helped by the Paris theologians; see Erasmus' *Responsio* 74, 76, and 91 above; and Allen Ep 2118:16–31. That Pio did insert additional scriptural proof-texts is evident from a comparison of his treatment of confession (Pio *Responsio* 9[r–v] and his MS version 7[r]), of paraphrases (*Responsio* 12[r] and MS 9[r]), of the respect due to church authorities (*Responsio* 23[v] and MS 17[r]), and so on. See *Responsio* nn 362 and 376 above and *Apologia* 140 below.

17 Erasmus later (156) suggests that the reason Pio's employees brought to his attention the teachings of theologians was because Pio was a layman, someone unacquainted with theological literature. Sepúlveda refuted this charge by pointing out that Pio was tutored in scholastic philosophy and theology by such noted scholars as Pietro Pomponazzi, Juan de Montesdoch, Graziano da Brescia OFM, the hermit Valerio, and others. He studied serious and difficult questions, especially the thorny enigmas posed by Duns Scotus. He learned the solutions offered by leadings Scotists and was wont to discuss and debate these questions with scholars both in private and in public. His keen, insightful, and eloquent answers used the terminology of Thomas Aquinas, Duns Scotus, and Alexander, but he would alter the wording to make his answers clearer, more pointed, and more intelligible. His answers thus won the admiration of all. He found time for such activities by shunning the idle princely pleasures of games, dogs, and romantic affairs. See Sepúlveda *Antapologia* ciii[r–v], Eii[r] / Solana 124–34. Pio's secretary Florido also denied that the Paris theologians had suggested to Pio phrases, proof-texts, and other things, and he insisted that Pio wrote his own response to Erasmus; see Florido *Libri tres* 264:19–35.

18 This accusation was implicitly denied by Sepúlveda; see his *Antapologia* cii[r], Di[v], Ei[r–v], Eiii[r] / Solana 122, 128, 132–5.

19 See *Adagia* I iii 41: *Fumos vendere* 'To sell smoke' LB II 128E–130A / ASD II-1 354–6 / CWE 31 270–2 and *Adagia* IV viii 83: *Fumus* 'Smoke' LB II 1141E; the expression here, *mag-*

public, so that they could both evade ill will themselves and use the splendour of his name to my disadvantage. Everything they do is managed under subterfuges of this sort, a fact better known to the world than I would like.

In addition, it is considered extremely invidious to write against the dead and, as the saying goes, 'to wrestle with ghosts.'[20] But what are you to do if the dead write against you, and bring against you, before the tribunal of the whole world, charges which it is impious to ignore? The proverb that one should not wrestle with ghosts might hold true if the dead did not contradict the proverb with their biting criticism. The bee has fled, but it has left its sting in the wound.[21] I have no one at whom I can strike back, but I do have a wound to heal.

This already invidious situation was worsened[22] by the fact that three days before he died, once the physicians had declared absolutely that he had no hope of life, Pio put on the most sacred habit of the Franciscans,[23] and in it was borne in solemn procession through the public thoroughfare and entombed in their most sacred monastery.[24] Consequently, there is a danger that if I write anything very harsh to defend myself against Pio, I will appear to be doing an injury to the entire Seraphic Order. So what am I to do? Should I be so in awe of his pious interment that by keeping silent[25] I allow myself to be framed on trumped-up capital charges by the people who are hiding in this Trojan horse?[26] How I wish that this too could be done with Christ's approval! He endured the Cross without retort, but he did not endure abusive charges of impiety in silence; rather, he made careful reply to those who accused him of having a demon and of casting out demons by the spirit of Beelzebub.[27]

* * * * *

nifici nominis fumus, recalls Lucan's allusion to Pompey in Bellum civile 1.135: 'stat magni nominis umbra.'

20 Adagia I ii 53: Cum larvis luctari 'To wrestle with ghosts' LB II 91B–C / ASD II-1 268 / CWE 31 194: 'Wrestling with ghosts is said of those who heap blame on the dead: nothing could be more unworthy of an honourable man.'

21 Adagia I i 5: Infixo aculeo fugere 'To flee after planting the dart' LB II 27B–F / ASD II-1 117– 18 / CWE 31 53–4: 'A proverbial metaphor; someone has uttered an accusation or committed some misdeed, and immediately removes himself, so as to escape being forced to uphold what he has said, or having the same treatment meted out to himself'; cf Allen Ep 2441:71 and 2443:346, where Erasmus applies this proverb to Pio, and see also n10 above for the 'Parthian dart.'

22 accessit 1531 accedit LB

23 See Introduction xc and 390 below.

24 See Introduction xciii–xcvii.

25 See Introduction xcvii–xcviii.

26 Adagia IV ii 1: Δουράτεος ἵππος 'Wooden horse' LB II 992A–E

27 Matt 12:24–8; Mark 3:22–7; Luke 11:15–22; John 7:20

I also wish that Pio had not himself gone to such lengths, whether in indulging his own bad temper or falling in with the bias of others, so that I could shift the entire opprobrium of my reply onto the others and very gladly have spared an Italian, a nobleman, a prince, and surely a supporter, even if a less than knowledgeable one, of good learning. For he really has been of help to learning by the very fact that a man of renown and of not exactly shabby circumstances followed the calling of scholarship. But since religious duty does not permit me to keep completely silent, I shall make what allowances I can for Pio's soul (for which I wish and pray the Lord's mercy), and reply selectively to certain points only, and, to the extent that it can be done, without affront to his good name. Although I thought that I had done this quite generously both in my letter and in the *Responsio* (and this is not my opinion only, but also that of a great many learned men),[28] nonetheless it looks as if there were some who convinced Pio otherwise. What is one to do? Human nature is such that whatever we say against others seems gentle to us, but what others say against us, even if extremely temperate, seems more bitter than vinegar. At this point some people are going to make an immediate protest: 'You have plainly pronounced a judgment against yourself!' I shall not contest this, nor deny that I am human; only let them admit that what I say is applicable to others equally as well.

It is said that no one is a good judge in his own case. I agree. But those who are critics of the books of another are themselves also pleading their own case. And these people should be given even less regard in this situation, because they are caught out in many passages in out-and-out slanders, in barefaced lies, in distortions of correct statements, in falsified quotations of passages they censure, in abridged quotation of what, were it quoted intact, would have a far different meaning, in uncontrolled tirades when they fail to understand what has been provided on their notecards, in dull witticisms whereby they present themselves most of all as objects of laughter, in repeated buffets and slaps,[29] particularly so since they are inciting these riots on the basis of a misunderstanding of notecards furnished by their servants, and all because they are seeking a braggart's renown[30] among the uneducated. My prefatory

* * * * *

28 There is no direct evidence of this in Erasmus' surviving correspondence.

29 *colaphos, et alapas*: words that occur most memorably in accounts of Christ's Passion, *colaphi* in Matt 26:67, *alapae* in John 18:22, 19:3, and both in Mark 14:65

30 *Thrasonicas glorias*, alluding to the proverbial braggart Thraso from Terence's play *Eunuchus*

index[31] will show clearly how these phenomena occur virtually everywhere in this book.

Indeed, through the entire argument his constant practice is this: an alarming assertion is made, my words are falsely quoted, assumptions are shamelessly drawn from these, and then comes a tirade of jokes and jeers at my expense just as if it were already proved that I attack what he is defending. There are some things in his tirade that are worth reading, though they do not apply to me at all, and yet I am nevertheless spat upon and pelted with witticisms just as if they did apply to me. I could put up with this if its occurrence were not so widespread.

I am sure that Pio wrote against me at Rome at the instigation of certain parties, and I know quite well who they are.[32] One is a Jew, a group of people with whom I have never been on good terms.[33] But they would never have been able to prevail upon Pio had Pio's cart not received a shove[34] from his desperate craving for fame.[35] This is a human enough failing, I ad-

* * * * *

31 This index was printed on sigs aa2[r-v]; it is translated above, 107–8.

32 The real target of Erasmus' complaint would seem to be Aleandro, but Erasmus wants it to appear that a whole group of Roman critics is arrayed against him. From the reports of others he knew the identity of two. In his letter of 16 February 1522, Jakob Zeigler reported to Erasmus from Rome that Zúñiga was denouncing Erasmus at several gatherings as the source of all evils; see Ep 1260:159–217, and Erasmus' response of 22 December 1522 in Ep 1330:45–60. Sancho Carranza de Miranda may also have joined in these criticisms during his visit to Rome at this time; see the letter of Juan Vergara from Brussels, 24 April 1522, Ep 1277:24–45 and Allen Ep 1277:24n. While it is possible that Zúñiga and Carranza might have urged Pio to write against Erasmus, Aleandro would seem a more likely candidate, if there was any 'instigator.' See Introduction xxxviii–xlii. Sepúlveda rejected this charge as unfounded. It implied that Pio did not act on his own judgment but on the exhortation of others, and hence that he was not a serious person concerned with the cause of religion. Given his poor health and the burden of grave affairs of state, Pio would not have undertaken to write against Erasmus unless he felt that Erasmus posed a serious threat; see Sepúlveda *Antapologia* Biii[r-v] / Solana 119.

33 Erasmus repeated this statement about Aleandro's ancestry in Ep 1166:93–4, Allen Epp 2414:15, 2578:31, ASD IX-1 150:717, and *Erasmi opuscula* 316–17:6–11. Aleandro's skill in Hebrew, which he learned from a Jewish refugee from Spain, Moses Perez, who later converted to Christianity, may have been the slender basis on which Aleandro's enemies later made this charge; see CEBR I 28. On Erasmus' negative attitudes towards Jews, see *Responsio* n168 above. On Eramsus' charge that Aleandro urged Pio to write against him, see Introduction xliii, cix–cxi; *Responsio* n41 above; and Heesakkers 'Erasmus' Suspicions.'

34 *Adagia* I vi 13: *Bene plaustrum perculit* 'He gave the cart a good shove downhill' LB II 226C–D / CWE 32 11–12: 'Commonly used, it is clear, of those who urge a man in a direction to which he was already tending of his own accord.'

35 Sepúlveda easily demolished this charge; see Introduction cxxiii. In making the charge Erasmus may have been repeating Louis de Berquin's accusation that Pio, 'in the way of an Italian, wanted to appear glorious in that he was able to find some fault in Erasmus'; see Allen Ep 2066:66–7.

mit, and one especially associated with the temperament of the nobly born; but it is ill bred to obtain one's renown from another's disgrace.

Whatever words they use to mitigate Pio's *Epistle*, calling it a suggestion, one that is gentle and friendly, in fact it is nothing but a fierce indictment tempered with a few pleasantries. One might call it a mixture of aconite and mead.[36] And yet I do not see how I could have handled his reputation more gently. Since I knew that Pio did not write the work by his own prowess,[37] I attributed whatever was liable to cause resentment to those who were exploiting Alberto's earnestness, while I assigned anything distinguished and conciliatory entirely to him. And yet, he took my considerable courtesy in a quite discourteous way, as if I had taken great pains not to miss any opportunity for insulting him. Consequently, he seems clearly to have been looking for an opportunity[38] to rage from a wagon,[39] as they say, against Erasmus.

* * * * *

36 See *Adagia* I viii 58: *Letale mulsum* 'A deadly honey-brew' LB II 321A–B / CWE 32 158

37 Translating *ipsius Marte*, based on the proverbial expression described in *Adagia* I vi 19: *Nostro marte* 'By our own prowess' LB II 228F–229B / CWE 32 15–16: 'Whenever we bring something to a conclusion with no outside help, by our own wits and such strength as we possess, we are said to do it by our own prowess or our own merits; also when something is done at our own risk.' While an element of sarcasm may be suspected in the expression *ipsius Marte*, because of Pio's well-known preference for peace and interest in quiet study as well as his lengthy career as a diplomat, nonetheless he was also known to be ready to resort to arms to defend his own or his patron's honour, he served with distinction at the battle of Concordia on 8 May 1511, and he secured from Julius II the important contingent of three hundred lancers who helped to defeat the Venetians at Olmo near Vicenza on 7 October 1513. The bronze statue atop his sepulchre depicts him reclining in antique Roman military garb (see illustration 388). See Semper 6–9; Alfonso Morselli 'Notizie e documenti sulla vita di Alberto Pio' in *Memorie storiche e documenti sulla Città e sull' antico Principato di Carpi* XI (Carpi 1931) 57–60; Guaitoli 'Sulla vita' 188, 207; Contini *Alberto III Pio* [45–7]; and Beaumont-Maillet 281. Erasmus' use of the epithet Mars when referring to Henry VIII's authorship of the anti-Lutheran works published under his name may have contained an element of flattery (Epp 1228:22–3, 1313:76–7, Allen Ep 2143:52–5). Henry liked to see himself as a heroic warrior king. The warrior image extended into the spiritual realm with his title 'defender of the Faith' and later his portrayal as the new David and Josiah. See eg John J. Scarisbrick *Henry VIII* (Berkeley, Calif 1968) 21–40; Lacey Baldwin Smith *Henry VIII: The Mask of Royalty* (Boston 1971) 150–62, 198–203; and Richard Rex *Henry VIII and the English Reformation* (New York 1993) 29, 173–5.

38 *ansam quaesisse*, literally 'to have been looking for a handle'; see *Adagia* I iv 4: *Ansam quaerere, et consimiles metaphorae* 'To look for a handle and similar metaphors' LB II 152C–F / ASD II-1 411–12 / CWE 31 321–2.

39 A variant of *Adagia* I vii 73: *De plaustro loqui* 'Wagon-language' LB II 290F–291B / CWE 32 110–11, referring to the liberty enjoyed by actors on the stage in ancient Greece whereby they could criticize with impunity not only leading citizens but even the gods. See also *Responsio* n223 above.

If what he wrote reflects his personal views, one must frankly admit what the situation indicates, that he was more ill in mind than in body. Now the minds of those whose death is imminent are commonly assailed by a desire to produce some notable achievement to pass on the memory of their name to posterity. His desire might have been acceptable if he had chosen a more praiseworthy theme, one more worthy of a nobleman. I know very well the capability of wicked *provocateurs* who market themselves in the guise of piety; but it was unbecoming for a prince, a scholar, a man seasoned not only by age but much more by experience, to entrust his reputation carelessly to those who were urging him on, to wield his pen so hatefully against one whose works he had not read,[40] and to construct from the notecards of untrained or careless servants, as if from coloured pieces of stone, a work not mosaic, but Momic.[41]

The bishop of Rochester wrote against Luther,[42] but he reports Luther's words reliably, and does not rely on snippets collected from wherever. Clichtove writes against Oecolampadius,[43] but he reproduces the man's entire work. Instead, the chosen model was Béda,[44] who had time to pass judgment though he had no time to investigate.

Pio admits that he pondered for a long time whether he should reply to me.[45] Indeed, certain learned friends in Paris[46] have told me in their letters[47] that Alberto, because of his illness, had altogether abandoned

* * * * *

40 See Introduction lxxvii and n271 below.
41 See *Adagia* I v 74: *Momo satisfacere, et similia* 'To satisfy Momus, and the like' LB II 210B–211C / ASD II-1 546–8 / CWE 31 448–50; of Momus: 'This god has the habit of producing nothing of his own, but staring at the works of the other gods with inquisitive eyes, and if anything is lacking or wrongly done, he criticizes it with the utmost freedom.'
42 John Fisher (1469–1535) wrote four major works against Luther: *The sermon ... agayn the ... doctryn of Martin luuther ...* (London: W. de Worde 1521?); *Assertionis Lutheranae confutatio ...* (Antwerp: M. Hillen 1523); *Defensio Regie assertionis contra Babylonicam captivitatem ...* (Cologne: P. Quentel 1525); and *Sacri sacerdotii defensio contra Lutherum ...* (Cologne: P. Quentel 1525), critical edition in *Corpus Catholicorum* IX ed Hermann Klein-Schmeink (Munster 1925). On Fisher's verbatim citations of Luther, see Edward Surtz *The Works and Days of John Fisher* (Cambridge, Mass 1967) 309, 313, 318.
43 *De sacramento Eucharistiae contra Oecolampadium opusculum ...* (Paris: S. de Colines 1527); the work of Johannes Oecolampadius which Josse Clichtove here refuted was *De genuina verborum Dei: 'hoc est corpus meum' etc. expositione* (Basel 1525). See Jean-Pierre Massaut, *Josse Clichtove, l'humanisme et la réforme du clerge* 2 vol, Bibliothèque de la Faculté de Philosophie et Lettres de l'Université de Liège 183) (Paris 1968) I 43, II 315–17.
44 On Noël Béda, see Erasmus *Responsio* n50 above; contrary to Erasmus' insinuations, Béda did on occasion (as in his 1526 *Annotationes* against Erasmus and Lefèvre d'Etaples) quote from his adversaries.
45 Pio *XXIII libri* 71^{v}N
46 The chief among them was probably Berquin; see Allen Ep 2066:60–79.
47 Eg Ep 1587:243–63.

the project of writing, but returned to his interrupted work at the urging of others. Now what sort of pillars of religion[48] shall we call those who drove a man who was ill beyond hope of recovery into making such tedious and pointless allegations? How much more suitable it would have been at that time for him to devote himself to meditating on a psalm, or on the passion of the Lord, or on the literature that contributes to a clear conscience and that, because of the promises of Christ and his unutterable kindness, either produces a contempt for this life or kindles a love for life in heaven! How I wish that the ones who undertook to persuade Pio to be clothed with the Franciscan habit as he lay dying had also persuaded him to beg of the Lord the spirit of Francis,[49] who, as is well known, was altogether averse to vainglory and to every sin of carping at others! As it is, in this work there is everywhere manifested such a zeal for carping that, although he was driven from the domain of Carpi,[50] nonetheless he can make quite a legitimate claim to the surname *Carpensis*.

But what did that amazing man intend? To refute the doctrines of certain Germans? Then he should have fielded a battle array of the kind of arguments with which they contend, and joined battle with them fiercely. But how could he have done this, for he did not know any of them, unless maybe he had heard something at dinner parties? Or was it that he wanted

* * * * *

48 *illa Religionis columina*, an ironic allusion to Gal 2:9: '*Iacobus, et Cephas, et Iohannes, qui videbantur columnae esse . . .*'

49 Erasmus elaborated these views on how to die properly and on adopting the spirit rather than the habit of St Francis of Assisi in his colloquy *Exequiae Seraphicae* (1531) in ASD I-3 688–99 / CWE 40 996–1032; and also in his *De praeparatione* LB V 1293B–1318D especially 1305E / ASD V-1 337–92 especially 366 / CWE 70 393–450 especially 420 (being garbed in a Franciscan habit and buried with religious rites does not prevent one from going to hell). The Council of Basel in 1444 condemned the opinion preached by Franciscans in the dioceses of Torino and Asti that anyone dying in the habit and who was professed as a Franciscan would not suffer more than one year of purgatory because St Francis by a special divine privilege descends to purgatory periodically to lead all the professed of his order with him into heaven; see Duplessy *Collectio judiciorum* I-2: *De novis erroribus* 242. On 10 April 1486 the theological faculty of the University of Paris formally condemned this opinion, now being advanced by Frater Johannes Mercator OFM, in the city of Besançon; (ibidem 319, proposition 11).

50 On 9 March 1525, while Pio was in Rome, the Spanish forces under Captain Juan Vargas entered Carpi without opposition from the few inhabitants who had not fled. The city and its domains were put under the control of Giovanni Andrea Spinola, the imperial commissar, and within a few days both the remaining and returning inhabitants swore fidelity to the emperor. The domain was declared confiscated by the emperor who eventually, on 8 May 1530, invested Alfonso d'Este with its land, titles, and rights; see Guaitoli 'Sulla vita' 263–5, 304–5.

to give me another piece of advice?[51] Then what need was there, to that end, for so much abuse, so many unamusing ironic remarks, so many slanderous misrepresentations, so many plain lies? But he was keen to prove a stubborn man wrong.[52] He should at least have read my works, reported them honestly, refuted the arguments that I put forward, and not merely tossed about some commonplaces. But irony is of little use, even for proving someone wrong.

If it was his plan to remove a temptation for those who are not steadfast and susceptible to heresy, it would have been better to find excuses for what could be excused, and to excerpt from my works different passages, ones that are openly opposed to condemned doctrines. This would have been better than distorting quotations in an attempt to make Erasmus seem to hold views he does not hold, the purpose being to prevent those who assign great weight to his opinions from being confirmed in their error!

But the people to whose wishes Pio has accommodated himself seem to have had this as their goal: to gain very considerable favour with the agents of divisiveness by misconstruing Erasmus, and yet not to provoke any attack on themselves by those hornets[53] whom they now fear instead of hate.[54] What a pious resolve!

If he thought that to have contended with Erasmus was a source of renown, he ought to have used his own weapons in the fight. As it is, what praise will he win? He let himself be forced to act the ungentlemanly role of an informer. He submitted to being decked out in others' plumage,[55] indiscriminately gathered, and to going before the public thus costumed. I admit that it is the prerogative of princes to have others write their letters and only to affix the signature themselves, but Pio speaks, through the entire work, as if he were waging the whole campaign on his own, though most people know what mercenaries he employed. I could myself name some

* * * * *

51 Pio *XXIII libri* 72ʳP: 'But I would like you to take my remarks as uttered not by an adversary but by a friend who is chatting companionably with you, one who is advising instead of disputing.'

52 'convincere pertinacem,' alluding to Pio *XXIII libri* 71ᵛN: 'in your attempt to defend these errors, you have introduced other new ones, as the stubborn (*pertinaces*) usually do.'

53 See *Adagia* I i 60: *Irritare crabones* 'To stir up hornets' LB II 50D–51B / ASD II-1 172 / CWE 31 105–6.

54 A variation, perhaps, on *Adagia* II ix 62: *Oderint dum metuant* 'Let them hate, so long as they fear' LB II 676D–F / CWE 34 113–14

55 See *Adagia* III vi 91: *Aesopicus graculus* LB II 866B / ASD II-6 386: 'He is called "The Aesopic Jackdaw" who usurps others property for himself, and advertises himself with others' goods.'

should circumstances require. Everyone knows Sepulvelus,[56] a learned man and a good Latinist.

In addition, he would have won more renown if the allegations he offered had been new ones. But when he only repeats, like Echo,[57] the allegations of others, what praise is due him, especially in this unpraiseworthy sort of activity? Only God knows how well the Seraphic fathers[58] took care of his soul; they surely took the worst possible care of his reputation. What do they really achieve by schemes like this except to expose their craftiness more and more every day? But I am sure that they will be reading this work of Pio in their refectories, swarming, as it is throughout, with countless false allegations, lies, tasteless witticisms, garrulous rubbish, and pointless arguments. They deserve this, for they nourish their bodies with no more wholesome food than their minds.

[REPLY TO THE MARGINAL NOTES][59]

But so as not to detain you, O reader, with trifles as unpleasant as they are unprofitable, I shall extract a few items from his marginal notes,[60] which are no less crazy than the rest of the book.

Whatever I brought up about things said or done in Rome, he declares that those who reported them were liars,[61] and I am inappropriately gullible.[62] He excised nothing from Sauermann's book.[63] He never spoke

* * * * *

56 That is, Sepúlveda; he comments on this misspelling of his name in *Antapologia* Diii[r] / Solana 130.

57 See Erasmus' colloquy *Echo* of 1526 in ASD I-3 555–8 / CWE 40 796–801.

58 The Franciscan friars in whose monastery Pio was buried

59 1531 UPPER MARGINS: '[Reply] to the Marginal Notes' on pages 14–47 [= 119–53 in this volume]

60 See Introduction c and *Responsio* n8 above.

61 Pio *XXIII libri* 49[r]q (on *Responsio* 15–16 above, 'a certain young man etc'): 'Whoever this was can have been a friend neither to me nor to you, and no gentleman, for he lied quite shamelessly'; 48[v]p: 'likewise, as to what you say about Sauermann, it was an out-and-out lie that was reported to you.'

62 Pio *XXIII libri* 48[v]o (on *Responsio* 13 above 'I shall not ... certain learned men there'): 'It is inappropriate for a man of learning to offer such ready ears to gossipmongers'; *XXIII libri* 52[r]u (on *Responsio* 38 above 'he left nothing untried to destroy me'): 'These were reported to you by the cunning of the heretics. You are very gullible.'

63 Pio *XXIII libri* 48[v]p (on *Responsio* 16 n71 above, Georg Sauermann): 'likewise, as to what you say about Sauermann, it was an out-and-out lie, that was reported to you. I have never laid eyes on Sauermann, so far as I know, nor did I see a word of his book. That is how far from the fact it is that anything was deleted from it at my instigation. I call Sauermann as witness. But it is amazing how you convinced yourself of such absurdities. Should I have done this, seeking vengeance, although I had never been offended by you?

A page from book 2 of Erasmus *Responsio*,
as reprinted in Pio *XIII libri* (Paris: Bade 1531) folio 70[r],
showing Pio's extensive marginal notes on the *Responsio*
Photograph courtesy of Kenneth J. Pennington and Ofoto, Kodak Company

of me save in a complimentary way.[64] The learned passed no unfavourable judgment on me.[65] The Bullbearer[66] made no hostile statement about me.[67] But were I to give the names of those who reported these things to me,[68] Alberto would not dare to call them liars, and were he to dare, he would gain nothing but a reputation for dishonesty. (Indeed, among these is one man to whom he defers in all things!) All Rome knows how cordially Alberto is wont to speak of Erasmus. On the contrary, though I pretended not to know this at the time, when Clement, at the urging of certain people, had intended some grand prelacy for me, Alberto, with his Achates, caused him to change his mind.[69] The people who wrote to me about this were present at the events they report, eyewitnesses,[70] not rumourmongers.

At the words 'what wisdom would it have been to draw my sword against a man [ie Luther] about whom I was really in doubt as to what he was up to, or by what spirit he was being led?' he says: 'Here you are admitting what we have said, that you were not initially displeased with his audacity.'[71] Can anyone be pleased with something on which he is not

* * * * *

Could I have acted out of rivalry although our lives are so unlike and your glory removes nothing from me? It must be that you think it was done out of malice, but to assume this without cause is most malicious'; see Erasmus *Responsio* nn 69 and 71 above.

64 See Erasmus *Responsio* 100 above and compare Pio's note on that passage: *XXIII libri* 65[r]e (on 'wanting in consistency ... always thought and spoken of me in a complimentary way etc'): 'How silly these words are! You talk as if it were inconsistent to speak in a complimentary way about an individual in public and to admonish or rebuke him in private, or to praise him in one sphere of activity and find fault with him in another. I wrote that I have spoken in a complimentary way about your scholarship, but that I have by no means approved of all you have written.'

65 Pio *XXIII libri* 48[v]o (on *Responsio* 13 above 'certain learned men there etc'): 'It is unsuitable for a man of learning to offer such ready ears to gossipmongers. I never heard such things stated about you at Rome.'

66 That is, Girolamo Aleandro; see *Responsio* n206 above.

67 Pio *XXIII libri* 52[v]i (on *Responsio* 42 above διπλωματοφόρος): 'Do not at all believe that the Bullbearer, whoever he may have been, babbled any such things. For in diplomacy it is not what can be said with meticulous accuracy that should be stated, but only what serves the purpose.'

68 See Introduction xlii–xliv.

69 Jacopo Sadoleto writing from Carpentras on 17 September 1530 reported that Clement VII had once thought of appointing Erasmus to a distinguished priestly office in Germany, but had been dissuaded from doing so by certain men whom Erasmus identified as Aleandro and Pio; see Allen Ep 2385:64–9, 68n; and Erasmus' response from Freiburg, 7 March 1531, Allen Ep 2443:262–350. Achates is Aeneas' trusty companion, *fidus Achates*, in the *Aeneid* passim; the allusion here is clearly to Aleandro.

70 Erasmus uses the term αὐτόπται, familiar from Luke 1:2; see Erasmus' comment on this point in LB VI 218D–219B / Reeve 150.

71 Pio *XXIII libri* 50[r]k (on *Responsio* 25 above)

ready to pass judgment? On the contrary, I did not conceal the fact that I was not pleased with his boldness, and for that reason I warned off those I could from becoming involved in the business. Alberto adds: 'But when you found out, why did you not draw the sword? Why do you still not draw it?'[72] Since he was reminded in this very letter how often and how fiercely I have unsheathed my pen against Luther and against the champions of his doctrines, was it not shameless to add 'you still not draw it'? And at the very time when Alberto was engaged in these labours, I sent the *Letter to Vulturius*[73] and the *Apology to the Churchmen of Strasbourg*.[74]

I had written: 'if you realized what trouble I could have caused, had I wished to put myself at the head of this movement, etc.' He draws the following conclusion: 'If you had influence enough to stir them up, you had no less influence for quelling them.'[75] What kind of inference is this? Is it as easy to harm as to heal? as easy to set in motion those who are eager as to recall those who are already charging forward? With silly arguments like these the master orator deals with Erasmus!

At my words 'if one who reminds priests of their duty, as opportunity presents itself, has clamoured at their honour, then I confess that I clamour frequently, but I am clamouring along with Jerome, with Cyprian, with Chrysostom, with Bernard,' Alberto makes the following note: 'If they did any such thing, it was done quite differently, and in a different epoch. But now, since piety has grown so cold that there is hardly any reverence towards priests, one should not have carelessly published the sort of remarks which would lead to the discarding of what little reverence remains.'[76] Thus Alberto, and he is quite correct. Of course they did it 'quite differently,' that is, more frankly and more satirically, in keeping with their prestige, and in an epoch in which it would have been better, because the population

* * * * *

72 Ibidem
73 Also known as the *Epistola contra quosdam, qui se falso jactant Evangelicos* (Freiburg: Johannes Faber Emmeus [1530]) ASD IX-1 283–309. 'Vulturius Neocomus' was Erasmus' attempt to put into humanistic Latin the name of Gerard (the Dutch *Gier* means 'vulture') Geldenhouwer of Nijmegen (Neocomus); see CEBR II 82–4 especially 84.
74 Also known as the *Responsio ad Epistolam apologeticam incerto autore proditam nisi quod Titulus, forte fictus, habebat: per ministros verbi, ecclesiae Argentoratensis* (Freiburg: Johannes Faber Emmeus [1530]) or as the *Epistola ad fratres inferioris Germaniae* ASD IX-1 329–425. Erasmus suspected that the churchmen of Strasbourg against whom he was writing were Wolfgang Capito and Caspar Hedio; see ibidem 331:65–6n.
75 Pio *XXIII libri* 50^{v}m (on *Responsio* 26 above 'Imagine ... that I had a considerable influence') where he quotes Erasmus' letter to him (Ep 1634:74–5, the passage quoted by Erasmus here)
76 Pio *XXIII libri* 51^{r}y (on *Responsio* 29–31 above)

was a mixture of pagans and Christians, to keep silent about the vices of Christians, in particular of priests and monks, whose welfare, at that time, depended on the public. Now they have so thoroughly fortified themselves with wealth, domains, citadels, troops, squads of devotees, leagues with monarchs, and other defences that they are awesome even to monarchs. But if it was their way of life that caused the popular good will towards them to grow cold, it was even more needful to point this out to them, so that they could regain lost favour through the reformation of their way of life. Furthermore, in this current epoch priests and monks rule and triumph. But my admonitions were given in a tranquil era, when no one conceived of this present age even in a dream.[77] Let the one who cares to do so read Jerome as he thunders against the vices of bishops, clergy, virgins, and monks, and he will declare that he is reading a Christian Juvenal.

Where I object that Pio himself wrote some rather blunt things against priests, theologians, and monks,[78] he writes in a note:[79] 'I have said that some are unworthy of that office, nor do I disown my statement; but you attack all without distinction, and you everywhere disparage the whole order.'[80] There are numerous passages in my writings that cannot even be twisted into applying to all priests, theologians, and monks. I affirm explicitly in countless passages that I am not attacking the order, and I warn the reader not to impute to the order the corruption of a few. I give lavish praise to some priests, theologians, and monks, by name. So what, then, could be more hateful than this lie, 'you everywhere disparage'? If he had said 'it is the order you disparage,' it would have been a single lie; but when he adds 'you everywhere disparage,' it is a double, and doubly shameless, lie, especially since he has seen hardly any of my works. He should at least have imitated the frank modesty of the lying Aeschines and added 'so far as I know.'[81] But Alberto seems to be more concerned about reverence for priests of whatever quality than about piety and Christ's glory.

* * * * *

77 See *Adagia* I iii 62: *Ne per somnium quidem* 'Not even in a dream' LB II 135D–136D / ASD
    II-1 370–2 / CWE 31 286–7.
78 MARGINAL NOTE: 'Lie'
79 At *Responsio* 30 above
80 Pio XXIII *libri* 51ʳz (on *Responsio* 30 above 'But why adopt a defence based on the ancients,
    since you, in your very book, made many frank criticisms of priests, theologians, and
    monks ...')
81 Aeschines, Athenian orator-politician (c 390–c 322 BC), opponent of Demosthenes in one
    of the great oratorical duels of antiquity; Erasmus was quite familiar with his extant
    speeches, as is evident from his many quotations from Aeschines in the *Adagia*. It is
    not clear precisely what passage of Aeschines Erasmus has in mind here; perhaps he is

Another instance quite similar to this is his saying[82] that I altogether condemn rituals everywhere, and that I fail to distinguish devout rituals from superstitious ones, although I do this explicitly and clearly in countless passages, as I shall prove quite clearly at the appropriate place.[83] Not even Luther condemns rituals altogether.[84]

No less a case of impudence is Pio's saying that I draw the sword of my pen against all monks.[85] And he repeats virtually[86] everything that he had set out so shamelessly in his catalogue, but he has not been able to prove any one of them so far. How well he has succeeded in this work the facts will show. At least he makes promises like a Thraso.[87]

'If,' he says, 'there is nothing erroneous in your writings, why have so many, of different estates, written books censuring you?'[88] I have never written that there is nothing erroneous in my books, but I do deny that they

* * * * *

thinking of Aeschines' expression ὡς ἐγὼ πυνθάνομαι 'as I learn' in *Against Timarchus* 99 and 173 and in *Against Ctesiphon* 217 and 225.

82 Pio *XXIII libri* 51ʳb (on *Responsio* 31 above 'I have always revered the rituals of the church etc'): 'If this is how you felt, you should have divided the genus and indicated which ones you judged superstitious and silly, and not condemned them as a whole, as you do everywhere. I, however, steadfastly maintain this: religion cannot long endure intact if rituals are eliminated from the church.'

83 Eg 209–23 below

84 For Luther's teachings on rituals, see *The Augsburg Confession* (25.6.1530) especially articles 15 and 24, in *The Book of Concord: The Confessions of the Evangelical Lutheran Church* trans and ed Theodore G. Tappert et al (Philadelphia 1959) 36–7, 56–61; Heinrich Bornkamm *Luther in Mid-Career 1521–1530* ed Karin Bornkamm trans E. Theodore Bachmann (Philadelphia 1983) 459–80; and Vilmos Vajta *Luther on Worship: An Interpretation* trans U.S. Leupold (Philadelphia 1958).

85 Pio *XXIII libri* 51ʳc (on *Responsio* 31 above 'Show me one monk etc'): 'How shamelessly you struggle to slip away! Is it not even worse to attack all [monks] in general, and to disparage their calling, than to accuse one only or a few? This is what you are doing, given that they are monks and it is their institutions your ridicule and criticize.'

86 *fere* LB *uere* 1531

87 See n30 above.

88 Pio *XXIII libri* 50ᵛs (on *Responsio* 29 above 'I am amazed that Alberto etc'): 'How dare you say this? If you have written nothing erroneous, it would be appropriate for you to say why so many learned men of different estates, of widely separated regions, of different languages, have recorded so many censures in their published books? If everything was correct and nothing was discordant or blameworthy, why have you published so many apologetic works? Why have so many doctors and preachers of the divine word, men who do not even know you, been moved to state, as you frequently complain, that you are in collusion with Luther, unless they had been provoked by your writings? In the case of most of these there is no basis for suspicion that they did so out of hostility towards you. Rather, it will become quite clearly apparent below what a remarkable lie you are telling about this.'

contain the things that Alberto, so savage in stating his case, says they do. I will not describe here the sort of people who have attacked me with pen or tongue. But to repel Pio's inference with one of my own, if there are so many impieties in my books, why have I been thanked for them by so many devout and learned men of the first rank? And among these there are Catholic princes, bishops, theologians, and monks. Why did so many people howl against Jerome? Because innovation, even the very best, offends most people; because envy and a thirst for fame goad many, and they see a wonderful shortcut to celebrity in attacking those who are already famous for their writings. The aulos-player told his student that, if he wished to obtain renown quickly in Greece, he should go to the most famous artists and win their regard.[89] But now an even shorter cut has been discovered: we attack a single famous person.

As to my words 'if sometimes I warn young men not to enter precipitously into an unknown manner of life, while they are virtually unknown to themselves,' just look at what he notes: 'But according to ancient practice, infants and small boys were given as oblates to the fathers of the monasteries[90] to take up a way of life to which they thought one should grow accustomed from infancy, etc.'[91] Who is denying that we should be accustomed to the very best from our tender years? I am talking about the obligation of vows. Or does he want young men to cast themselves into this thoughtlessly? If he does not want that, what is the point of a note that seems to criticize what I said? The households of bishops were once called monasteries, and in them was the school of letters and of piety.[92] Young men were often trained there, not for slavery, but for the freedom of the spirit. Pio is a native of Italy and a seasoned courtier; let him tell me why the young

* * * * *

89 Erasmus alludes here to Lucian *Harmonides* 2 (we are grateful to William J. McCarthy of the Greek and Latin Department of the Catholic University of America for locating this reference); Erasmus alludes to this story to similar effect in *Capita argumentorum contra morosos ac indoctos* LB VI ***3.

90 On this practice, see Maria Lahaye-Geusen, *Das Opfer der Kinder: Ein Beitrag zur Liturgie- und Sozialgeschichte des Mönchtums im Hohen Mittelalter* (Altenberg 1991); and the article by J. Dubois 'Oblato: nel monachesimo' in *Dizionario degli istituti di perfezione* (Rome 1980) VI 654–66.

91 Pio XXIII *libri* 51^{r}d (on *Responsio* 31–2 above)

92 Erasmus seems to be making reference to a tradition in the West that dates from the time of Augustine (354–430), who provided clerical formation and training in his episcopal palace and required candidates for ordination to spend a certain period of time there. This Augustinian system enjoyed an extensive diffusion. See James A. O'Donohoe *Tridentine Seminary Legislation: Its Sources and Its Formation* (Louvain 1957) 2–15 especially 3–4; and John Tracy Ellis *Essays in Seminary Education* (Notre Dame, Ind 1967) 3–22.

reject monasteries where discipline is regular, and hunt for those where it has collapsed.

There are many who have praised the monastic life,[93] I admit, but the old-time monks were quite different, and now most have nothing monastic about them but a habit and tedious rituals. Thus, in the old days monastic life was not a fish-trap, nor did such fish rule in it.

He admits that he has not seen my replies to Zúñiga. He says: 'What does it matter that you replied? Let them decide if your replies are satisfactory.'[94] On the contrary, it does not matter that they censured me, but what does matter is whether they were right to do so. Elsewhere, too, he admits that he had seen practically nothing of my works,[95] though he wrote against me nonetheless. 'Is that really necessary?' he says. It is indeed necessary for one who keeps on saying: 'There is nowhere Erasmus does not do this!' 'Erasmus condemns rituals everywhere.' How dare he say this when he has not read my works? Anyone who has observed him repeatedly stirring up great commotions on the basis of the incorrect, fragmentary, and misconstrued notes of his secretaries will admit that this was crucial.

He says: 'You approve of Luther's fierce temperament. You praise his learning.'[96] There are, in fact, many who do approve of his temperament, but I have always disapproved of ferocity. I did not think much of his learning at the beginning, but he has made progress in the course of controversy. Moreover, to approve is one thing, to praise is another; but the verb *praise* is handier for purposes of slander. He says: 'You never make clear of what your "true, evangelical piety" consists, though you exclude many things from it.'[97] What cheek to write this when he admits he has not read my works! In fact, everywhere I teach that true piety consists of the mortification of the emotions, stoutness of faith, alacrity of spirit, works of

* * * * *

93 Pio *XXIII libri* 51ᵛe (on *Responsio* 31 above 'if I censure monks etc'): 'The most holy and ancient fathers, Basil, Chrysostom, Jerome, Augustine, Gregory, and a great many others, have praised this sort of life in their books and made clear how salutary an opportunity the cloister provides.'

94 Pio *XXIII libri* 51ᵛi (on *Responsio* 32 above 'Although ... many others, and to Zúñiga etc'): 'What is it to me if you made reply? Let them decide if your replies were satisfactory. I had seen none of these, nor have I yet.'

95 See n271 below.

96 Pio *XXIII libri* 51ᵛk (on *Responsio* 32 above 'Is it any wonder ... in doubt?'): 'Luther revealed himself and immediately made clear what a monster [Horace *Odes* 1.22.13] he was bearing in his heart, and yet you not only approve his fierce temperament, but even praise his learning.'

97 MARGINAL NOTE: 'Lie.' Pio *XXIII libri* 51ᵛm (on *Responsio* 33–4 above 'evangelical piety'): 'nor yet have you ever made clear of what this authentic, evangelical piety of yours consists, though you exclude many things from it ...'

charity, contempt for the world, the hope of immortality to come. Somewhere I paint the complete monk in his true colours.[98]

Here he notes in passing that instead of the term ἐπαμφοτερίζειν [arguing on both sides] I later preferred ἀμφισβητεῖν [to stand aloof and waver],[99] whereas it was in the copy he sent to me that the latter word was added, as if as a correction of the former. He says that he does not understand what I mean by 'each side' except the Lutherans and the Catholics.[100] That is what I mean, I admit it. But is there, on the Catholic side, nothing to which one could object? As for the Lutherans, he himself admits that they have made some devout statements. What need was there for this slander? But I am passing over many items like this.

It is also a case of plain falsehood[101] to say that I condemn nothing in Luther save his rebellion, since I disagree with him about free will,[102] about the authority of the ancient doctors, about the self-evident character of Scripture, and about countless other points.[103]

He says: 'It is untrue that things human always deteriorate.'[104] What proof? 'Christ,' he says, 'has restored many things to a better state.' What, then, follows? I have pointed out the nature of the human condition: it is so prone to evil that even those elements which Christ restored with such care can deteriorate, and the best regulations turn into the worst models.

His note on my words 'and how do we know whether God sent such a beast (I am speaking of Luther) to reform our morals?' is a virtually senseless tirade. He says: 'It is ten years since Luther was condemned.

* * * * *

98 Perhaps *De comtemptu mundi* LB V 1250C–1262F; or Ep 858:548–80 (a 1529 addition) / Holborn 18–19; or Allen Ep 1887:25–64

99 Pio *XXIII libri* 51ᵛm (on *Responsio* 33 above 'If, in your opinion etc': 'But all the while you deny that you play the critic. But what is this approving and disapproving on each side etc? Nor yet have you ever made clear of what this authentic, evangelical piety of yours consists ...' On 'ἐπαμφοτερίζειν ... ἀμφισβητεῖν,' see *Responsio* nn 151 and 157 above.

100 Pio *XXIII libri* 51ᵛm (on *Responsio* 33 above 'on both sides'): 'nor is it possible to understand what you mean by "both sides," unless your are pitting Lutherans against Catholics.'

101 MARGINAL NOTE: 'Lie'

102 See *Responsio* 91 above.

103 Pio *XXIII libri* 51ᵛn (on *Responsio* 34 above 'So the charge ... etc'): 'From what premises follows what you are introducing here? Moreover, rebellion is an evil, surely, but heresy is an absolute evil, and you show that you are giving it your tacit approval when you disapprove of rebellion, but ignore heresy.

104 Pio *XXIII libri* 51ᵛp (on *Responsio* 35 above 'nature of human affairs ... always slip towards the worse': 'It is quite false that human affairs always slip towards the worse, and even if there are vicissitudes in certain areas, human affairs have, indeed, been exalted and exceedingly improved by the coming of Lord Jesus the Saviour, in fact sometimes they are restored and bettered from the worst possible condition.'

His madness has imperilled Germany and thrown the Christian world into confusion. And you are still wondering whether God sent him to reform our morals!'[105] (Elsewhere he pretends that I actually said that Luther was heaven-sent.)[106] He says: 'I suppose the upright morals of the Lutherans are, perhaps, your basis for thinking this, etc.'[107] Thus Pio. But more years have passed since we condemned the Turk, and yet we do admit that he was sent by God to reform our morals. We acknowledge that the Lord provoked Pharaoh against the Hebrews,[108] but we do not for that reason approve of tyranny, or of the *Allophyli*,[109] or Cyrus,[110] or Nabuchodonosor,[111] or Holophernes,[112] through all of whom God from time to time corrected the faults of his people. When Augustine writes that God led the church on through heresies to the knowledge of the Scriptures,[113] and through the savagery of persecutions to patience,[114] he is not expressing his approval of heretics or the slayers of Christians. And was not a lying spirit sent by the Lord to deceive people and king?[115] I ask you, O reader, what need was there to fill up the margins with such silly notes?

In order to make it seem that I did not fight it out with Luther, he says[116] that the *Diatribe* deals, not with a doctrine peculiar to Luther, but with one that is shared with many who are outside the church.[117] Is it then

* * * * *

105 Pio *XXIII libri* 52[r]r (on *Responsio* 36 above 'And how do we know etc'): 'Ah, what kind of a statement is this? And yet you do not want to be called into suspicion. Here I find your prudence wanting. It is ten years ...'

106 Pio *XXIII libri* 50[r]e (on *Responsio* 24 above 'But at the outset etc') : 'You yourself asserted this, for in your letter you had written thus: "Because, like many others, I thought Luther was a good man, sent from God to reform the evil ways of men" [Ep 1634:88–9].'

107 Pio *XXIII libri* 52[r]r

108 Exod 7:13, 14, 22; 8:19, and passim

109 Erasmus explains this term in a note on Jerome: '*Allophylo quodam hoste. Allophyllus* in Greek is *alienigena* (foreigner)' (HO 1 55[v]C; see also HO 1 80[r]B). This term for the enemies of Israel occurs only once in the Vulgate (Ps 55:1), but passim in the Old Latin Bible and in Christian Latin literature (see TLL SV).

110 Cyrus the Great of Persia, generally regarded as a friend to Israel because he allowed the exiles to return and build the Temple. Erasmus, however, may be thinking of Cyrus as one of a succession of foreign despots, as in Dan 13:65.

111 The King of Babylon, who ordered the deportation which led to the Babylonian Captivity

112 The Assyrian general slain by Judith

113 Eg in *De Genesi contra Manichaeos* 1.2 PL 34 173

114 Eg *De civitate Dei* 18.51 CCSL 48 649

115 2 Kings 22:23; 2 Chronicles 18:21

116 MARGINAL NOTE: 'Lie'

117 Pio *XXIII libri* 52[r]r (on *Responsio* 36 above 'And while I am fighting it out ...') 'If you had fought it out with him properly, no one would have attacked you. In fact, as to the *Diatribe*, the doctrine is not specific to the orthodox, but is shared by many who

the case that one who refutes the Arian doctrine that the Father alone is true God is not combatting Arius because that doctrine is shared by the Jews? If the *Diatribe* was ineffective against Luther, why was his reply to me so fierce? Why does he admit that I went for his throat? But why is it that Pio mentions only the *Diatribe*, whereas in my *Responsio* itself I report my having written the two books of the *Hyperaspistes*?[118] Does his unwillingness to read them mean that I did not write them? But he surely does not seem to have read even the very letter that he is answering, or, if he read it, he does not remember. If he is dissembling on this point, he is even more blameworthy. But who does not realize here the vast shiftiness of the man? He had read the *Diatribe*, he had been told about the *Hyperaspistes*, and he could have read them had he wished; but he prefers to allege one lie in excuse for another rather than admit his mistake. My *Diatribe* tears apart many teachings in addition to that on free will, as does the *Hyperaspistes* as well. Furthermore, since the theologians declare that Luther has nothing of his own, but has drawn everything from the works of Hus and Wycliffe,[119] then no one at all has written against Luther.

When I ask who had imposed on me the task of unsheathing my pen against Luther, his note replies[120] that it was two popes, Leo x and Adrian vi,

* * * * *

are outside the church [... *non proprium est orthodoxorum dogma, sed commune cum multis qui extra ecclesiam sunt*].' Subsequent notes by Pio (63ra quoted below in n245, and 64^rm quoted below in n254) seem to support Erasmus' interpretation of this confusing note as meaning that he failed to attack a specifically Lutheran doctrine, and perhaps Erasmus assumed that *orthodoxorum* was a typographical error for *Lutheranorum*. But on the face of it the note seems to mean, rather, that the doctrine of the *Diatribe* was not peculiarly orthodox, since many heterodox and non-Christians affirmed free will.

118 See *Responsio* 91 above. In his note on the *Responsio* (xxiii *libri* 63^rx), Pio states that when he wrote his *Responsio paraenetica* he had not seen Erasmus' *Hyperaspistes*, and even now (1530?) he has yet to read it.

119 The effort to identify Luther's views with those of Wycliffe and Hus which had earlier been condemned explicitly and implicitly at the Councils of Rome, Constance, and Basel, and the Fifth Lateran Council, was notably advanced at the Leipzig Debate (1519) by Johannes Eck and even incorporated into the bull against Luther, *Exsurge Domine*: 'with our own eyes we have seen and read many different errors, some of them already condemned by councils and the rulings of our predecessors, that also contain specifically the heresy of the Greeks and the Bohemian heresy.' On the conciliar condemnations, see Hefele-Leclercq vii-1 223–6, 233–4, 251–83, 298–350, 378; vii-2 675–6, 895–915; viii-1 421–2; for this text of *Exsurge Domine*, see *Causa Lutheri* ii 368. For an early work that sought to elucidate the connections between Luther and the Hussites at the Leipzig Debate, see Hieronymus Emser *De disputatione Lipsicensi, quantum ad Boemos obiter deflexa est (1519). A venatione Lutheriana aegocerotis assertio (1519)* ed Franz Xaver Thurnhofer, Corpus Catholicorum 4 (Munster 1921) 13–15, 36, 53, 66.

120 MARGINAL NOTE: 'Lie'

as is attested by their many letters, which, as he says,[121] I have seen fit to print.[122] There are two letters from Leo to me; in neither of them is there even a word about writing against anyone. Adrian's suggestion[123] came late, when I was in the course of explaining why I had not unsheathed my pen against Luther immediately, at the very outset. Thus the first allegation is plainly false, the second irrelevant [ἀποσδιόνυσον].

Ὁ Διὸς Κόρινθος [Korinthos, Son of Zeus],[124] he says, 'but what could be more silly?'[125] He is pointing out that something said elsewhere is repeated. In another place[126] he reprimands me quite abusively for this, saying that I did it on purpose to make the book longer, as if six words said twice create a huge book, or as if, should I fancy a large book, I would be short of words, or as if it were incorrect to have an occasional repetition. But I think that his intuition of my attitude is a projection of his own.

He says:[127] 'One ought to assert that supreme power has been entrusted to him [the pope] by the Lord.'[128] What delight he takes in the term 'supreme power'! I do not know what the expression means. The devout acknowledge a supreme pastor, and we do not read that supreme power has been assigned to the pope, but that supreme charity has been demanded of him. Oh, but about charity there is a great silence!

He says Eck was prudent to quote from my works the passages which support the authority of the pope.[129] So it must be imprudent to quote passages that run counter to this, and even to twist supportive passages so that they mean the opposite, a practice that this 'adviser' of mine engages in quite shamelessly. Moreover, it is not wavering to report the views of

* * * * *

121 Pio XXIII libri 52ʳt (on Responsio 37 above 'But who imposed upon me etc'): 'I say it was two popes; I will not mention the rest. You are not looking for a higher authority, are you? Leo x and Adrian vi, as many of their letters to you attest, letters you saw fit to publish.'

122 excudendas LB excidendas. These are Ep 338 (10 July 1515, first published 1516) and Ep 519 (26 January 1517, first published 1517). See Responsio 34 n158.

123 In Ep 1324:42–63, 105–113 Adrian vi exhorts Erasmus to write against Luther (1 December 1522), and in Ep 1338:27–32 he appeals to Erasmus for his advice (23 January 1523).

124 See Adagia ii i 50: Iovis Corinthus 'Corinthus son of Jove' LB ii 425A–426C / CWE 33 44–6: '... directed at those who always say or do the same things.'

125 Pio XXIII libri 52ʳy

126 Pio XXIII libri 67ᵛc

127 MARGINAL NOTE: 'Lie'

128 Pio XXIII libri 52ᵛb (on Responsio 40 above 'What contradictory views! etc'): 'It is not enough to assign the title of primacy to him, for other regions also rejoice in it. One ought ...'

129 Pio XXIII libri 52ᵛa (on Responsio 40 above 'But Johann Eck etc'): 'Eck acted prudently. What he did was skewer an enemy with his own weapons, the fate of waverers who are never consistent, but fight now with one sword and then with another, after they have thrown away the first'; see Responsio n194.

the ancients, nor does one waver henceforth and throughout if he expresses a doubt about the monarchy of the pope. But Pio, in order to be more forceful, loves to draw universal conclusions from specific premises, the claims of dialectic notwithstanding.

He says that I equate the pope's majesty not only with the rest of the bishops, but also with the lesser priests [*sacerdotes*].[130] What could be more uninformed that this silliness? I am quoting Jerome, and in his era only bishops were called priests [*sacerdotes*]. Jerome writes[131] that all presbyters are equal, but that for the avoidance of schism the episcopacy was established, to have one superior to the rest, not so much in dignity as in the right to ordain.

He says[132] that there is no one among those who support the pope who would not wish that his dignity be protected in the first place by divine safeguards.[133] If he had added 'who truly and devoutly support,' his statement could be acceptable. The divine safeguards are faith, charity, prayer, knowledge of the divine Scriptures, the sword of the Spirit, hope, and love of the heavenly life. What could be more craven than his toadying?[134] I speak of the conduct, not of the popes, but of their toadies.

He says[135] that complaints about the excessive number of the cardinals and their fanfare, far from being heard in the company of men of sense, are barely heard even in the company of the irresponsible.[136] What then has been achieved by the Pragmatic Sanction?[137] Or is there no complaint

* * * * *

130 Pio XXIII *libri* 52^{v}f (on *Responsio* 41 above 'words of Erasmus etc'): 'What a rhetorical manoeuvre this is for evading hostility! No one assigned you such importance that you could have been needed [*nemo tibi tantum tribuit uti deficere potueris*], although your wordlets [*verbula*] from time to time do equate the pope's majesty, not only with the rest of the bishops, but even with the lesser priests [*sacerdotes*], even if you do allow the pope primacy in honour.'

131 See nn 1084–6 below.

132 MARGINAL NOTE: 'Lie'

133 Pio XXIII *libri* 52^{v}h (on *Responsio* 42 above 'I wish that all Roman pontiffs etc'): 'Not even here can you restrain yourself from criticizing both the popes and those who support them. There is no one of these who would not wish that the pope's dignity be maintained in the first place by divine safeguards, although also by human safeguards, as much as may be required for its protection.'

134 MARGINAL NOTE: 'Toadying'

135 MARGINAL NOTE: 'Toadying and silliness'

136 Pio XXIII *libri* 53^{v}z (on *Responsio* 47 above 'There is no point etc'): 'I don't hear these complaints among men of sense, in fact scarcely among the hostile and frivolous. And you are disputing not about the title, but about the fact of cardinals.'

137 The Pragmatic Sanction of Bourges (1438) accepted the decree of the Council of Basel's twenty-third session (26 March 1436), *De numero et qualitate cardinalium* (see COD 501–4) and *Pragmatica Sanctio studiosis utilissima, cum Concordatis. Cosmae Guymier ... Commentarius ad Pragmaticam Sanctionem ...* (Lyon: Antonius Vincentius 1538) clxxxv^r–cxcii^r.

from the monasteries in Italy which are devoured by the cardinals?[138] But Alberto does not hear anything like this.

He denies that the actual work of the cardinals now attaches to the penitentiaries.[139] Well, I realize that this work requires a man of precise learning, one who is devout and zealous for the Lord's flock,[140] but, for all that, the leading figure in that group is, even today, a cardinal.[141]

He says[142] that the pope is superior to all kings, but is surpassed by all in <wealth>[143] and fanfare, 'though,' he says, 'it would be appropriate that the pope be superior to all in both respects.'[144] Is there, then, no king poorer than the pope? I will not mention fanfare. Many have lived in Rome who are in a position to give an opinion about this

* * * * *

138 On the practice of cardinals holding monasteries *in commendam*, see Barbara McClung Hallman *Italian Cardinals, Reform and the Church as Property* (Berkeley, Calif 1985) 39–42, 69. For the decree *Supernae dispositionis arbitrio* (5 May 1514) of the Fifth Lateran Council that sought to moderate some of these abuses, see COD 616–17.

139 Pio XXIII *libri* 53ᵛy (on *Responsio* 47 above 'For this work twelve cardinals were chosen etc'): 'So what if penitentiaries and cardinals were the same? But it is quite untrue that the actual work of the cardinals now attaches to the penitentiaries.'

140 In the colloquy *Adolescentis et Scorti* the young man Sophronius reports a sobering encounter with a prudent penitentiary in Rome (LB I 720A–B / ASD I-3 342:117–343:126 / CWE 39 385–6).

141 See *Responsio* n221 above. The origins of the office of cardinal were traced by Alvaro Pais OFM in his *De planctu ecclesiae desideratissimi libri duo* ... (Lyon: Johannes Clein 1571) 119ᵛ (book 2 art 16) (c 1320) to the assistance given to the bishop of Rome in his priestly functions by some of the local clergy. Thus by the time of Pope Marcellus (c 304) some fifteen cardinals were constituted to help with baptisms and funerals. According to Pope Benedict XII (1334–42) in his *Formulare penitentiariorum*, edited in Avignon c 1338, the cardinals of ancient Rome also assisted with confessions. But when the persecutions stopped and many more people came to Rome, the number of penitents so increased that other confessors had to be appointed, and eventually the cardinals themselves ceased hearing confessions, but one of their number, the major penitentiary, was put in charge of these confessors or minor penitentiaries. See Vincenzo Petra *De sacra poenitentiaria apostolica, pars prima* (Rome: Ex Typographia Josephi Nicolai de Martiis 1712) 60–1; and Filippo Tamburini 'Per la storia dei Cardinali Penitenzieri Maggiori e dell' Archivo della Penitenziera Apostolica. Il trattato "De antiquitate Cardinalis Poenitentiarii Maioris" de G.B. Coccino (†1641)' *Rivista di storia della chiesa in Italia* 36 (1982) 332–80 especially 335. For a listing of some of the grand penitentiaries at the time of Erasmus, see Hofmann *Forschungen* II 97–8.

142 MARGINAL NOTE: 'Toadying and silliness'

143 *opibus* LB (supplied from Pio's text) *om* 1531

144 Pio XXIII *libri* 53ᵛb (on *Responsio* 47 above 'I wish that the supreme pontiff etc'): 'What hostility towards the supreme pontiff you are stirring up with these words! But what you are saying is completely false. Although he far surpasses all kings in dignity, yet is he surpassed by all in wealth and fanfare, though it would be fitting that he excel all in both.'

point.[145] What king is carried on men's shoulders?[146] What king prof-
fers his feet to people to be kissed?[147] Who allows a triple crown to be
placed on him?[148] Who uses beadles to force the people to kneel?[149] Who
has countless cardinals, the equals of kings,[150] as worshippers in his ret-
inue? I will omit the rest, since I must rein in the horse that Pio keeps
on calling into the field![151] I am not saying all this because the pope is
unworthy of these honours, but for the purpose of rebuking this fellow's
shamelessness. Really, how is it fitting that the supreme vicar of Christ
should possess, as an individual, greater wealth than all kings? Forget about
wealth. Is it also fitting that the supreme teacher of piety should surpass

* * * * *

145 For a description of the liturgy and ceremonies of Renaissance Rome, see Charles L.
Stinger *The Renaissance in Rome* (Bloomington, Ind 1985) 46–59.

146 A reference to the pope's being transported in solemn procession, while seated on a
portable throne or *sella vectus*, ie by *sedia gestoria*. In outdoor processions the pope usually
rode on a white horse. For descriptions of papal processions of that time, see Agostino
Patrizi *Caeremoniale romanum* ed Cristoforo Marcello (Venice: Gregorius de Gregoriis
1516; repr Ridgewood, NJ 1965) book 1 sec 12 chs 1–6 especially L^v–LI^v.

147 Patrizi *Caeremoniale* CXX^v: 'All mortals ... ought ... to kiss his feet.' But only select
persons (eg emperors, kings, and princes, etc) were allowed to kiss also his hand and
mouth. By the time of Erasmus (and from at least 1404 onward), the papal slippers
were adorned with a cross that was kissed; see Giacomo Povyard *Dissertazione sopra
l'anteriorita del bacio dei piedi de' sommi pontifici all' introduzione della croce sulle loro scarpe o
sandali e sopra le diverse forme colori ed ornati di questa parte del vestiario pontificio negli antichi
monumenti sacri* (Rome 1807) 102–5. On the significance of this kissing, see Peter Burke
The Historical Anthropology of Early Modern Italy: Essays on Perception and Communication
(New York 1987) 173, 175. Polydore Vergil claimed that the practice of kissing the pope's
foot was taken from the ritual of kissing the Roman emperor's foot: see Beno Weiss
and Louis C. Pérez *Beginnings and Discoveries: Polydore Vergil's 'De Inventoribus Rerum':
An Unabridged Translation and Edition with Introduction, Notes and Glossary* Bibliotheca
Humanistica et Reformatorica 56 (Nieuwkoop 1997) 310.

148 Patrizi *Caeremoniale* CXLI^r. The Pope never used the tiara (a symbol of the greatest
priestly and imperial power and dignity) at mass or during the divine office, but reserved it for other most solemn occasions and specific feast days.

149 In a papal procession armed servants on horseback carried silver staffs; *parafrenarii* carried maces (*cum martellis*), cursors their rods (*cum virgis*) and silver staffs; the papal
bodyguard had wooden staffs. That these officials went beyond the tasks of protecting the pope and keeping his path open to forcing people to kneel before him is not
recorded in the standard papal ceremonial of the day; see Patrizi *Caeremoniale* L^r–LIII^r;
and Nelson H. Minnich 'Paride de Grassi's Diary of the Fifth Lateran Council' *Annuarium Historiae Conciliorum* 14 (1982) 370–460 especially 425, 445–7.

150 Because the pope and the emperor were considered to be at the top of the spiritual and
temporal hierarchies respectively, cardinals were equated with kings at the next level.

151 See *Adagia* I viii 82: *In planiciem equum* 'The horse to the plain' LB II 327F–328B / CWE 32
169: 'Used whenever a man is encouraged to do the thing he is best at and most enjoys
doing.'

all in fanfare? What sort of a champion of the papal dignity is this Pio? I am far more truly its supporter than he. But I gladly retreat from this topic.

Although he is strong in his assertions, he completely misunderstands[152] both my expressions and his own.[153] The term *sacrament*, in its strict sense, is applied to that through which, by divine institution, a peculiar grace is conferred. Now when Peter Lombard says that this sort of grace is not conferred through matrimony,[154] is he not clearly excluding matrimony from the number of sacraments properly so-called? Since Durandus expresses doubt about the fact of the matter, he must be expressing a doubt about the term as well. If, for example, one has a doubt about the Moors' rationality, he surely has a doubt about their humanity. 'But,' says he, 'the conferral of grace is not the rationale of a sacrament properly so-called.' Whatever is, then? Its being a sign of a sacred reality? How many sacraments do we have then? The sign of the cross will be a sacrament, all the vestments and gestures of the priest at mass will be sacraments properly so-called.

'Look,' says he, 'Peter Lombard lists matrimony among the seven sacraments.' Yes, he does list it, but[155] according to the non-technical usage of *sacrament* and according to received opinion. But as his discussion develops, he defines sacrament differently from Augustine and he separates matrimony from the sacraments properly so-called. This is the sort of 'theologian' with whom I am dealing, but, even so, he accuses me of heresy and dreadful crimes! And he is also the one who says there is no

* * * * *

152 MARGINAL NOTE: 'He carps hatefully at things he doesn't understand'
153 This paragraph and the next are a reply to Pio XXIII *libri* 54ʳh (on *Responsio* 49 above 'Everywhere I count matrimony ... but Peter Lombard etc'): 'No, you are arguing for the opposite doctrine. But what is this you are saying about Peter Lombard? In fact, in the fourth book of the *Sentences*, in the second distinction, he writes as follows: "Let us now go on to the sacraments of the New Law, which are baptism, confirmation, eucharist, penance, extreme unction, orders, marriage." Did you hear? He lists marriage among the rest of the sacraments properly so-called. And Durandus' doubt is not about whether it is truly a sacrament (for Paul states this clearly), but whether grace is conferred through it as in the others, a point about which Peter also seems to be unsure, although his reason is that it [marriage] is entered upon for a remedy, not for a ministry. But it is not the conferral of grace that seems to be the rationale of a sacrament properly so-called, but its being the sign of a sacred reality. Matrimony certainly is, even if no grace is conferred, although it is quite reasonable that it be conferred.'
154 See *Responsio* n235 above.
155 *sed* 1531 *om* LB

need to read my apologias![156] If he had read them, he would not have raved in quite this way, for Zúñiga suffered a notorious shipwreck on this very rock.[157]

He says[158] that it is not probable that in ancient times the practice was to hold the *synaxis* seven times a day, at least that which is properly called the *synaxis*.[159] This is not what I said, my friend, for I add 'with unconsecrated bread.' Why, then, does he rave that I have said something untrue and improbable? As it is, if one who denies the antiquity of the rituals of the sacraments is disparaging them, he must quarrel with other men than me, and he must erase from the Gospel the account of how Christ instituted this sacrament, lest someone realize that the ancient Eucharist was celebrated with different rituals.[160]

How suitable his bringing up here the business of associating with prostitutes and cutthroats, and of sleeping with another's wife![161] What a droll man! And yet, if anyone were to associate with prostitutes and cutthroats in an endeavour to draw them to Christ, would he not be accomplishing a more perfect work than if he were to sing a psalm, or take a nap,

* * * * *

156 See Pio *XXIII libri* 68ᵛ–69ʳF

157 See *Apologiae contra Stunicam* (1) LB IX 338D–E / ASD IX-2, 210–12:2784–6 especially 2784n.

158 MARGINAL NOTE: 'He doesn't understand his own words'

159 Pio *XXIII libri* 54ʳl (on *Responsio* 51 above 'Among the ancient Christians etc'): 'On the contrary, this is improbable, and it is false about what was properly a *synaxis* for the reception of the Eucharist, although they also held communal meals in church, according to Paul's "When you come together, etc" [1 Cor 11:20]. But Oecolampadius also uses this same passage to maintain that the true body of the Lord is not in the sacrament of the altar. Therefore you should not annotate these things which could deprive so divine a sacrament of any of its prestige. In addition, one who takes their antiquity away from ceremonies diminishes them considerably.' Note that 'seven times a day' does not occur in Pio's note; by using these words Erasmus is associating the *synaxis* with the prayers of the canonical hours, for which practice Ps 118:164, 'Seven times a day I have given praise to thee,' was a common proof-text. See *Responsio* n245 above.

160 *ritibus* LB *artibus* 1531

161 Pio *XXIII libri* 54ᵛq (on *Responsio* 53 above 'who cannot be corrupted ... seems stronger etc'): 'What an effective argument! You ought to have said "and is not corrupted," not "and cannot be corrupted." But if this is true, then it is even better to associate with prostitutes and cutthroats, and to sleep with another's wife, so long as you do not have intercourse, than to avoid these things and so escape even the occasion of sin [reading *peccandi* for *peccando*]. The precursor of the Lord thought differently, and withdrew to the desert in his tender years [*teneris sub annis*, from the hymn *Ut queant laxis*]. The sons of the prophets thought differently, as did the apostles, who left all things and followed the impoverished Lord, even though they did not withdraw at all from the work of evangelization, etc.'

maybe, in a Carthusian cell? He prefers the model of John [the Baptist]. Let the monks withdraw, and live like John the Baptist; no one will bother them. But Christ associated with sinners, publicans, and women of bad repute. Who was the more perfect, John or Christ?

I am talking about the regulations of the monks; he cites[162] the evangelical counsels to us,[163] leaping from horses to donkeys.[164] The counsels of Christ are not identical with the regulations of the monks. What I mean by the regulations of the monks is eating beans on some days, only eggs on others, meat sauce but not meat on others, being girt with a knotted rope, having open shoes, and countless other matters like these. I compare them to certain precepts of the Mosaic Law whereby that stubborn and dull people was kept within their enclosures.

He makes fun of me because I remarked upon his winning favour with bishops, theologians, and monks.[165] If he received nothing else, at least he won the prize of having been buried in the most sacred habit of the Franciscans,[166] of having been given a share with all the benefactors of the entire order,[167] and of having been born again better in a second baptism

* * * * *

162 MARGINAL NOTE: Ἀπροσδιόνυσον [Irrelevance]'

163 Pio XXIII libri 54^{v}r (on Responsio 53 above 'since the same ... Law of Moses etc'): 'But the rules of the monks embrace not only the Gospel's precepts, but its counsels as well. But they pertain to the fullness of perfection. Thus I really wonder how such things could slip from your learned mouth. But your hostility towards the monks is so great that it has twisted your judgment completely.'

164 See Adagia I vii 29: Ab equis ad asinos 'From horses to asses' LB II 273B–C / CWE 32 83: 'When a man turns aside from an honourable vocation to something less reputable ... It will also be suitable when someone has sunk from affluence to a humbler station.'

165 Pio XXIII libri 54^{v}s (on Responsio 53 above 'you are winning favour with bishops etc'): 'Of course, so that the former will load down me, a married man, and the sons I do not have, with priestly benefices, that the latter will honour me with the insignia of a bachelor or doctor, and that the monks may invite me to a sumptuous banquet of greens!'

166 See Introduction xc, xcii.

167 In Exequiae seraphicae it is said that one of the benefits obtained by those who make the Franciscan profession on their death-beds is that they are endowed with the merits of the entire order (LB I 868E / ASD I-3 690:145 / CWE 40 1006). Erasmus also comments on Pio's being garbed as a Franciscan in Allen Epp 2441:64–77, 2443:338–40, 2466:107–110, 2522:68–9, etc. In his will Pio wrote: 'I wish when death draws near that the care of me be granted to the friars of the said [Franciscan] order, who are to be with me day and night and to pray for me and console me spiritually, who are to clothe me, before my soul breathes forth completely, in the habit of the Friars Minor placed over my bare flesh, my night shirt having been removed, and gird me with a cord' (Pio Will 390 below). On the practice of being buried in religious garb, see Louis Gougand Dévotions et pratiques ascétiques du moyen âge Collection Pax 21 (Paris 1925) 129–42; C.M. Figueras 'Acerca del rito de la profésion monastica medieval "ad succurendum"' Liturgica [Monserrat] 2 (1958) 359–400; W. Brückner 'Sterben in Mönchsgewand. Zum Funk-

than he had been in the earlier one.[168] I hope he attained to this, and that I will also, along with him.

He asks why I neglect the laity and make such frequent attacks on priests and monks.[169] It is because we more often admonish those whom we consider of greater importance, those upon whose conduct the life of the people depends. And yet it is not unusual for me to criticize the conduct of princes, magistrates, soldiers, businessmen, artisans, and ordinary people. But what cheek he has, not to read my works while carping at everything.[170] He has obviously read the *Folly*.

Some years ago, because of the hypercritical pronouncements of certain parties, I revised the *Exomologesis*.[171] But he does not condescend to read anything, so great is his love of brawling. But – so as not to tarry over details – as to his statement that I wrote that the benefits of confession can be obtained short of confession in church, for example, if one should confess to God,[172] I do not know if this is in my book; it was certainly never in my mind.

* * * * *

tionswandel einer Totenkleidsitte,' in *Kontakte und Grenzen: Probleme der Volks-, Kultur- und Sozialforschung. Festschrift für Gerhard Heilfurth zum 60.Geburtstag* (Göttingen 1969) 259–77.

168 On religious vows as a second baptism, see E. Delaye 'La doctrine du second baptisme: les vœux de religion' *Dictionnaire du spiritualité ascetique et mystique* (Paris 1937) I cols 1229–30. The superiority of a Franciscan profession on the death bed to baptism is ridiculed in *Exequiae seraphicae* LB I 868D–E / ASD I-3 690:139–46 / CWE 40 104). For Luther's early critique of this Franciscan view of religious vows, see Heinz-Meinolf Stamm *Luthers Stellung zum Ordensleben* (Wiesbaden 1980) 115.

169 Pio *XXIII libri* 54ᵛu (on *Responsio* 54 above 'It is beneficial for the people etc'): 'δίς κράμβη [cabbage twice (served)]. But come now, outstanding sir. Why are you always concerned with priests and monks instead of with lay people whose whole life has gone astray? How broad a scope for speechmaking it would provide if you looked into all the states of life. For the precepts of Christ do not oblige clergy and monks alone, but all Christians. It is only fair, then, that you accuse all, or leave off.'

170 Erasmus makes this charge against Pio repeatedly, but he may be reacting here specifically to Pio's note *XXIII libri* 55ʳx (on *Responsio* 54 above 'But I . . . do not so much admonish . . .'): 'You are telling the truth: you do not admonish, but carp, tear, attack, and slander . . .'; and note 55ʳy (on *Responsio* 54 above 'Are a large number of Jerome's books . . . to be burned?'): 'Of course, your works are equal or similar to his! His advise, edify, teach; yours detract, disparage, and infect the minds of your readers.'

171 The *Exomologesis* was begun in November 1523, and at the press by February 1524 in order to be ready for the March fair at Frankfurt. A second edition with minor corrections was finished by September. In March 1530 the Froben press issued a new edition *recognita diligenter et aucta*. See Allen Ep 1426 intro.

172 Pio *XXIII libri* 57ᵛr (on *Responsio* 66 above 'That as much advantage etc'): 'In addition, you list the advantages in such a way as to assert that they can be obtained short of confession in church, as, for instance if one were to confess to God or to a sensible friend, as you urge.'

But, he says, I do not list among the benefits of confession the elimination of sins and the conferral of sacramental grace, which is the main point of the whole activity.[173] Certain matters are too obvious to require mention, but I do touch on this point more than once in passing. But who does not know that sins are forgiven through penance? Or what point was there in my teaching how to make a proper confession if confession is of no use for the elimination of sins?

Though I report others' views concerning the human imposition of the burden of confession on humanity, he twists this around against me.[174] He draws his conclusion from scattered remarks, but he does this with more craft than is necessary, more than the integrity of a man such as he really calls for.

I maintain that certain disadvantages cannot be avoided,[175] not altogether, but that they can be avoided to a great extent if only men proven in learning, maturity, and good conduct are employed for hearing confessions. If in time of plague the people always receive the Eucharist every Sunday, then those who are oppressed by more serious sins should confess at that time as well. The Dutchman explains what he thinks quite clearly,[176] but the Man from Carpi is all too glad to carp at everything.

I hold that confession should be scrupulously maintained just as if instituted by Christ, but that I am not yet certain whether it was instituted

* * * * *

173 Pio ibidem: 'But you overlook the benefit which is the main point of the activity, namely the elimination of sin and conferral of grace, which come from sacramental confession.'
174 Pio ibidem: 'And you argue that it cannot be proven that confession was instituted by Christ, and you express a doubt as to whether so great a burden could have been imposed on humanity by human ordinance, and, given that it has been imposed, whether it might be better to reject it because it is more of an obstacle than an aid to devotion. No one is so dull that he cannot draw the conclusions that would follow from this. Moreover, it makes no difference how widely scattered these statements are, since it is quite easy to collect them.'
175 This paragraph reacts to an earlier portion of the same marginal note: 'A sophistic reply, when one fends off a charge that has not been made, and ignores what is censured. I never wrote that you gather together the disadvantages of confession as if they arose from the thing itself, but that, whatever their origin, it was not right to collect them, both because they discourage people from the duty of confession, and because it is under attack by heretics at this time, and because the result of your collecting them is to show that although they originate from human failure, many of them cannot be avoided. For a penitent cannot have looked into whether the priest is sufficiently upright and learned to avoid slipping into these disadvantages, nor can a parish priest avoid the immediate danger of death when he is summoned to hear the confession of a plague-victim, and the other similar matters you mention under the "fifth disadvantage."'
176 Pio ibidem: 'Therefore there is no reason for you to think that you can, with these glib phrases, convince the sensible and devout reader that your *Exomologesis* is in any

by Christ himself, though I support this position and would like it to be established by the learned with irrefutable proofs.[177] What is the problem? Is this declaration of the Dutchman not sufficiently clear?

He says that confession has been received by the church among the sacraments that confer grace.[178] This calling confession a sacrament is a novelty. Those who discuss confession most skilfully term it 'the third part of a sacrament,' or 'a sacramental.' And yet he wants it to appear quite evident that confession was instituted by Christ, either by his very words, or through the apostles, or in another way. But if by 'in another way' he means by a decree of the church, the theologians say that the church is absolutely forbidden to institute any new sacrament.

Again,[179] it is plainly false that I have doubts about whether it would be more expedient to abandon confession because of its disadvantages or to keep it because of its advantages.[180] I report that such matters are argued by others;[181] under no circumstances do I express a doubt.

It is so aggravating to quarrel with a man who keeps on repeating either obvious falsehoods[182] or charges that have been clearly refuted in the very passage he is gnawing at. Where do I make it a matter of individual choice to confess or not to confess?[183] I report that this statement is being

* * * * *

way at all of benefit to the Catholic religion. Since the Dutchman does not dare to state openly there what he thinks, he so obscures the matter that it is impossible to grasp readily what he decides or means, unless the reader is extremely watchful and compares the statements to other passages in which you reveal your opinion more openly.'

177 In *Exomologesis* LB V 147A, in the course of reviewing opinions about the origin of confession, Erasmus mentions those who assert that it was instituted by humans, 'yet maintain it no less scrupulously than if it had been instituted from the mouth of Christ,' and a little later says 'I am, however, more inclined towards the group that believes that it was instituted by Christ.' In *Detectio praestigiarum* LB X 1570B / ASD IX-1 258:590–1 he asserts: 'I suspend my judgment, but in such a way that, in the meantime, I revere it as if Christ had instituted it [*quasi Christus instituisset*].'

178 Pio *XXIII libri* 58[r]s (on *Responsio* 68 above 'If I could prove etc'): 'Nor were all things written down, but they are adequately confirmed when they are either derived from what was written down, or have been promulgated by the decree of the church. Confession is of this sort. It is counted by the church among the sacraments that confer grace. This could not be done unless it had been instituted by Christ, either by his words, or through the apostles, or in another way.'

179 MARGINAL NOTE: 'Lie'

180 Pio *XXIII libri* 58[r]s (on *Responsio* 68 above 'If I could prove etc'): 'expressing a doubt whether it would be better that it be retained because of its advantages &c.'

181 See *Exomologesis* LB V 145B.

182 MARGINAL NOTE: 'Lie'

183 Pio *XXIII libri* 58[r]s (on *Responsio* 68 above 'If I could prove etc'): 'Since, after you list these [the disadvantages of confession], you make no determination that it is nonetheless

made by certain Lutherans.[184] As to my comment that he who confesses is more secure,[185] this was said, not on the basis of strong conviction, but in an argument against those who question the necessity of confession.[186] His repeating these allegations is all the more mean-spirited because I have refuted them quite plainly in my *Epistle*.

He gnaws at the particle *quasi* [as if]. 'What exists in a *quasi* way,' he says, 'does not exist.'[187] But I said this in answer to those who think that confession was introduced by ecclesiastical decree. So what poison is there in the particle *quasi*? It makes a distinction between the one instituting and the mode of institution. Who does not know that all things pious proceed from Christ? But when I say 'by Christ,' I mean by the words of Christ, in the same way as he instituted baptism and the Eucharist.

I had said that I adjusted the contents of the book in such a way that it could be of use even to those who question the necessity of confession. Look how he ridicules me here![188] 'Of course, it is in your control,' he says, 'when you publish books, to prescribe the sort of people who are to read them!'[189] He wants the whole business to appear irrelevant [ἀπροσδιόνυσον], and he makes sport of my 'wordlets.' (Indeed, he seems quite taken with this expression.) But what is the point of this snobbish sneer? Is it impious to wish to be of benefit to a greater number of people? Paul wrote to the Romans. Did he thereby forbid Greeks to read his letter? O sober and careful writer!

I point out that his proof-text from Moses, not contained in the earlier version of his book, was added in Paris. Though I remark on this phenomenon scarcely four times in the entire book, he says that I do it more

* * * * *

quite necessary to make confession, you are certainly leaving it to the reader to decide whatever he pleases about it.'

184 See *Exomologesis* LB V 145C–146A.

185 Pio *XXIII libri* 58va (on *Responsio* 70 above 'For the majority of them etc'): 'You subscribe to this position since you do not admit that it [confession] has validity on the basis of divine institution, even though, fearing hostility, you say that he who confesses is more secure'; Pio refers here to Erasmus' statement in *Exomologesis* LB V 145C.

186 MARGINAL NOTE: 'He is again going after things I have clearly refuted'

187 Pio *XXIII libri* 58^vc (on *Responsio* 70 above 'as if [*quasi*] instituted by Christ'): 'This "as if" is very important; what is as if gold is not gold, but brass; what is as if one is not one.'

188 MARGINAL NOTE: 'He hisses without justification'

189 Pio *XXIII libri* 58^ve (on *Responsio* 70 above 'Thus it was my desire etc'): after the words quoted here by Erasmus, the note continues: 'One cannot know why you make these irrelevant statements, unless it be to distract the reader's attention with these wordlets [*verbula*] from the focus of the question. But by design you ignore the fourth category of people, those who are convinced that it is of divine institution.'

than a hundred times![190] This is a quite shameless hyperbole. I am not finding fault with the fact of the addition, but with its having been tacked on
ineptly, and on the basis of someone else's initiative, and finally with the
fact that, though he went to the trouble to make additions, he did not make
the same effort to remove items that he ought to have removed.

Here he insists that I am distorting Scripture with my paraphrase,[191]
as if my paraphrase is the divine Scripture. He keeps on echoing the time-
worn allegations, nor does he see fit to read my replies to others concerning
this charge.[192] Indeed, with a view to my unpopularity, he even makes the
false charge[193] that I was the very first to expound Scripture by means of
paraphrase,[194] in spite of the fact that the paraphrase on Ecclesiastes of the
most saintly man Theodore, whom Jerome mentions, is extant;[195] in spite
of the fact that Arnobius' commentary on the psalms[196] is virtually nothing

* * * * *

190 Pio *XXIII libri* 60^{r}y (on *Responsio* 76 above 'Indeed, at this point etc'): 'So often you are
silly, or, to be more accurate, malicious. So what? Is this not appropriate when the book
has not yet been published? And yet more than one hundred times you make this objection about something which was done in barely two or three places!' For complaints
about later insertions at Paris, see *Responsio* 74, 76 above.

191 Pio *XXIII libri* 60^{r}z (on *Responsio* 76 above 'But the Translator etc'): 'I will answer all
these with only a wordlet. To make an addition to express the meaning more clearly is
far different from what amounts to distorting and altering the whole meaning by the
addition of many ideas of your own, as you do in the *Paraphrases*.'

192 Eg Ep 1381:441–50; Allen Ep 1721:3–23; Erasmus' responses to Béda's criticisms (*Divinationes ad notata Bedae, Responsio ad notulas Bediacas* and *Supputatio* LB IX 453–698; and
his responses to Cousturier *Apologia adversus Petrum Sutorem* LB IX 754C–D.

193 MARGINAL NOTE: 'Lie'

194 Pio *XXIII libri* 60^{r}x (on *Responsio* 76 above 'as if I were a new translator [*interpres*] of
the Scriptures'): 'New, I say, and the first in this type of publication in the case of the
Sacred Scriptures, nor yet an *interpres*, that is, a translator , but a paraphraser'; and see
Responsio nn 389 and 390 above.

195 That is, the work of Gregory Thaumaturgus; see Jerome *De viris illustribus* 65:1 'Theodore,
who was later named Gregory ...'; and 65:4 'He wrote a metaphrase of Ecclesiastes, a
short work, to be sure, but a very useful one ...' (Ceresa-Gastaldo ed 168, 170); see also
Jerome Ep 70.4.4 CSEL 54 706. In a note in his edition of *De viris illustribus* HO 1 143^v
Erasmus reports the presence of a copy of Gregory's *Metaphrase* in the library of the
Dominicans in Basel. In 1520 Oecolampadius published a Latin translation, *In Ecclesiasten Solomonis metaphrasis divi Gregorii Neocaesariensis episcopi, interprete Oecolampadio. Libellus hic brevis quidem est, sed valde vtilis vt ait diuus Hieronymus* (Augsburg: S. Grimm, M.
Wirsung 1520); see E. Staehelin 'Die Väterübersetzungen Oekolampads' *Schweizerische
theologische Zeitschrift* 33 (1916) 57–91 especially 58–9.

196 Erasmus refers to the *Commentarii in Psalmos* CCSL 25, no longer ascribed to Arnobius
Afer, as Erasmus had thought, but to Arnobius the Younger (fl 450); Erasmus published
the *editio princeps* of this work along with his own commentary on Ps 2 in *Arnobii opera*
(Basel: Froben 1522); the dedicatory letter to Adrian VI is published as Ep 1304 (see
Allen's introduction to this letter for additional information).

but a paraphrase, though it does not have that specific title. Who ever complained that they distorted Scripture? He also says that Titelmans did not write paraphrases of the psalms,[197] though the book exists to rebuke the fellow's shamelessness. And what about Jacques Lefèvre's continuous commentary on the Psalms,[198] what is it but a paraphrase? Finally, it is possible that something similar was done by even more authors whose works have not come down to us.

Here too I am foolish, according to my friend Pio. 'Preachers,' he says, 'who explain a passage quoted from Scripture by means of a long periphrasis are not paraphrasers when they quote the Latin.'[199] But when they translate into the vernacular with an extensive periphrasis, they definitely are paraphrasers! I am not talking about those who explain the lines of the particular characters by speaking as themselves in the style of commentaries, but about those who present a sort of version of Scripture in the persona of the one whose words they are relating, for example, of God, of Christ, of the Prophet, of the Apostle. Pio did not notice this, and yet the droll fellow

* * * * *

197 This is not what Pio said (see n217 below), nor could Pio have seen Titelmans' *Elucidatio in omnes psalmos* because Pio died on 7 January 1531 in Paris, while the first of many editions of Titelmans' *Elucidatio* was not published until later that year in Antwerp. Pio made his comments on the style of Titelmans' commentary in reference to either Titelmans' *Elucidatio in omnes epistolas apostolicas* (Antwerp 1528 and subsequent editions) or his *Collationes quinque super epistolam ad Romanos beati Pauli* (Antwerp 1529); see Pio *XXIII libri* 60ᵛi, quoted in n217 below. In his *Elucidatio in omnes psalmos* Titelmans placed the text of the psalm in the margin in one font, and in the text, set in another font, he provided a running paraphrase in which he spoke in his own or the psalmist's *persona*. Later on (146 below) Erasmus no longer argued from Titelmans' *Elucidatio* on the psalms, but from Titelmans' work on the Pauline epistles. See *Katholische Kontroverstheologen und Reformer des 16. Jahrhunderts* ed Wilbirgis Klaiber (Munster 1978) 284–5 nos 3092, 3100.

198 Although Lefèvre translated the Psalms into Latin (1509) and French (1525) and provided brief summaries of each psalm, explanatory notes, and accent marks for the proper pronunciation of the words, he does not seem to have provided for the Psalms, as Erasmus claimed, the running commentary that he provided for the Pauline Epistles (1512), the four Gospels (1522), and the Catholic Epistles (1527). In 1528 Lefèvre prepared 'explanations' of the Psalms for the royal children, but it is not clear if these were printed. See E. Amann 'Lefèvre d'Etaples, Jacques (Jacobus Faber Stapulensis)' DTC IX-1 cols 132–59 especially 157–8; *The Prefatory Epistles of Jacques Lefèvre d'Etaples and Related Texts* ed Eugene F. Rice JR (New York 1972) 468–77, 498–501; Philip Edgecombe Hughes *Lefèvre: Pioneer of Ecclesiastical Renewal in France* (Grand Rapids, Mich 1984) 154–62, 177 n22; and CEBR II 315–18.

199 Pio *XXIII libri* 60ʳb (on *Responsio* 77 above 'Do you not hear preachers etc'): 'Here you seem foolish. You are not really going to call preachers paraphrasers, since they state their interpretation only after the very words of Scripture have been quoted?'

accuses me of insanity, whereas he is the one who is insane, replying to matters he has not understood.

He says[200] that Leo x did not approve of my hard work on the New Testament, but only of my cleverness or elegance of language,[201] as if there were no brief of Leo's[202] in which one may read the following: 'we inferred ... what this new one would be, and how much it would benefit all who have at heart the progress of theology and of our orthodox faith.'[203] The brief has been published quite often,[204] and yet Pio dares to make this assertion, awarding himself the same licence in published books as he had allowed himself in table talk.

But Leo did not give his approval to the *Paraphrases* – nor was it suggested to him that he do so, for I was not worried about the *Paraphrases*! Domenico Cardinal Grimani certainly thanked me in a letter[205] for my *Paraphrase on the Epistle to the Romans*, and he was not the only one to do this.

Pio says: 'We granted that one might emend corruptions, but we did not grant that one might pervert through paraphrase.'[206] He is very fond of the word 'pervert,' as if every paraphrase is a perversion. And yet it is far more bold to alter a reading traditional in the usage of the church than to explain an obscure one by means of paraphrase. But what Pio has conceded, Cousturier, whom he praises, did not.[207]

* * * * *

200 MARGINAL NOTE: 'Lie'
201 Pio *XXIII libri* 60ʳd (on *Responsio* 77 above 'In fact, Leo x approved etc'): 'I, at least, will not admit this, since I was in Rome, and the pope was on very friendly terms with me. But if any such thing occurred, perhaps he approved of the elegance of the language and the versatility of your wit, though it is quite certain that he could not have approved of your rash endeavour, once he realized its true character.'
202 Published as Ep 864, requested by Erasmus through Antonio Pucci and others; see Allen Ep 864 intro.
203 Ep 864:9–12, Allen Ep 864 intro.
204 Allen Ep 864 intro reports that the letter 'was published in the second edition of Eramus' New Testament, and in all the other Froben editions published during his lifetime.'
205 While Erasmus' letters to Grimani (CEBR I 132–4) about the *Paraphrases* survive (Epp 710, 835, and 1017), Grimani's response – if he ever wrote one – does not; see Allen Ep 710 intro, 306n. In his *Apologia adversus Petrum Sutorem* (1525) Erasmus claims that Grimani gave thanks in a letter; see LB IX 801D.
206 Pio *XXIII libri* 60ʳc (on *Responsio* 77 above 'But if it is established etc'): 'Give this answer to yourself, not to me. This is rhetorical, sure enough, but not Christian. I conceded to you that corruptions could be emended, just as I proved that it is not appropriate to pervert sound texts with paraphrase.'
207 Eg *Apologia adversus Petrum Sutorem* LB IX 751F–752A, 755E–764A. See Bentley *Humanists and Holy Writ* 205–6.

Here again he brings forward an out-and-out misrepresentation,[208] that all of the ancients refrained from paraphrases.[209] He even takes virtually for granted the absurd position that it is one thing to explain an isolated section of Scripture by means of paraphrase and quite another to explain all of Scripture by means of continuous paraphrase, despite the fact that in both cases the Scripture itself remains intact. Moreover, I even distinguished the scriptural texts by indicating them in the margins. What difference does it make whether you read it in the margin or in your own book?

He is indulgent towards Juvencus,[210] Arator,[211] Prudentius,[212] Prosper,[213] and Nazianzen[214] on the grounds of poetic licence.[215] But who forced

* * * * *

208 MARGINAL NOTE: 'Lie'

209 Pio XXIII libri 60ʳf (on Responsio 78 above 'You say that no one has dared etc'): 'For a commentator interprets individual phrases separately as expressed in the author's own words. A paraphraser expounds his understanding of the meaning in continuous speech as if he himself were the author, making additions or deletions as he deems fit. The most holy Fathers, observing this, all refrained from paraphrase in expounding the Scriptures, but nonetheless published ever so many commentaries.'

210 See Responsio n389 above.

211 See Responsio n390 above.

212 The premier Christian Latin poet (348–c 405), whom Erasmus called 'that celebrated minstrel of Christian philosophy' (Psalmi LB V 421A). The editio princeps of Prudentius' opera omnia was printed at Deventer by Richardus Pafraet in 1492 (see J. Bergman De codicum Prudentianorum generibus et uirtute Sitzungsberichte der kaiserliche Akademie der Wissenschaften in Wien, Philosophisch-historische Klasse 157.5 [1908] 2 n1). The Aldine edition appeared in Poetae christiani veteres I (Prudentii, Prosperi, Ioannis Damasceni opera) in 1501. In 1524 Erasmus published, along with a commentary on the pseudo-Ovidian Nux, a commentary on two hymns of Prudentius, Cathemerinon 11 and 12, which he dedicated to Margaret Roper (Hymni varii LB V 1337–58 / CWE 29 173–218).

213 Prosper of Aquitaine (d shortly before 455); Pio and Erasmus are probably referring to his Epigrammata ex sententiis S. Augustini PL 51 497–532, a basic school text (described in Paul G. Gehl A Moral Art: Grammar, Society, and Culture in Trecento Florence [Ithaca, NY 1993] 137–43).

214 Gregory of Nazianzus (see n370 above). His poems had been published by Aldus in 1504 (Gregorii Nazianzeni carmina cum versione latina; the preface is reprinted in Aldo Manuzio Editore I 81–2); see Sister Agnes Clare Way in Catalogus Translationum et Commentariorum: Medieval and Renaissance Latin Translations and Commentaries II (Washington, DC 1971) 65–8. Erasmus had made a gift of a copy of this volume to Martin Lypsius (Ep 807). Erasmus cites Gregory's poem De Christi genealogia PG 37 479–87 in his annotation on Luke 3:23 LB VI 244C / Reeve 173.

215 Pio XXIII libri 60ᵛh (on Responsio 78–9 above 'Juvencus ... Arator ... etc'): 'You are leaving out Prudentius, Prosper, Cento, Nazianzenus, and very many others. But who is so stupid that he would say that the author Cento [sic], or any other, celebrating some sacred matters in verse, is a commentator on Scripture? Moreover, since, by your own testimony, giving delight and stirring the emotions are plainly no less a part of the poet's profession than being of benefit, poets therefore also enjoy great licence. But they cannot achieve this unless they add many things to the bare words of the text.'

them to handle the majesty of divine Scripture in verse? These are Aristotelian arguments!

Either he does not understand[216] what I said about Titelmans, or he pretends not to. He says that Titelmans' *Elucidations* are glosses, or notes, expressed in the author's own words, which Titelmans then explains in the same order at greater length.[217] In fact, except for the title, there is not the least bit of difference, except that he preferred to follow the more recent rather than the ancient authors, and he transcribed in the margin the entire passage, which I had only indicated succinctly for the reader so that he could read the paraphrases conscious of the context.

Again he has not understood what I said, and resorts to abuse nonetheless.[218] What I am saying is that the ancient commentators, particularly Origen[219] and Jerome, first quote the passage of Scripture; next they give a version of the text which is nothing but a paraphrase, for they speak in the persona of the one whose words they quote; and finally they proceed to a more lengthy comment. 'But,' says he, 'they do this after indicating the words of Scripture.' By indicating the passages in the margin I do the same thing as they do by giving quotations at intervals. Even in a paraphrase one may take a break. And after all this he repeats his absolutely false conclusion that no one, except me, has ever dared to undertake paraphrases![220]

At this point he admits that it is the task of the paraphraser to set forth the meaning of an author with greater clarity.[221] This is exactly what I

* * * * *

216 MARGINAL NOTE: 'He doesn't understand what he's criticizing'
217 Pio *XXIII libri* 60ᵛi (on *Responsio* 79 above 'Titelmans'): 'What are you saying about Titelmans' *Elucidations*? Clearly they are glosses or scholia in the author's own words, which he then expounds in the same sequence at greater length. But you have now gone around to all the corners of the arena, and have not found an escape: first commentators, second those who argue about difficult passages, next poets, and finally Titelmans' *Elucidations*, but they are all quite different from a paraphrastic reading. Indeed, by this line of inference what I had asserted is proven even more, that is, that no one ever, except you, had dared to treat Scripture in a paraphrase.'
218 Pio *XXIII libri* 60ᵛk (on *Responsio* 79 above 'How silly, as if etc'): 'Surely you are the one who is silly, since you undertake to prove that there is no difference between a commentator and a paraphraser, but you admit the opposite with your very words. For you say that commentators explain the meaning in their own words. But come, when they do this, do they not quote the words of Scripture from which they derive it? But a paraphraser does otherwise, since he manages his work without interruption, as you admit, as if he were the author. But commentators comment on passages that they quote separately.'
219 Origen of Alexandria (c 185–c 254); there is a detailed account of Origen's influence on Erasmus' paraphrases in André Godin *Erasme lecteur d'Origène* Travaux d'humanisme et renaissance 190 (Geneva 1982) 352–413.
220 See n217 above.
221 Pio *XXIII libri* 61ʳp (on *Responsio* 80 above 'As it is, if I were to add nothing etc'): Pio

was trying to do. In some places I do not succeed, but who succeeds every-where? The effort, at least, is praiseworthy, and allowance must be made for human frailty. If the use of a persona is a source of scandal, the same practice was, as I have said, followed by the Greek Theodore with Ecclesiastes, by Arnobius with the Psalms, by Jacques Lefèvre with the Psalms,[222] by Titelmans with the epistles of Paul,[223] and by Origen and Jerome, although intermittently, with the prophets.

What are the buffooneries, poetic fancies, and fables that I introduce into Scripture?[224] What portentous words! It is the nominalists and the realists who have imported these into Scripture. Indeed, Erasmus never used the expression 'fable of the Gospel,'[225] but some denouncer palmed this off on Pio, who would not have been taken in by him if the fastidious man had read my *Supputations.*[226]

Where do I quote Scripture in different words?[227] I use different words to explain it, and, granted, another's persona, the practice, as I have made clear, of the most esteemed men before me. But what will Pio make of those who, even though not explaining Scripture, quote it in different words? Cyprian quotes it one way, Tertullian another, Ambrose another, and Augustine another. Maybe he will say that there is no real difference between words that mean the same thing. Even this may not be the case everywhere, but I will let it pass. At least I make an honest effort to clarify the Apostle's meaning.

At this point he charges me with a crooked ploy.[228] He says: 'I did not write that there was no remedy at all, but that there was none at hand,

* * * * *

charges that Erasmus has gone beyond the scope of a paraphraser, 'whose sole task is, not to introduce new ideas, but only to expound in clearer language the author's meaning as gathered from his words.'
222 See n198 above.
223 *Elucidatio in omnes epistolas apostolicas* ed Klaiber (cited in n197) 284 no 3092
224 Pio *XXIII libri* 61ᵛq (on *Responsio* 81 above 'Aristotle and Averroes'): Pio defends the propriety of the use of philosophy and the arts in elucidating Scripture, and continues: 'But it is surely absolutely inappropriate to introduce poetic fancies, fables, and buffooneries in handling these things.'
225 See *Responsio* n271 above.
226 LB IX 654A–C (*Suppositio* no 132)
227 Pio *XXIII libri* 61ᵛs (on *Responsio* 81 above 'see if I have altered a single one etc'): 'You are being quite inappropriately ironic. In fact, you have changed them all, since you have introduced a new style that is unworthy of them.'
228 Pio *XXIII libri* 61ᵛx (on *Responsio* 81 above 'did you write this as a joke etc'): 'This too is a clever ploy, but crooked. I did not write that there was no remedy at all, but that there was none at hand, none ready, whereby [emending *quod* to *quo*] one fallen into error could be recalled, as when the words of Scripture are quoted line by line, but never in paraphrases. Accordingly, it could happen that many who do not go to the trouble of

none ready, etc.' We will, then, record here Pio's own words:[229] 'But in a paraphrase, once you have departed from the true meaning, there is nothing that can bring the reader who is relying on your faithfulness back to the true path, no gauge that he can keep checking to restore his critical judgment.' This is what he wrote. Is he not simply denying that a remedy exists? But he says this is not what he wrote, and maintains instead that he said only that there was no handy, ready remedy. Could anything more absurd than this argument be contrived? If a commentator goes wrong, there exists a means of getting oneself back on track. By what gauge? By the gauge of the Scripture, which one can examine more carefully. If a paraphrase goes wrong, there is no remedy. Why? Because all of Scripture has been annihilated! In fact, the entirety of Scripture remains in existence, intact and untouched. But there is no remedy ready at hand! In fact, there is one on the spot. That is why I noted the chapters and the *incipits* of passages in the margin, for the very purpose of making it easy for the reader to check the paraphrase against the original. Although I made this very statement in the passage he is gnawing on, he pretends that I did not, and croaks his tiresome complaints to me.

He says: 'Commentators put themselves forward as the servants of Scripture, paraphrasers as its partners and colleagues.'[230] I do not know what claim Theodore and Arnobius make. I declare often and clearly that I am a servant; I never claim to be a colleague. It is in this way, I suppose, that Aristotle used to reason!

No, it is Pio who does not realize what he is saying![231] I said that he made a fierce opening statement, but concludes feebly. Have I then lost track of what I said if I quote his words from the opening statement one way, those from his closing remarks in another?

* * * * *

comparing them [the paraphrases] with the texts could not be recalled from the error into which they have fallen.'

229 Pio *Responsio paraenetica* 12[r]I or Pio *XXIII libri* 7[r]I

230 Pio *XXIII libri* 61[v]c (on *Responsio* 83 above 'But it is nonetheless a work of piety etc'): Pio describes paraphrasers going wrong 'since they have not followed the authors whose persona they have put on with as great an attention and reverence as do commentators. I might say that, as a general rule, this is the difference: commentators claim to be the servants of their authors, but paraphrasers their collaborators and colleagues, wherefore it is by no means permissible to handle sacred Scripture in this genre.'

231 Pio *XXIII libri* 62[r]e (on *Responsio* 84 above '... leader of the entire Lutheran tempest etc'): 'You soon forgot what you just now said, namely that I had said that you had created a suspicion in many that you do not reject absolutely the Lutheran programme, whereas now you want me to have said that you were the instigator and cause of the entire Lutheran tempest.'

If he wished to appear forbearing in order to avoid offending me,[232] why is he more aggressive in presenting his case than in proving it? Nor am I arrogant, but he is deceptive. I have pointed this out to him, but he keeps to his old behaviour through this whole book.

What is his point here[233] but that all who desire church reform are heretics?

'Aha!' What is it that you've got? That I abjure all the doctrines of Luther?[234] I do not even know all of them! Why should I abjure doctrines to which I have never subscribed? Let others abjure!

So I thought that Luther was the antidote for our plagues.[235] Where did he read this? I am saying that this was the opinion of others; it is not my own view.[236] At least all the universities were my accomplices in the crime of silence at that time.[237]

I am not making excuses for Arius,[238] but I am criticizing his excommunication as being premature since the case was still unclear. What I say here about Arius, Jerome states about Tertullian,[239] and yet Jerome did not

* * * * *

232 Pio XXIII libri 62ʳg (on Responsio 85 above 'It is a suspicion etc'): 'You are finding fault with me because I was more temperate in my language in dealing with you than the situation requires. I did so to avoid offending you. But I could not prevent this, so vast is your pride.'

233 Pio XXIII libri 62ʳh (on Responsio 85 above 'But what are you calling a programme? etc'): 'All heretics always used this pretext, that they are teaching the truth, that they are bringing back true devotion and restoring what has fallen to ruin to its original sincerity. This was Arius' dodge, and that of the rest of the early heretics, of Wycliffe in the last century, and of Luther who was born from his tomb.'

234 Pio XXIII libri 62ʳi (on Responsio 85 above 'But if by programme you mean Luther's teachings ... falsehood'): 'Aha! In these words I've got most of what I was seeking, namely that you should declare your abomination and abjuration of all the doctrines of Luther. But if you really mean this, then remove from your writings everything that agrees with his views and seems to support them, and all will be well, and you will be at peace with all the orthodox.'

235 Pio XXIII libri 62ᵛs (on Responsio 90 above 'True enough. But at the beginning etc'): 'Ah, you thought that Luther was the antidote for plagues! And the universities again! But to the same argument, the same reply.'

236 However, see Ep 1634:88–9.

237 The universities of Cologne, Louvain, and Paris formally condemned some of Luther's positions; see Responsio n435 above.

238 Pio XXIII libri 63ʳt (on Responsio 90 above 'In the same way etc'): 'What is the point of this? But you would have ruptured yourself if you had not savaged the clergy according to your habit in this passage too, even though what you say is false. That conflagration arose because of the perversity and pride of Arius, as his blasphemies and stubbornness attest, for he rejected the most holy council's decrees to which he had earlier subscribed, and returned to his original impiety.'

239 De viris illustribus 53.4 (Ceresa-Gastaldo 150): 'He [Tertullian] was a presbyter of the church until the middle of his life. Afterwards he fell away to the teaching of Mon-

have to rupture himself.[240] And was it not the envy of the clergy that forced Chrysostom out?[241] But Pio is enraged as if it were a novel idea that envy arises among the clergy.

The debate about Arius went on for a long time before his condemnation. And Hilary is careful to explain his tardiness in writing against Arius.[242]

Yet again the *Diatribe* and the *Hyperaspistes* do not count against Luther at all:[243] 'Olive no kernel hath, nor nut no shell.'[244] Pio did not see the *Hyperaspistes*, but he surely could have seen it at Paris had he wanted. Why does he pass judgment on what he has not seen?

* * * * *

tanus as a result of the envy and abuse of the clergy of the Roman church'; cf Erasmus' note on this passage (HO 1 140): 'How well disposed Jerome is towards the talented Tertullian! He practically excuses his lapse into the Montanist faction by transferring the fault to the envy and abuse of the clergy. Arius too, who caused such vast disturbances, was provoked in the same way, nor did Origen escape envy, and Jerome barely survived.' In Allen Ep 2136:168–72 Erasmus says that his own experience 'is causing me to understand what I have heard, that Arius, Tertullian, Wycliffe, and many others departed from the communion of the church because of the envy of the clergy and the corruption of some monks, turning personal resentment into a public disaster for the church.'

240 'Rupture himself' is an allusion to Pio's expression *sed rupta tibi essent ilia* in his note XXIII *libri* 63^{r}t, quoted above in n238; see Virgil *Eclogue* 7.26 '*invidia rumpantur ut ilia Codro.*'

241 John Chrysostom was deprived of his bishopric and exiled partly as a result of the machinations of ecclesiastical enemies. Erasmus prefixed a *Vita divi Ioannis Chrysostomi*, based on book 10 of the *Historia tripartita* and Palladius' *Dialogue on the Life of John Chrysostom*, to his *Chrysostomi opera* of 1531 LB III 1332C–1347B, with the account of Chrysostom's being driven into exile on 1339A–1340F. Note Erasmus' remark about Chrysostom and his enemies: 'But if what Palladius tells about him [Chrysostom] in his *Dialogue* is as true as it is likely, and what he tells about Theophilus and the rest who followed his faction, then the bishops of our times are saints in comparison to them.'

242 See *Responsio* n447 above and Pio XXIII *libri* 63^{r}u.

243 Pio XXIII *libri* 63^{r}x (on *Responsio* 90–1 above 'in spite of the fact that I combated etc'): 'As you know, I have praised your effort on this subject, although in it you do not attack the specific doctrines of Luther in which he attacks the church, and the *Hyperaspistes* is on the same subject. Accordingly, you should not have listed it, unless it was by way of suggesting, by giving a lot of titles, that you had written a large variety of works against him. Moreover, it is quite beside the point that I had seen the *Hyperaspistes* when I wrote to you, because I had not yet read any of it.'

244 See *Adagia* I ix 73: *Nil intra est oleam, nil extra est in nuce duri* 'Olive no kernel hath, nor nut no shell' LB II 359F–360A / CWE 32 220–1: 'A proverbial line from the *Epistles* of Horace, which can be used against those who carry effrontery to such crazy lengths that they are not afraid to deny what is generally accepted and to maintain that manifest falsehoods are certainly true, who in fact will say anything rather than be seen to lose an argument.'

Pio escaped by making the distinction that the issue of free will is not a doctrine peculiar to Luther,[245] as if he does not virtually share the rest of his doctrines with Huss and Wycliffe! And that doctrine is so 'silly' that Augustine comes within a hair's breadth of saying the same thing, and he an orthodox figure of the first rank. 'It was rejected by the philosophers.' But not by all, nor had they read the Scriptures, which frequently seem to eliminate free will. As a matter of fact, even now there is no consensus about what freedom of the will means. The philosophers knew nothing of the distinction I have introduced, for they are, of course, ignorant of operative and cooperative grace.[246]

Again I am invited [to write against Luther] by the popes, though I have shown that this is untrue.[247] But now I am at least invited! He asks why I put off doing so. It is because I do not have the time because of slanders like these! In addition, even if I had written nothing at all against Luther, I would not now write a line, so excellent has been the recompense for my effort!

An Italian, he puts together a slander from the use of the term *crimen*, though what I mean by *crimen* is indictment.[248] If I am rightly to be blamed

* * * * *

245 Pio XXIII *libri* 63ʳa (on *Responsio* 92 above 'This is absolutely false etc'): 'You will prove it false if you can prove that you have refuted even one of the views specific to him. For his position on free will is so silly that it is rejected by all peoples and nations of whatever faith, by all philosophers, and, in fact, by common good sense.'

246 This distinction does not originate with Erasmus, but goes back to Augustine; see Alister E. McGrath *Iustitia Dei: A History of the Christian Doctrine of Justification* 2nd ed (Cambridge 1998) 27–8; *Justification by Faith: Lutherans and Catholics in Dialogue* ed H. George Anderson et al (Minneapolis 1985) VII 18–19 no 9. For an analysis of Erasmus' teaching on grace, see McSorley 283–93.

247 Pio XXIII *libri* 63ʳb (on *Responsio* 92 above 'In fact your friend the Bullbearer etc'): 'You keep on beating me on the head about something or other the "Bullbearer" said to you. But it is an established fact that this was urged on you by very many, and that you were invited to undertake it most graciously by the popes in their statements and letters. But let us grant that no one took this matter up with you then. Later on so very many have asked and are still asking. Why do you not do it, why are you still putting it off?' see 129–30 above. Erasmus was clearly invited in a letter from Adrian VI to write against Luther's teachings (Ep 1324:42–55, 90–6). In his response Erasmus claimed that he lacked the theological learning to do so properly, that his efforts would be misinterpreted, and that he lacked the influence attributed to him (Ep 1352:29–43, 159–67). Clement VII also urged Erasmus to write against Luther (Ep 1443B:8–12); Erasmus acknowledged this in a letter to Duke George of Saxony (Ep 1495:25–30). While Erasmus could and did deny that Leo X wrote to him on this topic, Leo's successors Adrian VI and Clement VII clearly had.

248 Pio XXIII *libri* 63ᵛd (on *Responsio* 92 above 'the University of Paris was the last ... remain under indictment'): 'Are you still carping at this renowned university? Surely you are making a display of your bitter feelings towards it. The other universities, though

because I did not write immediately, then the University of Paris is equally deserving of blame, for it was the last of all to publish articles. I am not carping at the renowned university, but the man from Carpi certainly[249] lives up to his name. I think, rather, that it redounds to the credit of the university that it was unwilling to rush into passing judgment.

What is the dialectician saying here?[250] If it was pointless for Erasmus to write against Luther after the censure of the pope, the lightning-bolt of the emperor, the findings of the universities, does it follow, then, that the king of England or the bishop of Rochester is being criticized?[251] It was pointless for me because I had no influence. How does this apply to so exceptional a king, so exceptional a bishop?

I see many squadrons and platoons of soldiers[252] who mind their own business and seek the glory of Christ with a sincere heart. Perhaps I know

* * * * *

silent, subscribed to the findings made by a few of them, yet you are branding them with infamy, saying that they still remain under indictment [*in crimine*], though the condemnation was made not only by a few universities but also by the edict of the supreme pontiff and by the decree of the holy senate.' Erasmus seems to have assumed that Pio took his expression *in crimine* to mean something stronger than 'under indictment,' perhaps 'guilty.' In classical idiom *crimen* could mean both *charge*, a meaning found mostly in forensic language, and *crime, sin, outrage*, the more popular usage. It is not clear what Pio's being Italian has to do with this. Is Erasmus implying that, being Italian, Pio is thinking of the more popular meaning of *crimen* that turns up in the Italian *crimine*, or that, although an Italian and, therefore, a Ciceronian, he is betraying an ignorance of Ciceronian legal usage in his misconstrual of Erasmus' use of *crimen*? As it is, Erasmus' charge here seems to lack any basis.

249 *utique* 1531 *ubique* LB
250 Pio *XXIII libri* 63ᵛf (on *Responsio* 93 above 'before the censures by the universities etc'): 'So you mean that there was no suitable time for you to write! Now did you say just now that you wrote the *Diatribe* and the two books of *Hyperaspistes* before or after Luther was condemned? But you are indulgent with yourself. You had forgotten them at once. And you do not see that by offering this excuse you are making a charge of imprudence and ineffectiveness against all who wrote against Luther either before or after the condemnation [emending *eundem natum* to *condemnatum*], in particular against the most glorious king of England, and against the most learned bishop of Rochester, who refuted Luther's assertions one by one with such learning and devotion.'
251 Henry VIII published three works against Luther: in 1521 *Epistola ... ad ... Saxoniae principes de coercenda ... Lutherana factione* and *Assertio septem sacramentorum adversus Martinum Lutherum*, and a response to Luther's letter of 1 September 1525, published in English in London in 1526 and in Latin in Cologne in 1527 as *Literarum quibus ... Henricus octavus ... respondit ad ... epistolam Martini Lutheri*. See Klaiber 135–6 nos 1457, 1461–2; and John J. Scarisbrick *Henry VIII* (Berkeley 1968) 110–15. On Fisher's writings, see n42 above.
252 Pio *XXIII libri* 63ᵛi (on *Responsio* 93 above 'If I had seen any band of men etc'): 'You saw platoons and battalions and thickly manned squadrons, and you may still see them if only your eyes would serve you rightly. But you chose to be a spectator of the battle rather than a warrior. But how I wish you had been a spectator only, and not also a supporter of the wrong side.'

some, but they keep quiet. I do not yet see the troops and battalions of which Pio brags.

I am not suggesting giving up, but I am offering a reminder of what should have been done in the first place.[253] As it is, I see many who shout against Luther, I see none reforming their lives. On the contrary, we observe some people getting worse.

Again he escapes by means of a distinction.[254] When you were bringing the charge that I had not written against Luther, you were making no distinction between his personal doctrine and what he has in common with others. No, that device was thought up later on, because there was no chink to escape through. And nonetheless he accuses me of lying because I said that I have unsheathed my pen against Luther!

He cannot endure that I be compared to Aristotle in any way,[255] as if I regarded myself as his equal in learning! 'But Aristotle,' he says, 'was compelled to disagree with earlier thinkers in order to vindicate philosophic truth.' True enough, but what a misrepresentation at the same time! Come now, if he could do this for philosophy's sake, cannot I do the same thing to clean up theology? But what theologian is there who does not somewhere disagree with his predecessors?

What's this I hear? Augustine's arguments in his dispute against Pelagius do not count?[256] Then he fights with him in vain. In fact, in the two books

* * * * *

253 Pio XXIII libri 64ʳk (on Responsio 94 above 'Nothing remains but for us etc'): 'An ineffective defence! So after weapons were drawn, writers ought to have held back and pursued the business with reasoned arguments. Cannot the opportune speech of an earnest man mitigate wrath, and calm disorders? "He controls souls with his words and eases hearts" [Aeneid 1:153].'

254 Pio XXIII libri 64ʳm (on Responsio 95 above 'Unless,' you say, 'I write against Luther again etc'): 'Why do you say "again"? Perhaps you thought I was nodding off here. My contention is that you have written nothing specifically against Luther; but you would rather have me say "Unless you write against Luther *again*." I, in fact, urged you to make an attack on topics specific [to Luther] once you had shown what you could do with this common one.' Adagia III ii 75: Reperire rimam 'To find a chink' LB II 764C–D / ASD II-5 145–6 / CWE 34 253

255 Pio XXIII libri 64ʳo (on Responsio 95 above 'If this is something ... with your Aristotle etc'): 'I did not think that you considered yourself so important that you compare yourself to Aristotle! But it was a necessity for him as he taught philosophy to condemn the mistakes of early philosophic writers in order to establish the truth. Your job was not like his, and accordingly you had no warrant for criticizing the ancient Fathers.'

256 Pio XXIII libri 64ʳp (on Responsio 95 above 'Neither the apostles ... nor Augustine etc'): 'Augustine supports Luther on this point as much as Luther supports the Catholic church, unless you are basing your claim on what he says when disputing against Pelagius, instead of on what he defines and teaches.'

to Abbot Valerius [sic][257] he gave his clear view when he was quite an old man. Of course, he gives nominal support to the freedom of the will, but he exaggerates the role of grace with arguments that leave scarcely any room for the freedom of the human will.

The one who withheld my letter to Pio[258] admitted in a letter to me that he did so on purpose, so that Pio would not seize upon a new opportunity from that document for venting his spleen.

[REPLY TO THE PREFACE][259]

So much for the scholia, a selective treatment. Now it remains for him to make good on the charges that he has repeatedly thrust into my face, which he has said, again and again, he would make clearer than sunlight. But he employs an extensive preface[260] in preparation for this. I shall answer some of its points selectively.

His discourse on the fact that many things occur by chance in human affairs, and that many things come out differently from what was intended,[261] is, I admit, as far as I am concerned, altogether true. For it was my impression that I had taken Alberto's dignity and the cordiality of our relationship as well into account in my *Responsio* in such a way that there would be in it nothing that could provoke his anger.[262] Some others, learned men, found this admirable.[263] And it does not matter that he terms it a friendly

* * * * *

257 Erasmus must mean the two books dedicated to Abbot Valentinus, that is, the *De gratia et libero arbitrio* PL 44 881–912 and *De correptione et gratia* PL 44 915–46 of the year 426, and not the two books of *De nuptiis et concupiscentia* dedicated to Count Valerius CSEL 42 209–319.

258 Pio *XXIII libri* 65^{v}k (on *Responsio* 101 above 'I wrote a letter to you etc'): 'It certainly was not delivered to me. I am amazed that it was not, if you sent it to some trustworthy friend. And yet I had been in Paris for many months more than a year before my *Epistola* was published. I would have resisted this being done all the more, by way of accommodating you, had I received the letter.' See *Responsio* n499 above.

259 1531 UPPER MARGINS: '[Reply] to the Preface' on pages 48–53 [= 153–8 in this volume]

260 Pio's *Praefatio* occupies *XXIII libri* 66^{v}A–72^{v}P.

261 Pio's theme (*XXIII libri* 66^{v}A–67^{r}A) when expressing his surprise that his interchange with Erasmus has come to this pass

262 In Allen Ep 2328:42 Erasmus insisted to Campeggio that he had responded *civiliter* to Pio. In Allen Ep 2443:332–5 he wrote to Sadoleto that he responded *civiliter et honorifice*, being very careful, insofar as the nature of the case allowed, that nothing slip out at which Pio might rightly take offence or which might harm his reputation. He repeats this claim more forcefully at 157 below. Pio did not agree with Erasmus' assessment of the tone and content of the *Responsio*; see Pio *XXIII libri* 67^{v}B–69^{v}I.

263 The identity of these men is not clear from the extant Erasmian correspondence.

and gentle suggestion,[264] for the point of his argument was this: to present a convincing case that I am the occasion, cause, instigator, and leader of this entire riot. What more hateful charge could be made?

I was much taken then with his courtesy in sending me a letter so thoughtful and obliging.[265] Far from taking offence at it, I even thought about having it published, except that I was put off by some rather hostile statements against me, though he attacked the doctrine of Luther with somewhat heavier artillery. I had begun a reply, and had already completed a couple of pages; but when Rome was occupied and the pope taken prisoner, I did not know where Alberto was living, nor was it safe to entrust anything to anyone. I heard that he was in Paris at the same time that I heard that he was thinking about publishing his letter.[266] I wrote to him then,[267] as I have said, not to ask him to refrain from publishing it, but to request that he soften certain rather acrimonious statements against me and to employ more effective weapons against the doctrines of Luther.

But it was uncivil of me not to send a copy to him.[268] (Though he did not send the printed book to me!) I did not think there was any need, but nonetheless I meant to send my book along with a letter, but I was not up to it because I was dead from work. For at that time I was slaving away on Augustine and Seneca almost to the point of being weary of life.[269] But nothing in this man's writing amazes me so much as his astonishing endeavour to misconstrue everything as proceeding from acrimony. This seems to me altogether foreign to that character which scholars, in their letters to me, have declared to be Pio's. Won't any unbiased reader of my *Responsio* be amazed at Pio's uttering such irrelevancies as these?

'And yet,' he says, 'as if vindicating yourself and exacting punishment from me, you have hurled countless darts at me, you have repaid my favour

* * * * *

264 Pio *XXIII libri* 67^{r}B: 'I advised you in a very friendly way'; 67^{v}B: 'my most friendly suggestion,' 'my kindly reply'

265 Pio had sent to Erasmus a manuscript copy of his *Responsio*; see Pio *XXIII libri* 69^{v}1–70^{r}1, and Erasmus' acknowledgment of his receipt of it in Allen Ep 2080:1.

266 Berquin in a letter from Paris of mid-October 1528 reported to Erasmus on Pio's activities there; cf Allen Ep 2066:60–79.

267 Allen Ep 2080; English translation in Introduction lxiv–lxv

268 Pio *XXIII libri* 67^{v}B: 'Behold, two years later suddenly another letter, much nastier than the earlier one, ... and printed this time. But neither before nor afterwards did you send it in any form to me. A boorish piece of business this, rather than urbane.'

269 Allen Ep 2157 intro gives a chronology of Erasmus' work on his edition of Augustine's *Opera omnia*, completed by 8 June 1529. Erasmus refers here to his second edition of Seneca (Basel: Froben March 1529); Allen Ep 2091 intro gives an account of this book. See also *Responsio* n33 above for Erasmus' health; for Erasmus' work on six other books at this same time, see Introduction lxvi

with as much vituperation and abuse as you could. I shall mention several instances of this. First you accuse me of stupidity and negligence on the grounds that I wrote against you without seeing your writings, but relying on the report of others who have a low opinion of you. You accuse me, in addition, of superfluity and irrelevance, declaring that I have mentioned nothing which has not been dealt with by others beforehand and about which you have made a thorough reply to them. In addition, you accuse me of being tardy, inconsistent, dull, and ignorant. Finally, to empty out your quiver, you allege that I am malevolent, cunning, and even, please God, insane. For you say that all my arguments are feeble and quite thin, that I am more forceful in the preliminaries and in the opening of my discourse than in the development of it, that I praise you only to criticize you more abusively, that I cite as statements of yours many things that you never wrote, and that, like a madman, I have attacked you when I really wanted to attack Luther, unless, cunningly and maliciously, I wanted to attack both of you, following the practice of those who, as the saying goes, wish to "whitewash two walls from the same pail,"' and so on.[270] These are his words.

Now what could be more unfair than to construe my words in this way? Where in my *Responsio* are these hateful remarks? I did say that he had not seen my writings, and he himself admits that this is true.[271] But I did not accuse him of stupidity for this reason. Rather, I was gentle with him, so

* * * * *

270 Pio XXIII *libri* 67[v]C

271 Pio XXIII *libri* 68[v]F: 'To admit it frankly, I had not read many of your works when I first wrote to you, not that I had a low regard for them, but because I had not found so much leisure.' Pio goes on (68[v]–69[r]F) to list the works of Erasmus with which he had been acquainted: 'First of all, ten years earlier, I glanced at some things in the book of the *Adagia*. I considered this to be very learned and useful, especially for literary studies, and I wish you had always involved yourself in the same sort of work and in similar subject matter. Next there soon appeared the *Moria*. I read it eagerly, attracted by its beginning, no less witty than elegant and learned. But when I had read the entire work, there were so many scandalous and annoying elements that they discouraged me from reading the other writings which you subsequently published. Even so, intrigued by its title, I read the whole book on confession carefully; the same with the *Diatribe*. I sampled your rendering of the New Testament in a few passages; the same with the *Paraphrases*; I had a glance at some other things in your books. I read nothing at all of your apologetical works, only a few things from the censures of Zúñiga, because he was in Rome, and your replies to them. Now you know what I have seen of your works and of those of others who have criticized you before I wrote to you. The rest were altogether unknown to me, so much so that I recognized neither the titles of the books nor the names of the authors whom you list.' Pio XXIII *libri* 84[r]F cites other Erasmian works by name, implying familiarity with their contents. See Introduction lxxvii n85.

that if my refutation seemed anywhere overly hostile, it would reflect badly, not on him, but on my denouncers. I denied his authorship of whatever was hateful and I left as his whatever was commendable. And I knew full well whom he was employing as his assistants for this,[272] but I pretended throughout not to know this for the sake of his good name.

I did write that a great many of his charges had been made by others and that I had refuted them. Was this an atrocity? He admits that he had seen none of my apologies.

Abusive charges of superfluity and irrelevance are nowhere to be found in my entire discourse. Rather, he made up this sort of thing so that he would have a theme for his display of temper. And where are there reproaches of tardiness, inconsistency, dullness, and ignorance, of malevolence and cunning? Where, finally, do I call him insane? This, in fact, is what he seems to be talking about: when he begs my forgiveness for attacking me 'too violently or too bitterly' because 'no one can even mention' Luther 'without glowing hot with anger and loosing the reins of self control,'[273] I remark that what he seems to be talking about is a condition similar to insanity, because people who are driven by illness attack anyone near them.[274] If it was Luther alone whom he was attacking fiercely, why does he beg my pardon? If an attack on Luther becomes an attack on me, that is the work of a madman. But I immediately provide a milder understanding of this very remark when I say: 'But perhaps I do not follow your meaning, etc.'[275]

I said that he uses thin arguments, that he is less convincing in his proof and conclusion than in his opening statement.[276] So? When dealing with someone who was making awful charges against me, charges that I thought and still think are absolutely inapplicable to me, should I have played the guilty party and said: 'Alberto, your opening statement is quite correct, and your conclusion is magnificent'?

I did indicate my regret that he was not better equipped against Luther. But really, what use was it that a layman, untrained in canon law, unacquainted with theological literature,[277] one weakened by a life of luxury in palaces,[278] should deal with the entire range of topics about Luther, and this

* * * * *

272 See nn 12–16 above.
273 Referring to Pio *Responsio paraenetica* 99^{v}N, see Erasmus *Responsio* 102 above.
274 See Erasmus *Responsio* 102 above.
275 Erasmus *Responsio* 102 above
276 Erasmus *Responsio* 84, 101 above
277 See Introduction xxiv, cxxii.
278 See n17 above.

in spite of the fact that there are in the world those who could do it both more appropriately and more effectively?

Although there is not a single word that is not respectful of him in my entire letter,[279] though his notes[280] are swarming with out-and-out abuse masked by no rhetorical artifice, though the rest of the book teems with even worse instances of the same thing, nonetheless the charming fellow writes: 'But what are you trying to do, Erasmus, attacking me in this way and pummelling me so fiercely? Believe me, you strive in vain. For although you are so insulting that you could enrage even the most patient of men, you will not succeed with your mordancy and scurrility in causing me to depart from my own character and become like you. I shall keep to my own style.'[281] These are Pio's words.

I approve his adding my letter to his book,[282] for it can answer for me. Although there is nothing ill-tempered in it, Pio concludes from it that I was upset by his 'suggestion' as I have never been before.[283] Ha ha ho! Yet elsewhere he says that my skin is so tender that a fly could draw blood.[284] I will not mention for the moment the hyperbole[285] whereby, when I said something twice, he says that I have repeated it more than one hundred times, no, countless times, more than a thousand, all for the purpose of producing a good-sized book, although he is the one who is constantly repeating himself.[286] I will not mention his taking out-and-out falsehoods for granted, for example, that I seem to be such a supporter of Luther that I maintain that 'the arguments of all who attack his doctrine are trifling and silly.'[287] My letter shows that I only wanted Pio to engage Luther with stronger arguments.

* * * * *

279 See Erasmus *Responsio* 101–3 above for veiled insults.
280 That is, Pio's marginal annotations on Erasmus' *Responsio* (Pio XXIII *libri* 47ʳ–66ʳ)
281 Pio XXIII *libri* 68ʳᴅ
282 Erasmus' *Responsio* along with Pio's marginal notes on it make up book 2 (47ʳ–66ʳ) of Pio XXIII *libri*.
283 Pio XXIII *libri* 71ʳ⁻ᵛᴍ: 'it seems far more reasonable and likely that you had done preparatory work on your reply to me far in advance, the one you brag you completed in six days … I had written to you almost three years earlier, and you were more upset by my letter than, perhaps, ever before, as your letter attests.'
284 Pio XXIII *libri* 71ᵛᴍ: 'particularly since your skin is so tender that a fly can scarcely touch it without drawing blood'
285 MARGINAL NOTE: 'Silly hyperbole'
286 Pio XXIII *libri* 68ᵛᴅ: 'finally asserting them again, you repeat inelegantly more than a thousand times the same things you had said before, so that at least by the very act of repetition you could fill a lot of pages and counterbalance a book with a book …'
287 Pio XXIII *libri* 69ʳꜰ: 'or you should go ahead and defend him [Luther] whom you seem so to support. For by maintaining that the arguments of all who attack his doctrine are trifling and silly, and not introducing better arguments yourself, you are, of course, supporting him indirectly.'

I do say that the writings of Prierias and Cajetan were less than success-ful,[288] not because they were uninformed, but because they accomplished nothing. Prierias was laughed at by everyone.[289] Cajetan was not even read, though he handled this business very acutely.[290] But I am talking about re-sults, not about erudition. But let him pretend that I said what he says I said about these two, even so I am not necessarily treating everyone's arguments as trifling and silly. How often he makes errors here in dialectic!

I think I have said enough in a selective way in response to only two books of his. I now come to the third book.

[FOLLY ACCUSED][291] [AGAINST BOOK 3][292]

He says that I have scattered some poison through all of my books, with the intention of harming every sort of reader.[293] First, how does he dare say this, when he admits that he has read very few things of mine?[294] He should at least have made exceptions of *Copia*,[295] *Pronuntiatio*,[296] and *Cicero-nianus*.[297] The *Enchiridion* has been in circulation for twenty-five years with-out anyone complaining about it.[298] No one has complained of *The Mercies of*

*****

288 Pio *XXIII libri* 69ʳF: 'But you say this about the arguments I brought against it [Luther's doctrine], and you also make the same statement about the very learned writings of Ca-jetan, the cardinal of St Sixtus, of Prierias, and of the rest'; see Erasmus' earlier remarks on Prierias in *Responsio* 26, 36, and 93 above, and on Cajetan in *Responsio* 42 and 95 above.

289 See Erasmus *Responsio* n177 above.

290 See Erasmus *Responsio* 93 above; and Ep 1225:215–19.

291 1531 UPPER MARGINS: 'Folly Accused' on pages 54–79 [= 158–81 in this volume]. Pio's book 3, entitled 'A List and Examination of Offensive Statements and Views in the *Moria*, by the same Alberto Pio,' occupies *XXIII libri* 73ʳA–84ʳP; in these notes references to line numbers in *Moria* are to ASD IV-3 67–194, while quotations from *Moria* are from the translation in CWE 27 83–153.

292 MARGINAL NOTE: 'Against book 3. Lie'

293 Pio *XXIII libri* 73ʳB: 'Some evil agency, envious of your renown and keen to do it public damage, seems to have driven you to this and to have taught you to scatter them [ab-surdities and flaws] in all your writings, so that nothing remains unblemished and not in some respect shabby; no, everything is tinged with some poison, with the express purpose of doing injury to every type of reader.'

294 Pio claimed that Erasmus' writings were almost infinite in number with new publica-tions issuing forth seemingly daily. See n271 above and Pio *XXIII libri* 68ᵛ–69ʳF.

295 *De copia* LB I 3A–110D / ASD I-6 / CWE 24 284–659, published in 1512, with revised edi-tions in 1514, 1526, and 1534

296 *De recta pronuntiatione* LB I 913A–968B / ASD I-4 11–103 / CWE 26 365–475, published in 1528, 1529

297 *Ciceronianus* LB I 973A–1026C / ASD I-2 599–710 / CWE 28, published in 1528, 1529, 1530

298 On the untroubled early reception of the *Enchiridion*, see John W. O'Malley CWE 66 Intro-duction xli–xlii and Charles Fantazzi CWE 66 Introductory Notes 3–7. The *Enchiridion* was first published in February 1503. By 1518 Edward Lee had criticized the anthropology of the *Enchiridion* and Erasmus responded in a letter to Maarten Lips of 9 May 1518;

God.[299] Not to mention the fact that what Pio calls poisons are exceedingly strong antidotes. Those things that dim the glory of Christ, that entomb true piety, are not poisons in his view, for he never complains of them; but statements against tyrannical pastors, against reliance on rituals, against the excessive wealth of the priests, these alone are poisons. Even though they are true, he wants them silenced,[300] lest the majesty of priests suffer any impairment. He has no worry about the majesty of Christ.

He says that my works have become so unpopular that they are rejected as soon as only their title-pages become known to the world.[301] Why, then, are they printed repeatedly, especially when they are rejected by the Lutherans most of all? In fact, the booksellers report to me that the allure of my name induces in a very great number of people an urge to buy.

Finally, after a long-winded series of complaints,[302] he takes the matter in hand and says he will demonstrate what he has repeatedly promised he would show, namely whether he is a forger or I am a liar.[303] Furthermore,

* * * * *

see Ep 843:690–9, point no 91. In 1520 Zúñiga addressed a work to Leo x in which he quoted thirteen passages from Erasmus' 1518 edition of the *Enchiridion* with its prefatory letter to Paul Volz. Following each of the quotations was a commentary by Zúñiga which pointed out the impieties, blasphemies, and scandalous nature of the passage. See Biblioteca Nazionale Vittore Emanuele III di Napoli, MS VII.B.14 *Contra Erasmum Retoredamum* [*sic*] Liber III 92ʳ–101ᵛ; and Henk Jan de Jonge 'Four Unpublished Letters on Erasmus from J.L. Stunica to Pope Leo x (1520)' in *Colloque Erasmien de Liège* (Paris 1987) 147–160 especially 147–4. Leo forbade Zúñiga to publish this work. But soon after the pope's death Zúñiga printed eleven of these offensive passages without commentary in his *Blasphemiae et impietates* Eivʳ–Fiʳ. Erasmus' claim that his *Enchiridion* was not criticized for twenty-five years is thus not accurate.

299 In its censures of 17 December 1527 the theological faculty at Paris cited for condemnation two statements from this work; Erasmus responded to each citation; see *Declarationes ad censuras Lutetiae vulgatas* LB IX 923C–924D. On the later popularity of this treatise in certain Italian circles, see Seidel Menchi 143–67.

300 Alluding, perhaps, to Pio's suggestion (*XXIII libri* 73ʳB) that Erasmus might have collected all his annotations which are offensive to piety into one volume, 'and then once that volume had been rejected, the rest could be given approval'; this is, in a sense, what Pio was pretending to do with his *XXXIII libri*, which is directed, according to the title-page, 'against passages of diverse works by Desiderius Erasmus of Rotterdam which he [. . .] thinks he [Erasmus] should review and revise.'

301 Pio *XXIII libri* 73ʳB: 'By these [defects] you have not only deprived them [your works] of authority, but also rendered them so hateful that they are rejected immediately by a great many people as soon as they see the title page.' Pio was describing the hostility towards Erasmus to be found at this time in certain conservative circles in Italy, Spain, and France; see Seidel Menchi 41–67; Marcel Bataillon *Erasme et l'Espagne: Recherches sur l'histoire spirituelle du XVIe siècle* (Paris 1937) 253–99; and James K. Farge *Orthodoxy and Reform in Early Modern France: The Faculty of Theology of Paris 1500–1543* (Leiden 1985) 176–96.

302 Probably referring to Pio's *Praefatio* (*XXIII libri* 66ᵛ–72ᵛ)

303 An allusion to the opening statement of Pio's book 3 (*XXIII libri* 73ʳA): 'And so, now that, with God's help, we have come to the end of your work [that is, Erasmus' *Responsio*] and

the charges he has promised to prove are heinous: I agree with Luther on many points, I condemn the ordinances of the church, I diminish the pope's majesty, I ridicule the sacred institution of monasticism, and I distort Scripture through paraphrases.[304]

After such striking and repeatedly brandished promises, the much-discussed *Folly* steps forward as a witness against me. 'Chronological order demands this,' because, he says, I got my start with *Folly*.[305] So, when the discussion is about the dogmas of faith, is it suitable that *Folly* have first place? And I did not get my start with it, for the *Enchiridion* had appeared much earlier, as had the *Preface to Lorenzo's Annotations on the New Testament*,[306] and many other things as well.

Although I entitle the work *Folly*, though I establish that its genre is burlesque,[307] though I provide Folly with attendants suitable to her,[308] Pio nonetheless requires that I present her speaking all things with wisdom, just as if I brought on the character of Peter or Paul. 'It is Folly who speaks,' he maintains, 'but she speaks from the mouth of Erasmus.'[309] When Gryllus

* * * * *

have examined those passages which could not be ignored in silence, the situation now requires that I prove by the evidence of your own words that I was neither a liar nor a slanderer in my hortatory epistle.' Erasmus is also alluding to earlier statements of Pio – 68[r]D: 'Now the truth of this will be clearly established by reviewing your passages, that is, whether I was forger or you are a liar'; 69[r]F: 'The following will prove (as I have said) whether I have been a liar or a forger in reporting these'; 72[r]o: 'that I may clear myself of the slur of being a liar or forger.'

304 These charges were made in Pio's *Responsio paraenetica*; they and some others are reviewed in *XXIII libri* 73[v]B.

305 Pio *XXIII libri* 73[v]C: 'But chronological order demands that *Moria* first be called into court, since in it you got your start in ranting and raving.' On the controversies surrounding the *Praise of Folly* see Myron P. Gilmore '*Apologiae*: Erasmus's Defenses of Folly' in *Essays on the Works of Erasmus* ed Richard L. DeMolen (New Haven, Conn 1978) 111–23. Written in 1509, the *Praise of Folly* was first published in Paris in 1511; see Clarence Miller Introduction ASD IV-3 14.

306 The *Enchiridion* was first published in February 1503, the preface to Valla's *Annotationes* in April 1505; see vander Haeghen I 79, II 66.

307 *genus ridicule*: Erasmus had established *Moria*'s generic context in his prefatory letter to Thomas More; see *Moria* 22–3 ASD IV-3 68 / CWE 27 83–4. Because the CWE translation of *Moria* does not have line numbers, in this volume line numbers are cited from the Latin edition of the text in ASD. It should be noted that the ASD edition numbers the lines of the text to 1000, and at lines 1001, 2001, and subsequently, begins the numbering again with 1. For clarity line numbers over 1000 are given in full in these notes; for example, in n312 below, line numbers 239–44 on page 138 of ASD IV-3 have been expanded to read 1239–44.

308 Listed by Folly in *Moria* 125–35 ASD IV-3 78–80 / CWE 27 89

309 Pio *Responsio paraenetica* 7[v]G: 'for although you substitute Folly herself as your *persona*, since, for all that, Folly speaks with your mouth, a great many accept what you impart there as not uttered by Folly, but by you, a man of renown and very great learning.'

argues that brute animals have more reason and more virtue than mankind, is he not speaking from the mouth of Plutarch?[310] But who has ever hailed Plutarch before the bar?

But what is it that Folly the Heretic says? She says that the arts were invented by the *daemon* Teuth, a feature found in Plato.[311] Now, of course, all the arts of mankind have been knocked flat, may God have mercy upon them!

What does Folly say about the grammarians?[312] She says that they would never continue in such pitiable labours unless Folly removed their awareness of hardship and turned their distress into pleasure. What, then, remains but for the grammarians to stone Erasmus?

But now he digresses to praise Aldus, whom, he says, I criticize and ridicule so very ungratefully.[313] *Folly* contains nothing about Aldus but these words: 'since my friend Aldus alone has brought out more than five.'[314]

* * * * *

Pio *Responsio paraenetica* 8ʳG: 'I hear "But it is Folly who uttered all this!" But she spoke with the mouth of Erasmus.'

310 The porcine interlocutor in Plutarch's dialogue *Bruta animalia ratione uti* (*That Irrational Animals Enjoy the Use of Reason*), cited in the prefatory letter to More (*Moria* 31–2 ASD IV-3 68:31–2 / CWE 27 83).

311 Pio *XXIII libri* 73ᵛC quotes *Moria* 717–27 ASD IV-3 110 / CWE 27 106–7, where Plato *Phaedrus* 274C–D is cited.

312 Pio *XXIII libri* 74ʳD quotes and comments on *Moria* 1239–44 ASD IV-3 138 / CWE 27 122.

313 Pio *XXIII libri* 74ʳD: 'In this section with great lack of gratitude you criticize and ridicule by name our friend Aldus, a man never adequately praised ... I mean Aldus, the premier restorer of good authors, the propagator of the Greek language, who by publishing all its best authors brought it out of darkness when it was almost lost because of antiquity and overwhelmed with barbarism. You criticize this most learned, excellent man, one who should be revered by all men of letters, because he produced a number of versions of his *Grammar*, as if you have not done the very same thing yourself.' Note that Aldus' *Institutiones grammaticae*, printed 9 March 1493, was dedicated to Pio (dedicatory preface and epilogue reprinted in *Aldo Manuzio Editore* 1 165–7). In the prefatory letter Aldus remarks on the long time he invested in producing his grammar: 'I have withheld my *Institutiones grammaticae* ... for so many years now and have compelled them to endure repeated revisons (as you too know very well, O Alberto Pio, glorious prince, whom we have taught, along with your noble brother Leonello, for six years and more with the greatest fidelity and care), that I seem not only to have obeyed Horace, who urges in the *Ars poetica* that publication not be rushed but be withheld until the ninth year, but to have invested even more time on this work than he commanded' (165); see Tiziana Plebani 'Omaggio ad Aldo Grammatico: Origine e tradizione degli insegnanti-stampatori' in S. Marcon and M. Zorzi eds *Aldo Manuzio e l'ambiente veneziano 1494–1515* (Venice 1994), 73–106 especially 89–100. The colloquy *Opulentia sordida*, inserted into the September 1531 edition of the *Colloquies*, is, according to d'Ascia *Erasmo e l'umanesimo* 88, a rejoinder to Pio's praise of Aldus that recounts the less attractive aspects of life at Manuzio's establishment.

314 *Moria* 1279–80 ASD IV-3 140 and n / CWE 27 123. This offending passage was not original, but inserted in the 1516 edition of the *Moria*; see Clarence Miller in ASD IV-3 46.

Extraordinary ridicule! Aldus himself told me that he had produced his grammar nine times. But where is the ingratitude? He says: 'Without the advantage of having lived with him you would surely never have become such a stylist. For you cannot deny that while at his house in Venice, when you were working for his press, you made a great deal of progress in both languages.'[315] Thus Pio.

Maybe he thinks that I learned Greek and Latin from Aldus! Not even Aldus himself, if he were alive, could hear this without a smile.[316] I will say no more about Aldus' learning. I loved the man when he lived, and I shall not denigrate him now that he is dead.[317] I can truly say this at least: when I came to Italy I knew Latin and Greek better than I do now. I had brought the collected raw material for the entire work [ie *Adagia*] with me from England to Venice, along with a load of the books, Greek ones particularly, in which I had written notes. It was my ambition, I admit it, that the work be published by a celebrated firm. Aldus was eager to have it. I lived in Aldus' quarters for about eight months.[318] The book was simultaneously written and printed, all within a few months.[319] Where was the time within this period to learn Greek and Latin? I had so much work to do that I scarcely had the leisure to clean my ears!

Aldus quite often expressed his amazement that I could compose so much extempore, and do so in the midst of the confusion and noise going on around me. I corrected the final proofs of my work only to the extent

* * * * *

By that time Aldus had published three revisions/enlargements of his grammar, which had been published first in 1493. Miller, who interprets this passage as referring to printed revised editions, proposes that *quinquies* (five times) should not be taken literally. Yet in the next sentence here Erasmus reports that Aldus told him that he had written (*scripsisse*) his grammar nine time (*novies*). Erasmus could not have spoken personally with Aldus after 1509, the last time he was in Italy. If his statement that Aldus told him that he had (re)written his grammar nine times has any truth, the 'five times' Folly refers to and the 'nine times' later reported by Erasmus would not refer to the successive printed editions of the grammar, but to the drafts that it went through before being printed, the ones he mentions in his dedicatory letter to Pio, quoted in the last note.

315 Pio XXIII libri 74ʳD.
316 Aldus died on 6 February 1515; see CEBR II 376.
317 For Erasmus' open criticism of Aldus' editions of classical authors, see Ep 1437:177–9, *Adagia* II iv 53 LB II 540E / CWE 33 218, II ix 76 LB II 679C / CWE 34 118, and IV ix 92 LB II 1143B.
318 Erasmus lived in the house of Aldus' father-in-law, Andrea Torresani, in the San Paternian district of Venice, in which house Aldus also lived; see the entries on Manuzio and Torresani in CEBR II 378, III 332.
319 In Venice Erasmus expanded his Paris 1507 edition of the *Adagia*, consisting of 841 proverbs, into the Aldine Venetian edition of 3260 sections, published in September 1508; see CEBR II 379 and Allen Ep 211 intro.

of making revisions if I wished to, for the work had a corrector apart from
me, a man named Seraphinus[320] hired for the purpose. Aldus read it after
I did. When I asked why he took on this job, he said 'I am studying at
the same time.' Aldus was of help to me by providing some manuscripts,
but he was not the only one to do so; Janus Lascaris,[321] Marcus Musurus,[322]
Giambattista Egnazio,[323] and Urbanus Rhegius[324] did the same. Is someone
who is there to see his own book through the press an employee?[325] And I
did not make a commitment to Aldus for any other work; it was rather the
case that the press worked for me.

But I shared his table. Yes, because I could not do otherwise, eager as I
was to finish the work as soon as possible. Moreover, the table and the entire
house belonged to Andrea Torresani of Asola.[326] Aldus was nothing but an
employee.[327] But that table begot a kidney-stone on me, a bane unknown
to me before,[328] and except that my impatience to finish the work held me

* * * * *

320 As yet this corrector remains unidentified; see Allen Ep 1482 n39.
321 Born in Constantinople into a branch of the imperial family in 1445, Lascaris emigrated
    to Italy, where he gained a reputation as a distinguished teacher of ancient Greek liter-
    ature. From 1504 to 1509 he served as the resident French ambassador in Venice where
    he frequented the Aldine Academy; see CEBR II 292–4.
322 Born c 1470 in Crete, Musurus came to Italy where he studied under Lascaris at Flo-
    rence. From 1493 to 1516 he made his career in Venice and neighbouring territories and
    was a collaborator of Aldus. Erasmus always spoke highly of him. See CEBR II 472–3.
323 Giambattista Cipelli (1478–1553) was born and educated in Venice. He opened a school
    there and was a member of the Aldine Academy. He remained a friend of Erasmus. See
    CEBR I 424–5.
324 Fra Urbano Valeriani OFM (c 1443–1524), after travelling extensively in the East, study-
    ing in Sicily, and tutoring the future Leo X, opened a private school in Venice. He was
    a member of the Aldine Academy and the author of the first Greek grammar intended
    for a Latin audience; see CEBR III 370–1.
325 After completing work on the *Adagia*, Erasmus remained in Venice to work for Aldus
    on the texts of Plautus, Terence, and Seneca. Erasmus was paid twenty ducats for his
    various services to the Aldine press (presumably Venetian ducats or *zecchini*, then worth
    56d sterling and 2.275 *livres tournois* [45s 6d *tournois*], so that this sum amounted to £4
    13s 4d or 45 *livres* 10 *sous tournois*; see CWE 1 314; CWE 12 650). See Lowry *World of
    Aldus* 99 and Allen I 13:4–8. Erasmus resented G.C. Scaliger's assertion that Erasmus
    had functioned as a mere corrector; see Allen Ep 2682:28–35.
326 Andrea Torresani of Asola (1451–1528), also known as Asulanus, was a printer in Venice;
    he merged his press with that of his son-in-law Aldo Manuzio in 1507; see CEBR III
    332–3.
327 Aldus and Torresani were unofficial partners from 1495 on. When Aldus married Tor-
    resani's daughter Maria, the assets of the two partners were fused and they remained
    on good terms until Aldus' death. Surviving records suggest that Aldus owned only 10
    per cent of the press, Andrea 40 per cent, and the Venetian patrician Pierfrancesco Bar-
    barigo 50 per cent. Aldus could have been considered merely an employee; see Lowry
    World of Aldus 83–6.
328 For Erasmus' special attack on the table of Aldus, see his colloquy *Opulentia sordida*
    (September 1531); for information on Erasmus' health see *Responsio* n33 above.

fast there, Janus Lascaris kept on inviting me most kindly to share his house and table. Also, my own funds were adequate for a two-year stay in Venice, even though I had received no additional funds from home. Since my health was in danger from the unfamiliar diet, I asked Aldus if Asulanus would be offended if I were permitted to prepare my meals in my own room. When he replied that he would not, I never afterwards shared Aldus' table.

He also says that Aldus was my master at that time, whereas he was not even my host.[329] In Rome I would not endure having cardinals as my masters, though their personalities were so congenial that you could not want better comrades, but Aldus was my boss, though I did not know it? Maybe Aldus considered me his servant, but he certainly made a place for me at his table, and admitted that he had learned a lot from me. He also tried to keep me there through the winter, even after the work was finished, for he wanted a bit of training in rhetoric.

But why is it that what offends Pio did not offend Aldus? He thought *Folly* worthy of his publishing house. Pio says: 'Doesn't respect keep you from calling such a man foolish and insane?'[330] What was Pio's state of mind when he wrote this? Is a man who repeatedly goes over the same material insane? No, he is diligent and enjoys his work. But how much credit should be given to Pio, who wants to be considered altogether good-natured and sympathetic even though he summons such melodramas forth from trifles? I have no debt to Italy in connection with my literary accomplishments; I wish my debt were very great indeed. There were men there from whom I could have learned, but there were such men also in England, in France, and in Germany. But I had no time when I was in Italy, for I had gone there only as a visitor.

But Folly criticizes poets as well.[331] Here the learned fellow expatiates on a very charming topic, in praise of poesy, and shows that poems are to be found even in Scripture.[332] But what has this to do with the point at hand?

Folly calls the legal profession asinine.[333] In fact, it is not she who calls it that; rather she reports that it is customarily so called by philosophers and

* * * * *

329 Pio XXIII *libri* 74^{r}D: 'Nor does respect keep you from calling foolish and insane a man so generous to scholars, so learned and devout, a saintly character, your host, and, whether you like it or not, your master.'

330 A paraphrase of Pio's statement quoted in the preceding note.

331 Pio XXIII *libri* 74^{r}E criticizes *Moria* 1288–94 ASD IV-3 140 / CWE 27 123.

332 Pio XXIII *libri* 74^r–74^{v}E, citing classical and biblical poets and poetry

333 Pio XXIII *libri* 74^{v}F quotes *Moria* 749–52 ASD IV-3 112 / CWE 27 107 and concludes: 'You could not have used more insolent language to demean the prudence of lawyers than to term their profession "asinine."' The more extended treatment of lawyers in the *Moria* occurs in lines 1348–53 ASD IV-3 142 / CWE 27 125.

even by some theologians,[334] who find the lawyers wanting in acuteness of dialectic and considered judgment. If this is so, why does such a wise fellow concoct a false charge against Folly?

Next, Folly criticizes the wranglings of certain dialecticians,[335] and now Erasmus is condemning all logic. Pio embarks on an excursus in its praise.[336]

Next she criticizes the philosophers, the researchers into nature, and their mathematical subtleties.[337] Pio takes this quite seriously, and provides a defence of sober philosophy,[338] as if Folly were making sport of any but overly curious philosophers, something which anyone can easily realize from the very tone of her speech.

Then we come to the theologians.[339] He says that Folly attacks them so insolently that no street-player or pimp could attack any condemned art more insolently. As if it were the prerogative of pimps to rage against condemned arts![340] But what is it that Folly says? That theologians are the 'bane of the human race, that they should be driven away completely from human society, that their factions should be banished to the Turks and Saracens to do battle with them.'[341] What is it you are saying under Pio's name, you shameless slanderer? Is it not the sophistic theologians whom Folly censures? If that were less than clear from the tone of her speech, Folly herself declares it quite plainly when she says: 'seeing that amongst the theologians themselves there are some with superior education who are sickened by these theological minutiae, which they look upon as frivolous. Others too think it a damnable form of sacrilege.'[342] 'Bane of the human race' are the words of the informer, not of Folly.[343]

* * * * *

334 See *Moria* 750 ASD IV-3 112 / CWE 27 107 where, in fact, only philosophers are cited as ridiculing the lawyers.

335 Pio *XXIII libri* 74ʳF criticizes excerpts from *Moria* 1354–5, 1357–8 ASD IV-3 142, 144 / CWE 27 142.

336 Pio *XXIII libri* 74ᵛ–75ʳG

337 Pio *XXIII libri* 75ʳG criticizes excerpts from *Moria* 1361–80 ASD IV-3 144 / CWE 27 125–6.

338 Pio *XXIII libri* 75ʳ–ᵛG twice

339 Theologians are discussed in *Moria* 1381–1523 ASD IV-3 144–58 / CWE 27 126–30; Pio's extensive critique of Folly's discussion of theologians is found in *XXIII libri* 75ᵛH–79ᵛN.

340 Note that Pio *XXIII libri* 76ʳI attributes this fierce criticism not to Folly but to Erasmus himself: 'Next you move on to the theologians. You attack their profession with such insolence and with language so aggressive that you could not be more energetic in denouncing hustlers, pimps, or the practitioners of any condemned art.' Pio's text contains the following anacoluthon: 'ut nec circulatorum, leonumue, aut cuiusuis damnatae artis professores magis criminari posses'; Erasmus' misinterpretation of this seems to be the basis for his comment, 'As if . . . arts!'

341 Pio *XXIII libri* 76ʳI, alluding to *Moria* 1470–3 ASD IV-3 154 / CWE 27 129

342 *Moria* 1479–81 ASD IV-3 154 / CWE 27 129

343 Erasmus is quite correct here.

But Folly wants to exile them to the Turks. Here too he is pleading against Folly with a false charge. Folly's words are as follows: 'And in my opinion Christians would show sense if they despatched these argumentative Scotists and pigheaded Ockhamists and undefeated Albertists along with the whole regiment of sophists to fight the Turks and Saracens.'[344] Do you hear? It is the sophists and the argumentative theologians she wants to have sent against the Turks, not to fight them with spears, but to refute them with their sophistries. Her point is that sophistries like these are ineffective against the enemies of the faith, and that they are superfluous among Christians. What has this to do with devout and responsible theologians? I think that this defence is adequate for Folly against all the arguments with which Pio here defends respectable and notorious theologians alike.[345] He has decided, it appears, to fawn upon all of them.

Next he is angry with Folly because she said that businessmen handle the basest commodity of all, that is, money, whereas money is an object worthy of the highest honour.[346] What was his next move here but to chant an encomium of wealth [τοῦ πλούτου ἐγκώμιον]? But what does all this have to do with what Pio had promised? In fact, Folly does not state even this so baldly, but is criticizing some businessmen who think they are better than everyone, although they handle the 'meanest sort of business.'

He defends soldiers who sell their souls for a bit of pay. Yes, the majority of them[347] are like this. I wish they were not brigands as well, a greater threat to their own people than to the enemy. The danger now is that we will not have any salaried soldiers.

Pio plays the slanderer everywhere, but nowhere more so than when he reaches the topic of monks. He says that Folly censures, not the vices of some monks, but the monastic way of life, mendicancy, fasting, psalmody,

* * * * *

344 *Moria* 1470–3 ASD IV-3 154 / CWE 27 129

345 In *XXIII libri* 76ʳ1–79ᵛN Pio defends theologians against what he summarizes as charges of 1 / novelty and barbarism in language, 2 / overly nice precision in theological discourse, and 3 / discussions of pointless questions.

346 Pio *XXIII libri* 79ᵛo criticizes *Moria* 1215–16: 'Most foolish of all, and the meanest, is the whole tribe of merchants, for they handle the meanest sort of business [*rem sordidissimam*]' (ASD IV-3 136 / CWE 27 121). Pio seems to think that this *res sordidissima* is *mercatura*, ie mercantile activity. It is Erasmus who suggests that Pio is thinking of money [*pecunia*].

347 Pio *XXIII libri* 79ᵛo defends soldiers against the charge of greed in *Moria* 1211 ASD IV-3 136 / CWE 27 121, saying among other things, 'the majority are men who undertake military service to protect justice, because of their nobility of soul and their love of virtue.'

and customs.[348] He was campaigning, I think, for the sacrosanct burial! Anyone who has heard Folly's harangue will declare that this is clearly untrue. In fact, in her opening statement on this topic, 'since most of them are a long way removed from religion,'[349] Folly attests plainly that she is speaking of the bad ones. These are, alas, an enormous throng, and I wish that were not the case. She does not condemn psalmody, but rather the fact that certain people think that thundering out in their churches psalms which they do not understand is a wonderful sacrifice to God.[350] Nor does Folly condemn mendicancy across the board, but rather the fact that there are scoundrels who are mendicants, who still, for that reason alone, wish to be regarded as apostolic.[351] For Folly speaks as follows: 'Many of them too make a good living out of their squalour and beggary, bellowing for bread from door to door, and indeed making a nuisance of themselves in every inn, carriage, or boat.'[352] You see that it is not any mendicancy but rascality in mendicancy that is condemned. Further, Folly manages her remarks concerning rules about clothing, diet, and things of this sort in such a way that she condemns reliance on these things, not their observance, for she adds: 'Even so, these trivialities make them feel not only superior to other men but also contemptuous of each other.'[353] She likewise censures reliance on rituals, not the rituals themselves, saying: 'Most of them rely so much on their ceremonies and petty man-made traditions that they suppose heaven alone will hardly be enough to reward merit such as theirs,' etc.[354] When you hear 'most of them,' you realize that not all are being censured. Nor yet does Christ call them Jews in the *Folly*[355] because they keep these observances, but because they are empty of charity, and have situated piety in these things after the practice of the Jews. I am so far from condemning the monastic order in this entire passage that I presented Folly out of character, speaking almost wisely.

He says that there is a consistent pattern: the apostles begged often, Christ begged a room to hold the Passover supper, and he implies that

* * * * *

348 Pio's criticism of Folly's treatment of monks covers *XXIII libri* 79ᵛP–80ʳQ; there he quotes excerpts from *Moria* 1524–76 ASD IV-3 160–2 / CWE 27 130–2. Again, it is Erasmus himself whom Pio accuses of saying these things.
349 *Moria* 1525–6
350 Alluding to *Moria* 1531–3
351 See *Moria* 1537
352 *Moria* 1533–5
353 Reacting to Pio's quotation of *Moria* 1544–5
354 *Moria* 1556–8
355 *Moria* 1569–70; responding to Pio's excerpt from *Moria* 1569–76

Christ and the apostles begged frequently, and that Paul was begging for money for the relief of the poor who were in Jerusalem.[356] Granting that all these are true, for that is not the point at issue now, what does it have to do with the kind of mendicancy that Folly is censuring? I am amazed that he left out the fact that Christ begged water from the Samaritan woman, that he invited himself to the house of Zachaeus, that even after the Resurrection he begged a meal from the disciples![357]

When Folly touches on kings, princes, and magnates of the court,[358] what is her goal but to show that their existence is more miserable than that of almost any human beings if they but recognized the burden they are carrying on their shoulders, but that, by Folly's ministrations, scarcely any men live more delightfully? Here the Aristotelian attempts to trap me with the question of whether I think the exercise of princely rule is a wicked thing.[359] No, I think it is an excellent thing. And I do not say that it should be avoided. I do say that there are many princes who would cast it off as a distressing burden if they were to look deeply into what is required of a good prince. And we read that certain enlightened monarchs have, in fact, done just this.[360]

Elsewhere, I think in the course of condemning studies, Folly says that all the most learned have been the least suited for governing.[361] Folly is not so stupid as to say a philosopher who is a prince is unfit for princely rule, but she says this: 'If power came to a philosophaster or someone enslaved to

* * * * *

356 Referring to Pio's defence of mendicancy (80[r–v]R), where he argues that their poverty proves that Christ and his apostles were mendicants, citing Mark 14:12–16 and parallels, and Romans 15:25–8.

357 John 4:7; Luke 19:5, 24:41; and John 21:5

358 Pio XXIII libri 80[v]s criticizes Moria 1675–8 and alludes to the rest of the treatment of the topic in Moria 1675–1738 ASD IV-3 168–70 / CWE 27 135–7, here, as always, ascribing the statements and views expressed to Erasmus.

359 Pio XXIII libri 80[v]s: 'I ask you, remarkable sir, whether you consider that to rule, to govern, to be in charge, is a wicked and blameworthy thing, or something good and praiseworthy.'

360 Among the more famous cases of rulers in that period who resigned their office to retire and practise a more intense spiritual life were Amadeus VIII (1383–1451), duke of Savoy (1391/1416–34), and Karl V (1500–58), king of Spain (1516–56) and emperor (1519–56); see Aenea Silvio Piccolomini Memoirs of a Renaissance Pope: The Commentaries of Pius II, an Abridgement trans Florence A. Gregg and ed Leona C. Gabel (New York 1959) 222; and Karl Brandi The Emperor Charles V: The Growth and Destiny of a World Empire trans C.V. Wedgewood (London 1939; repr 1965) 630–43. Wilhelm, prince of Anhalt-Zerbst, who adopted an austere Franciscan lifestyle, made a deep impression on the youthful Luther; see James MacKinnon Luther and the Reformation 4 vols (New York 1925; repr 1962) I 15.

361 Reacting to Pio's attack (XXIII libri 80[v]s) on Moria 496–514 ASD IV-3 98 / CWE 27 100

literature ...'[362] She says 'a philosophaster,' someone like Pio, not 'a philosopher.' Nor does that famous statement of Plato[363] count against me, for he specified a philosopher, not a grind in dialectics, physics, and mathematics, but one imbued with true opinions of right and wrong, who is possessed of a soul that is above all emotions. Folly does not disparage Marcus Antoninus, as Pio alleges.[364] This is what she says: 'And even admitted that he was good, he undoubtedly did more harm to Rome by leaving it such a son as his than he ever benefited it by his administration.'[365] She is accusing Commodus,[366] not his father. Someone else might not list Julius, the first to suppress Roman liberty by force of arms, among beneficial monarchs. I will say nothing about Octavian.[367] But what about Claudius, Caligula, and Nero, the sixth emperor, what about Hadrian and Julian, all of whom,[368] it is generally agreed, were quite learned?

Next, Folly calls human life a play.[369] What follows? Nothing, then, in this life is serious. I should have liked to have Democritus[370] here as my spokesman. But have not the most learned men declared that comedy is a mirror of human life? And do we not call comedies plays?

* * * * *

362 *Moria* 500–1; Clarence Miller in ASD IV-3 99:500n cites earlier uses of the term *philosophaster* by Augustine; Erasmus uses the coinage *theologaster* ('Pygmaei Theologastri') in *Apologia adversus Petrum Sutorem* LB IX 756F.

363 The celebrated passage in *Republic* 473C–D, cited by Pio (*XXIII libri* 80ᵛs) in connection with the satiric reference to it in *Moria* 497–8 ASD IV-3 98 / CWE 27 100; E. Surtz cites multiple and varied references to this passage by fifteenth- and sixteenth-century authors, including eight citations from the works of Erasmus in *The Complete Works of Thomas More* (New Haven, Conn 1965) IV 349.

364 Pio *XXIII libri* 80ᵛs attacks the handling of the philosopher Marcus Aurelius Antoninus, emperor AD 161–80, in *Moria* 505–9 ASD IV-3 100 / CWE 27 100.

365 *Moria* 507–9

366 Marcus Aurelius' infamous son, emperor AD 180–92

367 Pio *XXIII libri* 80ᵛs cites Julius, Augustus, and Tiberius as learned emperors under whom the empire flourished.

368 Gaius Caligula, emperor 37–41; Claudius, emperor 41–54; Nero, emperor 54–68; Hadrian, emperor 117–138; Julian the Apostate, emperor 361–3. Erasmus had edited Suetonius and the *Historiae Augustae scriptores* (Basel: Froben 1518); he sketches the various types of emperors in the dedicatory letter, Ep 586.

369 Pio *XXIII libri* 81ʳᵀ criticizes *Moria* 599 ASD IV-3 104 / CWE 27 103; see Clarence Miller's note on *Moria* 599 in ASD IV-3 104–5.

370 Democritus of Abdera (born between 460 and 457 BC), called in late antiquity 'the laughing philosopher'; see Clarence Miller's note on *Moria* 551 in ASD IV-3 103; and *Christiani matrimonii institutio* LB V 672A: 'Such were Heraclitus and Democritus; the first of these used to weep whenever he went out in public, the other used to laugh. It is unclear which was more mad than the other.' But Democritus is mentioned in the dedicatory letter of the *Moria* (line 15), where Thomas More is compared to him (ASD IV-3 67 / CWE 27 83).

But even more intolerable is her statement that human happiness is a matter not of reality but of opinion.[371] She spoke quite truly if you are thinking of the sort of happiness about which Folly is thinking. At this point Folly becomes, if it please the gods, an Academic,[372] and Luther is praised, because he called Erasmus a sceptic.[373] Here, of course, Luther's authority holds good. How ardent Pio's love for Erasmus, how seething his hatred for Luther!

Earlier he had said that Folly was not without justification in criticizing the behaviour of certain princes.[374] But now he says that the vices of princes and bishops should not be exposed, but concealed; instead, they should be admonished 'reverently and sparingly.'[375] As if Folly is speaking to the people at large, and not, on the contrary, to the learned, or as if she has published the secrets of evil bishops! She mentioned only matters that are common knowledge, matters that even the bishops could read with a smile. No prince or bishop has complained to me about the *Folly* so far; some have praised it.[376]

* * * * *

371 Pio XXIII *libri* 81[r]T paraphrases *Moria* 1096–8 ASD IV-3 130 / CWE 27 118.

372 Pio XXIII *libri* 81[r]T: 'But this view is Lucianic, not Christian, and Academic, not Stoic or Peripatetic.' Pio regularly associates Erasmus with Lucian (see especially n927 below). 'Academic' alludes to the scepticism and probabilism of the New or Middle Academy, best known from Cicero's *Academica*; see Charles B. Schmitt *Cicero Scepticus: A Study of the Influence of the Academica in the Renaissance* International Archives of the History of Ideas 52 (The Hague 1972). Erasmus describes the Academics in a scholion on Jerome: 'But this was common to all the Academies, to assert nothing, but to follow what was probable' (HO 1 104[v]C). Erasmus left an opening for this charge, for there is a favourable reference to the Academics in *Moria* 1098–1100 ASD IV-3 130 / CWE 27 118, and he urges in the *Methodus*: 'There are things ... about which the more refined course is to be uncertain and ἐπέχειν [suspend judgment] with the Academics than to make pronouncements' (LB V 134A / Holborn 297).

373 Pio XXIII *libri* 81[r]T: 'And Luther, in his argument with you on this point, is quite fair in his charge that you assert nothing, define nothing, but leave everything undecided. Thus you seem to subscribe to the mad and impious position of the Academics that there is nothing sure in life, but everything is based on opinion.' See Luther's reactions (*De servo arbitrio* WA 18 601:33–5; 603:2–7, 22–3; 605:27–34) to Erasmus' statement in *De libero arbitrio*: 'And I take so little pleasure in assertions that I will gladly seek refuge in Scepticism whenever this is allowed by the inviolable authority of Holy Scripture and the church's decrees' (LB IX 1215D / CWE 76 7).

374 Pio XXIII *libri* 80[v]s

375 Pio XXIII *libri* 81[r–v]v

376 In his *Responsio* Erasmus positively asserted that princes and leading church figures applauded his *Moria* and that Leo X had read it with pleasure (cf *Responsio* 58 above and his earlier statements in Epp 739:12–15, 749:13–18, and 967:204–7). While the archbishop of Brindisi, Girolamo Aleandro, may not have written to Erasmus directly to complain of *Moria*, he was probably among those prelates who censured it. In 1517 Erasmus reported to More that Jean Briselot, titular bishop of Beirut (an auxiliary bishop of Cam-

He says Folly censures popes, cardinals, and priests because they have plenty of wealth, as if they were behaving improperly, since Christ was poor.[377] In fact, Folly speaks with greater prudence than this thrice-wise Peripatetic. She is criticizing those who with their wealth, troops, and cannons imitate secular monarchs, and do not resemble Christ, whose vicars they are, in anything.[378] Pio considers valuable property, armaments, troops, sovereignty, and citadels necessary.[379] He makes no mention of those other matters that are the true properties of bishops. Furthermore, Folly makes a plain declaration that she is not speaking of all bishops and priests, tempering her discourse as follows: 'But it is not my purpose here to go into details of the lives of pontiff or priest. I do not want to look as though I am writing satire when I should be delivering an encomium, nor want anyone to suppose that when I praise evil princes I am censuring good ones.'[380]

Here too the droll fellow makes fun of me. He says: 'But I ask you, reverend teacher, what kind of syllogism is this? Christ was poor, and so the church ought to have no property, and therefore the possession of wealth by a priest is forbidden.'[381] It is outrageous to exchange jest for jest [ἀντιπαίζειν] against a dead man, though I do not think that these words were actually written by Pio. And why is the church brought up here in spite of the fact that Folly is talking about popes and bishops? But that enthymeme[382] held good up to the time of Jerome, when bishops who were eager to amass wealth used to have a bad reputation. How often, indeed, Paul castigates those who serve their bellies, and not Jesus Christ.[383] Could not an enthymeme be constructed from what he writes to Timothy (chapter 6): 'Since we have food and clothing, let us be content with what we have. For those who want to become wealthy fall into temptation and the

* * * * *

brai) and later archbishop of Oristano and thus primate of Sardinia, was openly criticizing his *Moria* (Ep 597:4–11).

377 Pio *XXIII libri* 81ᵛx criticizes *Moria* 1768–1840 ASD IV-3 172–6 / CWE 27 138–140.

378 *Moria* 1768–1829 ASD IV-3 172–3 / CWE 27 138–40; Listrius assumed that Erasmus had Julius II in mind (see ASD IV-3 175:1818n).

379 Pio *XXIII libri* 81ᵛx argues that nature and reason allow one to defend one's own property by force and in certain situations this is also allowed to clerics, especially when they are defending the church's goods. See n389 below and Erasmus' *Brevissima scholia* no 13.

380 *Moria* 1856–8 ASD IV-3 176 / CWE 27 140–1

381 Pio *XXIII libri* 81ᵛx

382 An enthymeme is a rhetorical syllogism with one premise suppressed; in this section Pio criticizes Erasmus' alleged positions from the point of view of formal logic.

383 Romans 16:18

devil's snare, etc.'[384] This is the sort of conclusion Jerome draws in the case of all clerics, that those who have something more than the Lord cannot have their lot with the Lord.[385]

These remarks of mine are extraneous, for I have never held that priests should be denied adequate support, even for a life-style of dignity. But it used to be the case that the possessions of bishops were expended for the needy, for widows, and for their deacons,[386] not for cavalry, infantry, castles, and cannon. In fact, Folly does not condemn even these things but, rather, the fact that it is these alone which are sought after. Let anyone who cares read the passage, and he will find that I am telling the truth.

Pio admits as probable the argument that 'whoever is a Christian ought to imitate the life of Christ in all respects,'[387] but he does not admit the argument that 'a bishop is the vicar of Christ, therefore he ought to imitate Christ.' He says: 'This argument is no more applicable to priests than it is to any Christians.'[388] So no more is required of bishops than of lay-folk?[389] Whoever heard doctrine like this? But when is Folly in danger of seeming not to have learned dialectic? Or is this enthymeme preposterous: 'The levites of the Old Covenant did not have a share among the tribes,[390] but they lived off the tithes; it is far more fitting that the clergy of the New Covenant do the same?' Or this one: 'Christ declared that it is impossible for a rich man to enter the Kingdom of Heaven.[391] Therefore it is much more of an impossibility that a rich man preside in the Kingdom of Heaven'; or: 'According to Paul, no soldier of God is entangled in the affairs of the world.[392] But bishops are the leaders of God's soldiery. Therefore they are bound to

* * * * *

384 1 Tim 6:8–9
385 Jerome Ep 52:5 CSEL 54 421
386 See Gratian *Decretum* pars 1, d 87–8 (Friedberg I cols 304–7). That these responsibilities were still considered part of a bishop's office is evident in the famous treatise of Giovanni Bertachini *De episcopo*, first published in Milan in 1511, a compilation of the current teachings and practice on the office and function of a bishop; see Johannes Bertachinus de Firmo *Tractatus de episcopo ecclesiastice facultati admodum conveniens* ... (Lyon: Vincentius de Portonariis de Tridino 1533) 78ʳ and 131ʳ, and M. Caravale 'Bertachini (Bertacchini), Giovanni' DBDI IX 441–2.
387 In fact, Pio XXIII *libri* 81ᵛx says: 'This, however, is not true in an unqualified way, though it may seem likely and probable.'
388 Pio XXIII *libri* 81ᵛx: 'But if this were true, it would be no more applicable to priests than to the rest of Christians.'
389 Note Pio's words in XXIII *libri* 81ᵛx: 'But if ... others are allowed to protect their property, and to repel force by force, surely priests will be allowed to do the same ... since this is nowhere found to be more restricted for them than for the rest.'
390 Num 18:20–1
391 Matt 19:23
392 2 Tim 2:4

be much less entangled in such affairs. But worldly wealth is most particularly to be classed among worldly affairs.' Likewise: 'Inasmuch as the pope is nearest to Christ in honour of his position, it is fitting that he be nearest to Christ in similarity of way of life'; or: 'Though he was wealthy, Christ became poor for our sakes.[393] Therefore it is not fitting that the vicar of Christ become rich from being poor'; also: 'Wealth, like thorns, chokes out the seed of the divine word.[394] Therefore wealth is a disadvantage to the chief steward of the word'; or: 'Christ's sheep cannot serve both God and Mammon,[395] much less, therefore, the shepherd of the Lord's flock.' Folly could have provided a hundred enthymemes, but Pio thinks that all of these are naive and unable to be brought into properly syllogistic form.[396] But he does like the enthymeme that goes 'The pope is the vicar of the impoverished Christ, and the successor of the impoverished Peter. Therefore it is fitting that the pope single-handedly surpass all kings in wealth and display.'

He says that Christ need not be imitated in everything he did or recommended, but that in many cases one must take into account time, place, the kind of activity, the rank of those involved, and their personalities.[397] How eager are these men to find excuses to depart from the Model! In fact, it is the duty of all Christians to imitate Christ according to their ability. But Folly does not understand the difference between precept and counsel, nor the difference between the things the Lord did to accomplish our redemption and what he did to teach by example the various sorts of men in their various callings, nor the difference between those things which are to be kept in spirit only and what is to be done externally.[398] No, she mixes everything up under a single Myconus.[399]

Who fails to see where all this is tending? He wants to relegate the counsels to those who follow the monastic calling, but in such a way, however, that they can have inner poverty but outer wealth. He wants popes and bishops to be rich as Croesus,[400] and warriors too. But from whom

* * * * *

393 2 Cor 8:9
394 Matt 13:22 and parallels
395 Matt 6:24, Luke 16:13
396 Pio wrote in XXIII libri 81ᵛx: 'To what suitable norm [*norma*] for syllogisms can these conclusions be reduced?' Erasmus read *forma* for *norma*.
397 Pio XXIII libri 81ᵛx.
398 Pio XXIII libri 81ᵛx criticizes Erasmus for failing to draw these distinctions.
399 See *Adagia* II iv 47: *Omnia sub unam Myconum* 'One Myconus to cover everything' LB II 538D / CWE 33 215; Erasmus quotes Strabo, 'used of those who bring under one heading things by nature separate.'
400 See *Adagia* I vi 74: *Croeso, Crasso ditior* 'As rich as Croesus or Crassus' LB II 251D–E / CWE 32 50–1.

shall we expect the evangelical counsels if they are not the particular concern of those who have succeeded specifically in the places of Christ and the apostles?

I have never declared without qualification that warfare is illicit for Christians,[401] though these wars that we have seen so far are absolutely pagan. Folly states that war is such an atrocity that war and Christ are mutually exclusive.[402] And this enthymeme is not preposterous: 'Christ never made war; in fact, he ordered Peter to put away his sword; therefore it is not fitting for bishops, the vicars of Christ, to make war.'[403] Likewise, 'Peter, in the role of the pope, is rebuked because he tried to protect the Lord's life with a sword; it is far less suitable for the successors of Peter to take up arms for wealth and dominion.'

Folly would not have been short of enthymemes, but her goal back then was different. When she comes to dispute at the Sorbonne,[404] then she will make those distinctions that Pio has made earlier here, and she will reach the conclusion that the model of poverty applies to monks alone.

Pio adds that we should fight more fiercely on behalf of the property of priests than the property of lay people, because clerical property has been procured for more compelling reasons and set aside for a better use.[405] So the priests are in danger of being denounced for being sluggards

* * * * *

401 Pio takes up the question of Christians and war in XXIII *libri* 81ᵛx, in particular the resorting to violence in defence of church property; he also devotes book 21 to Erasmus' views on war. For Erasmus' nuanced thinking in 1530 on the advisability or not of war with the Turks, see A.G. Weiler 'Einleitung (1986)' in the *Utilissima consultatio de bello Turcis inferendo* ... ASD V-3 3–28 especially 20–7. See also Robert P. Adams *The Better Part of Valor: More, Erasmus, Colet, and Vives on Humanism, War, and Peace 1496–1535* (Seattle 1962) eg 88–121, 296–7; Pierre Brachin 'Vox clamantis in deserto. Réflexions sur le pacifisme d'Erasme' in *Colloquia Erasmiana Turonensia* ed Jean-Claude Margolin 2 vols (Toronto 1972) I 247–75; J.-C. Margolin *Guerre et paix dans la pensée d'Erasme* (Paris 1973); and M.J. Heath 'Erasmus and the War against the Turks' in J.-C. Margolin ed *Acta conventus Neo-Latini Turonensis* (Paris 1980) 991–9.

402 *Moria* 1814–17 ASD IV-3 174 / CWE 27 139

403 Pio XXIII *libri* 81ᵛx had framed Erasmus' position: 'Christ did not make war; in fact, he stopped Peter when he drew his sword. Therefore he forbade Christians, especially priests, to make war'; and Pio observes: 'You are illogical, most eloquent Erasmus, if you think that you are arguing correctly when you say this.'

404 Cf *Moria* 1909–12: 'I ought to call on the spirit of Scotus (which is far thornier than any porcupine or hedgehog) to leave his precious Sorbonne and occupy my breast ... I only wish I could change my face and don a theologian's garb!' (ASD IV-3 179–80 / CWE 27 142).

405 Pio XXIII *libri* 81ᵛ–82ʳx

in defence of their property! But it is one thing to possess wealth, and quite another to be diligent in amassing it. Nor is possession a simple matter: one possesses wealth in such a way that it constitutes the greater part of his happiness; another possesses it as if he did not possess it. But these statements of mine in defence of Folly are beside the point, since she condemns neither war nor clerical wealth absolutely, but rather the fact that those who are concerned with war and wealth have nothing else about them that is worthy of the priesthood.

But so far Pio has dealt with trifles; now we will come to matters of life and death.[406] Folly condemns papal indulgences,[407] and this is one of the censured teachings of Luther, as is plain from the bull of Leo x.[408] Here Erasmus is certainly caught in a bear-hug,[409] and it will be clear from this issue which of us is a liar.[410]

In the first place, my reply, if I must defend whatever Folly says, is that she often sounds like a pagan when talking about the gods, and she praises many things that deserve criticism, for example, flattery and drunkenness, and, finally, herself. But, well and good, I do not refuse to stand trial with Folly in this matter. I will say only this by way of preface: the theologians have been unfavourable to Luther in no area less than in his condemnation of indulgences, so much so that the Parisians omitted this

* * * * *

406 Perhaps a sarcastic allusion to Pio's transition at *XXIII libri* 82ʳʏ: 'It remains to see what you write in the *Moria* about approved customs, about sacred rites and rituals, papal indulgences, and other things of this sort, and, finally, what you write about the whole of the Christian religion and about its Author, the Lord Jesus Christ.'

407 Pio *XXIII libri* 82ʳʏ, quoting *Moria* 970–2 and 979–82 ASD IV-3 122–4 / CWE 27 114; here Pio argues that Erasmus' treatment of indulgences in *Moria* proves that he provided the seeds for Luther's heresies and refutes Erasmus' claim that he agrees with Luther in none of his condemned doctrines.

408 *Exsurge Domine* of 15 June 1520 condemned several of Luther's statements related to indulgences; see eg nos 17–22 in *Causa Lutheri* I 378–80.

409 'tenetur Erasmus medius'; see *Adagia* I iv 96: *Medius teneris* 'You are held by the middle' LB II 180A–B / ASD II-1 472 / CWE 31 381.

410 Before taking up Erasmus' treatment of indulgences in *Moria*, Pio, referring to the earlier stage of their controversy, had remarked (*XXIII libri* 82ʳʏ): 'From this one will know whether I was deceitful or unfair in reporting these things in my letter, as you maintain while you shamelessly deny everything'; and after quoting the suspect passages from *Moria*, Pio exclaims, recalling the words of his *Responsio paraenetica* that Erasmus had so ridiculed (see *Responsio* nn 95 and 98 above): 'Who will dare to deny, when he hears these words, that Erasmus Lutherizes, or, rather, that Luther was Erasmusizing when he began his insanity? ... Who will not see ... that he gathered these, the seeds of his first heresy, from your gardens?'

article.[411] Only the author of the preface says that Luther writes many falsehoods about indulgences, but does not explain what the falsehoods are.[412] In fact, I myself heard the Carmelite theologian of Louvain, the one who promulgated[413] the bull of Leo, say before the people: 'We are not much concerned about indulgences either.'[414] Whatever is conducive to truth and to the salvation of the flock of Christ is not opposed to the authority of the pope. If indulgences are useless, and the people are taken in by deceptions like this, should we be providing for the advantage of the papacy or looking out for the salvation of countless men and women? It is quite well known what the theologians think about the sort of indulgences we have frequently seen offered for sale.[415]

And these indulgences that the decretals and the ancient charters mention were nothing other than a mitigation of a penance imposed by human agency.[416] They mean for it to be restricted, and not proclaimed except for things that are really necessary, for example, for building a basilica where none existed. Only later came those that reduce the punishment of purga-

* * * * *

411 This is true; see the Sorbonne article printed in Duplessy *Collectio judiciorum* 1-2 365–74 (15 April 1520), 374–9 (14 November 1523). See also D. Hempsall 'Martin Luther and the Sorbonne 1519–1521' *Bulletin of the Institute of Historical Research* 46 (1973) 28–40.

412 The Preface's long list of Luther's errors concludes: 'But in his ignorance he also says much that is mistaken about notable philosophic doctrines, as also about the church's jurisdiction and indulgences'; see Duplessy *Collectio judiciorum* 1-2 366. Noël Béda and Jacques Berthélemy compiled a list of Luther's errors that were condemned by the Sorbonne on 15 April 1521; see Pierre Imbart de la Tour *Les origines de la réforme* III: *L'Évangélisme (1521–1538)* (repr Geneva 1978) 205 n2; and Alexandre Clerval *Registre des procès-verbaux de la faculté de théologie de Paris* I 1501 á 1523 (Paris 1917) 280 n18, 285.

413 *evulgavit* LB *emulgavit* 1531

414 Nicolaas Baecham of Egmond (d 1526), assistant inquisitor for the Netherlands (1520–4), did defend Catholic teachings on indulgences; see CEBR I 81–3; and Epp 1299:60–4, 1301:140–8.

415 See n420 below.

416 On the origin and development of indulgences, see Paulus; Bernhard Poschmann *Der Ablass in Lichte der Bussgeschichte* (Bonn 1948); and Erwin Iserloh *The Theses Were Not Posted: Luther Between Reform and Reformation* trans Jared Wicks (Boston 1968) 3–17. On the practice of indulgences in the time of Erasmus, see Joseph Lortz *The Reformation in Germany* trans Ronald Walls 2 vols (London 1968) I 220–31; and Duffy 287–93. Erasmus erred in claiming that indulgences were originally merely the mitigation of a humanly imposed penance. In the eleventh century when indulgences first appeared they involved the remission both of external prescribed canonical penalties and of the corresponding punishment before God. Erasmus here voiced a later view of indulgences, one repeated in the fifth of Luther's ninety-five theses: *Papa non vult nec potest ullas penas remittere praeter eas, quas arbitrio vel suo vel canonum imposuit*; see WA I 233; Iserloh *Theses* 7–8; and Léon-E. Halkin 'La place des indulgences dans la pensée religieuse d'Erasme' *Bulletin de la Société de l'histoire du protestantisme français* 129 (1983) 143–54, repr as entry 10 in his *Erasme. Sa pensée et son comportement* (London 1988).

tory, that bid the angels to bear the souls redeemed for a price into heaven. The hawking of these to the people by some Dominicans[417] in a grossly obsequious, brazen, and, I might say, virtually blasphemous way provoked Luther to put forward some theses for disputation.[418] Prierias' answer to these was such that the conflagration blazed up.[419]

Even before the *Folly* ever so many people had objected, at least in part, to papal indulgences,[420] but Folly did not mention these at all. She attacked only imaginary pardons, and she designates as imaginary those that have been wrongfully obtained or improperly granted.[421] The pope grants them to the penitent and contrite, but the masses think that innocence is achieved by the payment of money, with no thought of amending their lives. The entire tenor of her speech proves that this is Folly's view, even the words which Pio excerpts to support his slander, for there she says: 'Take, for example, some merchant, or soldier [...] who believes he has only to give up a single tiny coin from his pile of plunder to purify once and for all the entire Lernean morass he has made of his life. All his perjury, lust, drunkenness, quarrels, killings, frauds, perfidy, and treachery he believes can somehow be paid off by agreement, and paid off in such a way that he is now free to start afresh on a new round of

* * * * *

417 Chief among them was Johann Tetzel with his famous sales pitch: 'As soon as your money clinks in the bowl, out of purgatory jumps the soul.' Once attacked by Luther, Tetzel's confrères rallied to his support; see Lortz *Reformation* (cited in n416 above) I 227; and Jared Wicks 'Roman Reactions to Luther: the First Year (1518)' *The Catholic Historical Review* 59 (1983) 521–62 especially 524.

418 For the famous ninety-five theses of 31 October 1517, see WA I 233–8. For a discussion of whether these theses were intended for debate, see eg Hans Volz 'Erzbischof Albrecht von Mainz und Martin Luthers 95 Thesen' *Jahrbuch der Hessischen kirchengeschichtlichen Vereinigung* 13 (1962) 187–228; Iserloh *Theses* (cited in n416 above) especially 76–97, where he lays out nine arguments against the posting of the theses for debate; and Heiko Augustinus Oberman *Masters of the Reformation: The Emergence of a New Intellectual Climate in Europe* trans Dennis Martin (Cambridge 1981) 148–50 n88.

419 His *In praesumptiosas Martini Lutheri Conclusiones de potestate Papae, Dialogus* (Rome 1518) is analysed in Wicks 'Roman Reactions' (cited in n417 above) 529–31.

420 For examples of such protests prior to the first publication of *Moria* in 1511, see the Sorbonne's 1482 rejection of the view that the payment of an alms by itself earned an indulgence for the dead (Paulus III 386). In 1518 the Sorbonne formally censured that opinion; see Clerval *Registre des procès-verbaux de la faculté de théologie de Paris* (cited in n412) I 232–8; and Duplessy *Collectio judiciorum* 1-2 355–6. For examples of protests against indulgences in Germany, see *Manifestations of Discontent in Germany on the Eve of the Reformation* trans and ed Gerald Strauss (Bloomington, Ind 1971) 38–9, 42, 47, 51, 57, etc.

421 *tantum incessit fictas condonationibus*, echoing *Moria* 970–1: 'Now what am I to say about those who are deluding themselves with imaginary pardons for their sins?' (ASD IV-3 122–4; CWE 27 114).

sin.'[422] These are Folly's words, and from them it is clear that Folly is not indiscriminately condemning indulgences, but is condemning those by whom they are improperly offered or obtained. Therefore Alberto's boast of victory at this point, as if he proved that Erasmus agrees with a condemned doctrine, is quite vain.[423] For not only is Luther's way of speaking of indulgences different from Folly's, but the substance of what he says is different as well.

The fact that different things are sought from different saints in particular regions[424] has nothing to do with the real business of the faith. Nor yet does Folly criticize this indiscriminately, but rather the fact that people seek of the saints not the things that are conducive to their salvation, but those that are suited to their desires.[425]

It is far from the case that the cult of the Virgin Mother of God[426] is anywhere censured, in humour or in earnest; rather, what is censured is the fact that the masses almost make more of her than of her Son,[427] and, if one takes into account the superstitions of some, the adverb 'almost' is redundant.

Folly is certainly not condemning votive offerings,[428] but she makes the joke that among them there has been none set up by one who had been a fool, and was made a wise man.[429] And Pio provides a serious reply to this drollery!

But he stimulates enormous hostility against me on the basis of Folly's conclusion that all of Christian life teems throughout with insanities of this sort.[430] What could be more clear? Why are we still looking for witnesses? 'You have heard the blasphemy!'[431] She calls the whole cult of the saints insanity. In fact, Folly is being framed. She labels as insane those who hasten

* * * * *

422 *Moria* 979–83 ASD IV-3 124 / CWE 27 114
423 Pio XXIII *libri* 82ᵗy: 'But, say there, you who maintain that ... there is no agreement between you and Luther in condemned doctrines. Come now, you deny the doctrine of the forgiveness of sins and the authenticity of indulgences. Are not these among the condemned ones?'
424 Pio XXIII *libri* 82ᵗy criticizes *Moria* 990–1 and 995–6 ASD IV-3 124 / CWE 27 114, 115.
425 See *Moria* 997–1007 ASD IV-3 124–6 / CWE 27 115.
426 This topic is taken up by Pio XXIII *libri* 82ᵗy, quoting *Moria* 995–6.
427 Paraphrasing *Moria* 996
428 Pio XXIII *libri* 82ᵗz and 82ᵛA quotes *Moria* 998–1002, 1009–116 ASD IV-3 126 / CWE 27 115.
429 *Moria* 999–1000
430 Pio XXIII *libri* 82ᵛz alludes to *Moria* 1014–15 ASD IV-3 126 / CWE 27 115.
431 A paraphrase/quotation of the words of Caiaphas in Matt 26:65, alluding, perhaps, to Pio XXIII *libri* 82ᵛz: 'These words so clearly demonstrate that all the practices which you reported ... are being mocked, criticized, and absolutely rejected by you that they need no special interpretation or exaggeration.'

to some locale on the assumption that the saint could not grant the same thing elsewhere, to those who consider the Mother more important than the Son, to those who seek from the saints things which should not be sought. Nor yet does Folly's conclusion have to do specifically with these issues, but with the many other matters which preceded them, the superstitious cult of the saints, magical prayers, and other similar matters. Our man Pio is ignoring these, not, as he repeatedly alleges, to prevent the book's growing too large, but to make room for false charges.[432]

By the Muses, has he not given us a list thus far of really outstanding heresies from the *Folly*?[433] But there remains the charge of blasphemy against Christ, because Folly says that Christ took delight in the foolish and hated the wise.[434] If she had first made a distinction here between the foolish of this world and the foolish before God, and between the wise of this world and the wise before God, he could admit that this is a thoroughly Christian point of view. And yet, when Folly is about to begin this section, she prays that her words will not cause offence.[435] Again, when she is going to quote skewed proof-texts from Scripture, she says by way of preface that she will be quoting inappositely [οὐδὲν πρὸς ἔπος].[436]

Now if anyone says that one should not make jokes with the words of sacred Scripture, I admit that it was inappropriate to do so, but this is something that many do without being slandered. Besides, I provided an ample reply on this point to Maarten van Dorp.[437] Since the Letter was attached to the *Folly*,[438] I am at a loss to know why Pio found it too much bother to read

* * * * *

432 Erasmus is referring to Pio's statement at the beginning of book 3 of *XXIII libri* (73ʳA): 'For I am not going to gather or discuss all of them [that is, the plainly suspect passages in Erasmus' books] (for that could hardly be done), since there is a countless number of them. They alone, collected together, would make an overlarge book.' In book 3 Pio indicates at various points (79ᵛO, 80ᵛS, 82ʳY) that he is omitting material that might be brought up for criticism.

433 In this paragraph Erasmus deals with Pio's criticisms (*XXIII libri* 83ʳB–83ᵛD) of *Moria* 2079–85, 2091–4, 2141–3, 2143–58, and 2176–91 ASD IV-3 186–90 / CWE 147–51, where, *inter alia*, Pio points out additions and deletions made by Erasmus to some of these passages in subsequent editions of the *Moria*.

434 Pio *XXIII libri* 83ᵛD implies a charge of blasphemy in connection with Folly's comparison of Christ's rejection of the wise to famous tyrants' suspicion of the very bright (*Moria* 2080–5); Pio observes: 'I could scarcely believe that Satan himself would have dared to make this comparison, it contains such an unspeakable blasphemy.'

435 See *Moria* 2158: 'Don't be put off by the words' (ASD IV-3 190 / CWE 27 150).

436 *Moria* 1889–90 ASD IV-3 178 / CWE 27 142

437 Ep 337 of May 1515, Erasmus' first apologia; see CWE 3 111–39, CWE 71 17–30, with references to secondary literature.

438 In the index of his works that he prepared for Hector Boece in 1530, Erasmus lists Ep 337 among his apologias as 'The Letter to the Theologian Maarten von Dorp which was

it before he wrote this. But what need is there for all that, since Folly herself carefully wards off every slander? Accordingly, the details, added for
effect, about Plato's cave[439] and about Julian and Porphyry[440] are, in fact,
nothing but feeble canards, and I am sure that Pio was less than proud of
himself when he wrote them, or, as I rather suspect, when he read them
after they were written by others. I am not really angry with him; rather, I
am sorry for the fellow, for he invested so much energy in a business that
is neither gentlemanly, nor handsome, nor learned, nor beneficial, nor pi-
ous/Pio's. I shall conclude, then, that Pio has so far proven none at all of the
many terrible charges that he had so often and so grandiloquently promised
to prove. Folly, at least, is forgiving if he has said anything foolish.

Now he promises[441] that he will point out horrific things from my
books,[442] this fellow, always fierce in opening, now too in his closing remarks, but sluggish and weak in providing proof. Just see if these things he
promises are not manifest falsehoods: there is virtually no book of mine in
which I do not disparage ceremonies, the holy traditions of the Fathers, and
monasticism; in which I do not write off the estimable order of theologians,
do not attack priests, do not tear apart the dignity of bishops and popes,
do not deprive certain sacraments of credibility and prestige, do not condemn the customs received by the church, deny the merit of good works,
do not denigrate the dogmas of religion, the practices of the church, the cult
of the saints, and the veneration of the Virgin Mother of God, do not condemn the taking of vows, and call many of the canonical Scriptures back
into doubt![443]

* * * * *

then appended to the *Moria* (Allen Ep 2283: 196–7); the letter was 'ordinarily reprinted
with the *Moria* during Erasmus' lifetime' (Clarence Miller in ASD IV-3 25), but it did not
invariably accompany the *Moria* (*pace* Allen Ep 337 intro), and appeared in no Italian
editions (Miller as quoted in CWE 71 142–3 n8).

439 Plato's cave is mentioned by Pio *XXIII libri* 83^{r}c, alluding to *Moria* 2176–9 ASD IV-3 190 /
CWE 27 150.

440 Pio had mentioned Julian and Porphyry earlier in his *Responsio paraenetica* 8^{r}G (see *Responsio* n284), and he repeats this allusion in *XXIII libri* 83^{r}c and 83^{v}D.

441 MARGINAL NOTE: 'He makes fierce promises'

442 Pio's final paragraph of book 3 (84^{r}F) is a transition to the rest of the *XXIII libri*. Pio
maintains that the erroneous opinions expressed in the *Moria* are to be found in the
rest of Erasmus' works: 'If only a slanderer could be protected by an excuse like "I was
declaiming when I said that, not offering instruction in earnest"! But you cannot even
allege this in your defence. You might have been able to if you had dealt with these
matters only in the *Moria*, but all your works teem with the same views. For everywhere,
whenever you get a chance, you go over again at greater length the matters you touched
upon in the *Moria*.'

443 A paraphrase of Pio *XXIII libri* 84^{r}F

What could be more impressive than this promise? If he can prove any of these charges I will confess that all of them are true! But it is surely an altogether tiresome business to argue with a man who has neither studied theology, nor involved himself in the Scriptures and the ancient doctors, nor ever, it seems, given any serious thought to real piety, and who, finally, provides abridged, skewed, and misunderstood quotations from my works because he has not read them. However much he denies this, the facts keep on crying out that it is absolutely true.

FASTING [444] [FROM BOOK 4] [445]

First he takes up my condemnation of the fasts enjoined by the leaders of the church and of abstinence from certain foods.[446] Where is this? In the *Methodus*.[447] Read it into the record:[448] 'The sabbath was established for the sake of man, man was not created for the sake of the sabbath. These things had, of course, been commanded by God, but he does not mean for them to be observed whenever the duty of charity must be fulfilled. What, then, will be the reply of the obscure little men who, because the eating of meat is forbidden, insist upon additional rules even more uninspired than these, with the consequence that the life of the whole person is endangered.'[449] So says the *Methodus*. What has this to do with the ordinances of the church?

* * * * *

444 1531 UPPER MARGINS: 'On Fasting' on pages 80–95 [= 181–92 in this volume]. The following section is the reply to Pio's book 4 of *XXIII libri* 84ʳ–93ᵛ, 'On Fasting and Dietary Observance.' It is by way of preface to this book that Pio describes (84ʳ⁻ᵛA) what will be his method throughout the rest of the *XXIII libri*, and clarifies the relationship of *XXIII libri* to his *Responsio paraenetica*; see Translator's Note cxlvi. At the beginning of book 4 Pio quotes ten excerpts (84ᵛB): three from the *Methodus*, three from *De esu carnium*, one from the *Supputatio*, and three from the *Annotationes in Novum Testamentum*.

445 MARGINAL NOTE: 'From book 4'

446 This is Erasmus' interpretation of Pio's sarcastic transition (*XXIII libri* 84ᵛB) to his presentation of Erasmus' passages on fasting: 'I shall begin with the rules for fasting, dietary observances, and the approved customs that have been introduced in the church. You deny that you disparage these, and insist that you do not object to them at all.'

447 Pio *XXIII libri* 84ᵛB quotes from the *Methodus* the following passages in this order: 1 / 'dietary rules, hatred for one's enemy, and war, which were permitted to the Jews, are not now permitted to Christians' (a paraphrase of the *Methodus* LB V 86F / Holborn 198:34–199:6; see 183 below); 2 / 'Christ cries out "It is not what enters the mouth that makes a man unclean," and you require of a Christian that he eat fish and cause damage to his health?' (*Methodus* LB V 107D / Holborn 240:34–241:2); 3 / 'What, then, will be the reply ... danger of death?' (*Methodus* LB V 107A / Holborn 240:1–3).

448 Erasmus' expression here, *Recita*, is the one Cicero employs when calling for documentary evidence to be read out in court; see Hugo Merguet *Lexikon zu den Reden des Cicero* (Jena 1877–84; repr Hildesheim 1962) IV 244.

449 Erasmus quotes directly from *Methodus* LB V 106F–107A / Holborn 239:33–240:3.

Are the popes obscure little men? Has the pope ever ordained or the church intended that the life of the whole person should be endangered because of regulations of human origin that have no public force? It continues: 'because it seemed good thus to mere humans, stupid, perhaps, and superstitious ones, who surely lacked any public authority.'[450] Both the content itself and the marginal note as well show plainly that this was not said against the practice of the church, but against the Carthusians.[451]

Pio also reports what follows: 'It is not what enters the mouth that makes a man unclean, yet you require of a Christian that he eat fish even with severe damage to his health? Indeed, do you even consider Christian (compared to yourselves) one who, compelled by reason of health, eats whatever food on whatever day, and gives thanks to God for it?'[452] Thus far the *Methodus*. Tell me, O Quarrelsome [φιλαίτιε] Censor: does the church demand of anyone that he eat fish and ruin his health? I do not think so, but even so certain superstitious little men do in fact demand it. In addition, when I say 'compelled by reason of health,' am I not condemning those who transgress the church's practices needlessly?

1[453] / In the *Letter to the Bishop of Basel*[454] I make an open and savage attack on those who are needlessly scornful of the church's dietary obser-

* * * * *

450 'because ... authority'] *Methodus* LB V 107A / Holborn 240:6–8

451 The marginal note is 'Listen to this, O Carthusian [*Audi Cartusiane*].' This marginal note does not appear in the early editions: see eg Basel: Johannes Froben, April 1519, p119 / sig H4ʳ [version I (1519A) in Holborn xv]; or Mainz: Johannes Schoeffer 1521, fol 67ʳ [version III (1520A) in Holborn xvi]. It does appear, however, in later editions: see eg Basel: Johannes Froben, June 1522, sig g2ʳ [version V (1522) in Holborn xvi]; and Köln: Hero Alopecius, December 1523, sig G4ʳ [version VI (1523) in Holborn xvi]. Pio may have used one of the earlier editions that lacked this marginal note. Hajo Holborn and Annemarie Holborn do not record marginal notes in their critical edition. The note would have appeared at 240:1–3. While Carthusians abstained from meat, their rule avoided any fanaticism and was prosaicly sane on questions of food; see David Mathew and Gervase Mathew *The Reformation and the Contemplative Life: A Study of the Conflict between the Carthusians and the State* (New York 1934) 284 n31; and n532 below.

452 *Methodus* LB V 107D; see n447 above.

453 Erasmus often assigns marginal numbers to the passages quoted by Pio; sometimes, as here, the numbers do not exactly correspond to the number, or sometimes to the order, of the passages in Pio.

454 *De esu carnium* (Basel: Froben 1522) ASD IX-1 19–50, addressed to Christoph von Utenheim, bishop of Basel (see CEBR III 361–2). On the disorders in Basel which prompted Erasmus to write this pamphlet, see ASD IX-1 3–5; the editor, C. Augustijn, lists fifteen Latin and four vernacular editions before 1536 (see 13–17). Pio *XXIII libri* 84ᵛB introduces this work in his collection of Erasmus' passages on fasting with the words: 'likewise, in the letter to the Bishop of Basel you endeavour, with great energy, in a careful and prolonged discourse, to prove that the law on dietary observance should be abolished'; he

vances. I do argue, however, that it would be beneficial, perhaps, in view of the temper of these times, that no one be bound to a particular type of food. If this were done, I think that there would be a minimal loss to Christian piety and a great gain for the unity of Christians.

2 / He reports this too: 'Dietary rules, hatred for one's enemy, and war, which were permitted the Jews, are not now permitted to Christians.'[455] But does not this remark attest that I am speaking about the Jewish observances? Is it we who abstain from pork and eels? The Jews used to believe that they were allowed to make war against any foreigners; we are not. He quotes from the Letter again where I add: 'because of this kind of trifling observance we observe people sicken and die, thinking that they are Christians, whereas they are Jews.'[456] Is what is being condemned here abstinence from meat, or rather the superstition of people who, on account of these things, cause their neighbour to sicken and even let him die, and all the while regard themselves as devout, whereas they are really Jews, that is, superstitious? Jewish dietary restrictions are forbidden us, and yet the Christian masses are a hair's breadth removed from them. Christians are not allowed to make war like the Jews, and yet we make war more criminally than they did. This is my opinion, and I have put it into words in more than one place, and yet it is in this passage that Pio has perceived a great impiety.

I did not write that our burden of dietary regulations and fasting is, in fact, more severe than that of the Jews, but that it seems to be. The theologians do not contradict this, but they devised the pious fiction that the burden of Christians is, admittedly, heavier in its regulations, but is made lighter by the increase of grace. And yet this fiction escaped Peter, since he said that the gentiles were free from the observance of the Law and gave his opinion that they should not impose on the brethren a heavy yoke that neither they nor their fathers had been able to bear.[457] Nor did Augustine

* * * * *

then quotes, with omissions and inaccuracies, *De esu carnium* ASD IX-1 46:824–8, 28:297–8, 295–6.

455 Erasmus is quoting Pio's summary paraphrase in *XXIII* 84^{V}B of *Methodus* LB V 86F / Holborn 198:34–199:6.

456 From *De esu carnium* ASD IX-1 46:826–8, quoted with omissions by Pio *XXIII libri* 84^{V}B; Erasmus here quotes Pio's version with one slight omission (*denique mori* for *et denique mori*; in ASD this reads (with no variants reported): *Et ob id genus obseruatiunculas videmus tot homines affligi, periclitari, denique mori, multos sibi stulte placere sibique christianos videri quum sint Iudaei* ('because of this kind of trifling observances we observe so many people sicken, become dangerously ill, and die, and many thinking that they are Christians, whereas they are Jews').

457 Acts 15:10; see Erasmus' paraphrase LB VII 727C–D.

see things this way when he wrote Epistle 108 to Januarius: 'And so the first thing I want you to hold,' he said, 'as the first principle of this discussion, is that our Lord Jesus Christ has placed us under his light yoke, as he himself says in the Gospel. Thus he has gathered together the society of his new people by means of sacraments, very few in number, quite easy of observance, and of surpassing meaning.'[458] You hear that here it is not grace that is being discussed, but regulations.

3 / Even in my annotations on the eleventh chapter of Matthew[459] I do not censure the constitutions promulgated by the authority of councils and received by the practice of all Christians, but I do criticize excessive prescriptions of human origin, not those of councils, but of bishops, princes, priors, and deans, and even those peculiar to the popes, like the *constitutiones camerae*.[460] If there is nothing excessive, nothing unfair, in these, then Erasmus' note was a waste of time.

At Rome Julius had established a rule that whoever obtained a bishopric should surrender his preferment.[461] What if he had none? It had to be

* * * * *

458 That is, Augustine Ep 54 (*Ad inquisitiones Ianuarii liber primus*) CSEL 34 159:4–9. In his edition of this letter, assigned number 108 in his *Secundus tomus operum divi Aurelii Augustini episcopi Hipponensis* (Basel: apud Io. Frobenium 1528) 359, Erasmus gives the readings *primo* and *leni*, as does CSEL; LB, however, reads *primum* and *levi*. After *nos subdidisse*, the Basel edition, and also CSEL, add *et sarcinae levi*, missing in the LB quotation. Pio *XXIII libri* 86ᵛE ignores this section of Augustine's letter to Ianuarius, but quotes from a later passage that defends Lenten fasts based on the authority of Moses, Elias, and Christ.

459 Erasmus is referring here to his lengthy annotation on Matt 11:28 LB VI 63B–65D / Reeve 53–6.

460 Rules governed the procedures and fees of the apostolic *camera*, the finance department of the Roman curia. For some of these rules, see Hofmann *Forschungen* II 155–62, 222–3.

461 This claim seems to have been based on a famous incident that occurred while Erasmus was visiting Rome. In a consistory of December 1509 Cardinal Guillaume Briçonnet publicly criticized the practice as being contrary to canon law, while Cardinal Marco Vigerio and Pope Julius II defended it as being beneficial to the apostolic see. See the report of the Venetian ambassador of 23 December 1509 in Roberto Cessi *Dispacci degli ambasciatori veneziani alla corte di Roma presso Giulio II, 25 giugno 1509–9 gennaio 1510* (Venice 1932) 207, as cited by Christine Shaw *Julius II: The Warrior Pope* (Oxford 1993) 165–6. In *Julius exclusus* Erasmus has Julius state that his predecessors had set down the rule that no one could be appointed a bishop until he had resigned his curial offices, and then he has Julius require a future bishop first to buy a curial office which he then must resign to become a bishop. See *Erasmi opuscula* 73–4:187–93; and Emmanuel Rodocanachi *Histoire de Rome: le pontificat de Jules ii (1503–1513)* (Paris 1928) 34 (no reference provided). In his life of Leo x Paolo Giovio claimed that Alexander VI and Julius II resorted to practices such as requiring rich and ambitious men to exchange benefices for offices; see his *Le vite di dicenove huomini illustri* (Venice: appresso Giovan Maria Bonelli 1561) 143ᵛ. In a ruling of 16 January 1505 Julius renewed the practices of his predecessors regarding the resignation of curial offices (he did not establish a new rule); see Hofmann *Forschun-*

surrendered, that is, he had to buy one only to surrender it at once. Who could not realize that this is unjust? As it is now, the observance of fasts, of dietary rules, and of feast days depends, to a great extent, on the bishops or on the custom of persons with no office.

4 / Pio piles up some passages from the *Annotations on the New Testament*; he does not cite the chapter but considers it sufficient to say 'in the same work.'[462] Still, I do not think that this is so much his fault as that of his quotation collectors, who were too lazy to indicate the book and chapter. One who condemns superstition among Christians is not condemning the regulations of the church. But is superstition not vast when adultery and drunkenness are considered a game, and a man who ate an egg or a bit of meat when he had, perhaps, a serious reason for eating them is thrown into jail as a heretic?

5 / He quotes[463] the passage on Colossians 2 differently from the way it is in my copy.[464] My words are as follows: 'And how I wish that some priests were not behaving in the same way in these times as well, neglecting those things which have more to do with devotion.'[465] When I say 'some priests' I am surely not condemning the church; when I add 'neglecting those things' I am not condemning abstinence from certain foods across the board, but wrong-headedness in this area. Moreover, I find the words 'it is contemptible that anyone should allow himself to be led about by human commandments like an ox by his nose, etc' neither in the first nor in the latest edition.[466] I do not know whence he took this, but under whatever

* * * * *

gen II 51 no 221. On the requirement of vacating a curial office upon promotion to the rank of bishop or cardinal, see Thomas Frenz *Die Kanzlei der Päpste der Hochrenaissance (1471–1527)* (Tübingen 1986) 193–4.

462 Pio *XXIII libri* 84ᵛB quotes three passages: the first is cited as 'in the annotations on the New Testament,' the second as 'ibidem,' and the third as 'in eodem volumine alio loco.' The first passage has defied location, but may be a paraphrase of that portion of Erasmus' lengthy annotation on Matt 11:28 that compares Christian superstition in dietary observance to Jewish observances LB VI 64c / Reeve 55. The second passage is from Erasmus' annotation on 1 Cor 8:8 (LB VI 704E / Reeve-Screech 482). The third is from Erasmus' annotation on Col 2:21, discussed in the next paragraph.

463 MARGINAL NOTE 'Deliberate misquotation'

464 Pio *XXIII libri* 84ᵛB quotes excerpts from Erasmus' annotation on Col 2:21 LB VI 892E / Reeve *Galatians to the Apocalypse* 642 with omissions and a distorting summary.

465 *Et utinam his* [LB VI 892E and Reeve *Galalatians to the Apocalypse* 642 add *quoque*] *temporibus non idem facerent Sacerdotes quidam, omissis iis quae propius ad pietatem pertinent* is rendered by Pio in *XXIII libri* 84ᵛB as *Quod et his temporibus faciunt sacerdotes ...'*

466 These words may be a garbled quotation from *Ennaratio in primum psalmum*: 'Are not some priests doing just this when they envelop their credulous flock in ritual? The people are led (by the nose, as it were) to depend on it, and thus they never grow towards the true teaching of Christ, but remain forever infants and children in Christ'

circumstance it was said, the statement is quite true, for no true Christian will be led about by the rules of just any men and never rise up to the freedom of the Spirit.

From these incorrect and distorted quotations he draws up some impudent and misleading propositions:[467]

1 / 'The dietary observances which the masses keep today (and how I wish they did keep them!) were not instituted by Christ.'[468] I think this could be true if you were to understand it as meaning 'by the words of Christ.'

2 / 'This observance was disapproved rather than approved.'[469] And I think that this is true, even if one associates Paul with Christ, however we interpret their statements. But there would be no need for these interpretations unless Christ and Paul seem to disapprove of dietary observances.

3 / 'Many illnesses and even death sometimes occur as the result of dietary observances and the Jewish superstition of certain people.'[470] No one will deny this.

4 / 'People are in no way bound to this observance.'[471] This is clearly false,[472] since I rebuke those who recklessly eat forbidden foods, and have done so in a number of passages.

5 / 'Those who, without necessary reason, eat meat on forbidden days sin against human law only.'[473] I judge this to be true providing one removes the distinction between directly and indirectly. He who violates the fast sins doubly. He who eats foods forbidden by men commits a single sin, and yet this is punished more severely.[474] It was in this connection that I said this.

* * * * *

(LB V 179E / ASD V-2 46:393–6 / CWE 63 23). Pio may have reworked these ideas into his distorted quotation from the annotation on Col 2:21, and one can only guess, as Erasmus does elsewhere, that Pio's notes had become scrambled.

467 The following formulation in the form of propositions does not come from Pio, but is Erasmus' abstraction from Pio's summary of Erasmus' views on *XXIII libri* 84^{V}B–85^r.

468 Pio *XXIII libri* 84^{V}B–85^r: 'Your arguments rest on these two foundations, one, that these things were not instituted by Christ ...'

469 Pio *XXIII libri* 84^{V}B–85^r: '... but were, on the contrary, abolished and rejected ...'

470 Pio *XXIII libri* 85^{r}B: 'these things are not only not beneficial, but even sometimes deadly for humans.'

471 Pio *XXIII libri* 85^{r}B: 'from these you conclude that Christians should not be compelled to keep these observances, since they are not subject to them.'

472 MARGINAL NOTE: 'Lie'

473 Pio *XXIII libri* 85^{r}B: 'you say in the Letter to the Bishop of Basel that, even if they eat meat on the proscribed days for no compelling cause, they do not sin except against human law.'

474 See the account of the barbaric punishment of one of the citizens of Basel given by Erasmus in his *In epistolam de delectu ciborum scholia* 17–30 ASD IX-1 66–7.

6 / It is not true that I draw the conclusion 'that it is appropriate that Christians be allowed to eat whatever foods they want, and that, in this connection, there ought to be no distinction between days.'[475] I am only arguing that it would probably be more beneficial, under contemporary circumstances, if the bishops would change this obligation to an exhortation. But I make no prescription or definition.

Pio's chiming in on these points is all the more impudent, since I have made a very careful reply to Béda about this very slander.[476] Pio wants to give the impression that he has read my *Supputatio*,[477] an occasional source for the content of his slanders. He did not look at the rest of it.

Next he provides a proof that fasting was instituted by Christ.[478] But I was talking about dietary restrictions, not about fasting. So let him go as far afield as he likes on his own choice of topics. It has nothing to do with me. His argument is based all the way through on the assumption that there is a natural connection between dietary restriction and fasting.

Now who does not know that fasting is recommended by Christ, and by the apostles too? And yet, when the issue is contempt for wealth and worldly dominion, then Pio teaches us that there is a great difference between counsel and precept.[479] He should have sought aid from that distinction here as well. In fact, I express my approval of fasting everywhere, so long as it is Christian! Rather, what Pio needed to prove is that fasting was in fact required on so very many days, and that in this matter humans can bind other humans under pain of hell. In addition, he should have proven that it was Christ who introduced into fasting superstition, anxiety, fear, over-reliance, and wrong-headedness, for it is these alone that Erasmus censures, whereas he praises Christian fasting in countless passages.

After a huge collection of texts in praise of fasting,[480] a practice of which no one disapproves, he vaunted over me like a victorious warrior. 'Why,' says he 'do you keep chattering that people sin only against human law if, even without a compelling reason, they fail to fast on the appointed

* * * * *

475 Pio XXIII *libri* 85ʳв: 'You claim that for these reasons it is appropriate that Christians should be free to eat whatever food they want, and that, in this matter, there ought to be no distinction between days, and so forth.'

476 The following of the Erasmian propositions formulated and criticized by Béda have to do with fasting: 111, 114, 118, 172–4; Erasmus' replies to these criticisms are in his *Supputatio* LB IX 644F–645C, 645F–648E, 681A–C.

477 Pio XXIII *libri* 84ᵛв quotes from the *Supputatio* on proposition 111 LB IX 645A.

478 Pio's argument on the antiquity of fasting and Christ's institution of it occupies XXIII *libri* 85ʳв–86ʳD.

479 See nn 398–9 above.

480 Pio XXIII *libri* 85ʳc–87ʳE quotes abundantly from the Scriptures and the Fathers, concluding on 86ᵛE with citations from Augustine, Basil, Ambrose, and John Chrysostom.

days and eat forbidden foods?'[481] He thinks that by human law I mean one 'which has been ordained on the basis of merely human understanding, with no divine inspiration.'[482] In fact, I call a law human whenever one human enacts for another a rule which was not stated in those writings whose authority no one can question. Whatever devout men do with good intention, it is likely that the Spirit is with them; but it is not necessary that we believe that all who make laws enjoy that inspiration of the Spirit on the basis of which the apostles wrote. I will not go into the fact that almost all the testimonia which he reports show that fasting has been recommended, but do not prove that it has been commanded. For not a single example or exhortation obliges one to fast under pain of hell. But granting that the apostles established some fasts, are all of them that are now required really of apostolic origin? I am talking about the fasts of our time. Christ enjoined vigils more often than fasting, and yet the church has done away with vigils,[483] but we push fasting as if it were of divine institution.

Listen to another premise. 'It is evident that anything that is never going to be carried out in practice is commanded or established to no effect.'[484] What is this I hear? No one would fast if there were not appointed fast days? Who appointed fast days for the apostles, who fasted voluntarily and almost every day? We are bidden to pray, and yet the church does not prescribe what we should pray for, how much we should pray, or on what days, unless, perhaps, we consider it enough to pray on Sundays. Almsgiving is commanded, but the days for almsgiving are not prescribed. Love of God is commanded, but no time for it is prescribed in the command. Bishops are commanded to study the Scripture, but no time is appointed for this. What has no fixed time appointed for it is to be done whenever there is an opportunity.

'What do you say,' he says, 'if the apostles ordained these things after the Lord ascended? What if they came a short time after the apostles?'[485] If Pio is unsure whether this took place, why does he triumph as if he has

* * * * *

481 Pio XXIII libri 86^{V}E
482 Ibidem
483 Erasmus must mean the popular congregational vigils of the patristic church; cf Modus orandi Deum LB V 1108E / CWE 70 168: 'Matins began with a prayer, and traces of the old custom survive today in the church.'
484 Pio XXIII libri 86^{r}D concluded that a divine command to fast can be found in Scripture, but that the days on which fasting is to be observed are not indicated; the words quoted verbatim here by Erasmus are Pio's transition to the next stage of his argument, that is, that the days appointed for fasting were part of an unwritten tradition.
485 Pio XXIII libri 86^{V}E: 'Granting you that Christ Jesus made no mention of these things, what do you say if the apostles enjoined these things after the Lord ascended, what

won? Furthermore, all are not forthwith obligated under pain of hell if it was the apostles who did or instituted something. But it is one thing to recommend by example or to urge people towards something, and quite another to command it under pain of hell. Christ recommended fasting, but where does he command it? But these observations of mine are beside the point.

Yet again, please observe his acuity as a disputant! In the *Letter to the Bishop of Basel* I argue the question of whether it would be better, in view of current circumstances, to change the rigid requirement of fasting to an exhortation, especially as affecting dietary restrictions.[486] He says that it is my view that both should be rejected and abolished.[487] In fact, I make the probable argument there that if my suggestions should be acted upon, more people would be fasting, and fasting more devoutly, than is now the case. I will pass over the rest of the fellow's silliness.

Lent is mentioned by the ancients;[488] but we do not read that the rest of the [fast] days were prescribed. St Augustine mentions fasting on Wednesday and Saturday among practices that were introduced by custom, not required by public regulation.[489] Otherwise there would have been the same observance everywhere, but there was one practice in Rome, another in Milan, another elsewhere. In fact, in the same city some would be fasting while others were eating twice a day. In this matter Augustine urges us to follow the bishop's practice. But he nowhere says what Pio asserts, that abstinence from meat is naturally linked with fasting.[490] But if we want to pay attention to the advice of St Jerome, who similarly urges frugality and temperance in food and in the eating of meat,[491] the first order of business would be abstinence from wine, which Jerome expressly forbids during fasting,[492]

* * * * *

if a short time after the apostles the leaders of the church did so, taught by the Holy Spirit?'

486 See *De esu carnium* passim, but especially lines 666–89, 917–47 ASD IX-1 41–2, 48–9.

487 Eg Pio *XXIII libri* 87ᵛG: 'But you are not afraid to propose that the laws of fasting and dietary observance should be abrogated by the leaders of the church, and that each should have a free choice as to what he pleases to do about these matters. This you urge at great length when writing to the bishop of Basel.'

488 Pio *XXIII libri* 86ᵛE–87ʳ cites Augustine, Gregory the Great, Origen, Chrysostom, Basil, and Jerome.

489 What follows is a summary account of issues raised in Augustine's Ep 54 CSEL 34 158–68, which is quoted several times by Pio in this book.

490 Pio argues for this connection with many patristic citations in *XXIII libri* 88ʳ⁻ᵛH.

491 Pio *XXIII libri* 88ʳ⁻ᵛH quotes Jerome passim in book 4; Erasmus probably has in mind texts like Ep 22.8 CSEL 54 154–7 and Ep 79.7 CSEL 55 95–6.

492 See Jerome Ep 100 (actually Jerome's translation of a paschal letter of Theophilius of Alexandria) sections 6 and 8 CSEL 55 219, 221.

and then from food cooked in spiced oil, from eggs, from dates, figs, snails, snakes, and finally from nourishing fish, oysters, eels, catfish, salmon, and the like.

Now, as if he were concluding, he waxes expansive in an epilogue with a review of my opinions[493] that is even more distorted that his earlier distorted list of them. He says:[494] 'You declare that holy fasting is a superstition.'[495] Erasmus never thought of this, not even in a dream,[496] but both in that pamphlet[497] and in many other places I censure people's superstition associated with fasting: for example, when they demand it from those for whom the church does not intend it; when people fast, contrary to the mind of the church, to the point of damaging their health, thus placing an obstacle before the real goal of fasting; when on a fast day they indulge in delicacies with greater abandon and load their stomachs more than they do on other days; when they fast only for fasting's sake and neglect those things for the sake of which fasting is appointed; when they fast from food but do not fast from drunkenness, from scurrility, gambling, malicious gossip, enmity, envy, and avarice, on the assumption that fasting alone amounts to a complete life of holiness. It is these and other abuses to which the majority of people are prone that I call superstition. Nowhere do I condemn Christian fasting, not in joke, not in earnest. Nor do I count among the superstitious those who, to mortify their bodies, abstain from delicate food, wine, soft beds, and extended sleep. But what has this to do with a universal obligation by which, human nature being what it is now, none are more afflicted than those who have no need to be? Accordingly, all the points that Pio heaps up here fall down flat since they rest on a rotten foundation.

He makes the assumption that I condemn the church's regulations on fasting, and again upon this rotten foundation he builds up his evil edifice.

He finally admits that it is not clear whether the lenten fast was established by Christ or by the apostles, although he has pronounced on this so often before.[498]

* * * * *

493 Erasmus refers to Pio's review of Erasmus' statements and views on *XXIII libri* 89ʳκ.

494 MARGINAL NOTE: 'Lie'

495 Quoted verbatim from Pio *XXIII libri* 89ʳκ; Pio goes on to argue against the view that fasting is superstitious on 89ʳκ–90ʳM.

496 See *Adagia* I iii 62: *Ne per somnium quidem* 'Not even in a dream' LB II 135D–136D / ASD II-1 370–2 / CWE 31 286–7.

497 Ie *De esu carnium*

498 *antea* 1531 *ante* LB. At various points in book 4 of *XXIII libri* Pio emphasizes Christ's fasting for forty days; on 86ᵛE he asserts that, even if the appointed days for fasting were established not by Christ but by the apostles, or even after the time of the apostles,

He explains to us how one should understand 'what enters the mouth does not make a man unclean.'[499] He says: 'The Lord is not giving permission there to eat foods forbidden in the Law.'[500] Yes, because of scandal, but he did prepare the minds of his disciples for ignoring that observance. An instance of this is Peter, who ate any foods with the gentiles. The proof is Paul, who always directs people away from dietary observance towards a zealous piety. What I really want to know is whether this supports the forbidding of foods. It does not. In fact, it counts against this rather than for it. And I have not said anything but this.

'The Jews,' he says, 'abstained, eg from things unclean.'[501] As if there were not too many Christians who abstain on the basis of a like attitude, for no other reason than because it has been so commanded by men. On this score it is the Jews who are the more deserving of forgiveness.

He closes with the statement that the view that no one is ever obligated to fast on the basis of the Law of the Gospel is false.[502] I have not said this anywhere. But no one is obligated on these or those days because it was so prescribed by Christ, a point he has himself made often from Augustine.[503] Dietary regulations can in no way be proven from the Gospel, nor from the letters of the apostles. Meat from sacrifice to idols is quite another matter, and is not relevant here.

He goes on to argue against my quotation of the text 'the kingdom of God is not food and drink.'[504] Therefore one who would gorge himself, who would drink until he vomits, would not be sinning![505] But the same thing

* * * * *

this was done at the instance of the Holy Spirit, and therefore the forty-day fast (ie Lent) is not a tradition of human origin.

499 Pio's discusses Matt 15:11 in *XXIII libri* 90ʳo–91ʳ.

500 This is the point Pio makes with support from quoted patristic exegesis in *XXIII libri* 90ᵛɴ–91ʳ.

501 Pio *XXIII libri* 91ʳɴ allows, following Titus 1:15, that for Christians 'all things are clean for the clean,' and that the Mosaic law which forbade the eating of unclean foods is abolished for them.

502 Pio ends this part of his argument (*XXIII libri* 91ʳɴ) by allowing that, while Christians are free of the Mosaic law, 'when you go on to infer that, therefore, one is never bound by the law of the Gospel to fast or to abstain from any type of food, this I deny you, for it does not follow from the facts.'

503 Pio *XXIII libri* 86ʳᴅ quotes Augustine's Ep 36.25: 'In the evangelical and apostolic literature [...] I see that fasting is commanded. But I do not find a command of the Lord or of the apostles determining on what days one should not and on what days one should fast' (CSEL 34 54:21–55:2); Pio returns to this text on *XXIII libri* 87ʳᴇ.

504 Pio *XXIII libri* 91ᵛᴘ–92ʳ takes up Erasmus' quotation of Rom 14:17 in *De esu carnium* 295–6 ASD IX-1 30.

505 Reacting, perhaps, to Pio's caricature in *XXIII libri* 93ʳs of Erasmus' position: 'What are you saying? If your view is to be approved, then "let us eat and drink, for tomorrow we die,"

can happen with fish, and in drinking water if one drinks to excess and injures his health. Paul gives his approval to every kind of food, taken with sobriety, and for the body's need, with thanksgiving; he does not approve of intoxication and drunkenness of any kind.

Pio stopped talking some time ago – this is shown by the completely different style.

He says that I am bringing back the sect of Epicurus, in spite of the fact that I everywhere praise sobriety! He says that I learned this from Jovinian, whom I have never read![506] What I write I learned from Paul. In fact, Jovinian did not speak out against the dietary rules prescribed by the church, since a universal rule did not yet exist; rather, he condemned those who abstained from more refined foods on their own initiative.

MONKS[507] [FROM BOOK 5][508]

Association of topics leads the fellow on from fasting to monks, who are always fasting, and from them, suitably enough, to rituals.

In the *Life of St Jerome* I wrote that the type of monk which that era knew was quite different from the type we see today.[509] And here he draws on the letter of Jerome to Eustochium[510] to prove to us that before the time of Jerome there were many swarms of monks in Egypt and in the deserts of Syria![511] Who is denying this? Only let him prove that those monks were the sort with which the whole world is now filled. In fact, I am not talking there about monasticism as a whole, but about that which Jerome embraced; it is that sort which I say was not like ours.

* * * * *

as Scripture says [Isa 22:13; 1 Cor 15:32]. Let us indulge the stomach and gluttony while we live, since fasting and continence are not meritorious, are not acceptable to God, for it was not he who established them, but human superstition and the tyranny of little men!'

506 MARGINAL NOTE: 'Stupid insults.' Pio charges in *XXIII libri* 93[r]s that Erasmus is reviving the doctrines of 'the Epicurus of the Christians,' Jerome's epithet for Jovinian in *Adversus Jovinianum* 1.1 PL 23 211A.

507 1531 UPPER MARGINS: 'On Monks' on pages 96–116 [= 192–209 in this volume]. Pio's book 5, entitled simply 'On Monks,' occupies *XXIII libri* 93[v]-103[v]. Pio opens this book with two quotations from the *Vita Hieronymi*, but later offers a collection of excerpts (see 198–207 below).

508 MARGINAL NOTE: 'From book 5.' The immediately following words allude to Pio's transitional remarks at the end of book 4 (*XXIII libri* 93[v]s).

509 Pio *XXIII libri* 93[v]s quotes *Vita Hieronymi* 309–20 *Erasmi opuscula* 145 / CWE 61 29 with many omissions and inversions, acknowledging that 'out of a concern for brevity I have gathered only the main points, but in your own words.'

510 Jerome Ep 22 CSEL 54 143–211; Pio *XXIII libri* 93[r]-94[r] quotes Ep 22.1–2 and 4–8 CSEL 54 197:16–198:7 and 198:14–200:9.

511 Pio's point on *XXIII libri* 94[r]B.

Even granting that what I said applies to all monks, did they have back then so many distinctions of garment, colours, cinctures, and footwear?[512] Did they have a rule with papal approval? Were there countless regulations? Were the bonds of their vows insoluble?[513] Were there the slavery, floggings, prisons, and even the right to execute which the Franciscans have without a danger of going against the rule? Were there so many anathemas of the popes, so many exemptions, privileges, and indults? They were called monks then, not because of their vows, but because of their withdrawal from the world;[514] they were fleeing the savagery of persecutors.

There were among them some who lived in a community under fathers and deans,[515] and they were not held to their duty by rebukes, blows, chains, and threats, but by good example, words of comfort, and fraternal admonitions. This is evident both from the very title of *father*, and from the very words of Jerome, who terms that way of life an alliance.[516] He says that, when 'they detect one who is not so fervent, they do not reproach him, but, pretending that they do not know this, they go to see him very often, and by beginning first themselves, they invite him to pray, rather than compelling him to pray.'[517] And this is the passage that Pio quotes in his own support. The prudent reader realizes how many aspects I could mention in which the monks of these times differ from the ancient monks.

And it is not the monks of this era that I am condemning. For I go on to remark about virgins, free in time past, but now enclosed behind bars:[518] 'I say this not because I condemn the common practice of the age, but because I grieve that Christian piety has deteriorated to the point that

* * * * *

512 Erasmus details and mocks these distinctions in his colloquy Ἰχθυοφαγία 'A Fish Diet'; see LB I 801E–802A / ASD I-3 523 1031–9 / CWE 40 705.

513 In a scholion on Jerome's letter to Rusticus Monachus Erasmus points out that Jerome makes no mention of the three solemn vows, but praises the monks' study and withdrawal to remote regions (HO 1 52D).

514 See Erasmus' note glossing the word *singularitas* in the edition of Jerome (HO 1 19ʳ): 'From this the Greeks took the term *monachus*, as if to say "alone" and "single." But the name was derived from the fact that once certain devout men, partly because they would not endure the blasphemies of the pagans against Christ, partly because they were harassed by the enemies of the Christian faith, fled into the vastness of the wilderness and lived there, some in communities, and these we call coenobites, others apart, and these we call anchorites.'

515 An allusion to Jerome Ep 22.35 CSEL 54 197–8

516 The term *confoederatio* occurs in Jerome Ep 22.32.1 CSEL 54 197:16, quoted by Pio XXIII libri 93ᵛA.

517 Jerome Ep 22.35.5 CSEL 54 199:10–12, quoted by Pio XXIII libri 94ʳA

518 Pio XXIII libri 93ᵛA quotes *Vita Hieronymi* 412–15 *Erasmi opuscula* 148–9 / CWE 61 31: 'This was far different from the situation of women today, who are shut up behind iron grillwork like untamed beasts, etc.'

the purity of virgins must be secured through the constraint provided by iron bars and prison walls.'[519]

So what is it that scandalizes Pio here? That I said that monasticism is something novel?[520] But I nowhere said this. When I say 'type' [*genus*][521] I am indicating its non-essential features. Unless, perhaps, he who has worn himself out with Aristotle understands 'type' to mean a *genus* which embraces *species* under it. In the same way a certain Dominican recently made the assertion that I am indefensible because somewhere I single out for reproach the base type of mass-priest, meaning the ignorant and those who offer mass only with a view to income.[522] He said: 'He is including all of them when he uses the term *genus*.' He had learned nothing but *genus* and *species*, so he deserves pardon. But Pio has read in Virgil 'this type of man' [*quod genus hominum*], etc,'[523] where a quality is understood, not the *species* of dialectic.

* * * * *

519 *Vita Hieronymi* 415–18 *Erasmi opuscula* 149 / CWE 61 31; Pio had broken off his quotation (see preceding note) just before the words Erasmus quotes here.

520 *novitium*; see Pio XXIII *libri* 93ᵛA, 94ʳB: 'First, so you can deprive monasticism of its antiquity ... and urge that it is something novel and artificial [*rem nouitiam et factitiam*] ... Therefore the monastic practice of our time is not novel [*nouitium*]'

521 In *Vita Hieronymi* 326–7 *Erasmi opuscula* 146 / CWE 61 29 Erasmus uses the expression 'that type of life.'

522 The identity of this Dominican has not been established and his criticism of Erasmus as here quoted could have been made in a sermon reported to Erasmus. Johannes Dietenberger, a German Dominican critic of Erasmus, defended the practice of priests receiving recompense for celebrating mass; see his *Fragstuck an alle Christglaubigen* (Frankfurt am Main 1529): 'Das vierzehent fragstuck: Ob sich auch gebuir, umb verdienst eyner gestifftenn oder ander presentz und nutzung mess auch in theutscher sprach zu halten' (sig o4ᵛ–p1ᵛ), but he makes no mention of Erasmus in this section. We are grateful to Hermann Immenkötter for having provided the text of his critical edition of this work for us to consult prior to its impending publication in *Corpus Catholicorum*. The other likely candidates as published authors of such a statement would seem to be Eustachius van Zichem OP and Ambrosius Storch (Pelargus) OP. The quotation cited is not, however, to be found in the 1531 edition of Eustachius' *Erasmi Roterodami [Enchiridion] Canonis quinti interpretatio* (Antwerp: Wilhem Vostesmann); see the modern edition introduced and annotated by Joseph Coppens in Verhandelingen van Koninklijke Academie voor Wetenschappen. Letteren en Schone Kunsten van Belgie. Klasse der Letteren 37 number 75 (Brussels 1975). Storch provided numerous critiques of Erasmus' writings and advised him on how to respond to the *Declarationes* of the Sorbonne during the years 1529–33 when they both resided in Freiburg im Breisgau. In his 1532 work *Conflictationcula Hieroprepii et Misoliturgi de ratione sacrificii Missae* published in his *Opuscula* (Cologne: Joannes Gymnicus 1534) 194–215, Storch defended the large number of priests celebrating masses (206). Unfortunately, we have been unable to consult his criticisms of Erasmus in his *Bellaria* (Cologne: H. Fuchs 1539); see Rummel *Catholic Critics* II 55–9.

523 *Aeneid* 1 539: 'quod genus hoc hominum? ...'

Of old, all those who had withdrawn were called monks, and every hermitage was called a monastery, and they called the solitary life monastic.[524] Augustine made such a withdrawal, but he did not, however, become a monk, at least the sort that exists now. So too the very many others who left this type of life would hardly have done so had they been bound by solemn vows as monks are now. To whom did the blessed Anthony[525] vow obedience? To whom did Jerome or Basil?[526] To whom did Hilarion?[527] To whom did Rufinus,[528] a very wealthy man and yet a monk? Not even Jerome himself would have been poor had the barbarians not plundered his homeland.[529] Nor did all the monks in Egypt and Syria have a common life. This is clear from the *Life of Hilarion*,[530] who visited the dwelling places of monks. If some of them had not owned property, the story of the stingy and the generous monks[531] would not be told. They owned gardens, vineyards, and farms from which those who wished collected rent. But what need is there to report these details from antiquity? Neither Benedict nor Francis founded the type of monasticism that exists now, nor were the Carthusians in their original foundation the way they are now.[532] They themselves know very well what I mean. And so, in this

* * * * *

524 Erasmus provides a thumbnail sketch of the varieties of monasticism in *De concordia* LB v 484A–C.

525 St Anthony Abbot (the Great, d c 355), the archetypal hermit, made famous by the influential biography by Athanasius

526 St Basil the Great (d 379); Erasmus mentions Basil's particular contribution to monastic life in *De concordia* LB v 484C; Erasmus had editorial oversight of an edition of Basil's works published by Froben in 1532 (see Allen Ep 2611 intro).

527 St Hilarion of Gaza (d 396), an early monastic figure whose *vita* was written by Jerome

528 Rufinus of Aquileia (d 410), Jerome's sometime friend and eventual antagonist (or victim)

529 See *Vita Hieronymi* 164–6 *Erasmi opuscula* 140 / CWE 61 25.

530 A reference to Jerome's *Vita Hilarionis* 15–17 AASS 9 October 51F–52A

531 Jerome *Vita Hilarionis* AASS 9 October 52A–B

532 In his study 'Carthusians during the Reformation Era: *Cartusia numquam deformata, reformari resistens*' *Catholic Historical Review* 81 (1995) 41–66 Dennis D. Martin provides a brief description of the evolution of the order. The major shifts from the practice of its original foundation would include the founding of houses in the environs of major cities (44) and the 'Copernican shift' identified with the activities of corresponding with leading humanists, lending out books, building lavish houses, and allowing lay visitors, even women (45). Martin suggests that Erasmus' urging Carthusians to restrict themselves to contemplation was related to the literary attacks on him by Cousturier (64–5). The Charterhouse of St Barbara in Cologne was noted for its contact with the outside world; see Gérald Chaix *Réforme et Contre-Réforme Catholiques: Recherches sur la chartreuse de Cologne au XVIe siècle* 3 vols, Analecta Cartusiana 80 (Salzburg 1981) I 48 (important visitors), 49 (alms to the poor), 62 and 82–94 (emphasis on the intellectual life), 73–5 (giving spiritual advice to women, notably Beguines), 94–9 (publishing

passage the antiquity of monasticism is not denied nor is its way of life criticized.

But they obeyed their fathers, he says.[533] I agree, but they did so willingly. They kept the vows of a coenobite.[534] Yes, but of their own free will. They were shabbily clothed.[535] But willingly. They did not retain anyone by force if he wanted to leave. Those who had withdrawn in time of persecution returned home once it had abated. If duty to parents or another serious reason called the monks back, the fathers themselves urged them to go.

But what about the solitaries,[536] the type of monk that, perhaps, Jerome and Rufinus were? To whom did they make their vows? Whom did they obey? To whose control did they hand over their possessions? And yet these monks were of an approved type, for Jerome condemns the third.[537]

In the case of the women for whom Augustine wrote the rule that men use now,[538] if they were unwilling to be obedient, they were thrown out of the community. That was the ultimate punishment. Far from being confined by iron bars, they could go out to the baths, but by twos and threes.[539]

So I am going to overlook both the insults and the irrelevant arguments that Pio hurls at me here. I do not begrudge the monks their antiquity.[540] If they like, they could take it all the way back to Adam,[541] who

* * * * *

on its own press works of spirituality and controversialist theology). Apart from his controversies with Cousturier and John Batmanson, Erasmus seems to have had good relations with the Carthusians. In 1526 he successfully appealed to the prior-general, the Dutchman Willem Bibaut (1521–36), to have Cousturier silenced; see Ep 1196:58n, Allen Ep 1687:50–147; the entry on Bibaut in CEBR I 145. See also Sigrun Haude 'The Silent Monks Speak Up: The Changing Identity of the Carthusians in the Fifteenth and Sixteenth Centuries' *Archiv für Reformationsgeschichte* 86 (1995) 124–40.

533 These are not Pio's very words, but an allusion to his summation in *XXIII libri* 94^{r}B of correspondences between the monks of Jerome's time and contemporary monks.

534 Pio *XXIII libri* 94^{r}B speaks of how the early monks lived a common life.

535 Pio *XXIII libri* 94^{r}B claims the early monks wore the prescribed garb.

536 *anachoretae*; Jerome discusses this class of monk in Ep 22.37 CSEL 54 200:10–201:4.

537 The undisciplined types whom Jerome calls the *remnuoth* in his account of them in Ep 22.34 CSEL 54 196:16–197:13, quoted in part by Pio *XXIII libri* 94^{v}C

538 In his edition of Augustine (10 volumes, Basel: Froben 1528–9) Erasmus wrote in his *Censura* on the *Regula tertia* (1 591): 'This rule is like Augustine both in its viewpoints and its style, and, indeed, declares its author by its devout and civilized humanity, although it is likely that it was written not for clerics but for women who were living, gathered in a community, under the direction of Augustine's sister.'

539 The *Regula tertia* (in Erasmus' edition 1 594) provides: 'And they should not go to the baths, or wherever they may need to go, except by twos and threes.'

540 Pio *XXIII libri* 93^{v}A charges Erasmus with writing 'first of all to deprive them of their antiquity (on which depends, to a great extent, the dignity and influence of any class).'

541 Pio *XXIII libri* 95^{r}D cites Elijah and Elisha as the first monks, and a little later (95^{v}F) argues that the holding of property in common existed in humanity's prelapsarian state.

lived as a solitary in paradise, in chastity, under the rule of God as abbot, and suffered the same punishment as the original Carthusians.[542]

Holy virgins used to make their vow solemnly.[543] I am not denying it. But they made only a vow of chastity. Otherwise they lived where they wished. They disposed of their possessions as they saw fit. But who has read that any of those who married after making this vow was thrown into chains? Who had their marriages voided? And yet Ambrose does complain that a great many of them had withdrawn from the bosom of the church.[544]

He imagines that I seethe with a remarkable hatred and malice towards the monks.[545] To prove this he alleges a distorted quotation from my

* * * * *

542 Because of his sin of disobedience Adam was expelled from paradise and made to earn his food by the sweat of his brow until he died (Gen 3:17–24). The parallel that Erasmus attempts to draw between Adam and the first Carthusians is difficult to understand. Perhaps he is referring to the fact that the first Carthusians initially abandoned their hermitage to follow their founder, St Bruno, to Rome, to which he had been called by Urban II in 1190, but then returned within the year to the Chartreuse mountains in Dauphiny to resume their solitary life. Was this advising of a pope, and other subsequent intellectual activities, seen as eating from the tree of knowledge that was forbidden to Adam in the Garden of Eden (Gen 2:17, 3:17)? Or perhaps Erasmus interprets the destruction of the early hermitage by an avalanche on 30 January 1132 as a punishment from God. A new hermitage, however, was soon erected near the original site. See Cyprien-Marie Boutrais *The History of the Great Chartreuse* trans E. Hassid (London 1934) 20–1, 27–9. The establishment of houses near urban centres (*tentations urbaines* to abandon the *appels au désert*) did not begin until the middle of the thirteenth century and continued, after a brief pause, in the fifteenth and early sixteenth centuries in France; see Gérald Chaix 'Les chartreuses françaises et la Réforme' in *Die Kartäuser und die Reformation: Internationaler Kongress vom 24. bis 27. August 1983* 2 vols, Analecta Cartusiana 108 (Salzburg 1984) 203–14 especially 204; and Sigrun Haude 'The Silent Monks Speak Up: The Changing Identity of the Carthusians in the Fifteenth and Sixteenth Centuries' *Archiv für Reformationsgeschichte* 86 (1995) 124–40. According to Martin 'Carthusians during the Reformation Era' (cited in n532 above), the rural character of the Carthusian houses nonetheless continued to predominate in the order (44). Polydore Vergil suggested that the significant change in the lifestyle of the Carthusians over time resulted from the growth of their wealth; see Beno Weiss and Louis C. Pérez *Beginnings and Discoveries: Polydore Vergil's 'De Inventoribus Rerum': An Unabridged Translation and Edition with Introduction, Notes and Glossary* Bibliotheca Humanistica et Reformatorica 56 (Nieuwkoop 1997) 395.

543 Pio *XXIII libri* 97ᵀH cites the account of the public vows of consecrated virgins described by Ambrose in *De institutione virginis* 17 PL 16 330D–334B.

544 Ambrose, bishop of Milan (d 397); Erasmus had published Ambrose's *Opera omnia* (Basel: Froben 1527). Pio *XXIII libri* 97ᵀH quotes from chapter 5 of the pseudo-Ambrosian *De lapsu virginis consecratae* PL 16 372A–373A.

545 Pio *XXIII libri* 98ᵛL: 'For so great is your spite, so great is your heart's uncontrolled malice towards the monastic order, that you give a contrary and distorted interpretation to everything so long as you can disparage the monastic order, as in the scholia on the letter to Eustochium [where] you interpret what Jerome relates against the virgins of various heretical factions, particularly the Manichees, as having been said against the

argumentum to the letter of Jerome to Euchtochium.[546] My actual words are as follows: 'And as a good teacher should, he not only points out the model that ought to be followed, but also discloses and reveals the tricks and disguises of the virgins, monks, and clerics who, under pretence of chastity, are slaves to the belly and avarice. Moreover, in his sketches of these people he makes sport of them rather more freely than the squeamish ears of some can bear. Thus he writes elsewhere that in Rome (it was in Rome that he wrote it) this little book was stoned.' What is there here to cause Pio to rave? He says: 'He did not write those words against the virgins and monks of the church, but against heretics.'[547] In fact,[548] whoever reads the letter will realize that Jerome is criticizing the way of life of the church's virgins and monks, the corrupt ones. Unless, perhaps, there were at that time no virgins whose lives were falling below their vows! Thus I can only wonder what came into Pio's mind here, unless, as I suspect, he had not even read this passage.

At last he comes to a really serious defence of the monks, and he reports some things from my books, as briefly, he says, as possible,[549] but in fact as slanderously as possible. He does so in a way that is quite annoying. He repeatedly fails to give citations so the passages could be checked. He also often gives the incorrect citations that were supplied to him. He frequently provides abridged quotations in cases where the preceding or following words preclude his slander. Finally, it happens fairly often that he fails to understand the passage. Now what could be a greater nuisance than to answer rubbish like this, particularly since I have already replied to most of it in published books. If he did not have time to read my apologias himself, he should at least have delegated this job to his employees, of whom he had quite a few.

So, first I will list the passages which he excerpted as being most obviously impious, so that by fitting these bits one to another he could put together a mosaic of slander.

* * * * *

devout virgins, monks, and clergy of the Catholic church, for you say: "He seems in this letter to direct the barbs of his satire against virgins and monks and clerics and to depict their way of life rather sharply [...] in their true colours etc" [CWE 61 192].'
546 That is, of Jerome's Ep 22, in HO 1 60ʳ / CWE 61 155
547 A paraphrase of Pio's words, quoted in n545 above.
548 MARGINAL NOTE: 'Lie'
549 See Pio *XXIII libri* 98ᵛL: 'You do the same thing in countless passages, but it is impossible to examine and run through all your countless statements, for there would be no limit to the book nor end to the toil'; and the passage from *XXIII libri* 99ʳM quoted in n554 below.

PASSAGES[550]

1 / From the *Supputatio* he quotes: 'What does "artificial religion" mean but one that is deceptive and contrived, as were the pagan religions that were human inventions?'[551] From this he draws the foolish conclusion that I am calling monasticism something artificial, whereas I am criticizing the improper use of this word by Béda, who terms the various types of monks 'artificial religions.' Thus he concludes his censure number 68 on the paraphrase of Luke: 'For he who attacks the status of holy artificial religions is a disciple of the heretics described in the earlier passages against Lefèvre, not a disciple of Christ.'[552] Thus Béda. But I draw the following conclusion:[553] If it would be absurd to speak of artificial devotion, then it would be absurd to speak of artificial religion, since devotion and the religion are the same thing. And Pio brags that these words are so obvious that they do not need anyone to refute them.[554] How are words obvious, poor fellow, that are misunderstood by the very one who is carping![555]

* * * * *

550 Pio *XXIII libri* 98ᵛ–99ʳ quotes fifteen passages from Erasmus' works: two from *Hieronymi opera*, two from *Supputatio*, eight from Ep 858 and *Enchiridion*, one from *Institutio principis christiani*, and two from the *Annotationes in Novum Testamentum*.

551 MARGINAL NOTE: 'Criticizes things he did not understand.' Pio *XXIII libri* 98ᵛL quotes two passages with slight changes and omissions from the *Supputatio* LB IX 615D, the first of which Erasmus quotes here from Pio's version. In his response to Carvajal Erasmus defended his statement that monasticism is not piety (LB V 65C / CWE 66 127), a defence he does not repeat in replying to Pio. See Erika Rummel '*Monachatus non est pietas*: Interpretations and Misinterpretations of a Dictum' in *Erasmus' Vision of the Church* ed Hilmar M. Pabel, Sixteenth Century Essays and Studies 23 (Kirksville, MO 1995) 41–55, especially 50.

552 Erasmus quotes from Béda's censure F on proposition 68 in his *Annotationum ... in Jacobum Fabrum Stapulensem libri duo et in Desiderium Erasmum Roterodamum liber unus ... in Paraphrases Erasmi super eadem quatuor Evanglia ...* (Cologne: Ex officina ... Petri Quentell 1526) 253ᵛ, reporting *sacrarum* for Béda's *sacrum*. 'Religion' here means, of course, the way of life of the religious orders, and by describing them as 'artificial' Béda no doubt meant devised and structured according to a particular heroic religious idea.

553 In the *Supputatio* LB IX 615D Erasmus had written: 'But what are the "artificial religions" Béda is telling me about? Is there some such thing as artifical devotion? For religion is the same as devotion. But what does "artificial religion" mean but one that is deceptive and contrived, as were the religions of the pagans that were human inventions? If he thinks this, as I believe he does, about the various orders of monks, he is damning with this so honorific name the very ones he wants to praise.'

554 Towards the conclusion of the Erasmian passages that he reviews in this section Pio observes (*XXIII libri* 99ʳM): 'I have reviewed as briefly as possible these few passages from the multitude that you have written against the order and way of life of religious. They are so clear that they do not require a commentator,' reading *enarratore* for *enerratore*, for which Erasmus substitutes *confutatore*.

555 MARGINAL NOTE: 'He doesn't understand what he's criticizing'

He says: 'You mean that monasticism is an artificial entity like the religions of the pagans, and that its way of life is not a work of devotion, but a kind of theatrical and ludicrous contrivance which should be done away with altogether.'[556] The more Pio develops this theme, the more precise he is in his reply, the more ridiculous is he to those who read my work, rather like Laches in Terence's *Eunuch* when he breaks into the house of Thais.[557] But Laches provoked the laughter of Pythias alone, while Pio serves himself up as a laughing-stock for the whole world. I am not angry with Pio, but embarrassed for him that such stuff is coming out under his name in printed books. I am indignant at the people who dumped texts they did not understand on a gullible and, it seems, vainglorious man. And yet he is the one who examines the individual words and syllables of my works and compares them down to the last detail.[558] He is the one who says that there is no need to read my apologias.[559] And yet if he had read them, he would not have served himself up as a laughing stock for all the learned.

2 / From the same work he excerpts the following: 'Since there is only one Christian religion, it is not far from blasphemy to speak of a variety of religions. This is really what it means to tear the seamless robe of Christ.[560] But now those who are called "religious" use the expression "we religious." And if anyone asks about a person who has not entered monastic life "is he religious?" they say "no, he's not." In the latter instance they are insulting to their neighbours, in the former instance they are arrogant.'[561] If there is one faith, one baptism,[562] one Spirit, one church, one God, would it not be the one who wishes to separate these who is blasphemous, would he not be the one who is rending the seamless robe of Christ? Christian religion includes all of these.

But by 'religions' they mean forms of religious life.[563] I know this, but which of the ancients dared to use the expression 'religions' when he was

* * * * *

556 Pio XXIII *libri* 99[r]M

557 There is some confusion about whether the name of the aged housebreaker is Laches or Demea; Erasmus assumes it is Laches, whose intention to barge into Thais' house is announced at *Eunuchus* 996; Pythias' laughter at his expense and Parmeno's occurs in lines 1002ff.

558 A sarcastic allusion to a remark by Pio in his earlier work; see Erasmus' *Responsio* 28 and n129 above.

559 See n549 above.

560 John 19:23

561 Pio XXIII *libri* 98[v]L quotes with some omissions from the *Supputatio* LB IX 615E; Erasmus here quotes Pio's version.

562 Eph 4:4–5

563 A summary of Pio's argument on XXIII *libri* 99[v]N

speaking of the association of Christians? Paul used the plural of *church*, but never of *faith*.[564] Let us allow them to call themselves 'religious'[565] (just as they call themselves 'dead to the world'!), but why do they say that others are not? But what has all this to do with the actual condition of monks, for this dispute is about a word incorrectly employed?

3 / He took the following curtailed quotation from the *Letter to Volz*: 'They live in leisure and are fed by the liberality of other people, possessing in common what without effort on their part has come to them (of wicked monks I say nothing for the present).' I compare monks to townspeople, and show that there is not so great a difference as the monks say there is: 'Monks live in leisure and are fed by the liberality of other people, possessing in common what has come to them without effort on their part (of wicked monks I say nothing for the present).'[566] It is the immediately following words that Pio left out: 'These,' that is, citizens, 'each according to his means, share what they have won by their own industry with those in need.'[567] I am not condemning monks because with the support of the generosity of others they live more grandly than the citizens themselves, but I am only comparing one set of people to the other. 'But,' he says, 'you deny that you are talking about the corrupt ones.'[568] I admit it. 'Therefore you are talking about the good ones.' I do not deny it, people being what they are at this time, but I am comparing good people to good people. For if I had been thinking of the corrupt ones, I would not have said 'fatten themselves up,' but 'wax wanton, indulging their sex and lust.'

'But,' he says, 'the monks do pray.' Yes, and devout layfolk pray even when they are not praying, when they are supporting wife and children by their labour, when they educate their families properly, when they deny themselves so they can share something with the needy. It is this sort of townsman I compare to the monk. In a similar way I compare a well-behaved city, one that obeys its bishops and magistrates, one in which, as Aristotle

* * * * *

564 A response to Pio's appeal to Pauline usage in *XXIII libri* 99^{v}–100^{r}o; see also *De concordia* LB V 475A.

565 A response to Pio's appeal in *XXIII libri* 100^{r}o for acknowledgement of degrees of perfection among Christians

566 Erasmus' prefatory letter addressed to Paul Volz, prefixed to the July 1518 edition of the *Enchiridion* (Ep 858 / Holborn 3–21 / CWE 66 8–23); Pio *XXIII libri* 98^{v}L quotes Ep 858:596–9 (Holborn 19:25–7).

567 Ep 858:599–600 / Holborn 19:27–8 / CWE 66 22

568 Pio *XXIII libri* 98^{v}L: 'If you did not make this statement about the corrupt ones, then it was about the upright ones'; words to the same effect are repeated on *XXIII libri* 100^{v}P.

says,[569] a mutual charity brings about a shared enjoyment of all things, to a large monastery. I compare them, I do not say that they are equal. I do not say that there is no difference between them, but not very much difference.

Moreover, the considerations on the basis of which some now aggrandize their vows, which they consider more important than baptism, are not articles of faith, but only the doctrines of the monks. Nor is their case helped much by the many commonplaces they mention about the marvellous way of life of the ancient monks and about the viciousness of worldlings, because just as that type of monk long ago disappeared, even so the condition of the world is now, in fact, quite different from what it was then. No pagan persecution rages now. Courtiers are not swollen with the haughtiness and over-refinement of the pagans. One may more readily be truly devout among layfolk than among monks. It is the monks and laypeople of these times that I compare, not their types of life.

4 / He quotes as being from the *Letter to Abbot Volz*[570] a passage that is really from the *Enchiridion*.[571] This is a shameful bit of thoughtlessness, since the remark would not apply to an abbot. The words are something like this: 'This I have written extempore here, as well as I could, to forward your pious enterprise. I did so with all the more alacrity because I was somewhat afraid that you might fall into the hands of that superstitious fraternity among the religious, etc.' What sacrilege is there if I advise a friend who does not know himself very well[572] not to throw himself rashly into a type of life from which he might later try to extricate himself in vain? Pio says:[573]

* * * * *

569 See n578 below; and see citations of Aristotle in *Adagia* I i 1: *Amicorum communia omnia* 'Between friends all things are common' LB II 13F–14F / ASD II-1 84–6 / CWE 31 29–30; apropos of the application of this adage to monastic life, see Erasmus' note on the term *coenobitae* in HO 1 61ᵛ / CWE 61 191: 'From κοινός meaning "common" and βίος meaning "life" or "way of life." There is a tradition that the practice derived from Pythagoras, who is well known for the adage: "Between friends all is common."' Pio (85ᵛF) criticizes what he takes to be Erasmus' derivation of communal life from the teachings of Pythagoras.

570 MARGINAL NOTE: 'Misleading citation.' Pio XXIII *libri* 98ᵛL introduces the quotation 'in the book to the same Volz.'

571 Pio XXIII *libri* 98ᵛL quotes from the *Enchiridion* LB V 65B / Holborn 134:33–6 / CWE 66 126 with some omissions and substitutions; Erasmus here quotes Pio's version. Pio quotes the same passage again in a fuller version in XXIII *libri* 100ᵛP.

572 See Ep 858:523–5 / Holborn 17:22–3 / CWE 66 20.

573 After paraphrasing John Chrysostom in XXIII *libri* 100ᵛQ–101ʳ, Pio continues at 101ʳQ: 'For you have heard with what praise the holy Father [Chrysostom] exalts the monk's vocation and urges the embracing of it, and how he expresses a higher regard for that way of life than for the common life of Christians who live in cities. But you are trying

'You are discouraging[574] entry into monastic life.' That is not true! I am discouraging doing so precipitously. The words of the *Enchiridion* that follow show this: 'I personally do not urge you to adopt it, nor yet do I urge you against it.'[575] What is it, then, that I am advising? That he not throw himself headlong into a community of wicked or superstitious monks. At that point I provide a portrait of them when I say: 'partly pursuing their own personal interests, and partly out of great zeal but not according to knowledge, [they] scour land and sea, and whenever they find anyone abandoning wicked courses and returning to a saner and better life, immediately attempt to thrust him into a monastic order by means of the most impudent urging and threats and cajoleries, as if Christianity did not exist outside the monk's cowl; and then, after filling his mind with mere quibbles and thorny problems that nobody could solve, they bind him to some petty observances of human origin.'[576] If Pio denies that there are anywhere monks like this, he will be refuted at once by a myriad of specimens. But if, in fact, there are, and alas there are all too many, then I was right to warn my friend to avoid monks like this, and, if he is drawn to monastic life, to join up with the really devout. For I go on to say: 'Associate with those in whom you have seen Christ's true image.'[577] You see, O reader, what is involved in constructing slander from excerpts.

5 / From the preface to Volz he alleges the following:[578] 'What else, I ask you, is a city but a great monastery?' and again: 'Why do we so closely confine the professed service of Christ, which he wished to be as wide open as possible?' and shortly after this: 'So they need not fear that the sect of Essenes[579] may not spread, in all this variety of men and minds,

* * * * *

to persuade your friend of the opposite, and discourage others from developing a zeal for that life, for you detest it.'

574 *Deterres* 1531 *Deterret* LB
575 LB V 65C / Holborn 135:9–10 / CWE 66 127
576 LB V 65B–C / Holborn 134:36–135:6 / CWE 66 127
577 LB V 66A / Holborn 135:12–3 / CWE 66 127
578 Pio XXIII *libri* 98ᵛL–99ʳM quotes Ep 858:591–4, and 519–21 / Holborn 19:22 and 17:16–19 / CWE 66 22, 20.
579 Erasmus uses 'Essenes' as a term of opprobrium in the letter to Volz (Ep 858:519); Pio XXIII *libri* 94ʳA–B quotes Jerome Ep 22.35, where they are said to be analogous to Christian monks on the basis of accounts of them by Philo and Jospehus. Erasmus discusses the Essenes in a note on Jerome's Ep 22: 'Here the text had been corrupted to read *esse nos debere* "we ought to be," instead of *Essenos* "the Essenes." The Essense were a people of the interior of Judaea situated towards the west. They were a new school of philosophers: they had no association with women, they lived on dates, they had contempt for money. Their numbers grew with the influx of refugees. No one was admitted, how-

which means that nothing is too absurd to seem attractive to many.' Here Pio says:[580] 'You reveal your hatred of the way of life of the ancient usages when you say: "Why do we so limit the types of commitment to Christ, whereas he wanted this to be completely open?"' How ill deserving of the name of Pio are they who furnished him with such uninformed rubbish! Do I really show my hatred for the monks' way of life there? In fact, I am criticizing certain arrogant monks who barely consider the rest of men to be Christians. My words are these: 'And yet because of some observance ... they claim so much sanctity for themselves that compared with them they consider other people hardly Christians. Why do we so closely confine the professed service of Christ, which he wished to be as wide open as possible?'[581] The haughtiness of some monks is quelled here, but the way of life is not condemned. For devout monks do not exalt themselves above any layperson; rather, they revere all who live devoutly in Christ, and place them before themselves. By 'Essenes' I mean the superstitious ones who have nothing of sanctity but an unchaste celibacy. It is the ones like this who most especially lie in wait for adolescents.

6 / From the same work he recalls the following: 'There will nowhere be a lack of naive youths whom they can lure into their snare, etc.'[582] The words he quotes here were said against superstitious monks who are furious when the true elements of piety are revealed because they are afraid that fewer tuna will swim into their fishponds. I am calming them down ironically. If their goal is only a throng of monks of whatever quality, then there will always be some they can entice with their wicked craft. What has this to do with devout monks who loathe such things?

7 / But he goes on weaving together the fragments supplied him by his staff, this from the fifth rule of the *Enchiridion* (they did not give Pio the citation):[583] 'It does not matter to me that they do not neglect vigils, fasts, silence, little prayers, and all other such observances. I shall not believe that they are in the Spirit until I see the fruits of the Spirit.' What am I getting at with these words? That those who fast or complete the prescribed

* * * * *

ever, who was conscious of any fault, even the slightest. This is pretty much what Pliny has to say, and it is also the description of Josephus and Philo' (HO 1 61ᵛ / CWE 61 191).
580 A paraphrase of Pio *XXIII libri* 98ᵛM–99ʳM
581 Ep 858:589–93 / Holborn 19:18–21 / CWE 66 22
582 Pio *XXIII libri* 99ʳM, quoting Ep 858:527–9
583 Pio *XXIII libri* 99ʳM introduces a paraphrase of the *Enchiridion* LB V 34E / Holborn 80:23–5 / CWE 66 77 with the words *Rursus ibidem*; Erasmus quotes Pio's paraphrase here.

prayers are not necessarily walking in the Spirit, if the fruits of the Spirit are lacking.[584] What are they? Joy, peace, and the rest. In fact, if the fruits of the flesh that I mention there are in fact present, what could be more true than this statement?

8 / He stitches on to this a statement that follows after an interval: 'I am not talking now about those monks,' etc.[585] But he leaves out what is said there: 'But those same monks should not be offended by my words, which are directed at vices, not men';[586] and shortly: 'It is no secret to me that there are many among them who, aided by their literary studies and gifts of intelligence, have had some experience of the mysteries of the spirit.'[587] These words make it clear that I am not speaking about all, but about the many who, even in the most disciplined religious life, either because of ignorance, or naivety, or superstition, assign a greater weight to rituals than to true devotion. For you will find a good number there who would rather die than alter their habits or eat meat, but you will not find so many who do not plot revenge when they are provoked, do not speak abusively of those they hate, who do good to those who deserve the contrary, and so on. I really wish that I were lying about all these points and that those who lay these false charges against me had a better case.

9 / He quotes from *The Christian Prince*:[588] 'Perhaps too, it would be to the state's advantage to limit the number of monasteries.' One who says 'perhaps' is not making an assertion. Nor do I say 'advantage' absolutely, but 'to the state's advantage,' for I am writing to a prince, not to a bishop. But what would be the loss to piety if there were a limit to the number of monasteries, especially since there are so few in which regular discipline thrives? It is about these especially that I am speaking, for there follows at once: 'especially for those whose lives have been far from blameless and who now fritter away lethargic lives in idleness.'[589] Pio left this part out since it did not support his slander.

* * * * *

584 Pio *XXIII libri* 101^{v}R observes that Erasmus cannot observe the fruits of the Spirit among the religious 'because you have become squint-eyed or blind from the darkness of hatred.'

585 Pio *XXIII libri* 99^{r}M quotes from *Enchiridion* LB V 35A / Holborn 81:6–7 / CWE 66 77 with those omissions (and more) indicated by Erasmus.

586 LB V 35A / Holborn 81:9–10 / CWE 66 77

587 LB V 35A / Holborn 81:11–3 / CWE 66 77–8

588 Pio *XXIII libri* 99^{r}M and 101^{v}s quotes the following and a bit more from *Institutio principis christiani* LB IV 598A / ASD IV-1 198:961–4 / CWE 27 267, but with the omissions which Erasmus indicates.

589 LB IV 598A / ASD IV-1 198:962–3 / CWE 27 267

'But,' he says, 'it should be our wish that there would be chanting to the Lord everywhere.'[590] If all are chanting, who will feed them? If the entire body is a tongue, where are the hands and the feet? I will not say anything now of the quality of the chanting of many of them and how pleasing to God it is!

'But Jerome,' so he says, 'wants everybody to be celibate.'[591] You see, whoever supplied this wrote down 'celibate' instead of 'continent,'[592] so you can realize that these are not Pio's words. Everywhere you can find ever so many who are celibate, but very few who are continent.

Is one who suggests a limit, so the commonwealth will not be burdened, condemning monasticism? Every day new monasteries are built, religious life degenerates, the crowds of wicked monks increase. What has this to do with Libanius' urging Julian to eliminate the monks' exemption from military service?[593] Where does Erasmus urge this? In fact, they them-

* * * * *

590 A summary of Pio's argument at XXIII libri 102[r]s

591 Pio XXIII libri 102[r]s: 'For Augustine, and Jerome, following Paul's view, desires that all men be celibate.'

592 In the Pauline passage referred to, 1 Cor 7:5–9, the operative words are in continentiam (verse 5) and non se continent (verse 9).

593 Libanius, the celebrated teacher of rhetoric of Antioch (314–c 393); Erasmus had published three short declamations of his in 1519 (ASD I-1 177–92). Libanius and Julian the Apostate were introduced into the argument by Pio XXIII libri 101[v]–102[r]s: 'But as to your statment in the book in which you instruct the Christian prince, saying that it would be to the state's advantage to limit the number of monasteries ... you borrowed this from the sophist Libanius who caused the emperor Julian to apostasize through his wicked advice. He convinced him of these points, and so Julian promulgated a law that monks should not be exempt from military service, decreeing that their withdrawal was disadvantageous to the state. The emperor Valens also followed this advice, and enacted as stated above.' Pio is referring here to a passage earlier in book 5 (94[r]vB) where, after telling of the vast numbers of monks dwelling at Nitria and elsewhere, he cites Valens' effort to coerce them into military service: 'For this reason not only the apostate Julian, but even the emperor Valens, a Christian, but (it seems) one less than devout, was moved to promulgate an edict compelling monks to do military service, because of which many of them were also beaten with clubs; but he paid the penalty of his great wickedness.' An enactment of Valens recalling to their civic duties idle men who have taken refuge among the monks survives in the Code of Justinian (Digesta 10.32.26). Accounts of Nitrian monks being clubbed to death pursuant to a law of Valens compelling them to do military service are found in the historical sources, notably Jerome's version of Eusebius' Chronicle for the year 375 (Griechische Christliche Schriftsteller, Eusebius VII ed Rudolf Helm and Ursula Treu [Berlin 1984] 248:1–4); and in Orosius' History CSEL 5 515:8–516:10, but there is no mention of Libanius and Julian. A possible connection of Valens' initiative with Julian might have been formed on the basis of a story found in Rufinus' Historia monachorum VII parts 2, 4 ed Eva Schulz-Flügel, Patristische Texte und Studien 34 (Berlin / New York 1990) 289–90. This story,

selves have no desire to be exempt and Pio wants them to make war in defence of their property even more fiercely than laymen.[594]

10 / Here he affixes like a centrepiece a quote from the *Annotations*,[595] but he does not give a citation. In fact it is on Matthew 23: 'Just as there are today among us those who make profession of their holiness before the masses with a variety of novel rituals and colours, etc.' 'You are comparing,' he says, 'monks to the Pharisees who were censured by Christ.'[596] I am comparing them, but the hypocritical ones! For the complete passage is as follows: 'Just as there are today among us those who make profession of their holiness before the masses with a variety of novel rituals and colours, although not all of them are consistent with their costume.' Pio left out this clause because it would preclude his slander. Oh, it was a contrivance of real Christian charity to drive an invalid to undertake this work! But my statement does not specifically pertain to monks.

11 / But then follow the horrific words: 'And I wish that this type of buffoon were not so common among Christians, etc.' But, again, the following words are: 'Not because devotion must not be manifested even in simplicity and cleanliness of toilette and clothing, but because superstition and ostentation should be everywhere absent.'[597] Pio left this out because it would not support his slander. What is it that I am censuring here but those hypocrites-about-town. You meet them everywhere, especially among my countrymen, but really thoughtful people gladly refrain from associating with them. Who fails to understand that this is friendly and polite counsel? In fact it is nothing but a collage of lies!

12 / He quotes from the scholia on Jerome, and does not give the citation. Though he does this all too often, I do not view it as having been done by Pio through guile, to avoid being found out; rather, I suspect that this failing is the result of the negligence of his note takers, whom he should have supervised. But what is it I say there? 'There were two theologians,

* * * * *

which tells of the activities of Egyptian monks on behalf of a brother who was being imprisoned because of his refusal of military service, is set 'in the times of Julian.' We have not been able to determine whether the Libanius-Julian-Valens connection is Pio's original surmise or a feature borrowed from a later historiographical tradition.

594 Pio *XXIII libri* 81^v–82^{r}x

595 Pio *XXIII libri* 99^{r}M quotes the annotation on Matt 23:5 LB VI 118C / Reeve 91–2, with the omission indicated by Erasmus.

596 A paraphrase of remarks by Pio *XXIII libri* 99^{r}M, 102^{r}S, 103^{v}Y. Erasmus makes no reference to Pio's further quotations from the same annotation.

597 *Annotationes* LB VI 118D / Reeve 92

one of them now a bishop, the other a member of that religious order which renders baptismal vows virtually irreligious.'[598] First he complains that I did not dare mention the name of the religious order.[599] (Should he not be praising my self-restraint instead?) What need was there to mention it? Surely the circumlocution does not point to the Dominicans who are, by and large, good fellows, and there are hardly any others less strict. If one wanted to conjecture, the Carthusians would have been the more likely guess.[600] At this point my friend Pio seizes the opportunity and digresses in an encomium of the Cherubic and Seraphic orders.[601] Let him go ahead and say that I had the Dominican order in mind, although it is untrue. Or, am I condemning the order just because I say that someone from that company was hostile towards me? I do not begrudge Pio a vast book, apparently his only goal in writing, but he could have achieved this without slandering me.

He makes the distinction between good monks and bad monks,[602] as if I do not do the same thing in countless places. But some of them cannot

* * * * *

598 We share Erasmus' frustration in not being able to locate this quotation in his scholia on Jerome, both in the original 1516 and in the revised 1524 editions published by Froben. In his scholia on Jerome Erasmus did attack an English Franciscan theologian, not because of his religious profession, but because he interpreted Latin terms in Jerome erroneously while disparaging Erasmus for being a mere grammarian and not a theologian; HO (1516) 4 4^{v}c and 6^{v}c; (1524) 3 11A–B and 15A. According to Rummel *Catholic Critics* I 118, this critic was Edmund Birkhead OFM (d 1518), who in 1513 became bishop of St Asaph (CEBR I 151). According to Richard Schoek (CEBR III 280), however, this critic was Henry Standish OFM, who succeeded Birkhead in 1518 as bishop of St Asaph. The solution to the problems surrounding the identification of the source of the quotation and of the first theologian there mentioned probably lies in recognizing that Pio gave a mistaken reference, misreading his notes and citing the scholia on Jerome instead of the annotations on the New Testament. In his annotation on 1 Cor 15:51 (added in 1522) LB VI 743B / Reeve-Screech 518, Erasmus makes a statement almost identical to that quoted by Pio. The English Franciscan theologian, later a bishop, there alluded to is Standish (d 1535). On his controversies with Erasmus, see Rummel *Catholic Critics* I 122–7; and Schoeck CEBR III 279–80.
599 Pio *XXIII libri* 102^r
600 Because this statement dates from 1516, the Carthusian adversary of Erasmus would appear to be John Batmanson. Thomas More seems never to have revealed to Erasmus the name of this opponent, whose identity has been suggested by David Knowles in *The Religious Orders in England* rev ed (Cambridge 1971) III 469 app 1. On Batmanson's criticisms, see Rummel *Catholic Critics* I 118–19. Erasmus' open controversy with a Carthusian dates from 1525, and this critic was Cousturier. On Batmanson, see CEBR I 99–100; on Cousturier, see CEBR I 353–3. Given the Carthusians' reputation as good religious, one may suspect that Erasmus had another order in mind.
601 Pio *XXIII libri* 102^{r}T assumes that Erasmus has in mind either the Dominicans or the Franciscans; a short encomium of them follows in *XXIII libri* 102^{r}T–102^{v}v; Erasmus gives a thumbnail sketch of his sufferings at the hands of Dominicans and Franciscans in Ep 2205.
602 Pio *XXIII libri* 102^v–103^{r}v

tolerate this; they want all to appear devout and holy! Next he must conclude, amplifying everything with his verbosity, but even so he has brought
forward nothing so far that proves that I condemn monasticism. In fact, I
could quote even more passages from my writings in which I praise monasticism splendidly. It would have been more effective to collect all those passages together if it was his intention to win favour for their orders.

But now hear the leaden dilemma. He says: 'If you approve, why do
you attack it with so many criticisms? But if you disapprove, how can you
say that you approve it?'[603] Nowhere do I express disapproval of monasticism, but I do disapprove of the morals of a great many individual monks.
What if he presses a parent who is rebuking his children with this sort of
dilemma? 'If you hate them, why are you maintaining them? If you love
them, why do you beat them so often?' This is the sort of triumph the
monks celebrate when they are prettily defended by Pio's troops!

[RITUALS] [FROM BOOK 6][604]

After his defence of the monks he quite suitably entered[605] upon an encomium of rituals, assuming quite impudently, but quite in keeping with
his practice, that I condemn rituals in general,[606] in spite of the fact that
I cry out everywhere to the point of hoarseness[607] that I do not condemn
the rituals appointed by the church, but only superstitious rituals of human origin, or those people who place their faith in them and those who go
along with this. I have made reply to slanderers so often about this matter
that it is superfluous to repeat the same points. In fact, what kind of Christian earnestness is it, really, to construct so hateful a slander against one's
neighbour from fragments excerpted piecemeal, neither condescending to

* * * * *

603 Pio's words (*XXIII libri* 103ᵛʏ) '*Si igitur probas, cur tot maledictis insequeris? Si improbas,
    quomodo dicis te illud probare?*' are altered slightly by Erasmus to '*Si probas, cur tot male-
    dictis insectaris? Sin improbas, quomodo te dicis illud probare?*' According to Erasmus, 'A
    dilemma is a horned syllogism, when two propositions are presented in such a way
    that one must run onto one of them' (HO 3 113ᵛD).
604 1531 UPPER MARGINS: 'On Rituals' on pages 116–23 [= 209–15 in this volume]. A note
    in the left margin reads: 'From book 6'; Pio's book 6 (*XXIII libri* 103ᵛA–120ʳz) is entitled
    'An Assertion of the Rituals Received by the Church'; throughout this section the term
    ceremoniae will be translated 'rituals.'
605 An allusion to Pio's transitional sentence at the beginning of book 6 (*XXIII libri* 103ᵛA)
606 MARGINAL NOTE: 'Lie.' Pio *XXIII libri* 104ʳB gives a long list of ritual practices and ob-
    jects and concludes: 'all of these, included under the generic term "ritual," you attack,
    ridicule, and criticize, both in general and in particular, in the most offensive language.'
607 See *Adagia* IV 1 70: *Usque ad ravim* 'Unto hoarseness' LB II 985D–E.

examine statements which preclude the slander nor reading my defence. He derived his obviously false opinion from the following passages.

PASSAGES[608]

1 / In the *Methodus* I wrote: 'Go through the whole New Testament, and you will not find anywhere a precept which has to do with rituals.'[609] But he presses me with my own definition, that any external activity pertaining to the divine worship or to religion is a ritual.[610] If, he says, one joins this definition to the words of the *Methodus*, does it not become obvious that Erasmus condemns all the rites of the church?[611] Not at all, renowned sir, for what is not prescribed in the Gospel is not thereupon condemned, though Christ did teach those things that are best and most conducive to devotion. But it was Christ, says Pio, who instituted baptism and the Eucharist; these are rituals; therefore the *Methodus* lies.[612] Not at all! But granting that sacraments are correctly termed rituals because they are accomplished by means of outward rites, it was Aristotle who taught that words are to be interpreted according to their material subject.[613] The *Methodus* is not talking of the sacraments about which doubt is forbidden, nor should even Pio suspect that I am so mad as to deny what is clearer than the sun, something of which I express approval in countless passages in my works. Furthermore, the very progression of the statement makes clear the matters about which I am expressing my opinion, for it continues immediately: 'Where is there a single word about food or clothing, where any mention of fasting or the like?'[614]

But what Pio wove in here follows after a large interval: 'Yet we, practically deaf to all this, burden with worse than Judaic rules those who have been set free by the blood of Christ.'[615] This passage, as was shown above, begins from the peculiar superstition of certain men who force a brother

* * * * *

608 Pio *XXIII libri* 103ᵛ-104ʳA quotes the seven passages Erasmus discusses here: one from *Methodus*, one from the scholia on Jerome, three from *De esu carnium*, one from the letter to Volz, and one from the *Annotationes*.
609 LB v 106F / Holborn 239:23–4
610 Pio *XXIII libri* 104ʳB: 'it is plain that we will have no argument about the term "rituals," since under their name you have included whatever is done or performed externally for the celebration of divine worship or for the sake of religion.'
611 A paraphrase of Pio *XXIII libri* 104ʳB
612 Ibidem
613 For Aristotle's teachings on how properly to define a term, see eg his *Posterior Analytics* 1.4 and *Topica* 1.5 Loeb 43–7, 281–9.
614 From *Methodus* LB v 106F / Holborn 239:24–5
615 From *Methodus* LB v 107E / Holborn 241:8–10, quoted by Pio *XXIII libri* 103ᵛA

into peril of his life because of their fanaticism about meat eating or fasting. There is no mention at all there of the regulations handed down by the public authority of the church.

2 / The second passage is quoted from my scholia in the first volume of Jerome.[616] I have not been able to determine the passage so far, but only that it was quoted incorrectly (*ad unum* instead of *ad vivum* and *voluimus* instead of *volumus*).[617] My words are as follows: 'We have lost a large part of the earth in our desire to exact exhaustive observance of all rituals.' I have no doubt but that this was said about the Greek schismatics, the Muscovites, and the Bohemians,[618] and any others of this sort who, because

* * * * *

616 Cited by Pio *XXIII libri* 103ᵛA as *et in scholiis in Hieronymum Tomi primi*, it is from a note on the Irenaeus section of *De viris illustribus* HO [1516] 1 139ᵛ, [1524] 1 310; Erasmus' speculation as to what he had written is correct regarding *vivum*, but wrong on *volumus*. Instead of *exigi volumus* he had written *exigimus*. A more extensive quotation from this note was given by Zúñiga in *Blasphemiae et impietates* DIVʳ and in the manuscript version, Biblioteca Nazionale Vittore Emanuele II di Napoli, MS VII.B.14 Liber II no xxii 59ʳ⁻ᵛ.

617 Pio's text (103ᵛA) misquotes the passage found in HO 1 139ᵛ '*dum omnes cerimonias ad vivum exigimus*' to read *dum omnes ceremonias ad unum exigi voluimus*. The crucial difference is Pio's *ad unum* for Erasmus' *ad vivum*. The phrase *ad unum* is a Latin idiom for 'to the last one' or 'to the last man.' Erasmus used *ad vivum* thinking of *Adagia* II iv 13: *Ad vivum resecare* 'To cut back to the quick' (LB II 527A–B). *Ceremonias/cerimonias* is an orthographical variant; the other differences are due to Pio's paraphrasing of Erasmus' words.

618 Erasmus is incorrect in this speculation. Jerome's comments are in regard to Irenaeus' opposition to the insistence of Pope Victor that the eastern church follow Rome's way of calculating the day for celebrating Easter; HO (1516) 1 127R, (1524) 1 282D. Erasmus' claim that this comment was directed at the Greeks and Bohemians ironically agrees with Zúñiga's similar opinion: 'Apparet per haec verba nolle Erasmum, ut haeretici aut scismatici ab ecclesia rescindantur, aut graecis in hac parte [illegible], aut quod verius puto hussitis ac Bohemis': from the unabridged manuscript version of his *Blasphemiae et impietates* (cited in n616) fol 59ᵛ.
The dispute between Latins and Greeks (and Russians) over using leavened or unleavened bread for the Eucharist was resolved at the Council of Florence on 6 July 1439 by a decree allowing both; see COD 527; Joseph Gill *The Council of Florence* (Cambridge 1959) 266–7, 275; and Joseph B. Koncervicius *Russia's Attitude towards Union with Rome (9th–16th Centuries)* (Washington 1927) 66–7. The Latin church did *not* cut off the Orthodox because of their use of leavened bread, as claimed by Erasmus; rather, the Orthodox churches objected to the Latin church's use of unleavened bread. On the theological significance attributed to leavened or unleavened bread, see John Meyendorff *Byzantine Theology: Historical Trends and Doctrinal Themes* (New York 1974) 204–5. A report on the errors of the Russians in Poland-Lithuania presented to Leo X in 1514 does not list the use of yeast and milk (ie yogurt) in the Eucharist but does cite other differences; see Odorico Rainaldi *Annales ecclesiastici* 12 (Lucca: Typis Leonardi Venturini 1755) 85–6, for the year 1514 nos 72–8; and Petro B.T. Bilaniuk *The Fifth Lateran Council (1512–1517) and the Eastern Churches* (Toronto 1975) 91–4, 115–20. Against the wishes of Eugenius IV the Council of Basel in 1437 agreed to allow the Bohemians to receive the Eucharist under both species provided they recognized also the validity of reception under only one

they consecrate with leavened bread and milk, because they receive the Eucharist under both species, or something else of this sort, have been cut off
from the Roman church. But one who thinks some relaxation in the case of
like rituals is better than rending the unity of the church does not condemn
all rituals, now, does he?

3 / He quotes the third passage from the *Letter to the Bishop of Basel*:[619]
'But if devotion is not imperilled by rituals, why is it that Paul everywhere
fights with such energy against the rituals of the Mosaic law?'[620] Here Pio
gives us a lesson in dialectics, that is, that it is incorrect to conclude a universal from a particular, or even a particular from a particular. He condemned Mosaic rituals, therefore he condemned all rituals? He condemned
Mosaic rituals, therefore he condemned the church's rituals?[621] I am not saying, O man most acute, that he condemned our rituals, but that he fought
against those which were gaining strength against the grace of Christ. And
how could Paul condemn these rituals of ours which he did not know? Superstitious Jews were prescribing rituals for the gentiles as if they were
necessary for salvation. Whoever heard in the times of Paul that on such
and such a day one must fast under pain of hell, that one must abstain from
eggs and meat under pain of hell?

4 / There follows after a space: 'If no one resists rituals of this kind,
gospel liberty is clearly finished.'[622] Here too he leaves out the words that
prevent slander. I wrote: 'rituals of this kind which are always swelling in
number.' Why was Paul fighting[623] against Judaic rituals? Because they were
evil by nature? Not at all. It was because they were eclipsing the Gospel's
grace through human superstition. But if the same thing is happening today
through the church's rituals of merely human origin, I am not saying that
these should be abolished, but reduced. In these matters if we go along with
the superstitious tendencies of certain people, there will be no limit nor end
to rituals, and for that reason I added 'always swelling in number.' Now if
Christ is praised, and rightly, because he gathered together the church by
imposing on us an easy yoke, with very few rituals, and those most easy

* * * * *

species; see Hefele-Leclercq VII-2 858–67, 907–16. Leo X in 1513 was willing to accept
this arrangement; see ibidem VIII-1 410.
619 See n454 above.
620 *De esu carnium* 807–9 LB IX 1211F / ASD IX-1 45, quoted by Pio XXIII *libri* 103ᵛ–104ʳA. For
a similar statement, see Ep 858:470–2.
621 Pio XXIII *libri* 108ʳI
622 *De esu carnium* 821–3 LB IX 1212A / ASD IX-1 46, quoted by Pio XXIII *libri* 104ʳA with the
omission noted here by Erasmus
623 *pugnabat* 1531 *pugnat* LB

to observe,[624] is there any inconsistency if the servants in imitation of their Lord are temperate in the rituals with which they burden the Lord's flock?

But I am not dealing there with rituals, but with the superstition of men who for the sake of rituals violate the greatest precepts of God, who punish the violation of ritual observances more severely than adultery, or slander. My intervening words are these: 'If rituals are introduced intemperately, they overwhelm gospel freedom. To place one's trust in them, as the common run of people usually do, is the bane of true devotion. To denounce one's brother on account of rituals is poison to evangelical religion.'[625] And shortly after: 'I disapprove of the defiance of those scoffers who eat meat in public as if to make fun of the general practice, but I condemn them only as wanting instruction, admonition, and, if the circumstances demand it, rebuke, but not as needing to be denounced to the magistrates as if they were guilty of parricide.'[626] I wrote this against the people at Basel,[627] who were then beginning publicly to eat eggs and meat during Lent. My whole letter combats the heedless violators of ritual observances, and yet it is from it that they are excerpting and distorting fragments in their effort to prove that I condemn all the rituals of the church. There is no method of criticism more ungentlemanly than to attack a person with vicious slanders on the basis of snippets excerpted from here and there, and distorted, and further twisted by malicious misconstruction. But however many such slanders there are, it is the Lord who will surely judge us all. Books like this seem to me to contribute to nothing but the rending of Christian concord. Satan is abundantly crafty [πολυτεχνότατος], but he is never more harmful than when he creeps in under the guise of devotion.

5 / I have not yet found the fifth quotation: 'We say, in keeping with the opinion of Paul, that rituals were established as a temporary measure.'[628] I do find this: 'Paul cries out that the rituals of the Law were established as a temporary measure, that they ought to be rejected in the face of the vitality of the Gospel.'[629] You will say: how does this apply to our rituals? At least to the extent that rituals of human origin should give way whenever the piety of the Gospel is imperilled on their account.

* * * * *

624 An allusion to Augustine's use of Matt 11:30 in Ep 54; see n638 below.
625 *De esu carnium* 809–11 LB IX 1211F / ASD IX-1 45
626 *De esu carnium* 815–18 LB IX 1211F–1212A / ASD IX-1 46
627 For the disturbances in Basel that were the occasion for writing *De esu carnium*, see Augustijn's introduction in ASD IX-1 3–6.
628 Pio XXIII *libri* 104ʳA
629 *De esu carnium* 891–2 LB IX 1213C / ASD IX-1 48

6 / I have already dealt with what he quotes from the *Letter to Volz*:[630] 'If only a moderate danger to religion lurked in ceremonies, Paul would not vent his indignation upon them so vigorously in all his epistles.' The danger lies hidden in them, not from their own nature, but because of human superstition, or hypocrisy, just as Paul said 'the letter kills'[631] and 'knowledge puffs up,'[632] condemning not the letter of the Law, nor knowledge of it, but human frailty.

7 / His seventh quotation, from the *Annotations*, is one on the first chapter of the Letter to Titus: 'You see how Paul is fierce in his anger at no one more than at those who were striving to do away with Christ by means of Mosaic rituals and under pretext of these were setting up a tyranny over the Christian people.'[633] Is there anything impious here? He will say: you are poking fun at the rituals of the church. If there are now among Christians any like those false apostles, they are even more deserving of denunciation.

These are the terrifying passages, fragments collected from everywhere through hired labour. From these it is plain, as none can gainsay, that I am for the abolition of all, each, and every ritual of the church. And Pio includes among rituals even the fact that Christ was bound with ropes, and scourged with whips, and many like things.[634] And because at the Ascension of Jesus two angels appeared in white garments,[635] he draws the convincing conclusion that my statement is plainly false, namely that in the New Testament Christ instituted no ritual at all.[636] This is how the children of the Aristotelians draw conclusions [οὕτως περαίνονται οἱ τῶν Ἀριστοτελικῶν παῖδες]!!

But he admits that he has been 'more critical' in reporting these *testimonia*.[637] Oh he has, indeed, just as he is everywhere most critical, and these *testimonia* were very much to the point! Although Augustine teaches that Christ assembled his church by means of very few rituals, ones quite easy of observance,[638] Pio asserts that very many rituals were instituted by

* * * * *

630 Ep 858:470–2, quoted by Pio *XXIII libri* 104ʳA. See 212 above.
631 2 Cor 3:6
632 1 Cor 8:1
633 From the annotation on Titus 1:15 LB VI 968F / Reeve *Galatians to Apocalypse* 697, quoted by Pio *XXIII libri* 104ʳA
634 Pio *XXIII libri* 105ᵛC
635 Pio *XXIII libri* 105ᵛE
636 Pio *XXIII libri* 106ʳE
637 Pio *XXIII libri* 105ᵛE: 'I have been, perhaps, more critical in reviewing these than would have seemed permissible to some.'
638 Ep 54.1 (*Ad inquisitiones Ianuarii liber primus*) CSEL 34 8–9

the Law of the Gospel.[639] But he makes no distinction between types, historical events, and the rituals whose observance was prescribed. Even so, Pio attacks inauthentic and hypocritical rituals,[640] agreeing with me here, except that he should have added excessive rituals, and rituals that are destructive, because of human failing, of authentic devotion. As it is, what Pio says only once, I keep insisting upon in ever so many places. But, nonetheless, Erasmus is the foe of rituals, Pio their defender.

FEAST DAYS[641]

After making it wonderfully evident that I condemn all rituals, he makes it equally evident that I disapprove of feast days.

He alleges a passage from my scholia on Jerome, and fails to give the citation in spite of the fact that there are four volumes, and the arrangement is not the same in the two editions.[642] 'At the time of Paul the church knew no feast days, etc.'[643] Pio admits the same,[644] because Christians did not have churches, and they would not have been allowed to celebrate feast days with novel rites, I mean feasts that were prescribed by the church. But gradually the church established for herself feast days for solemn celebration.[645] I do not deny this, but it does not seem probable to me that this took place in Paul's time. But even granting that this is untrue, am I therefore attacking feast days?

He brings forward something more severe from the *Annotations on the New Testament*, and, as usual, he does not give a citation, but it is, in fact, on

* * * * *

639 Pio *XXIII libri* 106ᵛH–I
640 Pio *XXIII libri* 106ᵛ–107ʳI
641 1531 UPPER MARGINS: 'On Feast Days' on pages 124–5 [= 215–17 in this volume]. Erasmus neglects Pio's formal 'Refutation of the Arguments of Erasmus against the Rituals Received by the Church' that occupies 107ᵛ–109ᵛI of *XXIII liber*, and responds to the first of Pio's discussions of particular aspects of ritual, the feast days of the liturgical year (109ᵛN–112ᵛ[Q]); Erasmus takes up the two passages Pio quotes at the beginning of this discussion, but ignores the three subsequently quoted.
642 The edition consisted of a total of nine parts or volumes, but Erasmus had responsibility for the first four; in the revised edition of 1524–6 volumes I, II, and III correspond to volumes I, III, and IV of the first edition (see CWE 61 xxii–xxx).
643 Pio *XXIII libri* 109ᵛN quotes from a scholion of Erasmus on Jerome's Ep 121 (HO 4 79ʳB); that Pio quoted from the 1516 edition is evident from his failure to include the word *opinor* (between *dies* and *nesciebat*) added in the revised 1524 edition HO 3 178C. Erasmus makes a similar statement in his annotations on Rom 14:6 LB VI 641C / Reeve-Screech 427 / CWE 56 374.
644 Pio *XXIII libri* 109ᵛ–110ʳN
645 Pio *XXIII libri* 110ʳO

the first Epistle to the Corinthians, chapter 16.[646] There, when Paul says 'I shall remain at Ephesus until Pentecost,' I only wonder whether he means the feast day or only the period of time, that is, until the fiftieth day. This, as I show there, is the interpretation of Ambrose and Theophylact,[647] but I leave final judgment to the reader.

But I add: 'None of the ancients, so far as I recall, makes any mention of the feast.' Here he had to bring forward one hundred authors who speak of feast days.[648] But I wrote 'none of the ancient commentators,'[649] meaning those who comment on this passage of Paul. That is how my words are quoted with a view to slander. Here he howls as if bitten by a gadfly, here he lays about with his scourge, here he stops just short of acting like a madman, even though he fails to understand what I am saying. Who is so stupid that he does not know that Jerome, Augustine, and Ambrose speak of feast days? How vast a field is open for Pio's rhetorical gallop, but he goes, as the proverb has it, beyond the olives.[650] He challenges me by name[651] to say, if Paul kept the feast of Pentecost at Ephesus, whether he kept the feast of the Jews or the feast of the Christians; if not that of the Jews, then that of the Christians.[652] Here is my reply: first Pio takes as something admitted a matter about which I am not sure. I do not care that Ambrose mentions the feast day in passing,[653] since when explaining this passage he is making another sort of comment. But if he celebrated Pentecost, he celebrated the feast of the Jews. But, says Pio, there were no Jews at Ephesus. On the contrary, there were synagogues of Jews through all Asia.

* * * * *

646 Pio *XXIII libri* 109ᵛN quotes selectively from the annotation on 1 Cor 16:8 LB VI 745E–F / Reeve-Screech 520–1.
647 Not Ambrose, but Ambrosiaster CSEL 81 2:189–90 and Theophylact PG 124 788C, both cited by Erasmus LB VI 745E–F / Reeve-Screech 521
648 Pio *XXIII libri* 110ʳ–111ᵛo cites Augustine (six times), Ambrose (three times), Jerome (four times), Origen (twice), Bede, the *Glossa ordinaria*, John Chrysostom, and Origen again.
649 Erasmus' text reads *Certe nullus veterum interpretum, quantum equidem memimi*; Pio had omitted the word *interpretum*.
650 *Adagia* II ii 10: *Extra oleas* 'He runs beyond the olive trees' LB II 451F / CWE 33 82–3: 'when a man oversteps the prescribed limits, or does or says irrelevant things which have no connection with the matter in hand'
651 Pio *XXIII libri* 110ʳo: 'But say, I beg, O Desiderius, for I address you at this point, which feast of Pentecost did he celebrate with them?'
652 A severe compression of Pio's argument in *XXIII libri* 110ʳo: 'You will surely not say that of the Jews, since those [at Ephesus] converted to Christ were gentiles. And so it was the feast of Christians. Therefore it was already celebrated then by Christians.'
653 Pio *XXIII libri* 110ʳo cites Ambrose *Idem quoque in lucem [...] ait*, ie Ambrose *Expositio evangelii secundum Lucam* 8.25 CCSL 14 30:284–6.

But what need is there to declare which he celebrated, since he writes only that he wished to remain until that day? Moreover, since Augustine[654] is unsure whether the Lord's Day and the rest of the feasts, in particular two of them, were established by the apostles or by the public authority of the entire church, what danger is there if I query whether the feast of Pentecost had been appointed for Christians at the time of Paul?

CHANT[655]

He continues: 'After expressing your abomination for the observance of feasts,' he says, 'you also attack the very practice of singing the divine praises in as many passages as you can.'[656] Who could believe that Pio himself wrote, or even read, such obvious slanders? Where do I express an abomination for the feast days of Christians? Where do I attack the chant of the church?

1 / I wrote 'in the *Annotations.*'[657] Where? Search for it, for he does not cite the passage. It is on chapter 14 of 1 Corinthians. There I condemn only the practice of perhaps some people who either do nothing but sing, in spite of the fact that Paul enjoins the presence of a prophet whom he places before all others,[658] or sing an inappropriate sort of chant by which the words are obscured, whereas chant, that is, intoned delivery, was devised to the end that the words might more effectively flow into the minds of men. And Augustine himself declares that in his time certain bishops did not allow any chanting, but only recitation.[659] And if it is so angelic a thing

* * * * *

654 Ep 54.1 (*Ad inquisitiones Ianuarii liber primus*) CSEL 34 159
655 1531 UPPER MARGINS: 'On Chant' on pages 116–29 [= 217–21 in this volume]. Pio's treatment of chant is in XXIII *libri* 112ᵛR–117ᵛ[v]; Erasmus takes up the three passages that Pio quotes in this connection. Erasmus' views on music have been studied by Jean-Claude Margolin *Erasme et la musique* (Paris 1965); Clement A. Miller 'Erasmus on Music' *The Musical Quarterly* 52 (1966) 332–49; and Helmut Fleinghaus *Die Musikanschauung des Erasmus von Rotterdam* Kölner Beiträge zur Musikforschung 135 (Regensburg 1984). Margolin focuses briefly on Erasmus' debate with Pio in a later study in *Recherches Erasmiennes* Travaux d'humanisme et renaissance 105 (Geneva 1969) 93–5.
656 MARGINAL NOTE: 'Lie.' Pio XXIII *libri* 112ᵛR
657 Pio XXIII *libri* 112ᵛR quotes excerpts from the annotation on 1 Cor 14:19 LB VI 731C–732C / Reeve-Screech 507–8, introducing the quotation *ut illud in Annotationibus.*
658 See 1 Cor 14:29; Erasmus concludes his note on 1 Cor 14:19: 'Let there be heard before all the voice of a prophet which is to rebuke the conscience of the wicked, to console the downcast, to arouse the drowsy, to utter the mysteries of the divine Spirit' (LB VI 732C / Reeve-Screech 508).
659 Erasmus alludes to the anecdote about Athanasius' provisions for chanting psalms in *Confessiones* 10.33.50 CCSL 27 181–2.

that monks always be singing in the churches, why is it that Paul summons to a more outstanding charism, that is, the elucidation of the Scriptures?[660] Why are monks told to work with their hands?[661] Why does Gregory, in the *Decretum*, express anger that priests, whose duty is to expound the Gospel, are singing in the churches?[662] Why did Benedict and Francis not mandate it in their rules? And all this especially now that the commons do not understand the words of the singers? In addition, since Augustine admits that he sins if he is more taken with the singing than with the meaning of the words,[663] what is to be said of the people who hear nothing but the din of voices? What does Paul mean when he introduces the simile of a cymbal tinkling without profit?[664]

2 / But I say there:[665] 'It is more holy to provide for one's children and wife than to listen in church to incomprehensible bits of chant.'[666] These are the words of his unreliable source, not mine. I speak as follows: 'And what is more burdensome still, the priests are obligated to these things by bonds virtually more confining than those whereby they are bound to the teachings of Christ. The people are compelled to listen to these things, driven from the work by which they feed their wives and children. And what could be more holy than that work?'[667] This applies particularly to the English, for I was living there at that time.[668] Although among them there is a constant tuning of voices which does not allow the words to be understood, nonetheless the people are told to hear the chants of matins, and of all the hours, under pain of hell. Now which is more holy, by the work of one's hands to help one's children who are imperilled by hunger, or to listen all

* * * * *

660 See Erasmus' note on 1 Cor 14:19 LB VI 731C–D / Reeve-Screech 507.

661 See eg *Benedicti regula* chapter 48 'De opera manuum cotidiana' in CSEL 75 114–19.

662 See the decree of Gregory the Great in Gratian's *Decretum* pars 1 d 92 cap 2 (Friedberg I col 317) translated and annotated by John Gilchrist in *The Collection in Seventy-Four Titles: A Canon Law Manual of the Gregorian Reform* (Toronto 1980) 203–4.

663 *Confessiones* 10.33.50 CCSL 27 182

664 1 Cor 13:1

665 MARGINAL NOTE: 'Misleading citation'

666 Pio *XXIII libri* 112ᵛR

667 From the annotation on 1 Cor 14:19 LB VI 731E / Reeve-Screech 507

668 Erasmus visited or resided in England during the years 1499, 1505–6, 1509–14, 1515, and 1517. On Erasmus' visits to England and contacts with English monasticism, see David Knowles *The Religious Orders in England* III: *The Tudor Age* (Cambridge 1959) 151–4. In his annotation on 1 Corinthians Erasmus refers to the musical practices in monasteries, including those of the Benedictines, but he cites none by name (LB VI 732B / Reeve-Screech 508). See also Roger Bowers 'To Chorus from Quartet: The Performing Resource for English Church Polyphony c 1390–1559' in *English Choral Practice, 1400–1650* ed John Morehen (New York 1995)

day long to chant which is not understood, and, therefore, without profit? He is going to say: therefore you are doing away with chant? No. In fact, I say there: 'Certainly let the churches have the traditional chanting, but in moderation.'

Pio condemns the use of polyphonic music in church.[669] I censure either excessive music, or music not suited to church. But, says he, you are condemning what has been received. Pio is doing the same thing, for in the papal chapel this sort of warbling is heard, but hardly ever a sermon.[670] This warbling is heard also in the court of the emperor, and of other kings.[671] Is Pio not afraid to condemn this?[672] And what I am condemning has not been universally received, but has come into practice gradually and unnoticed, like so much else.

* * * * *

669 *Pius damnat ... Musicam* πολύφωνον; Pio *XXIII libri* 116^{V}τ condemns specifically 'theatrical singing, effete and impure songs' (*theatrales cantus, mollesque et impurae cantilenae*); earlier, in his *Epistola paraenetica* 78^{r}z or *XXIII libri* 37^{r}z, Pio had expressed his distaste for *cantus perfractus* (see *Responsio* nn 144 and 251 above). The theological faculty of the University of Paris in 1498 condemned 'lascivious and theatrical songs' in churches; see Duplessy *Collectio judiciorum* 1-2 341 (proposition 6).

670 On the music of the papal chapel, see Franz Xaver Haberl *Die römische 'Schola cantorum' und die papstlichen Kapellsanger bis zur Mitte des 16. Jahrhunderts* Bausteine fur Musikgeschichte 3 (Leipzig 1888) 42 (on polyphonic music that became the new style under Sixtus IV, 1471–84), 47–8, 61–3 (on the papal chapel under Julius II, 1503–13, during whose pontificate Erasmus visited Rome); and Richard J. Sherr *The Papal Chapel c 1492–1513 and Its Polyphonic Sources* (Dissertation, Princeton University 1975) especially 93–4 (where the papal master of ceremonies, Paride de Grassi, complained that during the mass on Saturday of Easter week 1507 it was difficult to understand the words being sung in polyphonic style) and 91 (where the celebrant had to wait until the musicians had finished).

That Erasmus has grossly misinterpreted the quantity and quality of sermons delivered in the papal chapel has been established by John W. O'Malley *Praise and Blame in Renaissance Rome: Rhetoric, Doctrine, and Reform in the Sacred Orators of the Papal Court, c. 1450–1521* (Durham, NC 1979) especially 29–31, 114. Even the infamous Ciceronian Latin sermon that Erasmus had heard in Rome and later vilified in his *Ciceronianus*, which some scholars attribute to the noted curial orator Tommaso Inghirami, has been reevaluated; see L. Gualdo Rosa 'Ciceroniano o cristiano? A proposito dell' orazione "De morte Christi" di Tommaso Fedra Inghirami' *Humanistica Lovaniensia* 34 (1985) 52–64. A good example of inappropriate Ciceronian oratory is found in Pietro Alcionio's oration on Pentecost Sunday (4 June 1525) before Clement VII; see Kenneth Gouwens 'Ciceronianism and Collective Identity: Defining the Boundaries of the Roman Academy, 1525' *Journal of Medieval and Renaissance Studies* 23 (1993) 173–95 especially 177–9.

671 See Louise E. Cuyler *The Emperor Maximilian and Music* (London 1973) especially 225–6 (on the elaborate polyphonic settings of some of the music composed for Maximilian's court). On Henry VIII (1509–47) and his love of Renaissance sacred music for the royal chapel, see J.J. Scarisbrick *Henry VIII* (Berkeley, Calif 1968) 15.

672 Using the punctuation of 1531, which provides an interrogation mark, against LB, which provides a full stop

3 / But he also reports the following, in his own words, not in mine:[673] 'Also,' he says, 'when instructing a Christian prince, you try to persuade him that it is better for him to tend to his own obligations than have his time taken up with listening to the chants of the priests.'[674] (A solecism was inserted here, for who was it who said *conatur suadere*?)[675] But what is it that Pio means, that a prince should be urged to listen to the chants of a priest rather than to meet his own obligations? Is listening to chants, then, in itself a heroic work, especially the kind that princes usually hear? If he had said this about praying, or listening to a sermon, the statement he alleges could have had some semblance of the truth. As it is, I said nothing like this. Rather, I criticized courtly princes who now make it a practice to say the prayers of the hours, for which, unlike priests, they have no obligation, and under this pretext fend off those who wish to approach them.[676] But imagine that I am speaking about praying and listening to a sermon, if the public tranquillity is endangered, or if the innocent are being oppressed, should a prince come to the relief of those whom it is his duty to aid, or neglect them and listen to the sermon? I mean under a circumstance in which he cannot do both. A Pharisee goes to the aid of a sheep fallen into a pit on the Sabbath[677] on which God enjoined rest, and a prince, because of a sermon or the mass, does not come to the aid of his country?

Now, so as to appear witty as well, Pio pretends that I am violently scandalized by ecclesiastical chant, and asks what I would prefer to hear in church, the din of mimes or actors presenting a comedy, etc.[678] What could be more dull than these jokes? What is heard in the heavenly church? No voice, but the yearnings of the devout spirits. Let a reader be heard, but

* * * * *

673 MARGINAL NOTE: 'Misleading citation'
674 Pio *XXIII libri* 112ᵛR is referring probably to statements in the dedicatory letter of the *Paraphrasis in secundum Ioannem* addressed to Ferdinand of Austria (Ep 1333:300–25), though Erasmus does not mention chanting explicitly there, but only the canonical hours. Similar views are expressed in his *Institutio principis christiani* LB IV 567C / ASD IV-1 147:338–41 / CWE 27 216; in the *Exomologesis* LB V 163D; and in the *Modus orandi Deum* LB V 1128E, 1129E–30D / ASD V-1 167–70:614–21, 657–707 / CWE 70 217–21.
675 Pio's text reads '*illi suadere conaris quod praestaret etc.*' Erasmus' objection is to the collocation *conatur suadere*; correct usage, according to Valla, would have been *conatur persuadere* or *suadet*; see *Paraphrases in Elegantias Vallae* LB I 1105D / ASD I-4 290–1. Pio used *persuadere coneris* (*XXIII libri* 57ᵛr) and in Erasmus' *Apologia* we find *conatur persuadere* (1118C) and *persuadere conantur* (1190B).
676 Ep 1333:300–25; *Institutio principis christiani* LB IV 567C / ASD IV-1 147:338–41 / CWE 27 216; and perhaps *Responsio ad notulas Bedaicas* LB IX 481C–D
677 Alluding to Matt 12:11; both 1531 and LB inexplicably read *montem* instead of *ovem*.
678 Pio *XXIII libri* 115ᵛs

let there be a prophet present to interpret; let a canticle be recited, but in a devout way. But let the voice of the prophet take the first place, and silent, mental prayer.

CANDLES[679]

Here the craftsman surpasses himself. He asks in amazement how I dared write that Jerome seems to have held the view that it was superstitious to light candles in the daylight in honour of the martyrs.[680] So let Jerome's own words be read out from the book he wrote against Vigilantius. He says: 'We do not light candles in broad daylight, as in your groundless slander, but to abate the night-time darkness by means of this comfort, and to keep watch for the light, so that we will not be blind and asleep with you in the darkness. But if any people of the world, or, for that matter, devout women, do this in honour of the martyrs as a result of their ignorance or naivety, people of whom we can truly say, "I admit, they have a zeal for God, but not according to knowledge," what harm is that to you?'[681] Thus far Jerome. He says that candles are not lighted in honour of the martyrs. He calls the people of the world or the devout women who do so in honour of the martyrs naive, possessed of a zeal for God, but not according to knowledge. What could be more clear than these words of Jerome?

But Jerome immediately adds that the apostles were rebuked when they murmured about the loss of the ointment![682] I reply: this event was unique, one full of mystery, and not to be applied as an example. Jerome says: 'And whoever light candles have a reward according to their faith, as the Apostle says, "Let each abound in his own understanding."'[683] What is defended by these words but a pious, if somewhat superstitious, proclivity in a non-essential matter? Thus it becomes a matter of free choice for each to light candles or not. The same goes for his remark in the same passage that in the churches of the East candles are customarily lighted when the

* * * * *

679 1531 UPPER MARGINS: 'On Candles' on pages 130–5 [= 221–2 in this volume]. Pio's discussion of candles in *XXIII libri* 117^{v}x–119^v[y] is targeted at the one passage that Erasmus discusses here.

680 Pio *XXIII libri* 117^{v}x quotes from Erasmus on Jerome: *Quod ais in scholiis Hieronymi*; this is a quotation from the Antidotus to the *Contra Vigilantium* HO [1516] 3 58^{r}B, [1524] 2 129A. Pio urges that this is a misrepresentation, and that Jerome's view to the contrary is stated in so many passages that he is amazed (*stupescam*) that Erasmus wrote what he did.

681 Jerome *Contra Vigilantium* 7 PL 23 345B

682 This is the passage from *Contra Vigilantium* 7 PL 23 345B–346A that Pio *XXIII libri* 117^{v}x–118^r actually quotes.

683 Jerome *Contra Vigilantium* 7 PL 23 345C, quoting Rom 14:5

Gospel is read;[684] one may conclude from this that this was not the practice in the churches of the West.

And here Pio scoffs: 'Aha,' he says, 'you read those words, and you say that Jerome judged that candles should not be lighted, etc,' and shortly: 'Did you not hear him saying the complete opposite?'[685] I did not say that this was Jerome's judgment, but that he *seems* to have held the view that it was superstitious. If he says the complete opposite, then he contradicts himself. And, in fact, I am so far from condemning a temperate use of candles that I make excuse for Jerome there in these words: 'But it appears that practice was more tolerated than approved of in that period, and the passage of time changes many things.'[686] With these words I protect Jerome, so that he does not appear self-contradictory, and I express my approval of a temperate use of candles. What place, then, is there for this fellow's wranglings, abuse, and insults with which he ungrudgingly [ἀφθόνως] fills up this entire work.

He says: 'A lighted candle has mystical meaning.'[687] So let the one who wishes go ahead and bring in something extraneous, but let all be zealous to provide what is really mystical; otherwise the lighting of candles is in itself superstitious.

[THE EUCHARIST]
Next some one of Pio's 'stitchers' sewed on a fragmentary observation on the Eucharist[688] that is quite irrelevant, though of a kindred style. It begins thus: 'In addition, you say that one can conclude from that statement of Jerome, "Why they do not enter the churches, etc," that of old everyone who wished used to receive the Body of Christ in his own home.' Where are those words of Jerome; where does Erasmus draw this conclusion?

* * * * *

684 Jerome *Contra Vigilantium* 7 PL 23 346A, quoted by Pio *XXIII libri* 118ʳx
685 Pio *XXIII libri* 118ʳx
686 From the Antidotus to the *Contra Vigilantium* HO 3 58ʳB
687 A brief allusion to Pio's extended discussion of fire as a religious symbol (*XXIII libri* 118ᵛ[Y])
688 Pio concludes book 6 of *XXIII libri* with an attack on a scholion in Erasmus' edition of Jerome Ep 69 (119ᵛ-120ʳz): 'You say, in addition, that it can be concluded from the statement of Jerome "wherefore they do not enter the churches, etc" that it was the custom of old that each one who wished received the body of Christ at his home. But it is not apparent how you draw this conclusion, for Jerome's words do not at all suggest this, and your view is clearly quite false.' The note to which Pio refers, on Ep 49.15.6 CSEL 54 377:12, reads: 'why they do not enter, etc. From this passage we have concluded that it was the custom of old for each one who wished to receive the Body of Christ at his home. Paul seems to suggest the same thing, when he says: "Do you not have homes for eating"' (HO [1516] 3 51ʳB, and [1524] 2 113B).

It is absolutely clear from Chrysostom and Tertullian[689] that of old it was permitted that any who wished might take home the bread that had been consecrated by the priest and given to him, and consume it there. But he says that I am talking about the *synaxis* which took place with bread that was only blessed, not consecrated, by the priest.[690] Even if this is true, what is his complaint? For it is certain that such a *synaxis* was customarily held in private by many frequently in the course of the day. But what has this to do with the sacrament of the Eucharist? Whoever tacked on this appendix was clearly delusional and does not deserve a reply.

THE DECORATION OF CHURCHES [FROM BOOK 7][691]

He quotes from a dialogue 'For the Sake of Pilgrimage.' Here it is at once apparent what sort of help Pio employed, for they could not even quote the title correctly! But what is it Erasmus says in the dialogue 'Pilgrimage for the Sake of Religion'? Hear, O reader, a quotation that outdoes any false informer.[692] Someone says there: 'But seriously, I wonder sometimes what

* * * * *

689 The ancient evidence for reservation and reception of the Eucharist at home is reviewed by Nathan Mitchell *Cult and Controversy: The Worship of the Eucharist outside Mass* (New York 1982) 10–43.

690 Pio *XXIII libri* 119ᵛz alludes to Erasmus' point that the *synaxis* was not always celebrated with consecrated bread; see *Responsio* nn 244 and 245 above. There is a clash of meanings here. Erasmus generally uses *synaxis* to refer to gatherings for the liturgy of the canonical hours, and only occasionally to refer to the Eucharist; Pio seems here to restrict the meaning of *synaxis* to the Eucharist; compare Thomas Aquinas *Summa theologiae* 3.73.4, on the name of the Eucharist: 'It has another meaning in regard to the present reality, sc the unity of the Church unto which people are gathered by this sacrament. With this meaning it is called communion or *synaxis*' (Ottawa 1941–5) IV 2925a1–2925b2.

691 1531 UPPER MARGINS: 'On the Decoration of Churches' on pages 134–41 [= 223–9 in this volume]; and MARGINAL NOTE: 'From book 7.' Erasmus is replying to Pio's book 7 in *XXIII libri* 120ᵛA–133ᵛz, of which this is the title; the book's opening words provide an outline of its topics: 'But you were not content with this. No, your habitual temper of mind drives you on to a detailed attack on the embellishment of churches and their decoration and construction, even though you had already abundantly denounced such things and expressed your bitter dislike for rituals in a general way. But you would still not have played the role of a very Momus among priests and ceremonies if you did not covertly criticize this devout and praiseworthy embellishment as well. Moreover, lest anything be left untouched, you also had to criticize and condemn the church's last rites and works of devotion that are provided in funerals, stooping to criticize even the ringing of bells, as will be reported below.' On 120ᵛA and 132ʳy Pio quotes passages from the colloquies *Peregrinatio, Convivium religiosum*, and *Funus* that Erasmus discusses here.

692 MARGINAL NOTE: 'Abridged quotation'

possible excuse there could be for those who spend so much money on building, decorating, and enriching churches that there's simply no limit to it.'[693] What is being criticized here but excessive expenditures on the decoration of churches? But his supplier left out the words that follow immediately: 'Granted that the sacred vestments and vessels of the church must have a dignity appropriate to their liturgical use; and I want the building to have grandeur.'[694] That supplier left these out, and attached to the earlier words what follows: 'But what is the use of so many baptisteries, candelabra, gold statues? What is the good of the vastly expensive organs? (We are not content with a single pair, either.) What is the good of that costly musical neighing when meanwhile our brothers and sisters, Christ's living temples, waste away from hunger?'[695] You hear of golden baptisteries and monstrously excessive expenditures, of the neglect of Christians dying of hunger, you hear a due dignity and even majesty allowed to the churches, and yet the speaker who follows next in the colloquy modifies this statement, saying as follows: 'Every decent, sensible man favours moderation in these matters, of course. But since the fault springs from excessive devotion, it merits applause, especially when one thinks of the opposite vice in those who rob churches of their wealth. These gifts are generally given by kings and potentates, and would be worse spent on gambling and war. And removal of anything from there is, in the first place, regarded as sacrilege; next, those who are regular contributors stop their giving; above all, men are incited to robbery. Hence churchmen are custodians of these things rather than owners of them. In short, I would rather see a church abounding in sacred furnishings than bare and dirty, as some are, and more like stables than churches.'[696]

What is said in *The Godly Feast*? 'Therefore I do not see how they can be exempt from a capital charge who with excessive expenditure build or decorate monasteries or churches, when all the while so many living temples of Christ are in peril from hunger.'[697] Here too the supplier substituted 'do not see how they can be exempt' for what had been written, 'hardly see how they can be exempt.'[698] Oh yes, this is what it is to write in a brotherly

* * * * *

693 *Peregrinatio religionis ergo* LB I 784E / ASD I-3 489:704–7 / CWE 40 644
694 Ibidem LB I 784E / ASD I-3 489:707–8 / CWE 40 644
695 Ibidem LB I 784E–785A / ASD I-3 489:708–12 / CWE 40 644, quoted by Pio XXIII *libri* 120ᵛA
696 Ibidem LB I 785A / ASD I-3 489:713–490:721 / CWE 40 644–5
697 *Convivium religiosum* LB I 684F / ASD I-3 257:787–8 / CWE 39 198–9
698 Pio XXIII *libri* 120ᵛA substitutes *non … excusari posse* for *vix … excusari posse* of LB I 684F / ASD I-3 257:787–8 / CWE 39 198–9

way![699] But what is being censured here but excessive expenditures while the dire necessity of one's neighbours is neglected. And at this point Pio spits in my face. 'This statement,' he says, 'is not yours, but was taken from the foul criminal Judas, the betrayer of the Lord, etc.'[700] If it was Pio who wrote this bitter remark, I implore the Lord's mercy for him. But I would rather believe that my quarrel is with the wicked supplier. You, O reader, appreciate what he would deserve to hear were he alive. These were the words of the apostles,[701] but Judas' attitude was very different, since he had no care for the needy, but was a thief.[702] As for me, what loss do I suffer because of expenditures on churches? But whose were the words rebuking the Pharisees because they taught sons to pay the *corban* while neglecting their parents?[703] Whose are the famous words: 'Tell us, O priests, what does gold accomplish in the temple?'[704] 'A pagan poet's,' he will say. Yes, but the words were approved by Christian doctors.

He admits that there were saintly men who condemned wealth in churches,[705] one of whom was St Jerome. And the same Jerome praises Nepotianus because he took care of the decoration of the basilica, as Pio says,[706] but with what words? 'He was, then, anxious,' he says, 'that the altar gleam, that the walls were free of soot, that the floors were washed, that the porter was constant at the entry, <that>[707] there were always hangings in the doors, that the sanctuary [*sacrarium*] was clean, that the vessels were shining, and his devout concern for all the rituals.' Is there any mention here

* * * * *

699 MARGINAL NOTE: 'Quotes deceptively'; *fraterne* could also be read as 'like a friar.'
700 Pio XXIII *libri* 120[V]A, alluding to John 12:4–6
701 Erasmus counters with an allusion to Mark 14:4–5.
702 John 12:6
703 The words of Jesus, in Mark 7:11–13; in his annotation on this passage Erasmus observes: 'Here he explains what is rather obscure in Matthew [15:5], that some were being taught by the pharisees to neglect their parents and give their money to the temple' (LB VI 179F / Reeve 131). In a note on the term *corban* in Jerome's Ep 123 HO 1 39[V]C Erasmus says that the Hebrew term 'signifies not any gift, but specifically what is dedicated to the temple. Wherefore they also called the place in which gifts like this were kept *corbana* [Matt 27:6]'; for a contemporary account of the term see the article on *korban* in Gerhard Kittel ed *Theological Dictionary of the New Testament* trans and ed by Geoffrey W. Bromiley (Grand Rapids, Mich 1964–76) III 860–6.
704 Persius *Satires* 2.69
705 Pio XXIII *libri* 126[r]I: 'But if you continue to argue, saying that it is evident from certain statements of the holy Fathers that one should not expend great effort on decorating temples of this kind, but should toil instead on the spiritual temple, and that, in fact, they sometimes seem to prefer the works of mercy to this sort of effort ...'
706 Pio XXIII *libri* 109[V]M quotes from Jerome's Ep 60, the so-called *Epitaphium Nepotiani*, section 12 CSEL 54 563:18–564:2.
707 *si* LB *om* 1531

of golden candelabra, of jewels as valuable as whole cities? Pio replies here that the words of the saints should not be taken too seriously when they are engaged in a dispute.[708] (What a window he has opened for appeal from the authority of the ancient doctors!) In fact they do not dispute against luxurious wealth in the churches, they make pronouncements against it. But what is this to me, for I do not condemn even excessive wealth in churches in an unqualified way, but only so that the needy are not neglected on this account?

'But,' he says, 'all worship is to be preferred to almsgiving.'[709] Why, then, does Christ inveigh against the Pharisees who were teaching that one should contribute to the treasury by defrauding one's parents?[710] That was not an act of worship, but enriching churches or monasteries with more than regal wealth is an act of worship? Was not the observance of the sabbath an act of worship? Yet Christ does not rebuke its being violated because of an ox or an ass that had fallen into a pit,[711] and he proclaims with a loud voice that the sabbath exists on account of man, and not man on account of the sabbath.[712] But for what purpose are churches built, but that Christians may assemble there? 'But,' he says, 'we read: "You shall not build a house for me"[713] and "My house shall be called a house of prayer."'[714] But Paul proclaims: 'God does not dwell in buildings made by hand,'[715] and Jerome does not admit the precedent of the sumptuous temple at Jerusalem unless at the same time we also accept the rest of Judaism.[716]

Let us imagine a case in which your neighbour is in distress, and can hope for help from none but you.[717] Which will you do? Will you give help to a member of Christ, or will you donate that money for the needless

* * * * *

708 Pio's statement (*XXIII libri* 126[r]I) is, of course, more nuanced.

709 A point Pio makes early (*XXIII libri* 121[r]B): 'And so almsgiving and the rest of the works of mercy are never to be preferred, but should be placed far below things that have specifically to do with worship.' In his scholion on Jerome's Ep 108 Erasmus notes: 'Therefore Jerome does not condemn the elegant structure of churches, he does not reprove what is adorned and clean, but he denies that the culmination of virtue lies in these. He prefers alms that are donated to living temples of God, that is, to the poor of Christ' (HO 1 80[v]C).

710 See n703 above re Mark 7:9–13.

711 Luke 14:5

712 Mark 2:27

713 Pio *XXIII libri* 124[r]F: 'For the Lord said to David through the prophet Nathan: "You are not going to build a house for me to live in, are you?" [2 Sam 7:5]'; Erasmus changes *numquid* to *non* and omits the question mark.

714 Isa 56:7, repeated in Matt 21:13, Mark 11:17, and Luke 19:46, quoted by Pio *XXIII libri* 124[v]F in conjunction with John 2:15, 16

715 MARGINAL NOTE: 'Act 17' (ie Acts 17:24)

716 Ep 52.10.1, 3 CSEL 54 431–3; Pio, in fact, deals with just this objection in *XXIII libri* 129[v]R–130[r]S.

717 Echoing, perhaps, the more nuanced hypothetical case proposed by Pio *XXIII libri* 127[r]L

decoration of a church? This is what we are talking about. Pio allows that one should come to the aid of those in distress.[718] But where now is that worship, so very preferable to almsgiving? In such a case I think that the life of one's neighbour is to be preferred to even necessary expenditures. If my neighbour dies, he cannot be restored. A church can be repaired [*succurri*], and even without churches worship goes on. He is maintaining the same thing as I, and yet I am the blasphemer appropriating the words of Judas, while Pio is the pious one who defends the church.[719]

But the same fellow maintains that, no matter how much wealth we donate to the churches, more is due them.[720] Indeed, some of those who are very shrewd insist that, even though a hundred thousand people were in danger of starvation, it is not permissible to remove anything from the treasuries of the churches. It is for people that churches were first established, as a more convenient place for them to carry out the works of worship, though each one can do this in his own home.

But at this point Pio does not hesitate to compare the mystical anointing[721] of the Lord himself with the ambition of certain grandees who pass on their names to posterity by means of excessively expensive building projects.[722] The Lord did not wish this precedent to be applied to the many when he said: 'The poor you have with you always, but you will not have me always.'[723] He has departed, and it is his wish to be adorned in his members, and to be worshipped in spirit.[724]

Nor is it an act of worship[725] to build churches that are costly beyond measure; rather, the real act of worship is carried on in our souls. But if whatever is done externally for the honour of God[726] is an act of worship,

* * * * *

718 Pio *XXIII libri* 126ᵛK, 131ʳT
719 Pio *XXIII libri* 120ʳA
720 Referring perhaps to Pio *XXIII libri* 128ᵛO: 'even though in other things a limit must be applied, only in loving and glorifying God is none to be observed'; or to *XXIII libri* 124ᵛF: 'So let them cease, let them stop yelping and criticizing divine worship, something that can never be provided with as much honour and splendour as it deserves'; or to his lament at the contemporary neglect of church buildings in *XXIII libri* 131ʳ-ᵛv
721 Matt 26:6–13, Mark 14:3–9, John 12:3–8, discussed by Pio *XXIII libri* 126ᵛK–127ᵛM in the light of commentary by Theophylact and John Chrysostom
722 Pio *XXIII libri* 130ᵛT
723 Matt 26:11, Mark 14:7, John 12:8, a phrase central to Pio's argument against the priority of almsgiving
724 John 4:23–4
725 *opus latriae*, probably reacting to Pio's statement (*XXIII libri* 130ᵛs) that 'every act of worship [*omne opus latriae*] is more estimable by its very nature than any works of mercy'
726 A favourite Erasmian formulation; see n610 below. For a fuller treatment of Erasmus' views on worship or external cult, see eg his comments in *Modus orandi Deum* 577–95 LB V 1128C–D / ASD V-1 166 / CWE 70 216–17 and in *De puritate tabernaculi* LB V 308C–D.

then every action of the devout is worship, according to Paul's statement in 1 Corinthians 10: 'Do all things for the glory of God.'[727] The one who decorates a church decorates it for the honour of God, and the one who helps out his neighbour because of a devout disposition, in whose honour does he do it? Christ did not say whatever you spend on churches, you spend on me, but 'whatever you have done for one of these little ones, you have done for me.'[728]

But in Pio's view decorating a church with non-essentials is worship, but to come to the aid of your neighbour is a matter not of precept, but of counsel, unless he is in extreme necessity.[729] What angel[730] taught him that doing good to one's neighbour in need is a matter of counsel, whereas hardly anything else is more forcefully commanded? The Lord said of continence: 'Let him accept it who can';[731] he said of almsgiving: 'Give to everyone who asks you.'[732] John cries out: 'He who possesses the wealth of this world, and sees his brother suffering necessity, and closes his heart to him, how is God's love in him?'[733]

You will say: 'He says "necessity," granted, but he does not say "extreme."' In fact, in Greek it is χρεῖαν ἔχοντα, that is, 'needy' or 'having need.' Even if the word 'necessity' were there, we often say 'necessary' especially where there is particular need. Ultimate necessity can be imagined in disputation, but in reality it cannot be ascertained. But finally at this point, even according to these people, almsgiving becomes a matter of precept. But they maintain that it is to be given in such a way that one may demand it back if ever the needy man has a capacity for returning it!

To conclude. His citation of the luxurious temple of Solomon, that it was a figure of the church adorned with spiritual wealth,[734] misses my point. Nor is help provided by those authorities[735] who maintain that the service of the woman to the Lord's person was preferred to ordinary almsgiving. Nor yet is the building of a church worship properly so-called. Many build

* * * * *

727 1 Cor 10:31
728 Matt 25:40
729 Pio *XXIII libri* 131^rT
730 A barbed allusion to Gal 1:8 and a rejoinder to Pio *XXIII libri* 125^rF
731 Matt 19:12
732 Matt 5:42
733 MARGINAL NOTE: '1 John 3' (ie 1 John 3:17)
734 Alluding perhaps to Pio *XXIII libri* 122^v–123^rD, where Pio cites John's vision of the ornate New Jerusalem, a figure of the church (Rev 21:10–27); other references to Solomon's temple in this connection occur in *XXIII libri* 123^v–25^rF, 128^vO.
735 In his discussion of the anointing of Christ, Pio *XXIII libri* 126^vK–28^rM quotes Theophylact on Matt 26:8–11 and Mark 14:6–9, and John Chrysostom's Homily 80 on Matthew.

to the glory of their own name. But, so long as the mind is devout, worship is nothing other than an external action which looks to the service of God. Acts of worship, more strictly speaking, are scripture readings, genuflections, and other acts of this sort. But anyone would interrupt these if his neighbour were suffering grave necessity.

But these remarks of mine are beside the point. I have criticized only excessive expenditures on churches, and that when done to the neglect of neighbours at risk from starvation or thirst. Although Pio has to agree with me and admit that in this case almsgiving is to be preferred, nevertheless he quarrels with me in a lengthy disputation as if I disparage the decoration of churches, whereas in the very passage he quotes I make allowance for the dignity and even the majesty of churches. But what could be more ridiculous than to consider churches inadequately decorated unless in gems and gold they match the wealth of many kings? I will leave out now the fellow's garrulous complaints, abuse, buffets, and slaps.[736] He treats me throughout just as if I were directing all my mental energy to the end that I might abolish all the rites of the church. But I frankly express my approval of them in countless passages, and nowhere condemn them, though sometimes I do criticize the superstition or excessive reliance or hypocrisy connected with them.

ELABORATE FUNERALS[737]

He quotes a terrifying passage from 'The Funeral,'[738] and cautions me to listen well to my own words, this after a scary introductory remark to the effect that in this passage I make fun of everything.[739] At least I do not make fun of Cornelius' making confession to his parish priest in good time, when the fever was first stealing upon him, and his receiving the Body of the Lord, which he received from the same priest at home as he lay dying, of his lack of concern about his burial because of his hope in the resurrection, of his preferring to place his hope of salvation in the passion

* * * * *

736 See n29 above.
737 1531 UPPER MARGINS: 'On Elaborate Funerals' [De Funerum Pompa] on pages 143–5 [= 229–32 in this volume]. This topic is discussed by Pio XXIII libri 132^{r}Y–133^r.
738 Pio XXIII libri 132^{r}Y, quoting Funus LB I 816B / ASD I-3 549:434–7 / CWE 40 777
739 Pio XXIII libri 132^{r}Y: 'And I ask you to pay attention to your own words, and judge for yourself just how devout they are.' Pio opens this section with the words: 'It remains to see what things you carp at in the funerals and obsequies that are provided for the deceased, all of which usually take place in churches, seeing that in the dialogue entitled Funus you make fun of everything and condemn and carp wantonly.' Funus opens with an account of the excesses at the death and funeral of the wealthy George Balearicus.

of Christ and the mercy of God rather than in indulgences purchased from the popes.

But do let us listen to my words:[740] 'Shortly mention was made of tolling the bells, of thirty-day[741] and anniversary masses,[742] of a brief,[743] and the purchase of the community of merits.'[744] This was said about the parish priest, one such as a great many are, who brought to the dying man no consolation based on the goodness of Christ or the hope of life in heaven, but who, with a view to his revenues, talked about the funeral and place of burial, then about the tolling of the bells, thirty-day masses, a brief, and the sharing of merits, all of which parish priests generally sell, especially to those on the point of death. It would be another matter if out of pity these were provided for all Christians.

But what answer does Cornelius make? 'Pastor, I will be none the worse off if no bell tolls; or if you deem me worthy of one burial service, that will be more than enough. I do not desire to buy up someone's prayers or deprive anyone of his merits. There is sufficient abundance of merits in Christ, and I have faith that the prayers and merits of the whole church will benefit me, if only I am a true member of it. In two "briefs" I rest my entire hope. One is the fact that the Lord Jesus, the chief shepherd, took away my sins, nailing them to the cross. The other, that which he signed and sealed with his own sacred blood, by which he assured us of eternal salvation if we place our whole trust in him. For far be it from me that, equipped with

* * * * *

740 LB I 816B / ASD I-3 549:434–5 / CWE 40 779
741 A trental was a series of thirty masses said for the benefit of the dead, based on a practice mentioned in the *Dialogues* of Gregory the Great; see his *Dialogi* 4.57.14–16 (SC 265 192–4 and 194n); Joseph A. Jungmann *The Mass of the Roman Rite: Its Origins and Development* trans Francis A. Brunner (repr Westminster, Md 1986) I 130; Francis Clark *Eucharistic Sacrifice and the Reformation* 2nd ed (Chulmleigh, Devon 1981) 57–9; and Duffy 293–4. Pio provided in his will for the celebration of the *septimae, tricenarius, et anniversarius* masses for the repose of his soul (Pio *Will* 390).
742 Upon payment of a large fee (property or money), the donor was promised that henceforth until the end of time on a specified day (usually the date of his death) a mass would be celebrated for his soul and intentions. On this practice among Franciscans, see John Moorman *A History of the Franciscan Order from Its Origins to the Year 1517* (Oxford 1968) 354–5; and Paul L. Nyhus *The Franciscans in South Germany, 1400–1530: Reform and Revolution* Transactions of the American Philosophical Society new series 65 part 8 (1975) 1–47 especially 7, 14, 40. See also Philippe Ariès *The Hour of Our Death* trans Helen Weaver (New York 1981) 180.
743 A papal document granting a plenary indulgence at the hour of death; see Paulus II 149–59.
744 From the eighth century on in the West there developed, especially among religious orders, a notion of forming a common fund of masses and prayers that could be drawn upon by those who were granted a share; see Ariès *Hour of Our Death* (cited in n742 above) 160.

merits and briefs, I should summon my Lord to court with his servant, certain as I am that in his sight shall no man living be justified! For my part, I appeal from his justice to his mercy, since it is boundless and inexpressible.'[745] These are the words of Cornelius. Who ever heard a more impious speech?

But according to him I am criticizing the remission of sins imparted by the church.[746] How does this apply since Cornelius made confession and was absolved? But Cornelius talks about the bulls for sale which certain men with a view to profit have begun to shove into the faces of the dying and to din reliance on them into their ears. Even so, I do not condemn these briefs absolutely, but I am criticizing talk of them alone when other matters are more important. I know what the theologians say in the schools about the treasures of the church[747] and about their steward, the Roman pontiff, and I am not unaware of what these same man say in their private conversations, but I will not move this Camarina.[748] Cornelius did not leave out the almsgiving Pio mentions,[749] for he speaks as follows to the one who was asking him whether, as he was dying, he had bequeathed anything to monasteries or the poor: 'I have shared my modest fortune as I could afford. Now, as I am handing over possession of it to others, so I hand over the spending of it. And I am confident that my family will spend it better than I myself have done.'[750] From this it is clear that Cornelius bestowed as much as he could on the poor in the course of his life. Even the monks admit that it is safer to give alms while alive than to bequeath them in one's will.

If I should care to recall here the outrageous income of certain priests and monks and their bare-faced mockery of the dying, it would be quite clear that these matters were not spoken of unadvisedly in the colloquy. To be sure, Pio's discourse[751] on the honour of funerals and on prayers for the dead has nothing to do with me; its only purpose is to fill out his book. Finally, I have no need to defend whatever Cornelius said, just as I do not condemn all that George said or did. I set out a twofold example of dying, one simple and the other elaborate, and leave judgment to the reader.

* * * * *

745 *Funus* LB I 816B–C / ASD I-3 549–50:435–49 / CWE 40 777
746 Pio *XXIII libri* 132^rY: 'Are you not branding the Catholic church with the charge that it sells the remission of sins, whereas it appoints for the penitent almsgiving and prayers in consideration of which it forgives their sins.'
747 Pio *XXIII libri* 132^vY says that the church 'generously offers to the Most High the spiritual treasures stored up with her to buy off the punishments of her children.' For a study of the development of a theology of the treasury of the church, see Paulus II 184–206.
748 See *Responsio* n82 above.
749 Pio *XXIII libri* 132^rY, quoted in n746 above.
750 *Funus* LB I 815F–816A / ASD I-3 549:416–18 / CWE 40 777
751 On *XXIII libri* 132^vY–133^r, where, in addition to scriptural texts, Pio provides multiple quotations from Augustine and Jerome, and cites the Pseudo-Dionysius and Ambrose.

BELLS[752]

Next comes a defence of the bells that I do not anywhere condemn. I criticize only the parish priest's eagerness for gain.[753] He wished to sell the dying man the bells' din and clanging, something that should be provided for all Christians if they are of such great benefit.[754] Then, too, if the sound of the bells summons us to pray for the dead, why are fifteen bells sounded at the funerals of the wealthy when just one would be enough for this? Moreover, inasmuch as I presented the example of a man dying with no pretence at all, who did not bother to purchase that din of a lot of bells whereby the funerals of the wealthy are customarily ennobled, what place was there here, I want to know, for the tasteless jokes with which he begins? 'Did the sound of bells drown you out, or distract you from your note taking, that you should be so unmerciful to them? Your ears would, of course, have been less troubled if you had dwelt in Muslim rather than Christian lands. O delicate Desiderius, was the use of bells really introduced without good reason, etc?'[755] What could be more raving mad than these jokes? And yet it is with witticisms of this sort that he beguiled himself when he was on the point of death.

THE CULT OF THE SAINTS [FROM BOOK 8][756]

Here too the supplier begins with a huge lie.[757] Neither in the *Enchiridion* nor anywhere else, either in jest or in earnest, do I criticize the veneration of the holy cross, of the Virgin Mother of God, or of the image of

* * * * *

752 Pio *XXIII libri* 133ᵣz–133ᵛ provides a defence of the use of bells prompted by Erasmus' references to bells in the *Funus*.
753 *Funus* LB I 816B / ASD I-3 549:434–6 / CWE 40 777
754 Pio *XXIII libri* 133ᵛz says that bells came to be associated with rites for the dead 'to alert and forewarn their friends, for they are like the voices of the dead crying out and imploring their aid in providing those rites, as if they were saying "At least you, my friends, have pity on me!" [Job 19:21]. They also allay their grief by their happy sound which declares that the deceased has not perished, but is only asleep in the Lord and will rise again on the day of the Lord's coming.'
755 Pio *XXIII libri* 133ᵣz
756 MARGINAL NOTE: 'From book 8'; 1531 UPPER MARGINS: 'On the Cult of the Saints' on pages 146–50 [= 232–9 in this volume]. This heading is imprecise; Pio's book 8 (*XXIII libri* 133ᵛA–147ᵛz) is entitled 'On the Veneration of Images of the Saints, Male and Female.' Pio cites (134ᵣA) the treatment of images in his *Responsio paraenetica* (78ᵛA–86ᵣ, reprinted in *XXIII libri* 37ᵣA–38ᵣ) as providing basic principles for his argument, and quotes only the two passages mentioned here by Eramus. For the relevant passages, see G. Scavizzi *Arte e architettura sacra: cronache e documenti sulla controversia tra riformati e cattolici (1500–1550)* (Reggio Calabria 1982) 154–86, 202–13.
757 MARGINAL NOTE: 'Lie'

Christ,[758] but I do censure the superstition of certain lower-class people in connection with them. He did not quote the passage there, for it would refute what he says, but only its conclusion. It goes: 'I am ashamed to report with what superstition so many of them observe silly little rituals instituted by men, etc.'[759] You hear primarily a criticism of the superstition of certain people, not of the cult itself. But devout ears cannot endure[760] what is quoted from the annotations on the Epistle to the Hebrews, chapter 11:[761] 'whereas nowadays it has been discovered on the basis of supernatural philosophy that a wooden statue is adored with the same adoration as the eternal Trinity.' Here[762] Pio errs twice. First, he thinks that it is my view that any image you like is to be adored with *latria*,[763] whereas I am speaking of an image of the Holy Trinity. Secondly, he thinks that I approve the statement I report, whereas I disapprove of the fabrication of Scotus,[764] who

* * * * *

758 Paraphrasing Pio's opening charge (*XXIII libri* 133ᵛA): 'For in the *Enchiridion* you express at some length your dislike for the honour given to the sign (*signo*) of the most holy cross and to statues (*simulacris*) of the Lord Jesus and of his most holy mother and the saints. At the end of this business you burst into the following words: 'I am ashamed ...'

759 Quoted by Pio *XXIII libri* 133ᵛA from the *Enchiridion* LB V 33A / Holborn 77:15–17 / CWE 66 74. Pio's quotation alters the text somewhat, reading *observant* for *observent*, *has caeremonialas* for *caerimoniolas quasdam*, and omitting *non ... animo* from *ab homunculis, non tamen hoc animo institutas*. However, Erasmus here quotes the text as given by Pio, with the single exception that he alters *homunculis* to *hominibus*.

760 Alluding, perhaps, to Pio's expression of outrage, quoted in n763 below.

761 From Erasmus' annotation on Heb 11:21: 'And adored the top of his rod': ... To such an extent did they recoil at that time from adoring any created things, reserving this honour for God alone, whereas nowadays it has been discovered on the basis of supernatural philosophy that a wooden statue is adored with the same adoration as the eternal Trinity. Nor is there anyone whose ears [*aures*; LB unaccountably has *auctores*] this remark could scandalize, since on all sides so many statements are made that are scandalous to devout ears [*offensiva piarum aurium*]' (LB VI 1015D / Reeve *Galatians to the Apocalypse* 729).

762 MARGINAL NOTE: 'Criticizes things he doesn't understand'

763 Pio *XXIII libri* 134ʳB: 'The soul of any Christian is absolutely shocked and his ears cannot endure to hear such an outrage, that anything shaped by hands or produced by nature should be adored with the worship of *latria*.' Pio (134ᵛC) defines *latria* as follows: 'The worship of *latria* is the supreme honour and greatest reverence which must be shown only to that supreme and paramount Lord. This worship can be attained barely, to some small degree, only by the mind, more when it is struck with amazed contemplation of God's loftiness, our lowliness, and our indebtedness to God, than when it is confident it can draw near to the worship of God.'

764 Erasmus' identification of Scotus as the author of this thesis is problematic; see Minnich 'Debate on Images' 396 n40. When first mentioned, in the annotation on Heb 11:21 (1522; see Reeve *Galatians to the Apocalypse* 729), the thesis bears no attribution; but attribution to Scotus in the same words as used here (*Scoti commentum*) is found in a letter to Sadoleto dated 7 March 1531 (Allen Ep 2443:23). That Erasmus himself may have been unsure about this attribution is, perhaps, suggested by his words below: 'If

maintains this because the action of one moving towards an end through a means is one and the same, so that one who adores God through an image does not adore twice, but in a way different from one who does so without a statue.[765]

But I disapprove not because I think that what he says is in no sense true, but because the words are jarring to devout ears.[766] Since Pio admits this, and says that no sign is adored with the same adoration as the thing signified,[767] why does he wrangle with me, why does he hurl so much abuse at me, whom he does not understand? If Scotus does not, in fact, maintain this,[768] he should say only that I am wrong to this extent. If he does maintain it, Pio should quarrel with him, and not with me who censure Scotus' words as offensive to devout ears. It would be more authentically Christian to abstain altogether from words like 'adore' in connection with a wooden

* * * * *

Scotus does not, in fact, maintain this, he should say only that I am wrong to this extent. If he does maintain it, Pio should quarrel with him.' Pio himself does not mention Scotus in connection with this statement, but turns it back on Erasmus in his discussion in *XXIII libri* 146ᵛγ: '"On the basis of supernatural philosophy" is right, I tell you, Erasmus, because it was on the basis of sacred theology, on the basis of instructions and precedents taken from Scripture, on the basis of statements of the holy Fathers, on the basis of the authority of holy church, not today, but in antiquity, from the times of Christ himself and of the apostles, that the practice was introduced of adoring sacred images, not, as in your crafty and slanderous fantasy [*ut tu vafre ac calumniossisime commentaris*], in the way that "the eternal Trinity is adored," as was shown above, but with a suitable, reasonable cult, assigning the honour not to the images, and not stopping short with them, but to that deity to which the image is referred.'

765 Erasmus seems to be objecting to Pio's use of the verb *adorare* in this connection, but Pio makes careful distinction between *adoratio* and *adoratio* with *latria*, eg *XXIII libri* 134ᵛc: 'We bend the knee also to profane men, to kings and others who hold office in the state. We bend the knee to priests, children bend the knee to parents, and servants to their masters. But no one is said on that account to be adoring them with *latria*, though there is a sort of real adoration.' Pio cites a series of cases from Scripture of adoration of entities other than God, concluding: 'These, to be sure, are indications of veneration and honour, but they are not adoration with *latria*. Therefore, when we bend the knee to images and statues, we should neither be said to be adoring with *latria* nor to be assigning that honour to insensate objects, but to the almighty God signified by them.'

766 Erasmus is alluding to his own ironic statement to the contrary in the annotation on Heb 11:21 quoted in n761 above.

767 Pio *XXIII libri* 134ʳ–ᵛB, 135ʳ–ᵛE; eg 135ʳ: 'Although, as an image, it is to be venerated with due reverence and gesture because of what it represents, yet it must not be worshipped with the same adoration as if it should be adored along with the thing it represents.'

768 See n764 above; Pio, a student of Scotus, nowhere denies the ascription of this view to him, but he does say that as an account of Christians' veneration of images it is *falsissimum et calumniosissimum* (*XXIII libri* 135ʳ⁻ᵛ[E]).

statue or a painted image. It would have been better to say that we adore God or venerate the saints through the occasion provided by an image. Nor is it relevant here if kings want their standards to be adored.[769] God and the saints do not proceed on the basis of human desires. Princes do many things in a very human way.

What praise is due to those who serve up a nobleman to the learned as an object of ridicule? And it is on this foundation he relies as he enters upon a lengthy disputation supplied to him by some monk or theologian. The very style proves this as does the subtlety of the distinctions and arguments. A theologian, I think, had prepared this against the Zwinglian image smashers.[770] He turns the full force of it against me, although he is using the same arguments that I usually use against the attackers of icons [εἰκονόμαχοι]. Nowhere do I condemn the veneration of images except when it is superstitious, and even so the gloomy joker deals with me just as if with deep sighs I were condemning 'the practice of Christians in the veneration of images,' 'not only as foolish and superstitious, but even as sacrilegious.'[771] Who would not feel sorry for such great insanity as this? In one passage he calls upon me by name[772] as follows: 'For it is not you, Erasmus, who are the first attacker of images, nor is it a novelty that we are hearing in our times that images should not be venerated. For this controversy was begun in antiquity by heretics.' Let him go ahead and pretend what is not the case, that is, that anywhere I maintained this. But why does he prefer to mention the ancients to us over Zwingli, Oecolampadius,[773] and ever so many others? Oh, he is a prudent man and preferred a quarrel with Erasmus to calling down upon himself the pens of Germany, that is to say, he preferred to make one man a heretic rather than to refute many heretics, if, in fact, this is really part of the Catholic faith.

* * * * *

769 Pio does not make this point; perhaps Erasmus is alluding to Pio's statement that kings and civic officials are honoured with a bent knee, quoted in n765, or to the imperial statues in Antioch (*XXIII libri* 135^{r}D).

770 Eramus has in mind the scholastic argumentation in Pio's set-piece refutation of Erasmus' supposed critique of the cult of images in *XXIII libri* 138^{r}1–146^{v}Y. For Pio's explicit references to the German iconoclasts, see *Responsio paraenetica* 37^{r}z–38^{r}A; for his implicit and explicit references to Erasmus, see his *XXIII libri* 133^{v}A–137^{r}G, 143^{v}Y, 146^{v}Y. On the iconoclasm of Zwingli's followers, see eg Charles Garside *Zwingli and the Arts* (New Haven, Conn 1966) 106–28; Carlos N.M. Eire *War against the Idols: The Reformation of Worship from Erasmus to Calvin* (Cambridge 1986) 79–83, 92–4, 105–65; and Lee Palmer Wandel *Voracious Idols and Violent Hands: Iconoclasm in Reformation Zurich, Strasbourg, and Basel* (Cambridge 1995) especially 149–89 (Basel).

771 An almost verbatim quote from Pio *XXIII libri* 137^{r}G

772 Pio *XXIII libri* 143^{v}T

773 See CEBR III 24–7; and Eire *War against the Idols* (cited in n770 above) 114–19.

VENERATION OF RELICS[774]

They quote from the scholia on Jerome, but as usual it is a misquotation: 'Now we make a display of the kerchiefs or sandals of the saints, and they regard it a kind of sacrilege to handle the relics of the saints without sufficient reverence, for example, the girdle of Anne, the comb of George, or the shoe of Thomas. These relics in our reverence we also kiss, etc.'[775] What a terrifying passage, but one that was misunderstood![776] There,[777] and else-

* * * * *

774 1531 UPPER MARGINS: 'On Relics' on pages 150–4 [= 236–9 in this volume]. Pio's book 9 (*XXIII libri* 147^{r}A–172^{v}K) is entitled 'On the worship of the saints and the veneration of relics; also on the praises, titles, and innocence of the Virgin Mother of God; and on their being enrolled in the catalogue of saints according to canonical custom.' After a preliminary denunciation of Erasmus' scandalous views on these subjects (147^{r}A–148^v), Pio sketches (148^{v}C) the arrangement of book 9: 'First we will discuss the veneration of the saints in general and the cult of their relics, for it is because of the saints that we honour their works and relics; second, the particular veneration of the most glorious Virgin, exalted above all the saints, for in this way we will maintain the due progress of the discussion from lesser to greater matters; finally, we will also deal with the practice of canonization, thus completing the whole treatment of this topic.' Pio opens this book with long quotations from the *Enchiridion* and from the colloquies *Naufragium* and *Peregrinatio religionis ergo*, later introducing additional passages from the scholia on Jerome, *Enchiridion*, *Naufragium*, the *Annotationes*, *Apotheosis Capnionis*, and the colloquy Ἰχθυοφαγία. Erasmus, however, chooses to deal first with Pio's discussion of relics, and then returns to the topic treated at the beginning of book 9, his reply being divided into four parts: veneration of relics, cult of the saints, veneration of the virgin Mother of God, and canonization of saints.

775 The bulk of this quotation from Pio *XXIII libri* 152^{r}H ('For about their veneration [of relics] you say as follows in the scholia on Jerome ...') is taken from Erasmus' preface to the third series of documents in volume 2 of his edition of Jerome's *Opera omnia*, where it appears as follows: 'It is deemed a kind of sacrilege to handle the relics of the saints without sufficient reverence, for example the girdle of Anne, or the chalice of Edmund, the comb of George, or the shoe of Thomas. These relics in our reverence we also kiss or apply to our eyes. How much veneration therefore is owed to the writings of so saintly a man, which offer us not his cap or headband but the living and still breathing image of his mind!' (HO 2 190^v / CWE 61 91). Pio has added to the beginning: 'now we make a display of the kerchiefs or sandals of the saints'; and also adds an enclitic *-que* to *sacrilegii*, alters *habetur* to *habent*, substitutes *annulum* for *cingulum*, omits *aut calicem Edmondi*, and, of course, truncates the passage at *exosculamur*. In his reply here Erasmus quotes Pio's version except that he gives the original *cingulum* for Pio's *annulum*.

776 MARGINAL NOTE: 'Criticizes things he doesn't understand'

777 Though Erasmus virtually repeats Pio's quotation from HO 2, he seems to think that Pio has in mind a similar invective against the abuse of relics in the annotation on Matt 23:5 LB VI 118E–F / Reeve 92. There Erasmus refers to Jerome's remark on superstitious women in his *Commentarius in Matthaeum* 4:96–100 CCSL 77 212, which he mentions here, and goes on to complain of a large catalogue of questionable practices in which the shoe of Thomas of Canterbury is the only item in common with the catalogue from HO 2 190^v. (Erasmus also alludes to this criticism by Jerome in his *Declarationes ad censuras Lutetiae* LB IX 863A.)

where as well, Jerome is criticizing the superstition of silly women who used to carry around the wood of the cross. On the basis of this I reason that, if it is superstitious to carry the wood[778] of the cross, what is to be said of us who not only venerate lowly or even fake relics of the saints, something pardonable in a devout disposition, but also display them and force them on people for gain? If our behaviour in making a display of even undignified items is pious, then Jerome is impious in calling the behaviour of the women a superstition. But the shrewd friar-fraternity ignores this. You see, for some time now there is not much here that is Alberto's, for an exchange has taken place: they gave Alberto a cowl, he provided them with his name as a kind of cosmetic.

They quote from the colloquy 'A Pilgrimage for Religion's Sake': 'What, do these brutes want us to kiss all good men's shoes? Why not, in the same fashion, hold out spittle and other excrements to be kissed?'[779] These words are there, but if it is fair that I defend at my peril whatever is said in the colloquies by whatever character, in jest or in earnest, then a stern law has been laid down for me, especially since in the colloquies I have some interlocutors who are devout and some impious, learned and unlearned, foolish and wise. In the colloquy these are the words of Gratian, whose travelling companion I was, Ogygius being my name in the dialogue.[780] Now Gratian was sympathetic to the teachings of Wycliffe,[781] but what I have him saying there is not so very absurd, for he was not condemning the pious disposition of one who would venerate such things. And is my adversary here really approving our kissing the foul leavings of bodies?[782] Then what difference does it make whether it is the saints' spittle or urine we kiss or a kerchief dirty with phlegm or a shoe rotten with filth and maybe stained with urine?

* * * * *

778 *gestare lignum* 1531 *gestare signum* LB
779 *Peregrinatio religionis ergo* LB I 786D–E / ASD I-3 492:816–18 / CWE 40 648, quoted by Pio *XXIII libri* 152ʳH
780 Preserved Smith *A Key to the Colloquies of Erasmus* Harvard Theological Studies 13 (Cambridge, Mass 1927) 40, where he describes Ogygius as 'the "primaeval" champion of an obsolete form of piety.'
781 For an earlier description of Gratian, see *Peregrinatio* ASD I-3 488:648 / CWE 40 643, where he is depicted not as being a Wycliffite but as having read Wycliffe's books. On 'Gratianus Pullus' as John Colet and views of his that could seem similar to those of the Lollards, see Smith *A Key to the Colloquies* (cited in n780 above) 40; and John B. Gleason *John Colet* (Berkeley, Calif 1989) 253–5.
782 *sordidas . . . reliquias* can mean both 'foul leavings' and 'base relics'; Erasmus alludes here to Pio's treatment in *XXIII libri* 153ʳH of Gratian's words 'other excrements' (*alia corporis excrementa*), where he distinguishes the *varia . . . excrementorum corporis genera*.

But what is it Ogygius says there? 'I felt sorry for the old man and cheered him up with a bit of money, poor fellow.'[783] Now this is Erasmus' persona, under the name of Ogygius. Menedemus answers him: 'If soles of shoes were kept as evidence of a simple life, I wouldn't object, but I consider it shameless to push soles, shoes, and drawers at one to be kissed.'[784] Do you hear, O slanderer? It is not the pious disposition of the people who venerate these things that is being condemned, but the attitude of those who thrust them on others as objects of adoration, for kissing is a type of adoration. Ogygius, that is, Erasmus, answers him: 'I won't pretend it wouldn't be better to leave those things undone.'[785] What things would it be better not take place? The veneration of the relics of the saints? That is not what Ogygius says, but the shameless pushing, especially when it is for a profit, of such base relics as these, which are often fakes anyway. But there Ogygius does not even attack this, but goes on: 'But from what cannot be amended at a stroke I am accustomed to take whatever good there is';[786] and to the rest of his praise of St Thomas he adds that the shoe of a dead man was supporting a hostel for beggars.[787]

From this crumbling foundation the brawler goes on to a proof that it was appropriate that the veneration of the relics of the saints was introduced by the church,[788] as if I maintained that it is an impiety to give any honour to the relics of the saints. Thus it makes no sense to answer the stupid arguments of these men, or their raving abuse. They say that I am recalling Eunomius, who had long been consumed by the flames, back from the lower world, and summoning up Vigilantius from the mouth of hell,[789] as I strive to defend their fanatically impious doctrines which had been forgotten now for so many centuries. These things savour of the stupidity of

* * * * *

783 *Peregrinatio* LB I 786E / ASD I-3 492:819 / CWE 40 648
784 *Peregrinatio* ibidem 821–2, quoted by Pio *XXIII libri* 152ʳH
785 *Peregrinatio* ibidem 823, quoted by Pio *XXIII libri* 152ʳH
786 *Peregrinatio* LB I 786E–787A / ASD I-3 492:824 / CWE 40 648
787 *Peregrinatio* LB I 787A / ASD I-3 493:836 / CWE 40 649
788 After the quotations from *Peregrinatio*, Pio asserts (*XXIII libri* 152ʳH–152ᵛ): 'But it is quite clear that venerating them [relics] is altogether suitable, and that this practice was an appropriate development in the church'; and he then proceeds to give an elaborate, richly documented apologia for the cult of relics.
789 A paraphrase of Pio's invective in *XXIII libri* 155ʳL. Eunomius, the great Neo-Arian leader (d c 394), is identified as the inventor of the heresy of rejecting veneration of the relics of martyrs by Jerome in his *Contra Vigilantium*, the source of some of Pio's arguments in this section. Jerome accused Vigilantius, an Aquitanian cleric against whom he wrote the treatise in 406, of holding the same heterodox view; see *Contra Vigilantium* 8 PL 23 347A.

the friary! When would Pio have written such stuff, if he were of sound mind in a sound body?[790] He goes so far as to say that no one can give more honour to relics than is their due,[791] although I have seen ragged kerchiefs which still bore traces of snivel, or maybe even of sweat from armpit or groin,[792] adored by monks with the same ceremonies as the Body of Christ, not to mention the countless stories about the superstitious behaviour of the masses. But Pio has always lived among the pious and has neither seen nor heard any of these. It was of no concern to him, I suppose, intent as he was on other matters.

CULT OF THE SAINTS [FROM BOOK 9][793]

To demonstrate that I detest the cult of the saints[794] they quote a snippet from the *Enchiridion*: 'In short, in this way we have appointed certain saints to preside over all things we fear or desire. These differ with each nation.'[795] I clearly give my approval to the cult of the saints in the *Enchiridion*,[796] but I disapprove of it when it is superstitious, a malady with which my countrymen are particularly infected. But the fellow who is wrangling with me denies that this view was asserted by any of the Fathers or advanced by any authoritative writer, and so it is clear that it was invented by me alone out of my zeal for abusive criticism.[797]

Hear the contrary, you stupid babbler, whoever you are, who understand neither my words nor your own statement. Did I claim authoritative

* * * * *

790 Alluding to Juvenal 10.356

791 Pio opens the section on relics with the following words (*XXIII libri* 152^{r}H): 'It remains for us to show that in the veneration of their sacred relics no undue honour is bestowed and nothing is given that is excessive.'

792 This reference to disgusting relics recalls the experience of Ogygius and Gratian at Canterbury (reported in *Peregrinatio* LB I 785 / ASD I-3 491:763–6 / CWE 40 647) who were shown 'some linen rags, many of them still showing traces of snivel. With these, they say, the holy man wiped the sweat from his face or neck, the dirt from his nose, or whatever other kinds of filth human bodies have.'

793 1531 UPPER MARGINS: 'Cult of the Saints' on pages 154–9 [= 239–43 in this volume]; with a note in the left margin: 'From book 9.' Here Erasmus returns to the topic with which Pio had begun book 9.

794 In fact, Erasmus is misrepresenting Pio's introduction of the quotation from the *Enchiridion XXIII libri* 155^{r}M: 'But it must not be overlooked that you do not so much loathe or ridicule the cult afforded in the veneration of the saints as the fact that different saints are invoked in different places to obtain different favours.'

795 *Enchiridion* LB V 26F / Holborn 66:14–16 / CWE 66 64, quoted by Pio *XXIII libri* 155^{r}M

796 For examples of Erasmus' conditional endorsement in the *Enchiridion*, see LB V 27B–C / Holborn 66:37–67:20 / CWE 66 64; and LB V 31D–E / CWE 66 72.

797 Pio *XXIII libri* 155^{r}M

figures as the source for what I maintain arises from the superstition of the ignorant masses? Furthermore, so that it will be clearer that I am speaking there not of a Christian cult but of a superstitious and pagan one, I will point out the drift of the passage. In the fourth canon (you see they suppress the passage, pouring out blackness like an octopus, so that they will not be caught out) I maintain that Christ must be considered the goal in every Christian activity.[798] Eventually I put forward these instances: 'You take food in order to be healthy of body. If you wish to be healthy of body so that you may have sufficient strength for holy studies and holy vigils, you have attained your goal. But if you take care of your health in order not to become less attractive and less capable of satisfying your lustful desires, then you have fallen away from Christ and fashioned another god for yourself.'[799] There follows immediately: 'There are those who honour certain saints with certain specific ceremonies. One person greets Christopher each day, but only if he sees his image. To what end? Evidently because he is convinced that thus he will be preserved from a bad death that day.'[800] Thus far the *Enchiridion*. Even this very fact is superstitious: they promise themselves as a result of seeing an image of Christopher what ought not to be promised even as a result of seeing the Body of Christ[801] unless they straighten out their lives. This too is not without superstition, that they term a bad death not one that occurs when one is in state of sin, but one that is sudden and unforeseen. A third superstition lies hidden here. The ignorant masses do not seek these bodily goods from Christ through the intercession of the saints, but from the saints themselves as if they were the source. They will argue: 'It is not the church that does these things.' I agree. Nor does the church assign specific departments to specific saints, but this is done by the superstition of the ignorant crowd, while certain priests, who hold revenue more dear than the glory of Christ, look the other way.

* * * * *

798 *Enchiridion* LB V 25A–B / Holborn 63:9–10 / CWE 66 61
799 *Enchiridion* LB V 26E / Holborn 65:35–66:3 / CWE 66 63
800 *Enchiridion* LB V 26E / Holborn 66:4–7 / CWE 66 63; on the image of St Christopher, see CWE 66 295 n14.
801 See Adolph Franz *Die Messe im deutschen Mittelalter: Beiträge zur Geschichte der Liturgie und des religiosen Volkslebens* (Freiburg-im-Breisgau 1902; repr 1963) 102–5; he quotes (103 n3) from a fifteenth-century manuscript from Graz: 'The merits of seeing the Body of Christ are the following, according to St Augustine [*sic*] *On the Truth and Wisdom of God*: that on that day necessary sustenance is provided, frivolous conversations are avoided, unintended oaths are avoided, eyesight is preserved, sudden death does not assail on that day, [...] if anyone dies suddenly without communion, he is regarded as having received communion'; see the vernacular prayer at the time of the elevation of the host, quoted by Clark *Eucharistic Sacrifice* (cited in n740 above) 555.

There remains a fourth superstition. They seek these external goods from the saints, not with Christ as their goal, but rather the desires of the flesh. If criticism is due to one who seeks a long life from Christ so that he may longer enjoy the delights of this world, then an even more just criticism is due to him who seeks the same,[802] with a like intention, from Christopher or Barbara.[803]

But so that no one will think that these points have been twisted by rhetorical artifice, the passage continues: 'If this sort of piety is not turned from mere consideration of material advantages or disadvantages and redirected towards Christ, then, far from being Christian, it is not much removed from the superstition of those who pledged a tenth part of their substance to Hercules so that they might become rich or a cock to Aesculapius in order to recover from some sickness, etc.'[804] And I return at once to the worship of Christ: 'You pray to God that you may not suffer a premature death rather than pray that he grant you a better frame of mind so that wherever death overtakes you it will not find you unprepared. You give no thought to amending your life, yet you ask God that you may not die. What is the object of your prayer? Obviously that you may go on sinning as long as possible, etc.'[805] You see, you absolutely shameless slanderer, that here I employ a euphemism when I call the superstitious temper of the masses 'piety,' but I do warn that they should look to Christ in all things. Does this constitute abomination or ridicule of the cult of the saints? By the same line of reasoning I ridicule the cult of God!

From these remarks I think it is clear enough that there is nothing more stupid than denunciations of this sort, which are fabricated from misunderstood notecards and the gossip of the friars. But this will be even more apparent from the words that follow there. They run as follows: 'There will be objections at this point from certain petty religious who consider piety a source of gain[806] and, as the same [Apostle] said, "by sweet and flattering words seduce the hearts of the simple-minded, serving their own appetites rather than Jesus Christ." "Therefore," they will say, "do you forbid the veneration of the saints, in whom God is honoured?" I do not condemn those who do these things out of a naive superstition so much as I do those who pursue them for their own gain. They glorify practices that

* * * * *

802 *eadem* 1531 *eam* LB
803 See CWE 66 285 n16.
804 *Enchiridion* LB V 26F–27A / Holborn 66:18–23 / CWE 66 64
805 *Enchiridion* LB V 27A / Holborn 66:26–30 / CWE 66 64
806 1 Tim 6:5

are barely admissible as if they were the paragon of consummate piety, and they encourage for their own advantage the igorance of the masses, which even I do not entirely condemn, but I will not allow them to lend great importance to things that are merely indifferent or consider matters of little value as being very important. I shall commend them for asking their patron Rocco to keep their lives from harm if they consecrate that life to Christ, etc.'[807] This is what Erasmus says in the *Enchiridion.* You deaf asp,[808] can you not hear that in this passage, far from the devout and correct worship of the saints being condemned, the invocation of the saints, even when it is not wholly free from superstition, is generously tolerated? The most important worship of the saints is, however, imitation, but with discretion.

They rush back to the *Colloquies,* a work designed to bring boys to learn Latin through the enticement of pleasure, as if through play. But one must humour the ill-tempered. What is in 'The Shipwreck'?[809] One finds portrayed there a variety of dispositions of ordinary people; each usually reveals his character in a crisis like this. I do not necessarily approve of whatever is done there, nor, again, do I disapprove. If I approve of everything, then there is the mention there of sacramental confession. If I disapprove of everything, then the invocation of God has been condemned. But let me not waste a lot of time in ludicrous replies to all their silliness, especially since in the *Liturgy of the Virgin* I both show my approval of the cult of the saints and show how they should ideally be venerated.[810] They set out arguments advanced in the books of the Lutherans,[811] and say to me: 'Maybe I meant to scratch your itchy back with these.'[812] Who does not realize that this is a witticism of the

* * * * *

807 *Enchiridion* LB V 27B–C / Holborn 66:32–67:7 / CWE 66 64. One variant occurs: LB V reads *innocentium,* Erasmus here gives *simplicium.*

808 Ps 57:5

809 Pio *XXIII libri* 156^{v}N had referred, though without naming the work, to several details from the colloquy *Naufragium* LB I 713A–B / ASD I-3 328:91–4, 98, 103; 329:126 / CWE 355, 356.

810 Erasmus refers here to his *Virginis Matris Lauretum cultae liturgia,* a votive mass in honour of the Virgin of Loreto, published in 1523; see eg LB V 1327A–B, 1329A–32F, 1333F / ASD V-1 97:7–13, 99:85–6, 100:91–105, 107:356–69 / CWE 69 86–7, 91–2, 104. Erasmus' emphasis is on praising and imitating Mary's virtues as wife and mother.

811 Pio presents a series of arguments against the intercession of the saints in *XXIII libri* 156^{v}N–157^r, but refutes them in detail in 157^r[0]–162^{v}R.

812 Pio *XXIII libri* 157^r[0]: 'Maybe I meant to scratch your itching back with these, but there is no reason for you to be glad, for arguments like these fall far short of delivering what they promise at first glance.'

lice-infested? But this was their plan: 'We will not provoke the Lutherans against our order for fear we may hear things we would rather not, but if we rave against Erasmus, we will win considerable favour even with them.' You are, however, fooling no one. The world is now too familiar with your low tricks, and therefore you are headed for destruction, sure to perish unless you recall your lives to their original sincerity. And so I pass over this discussion, for there is not a word there that pertains to me.

VENERATION OF THE VIRGIN MOTHER OF GOD[813]

Here there is a real outcry. 'O heaven, O earth, O seas of Neptune!'[814] An excerpt from 'The Shipwreck'[815] is brought forward in support of the slander. My *Paean* and *Prayer*,[816] now so often printed,[817] show how highly I regard the most holy Virgin, as does the *Liturgy* and the sermon[818] that I published against the Virgin's detractors. I have always supported the opinion

* * * * *

813 1531 UPPER MARGINS: 'Cult of the Mother of God' on pages 160–73 [= 243–55 in this volume]. This section is a reply to Pio *XXIII libri* 162^{v}s–171^{r}G. All of Erasmus' major statements on Mary and her cult have been collected and studied by Joaquín María Alonso *Erasmi corpus mariologicum* 2 vols, Marian Library Studies new series 11–12 (Dayton, Ohio 1979–80); selections from this section of Pio's work are given on II 399–411 and from Erasmus' reply on II 412–20, with notes. See also Alonso's study of this phase of Erasmus' mariology in II 394–7 and the works cited there.

814 Quoted from Terence *Adelphoe* 790

815 Pio *XXIII libri* 162^{v}s quotes from the colloquy *Naufragium* LB I 713A / ASD I-3 327:71–4 / CWE 39 355; cf *Declarationes ad censuras Lutetiae vulgatus* LB IX 942C–943A.

816 Erasmus' *Paean Virgini Matri* LB V 1227C–1234C / CWE 69 20–38, and his *Obsecratio ad Virginem Mariam* LB V 1233D–1240A / CWE 69 41–54, both dedicated to Anna de Veere and published in Erasmus' *Lucubratiunculae* (Antwerp: Martens 1503). Erasmus' statement here notwithstanding, he early disassociated himself from these compositions. In a letter accompanying the gift of his *Lucubratiunculae* to John Colet in 1504, Erasmus says of much of the contents 'as for all the rest, I wrote them almost against the grain, especially the *Paean* and *Obsecratio*; this task was discharged in deference to the wishes of my friend Batt and the sentiments of Anna, princess of Veere' (Ep 181:60–2); and in the *Catalogus* of his works addressed to Johann von Botzheim in 1523, Erasmus observes of these works: 'and besides that, two prayers to the Virgin Mother, written to please the boy's mother, Anna, lady of Veere, in a childish style designed to suit her feelings rather than my judgment' (Ep 1341A:742–4).

817 See vander Haeghen I 131, 136, who lists seven editions of the *Obsecratio ad Virginem Mariam* (1503–24) and nine editions of the *Paean Virgini Matri* (1503–25).

818 *Virginis Matris Lauretum cultae liturgia* (1523), with its sermon (*concio*) which Erasmus added in the second edition of 1525. See n810 above. Léon-E. Halkin (ASD V-1 90) acknowledges that Erasmus may have wanted to prove his orthodoxy by composing this liturgy, but holds that he probably also wanted to display his skills in another literary genre, namely, liturgical texts.

that she is free of all sin, even original sin,[819] though this is not properly called a sin[820] except in the case of the first parents.

But let us listen to 'The Shipwreck.' It says: 'the sailors singing *Salve Regina*, praying to the Virgin Mother, calling her Star of the Sea, Queen of Heaven, Mistress of the World, Gate of Salvation, flattering her with many other titles the Sacred Scriptures nowhere assign to her.'[821] 'You are cracking jokes there,' says Pio. No, I am instead ridiculing,[822] not the most holy Virgin, but rather the stupid sailors who did not think of Christ in their pressing danger, but only of his Mother. Nor do they ask her[823] to intercede with her Son, but to show them her Son after this life.[824] And they flatter her with novel titles, as if, being a woman, she could be cajoled with such flattery. And it adds 'with other titles,'[825] probably even more superstitious ones which it does not report for that reason.

Now who is denying that the blessed Virgin deserves the most splendid titles? But not all are equally suitable to her. If this world is a sea in which we sail while we live in the body, if there is a star, a *Cynosura* [Ursa Major] or *Helice* [Ursa Minor],[826] from which one should not turn one's eyes,

* * * * *

819 Erasmus is responding to Pio's accusation (*XXIII libri* 163ʳs, 167ᵛD–171ʳG) that he denies Mary's freedom from original and actual sin. For examples of Erasmus' varying statements on this question, see Epp 1126:327–37 and 1196:58–67 and *Annotationes* LB VI 69F–70F, 236D–237E (favouring Mary's sinlessness), but also LB VI 696C–D (the church's teaching is not clear enough). See also *Supputatio* LB IX 569E–570B (Erasmus' position); and *Apologia adversus monachos* LB IX 1036C (for an acceptable argument) but 1051D (for an unacceptable one). See also José Ignacio Tellechea Idígoras 'María en los escritos de Erasmo' in *Quincena Semana Española de Teología* (Madrid 1956) 291–325; and Léon-E. Halkin 'La mariologie d'Erasme' *Archiv für Reformationsgeschichte* 68 (1977) 32–54.

820 *Annotationes* LB VI 70E / Reeve 59

821 *Naufragium* LB I 713A / ASD I-3 327:71–4 / CWE 39 355

822 Translating: *Rides, inquit Pius. Immo etiam derideo* ...(emphasis supplied). In *XXIII libri* 163ʳs Pio says that the whole dialogue has no other purpose than 'to excite laughter and make sport of the Virgin and all the saints,' and describes one of Erasmus' witticisms as 'a buffoonery designed to make boys laugh'; but he also complains of Erasmus' *derisoria verba*, his *inrisorium* directed against Mary, and delivers an impassioned expression of righteous indignation to which Erasmus will refer later (248 below).

823 *illam* 1531 *illum* LB

824 The concluding request of the antiphon *Salve Regina* is: 'and after this exile show us the blessed fruit of your womb, Jesus'; among Erasmus' never published juvenilia (tentatively dated spring 1499) is a paraphrase in elegiac distichs of this most famous of Marian antiphons (CWE 85 338–9 and nn). On the situation of the *Salve Regina* in Reformation disputes, see Jose Maria Canal *Salve Regina Misericordiae* (Rome 1963) 109–10.

825 *Naufragium* LB I 713A / ASD I-3 327:71–4 / CWE 39 355. On the abundance of titles assigned to the Virgin Mary by Erasmus himself in his youthful writings, see Alonso (cited in n813 above) II 481–4.

826 That is, the constellations Ursa Major and Ursa Minor, also known as the Big and Little Dipper; Ursa Minor contains the North or Pole Star, a key directional guide.

if eternal life is the harbour, then it is, rather, Christ who is the Star of the Sea, he who said: 'I am the way, the truth, and the life';[827] 'he who follows me does not walk in darkness.'[828] But if we turn our gaze from Christ, it was said to the apostles: 'You are the light of the world';[829] and Paul says: 'You be imitators of me, as I am of Jesus Christ.'[830]

Now I know very well that these can all be defended by means of an accommodating interpretation,[831] but one can with the same accommodation just as readily defend also those prayers turned out by the printers which are hung up in the churches, in which whatever could be sought from God – hatred of sins, true confession, sincere and constant charity, hope, and, finally, eternal life – are sought from the blessed Virgin, for she who bore him who bestows all things does give to some extent, or again, she by whose intercession things are obtained does give in a way.[832] But, O man of renown, it is not safe to depart very far from the Cynosura of divine Scripture, and there is a danger that these accommodating interpretations will palm off superstition on us in place of religion, especially when we see the ignorant throng so inclined in this direction. Let human enthusiasm devise countless titles, none is more pleasing to the Mother than that whereby her Son is most glorified.

* * * * *

827 John 14:6
828 John 8:12
829 Matt 5:14
830 1 Cor 11:1
831 Pio XXIII libri 163^{v}s–164^r reviews titles given to the Virgin in the New Testament, and then from the Old Testament a more problematic series of titles that were applied to the Virgin 'mystically.'
832 On the practice of *ex voto* writings, paintings, and figurines placed in churches near tombs of saints and sacred images to serve as tokens of prayers of praise and thanksgiving for favours received, see 'Voto o Tabella o Tavoletta votiva' in Gaetano Moroni *Dizionario di Erudizone storico-ecclesiastica* 103 vols plus 6 vols of indexes (Venice 1840–79) CIII (1861) 133–44. For examples of printed woodcuts and metal engravings of Mary accompanied by prayers, wee Wilhelm L. Schreiber *Handbuch der Holz- und Metallschnitte des XV Jahrhunderts*, 2nd ed 11 vols (Stuttgart 1926–30; repr Nendeln 1969) II nos 1009, 1012, 1039a, 1044, 1111, 1152, all photographically reproduced in Richard S. Field, ed *The Illustrated Bartsch 164* (Supplement), *German Single-leaf Woodcuts before 1500* (New York 1992) 18, 20, 60, 69, 143, 194. One of the most famous instances of a similar practice from the time of Erasmus took place annually in the church of Sant' Agostino in Rome on 26 July, when poems written by leading humanists that celebrated the virtues of the saint or gave thanks for favours received were affixed to the statue of St Anne. The poems of Giano Francesco Vitale and Gaius Silvanus giving thanks for Pio's recovery from illness can be seen as examples of such appended prayers. See D'Amico *Renaissance Humanism* 108–9; Virginia Anne Bonito 'The Saint Anne Altar in Sant' Agostino: Restoration and Interpretation' *Burlington Magazine* 124 (1982) 268–80; and Introduction n28, where it is noted that Vitale's poems are called 'voti.'

But in 'The Shipwreck' one answers: 'What has she to do with the sea? She never went voyaging, I believe.'[833] A second replies: 'Formerly Venus was protectress of sailors, because she was believed to have been born of the sea. Since she gave up guarding them, the Virgin Mother has succeeded this mother who was not a virgin.'[834] And another says: 'You're droll.'[835] And so there is irony at work. Against whom? Not against the Virgin, but against the sailors who were earlier accustomed to invoke Venus, but now invoke Mary. This might be more acceptable, but there is not much improvement if they make their invocation in a spirit like that of most sailors, for there is no class of men more superstitious. But who placed the Virgin Mother of God in charge of the sea unless it was a pious, but superstitious, enthusiasm? Christ said: 'To me has been given all power in heaven and on earth.'[836]

Finally the colloquy presents certain people flattering the sea itself just as the sailors were flattering the Virgin Mary: 'O most merciful sea, O most kind sea, O most splendid sea, O most lovely sea, grow mild, save us!'[837] And these are not fictions, they really happen. Denouncing these things strengthens Christian piety; it does not diminish the Virgin's dignity. In addition, just as these people bellow whenever they encounter something that seems to diminish the pope's visible majesty, although, in fact, it advances his true dignity, and likewise whenever they find something that takes anything away from the external cult of the Mother, although, in fact, all the Mother's glory is situated in her Son, even so do those who love the Lord Jesus with all their hearts sometimes grow angry when they see Christ's glory obscured under pretext of piety by the ignorant with the connivance of bishops and priests. In fact, I replied some time ago to the Spanish monks about this slander,[838] though this brawler pretends not to know it.

He excerpts, for purposes of slander, certain snippets from the annotations on the twelfth chapter of Matthew,[839] but if anyone will read the whole annotation, he will see the wicked craftiness of this slanderer. For I

* * * * *

833 *Naufragium* LB I 713B / ASD I-3 327:75 / CWE 39 355. It is this line that Pio describes (163ʳs) as 'a bit of buffoonery designed to make boys laugh, and perhaps not an inappropriate line if it had been said in some entertaining comedy about Diana who dwells in the groves, or about Vesta.'
834 Ibidem lines 76–7
835 Ibidem line 78
836 Matt 28:18
837 *Naufragium* LB I 713B / ASD I-3 327:82–4 / CWE 39 355
838 *Apologia adversus monachos* LB IX 1084B–1087B especially 1087B
839 That is, on Matt 12:48 LB VI 69F–70F / Reeve 58–9; Pio gives a severely condensed version

begin as follows: 'But how I wish that all of the Christian people were so given over to the cult of the most blessed Mary that they would imitate her virtues with all their zeal. Otherwise, either the contemporary public assigns her more than enough because of some devout enthusiasm, or Chrysostom assigns her less than enough when he expounds this passage.'[840]

Here he shouts that I am twisting the testimony of the holy Fathers to insult the Virgin.[841] In fact, I quote Chrysostom verbatim there. He says: 'Consider, then, the importunity both of his mother and of his brethren. For although they should have gone in and listened along with the crowd, or at least awaited the conclusion of the speech outside and then approached, they were prompted by a kind of pretention and ostentation and called him out in the presence of all.'[842] But, he says, the Aldine edition spares the Mother.[843] Of course, it was 'corrected,'[844] as has been done in the case of very many of the Doctors of the church, but the Greek is otherwise. Indeed, Theophylact, who is hesitant to depart from Chrysostom's footsteps, writes about this very passage as follows: 'The Mother wanted to show something all too human, that she had power over her Son. For she did not yet realize anything remarkable about him, and on that account wanted to call him to herself while he was still speaking, as if she were a bit pretentious because so great a son was subject to her'; and shortly after he says: 'He did not say this to hurt his Mother, but to correct her character,

* * * * *

of this annotation with many elipses and some paraphrase in *XXIII libri* 162ᵛs, and later returns to a detailed criticism of it (167ᵛD–171ʳG).

840 *Annotationes* LB VI 69F / Reeve 58; Pio does quote these words later (169ʳF).

841 Pio *XXIII libri* 168ʳ–ᵛD quotes texts from Origen, Augustine, Cyril of Alexandria, Hilary of Poitiers, and Ambrose that assert Mary's freedom from sin, and concludes (168ᵛE): 'Although these testimonies of the holy Fathers are so very explicit in their assertion of the freedom of the glorious Virgin from every fault, you still dared to assign to her not only the taint of original sin but also the commission of actual sin, and you did not blush to so assert with sacrilegious boldness that she was ambitious and weak in faith in the opinion of Chrysostom and Augustine. If they had said this, they would have been contradicting themselves.'

842 *Annotationes* LB VI 69F / Reeve 58–9, quoting John Chrysostom *In Matthaeum* homily 44.1 PG 57 465

843 Pio does not mention a specific edition, but with the words of *XXIII libri* 169ʳF, 'Chrysostom in another version [*in altera editione*] also ascribes this importunity to his brethren rather than to his Mother, as not even you deny, thus sparing his Mother,' he is reminding Erasmus of his annotation on Matt 12:48: 'But in another version [*in altera editione*], for the codices vary, a homily with the same number ascribes this domineering importunity to his brethren in particular, thus sparing his Mother.' See Reeve 59 for this quotation and the various versions of this annotation.

844 In *Responsio ad annotationes Lei* Erasmus specifically criticizes the Aldine edition of Chrysostom LB IX 141D.

which was eager for glory and merely human'; and a little later: 'Therefore, once he had healed the disease of vainglory, he was once again obedient to his Mother.'[845] There I also quote from Chrysostom on John: 'And perhaps she was affected with some merely human yearning, as were his brethren also.'[846] (Here the brawler says that I left out the adverb 'perhaps.' Possibly it was left out on the denouncer's notecard, but it was not left out in my *Annotations*.)[847] And shortly: 'For they did not yet have a due regard for him, but in the way of mothers Mary thought that she had a right to complete command over her Son, whereas it was required that she worship and revere him as her Lord.'[848] Theophylact too says that his Mother was reproached or rebuked by Christ.[849] In addition to these is the point I allege from the view of Augustine,[850] who assigns to Mary a weakness of faith at the death of Christ, though a less severe one than the disciples. And do I not indicate this plainly in my annotation on chapter 2 of Luke[851] where I quote Origen,[852] who plainly assigns some sin of action to Mary? Nor does Theophylact[853] keep silence about Mary's scandal at the death of her son.

But if it is certain that it was the sacred authors who wrote and thought this, then where is the blasphemer who attempts to prop up his impious and deranged view with the testimonies of saintly men? And he raves and shouts:[854] 'O manners, O laws, O zeal, O devotion, where have you gone? How can it be that these things go unheard and ignored by these princes and leaders of the church?'[855] I too would like to shout: 'O Aesculapius,

* * * * *

845 On Matt 12:46–50 PG 123 276C–D
846 John Chrysostom *In Ioannem* homily 21.2 PG 29 130, quoted in the annotation on Matt 12:48 LB VI 70D / Reeve 59
847 Pio *XXIII libri* 169ʳᴇ: 'Chrysostom, however, does not ascribe these feelings to her absolutely, but says "perhaps," which you left out [*quod tu praetermisisti*].' *Praetermittere* could mean 'omit,' as Erasmus clearly understands it here, or merely 'overlook'; in any case, Erasmus did not omit the word in his annotation.
848 PG 29 130–1, also quoted in the annotation on Matt 12:48 LB VI 70E / Reeve 59
849 On Matt 12:47–50 PG 123 276D
850 This is, in fact, Pseudo-Augustine (ie Ambrosiaster) *Quaestiones veteris et novi testamenti* 77 (73).2 CSEL 50 131:12–25, cited in the annotation on Matt 12:48 LB VI 70E / Reeve 59 and paraphrased in the annotation on Luke 2:35 LB VI 236D–E / Reeve 166.
851 On the prophecy of Simeon at Luke 2:35: 'And your own soul a sword shall pierce'; *Annotationes* LB VI 236C–237F / Reeve 166–7.
852 *In Lucam homiliae* 17.6 SC 87 258, an addition made to this annotation (LB VI 236F) in the edition of 1527 (see Reeve 167)
853 On Luke 2:35 PG 123 732, a text subsequently mentioned in the 1535 version of the annotation on Luke 2:35 (Reeve 166)
854 MARGINAL NOTE: 'He raves without cause'
855 Referring back to Pio's indignant outburst at the beginning of this section (*XXIII libri* 163ʳs)

O Hippocrates,[856] O mighty art of medicine, where have you gone, where are you loafing, that such revolting persons rave so with impunity?' I have proven that I nowhere ridicule the sainted Virgin, but only human superstition. To exactly which leaders of the church is this madman, then, making his appeal? To Chrysostom, Augustine, Theophylact, and Origen?[857] In that very passage which he quotes[858] I nowhere subscribe to their view; rather, I declare plainly that in this matter I am not inserting my own opinion, and I profess that I favour the view of those who describe Mary as free from every sin.[859]

Also there could be an appetite for renown or a weakness of faith that would not be sinful, a matter of error or ignorance; but not in the minds of those who assign to Mary a complete revelation of and absolute faith in the divinity of Christ from the moment of conception! If they want to hold on to this view tenaciously, they would have an easy escape if they were to say that Chrysostom and the rest erred in this matter, as in many others. For when it comes to original sin, all the ancients exempt Christ alone from every sin, nor are there lacking today those who maintain publicly in their churches that Mary was not free from original sin,[860] though I express the opposite view in my books.

But to return to the *Colloquies*, from which serious men select things said in jest for reproach in earnest. In 'A Fish Diet' someone says: 'How many there are who put their trust in the Virgin Mother's protection, or Christopher's, rather than that of Christ himself! They venerate the Mother with images, candles, and canticles. Christ they offend recklessly by their wicked

* * * * *

856 Aesculapius (Latin form of the Greek Asklepios), son of Apollo and Coronis, a prominent deity of healing and medicine whose shrines were renowned both for miraculous cures and the expertise of their resident physicians; Hippocrates (fourth century BC), the most celebrated ancient physician, gave his name, if nothing more, to the Hippocratean corpus of medical texts.
857 Using the interrogation point of 1531 against the full stop provided in LB
858 That is, the annotation on Matt 12:48, which Pio quotes piecemeal at XXIII *libri* 162ᵛs, 169ʳE
859 Alluding to the annotation on Matt 12:48 LB VI 70F / Reeve 58–9: 'I am not inserting my own opinion in this matter ... Thus far the words of Chrysostom. <1522 Augustine seems to assign to her some little lack of faith.> I assume that she was altogether free of original sin, but he seems to assign her some so-called actual sin, unless, perhaps, that attachment which he assigns to her is free of all blame. <1522 And that is rather what I think our view should be. For as there can be an ignorance without culpability, so can both doubt and maternal attachment be without culpability.>'
860 Principally the Dominican followers of Thomas Aquinas; see William A. Hinnebusch *The History of the Dominican Order: Intellectual and Cultural Life to 1500* (New York 1973) 171–80. See also Farge *Orthodoxy* (cited above n301) 129 n73 (1), 163.

life. [. . .] And they think the Virgin will help them because at nightfall they sing a hymn they do not understand, *Salve Regina*.'[861] Tell me, O pitiable brawler, what is here that is impious? Are there no such people? How I wish that there were not! Or do you believe we should approve of those who, placing their reliance on a song, think they have the good will of the Mother, although by their whole way of life they are intensely offensive to Christ? I cannot figure out why they thought they should quote this passage.

He says I criticize the practice whereby preachers implore the Blessed Virgin in place of Christ, calling her 'the Font of Grace.'[862] In fact, I do not criticize, but suggest in a very restrained way what is more in keeping with both the authority of Scripture and the custom of the primitive church. 'But of old,' he says, 'it was necessary to celebrate Christ; but now that he is well enough known, it is appropriate to celebrate his Mother.'[863] As if there were no opportunity to celebrate the Mother, or as if Christ has not been practically abolished from men's minds by judaizers!

'But at the end,' he says, 'they add "Pray for us sinners."'[864] Maybe this is done among the Italians;[865] among us each salutes the Virgin in

* * * * *

861 Ἰχθυοφαγία 1462–3 LB I 808F / ASD I-3 535:1462–3 / CWE 40 719; Pio quotes this passage in a somewhat garbled form at *XXIII libri* 166ʳA.

862 Erasmus complains of this widespread practice in *Modus orandi Deum* LB V 1132C / CWE 70 255 and *Ecclesiastes* 2 LB V 873C–874A: 'a great many, upon completing the exordium, hail the most blessed Mother of Christ, and this with greater religious fervour than they invoke Christ or his Spirit, calling her "Font of All Grace"'; for an example of this practice, see the sermon of Stefano Taleazzi (1515) at the Fifth Lateran Council (Mansi, *Conciliorum amplissima collectio* 32 col 918A–B). Erasmus reports among the alleged erroneous propositions found by Parisian theologians in a work of Berquin 'that it is unsuitable for the Blessed Virgin to be invoked in sermons instead of the Holy Spirit, and that it is not suitable that she be called the Source of All Grace' (Allen Ep 2188:101–3).

863 A summary of Pio's argument on *XXIII libri* 166ᵛB, where he admits the point made in Erasmus' Ἰχθυοφαγία that the early Christian practice was to invoke Christ, not Mary, at the beginning of sermons, but argues that, whereas this was necessary then, to make the name of Christ known to the many pagans who came to hear the sermons of the Fathers, later it was needful to invoke his Mother as an advocate with Christ so that her name would become better known and Christ would be the more glorified in the praises of his Mother. Pio also argues that the invocation of Mary is a sign of humility and a blow against heretics.

864 Pio *XXIII libri* 166ʳ⁻ᵛB defends preachers' beginning their sermons with the *Ave Maria*, describing the prayer thus: 'For after the words of the salutation in the Gospel have been spoken and the most holy Lord, ie Jesus, has been celebrated and declared blessed under the name of Fruit of her womb, one says "Holy Mary, Mother of God, pray for us sinners."'

865 While often criticizing the insertion of the *Ave Maria* into the exordium of sermons, Erasmus in *Ecclesiastes* 2 (LB V 873D) seems to favour adding on such occasions a petition after the salutations of Gabriel and Elizabeth. In Italy a petition was gradually added between the thirteenth and early sixteenth centuries. Various vernacular versions tend

silence. But if she is the Font of Grace, what need is there to say to her 'Pray for us'? It is unlikely that this custom was introduced by serious men, but by some bungler who, in keeping with what he had learned in the poets, that invocation should follow the statement of theme, substituted Mary for the Muse.[866] I have listened to a Franciscan[867] who, through all of Lent, proclaimed the praises of the Virgin at the beginning of his sermons with three headings, a devout practice, but that was not the place for them.[868] There is, moreover, an unspoken bit of fawning also at work here, because the female sex, as a result of some human craving, loves to hear the praise of a woman rather than of Christ. But this craving should have been chastened, so that after the praises were given, he would have urged his hearers to imitate her, and rebuked those who boast of the loftiness of the Virgin while by the impiety of their lives they disgrace both her and her Son.

* * * * *

to conclude with a petition like 'pray for us sinners'; eg Bernardino of Siena records in 1444 the addition of the words 'pray for us sinners.' The complete, final text of the *Ave*, including 'now and at the hour of our death,' appears in print first in the Camaldolese breviary of 1514, but does not appear in a document sanctioned for the Roman church until the breviary of Pius v (1568). By the end of the fifteenth century a conclusion close to the one at issue here was in use and known as the 'prayer of Pope Alexander vi.' The present formula probably came from Italy, but a similar wording was also used in France. See Herbert Thurston 'Hail Mary' *Catholic Encyclopedia* vii (New York 1910) 110–12; and John Hennig 'Ave Maria' *Dictionary of the Middle Ages* ed Joseph R. Strayer 12 vols (New York 1983) ii 13–14.

866 Insertion of an invocation, commonly to the Holy Spirit or to the Virgin Mary (the *Ave, Maria* being the most popular), was mandated by medieval preaching manuals, the *artes praedicandi*. As to the substitution of Mary for the Muse, note the opening of Valla's *Encomium sancti Thomae Aquinatis* trans M. Esther Hanley, in Leonard A. Kennedy, ed *Renaissance Philosophy: New Translations* (The Hague 1973) 17–18, where he is emulating the pagans in this practice: 'That is why it is my pleasant duty today to imitate that excellent practice of theirs ... and in the customary manner to invoke the most holy ever-virgin Mother of God, greeting her in the angel's words, "Hail, Mary, full of grace ..."'; see also Erasmus *Ecclesiastes* 2 LB v 873C.

867 Erasmus was a keen observer of contemporary preaching, pointing out its failings and suggesting remedies, especially in his major work *Ecclesiastes* (1535); see Robert G. Kleinhans *Erasmus' Doctrine of Preaching: A Study of 'Ecclesiastes, sive de ratione concionandi'* (Dissertation, Princeton Theological Seminary 1968) 15, 67–73; and John W. O'Malley 'Erasmus and the History of Sacred Rhetoric: The *Ecclesiastes* of 1535' *Erasmus of Rotterdam Society Yearbook* 5 (1985) 1–29 especially 14.

868 See *Ecclesiastes* 2 LB v 862D, where, in the course of detailing vices of the exordium, Erasmus reports: 'I myself listened to a certain man who used the same exordium every day through all of Lent. The Virgin Mother was praised with three headings, eg that she is holy, that she is a virgin, that she is prophetic ... It was pious to extol with praise the worshipful Mother of Christ, but what had this to do with a penitential season?'

Now the letter of the Blessed Virgin to Glaucoplutus[869] is entered in evidence, as being unworthy[870] of the Mother of Christ. Of course there is a real danger that someone may think that it was written in earnest by the Blessed Virgin! It is a spoof, but one aimed against those who publicly condemn the invocation of the saints and have thrown all images out of the churches. They will say that 'this should not have been done with humour.' I answer that this was rather better than presenting a serious plea with the kinds of argument and abuse with which this bulwark of the church pleads.

The Germans who know Greek realize that Glaucoplutus stands for Zwingli.[871] And I do not bring in as a character there the real Mother of Christ, but a statue which is named from a stone,[872] and, indeed, she writes not from heaven, but from her stony house.[873] Still, many revere this statue as if it were sentient. But what does the Stone Virgin write to him? 'Know that I am deeply grateful to you, a follower of Luther, for busily persuading people that the invocation of the saints is useless, etc.'[874] What a blasphemy! The Virgin approves his condemnation of the invocation of the saints. In fact, the invocation of the saints is not an article of faith.[875] But what follows? Mention is made of the intolerable superstition of people who demand everything of the Virgin, as if she could command her Son who reigns with the Father and he cannot deny her anything she bids since he is not yet of age. Indeed, this is what Béda the thrice-theologian[876] teaches us,[877] because

* * * * *

869 This letter is given in the colloquy *Peregrinatio religionis ergo* LB I 775B–776B / ASD I-3 472:79–424:129 / CWE 40 624–8. It is mentioned by Pio *XXIII libri* 158ʳB, and discussed at 162ᵛs–163ʳ.

870 Pio *XXIII libri* 162ᵛs introduces a quotation from the beginning of the letter with the words: 'In addition, how unworthy is the fictional letter from the Virgin Mother of God to Glaucoplutus.'

871 Glaucoplutus is the Latinized Greek form of Zwingli's forename 'Ulrich,' which in German sounds like 'owl rich.' See CWE 40 624 23n.

872 Mary is so described in *Peregrinatio* LB I 775A / ASD I-3 472 / CWE 40 624: 'Which Mary? The one called Mary a Lapide.' The statue was called 'Mariastein'; see E. Gutmann *Die Colloquia familiaria des Erasmus von Rotterdam* (Basel 1968) 70 n28; M. Fürst 'Erasmus von Rotterdam und Mariastein' in *Jahrbuch für Solothurnische Geschichte* 47 (1974) 277–83; and ASD I-3 472 n57.

873 'Mary' concludes her letter: 'From our stony house . . .,' probably a reference to the stone statue rather than to the stone church in Mariastein (compare Job 4:19). See *Peregrinatio* LB I 776B / ASD I-3 474:127–8 / CWE 40 628.

874 *Peregrinatio* LB I 775B / ASD I-3 473:80–1 / CWE 40 624–5

875 That is, as opposed to the communion of saints mentioned in the creeds. The first conciliar teaching approving the invocation of saints comes from the twenty-fifth session (3 December 1563) of the Council of Trent; see COD 774:32–775:6.

876 *Bedda ter Theologus*, the multiplicative *ter* being used for hyperbole (cf Horace *Odes carm* 2.14.7–8: 'ter amplum / Geryonem'), and perhaps by analogy to Hermes Trismegistus, as a jab at the mystique of establishment theology.

877 *nos* 1531 *om* LB

in church one sings 'show thyself to be a mother,'[878] that is, 'command your Son, with maternal authority, to hear our prayers.'[879] But at the wedding feast the Son did not accept his Mother's command about the wine[880] because human agency could have provided it. And Augustine says 'you have not given birth to that in me which does miracles,'[881] and nonetheless, when we are asking of him who reigns with the Father the gift of faith, sound understanding, eternal life, and forgiveness of sins, favours which God alone bestows, they want his Mother to have the right to command God the Son. For she goes on in the colloquy: 'They demanded everything from me alone, as if my Son were always a baby (because he is carved and painted as such at my bosom), still needing his mother's consent and not daring to deny a person's prayer; fearful, that is, that if he did deny the petitioner something, I for my part would refuse him the breast, etc.'[882] This is silly, I admit, but what Béda maintains is no less silly. If the Virgin can command God the Son, what need is there that we be suppliants before God?

It is against these superstitions, these blasphemies, that one should have cried out 'O manners! O laws!' Our salvation depends on Christ,[883] to

* * * * *

878 'Monstra te esse matrem,' the first verse of the fourth stanza of the very popular Marian hymn *Ave maris stella*; see Heinrich Lausberg *Der Hymnus 'Ave maris stella'* Abhandlungen der Rheinisch-Westfälischen Akademie der Wissenschaften 61 (Opladen 1976) 69–70.

879 In his critique of Erasmus' paraphrase of Luke 2:51 ('and he was subject to them') Béda attacks the proposition, which he imputes to Erasmus, that 'Jesus owed obedience to none but his heavenly Father,' urging: 'The will of God the Father is that children obey their parents' just bidding, for the law of nature is so unalterable that its power is not made void even in heaven. For this reason even now the church is not afraid to say in prayer to the Mother of Mercy: "Show that you are a mother," as if to say along with one of the saints as one posssessed of a great confidence: "With a mother's right command the Redeemer." Surely as a human Christ owes his mother obedience even now': *Annotationum . . . in Jacobum Fabrum Stapulensem libri duo: Et in Desiderium Erasmum Roterodamum liber unus* (Cologne: Petrus Quentel 1526) 250ᵛv (censure on proposition 59). Béda developed this theme further in his criticism of Erasmus' paraphrase of John 2:3–4 in his censure on propositions 81–3 (ibidem 259ᵛx–260ᵛz). Erasmus' reaction to this in his *Supputatio* was considerably more violent than what we read here: 'He proves this with a clever quotation. Because the church sings "show thyself to be a mother," as if to say "with a mother's right command the Redeemer." If this does not savour of blasphemy, I do not know what blasphemy is' (LB IX 608C). His answer to Béda on this question is further developed in his rejoinder to Béda's censure of propositions 81–3, *Supputatio* LB IX 625D–629E. Erasmus commented on this text in his annotation on Luke 2:51 LB VI 239D–E / Reeve 169.

880 John 2:3

881 Augustine *Tractatus in Iohannis evangelium* 8.9.12 CCSL 36 87

882 *Peregrinatio* LB I 775B–C / ASD I-3 473:82–6 / CWE 40 625

883 Erasmus is fighting a straw soldier here, since Pio explicitly states the same: see Pio *XXIII libri* 165ᵛx and 167ʳc.

whom even the Virgin herself owes her salvation. We are bidden to boast in Christ, not in the Blessed Virgin. We will take up the question of invoking her elsewhere.[884] I am going to omit their proofs, for they do not apply to me. They are amazingly thin in these proofs, though very savage in making opening statements and in giving abuse.

Finally he advises that if there are any so foolish as to assign greater importance to the Mother than to the Son, these should be rebuked by church officials.[885] But I do the very same thing he wants them to do, and I am called a blasphemer! 'You do it,' he says, 'but in jest.'[886] But elsewhere I do it in sober earnest.

For the rest, to conclude the topic with an epilogue, I do not make fun of the Virgin, but of the monstrosity of human superstition, and I claim his due glory for Christ, nor do I lie in saying that Chrysostom and Augustine attributed to her a human craving; I report their words honestly. Nor do I agree with them; rather, I leave judgment to others, though I am more inclined to the school that is more generous to the Virgin, for I am a supporter of her honour even if I do not support human superstition. Since these points are as clear as can be, I must excuse his unsound mind for the conclusions, corollaries, and abuse which are heaped up against me here, and I pray for a sounder mind for his suppliers. I am similarly leaving out the many points that they failed to understand.

He had come as far as proving that the most holy Virgin was free of original sin, when the printer interrupts to say that thus far this glorious work had been revised by Alberto, its author, but he was prevented by death from revising what follows, and it was delivered in a copy made by members of his household.[887] But nothing is achieved by this ruse. He did not write on his own even the earlier book which he sent to me, so I am far from believing that he completed this work. If he did the copying with his own hand, I am amazed at the labour; if he revised the

* * * * *

884 Erasmus seems to have forgotten to return to this topic in the remainder of his response to Pio, unless perhaps he is referring to section 282 below.

885 Pio *XXIII libri* 167^{v}c

886 These are not Pio's words. He cautions instead (*XXIII libri* 167^{v}c) that the appropriate medicine should be used to cure the sick, because too harsh a medicine can cause even greater illness, thus warning of the dangers to popular faith if these abuses are corrected imprudently.

887 The following note appears in the middle of folio 171^r, its presence signalled by a substantial indention from the left margin: 'Thus far the author, Alberto Pio, himself reread and revised his work. Prevented by untimely death, he was unable to revise what follows, but he wrote all of it, and it was delivered to us in a copy made from the original by the reliable skill of members of his household'; see Introduction lxxix.

materials compiled by others, it is amazing that he left so many grammatical errors.

CANONIZATION OF SAINTS[888]

I do not think that it is worthwhile to reply on this point, for all I do is point out that some are considered saints who have not been entered in the canon by the Roman pontiff.[889] Here he criticizes me for imprecision in speech because I designated the *civitas* of Siena with the noun *urbs*,[890] as, in fact, I did. A serious slip in the Catholic faith!

SCHOLASTIC THEOLOGY [FROM BOOK 10][891]

Nowhere has he been more assiduous in quoting,[892] or, as he likes to say, more meticulous, in the hope of inciting all theologians against me. In fact,

* * * * *

888 Pio's handling of this, the final topic of book 9, is found in *XXIII libri* 171ʳH–172ᵛK. After identifying what he takes to be Erasmus' implied criticism of the process of canonization, Pio sketches the history of the cult of saints arguing (*XXIII libri* 172ʳK) that the process of canonization was introduced to curb abuses in the identification and cult of saints, abuses which gave rise to the proverbial saying that many bodies of saints are venerated on earth while their souls are punished in hell.

889 Referring to the text from the colloquy *Apotheosis Capnionis* 'Apotheosis of Reuchlin' LB I 692A–B / ASD I-3 272:178–9, 181–2 / CWE 39 250–1, as quoted by Pio (171ʳH): 'Who canonized . . . St Jerome? or Paul? or the Virgin Mother? Whose memory is more sacred among all the devout . . . or Catherine of Siena, whom Pius II is said to have canonized to please an order and a city [*in gratiam ordinis et urbis*]?'

890 Pio *XXIII libri* 172ᵛK points out that St Catherine was a citizen of Siena (*civis Senensis*), taking *urbs* to refer to Rome, or 'if by *urbs* you mean Siena, you are speaking incorrectly.' St Catherine died in Rome in 1380 and was canonized there in 1461 by Pius II on a petition from the citizens of Siena; see his *Commentarii rerum memorabilium que temporibus suis contigerunt* 2 vols, ed Adriano van Heck, Studi e Testi 312–13 (Città del Vaticano 1984) I 341 (book 5 no 24).

891 MARGINAL NOTE: 'From book 10'; 1531 UPPER MARGINS: 'Scholastic Theology' on pages 173–81 [= 255–65 in this volume]. Pio's book 10 (*XXIII libri* 172ᵛL–176ᵛs) is entitled 'On Theologians' Innovations and on Scholastic Theology.' Early on (173ᵛM) Pio refers the reader back to his earlier treatment of Erasmus' views on this topic in connection with the *Moria* (*XXIII libri* 76ʳI–79ᵛN): 'I can reply to these passages of yours with rather less trouble since earlier, when dealing with your statements in the *Moria*, I developed quite a few ideas relevant to this topic and need not repeat them here.'

892 Pio opens book 10 (*XXIII libri* 172ᵛL–173ʳ) with thirteen quotations from Erasmus' works – seven from the scholia on Jerome, one from *Methodus*, two from *Enchiridion*, two from *Annotationes*, and one from the preface to Hilary – plus one vague reference to his treatment of Thomas and Scotus. Erasmus takes these up in order, combining Pio's third and fourth quotations as his number three. Pio insists here, as usual, that he is providing only a sampling: 'I am not going to bring in every passage in which you attack all these, since the passages are so numerous that a huge book would scarcely contain them.'

I express my approval of scholastic theology, one that is tempered with a sober admixture of dialectic and philosophy, in so many places that it is superfluous to reply to these[893] complaints.

1 / What do good theologians care if I value Jerome and authors like him more highly than some primer-writers and compilers for whom not even the theologians of Paris have much regard?[894]

2 / What is my complaint in the scholia of volume II[895] except that too much human philosophy is applied in scholastic theology? This is a criticism of Thomas, but most particularly of Scotus. The correctness of this view is clearly shown by the fact that now a good many schools commit themselves neither to Thomas nor to Scotus, but are shifting to a more sober study of Scripture.[896]

3 / In a scholion in volume III,[897] on the basis of a remark of Jerome, I criticize the theologian who prattles nothing but Aristotle. What has this

* * * * *

Unless otherwise indicated, the text of these passages translated in the notes will be the versions given by Pio, with only significant divergences from Erasmus' originals noted.

893 HAS 1531 HUJUS LB

894 Erasmus deals with Pio's first quotation, from his preface to his scholia on *De viris illustribus* HO 1 138ʳ where he complains of the negligence of bishops and theologians who have allowed the work of the Fathers to be lost 'so that their place would be taken by this base sort of writer, writers of *sententiae*, little primers, *fasciculi*, and *specula*, and, O God immortal, people whose very mention nauseates noble and well-born minds.'

895 Erasmus' parodic response to Pio's second quotation: 'But of theology you speak in the scholia of volume 2 as follows: "The subject matter now treated far and wide in the schools is so tightly packed with Aristotelian dogmatism and sophistical nonsense (not to say contaminated by vain fancies) and so enveloped in the labyrinths of vain and trivial questions that Jerome himself, should he return to life, or even Paul, would be considered utterly ignorant of theology in that milieu"'; this passage comes from Erasmus' preface to volume 2 of Jerome HO 2 2ʳ / CWE 61 68.

896 On the curricular reforms in German universities during the years 1515–35, see James H. Overfield *Humanism and Scholasticism in Late Medieval Germany* (Princeton 1984) 298–327. On Spain, see Andres II 5–76, 101–6, 347–86; on France, see James K. Farge *Orthodoxy and Reform in Early Reformation France: The Faculty of Theology of Paris 1500–1543* (Leiden 1985); and Farge *Le parti conservateur au XVIe siècle: Université et Parlement de Paris à l'epoque de la Renaissance et de la Réforme* (Paris 1992); on Louvain, see Jeremy H. Bentley 'New Testament Scholarship at Louvain in the Early Sixteenth Century' *Studies in Medieval and Renaissance History* new series 2 (1979) 51–79; on English universities, see James Kelsey McConica *English Humanists and Reformation Politics: Under Henry VIII and Edward VI* (Oxford 1965) 76–105 especially 82–3. Italian universities, however, successfully resisted introducing more humanistic theology and continued to teach Thomistic and Scotist scholastic theology; see Grendler *Universities* 389–92. See also Rummel *Humanist-Scholastic Debate* especially 63–125.

897 Erasmus' reaction to Pio's quotation from two scholia from HO 3: 'But in the third volume you say again about theologians: "Read this, O theologians who babble nothing but Aristotle." And again, in the same place: Jerome "is indicating that all heresies were

to do with theologians in general? Moreover, Jerome was not the only one who said that a large number of heresies have sprung from philosophy.[898] But this fact alone does not condemn all of philosophy, for otherwise one would have to condemn Sacred Scripture as well. As it is, my very words – 'But now some of these theologians' – make it plain that I am criticizing only some theologians, not all of them, although the use of *modo*, instead of *nunc*, is not mine;[899] this is just another example of how they are pleased to corrupt everything I say.

4 / What I say is absolutely true, but not all theologians embrace a theology that has been 'tainted or, rather, overwhelmed by Aristotelian rules.'[900]

5 / If in the books of the scholastic theologians, for example, Scotus, Albert the Great, and the like, there are no human fabrications, then I have spoken incorrectly.[901]

6 / What is the outrage if I have expressed the opinion that there should be an admixture of Platonic philosophy, as being the more adaptable to divine mysteries?[902]

* * * * *

born from the writings of the philosophers and that the purity of Christian wisdom is not aided, but defiled, by their teachings. But now some of these theologians think that nothing in Scripture can be understood unless you waste a good part of your life studying Aristotle"' (HO 3 67^v and 117^{v}D, the second partly misquoted as Erasmus indicates below).

898 Indeed, the text of Jerome's Ep 133.2.1, on which the second scholion at issue here comments, is: 'One of our own people put it very well, "The philosophers are the heretics' patriarchs"' (CSEL 61 242:1–2); Jerome is referring to Tertullian who wrote in *Adversus Hermogenem* 8.3 'haereticorum patriarchae philosophi' (CCSL 1 404). Erasmus remarks in the same scholion: 'For they will not be able to escape by saying that this was said by someone else, not by Jerome. Surely when he approves of another's view he is making it his own.' In his preface to Hilary Erasmus comments on Tertullian's views of philosophy (Ep 1334:185–8).

899 Pio had, in fact, drastically altered this part of the scholion. Erasmus had written: 'But why do they not call Jerome into court, these men, some theologians who think nothing can be understood ... (*At cur non in ius vocant Hieronymum isti quidam theologi, qui nihil putant intelligi ...*)'; Pio altered it to read: 'But now some of these theologians think that nothing can be understood ... (*At modo isti quidam theologi nihil putant intelligi ...*).'

900 Replying to Pio's quotation from Erasmus' antidotus to Jerome Ep 78 HO 4 23^{v}D: 'I have no doubt that if Jerome had seen this theology that is everywhere tainted, or rather overwhelmed, by Aristotelian maxims and rules, he would have thundered at this disgrace quite satirically.'

901 Replying to Pio's quotation from Erasmus' *Methodus* LB V 82A–B / Holborn 189:34–6: 'What you hear in the books of the scholastic theologians are human fabrications. The more closely you inspect them, the more they fade away like dreams.'

902 Replying to Pio's garbled quotation from the *Enchiridion*: 'And in the *Enchiridion*: Our theologians (you say), content with our Aristotle, banish the Platonists and Pythagoreans from the schools. They either practically despise allegory or treat it very coolly.'

7 / In 1 Timothy 1,[903] Paul advises his son to avoid *mataeologia* [vain talk].[904] I take this occasion to warn the theologians of our[905] time that they should do theology in such a way that they do not fall into *mataeologia*, and it is there I mention the countless questions that are pointless, or even harmful, or that, because of their difficulty, surely keep people from things that are more important.

8 / Next, I criticize certain Franciscans who waste their entire lives on the subtleties of Scotus, a practice which the most approved men of that order also condemn.[906]

9 / What theologian is there who does not condemn many of Scotus' subtleties?[907] What school is there that does not today direct itself towards more sober disputations?

* * * * *

Erasmus' topic in the section of the *Enchiridion* from which Pio excerpted these words is the allegorical interpretation of Scripture, and he says: 'Theologians of the present day either practically despise allegory or treat it very coolly. In the art of subtle distinction they are equal or even superior to ancient commentators, but in the treatment of this subject they cannot even compare with them, for two reasons, as far as I can determine. First, mystical exegesis cannot fail to fall flat if it is not seasoned with the powers of eloquence and a certain gracefulness of style, in which the ancient achieved an excellence that we cannot even approach. The second reason is that they are satisfied with Aristotle alone and banish the Platonists and Pythagoreans from the schools. But Augustine preferred the latter two' (LB V 29F–30A / Holborn 71:27–35 / CWE 66 69).

903 Erasmus responds to Pio's quotation from his annotation on 1 Tim 1:6 LB VI 926D / Reeve *Galatians to the Apocalypse* 662: 'Likewise, in the Annotations: "Accordingly we must be very careful not to pursue theology in such a way that we lapse into *mataeologia* [. . .] For what good is it to know how many ways sin can be understood, whether it is a privation or a stain sticking to the soul," and so on, for I am leaving out the rest of the questions you list.'

904 A favourite insulting word of Erasmus, derived from 1 Tim 1:6, to describe corrupt theology; see H.J. de Jonge in ASD IX-2 229 n52.

905 *nostri* LB *nostris* 1531

906 Erasmus' corrective to Pio's immediately subsequent quotation of a later segment of the annotation on 1 Tim 1:6 LB VI 928C / Reeve *Galatians to the Apocalypse* 663: '. . . the additional remarks in which you condemn the mingling of theology and philosophy. It is about this last point that you conclude with the following: "And it is on this theology that is so far from simple that they spend their entire lives, these men who profess a simple, apostolic life, who by their very nickname [*fratres minores*] profess the height of modesty. In the teaching of this sort of philosophy they raise up their crests ... and blather nothing but Averroes and Aristotle. They recoil from the touch of money, but spend their whole lives on books of a blasphemous philosopher."'

907 Responding to Pio's quotation from the same annotation (LB VI 928D / Reeve *Galatians to the Apocalypse* 664): 'And next you criticize Johannes Scotus quite irresponsibly, saying: "It is by means of labyrinthine arguments like these that Basilides, Valentinus, and Marcion used to snare the minds of the naive."'

10 / And it is not[908] scholastic theology that is censured there without qualification, but a mange, a chronic itching for disputation, a characteristic of sophists.[909]

11 / The supplier had incorrectly copied out a misunderstood passage from the preface to Hilary[910] in this way: 'In this regard certain men were so lacking in moderation that, after defining everything in theology, they contrived for those who are no more than men a new status of divinity that aroused savage commotions in the world.'[911] That is how they quote it. But the passage, in fact, runs as follows: 'They contrived for those who are no more than men a new status of divinity, and this has aroused more questions and greater commotion in the world than the Arians in their foolishness once did. But certain pundits [rabini] on some occasions are ashamed to have no rejoinder to make.'[912] You see how great a difference there is between my words and their quotation. It is an amazing earnestness that takes from ignorant informers passages incorrectly copied out and misunderstood, and then uses them to prosecute a savage quarrel with one's neighbour! But how am

* * * * *

908 *Nec* 1531 *Ne* LB
909 Reacting to Pio's quotation XXIII *libri* 173ʳL), which he identifies as coming from the Annotations: 'And again, in the same [annotations]: "For it is customary to call the schools *diatribae* perhaps because, just as mangy sheep, when they scratch themselves, fill the healthy sheep with disease, so these men corrupt others when they scratch themselves on them."' We have not been able to locate this passage in Erasmus' works. He attacks the urge for disputation in annotations on 1 Tim 1:6 LB VI 926D–928E / Reeve *Galatians to the Apocalypse* 662–5, but notably in a comment on 1 Tim 6:4 LB VI 944C / Reeve *Galatians to the Apocalypse* 680 he says that this text applies especially to sophistic theologians, citing the few theologians who are 'unworthy of a most upright profession, itching with a chronic mange of silly *quaestiones.*'
910 The preface to Erasmus' edition of the works of Hilary of Poitiers (*Hilarii opera* [Basel: Froben 1523]), also printed as Ep 1334, with useful notes of the passages cited for criticism by the theology faculty of Paris and the Spanish monks.
911 MARGINAL NOTE: 'Deceptive quotation of things not understood.' Erasmus is replying to Pio's faulty quotation from Erasmus' preface to Hilary (Allen Ep 1334:223–7): 'Atque hac in re adeo modum nesciere quidam, ut posteaquam nihil non definierant de rebus divinis, nova [novam *Eras*] etiam in his, quae [qui *Eras*] nihil aliud sunt quam hominum [homines *Eras*], theothita [θεότητα *Eras*] commenti sunt, quae ... atroces [atrociores *Eras*] tumultus excitarunt [excitauit *Eras*] orbi ... Sed pudet rabbinos [+ quosdam *Eras*] alicubi non habere quod respondeant.'
912 Erasmus corrects Pio's errors, quoting Ep 1334:240–4; 'rabbis' (*rabini*) is a favourite Erasmian expression for scholastic theologians (see eg Allen Ep 2448:44–7), based, perhaps, on Matt 23:7 (see *Methodus* LB V 101E, 138C / Holborn 229:26, 305:25; and *Apologia adversus monachos Hispanos* LB IX 1090A); see also Pio XXIII *libri* 176ᵛv where Pio says: 'You call the theolgians *rabini* because they cannot tolerate awful monstrosities, because they raise the cry when they see wolves or foxes lurking in the sheepfolds or barnyards of the Lord.'

I at fault here if I remind teachers that it is not necessary to answer all of the idle questions of everyone? If there are none who put forward curious and idle questions, if there are none who define matters that ought not to be defined, my warning was in vain. But, alas, there is all too great a crowd of such people. If anyone wants examples of this, he should read my annotation on the first chapter of the first Epistle to Timothy.[913]

12 / In the *Enchiridion*[914] I spoke out, not against theology, but against its corrupters and against some superstitious monks who overwhelm Christian freedom with their regulations. Criticism of its vices does not harm, but helps, a profession.

13 / He says I treat Thomas and Scotus as if they were cattle and fools.[915] No, that is how Pio, neither a theologian nor a priest,[916] treats me! The fellow is disgruntled because Folly prays in jest that the soul of Scotus might travel into her body.[917] Thomas I treat with respect[918] in many passages, even though occasionally I disagree with him.

On the basis of these mutilated and misunderstood passages he takes it for granted that I condemn all of scholastic theology absolutely,[919] that the

* * * * *

913 On 1 Tim 1:6 LB VI 926D–928E / Reeve *Galatians to the Apocalypse* 662–5, where one can see at a glance how Erasmus increased his list of pernicious theological questions over the years.

914 Reacting to a text that Pio *XXIII libri* 173^{r}L presents as quoted from the *Enchiridion*: 'In the *Enchiridion* too, after you have raved as much as you wanted against them, you conclude your speech with these words: "I wish Christ would eventually awaken and free his people from this Judaism and this tyranny, unless, perhaps, he redeemed us by his blood only to be enslaved to monstrosities like these."' In fact, the text is not from the *Enchiridion*, but from the annotation on 1 Cor 15:51 LB VI 743E–F / Reeve-Screech 519. Pio later (176^{v}v) says that the quotation is 'in the Annotations on the New Testament'; see n937 below. In the *Enchiridion* Erasmus proposed a 'new' type of theology based on *pietas*, a theology unlike contemporary contentious scholasticism; see John W. O'Malley in CWE 66 xl–xliii.

915 Pio's concluding remark (*XXIII libri* 173^{r}L) to the series of quotations: 'I am omitting the rest of what you have had to say against theologians and theology, treating St Thomas and Johannes Scotus as if they were cattle or fools.'

916 See Introduction xxiv–xxv, cxxii; and *Responsio* 27 n125, 86 n427 above.

917 *Moria* LB IV 489A / ASD IV-3 179:910 / CWE 27 142

918 See Christian Dolfen *Die Stellung des Erasmus von Rotterdam zur scholastischen Methode* (Osnabruck 1936) 86–8; and J.-P. Massaut 'Erasme et Saint Thomas' *Colloquia Erasmiana Turonensia* ed Jean-Claude Margolin 2 vols (Toronto 1972) II 581–611.

919 Pio *XXIII libri* 174^{v}P–175^r: 'But you will, perhaps, say: "I am not so much condemning the mixing of philosophy with theology as the method of dealing with it developed by the moderns, by those commentators on the *Sentences* [*sententiarii*] who complicate everything with thorny and fretful *quaestiones*. For this style of theology is alien and different from that employed by the ancient Fathers."'

name of Aristotle is amazingly hateful to me,[920] that I am of the opinion that philosophy should be separated altogether from the study of theology. In fact, in my writings I combat those who are of this opinion. Indeed in the *Methodus*, which Pio quotes repeatedly,[921] I maintain that the liberal arts should be soberly applied to theology, following St Augustine in this.[922] But Augustine does not mention the eight books of the *Physics* to which Pio assigns such great importance, inasmuch as he is quite an expert in such things,[923] though I nowhere find fault with them.[924]

And the deceptive fellow throws up to me Plautus and Diomedes,[925] as if I have never looked at Aristotle, or do not quote Aristotle in my writings more often than Pio himself, who for many years has been a dinner-party Peripatetic.[926]

When I say somewhere, 'I ask you, by the shades of Aristotle,' he turns the text back on me, saying 'I ask you, by the ghost of Lucian,'[927] just as if in all my writings it is not Christ whom I sincerely profess.

* * * * *

920 Pio *XXIII libri* 174^{v}o apostrophizes Erasmus: 'O you who are so hostile to the name of Aristotle ...'; Pio (174^{r}o–174^v) provides an extravagant defence of Aristotle, calling him 'the greatest [of the philosophers] and nearly godlike.'

921 Pio *XXIII libri* 173^{r}L provides multiple quotations from *Methodus*, and makes other indirect references to ideas stated in it.

922 See *Methodus*, eg LB V 79C–83A / Holborn 184–91.

923 Erasmus refers here to Pio's citation in *XXIII libri* 175^{v}P of the superiority of Aristotle's natural philosophy to that of Pliny and Seneca as an illustration of the superiority of the scholastic method even in the sciences, and takes an opportunity to snipe at Pio's philosophical background. Of course, the Aldine edition of Aristotle's *Physics* (1497) had been dedicated to Pio (*Aldo Manuzio Editore* I 14–17).

924 But see *Methodus* LB V 80B / Holborn 186.

925 Plautus, the great Roman comedian (d after 184 BC), and Diomedes (late fourth century AD), the grammarian to whom Erasmus assigns first place among the ancient Latin grammarians (*De ratione studii* LB I 521C–D / CWE 24 667–8), are both mentioned by Pio (*XXIII libri* 174^{v}o) in a sarcastic apostrophe to Erasmus: 'But you who are so hostile to the name of Aristotle, just make believe that what Aristotle produced came not from him, but from Plautus or Prisican or Diomedes, or, if you prefer, from one of the saints, or, if you would rather, was divinely revealed.'

926 See Introduction xxviii; Erasmus' epithet for Pio here is reminiscent of *Adagia* IV ix 70: *Ad vinum diserti* 'Eloquent over wine' LB II 1158F–1159A.

927 Erasmus refers here to a marginal comment (*XXIII libri* 62^{r}m) in which Pio, replying to *Responsio* 86 n426 above, notes: 'Oh, but I too adjure you, man most distinguished, by the shades of Lucian ...' Lucian of Samosata (second century AD) is known best for his cynically satiric dialogues; Erasmus and More translated selected works of his into Latin; see Rummel *Erasmus as a Translator* 49–69. Pio elsewhere, apropos of *Moria*, identifies Lucian as a principal source of Erasmus' world view (*XXIII libri* 73^{v}c): 'But you have learned these things from your Lucian, who with drolleries like these and his blather discourages mortals from every pursuit, not only of studies, but of all the arts and professions, and every style of life'; see also n372 above. Pio's use of Lucian against Erasmus

He accuses me of having called Aristotle blasphemous.[928] How could I have done this, since Aristotle antedates Christ? But I do call Averroes a blasphemer.[929] He was born after Christ and detests his teachings, yet I know Franciscans who have learned Averroes verbatim.[930]

Pio conjures up a double theology[931] for us, that of the ancients, serene, to be sure, but unarmed, that of the moderns, armed with shield and helmet. The first is useful for exhortation, but only among believers; the latter is just right for refuting pagans and heretics. In fact, those ancients carried on a constant struggle with pagans and heretics. Our theologians only shadow-box [σκιαμαχοῦσιν] in the schools and write *articuli*. What success could they have against the pagans when they base their arguments on the scholastic foundations of the Thomists and Scotists, the realists and nominalists? They use *fasciculi* in their fights more than the Scriptures.

But the easygoing fellow, so that there will be plenty of time for learning Averroes, declares that in the divine Scripture there are very many things that 'contribute less to the substance of the faith,' like the deeds of the kings, 'the sayings of the prophets, and other similar matters,' of which one can be ignorant and still be a theologian nonetheless.[932] I would surely

* * * * *

is described by Christiane Lauvergnat-Gagnière *Lucien de Samosate et le Lucianisme en France au XVIe siècle: Athéisme et polémique* (Geneva 1988) 141–3.

928 Pio *XXIII libri* 174ᵛo: 'But you call him a blasphemous philosopher, and thus give clear proof that you were not very well trained in his teaching. Quote a passage, if you can, in which Aristotle blasphemes ... Aristotle is as remote from blasphemy as you are from his teaching'; see Erasmus *Psalmi* LB v 253C / CWE 63 196 (errors of Aristotle).

929 Erasmus is referring to his denunciation of perverse theology in the annotation on 1 Tim 1:6 where he exclaims: 'The mouth has been consecrated to the gospel, and it chatters nothing but Averroes and Aristotle. The contact with money is religiously avoided, but one's whole life is spent on the books of a blasphemous philosopher' (LB vi 928C / Reeve *Galatians to the Apocalypse* 664).

930 In the faculty of arts at Paris the works of Averroes furnished the traditional commentary on the texts of Aristotle that were the basis of a university education; see Augustin Renaudet *Préréforme et humanisme à Paris pendant les premières guerres d'Italie (1494–1517)* 2nd ed rev (Paris 1953; repr Geneva 1981) 59–60, 92.

931 Summarizing Pio *XXIII libri* 174ᵛo–75ᵛq. On Erasmus' criticism of theologians who collect – apart from their context and the author's intention – statements of opinion (*articuli*) on which they pronounce determinations as to their truth or falsity, see Charles G. Nauert 'The Articular Disease: Erasmus' Charge That the Theologians Have Let the Church Down' *Mediaevalia* 22 (1999) 9–27; and his ' "A Remarkably Supercilious and Touchy Lot": Erasmus on the Scholastic Theologians' *Erasmus of Rotterdam Society Yearbook* 22 (2002) 37–56.

932 Pio *XXIII libri* 175ᵛq: 'It is absolutely obvious that in the Scriptures there are a good many things the knowledge of which is not very important (although they are well worth knowing and quite useful, provided that they are correctly understood) for they supply rather less of the real substance of the faith; for example, a great many of the

rather have this statement come from the mouth of someone else, and not from mine. If any such thing had been reported to him from my works, how he would have cried out that I am disparaging the Scriptures, in which, as he maintains elsewhere, not a tittle is without its mystery![933] Nonetheless, he concludes that the complete theologian will be he who has joined the liberal arts to a precise knowledge of Scripture.[934] I asserted this same thing before him, so why does he quarrel with me?

He admits that certain theologians deal with sophistic questions, but he says that the method is not to be blamed on that account.[935] He agrees with me, and yet he quarrels with me as if with an opponent. He will say: 'You should have expressed yourself on this point more clearly, and not have criticized scholastic theology in general, but rather the practices of certain theologians.' I say this quite clearly in a hundred passages, but these passages were not reported to Pio by his note gatherers.

There remains an extremely difficult passage which those poison gatherers[936] provided to him a bit late. It is 'in the *Annotations on the New Testament*.'[937] A fine citation! He could have said 'in chapter 7 of Matthew,' but the octopus was afraid of being caught out. I have not so far been able to find the passage, and so I have no suitable response. The words he quotes are as follows: 'How I wish that eventually Christ would awaken, and free his people from this Judaism and this tyranny, unless, perhaps, he redeemed us so that we would be slaves to monstrosities like this, etc.'

* * * * *

historical events which are reported there, as well as many statements of the prophets, and other similar things.'

933 Pio had made this point in his earlier work (*Responsio paraenetica* 19ʳ1-12ᵛ); see the quotation in *Responsio* n375 above, and restated in n409; in *XXIII libri* 61ᵛz he states: 'This does not alter the fact that certain mysteries lie hidden in the words and letters in the original language in which the divine Spirit produced Scripture, and that a great many of these mysteries, although not all, were able to be preserved in the language into which it was translated under the inspiration of the same Spirit. It was necessary to translate Scripture so that it could spread more widely, but this was done with such restraint that, when these Scriptures were translated under the inspiration of the same Spirit, because only the words were changed or rearranged, a pure and authentic integrity of meaning linked with the holiness of the mysteries endured.'

934 Pio *XXIII libri* 175ᵛQ

935 Pio *XXIII libri* 176ʳ–ᵛs

936 Erasmus uses the term φαρμοκολόγοι, perhaps an Erasmian neologism; 'collectors of poisons' is the meaning suggested by C.L. Heesakkers in his article 'Argumentatio a persona in Erasmus' Second Apology against Alberto Pio' in J. Sperna Weiland and W.Th.M. Frijhoff eds *Erasmus of Rotterdam: The Man and the Scholar* (Leiden 1988) 79–87 especially 83.

937 Pio *XXIII libri* 176ᵛv quotes from the annotation on 1 Cor 15:51 LB VI 743E–F / Reeve-Screech 519; see n914 above.

He takes this as having been said against the entire profession of theologians. But who could not realize that this was said against superstitious and tyrannical monks? Nonetheless, just as if it were established that I said this against all theologians, with the purpose of removing their entire profession altogether from human affairs, he thunders at me in tragic diction: 'With these words you have shown all too clearly your raw hostility and uncontrollable hatred for them. Indeed, what more savage and monstrous statement or wish could be made against one's most ferocious enemies, against the bane of humanity, than for the entire group you are attacking to be abolished and done away with altogether? And you are wishing for the ruin of this very order of theologians, blameless and deserving of reverence both because of the dignity of their profession and the innocence of their lives. And yet it is this group that has honoured you and admitted you to membership.[938] (If I were not afraid of offending you, I would say that this was done more on account of their kindness and generosity than because of your merits.) But it is a peculiar characteristic of yours, that you are more likely to attack those in whose company you have enlisted. You pray that the order of theologians may be destroyed by Christ, clearly so that the heretics may have even greater freedom to rage unchecked.'[939]

He screams a good many things like this in his demented way against the Subverter of the Entire Profession of Theologians. How great an illness gripped him for him to write such stuff! May the Lord forgive him, and straighten out those who drove the sick man to this extreme. If Pio did not learn theology,[940] at least he spent some time with the teachers of rhetoric.

What could possibly bring me to this monstrous audacity, to wish to have the whole profession of theologians done away with?!? And as it is, even granting that this were my wish, what hope had I of accomplishing it?

* * * * *

938 Probably a reference to Erasmus' having received a doctorate in theology from the University of Turin on 4 September 1506 and having been appointed to the theological faculty at Louvain on 30 August 1517 by the rector, Jan Calaber, a professor of medicine; Erasmus was apparently never elected to admission into the 'strict college' of the faculty of theology. See Allen Ep 200:8n; Grendler 'How to Get a Degree' 40–69; Grendler *Universities* 384–5; and Marcel A. Nauwelaerts 'Erasme à Louvain: Ephémérides d'un séjour de 1517 à 1521' in *Scrinium Erasmianum* ed Joseph Coppens (Leiden 1969) I 3–24 especially 7.

939 Pio XXIII *libri* 176ᵛv; Erasmus tacitly corrects *odium minus patefecisti* of Pio's text to *odium nimis patefecisti*; he prudently truncates the excerpt he gives here, for the rest of Pio's statement has an uncomfortable, *ad hominem* tone: 'so that the arrogant and headstrong will feel free to write whatever comes into their mouths, so long as it is somewhat witty, as much about the sacred matters of theology (which they do not understand) as about things profane . . .'

940 But, of course, he had; see Introduction xxiv–xxv, cxxii.

Finally, what sense does it make that I would wish the absolute elimination of the profession for the advancement of whose studies I have expended so much toil? I do not think that there is any theologian who would believe what Pio asserts so fiercely in his effort to curry favour with the theologians. Now does one who strives for the improvement of a profession really desire that profession's abolition?[941]

ON THE AUTHORSHIP OF THE SCRIPTURES
[FROM BOOK 11][942]

In this area Pio brings forward nothing about which a thorough reply has not been made to the Zúñigas, Bédas,[943] and Spanish monks. He gathered up only the raw material for brawling with Erasmus and ignores the authorities I have followed in these areas and the arguments I have used. He should have reviewed and refuted these before he descended to the deceitful abuse and captious wrangling with which he flogs Erasmus mercilessly.

Jerome in some places makes a declaration, in others expresses his doubts, about the Second Epistle of Peter, the Epistle of James, the Epistle to the Hebrews, and the two Epistles of John.[944] Indeed, as to authorship, he calls the two later books of Esdras fantasies,[945] makes fun of the story of Bel and the Dragon, denounces the history of Susanna as a fiction,[946] and

* * * * *

941 Using the interrogation mark of 1531 against the full stop of LB
942 1531 UPPER MARGINS: 'On the Authorship of Scripture' on pages 182–8 [= 265–72 in this volume]. Pio's book 11 is entitled 'On the Sacred Scriptures and their Authorship.' After citing twenty-four passages from Erasmus' works (*XXIII libri* 177^{r}A–177^v), three from *Methodus*, six from the scholia on Jerome, ten from the *Annotationes in Novum Testamentum*, and five from *Supputatio*, Pio takes up the issues of the uniform authority of Scripture (177^{v}C–178^r), the creed (178^{r}D–178^{v}F), the authorship of various books of Scripture (178^{v}F–179^{v}I), alleged dissembling by Paul and Christ (179^{v}K–180^{v}M), and the use of figures and allegories in Scripture (180^{v}N–181^{v}P).
943 *Beddis* 1531 *Beddeis* LB
944 Notably in *De viris illustribus* 1.3 (2 Peter), 2.2 (James), 5.10–11 (Hebrews), 9.5 (2 John and 3 John), (Ceresa-Gastaldo 72, 74, 84–6, 94); Erasmus is alleging Jerome in defence of his own doubts or reports of doubts of others about the authenticity of Hebrews, the epistles of James, Peter, and John, and Revelation, which Pio catalogues in the series of quotations which opens book 11. In his argument Pio lays down the rule that 'he who expresses a doubt about only one little word of Scripture is asserting that the whole of it is lacking in authority' (*XXIII libri* 178^{r}C).
945 See Jerome's prologue to his translation of Esdras: 'Neither should anyone be disturbed that we have produced a single book nor should he delight in the fantasies of the apocrypha of the third and fourth book' (*Biblia sacra vulgata* 638:18–19).
946 Erasmus errs here if he has in mind Jerome's prologue to his translation of Daniel, for it is not Jerome who makes fun and denounces, but a rabbi whom Jerome cites: 'I myself heard a certain one of the teachers of the Jews, when he was making fun of the history

marks many other items with obeli and asterisks,[947] he so great a figure and of such great authority.

Nowadays all are read without distinction.[948] But when did a council explicitly pronounce that all of them are of equal authority, that they belong to the authors under whose names they circulate?[949] What if in this matter the Fathers have indulged the sentiments of the devout? What if they received certain items as hagiographa?[950] Indeed, nowadays what is *not* read in churches? In fact, there is not an identical format for all the texts that originate with the Holy Spirit. The Spirit displays its mysteries

* * * * *

of Susanna and saying that it was a fiction composed by some Greek ... he was carping at ... the dragon slain by a snack of pitch or the detection of the contrivances of the priests of Bel' (*Biblia sacra vulgata* 1341–2:23–5, 31, 33–4).

947 For Jerome's use of these critical signs, see eg his prologue to the 'Gallican' Psalter (*Biblia sacra vulgata* 767:9–14), and see his use of *obleli* in Daniel 3:24–90, 13 (Susanna), and 14 (priests of Bel, the dragon).

948 Here Erasmus reacts to Pio's criticism of part of the statement quoted from *Methodus*: 'With me, to be sure, Isaiah has more weight than Judith or Esther, the Gospel of Matthew more than the Apocalypse ascribed to John, the epistles of Paul to the Romans and Corinthians than [...] to the Hebrews' (LB V 92C–D / Holborn 211). It is surprising that Pio failed to quote the more general statement of Erasmus' position that immediately precedes these words: 'Nor would it, perhaps, be ridiculous to establish some ranking of authority in the sacred books as well, something which Augustine did not fear to do. For the first place is due to those books about which the ancients never had a doubt.' Pio *XXIII libri* 178ᵛc denounces this view: 'Since the very character and law of things divine decree that all [ie the books of Scripture] must be of the same weight and authority, because they proceeded from the same author, I mean the divine Spirit ...'

949 In the bull on union with the Copts, *Cantate Domino* (4 February 1442), of the Council of Ferrara-Florence-Rome, which repeated the Constitutions of Nicosia (1340), the canonical books of the Bible were clearly stated. This list included the second Epistle of Peter, the Epistle of James, the Epistle to the Hebrews, three epistles of John, Esdras and Nehemiah (= 2 Esdras), Daniel (presumably including chapter 13 on Susanna and chapter 14 on the statue of Bel and the Dragon). Hebrews is listed as one of Paul's fourteen epistles. See COD 572:14–39. Erasmus seems to have been unaware of this decree of Florence; see *Supputatio* LB IX 594D–595E, *Apologia adversus monachos* LB IX 1079D–E, *Responsio ad notulas Bedaicas* LB IX 713D–E, and *Declarationes ad censuras Lutetiae vulgatas* LB IX 865A–866B, where ignorance of this decree by the Paris theological faculty, Erasmus, and Tommaso de Vio is implied. The Paris theologians cite the Council of Nicaea (325), the Council of Laodicea (363), the third Council of Carthage (397), and a council under Pope Gelasius (492–6), but not the Council of Ferrara-Florence-Rome (1438–45). On ignorance of the decree of Florence (which was missing from Pierre Crabbe's 1538 edition), even among some bishops at the Council of Trent in 1546 until the cardinal-president had the original bull brought to Trent from Rome, see Hubert Jedin *A History of the Council of Trent* 2 vols only trans Ernest Graf (London 1957–61) II 65–6.

950 Hagiographa is, in Jerome's account, the third Jewish category of the books of scripture: Law, Prophets, Hagiographa; see Jerome's prologue to his translation of Daniel (*Biblia sacra vulgata* 1342:46–8).

more plainly in some and more darkly and obscurely in others. Accordingly, when theologians argue about the principal articles of the faith, they do not quote very much from Ecclesiasticus or from the Apocalypse, but a great deal from Paul.

[THE CREED]⁹⁵¹

What need is there for endless replies about the Apostles' Creed?⁹⁵² I ascribe to it an apostolic gravity and spirit.⁹⁵³ I have no doubt at all that the apostles taught those things; my only doubt is whether they were published by them in writing.⁹⁵⁴ This is not very likely for a number of reasons, in particular since the creed was not in ancient times the same everywhere, as I have shown quite clearly.⁹⁵⁵ But who would dare add, delete, or change

* * * * *

951 MARGINAL NOTE: 'The Creed.' A response to Pio XXIII libri 178ʳD–178ᵛE. The creed comes up in the course of a discussion of Erasmus' views on Scripture because the Apostles' Creed is mentioned in the passage from the Methodus with which Pio had begun his collection of Erasmian passages at the beginning of this book (177ʳA), the conclusion of which he repeats at the beginning of this section. After these (178ʳD), 'among the first I place the creed promulgated [...] at the Council of Nicaea, which is popularly called "the Apostles' Creed."' The quotation is garbled in that it omits intervening material ('... to the Hebrews. The place after these is held by certain items entrusted to us, as it were, by hand, that either come down to us from the apostles themselves or at least from those who were near to the era of the apostles. In this category, among the first ...') and omits the proviso ('promulgated, unless I am mistaken, at the Council of Nicaea') that Erasmus had added in the 1520 edition of the Methodus LB V 92C–D / Holborn 211:20–1.
952 Erasmus had repeatedly to defend his views on the Apostles' Creed. The flashpoint seems to have been an obiter dictum in his preface to the reader in the Paraphrasis in Matthaeum: 'from the creed, which, whether it was produced by the apostles I do not know, yet surely it exhibits an apostolic majesty and simplicity' (LB VII f**3ᵛ). This was seized upon by Béda. Erasmus produced several replies to him (LB IX 457D–458D, 497C, 554C–557C), and later to the criticisms of the theology faculty of Paris (LB IX 868C–870F), and to the Spanish monks (LB IX 1080B–D), all the replies having much in common in terms of types of arguments, texts cited, etc. A brief overview of fifteenth- and sixteenth-century controversies (Lorenzo Valla, Reginald Pecock, Erasmus) about the Apostles' Creed is provided by J. de Ghellinck Patristique et moyen âge: études d'histoire littéraire et doctrinale I: Les recherches sur les origines du Symbole des Apôtres (Gembloux 1946) 18–21.
953 After mentioning this creed, Erasmus had continued: 'this is, I think, because, it displays an apostolic gravity, sobriety, and terseness as well; and how I wish our faith had been content with it!' (Methodus LB IX 556B and 1080C / Holborn 211:22–4).
954 See earlier statements of this difficulty in Divinationes ad notata Bedae LB IX 458C; Elenchus in N. Bedae censuras LB IX 497C; Supputatio LB IX 555B; and Declarationes ad censuras Lutetiae vulgatas LB IX 868E.
955 See Divinationes ad notata Bedae LB IX 457E–F; and Declarationes ad censuras Lutetiae vulgatas LB IX 868E–869B.

anything in a document published by the authority of the entire apostolic senate, since to dare to do this in the case of the Gospel, which some individual among them wrote single-handed, is more than a sacrilege?

Even so, Lorenzo Valla maintains in his *Antidote to Nicholas v*,[956] with surely sound arguments, that this creed, which because of its very great authority is called apostolic, was promulgated at the Council of Nicaea.[957] He shows that in the *Decretum*, distinction [1]5, the chapter entitled 'Canons,'[958] the text should read 'under him also the holy Fathers from all the world who met at the Council of Nicaea, the second after that of the apostles, promulgated the creed,' and he says that he found this version in an old manuscript of Isidore.[959] You may, thus, understand that the creed which is now called the Apostles' Creed was composed at the First Council of Nicaea, which was the second after the Council of the Apostles.

In addition, Valla thinks that the Council of Nicaea composed the creed only as far as the words 'and of his kingdom there shall be no end.'[960] The rest was added at the Council of Constantinople at which Macedonius, who denied the divine nature of the Holy Spirit, was condemned.[961]

* * * * *

956 *Antidoti in Pogium, ad Nicolaum Quintum Pontificem Maximum* book 4 (c 1451–3) in *Opera omnia* (Basel: per Henricum Petri 1540) I 253–366. On the *Antidotum*, see Mario Fois *Il pensiero cristiano di Lorenzo nel quadro storico-culturale del suo ambiente* (Rome 1969) 359–82, 653.

957 Erasmus erred in claiming that Valla held that the Council of Nicaea had composed the Apostles' Creed; see Fois *Il pensiero cristiano* (cited in n956) 372.

958 *Decretum* pars 1 d 15 cap 1 (Friedberg I 34): 'Sub hoc [Constantino] etiam sancti Patres in concilio Niceno de omni orbe terrarum convenientes iuxta fidem evangelicam et apostolicam secundum post apostolos simbolum tradiderunt.'

959 *Antidoti in Pogium* (cited in n956) I 359–60; Poggio reports that he first conjectured that *secundum* in the text quoted by Gratian from Isidore (*Etymologiae* 6.16.4) should be *secundo* and immediately found support for this reading in a manuscript of the *Etymologiae*; the critical edition of the *Etymologiae* by Wallace Martin Lindsay (Oxford 1911) I 234 shows no such variant.

960 *Antidoti in Pogium* (cited in n956) I 360; for Erasmus' views on the original conclusion of this creed (sc *Credo in Spiritum sanctum ...*) and subsequent additions to it, see *Declarationes ad censuras Lutetiae vulgatas* LB IX 870A.

961 As Fois points out in *Il pensiero cristiano* (cited in n956) 368–9, Valla has erroneously assumed that the creed said at mass (the Nicene-Constantinopolitan Creed) is the same as the Apostles' Creed. According to Valla, the Council of Constantinople (381) must have added the section on the Holy Spirit because prior to that council the bishop of Constantinople, Macedonius (d c 362), was not considered heterodox even though he held mistaken views about the Holy Spirit (*Antidoti in Pogium* [cited in n956] I 360). The semi-Arian Macedonius was deposed at the Arian Council of Constantinople in 360. Because this section of the creed was thus added later, it could not have been composed by one of the apostles.

But what creed was it that was composed at the Council of Nicaea? The one that is chanted at mass?[962] But in it there is a declaration about the Holy Spirit, that along with the Father he is 'adored and glorified.' Then what need was there for another council against Macedonius? The Athanasian Creed is different from these. Moreover, a different version of the creed is given in the book of the early canons,[963] so if we want this particular version to have weight, then at the same time all the regulations that are transmitted there under the name of the apostles would be valid, and valid as well would be the church order which goes under the name of Peter.[964] But I would rather not argue about Lorenzo's opinion. This much is clear, from very many weighty arguments, that the creed which is called the Apostles' Creed was not published in written form by the apostles, much less were its individual articles supplied by individual apostles.

Oh, but that is what Augustine says![965] Yes, so does John Gerson who follows him![966] Yet I have found through careful examination that those

* * * * *

962 Pio *XXIII libri* 178ʳᴅ refers to the creed drafted at the First Council of Nicaea against Arius: 'it is the one called the Nicene Creed, which also from those times was constantly chanted in the churches, not the one which is called the Apostles' Creed.' In *Explanatio symboli* Erasmus distinguishes three creeds: the Apostles' Creed (LB V 1136ᴅ / ASD V-1 210:127–8 / CWE 70 241); the *Symbolum missae*, that is, the Creed used at mass (LB V 1178ᴇ / ASD V-1 290:458 / CWE 70 347), the so-called *Symbolum Niceno-Constantinopolitanum* or *Symbolum Constantinopolitanum*; and the *Symbolum Athanasii* (LB V 1178ᴇ / ASD V-1 290:465 / CWE 70 347), the pseudo-Athanasian Creed, often called, from its opening words, the *Quicumque vult*.

963 Probably a reference to the creed contained in the *Constitutiones Apostolorum* 4.11.1–10; see *Les constitutions apostoliques* ed Marcel Metzer 3 vols, Sources chrétiennes 320, 329, 336 (Paris 1985–7) II 322–7.

964 The regulations contained in the *Constitutiones Apostolorum* were supposedly promulgated at a meeting of the apostles (Metzer ed I 45). To St Peter is attributed the ritual for episcopal ordination (8.4.2); see ibidem III 140–1.

965 Pio *XXIII libri* 178ʳᴅ: 'about this [creed] you doubt whether it was of apostolic origin, though you indeed admit that it did proceed from apostolic men, nor in this matter do you yield to the judgment of Augustine who bears witness that it was handed down by the apostles themselves; you say that in this matter he was overly credulous and exploited the pietistic belief of the people'; Pio is referring here to Erasmus' remark on Augustine's credulity in the *Supputatio* LB IX 555ꜰ.

966 Eg in his 'A.B.C. des simples gens' Jean Gerson describes the creed as consisting of twelve articles of faith according to the twelve apostles; see his *Œuvres complètes* ed Palémon Glorieux VII: *L'œuvre française 292–339* (Paris 1966) 155 no 310. He was repeating a traditional teaching in this regard. For a study that shows how the two traditional versions of the Apostles' Creed were subdivided, and how these subsections were assigned variously to the different apostles, see J. Gordon 'The Articles of the Creed and the Apostles' *Speculum* 40 (1965) 634–40. For a study that distinguishes fifteen different traditions (Gerson is not cited) on how the articles were assigned to various apostles,

sermons[967] contain nothing of Augustine's. In his book to the catechumens he does not call this creed the Apostles', but says that it was put together from the Scriptures by the holy Fathers.[968]

Though I have made these and other points by way of reply in published books,[969] either Pio did not read them, or, more criminal still, he pretends not to have, and intones again the old song of slander so as to have an opportunity for the savage abuse in which the brawler, whoever he is, seems to find here such amazing pleasure.

But I admit, as he says, that I am going to have doubts about the authorship of the Epistle to the Hebrews. I do not deny this, but I do add 'until I see the church's express judgment on this point.'[970] If he should say that the Spirit is the author, I will answer that he is playing the sophist, for I am talking about the human author. We have doubts of this sort about the author of the book of Job[971] and the books of Kings,[972] but their authority is not shaken on this account. We have doubts about the authorship of the Gospel according to Mark,[973] but its authority is nonetheless universally

* * * * *

see C.F. Butler 'The Apostles and the Creed' *Speculum* 28 (1953) 335–9. Gerson is cited in other apologias with reference to the Apostles' Creed (eg *Divinationes ad notata Bedae* LB IX 457F–458A, *Supputatio* LB IX 556A).

967 The sermons are the pseudo-Augustinian sermons 240 and 241 PL 39 2188–91, which assign the contribution of the articles of the creed to specific apostles; for a similar denial of the authenticity of these sermons, see *Supputatio* LB IX 555F, *Declarationes ad censuras Lutetiae vulgatas* LB IX 870C, *Apologia adversus monachos* LB IX 1080C.

968 Augustine *Ad catechumenos* 1.1 PL 40 630; this, with three pseudo-Augustinian sermons on the creed (PL 40 637–8) found with it in the manuscript and early printings, formed the *De symbolo libri quatuor*, or, as Erasmus calls it in *Supputatio* LB IX 555D *Ad catechumenos libri quatuor*, or in *Declarationes ad censuras Lutetiae vulgatas* LB IX 870C the *Liber de symbolo*; this same text is introduced into the argument in *Supputatio* LB IX 555D and *Apologia adversus monachos* LB IX 1080D.

969 See n952 above.

970 Pio *XXIII libri* 177ʳA had quoted Erasmus' statement in the *Supputatio*: 'I think that in all centuries there have been learned men who had doubts about the author of [this] Epistle to the Hebrews; I too, to admit it candidly, still have doubts, will have doubts until I see the church's express judgment on this point' (LB IX 595C). The *casus belli* about the Epistle to the Hebrews seems to have been the passage in Erasmus' dedicatory letter to Henry VIII in the paraphrase on Luke (1523) (Ep 1381:103–5, and 16n), though Erasmus had reported doubts about Pauline authorship earlier in his preface to the paraphrase on the Epistle of James (1520), dedicated to Cardinal Matthäus Schiner (Ep 1171:9–12). On the Pauline authorship of this epistle, see also Pio's comments in *XXIII libri* 179ʳG and Erasmus' statements in his *Declarationes ad censuras Lutetiae vulgatas* LB IX 864A, 864E–F, 865A–866B, 866D.

971 See *Supputatio* LB IX 554F–555A; *Declarationes ad censuras* LB IX 863F–864A; and *Apologia adversus monachos* LB IX 1080C.

972 *Regum* LB *Regnorum* 1531

973 *Declarationes ad censuras* LB IX 866E; *Apologia adversus monachos* LB IX 1080C

sacrosanct. Finally, why is it that we make such a commotion about the question of authorship, since the writings of the apostles do not have such great authority because they came from them, but because they have been approved by the consensus of the church.

I have made reply about the Apocalypse[974] too often now. I mention there the opinions of various people, but I profess candidly that I submit my view to the judgment of the church. If I did not find the church's authority compelling, I would assert plainly that it is not the work of John the Evangelist. As it is, I receive it with the same attitude with which the universal church has received it.

He says: 'But these matters ought not to have been mentioned, because of the frail.'[975] Then, even more the books from which I took these points ought not to have been printed so often and published promiscuously. I thought it best to warn the reader because of certain people who, since they have read nothing of the ancients, have on their lips only two words, 'the church' and 'the Holy Spirit,' and under this pretext want whatever is read or chanted in the churches, however faulty and corrupt, to be considered the irrefragable Scripture. How I wish that the church's ancient bishops and princes had been a bit more vigilant in the beginning, and then we would not be tormented with so many difficulties pertaining to the letter.

But Pio is everywhere vigilant in the craft of slander. When talking about Eusebius I say: 'He cites a certain Gaius, an orthodox author,' and this is the way Pio reports it: 'And you cite Gaius, an orthodox author, as you say.'[976] No, it is not I who cite him, amazing sir, but the weighty author Eusebius, nor is it Erasmus who calls him orthodox, but Eusebius. It would be silly to go on with this sort of thing, but he everywhere employs this

* * * * *

974 Erasmus' response to Pio's extended series of combined quotations from and paraphrases of Erasmus' annotation on Rev 22:12 LB VI 1123E–F / Reeve *Galatians to the Apocalypse* 782; for Erasmus' replies, see also eg *Declarationes ad censuras Lutetiae vulgatas* LB IX 864A, 867A–868B.

975 Paraphrasing Pio *XXIII libri* 178ʳc

976 Erasmus quotes from his annotation on Rev 22:20 LB VI 1124F / Reeve *Galatians to the Apocalypse* 782: 'Eusebius ... quotes a certain Gaius, an orthodox author' and from Pio *XXIII libri* 177ʳA: 'Next, pursuing the matter at great length, you strive to prove that it [the Apocalypse] is not John's, and you cite Gaius, an orthodox author, as you say, who attributes this work to a heretic, Cerinthus, and Eusebius who suspects that it belongs to a certain John the Presbyter.' Erasmus is correct in that Pio did not report his statement correctly, but Erasmus seems to have garbled Eusebius, for it is not Gaius who is quoted in *Historia ecclesiastica* 7.25.2 SC 41 205, but Denis of Alexandria, with whom earlier (6.40.9; 7.11.22, 23) a Gaius is associated.

kind of straightforwardness, the guileless fellow, a man free of all hatred and vainglory.

The remaining passages are not worth entering into the record. He adds this conclusion: 'Let these suffice from the many things you have written about the Scriptures which you call into doubt.'[977] Now really, does one who has doubts about their human author call the Scriptures themselves into doubt? I suppose this is how Aristotle taught him to draw his conclusions.

THE AUTHORS OF THE SCRIPTURES[978]

I would rather grind in a mill[979] than answer this captious Thraso,[980] who makes the most slanderous pronouncements on the basis of the fragments supplied to him. Somewhere I credit Paul with a pious cunning in argumentation,[981] that is true, but I also credit him with a holy pride.[982] What I said in praise of Paul, this craftsman twists around into an insult. Paul often changes his tone so as to draw all to Christ, and he says one thing, keeps silence about another, suggests one thing, and enjoins another. Although Origen, Chrysostom, and Jerome along with them frequently point this out to Paul's credit, my friend the brawler takes this as blasphemy, and as an Aristotelian draws the following conclusion: 'Inasmuch as Paul speaks from the mouth of the Divine Spirit (for he is only the Spirit's herald), the Divine Spirit is cunning, slippery, deceitful, a twister of Scripture, and a dissembler. But even to think this is blasphemy.'[983] Who cannot see that this brawler has been driven to slander by a demented zeal for renown?

* * * * *

977 The conclusion of Pio's series of quotations from Erasmus in XXIII *libri* 177^{v}A
978 1531 UPPER MARGINS 'On the Authors of Scripture' on pages 188–91 [= 272–5 in this volume]. This heading also interrupts Erasmus' text at this point. It corresponds to a paragraph break in Pio's series of quotations on XXIII *libri* 177^{v}A–B, but does not otherwise make much sense.
979 Providing power for a mill in the place of a mule or an ox was a proverbial punishment for Roman slaves.
980 See n30 above.
981 In a series of quotations about Paul, Pio XXIII *libri* 177^{r}B quotes from the *Supputatio* LB IX 679E: '"It was a pious cunning whereby ... among the Athenians ... he twisted the inscription of an altar dedicated to demons into an argument for the Gospel" and you repeat this in countless other passages when reporting the words of the inscription'; see especially the annotation on Acts 17:23 LB VI 501E / Reeve-Screech 311; Pio also quotes from Erasmus' long annotation on 1 Cor 7:39 LB VI 700E–F / Reeve-Screech 478: 'But it is a peculiarity of Paul [Erasmus had written 'peculiar to Paul'] to twist everything to the business of the Gospel by a pious Christian cunning.' See also LB VI 501E / Reeve-Screech 311 (devout cunning); LB VI 815E / Reeve *Galatians to the Apocalypse* 579 (mistranslates Bible to make a point); and *Apologia ad Fabrum* LB IX 51A (twists words).
982 Eg *Annotationes* LB VI 648E / Reeve-Screech 431 (proud in Christ)
983 Quoted from XXIII *libri* 179^{v}K

I cite Jerome as my authority concerning the inscription on the altar at Athens.[984] Why, then, does Pio not quarrel with him? Jerome was even more bold in saying that scriptural *testimonia* which are not in disagreement in their own contexts are in conflict in Paul.[985] Is one who pretends necessarily a liar, or one who says one thing and does another? Is whoever misleads a criminal? When Paul says, so as to avoid provoking the false apostles, 'one says I am of Cephas, another I am Apollo's, another I am Paul's,'[986] is he not thinking one thing, and saying another? When he makes himself out to be a fool[987] for the Corinthians, is he not thinking something different from what he is saying? Again, when he says 'although unskilled in speech,'[988] does Jerome not expound this as a dissembling statement, since Paul was a master craftsman of argumentation?[989] Does the use of irony not mean saying something different from what one feels? Does the use of hyperbole not mean stating something different from what one feels? Is one who accommodates Scripture to the present circumstance not twisting it in a way? Again, is one not being deceptive who says what he knows will be understood by his hearers in another sense for the time being? When Christ said 'destroy this temple, and I will raise it up in three days,'[990] he wanted this statement, misunderstood as it was, to stick fast in the hearts of the apostles, so that when they remembered it later they would understand. But Pio treats all these as lies, and whatever is worse than lies. Now what could be more bothersome than to reply to wordy, impudent, and ignorant trifles, especially when they have already been answered so often?[991]

In addition, his saying that I impute forgetfulness to the apostles in the Scriptures is an impudent lie.[992] I show Jerome reporting the view of

* * * * *

984 In the annotation on Acts 17:23 LB VI 501E / Reeve-Screech 311: 'Here too Jerome points out that Paul employed a kind of pious cunning . . .etc,' referring to Jerome's comment on Titus 1:12f PL 26 572C–73A.

985 Erasmus repeats here in the same language a point he had made in his *Apologia ad Fabrum* LB IX 51A–B; the source of his reference to biblical *testimonia* is obscure; he may have a confused recollection of Jerome's comments on Titus 1:12 PL 26 571–4, where Jerome describes Paul's quoting pagan poets out of context.

986 1 Cor 1:12, 3:4

987 2 Cor 11:16, 12:11

988 2 Cor 11:6

989 Erasmus here paraphrases his own annotation on 2 Cor 11:6 LB VI 788E / Reeve-Screech 557. For an example given by Erasmus of Jerome's pointing out the clever argumentation used by Paul, see LB VI 815D. Erasmus in his scholia noted how Jerome felt Paul had engaged in dissimulation and feigning; see his scholia in the letter of Augustine to Jerome HO 3 150ᵛC, and of Jerome to Marcella HO 4 42ᵛD.

990 John 2:19

991 *toties* 1531 *om* LB

992 MARGINAL NOTE: 'Lie.' Pio XXIII *libri* 177ᵛB attempts a summary of Erasmus' annotation on Matt 2:6 LB VI 12E–14E especially 12E–13D / Reeve 13–15: 'And in the same work, in

some who thought that the Evangelist put down one name for another as a result of a lapse of memory, and Jerome does not reject this view as impious.[993] I show that Augustine seems to impute something like this to the evangelists in the matter of the order in which events took place.[994] It is not I who impute this to them; I only say that, even if there were something like this, particularly in those matters upon which the whole of the Gospel truth does not depend, the authority of all Scripture is not undermined,[995] so far as I am concerned.

An even more impudent lie is his statement that I wrote that Christ did not know the Day of Judgment in the same way other men did not.[996] But because it was not appropriate in the paraphrase to give an explanation in the persona of Christ, I report the words of the Evangelist in the sense in which he wrote.[997] But in the annotations I explain this text in three places.[998]

* * * * *

another passage, you say that the apostles could have suffered frequent lapses of memory in marshalling the scripture texts and testimonia.' Erasmus' defence here was anticipated by his more elaborate apologia for his views on this topic in *Apologia adversus monachos* LB IX 1070C–1073B.

993 In the annotation on Matt 2:6 LB VI 12F / Reeve 13 Erasmus quotes Jerome *In Micheam* 2.5.76–86 CCSL 76 481–2. Erasmus comments (LB VI 13B–C): 'Indeed the second defence [of discrepancies in New Testament quotations of Old Testament passages when compared with the Old Testament texts] that Jerome presents under the name of other people is somewhat more severe, namely that the evangelist suffered a lapse of memory and often disagrees with the Hebrew words and sometimes with its meaning. And yet he reports this view, while ascribing it to others, in such a way that he does not attack it as impious.'

994 Referring to his annotation on Matt 27:8 LB VI 140D–E / Reeve 106–7, where he summarizes Augustine *De consensu evangelistarum* 3.7.29–31 CSEL 43 304–8 and quotes *Quaestionum XVII in Evangelium secundum Matthaeum* 14:11–13 CCSL 44B 138.

995 Erasmus is paraphrasing his own comment in the annotation on Matt 2:6 LB VI 13D / Reeve 14.

996 MARGINAL NOTE: 'Lie.' Erasmus is reacting to Pio's quotation in *XXIII libri* 177^{v}B: 'And in the prologue to the *Supputatio* [LB IX 445A] you say as follows about the Lord Jesus Christ: Christ replied to his disciples "that neither the angels nor the Son knew the day of Judgment [. . .] and allowed them for the time being to be mistaken so that they believed that he did not know the day in the way that men did not know it."'

997 Referring to his paraphrase of Matt 24:36 LB VII 126D–127A: 'It is enough for you then to know the signs that portend my coming, so it will not catch you off guard. But it is not for you to ask to know precisely the day or hour at which the Son is to come, since it has not been given even to the angels of heaven to know these things, and besides, not even the Son of Man knows. The Father has kept this for himself alone'; or of Mark 13:32 LB VII 256E / CWE 49 155, both of which are quoted in the prologue to the *Supputatio* LB IX 443E–F; the impropriety of dealing with this exegetical problem in a paraphrase had been indicated to Béda LB IX 444E–F.

998 Referring, perhaps, to his comments on the parallel passages in which the text at issue occurs: Matt 24:36 LB VI 126F–127D / Reeve 97–8; Mark 13:32 LB VI 202F / Reeve 142; Acts 1:7 LB VI 436D–E / Reeve-Screech 274

On both of these points there is my crystal-clear reply against the denunciations of Béda.[999] At least Pio should have avoided a rock that had been pointed out to him, the one on which Béda made such a shameful wreck. But although it is too much trouble for Pio to read this, he is not embarrassed to thrust at his neighbour dangerous lies based on gossip. May the Lord forgive him all his sins. If he were alive, he would be treated as he deserves!

THE HOLY TRINITY [FROM BOOK 12][1000]

I have made answer to these slanders all too frequently, to Lee,[1001] to Zúñiga,[1002] to Béda,[1003] to the Tyrologus,[1004] to the Spanish,[1005] and when

* * * * *

999 Prologue to the *Supputatio* LB IX 443E–445B; *Supputatio* LB IX 548C–550A

1000 1531 UPPER MARGINS: 'On the Trinity' ('De Trinitate' as opposed to 'De Triade sacra' of the heading in the text) on pages 192–209 [= 275–90 in this volume]. Erasmus now takes up Pio's book 12 (*XXIII libri* 181ᵛQ–186ᵛY), entitled 'On the Mystery of the Trinity, and the Doctrine of the Arians which Erasmus Seems to Defend in Certain Passages.' Pio here departs from his usual pattern by not quoting a series of passages from Erasmus' works at the very beginning of the book, but instead using a structure of alternating quotation and discussion. He cites four passages from the scholia on Jerome, two from the preface to Hilary, one from the *Annotationes*, and one from the *Supputatio*. Pio had anticipated this topic in book 11 (179ʳG) with a reference to 'the Arian impiety ... towards which you seem to be wholly partial, in your words, if not, perhaps, in your meaning, as we will make clear later.'

1001 Eg *Responsio ad annotationes Lei* LB IX 171–176E, 183A–189A, 229A–D, 231A–232A, 251D–F, 252E–254D, 275B–278E

1002 Eg *Apologiae contra Stunicam* (1) LB IX 309D–311C, 315A–317D, 351F–353E / ASD IX-2 124–30, 140–6, 252–8

1003 Eg *Supputatio* LB IX 570C–572B, 583F–585D; *Responsio ad notulas Bedaicas* 717D–F

1004 A name fashioned out of the Latin word *tiro* 'a beginner' and the Greek *logos* 'a word, speech, discourse,' suggesting that the person is a novice at the Latin and Greek languages. Arguments can be made for identifying *Tyrologus* as either Cousturier, Titelmans, or Carranza. Cousturier (Sutor) was lampooned by Erasmus for his ineptitude in classical languages, eg *Apologia adversus Petrum Sutorem* LB IX 741D–F and 804C. Erasmus also attacked him as a particularly stupid theologian and a disgrace to the Sorbonne, eg LB IX 740D, 795D. By referring to him as *tyrologus* instead of *theologus*, Erasmus may also have been retaliating for Cousturier's attacks on Erasmus as a *theologaster* and *latinisator*; see Rummel *Catholic Critics* II 62. In his *Appendix respondens ad Sutorem* Erasmus briefly treated problems related to the hidden divinity of Christ; see LB IX 805C–807B. For more on Cousturier, see CEBR I 352–3; and Rummel *Catholic Critics* II especially 61–73. Erasmus also criticized Titelmans' abilities in Latin and Greek (Allen Ep 2807:13–15), and Titelmans was the recipient of the *Responsio ad collationes*, in which Erasmus treats Christ's humanity (LB IX 998C–F) and his divinity (1002B–1003C). Carranza (see CEBR I 273–4) presented himself as unskilled in writing (see Rummel *Catholic Critics* I 158), and most of the scriptural passages taken up in Erasmus' controversy with him (*Apologia ad Caranzam* LB IX 401A–413F) have to do with the divinity of Christ.

1005 Eg *Apologia adversus monachos* LB IX 1023E–24E, 1029E–1054A

annotating the Epistle of John.[1006] This most captious of triflers is deaf to all these, and on the basis of misunderstood scraps supplied by his secretaries he makes me out to be the patron and champion of the Arian heresy,[1007] and he rages at me with such abuse that, if he were alive, he would hear what he deserves.

In the third volume of Jerome I say that the Arian disturbance was more correctly a schism or faction than a heresy.[1008] I am not denying that it was a heresy. But at the beginning, when the emperor was an Arian, the pope an Arian, when at the council at which the Arian position was approved there were more bishops in session than at the one at which it was condemned, it was then more correctly termed a faction than a heresy. It was a heresy before God, but among men there was uncertainty, for the public declaration of the church had not yet been heard.

In the second passage[1009] it is not the Arians I am talking about, but the Origenists, who sprouted up in the era of Jerome. At that time virtually everybody was setting forth his faith in creeds, but these would have achieved nothing had not the influence of the rulers come into play and eventually taken away the churches from the Arian bishops, allowing them not even private assemblies. If these matters are clearly reported in ecclesiastical history, why is Pio quarrelling with me?

* * * * *

1006 The annotation on 1 John 5:7 LB VI 1079C–1081F / Reeve *Galatians to the Apocalypse* 768–71 was steadily revised to keep up with the controversies about it (see Rummel *Erasmus' Annotations* 132–4); most of what Pio complains of here Erasmus had added verbatim to the 1527 edition of his *Annotationes* from his note on 1 John 5:7 in *Apologiae contra Stunicam* (1) LB IX 351F–353D / ASD IX-2 252–8, with excellent notes by H.J. de Jonge.

1007 The opening sentences of book 12 of *XXIII libri* are: 'But it is right that we have a look also at what you have written to defend Arian doctrine against the view of our authors. You appear so to support it that it does not seem to have had a like patron in the nine hundred years since Arius was condemned'; later (183ᵛT) Pio addressed Erasmus 'the patron of Arian doctrine' and declaimed (185ʳY): 'who is so blind or mad that he does not see that in these words you want nothing else than for the error of the Arians to come back to life?'

1008 A reply to Pio's summary of a statement of Erasmus in *XXIII libri* 181ᵛP: 'For you say in the scholia on Jerome in the third volume that the error of the Arians was more correctly a faction or schism than a heresy, because the opposition was almost equal to our people in number and superior to them in eloquence and learning'; this statement is found in Erasmus' preface (HO 3 1ᵛ).

1009 Reacting to Pio's quotation (*XXIII libri* 182ʳQ), also from the preface to the third volume of Jerome (HO 3 1ᵛ): 'But again, in the same volume in another place you say as follows: "Against [the evil doctrine of Arius] which had been reborn in the Origenists the struggle was prosecuted fiercely with councils, decrees, and countless creeds, hardly the most suitable instruments, in my view, for suppressing heresy unless the issue is one of authority. Otherwise, the greater the number of doctrines, the more abundant the material for heresies. And the Christian faith was never more authentic and pure than when the world was content with that shortest creed."'

If I had in any way supported the insanity of the Arians, why would I express my hatred for their position in countless passages[1010] and declare forcefully and expressly that those who do not make profession among Christians are not to be regarded as such? I reviewed these passages in my *Apologia to the Spanish Monks*.[1011] But Pio says that it is not important to have read my writings, yet how important it is, in fact, will soon be plain.

In the scholia I criticize the argument which Jerome uses in the dispute as not being sufficiently effective against the stubborn Arians.[1012] They agreed that the Son is God and a great God, and they interpret the text of Paul, 'Who is blessed above all,'[1013] as being about the Son. But, says he, why are you pointing this out, unless you desire a return of the Arian heresy?[1014] I am, in fact, pointing this out to keep the Doctors from fighting heretics with weapons like these which are easily fended off. This is an effort to strengthen the church's doctrines, not an attack on them.

About the matter of the Threefold Testimony[1015] I have already given a careful answer, and more than once,[1016] so it is silly to repeat it here.

* * * * *

1010 See James D. Tracy 'Erasmus and the Arians: Remarks on the *Consensus Ecclesiae*' *Catholic Historical Review* 67 (1981) 1–10 especially 8, where an evolution in Erasmus' stance towards the Arians is noted.

1011 Erasmus presents eighty passages on trinitarian theology from a wide selection of his works (*Apologia adversus monachos* LB IX 1023E–1029B).

1012 Pio *XXIII libri* 182^{r}Q goes on to quote from Erasmus' scholia on *Orthodoxi et Luciferiani dialogus* HO 3 67^{v}D: 'And again you say in the same place about Jerome's expression *true God*, "That is, created by no one, whereas according to the Arians the Son was created by the Father and the Holy Spirit created by the created Son."' He then adds what appears to be a continuation of this passage, but is not. We have not been able to locate the passage Pio quotes, but it reflects ideas like those in Erasmus' annotation on Rom 9:5 LB VI 610D / Reeve-Screech 392 / CWE 56 243, and, more precisely, in *Apologia adversus monachos* LB IX 1044B: '"But I do not see what effect this text really has against the Arians, who did not deny that the Son of God is God; rather, they even acknowledged him to be a great God and God blessed over all [Rom 9:5]. But they thought that the Father is called God in some particular way in which the Son and the Spirit are not. Paul is not dealing here with the question of what Christ was, but of how he behaved, providing us with a model, etc."'

1013 Rom 9:5

1014 Pio *XXIII libri* 182^{v}R: 'But, O man most learned, I surely cannot guess what you had in mind when your wrote this. For if it were your purpose to defend Arian doctrine because you thought it true, you would have to be considered far more impious than Luther, indeed, absolutely blasphemous as well!'

1015 The so-called Johannine Comma of 1 John 5:7-8; Pio *XXIII libri* 182^{r}Q summarizes Erasmus' annotation on this text (LB VI 1079B–81F / Reeve *Galatians to the Apocalypse* 768–71), quoting his criticism there of Jerome (LB VI 1079E) and Erasmus' account of the Arian interpretation of the passage and his insistence on the need for rational argument to refute the Arians (LB VI 1080C).

1016 Eg *Responsio ad annotationes Lei* LB IX 275B–278E; *Apologiae contra Stunicam* (1) LB IX 351F–353E / ASD IX-2 252–8; and *Apologia adversus monachos* LB IX 1026D, 1031D–1032A

But who could endure my writing in the preface to Hilary like this: 'But we have rushed to such extremes of boldness that we have no scruples about dictating to the Son how he ought to have honoured his mother. We dare to call the Holy Spirit [. . .] God, [. . .] which the ancients did not dare to do'?[1017] At this point Pio, if he were alive, would surely blush, even though he was the most impudent of men. What is more disgraceful than to wreck on the same rock on which the Spanish monks had crashed?[1018] I am not defending the Arians against Hilary there, but showing the reticence of the ancients in making declarations about the divine persons. But Pio punches my face, saying: 'By this lengthy speech on behalf of the Arians you are taking a stand against the opinion and book of the very author who was being published as edited and emended by you. Let anyone who can stand the toil read the entire passage which we have reported in summary, and he will see clearer than light your eager zeal for defending the Arians.'[1019] He adds as the conclusion: 'But only about these statements you have made for the Arian doctrine.'[1020] To save the labour of replying to these charges, I will repeat here what I wrote in my *Apology to the Spanish Monks.*[1021]

The passages presented above make it immediately clear how they have been shamelessly distorted for the purposes of slander. The passages they pile up have no other goal, in their proper contexts, than to show the religious awe of the ancients in speaking about things ineffable, and the irreligious rashness of some moderns in doing so. If Hilary somewhere in his twelve books declares that the Spirit is to be adored, if he assigns to the Spirit the

* * * * *

1017 MARGINAL NOTE: 'He quotes things he doesn't understand.' Pio *XXIII libri* 182ᵛʀ quotes Ep 1334:473–7, with notable omissions in lines 475–7 (indicated by square brackets): 'We dare to call the Holy Spirit [true] God, [proceeding from the Father and the Son], which the ancients did not dare to do.' These same omissions occur in the quotation of these lines given by the Spanish monks (LB IX 1050D). Erasmus quotes the text here as given by Pio, changing the initial *Qui* to *Quin.* In the 1535 edition of Hilary the words *ausi non sunt* were changed to *aliquandiu non videntur ausi* 'sometimes do not seem to have dared' (Allen v 182 n446).

1018 *Apologia adversus monachos* LB IX 1028E, 1050C–1054A; *Adagia* I v 8: *Iterum eundem ad lapidem offendere* 'To stumble twice over the same stone' LB II 185C–186A / ASD II-1 484–5 / CWE 31 392

1019 Pio *XXIII libri* 182ᵛʀ

1020 This is an imprecise quotation of the final sentence of book 12 of *XXIII libri* 186ʳʏ: 'For it was not our intention to produce a book on the godhead of the divine Trinity, but only to reply to these statements you have made in defence of the Arian doctrine.'

1021 *Apologia adversus monachos* LB IX 1050D–1052B. Except for the insertion of *ita* (1174C:2 *ita loquitur*) and a reversal of word order (1174C:5 *faceret rem*) the section 1172E:12–1174C:12 is a verbatim repetition of 1050D:6–1052B:5, the reply to the Spanish monks' objection to a series of passages from the preface to Hilary concerning the Holy Spirit.

Title-page of Pio *XXIII libri* (Paris: Bade 1531)
By permission of the Biblioteca Apostolica Vaticana

name of God, then I will admit a lapse in memory on my part, since there is no controversy about fact. And it was to avoid being charged with a lapse of memory that I had inserted 'so far as I know,'[1022] but they left out these words because they do not support their slander. You recognize, O reader, Christian charity at work! But I would prefer that this be blamed, not on them, but on the libellous book from which, it is plain, these things were quoted. Again, if the epithet 'God' is thus clearly added to the name of the Holy Spirit in the canonical books, as frequently it is added in the gospels and epistles to the name of the Father, but rarely to the name of the Son, then I will admit a lapse of memory. But I attribute the ancients' reserve in pronouncing about things divine to their awe, that is, their reverence for ineffable mysteries. For they were awaiting a more careful examination of Scripture and the authority of the church. The words which follow there make this clear. Later authors, relying on more precise studies of the sacred books and on the authority of the ancients, had no apprehension about calling the Son of God 'true God,' and offering their prayers to him.

I speak along the same lines in my preface to Hilary: 'St Hilary at the end of the twelfth book does not dare to make any pronouncement about the Holy Spirit except that he is the Spirit of God – and it was unlikely that he would have dared to say this unless he had read it in Paul: he does not dare to use the word "creature" because he has not read it anywhere in Holy Scripture. This kind of profession would not be sufficient in this age because the needful diligence of the ancient Fathers has been very instructive for us, but we are carried far beyond what is needful.'[1023] They truncate what I say a little later: 'Though in Scripture the name of God is several times assigned to the Son, yet nowhere is it explicitly assigned to the Holy Spirit,' for these words follow: 'It should be acknowledged, however, that the devout probing of the orthodox later ascertained with sufficient proof from Holy Scripture that whatever was attributed to the Son was appropriate for the Holy Spirit.'[1024] You see, O reader, the craftiness of the holy inquisition about which Erasmus holds an incorrect

* * * * *

1022 From *Apologia adversus monachos* LB IX 1050E, but these two words do not appear in the passages of the preface to Hilary where Erasmus discusses Hilary's silence concerning the Holy Spirit. Perhaps he was thinking of Hilary's disclaimer in the sentence: 'Indeed nowhere does he write that the Holy Spirit must be adored and nowhere does he assign the word God to him, save that in one or two passages in *De synodis* he refers to those who dared to call the Father, Son, and Holy Spirit three Gods as condemned' (Ep 1334:440–3).
1023 Ep 1334:403–9
1024 Ep 1334:484–6

view.[1025] Why did they not add this clause? Because it absolutely precluded slander! They were not, therefore, inquiring after the truth, but grasping for material for false charges. It is with this kind of skill that malicious slander stitches together its patchwork.

With a like contrivance they also report this passage: 'The Father is most frequently called God, the Son several times, the Holy Spirit explicitly never.'[1026] For what follows is: 'And these remarks of mine are not meant to call into question what has been handed down from Holy Scripture by the authority of the orthodox Fathers, but to show how much reverential reluctance the ancients had in making pronouncements in theology.'[1027] Now what kind of artifice is this, to collect proof of heretical opinion from the very page on which there is so much that contradicts this same opinion? This is not the assiduity of bees, not the sincerity of charity which does not suspect evil, but rather the craft of spiders who suck out poison from health-giving flowers, or rather, to speak the truth, turn curative nectar into poison. Indeed, if one may utter the plain truth, it is the craft of him who among the Greeks has his name from slandering.[1028]

I have no doubt but that their remarkable shamelessness is apparent to all. But listen to something that is even much more criminal than these. I criticize the rashness of certain moderns in making declarations about things divine with the following words: 'We have rushed to such extremes of boldness that we have no scruples about dictating to the Son how he ought to have honoured his mother. We dare to call the Holy Spirit [. . .] God, [. . .] which the ancients did not dare to do.'[1029] This is what Erasmus said. Who could endure the blasphemy of these words? But hear, O reader, their utterly shameless craft. The occasion for this statement was as follows: I was saying that those ancients who, because of a kind of awe, were more sparing in their statements about the Holy Trinity, worshipped that Trinity in a more holy manner than we who hold more precise beliefs about it. Here I criticize in passing the irreligious boldness of certain people who, in their attempts to prove that the Virgin

* * * * *

1025 The fourth heading of the Spanish critique of Erasmus' works is 'Against the Holy Inquisition into Heretics,' with Erasmus' reply at *Apologia adversus monachos* LB IX 1054B–1060F.

1026 Ep 1334:466–8

1027 Ep 1334:469–72

1028 That is, the devil, διάβολος 'slanderer'; see eg *Christiani matrimonii institutio* (LB V 622A): 'The devil has his name from slander.' Note the use of the same conceit at the end of the *Responsio ad annotationes Lei* LB IX 284B.

1029 Ep 1334:473–7; note that Erasmus quotes the text with the same omissions as the Spanish monks.

Mother was conceived without the stain of original sin, employ this argument: 'He owed the highest honour to his Mother, and therefore it was his duty to provide that she be so conceived, and what he owed he could bring about, and so he did.' There is a like boldness at work in the fact that certain people maintain that Christ, even enthroned in heaven, is subject to the command of his Mother, and that it is on that account that the church sings 'Show thyself to be a Mother,'[1030] that is, 'Tell your Son to do what we ask.' But this, I think, does not particularly disturb them. What does disturb them is my adding: 'We dare to call the Holy Spirit God, which the ancients did not dare to do, etc.'[1031] I intended what went before to seem rash, and I appended this as a specimen of similar rashness.

O hateful words! Who could endure them? Who would not loathe such shameless malice in slander? I scarcely think that Satan would dare anything quite like this. How malicious the contrivance whereby these fellows, dead to the world, have stitched together horrifying statements! But the statement would be pious if all of it had been quoted. For it continues immediately: 'but on the other hand we have no scruples about driving him repeatedly out of the temple of our soul by our evil deeds, just as if it were our belief that the Holy Spirit is nothing more than a meaningless name. By the same token the majority of the ancients who revered the Son with the greatest devotion nevertheless feared to use the term ὁμοούσιον [of the same essence], etc.'[1032] With what am I finding fault here? Is it the fact that we dare to call the Holy Spirit God? Not at all, for I express approval of this in the same preface. I label as a defect the fact that, although we are more fully informed about the nature of the Holy Spirit and the Son, even so we are inferior in worship to those who were less informed, in spite of the fact that with the increase in knowledge there ought to have been an increase in veneration of majesty. What I am saying is so clear that it should be apparent, as they say, even to a blind man.[1033]

What, then, did they think when they were denouncing these statements to create a suspicion of Arian heresy? I will clear this up briefly. Unless they are plain blockheads, they either thought that there would be no one who would read my works, or they considered the judges blockheads.

* * * * *

1030 See n879 above.
1031 Ep 1334:475–7
1032 Ep 1334:477-81; Erasmus explains the Greek term at Ep 1334:436–7: 'the Greeks use the term ὁμοούσιον (that is, equal in power, wisdom, goodness, eternity, immortality, and all other attributes).'
1033 Compare Ep 1334: 384–93, 469–73; see *Adagia* I viii 93: *Vel caeco appareat* 'A blind man might see that' LB II 331B–E / ASD II-2 315–16 / CWE 32 174–5.

If anyone were to twist words excerpted from the Pauline epistles in this way for purposes of slander, well, in 1 Corinthians 8 he says: 'There are many gods and many lords.'[1034] This is against the first commandment of the Decalogue. In Romans 3 he writes: 'Let us do evil, that good may come about.'[1035] This is contrary to correct doctrine. Likewise in 1 Corinthians 15 he says of himself and the rest of the Christians: 'We are more pitiable than all other men.'[1036] This is contrary to the statement in the Gospel: 'Blessed are they who have not seen and yet believe.'[1037] Would not anyone who would try this be doing something everyone would consider absolutely absurd? No one is so shameless that he would not agree. And yet, among the instances which I have pointed out above there are some that display no more sense or shame than these.

But if they are carping about the expression 'we dare,' because it seems to have a bad ring to it, I will answer them with that text from the Canon of the mass: 'Admonished by saving commands and informed by divine instruction, we dare to say "Our Father, etc."'[1038]

If Pio had not found it too much trouble to read this, he would not have brought such disgrace upon himself by attacking a friend with such a stupid slander. If anyone were to speak like this, 'We dare to call the Father the Creator of all things, the Governor of the World; we make bold to call the Son its Redeemer; we make bold to call the Holy Spirit the Giver of Charisms,'[1039] and did not add anything else, would his words not seem impious? But if you add, 'and yet, though sons, we neglect the commands of our Father; creatures, we scorn our Creator; ungrateful to our Redeemer, we willingly slip back under the power of Satan; we push away from us the Spirit's abundant gifts by the impurity of our lives,' is it not now a pious one? Yet it was a statement like this that Pio attacks, raving on with savage abuse even though he did not understand it.

From all this it is clear what it means not to read what you are criticizing, what it means to put together a dangerous slander from the notecards of your employees, what it means to weave together a patchwork accusation from small fragments broken off from here and there. Furthermore, it

* * * * *

1034 1 Cor 8:5
1035 Rom 3:8
1036 1 Cor 15:19
1037 John 20:29
1038 The preamble to the Lord's Prayer, not strictly speaking part of the canon, but following immediately upon it
1039 Ep 1334:475–9

is silly for him to urge the reader to examine the thoroughgoing purport of a passage that he himself did not think worth reading.[1040] If he had read it, I think the fellow would have been ashamed to vomit forth abuse like this at his undeserving neighbour.

Meanwhile, Pio wonders whether I really agree with Arius, or wanted, as a display of cleverness, to present a defence of a bad case, though I do not succeed, however remarkable my attempt.[1041] What would this trifler deserve to hear, if he were alive? But he slipped away from my talons, and I hope he also slips away from the judgment of God. I made a list for him of ever so many passages[1042] in which I clearly, forcefully, ardently profess the same faith in the Holy Trinity as does Holy Church, yet here he doubts whether I am sincere in my defence of the Arians, or am doing it as a display of cleverness!

But he ascribes my lack of success in my endeavour to the fact that I have neglected Aristotle,[1043] as if the only one to have read Aristotle were he, who never was a philosopher, as those who know the man assert, except at dinner-parties,[1044] or as if I did not quote Aristotle more frequently in my works than he who for all these years has just talked Aristotle.[1045]

But will Aristotle, may it please the gods, extricate us from this labyrinth? St Jerome asserts in the course of his dispute against the Luciferians that this heresy had been supported by the philosophy of Aristotle.

* * * * *

1040 Pio XXIII libri 182ᵛʀ: 'Let anyone who will not find it too troubling read the uninterrupted text of the passage, a summary of which we have reported, and he will see clearer than any light your eager zeal to defend the Arians.'

1041 Summarizing Pio XXIII libri 182ᵛʀ

1042 Erasmus refers to the list of passages where he clearly states an orthodox Trinitarian theology published earlier in his controversy with the Spanish monks; see n1011 above.

1043 Pio XXIII libri 182ᵛʀ remarks that Erasmus has been thrust into labyrinthine confusion because of his 'disdain for philosophy and more exacting studies. If you were equally as well versed in them as in less difficult studies and your nimble wit had not fled from the study of Aristotle, you would not be ignorant that it is a necessary conclusion that, etc.'

1044 See Introduction xxiv–xxv, xxxviii, cxxii.

1045 Pio did not just talk about Aristotle. He actively supported the publication of Aristotle's works in Greek by Aldus and their retranslation into Latin by Sepúlveda. As a youth Pio was tutored by some of the leading scholastic philosophers of his day, and his contemporaries praised his abilities in philosophical discourse and argumentation. In his writings against Erasmus he often defended Aristotle and his followers. For his part, Erasmus showed great disdain for scholastic philosophy as practised in his day. Even if he did mention Aristotle in his writings (LeClerc provides at least twenty-four references: LB X a7ᵛ–8ʳ), Erasmus often criticized the adulation given to him. Only after Pio's death did Erasmus publish an edition of Aristotle's works; see vander Haeghen II 10.

He says: 'There is the additional fact that the Arian heresy fits better with worldly wisdom, and derives the streams of its arguments from the springs of Aristotle.'[1046] Jerome thinks that the Arians were invincible because they were equipped with Aristotelian assistance, and Pio will free me from my confusion with the same assistance!

But let us see how Pio extricates us with his Aristotle. 'Whoever is true God,' he says, 'is God by nature.' Someone might object that the Son is called the True Vine,[1047] and is, therefore, a vine by nature; but I do not want to play the sophist, so I will let it pass. 'But Scripture calls the Son true God. Therefore, he was not created.'[1048] What if I deny his minor premise, how is he going to prevail? I am not now dealing with reality, but with Aristotle. If he does not prove it, what is his conclusion? He could have quoted the text of John, 'That they may know you the only true God, and the one whom you have sent, Jesus Christ,'[1049] and it will have force, if we accept Augustine's interpretation, 'That they may know that you and the one whom you have sent, Jesus Christ, to be the only true God.'[1050] But where in all this is the Holy Spirit?

But why did Pio add: 'if he were only called God in the way that the "many gods, many lords" are so called'?[1051] Who proclaimed of Christ what Paul said of the demons of the pagans? He should have quoted the passage that Christ quotes in the Gospel: 'I have said, you are gods, and sons of the Most High.'[1052] (With this passage the Lord is protecting himself against the slander of the Jews, that he had called himself the Son of God. But Pio was acting in another kind of arena here.)

The Aristotelian continues: 'If [. . .]' the Son is 'of the same nature with the Father, [. . .] then he is the same *individuum*.'[1053] I might make the substitution: 'Peter is of the same nature with John, therefore he is the same

* * * * *

1046 *Dialogus contra Luciferianos* 11 PL 23 166B

1047 John 15:1

1048 Erasmus gives here in a very compressed and distorted syllogistic version part of Pio's more elaborate argument in XXIII *libri* 182ᵛR.

1049 John 17:3

1050 See eg Augustine Ep 238.22 CSEL 57 550:23–551:1 and *De spiritu et littera* 22.37 CSEL 60 190:16–17; Erasmus added this Augustinian paraphrase to his annotation on John 17:3 LB VI 405F in the edition of 1535 (see Reeve 262).

1051 Erasmus is taking up Pio's allusion in XXIII *libri* 182ᵛR allusion to 1 Cor 8:5: 'Now because he is true God, he is God by nature and essence, as a true man is man by nature, even as, also, true gold is gold by essence, not by likeness or colour only. For the Son would not be named true God if he were only called God in the way that the "many gods, many lords" are so called.'

1052 Ps 81:6, part of which Jesus quotes at John 10:34

1053 Pio XXIII *libri* 182ᵛR

individuum.'[1054] I read in the orthodox writers that the Son is the same with the Father, but I do not read that he is the same *individuum*. I profess that there is the same undivided nature. I do not know if it is permissible to call three persons really distinct from one another an *individuum*, unless you add 'in essence,' for the persons differ one from another in number.

He goes on: But 'since the divine nature is absolutely simple, it cannot be shared by many *individua*.'[1055] And here some theologian added: 'And it is so united that it cannot be distributed to very many *individua*.' (You observe, reader, the jargon [διάλεκτον] at work.) And shortly: 'The true God, then, is thus one, and this opposes the existence of a plurality. For just as that which is what it is by nature is opposed to that which exists by participation, so too that which is by nature a unity resists multiplication, etc.'[1056] Let anyone who wishes now go ahead and assert that Alberto wrote all this!

Has the knot been untied now with the aid of Aristotle? No, it has not been untied, because nowhere in Scripture is Christ called 'true God.'[1057] If this had been the case, the Arians would have surrendered. But what Pio the Aristotelian has attempted in vain has already been achieved in published books by Erasmus the Grammarian. (Pio, who seems never to have opened a single volume of divine Scripture, often pokes fun at me under the title Grammarian.) For what I wrote is that the obstinacy of the Arians cannot be convincingly quelled from Scripture without the addition of

* * * * *

1054 In the philosophical-theological system of the Scotists, which Pio followed, the term *individuum* can mean the supreme genus that is not divisible by differences, something that is one in number and cannot be divided into many. In an equivocal sense it can also refer to a most special species that can be divided through individual or material differences and thus become particularized. See Mariano Fernández García *Lexicon Scholasticum philosophico-theologicum in quo termini, definitiones, distinctiones et effata a Joanne Duns Scoto exponuntur, declarantur* (Hildesheim 1988) 340 (*Individuum*). When this term is applied to God, he is thus the supreme genus, one in number, and because absolutely simple (a pure spirit) cannot be divided by material differences. Pio's argument at *XXIII libri* 182^{V}R is that because the Son is of the same nature, substance, and essence as the Father, he is therefore not created but true God; and since he is of the same nature, he is therefore the same *individuum*, God. For a study of the Trinitarian teaching of Scotus, see Friedrich Wetter *Die Trinitatslehre des Johannes Duns Scotus* (Münster 1967). Erasmus is here either betraying a remarkable lack of comprehension of Scotist theology, which he claimed to have studied for years in Paris, or he is attempting to lampoon it.

1055 Erasmus is careless in his quotation of Pio *XXIII libri* 182^{V}R here, substituting *plurimis* for *pluribus*, ie 'very many individuals' for 'a plurality [that is, more than one] of individuals.'

1056 Quoting, with some imprecision, Pio *XXIII libri* 183^{r}s

1057 Pio's argument in *XXIII libri* 182^{V}RS–183^r depends on acceptance of the condition 'if we acknowledge that he is true God,' and Pio asserts 'but Scripture proclaims that he is true God.'

argumentation based on reason,[1058] and I showed that this must be derived from the absolutely simple simplicity of the divine nature. It is beyond controversy that Scripture calls the Son the Only-begotten of God, as in the first chapter of John: 'We have beheld his glory, as of the Only-begotten of the Father.'[1059] Again, in chapter 3: 'God so loved the world that he gave his only-begotten Son.'[1060] He is not, then, a son by grace, like the rest of the saints. Therefore he is Son by nature, and if by nature, then he was born of the substance of the Father. But since that nature is most simply simple, it cannot be shared by division. Therefore it should be the same in number in Father and in Son.

But Pio did not read this, though it would have been better if he had, so that he would not be both attacking me undeservedly and serving himself up to the learned as a laughing stock. But what need was there of Aristotle to demonstrate the simplicity of the divine nature, as if this could not be better learned from Plato?[1061] As it is, since Pio's argumentation is based on this foundation – that Scripture proclaims the Son true God,[1062] and this cannot be proven to the very obstinate – then the rest of his structure also totters.

When I read Pio's complaints about such trivial matters rendered in tragic style, it seems to me that I am dealing with a petty and quarrelsome woman who seizes upon everything as a subject for contention. I wrote somewhere that Jerome 'is violent, less than bashful, often variable, and not very consistent.'[1063] What I wrote is true, and I have proved it, and could prove it with even more arguments if the situation required, even though in a good cause I take his authority very seriously. Here Pio spits in my face: 'You, you utter this about St Jerome, a man most learned and holy, to whom you would not be worthy to offer even a chamber

* * * * *

1058 That is, *ratiocinatio*, the term Erasmus used in *Apologiae contra Stunicam* (1) LB IX 353C / ASD IX-2 258:1528 and the annotation on 1 John 5:7: 'I surely do not see how what is denied by the Arians can be proven except by argument based on reason (*sine ratiocinatione*)' (LB VI 1080C / Reeve *Galatians to the Apocalypse* 768–9), quoted earlier in Pio XXIII *libri* 182ᵛQ and 184ʳv. De Jonge glosses *ratiocinatio*: 'By theoretical, speculative reasoning, as opposed to demonstration by reference to biblical proof-texts' (ASD IX-2 259 n528).

1059 John 1:14

1060 John 3:16

1061 In the *Enchiridion* Erasmus advocated following the Platonists 'because in much of their thinking as well as in their mode of expression they are the closest to the spirit of the prophets and of the gospel' (LB V 7F / Holborn 32:25–8 / CWE 66 33).

1062 See n1057 above.

1063 *Apologiae contra Stunicam* (1) LB IX 352C–D / ASD IX-2 254:1477–8 and repeated in the annotation on 1 John 5:7; for Pio's source for the quotation, see n1064.

pot!'[1064] Who would not laugh at this? Did Jerome, an effeminate fellow given to luxury, have a pretty boy to provide him with a chamber pot at the snap of his fingers?[1065] But joking aside, is whoever disagrees with an author or criticizes something in him thereupon insisting that he be considered that author's superior? Aulus Gellius criticizes something in Cicero,[1066] but no one spits in his face as if he were ranking himself higher than Marcus Tullius.

I maintain that I have provided sufficient answers about the passage on the Threefold Testimony.[1067] Let him read who wants to. Pio did not want to read them because he preferred to brawl with Erasmus. He teaches us that certain things are called one by nature, certain things by accident.[1068] Peter and John, he says, are one, because both are men. Alexander and Bucephalus are not one, except in that both were white.[1069] Even granting that

* * * * *

1064 Erasmus is reacting to Pio's paraphrase and criticism in XXIII libri 183^{V}T of the harsh words about Jerome in Erasmus' annotation on 1 John 5:7 LB IX 1079E / Reeve Galatians to the Apocalypse 769: 'You are so stubborn in your endeavour to prove this that you are not afraid to attack Jerome quite insolently with a view to diminishing his authority, not blushing to say that he is often violent, lacking in restraint, often variable and inconsistent. You, you utter this about St Jerome, a man most learned and holy, to whom you would not be worthy to offer even a chamber pot or untie his shoelace.' Pio is combining the adage 'Unworthy to hold out a chamber pot to him' (see Adagia I v 94 LB II 217D–218A / CWE 31 465–6) with an allusion to Scripture (Mark 1:7, John 1:27); note that Erasmus concludes Adagia I v 94 with the remark: 'There is a very common saying used in our own day: "He is not fit to take off that man's shoes." This adage is used in the Gospel by John the Baptist' (CWE 31 466).

1065 Erasmus takes up Pio's chamber-pot reference, elaborating it a bit by a telescoped allusion to Petronius (a locus not mentioned in Adagia I v 94). In Satyrica 27, early in the set of fragments known as 'The Dinner of Trimalchio,' Trimalchio is spied playing ball with his effeminate slave boys, attended by 'two eunuchs standing on opposite sides of the circle. One of them was holding a silver chamber pot, and the other was keeping a record of the number of balls'; eventually 'Trimalchio snapped his fingers, at which signal the eunuch put the chamber pot under him while he went on playing. When his bladder was unburdened, he called for water for his hands, and wiped his slightly moistened fingers in the hair of a boy.'

1066 Aulus Gellius (2nd century AD) was an admirer of Cicero, but does criticize him in his Noctes Atticae, eg 1.3.12, 15.6; Erasmus may, however, have in mind what he mistakenly reported as a criticism of Cicero by Gellius (Allen Ep 2172:41–2 and n).

1067 Pio XXIII libri 182^{V}R–183^{V}T takes up Erasmus' annotation on 1 John 5:7–8; for Erasmus' earlier replies on this point, see n1015 above.

1068 Pio XXIII libri 183^{V}T–184^r, part of his attack on Erasmus' view that the text of 1 John 5:8, 'and these three are one,' was an orthodox interpolation.

1069 Referring to Pio XXIII libri 183^{V}T; but Pio uses the names Erasmus and Albertus in his first example, an association that Erasmus, for whatever reason, chose to overlook. Bucephalus was the name of Alexander the Great's horse.

some people talk in this way,[1070] is that rationality which makes Peter a man a single indivisible thing? Or is it the same in number as the rationality which makes John a man? Is the soul of both also the same indivisible entity? For the human soul is the form of man.

But who was it who said that a white horse is the same as a white man? And is the whiteness of the horse the same in number as the whiteness of the man? Did he learn these subtleties from Aristotle? Why did he not expound the passage in John 17 'that they may be one even as we are one'?[1071] Are all the pious[1072] one by an indivisible nature, like the three Persons? When it is said in Acts 'and they had one heart and one mind,'[1073] what is meant but an agreement of minds? But I must stop making sport of a dead man.

He charges that I reject the dogma of John the Evangelist about the divinity of the Son.[1074] I suspect that Pio thinks *dogma* comes from *docendo* [teaching],[1075] whereas it is a Greek word derived from the verb δοκεῖν [παρὰ τὸ δοκεῖν], which means to have an opinion. Who is it that calls the divine Scripture dogmas?[1076] I speak of the dogmas of those who produced the many creeds, creeds in which it comes about, from time to time, that while

* * * * *

1070 Pio (*XXIII libri* 182ᵛʀ, 183ʳ⁻ᵛᴛ) makes an extensive appeal to ordinary usage as a norm for interpreting 1 John 5:7–8: 'For this is the way we speak, and it is the customary way to make declarations ... For unless we speak by fixed rules, which are derived for the most part from usage, nothing can be uttered in word or writing whose meaning may be regarded as sure.'

1071 John 17:21

1072 Of course, this also means 'all the Pios.'

1073 Acts 4:32; both John 17:21 and Acts 4:32 are quoted in the annotation on 1 John 5:7 LB VI 1981D and E / Reeve *Galatians to the Apocalypse* 770–1.

1074 Pio *XXIII libri* 182ᵛʀ, 184ʳ–185ʳ asserts Johannine authorship of the problematic words in 1 John 5:8, 'and the three are one' (see nn 1015 and 1068 above) and returns to the depreciation of dogmatic statements in the preface to volume 3 of the edition of Jerome, where Erasmus states: 'And never was the Christian faith more sincere and holy than when the world was content with that most brief creed' (HO 3 1ᵛ). Pio provides a brief history of the causes and development of elaboration in dogmatic pronouncements, including the point that John was compelled to write his Gospel, a book of Scripture, to combat the dogmas of Cerinthus and the Ebionites, a detail borrowed from Erasmus' preface to Hilary (Ep 1334:154–6).

1075 Pio *XXIII libri* 182ᵛʀ, 185ʳx: 'For the promulgation of dogmas, the revelation of things divine, does not provide the material for heresies, but instead eliminates and eradicates those that already exist, and is a precaution against the appearance of new ones, since it both teaches [*doceat*] and instructs in what should be thought.'

1076 This reference is obscure; Erasmus may be alluding to *XXIII libri* 182ᵛʀ, 185ʳ, where Pio misquotes 2 Tim 3:16, giving *Omnis doctrina divinitus revelata*, ie 'All doctrine divinely revealed' instead of *Omnis scriptura* 'All Scripture.'

they are trying to close all loopholes for heretics, they themselves fall into the pit that gapes behind them. I have shown that this happened to Hilary, and I could show, if I liked, that it has happened to other great Doctors of the church.[1077]

But so that he may prove that the ancients assigned the epithet 'God' to the Holy Spirit, he brings forward Chrysostom, Didymus, and Ambrose,[1078] by whose time the declaration against Macedonius[1079] had already been made. 'Who are more ancient?' he says.[1080] Origen and Tertullian. But far be it from me to think that there was no one at all who did this. I was talking about published books. Certainly Hilary does not do this in all of his *On the Trinity*.

Pio moves on to the arguments by which he means to prove that Christ [*sic*][1081] is true God. I accept these, but it was the explicit appellation that I meant. But it is tiresome to deal any longer with someone who has not read my works, but takes this opportunity for a quarrel from the notecards of his secretaries, all as if others had not quarrelled with Erasmus on this point earlier, to their own very considerable disgrace.

THE AUTHORITY OF PRIESTS AND BISHOPS [FROM BOOK 13][1082]

1 / First, Pio's using the term *sacerdotes* instead of *presbyteri*[1083]should be blamed on the authors of the notecards. For I do not recall that anyone at the time of Jerome used the title *sacerdos* for anyone who was not a bishop.

* * * * *

1077 In the preface to Hilary Erasmus makes this point briefly about Tertullian (of whom he uses the 'pit' imagery), Jerome, Montanus, and Augustine (Ep 1334:504–18), and in detail about Hilary (Ep 1334:521–603).

1078 Pio *XXIII libri* 185ʳʏ–186ʳ returns to the preface to Hilary: 'We dare to call the Holy Spirit true God ... which the ancients did not dare to do' (Ep 1334:475–7), citing John Chrysostom, Didymus the Blind of Alexandria (d 398), Ambrose, Hilary, and Augustine.

1079 See n961 above.

1080 It is unclear whether this is provoked by Pio's question to Erasmus at *XXIII libri* 185ʳʏ, 'Which ancients are you talking about?' or his rhetorical question (185ᵛ), 'Who more ancient, who greater, who more sublime can be adduced than John the Evangelist?'

1081 *Christum* is the reading of both 1531 and LB, but it must be an error, going back, perhaps, to a slip of the pen on the part of Erasmus, for the arguments assembled by Pio *XIII libri* 185ᵛ–186ʳ are all aimed at proving that the Holy Spirit is true God.

1082 MARGINAL NOTE: 'From book 13'; 1531 UPPER MARGINS: 'On the Authority of Priests' on pages 210–19 [= 290–8 in this volume]. Pio's book 13 (*XXIII libri* 186ʳ–194ʳ) is entitled 'On the Authority and Functions of Priests and Bishops,' and opens with the series of thirteen passages which Erasmus takes up here.

1083 Pio *XXIII libri* 186ʳA introduces his first quotation: 'Now it is appropriate after this that we look also at certain of the things you have written about priests [*sacerdotes*] and

Let us forgive him this error, although so savage a detractor does not really deserve forgiveness. Since Jerome proves from Scripture in so many passages that at the time of Paul presbyters and bishops were the same (for example, in the Letter to Oceanus,[1084] in the Letter to Evagrius,[1085] in his commentary on the Epistle to Titus,[1086] chapter 1, and in many other passages), why is it that Pio argues with me about this point?

2 / But elsewhere in the scholia (of course, it is on the Letter to Evagrius) I say Jerome 'does not think that a bishop is more important than any presbyter, except that he has the right to ordain.'[1087] What is Erasmus' fault here? He is not the one who thinks this! He is only pointing out Jerome's view. But if Pio refuses to accept that Jerome wrote this, then we must bring forward exhibits and witnesses.

These are his words in the Letter to Evagrius: 'Now the selection of one man' from among the presbyters 'to be placed over the rest came about as a remedy for schism, so the church of Christ would not be broken up by each claiming this for himself. For at Alexandria too, from Mark the Evangelist down to the bishops Heracles and Dionysius, the presbyters always named as bishop one who had been chosen from among them and advanced to a higher rank, just as an army might make a general, or deacons would choose one of their number whom they knew to be industrious and call him "Archdeacon."[1088] For what does the bishop do, with the exception of ordination, that the presbyter cannot do? Nor is the church of the city of Rome to be considered one thing, the church of the entire world something else.'

* * * * *

bishops [*episcopi*]. For you say in the scholia of volume 4 of the works of Jerome: "Of old only bishops were called *sacerdotes*, and the same were called *episcopi*. Then when there began to be more of them, the chief of all of them was called *sacerdos* and *episcopus*, and the rest *presbyteri*."' Pio (186ᵛA) seems to be concerned that a real distinction be maintained between priests (*sacerdotes*) and bishops (*episcopi*), a distinction that Erasmus, citing Jerome, seeks to blur. Our review of the 1516 scholia in volume 4 did not turn up this quotation, but in his scholia on Ep 69 (HO 2 195ᵛD) Erasmus does state a similar idea: 'He [Jerome] took care that not even once did he call a priest someone who was not a bishop, for in that century they were not called priests unless they were bishops. I wonder why the same work does not call him [Jerome] cardinal, since in those times that term was as yet unknown.'

1084 Ep 69.3.4 CSEL 54 683:19–21

1085 Ep 146.1.1, 2 CSEL 56 308:5–6, 309:4–5; this letter was not in fact addressed to Evagrius, as Erasmus has it, but to Evangelus.

1086 On Titus 1:5 CCSL 77C 13–16, with many scriptural citations

1087 Reply to Pio's quotation from Erasmus' *antidotus* to Ep 146 HO 3 150ᵛC. Note that in Erasmus' original and in Pio's quotation the text runs 'more important than any *sacerdos*'; Erasmus has fitted the quotation to this immediate context by substituting *presbyter* for *sacerdos*.

1088 Erasmus quotes Ep 146.1.5–6, 7 CSEL 56 310:5–14, 17; 310:19–311:3.

And shortly: 'If it is authority that is sought, the world is greater than a city.' With these words Jerome shows that the bishop of Rome is no more the superior of his presbyters than any bishop in a remote town, and that if Roman practice should differ, then what is done in the whole Christian world is what ought to be observed rather than what is practised in a single city. He continues: the bishop of any place 'has the same dignity and the same priesthood. It is the power of wealth and the lowliness of poverty that make a bishop either higher or lower. Otherwise, they are all successors of the apostles.' This holy man frankly declares that the bishop of Rome is not higher than the rest of the bishops by virtue of his priesthood, but only by virtue of his wealth.

His meaning, moreover, is that the presbyters in any church are equal to their bishops, except in dignity and in the right of ordination granted to them. He makes the same point on chapter 1 of the Epistle to Titus: 'The foregoing,' he says, 'to show that among the ancients presbyters were the same as bishops, and that gradually, to pluck out the sprouts of dissension, the entire pastoral concern was transferred to one man. So, just as presbyters know that it is by the custom of the church that they are subject to the one placed over them, so too the bishops should know that it is by custom rather than by the Lord's authentic dispensation that they are greater than the presbyters, and that they ought to share governance of the church, etc.'[1089] Let others judge what Jerome means by these words. I am so far from inserting my own judgment that I attempt to explain away this passage in an *antidotus*, and do so in even greater detail in the second edition.[1090] But the poison-gatherers did not supply passages like that, to Pio's great disgrace.

3 / What wrong have I done if at the opportunity provided by the divine Scriptures I remind bishops that, punctilious as they are in demanding honour from their charges, they should in turn be mindful of their obligations to the Lord's flock?

* * * * *

1089 Erasmus quotes with imprecision Jerome's commentary on Titus 1:5 CCSL 77C 15.
1090 See n1083 above. Erasmus added in the 1524 Froben edition a number of comments to his scholia on the letter to Evagrius in the earlier 1516 Froben edition: eg in the first edition he stated: 'Nor did he think any bishop to be less than another, except insofar as he was superior to him in humility' (HO 3 150ᵛC); he expanded this text in Froben 1524 to read: 'Nor did he think any bishop to be less because he was more humble, or greater because he was wealthier' (HO 2 335ʳA). To explain the superiority of bishops over priests, Erasmus adds in this revised edition: 'And somewhere he [Jerome] says that presbyters succeeded to the position of vicar of the apostles, [while] a bishop to the place of Christ. In this regard therefore presbyters and bishops are equal, because they are everywhere preferred to deacons' (ibidem).

4 / They also quote this: 'Where has it been considered acceptable that bishops have secular jurisdiction?' By adding the interrogation mark[1091] they have made it a complete sentence, as if it was my view that this had not been considered acceptable anywhere, whereas it is considered quite acceptable almost everywhere, in France, in both Germanies, in Italy, in Spain.[1092] The very absurdity of the matter could have suggested to the fellow that this is not my view. I have not yet found the passage,[1093] but my guess is that I wrote: 'Where it has been considered acceptable, etc, care must be taken that they be learned and devout pastors,' or something like that. A curse on these drunken suppliers who cause me so much trouble! And yet, taken in by their mistake, the pitiful old man raves at me, shouting: 'O destructive, seditious, blasphemous statement, full of scandal and riot, etc!'[1094]

I forgive the deceased, and would gladly find excuse for him altogether, if I could, but a man of prudence should not rave so venomously at his neighbour because of the shitty notes[1095] of the friars and his servants. How could I properly ask where it has been considered acceptable when there is virtually nowhere where it has not? And at this point, of course, he gives a disputation against Erasmus,[1096] as if it were my view that having possessions is inappropriate for priests or monks. It is one thing to have possessions, another to have secular jurisdiction, which even abbots and

* * * * *

1091 A question mark concludes the quotation when it is first given (*XXIII libri* 186ᵛA), and when Pio returns to the passage (*XXIII libri* 192ᵛM), he says 'As to your asking where has it been acceptable that bishops have secular jurisdiction ...'

1092 Examples of such secular jurisdiction can be found: in France, with the bishoprics of Albi, Beauvais, Chalons, Langres, Reims, and Rodez, see Pierre Imbart de la Tour *Les origines de la Réforme* I: *La France moderne* (Paris 1905) 361–6; in both Germanies, with Liège, Utrecht, Münster, Mainz, Wurzburg, and Salzburg, see Joseph Lortz *The Reformation in Germany* trans Ronald Walls (New York 1968) I 93–7; in Italy, with Trent, Brixen, Aquileia, and the Papal States, see Denys Hay *The Church in Italy in the Fifteenth Century* (Cambridge 1977) 29, and Paolo Prodi *Il sovrano pontifice: un corpo e due anime: la monarchia papale nella prima eta moderna* (Bologna 1982); in Spain, where bishops often exercised secular jurisdiction over their see city and other territories, as was the case with Palencia, Siguenza, Santiago de Compostela, Zamora, Tuy, etc, see Tarsicio de Azcona *La eleccion y reforma del episcopado espanol en tiempo de los reyes catolicos* (Madrid 1960) 36–53.

1093 The passage is from Erasmus' annotation on 1 Tim 3:2 where he describes the primary work of bishops and complains that 'now many gladly delegate it to others, however low-down they may be. Where it is considered acceptable that bishops have secular jurisdiction, they ought at least to lend lustre to the ministry of preaching the Gospel by preaching on major feasts' (LB II 934D–E / Reeve *Galatians to the Apocalypse* 671).

1094 MARGINAL NOTE: 'He raves without cause.' Erasmus truncates this outburst of Pio *XXIII libri* 192ᵛM, which continues: '... that gives an opening to the greedy and impious to invade and plunder the entire domain of the church.'

1095 *cacatis schedis*; cf Catullus 36.1: *Annales Volusi, cacata carta.*

1096 Pio *XXIII libri* 192ᵛM–193ᵛO

canons have now. If Pio were alive, I would lament his bad luck; as it is, I pray God to extend to him his mercy.

5 / Alberto and I almost agree about burial.[1097] He would prefer that corpses not be buried within the walls of a church.[1098] As it is, in certain cities hardly a day passes without a burial in church. He thinks it would be more appropriate if priests would take nothing in return for burial. However, as to his remarking that no one has the right to bury anyone in another's land, citing to us the example of Abraham, who bought a place for burial,[1099] I say in reply that cemeteries are public plots set aside for the burial of the dead, just as churches are open to all for prayer. The example of Abraham has no relevance here.

6 / The sixth passage[1100] is in the annotations on the First Epistle[1101] to Timothy, chapter 5. My words are as follows: 'This passage should be noted by those who with such great outcry and such great tyranny exact tithes, and more than tithes, even from the poorest of the laity, and do not understand that, though the Law forbids that anyone muzzle the ox, it means the ox who threshes, that is who teaches, warns, exhorts, offers mass, and performs the rest of the duties of a priest. What has this to do with certain priests who pass their entire lives not only at leisure, but with the luxuries of Sardanapalus?[1102] But it is one thing to work with mouth unmuzzled, and quite another to fill your coffers with the blood of the poor.' Here Pio deals with me as if I maintain that tithes should not be given to the priests, and elaborates this point as if he were fighting with me,[1103] whereas he is fighting with a shadow.

My admonition is addressed to certain priests who are excessively aggressive and unbending in demanding tithes and are not content even with these. They do not allow the consciences of the layfolk to be at rest in this

* * * * *

1097 Reply to Pio's quotation from the annotation on 1 Cor 9:18, where Erasmus wrote: 'Nothing is free in church nowadays, not even burial, although Augustine in letter 64 does not want the rites of the dead to be sold' (LB VI 708D / Reeve-Screech 485, citing Augustine Ep 22.6 CSEL 34 59:5); Pio provides a critique of Erasmus' note on 1 Cor 9:18 on 189^r–^{v}F.

1098 Pio XXIII libri 189^{r}vF. Pio, nonetheless, provided in his will (see Pio Will 398 below) for the burial of his son Francesco and grandfather Alberto II Pio in churches in Carpi.

1099 Pio's point in XXIII libri 189^{r}F, where he cites Genesis 23:2–19

1100 Pio's abridged quotation from the annotation on 1 Tim 5:18 LB VI 942E / Reeve Galatians to the Apocalypse 678, which Erasmus quotes here at greater length

1101 1531 gave the citation as Epistolae ad Timotheum; the number of the epistle was added in LB.

1102 Assyrian king emblematic of luxury and excess

1103 Pio XXIII libri 189^{v}G–192^v gives a detailed defence of tithes and the immunity of the clergy from taxation.

respect, and demand additional tithes on their annual income, which is frequently unreliable. This is particularly the case with the English.[1104] Secondly, it is addressed to those who are plundering the pitiable folk whom the pastor ought to be aiding from his own resources. In addition, it is addressed to those who maintain their lechery and luxury with tithes. It is addressed, finally, to those who are not mindful of their duty in any respect, and neither teach, nor admonish, nor console, nor protect, nor light the way before their flocks. I end the annotation with this conclusion: 'And my point in saying this would be, not that the laity should be less prompt in kindness towards their priests (if they are good priests, no adequate thanks can be given), but that we, in our turn, should be mindful of our duty.' Is there a word here that says that tithes are not owed to priests?

7 / This is reported,[1105] but the citation is not given. It is, in fact, on 1 Corinthians, chapter 9: 'But today what a burden it is to endure certain men, tyrants more truly[1106] than bishops, of whom the greatest is proportionately the most grievous burden upon the people, etc.' I criticize certain ones, not all.

8 / This too is reported[1107] from the annotations on the Second Epistle to the Corinthians, chapter 6: 'It should be noted how this great apostle humbles himself, even before those who had sinned, whereas now men quite unlike him in their amazing haughtiness babble nothing but mandates, [. . .], excommunications, and anathemas, etc.' It is not the authority of bishops against the shameless and intractable that I am condemning, but those bishops who babble nothing else, and neither teach well, nor live decently, nor admonish like a father. The throng of these is, alas, all too great.

9 / This is quoted from the annotations on chapter 4 of the Second Epistle to Timothy: 'O apostolic accoutrements, a cloak to protect from the rain,

* * * * *

1104 The great price rise beginning in England in the second decade of the sixteenth century affected the cost of goods and services and added an element of uncertainty when calculating the tithe, the one-tenth church tax that in some situations was no longer based on agricultural products in kind but calculated on their cash equivalent or on a salary that changed from year to year ('personal tithes'). Disputes over what was properly owed could and did become the matter for legal suits and sermons. See John A.F. Thomson 'Tithe Disputes in Late Medieval London' *English Historical Review* 78 (1963) 1–17; Peter Heath *The English Parish Clergy on the Eve of the Reformation* (London 1969) 148–52; and Susan Brigden *London and the Reformation* (Oxford 1989) 49–50.

1105 Pio's somewhat imprecise quotation in *XXIII libri* 186ᵛA from the annotation on 1 Cor 9:6 LB VI 706E / Reeve-Screech 484, which Erasmus quotes more accurately here

1106 *uerius* 1531 *versus* LB

1107 Pio's quotation in *XXIII libri* 186ᵛA from the annotation on 2 Cor 6:1 LB VI 769E / Reeve-Screech 539–40, also quoted here; Pio gives *peccaverunt* for *peccaverant*; Erasmus himself omits the word *minas* 'threats.' Pio discusses this passage in detail on 188ʳE–189ʳ.

and some books, surely holy ones! Nowadays what a quantity of horses, of cannon, and of other things which one would not want to mention? etc.'[1108] Even if Erasmus keeps silent about these last matters, nonetheless the world has long been sighing over the boundless luxury and arrogance of certain bishops.

10 / The tenth passage is on chapter 17 of Matthew,[1109] where I examine a comment by Jerome and give it a positive interpretation as being not anticlerical, but on the clergy's behalf.[1110] Whoever likes can read the passage, and he will find that this is so.

11 / He quotes[1111] from the annotations on 1 Peter, chapter 5: 'The teaching of the Prince of the Apostles should be inscribed in all <palaces> of all bishops, <even> in letters of gold. He says: "Feed your flock, do not oppress it, do not plunder it; and do so, not under compulsion as a matter of duty, but from a sincere love, like fathers, and not for the sake of foul gain" (it is as if he had a presentiment that a plague would arise for the church from this), "and finally, do not rule over them like kings, but feed them with your good example, surpass them in kindnesses."' What offended Pio here? Nothing, I think. But what follows is hard: 'Nowadays the common run of bishops hear nothing from their learned yes-men but dominion, jurisdiction, swords, keys, powers; and from this has developed in some a haughtiness worse than that of kings, a savagery beyond that of tyrants, etc.' In the first bit I am criticizing not bishops, but their learned flatterers, a common type among experts in pontifical law. In the second I censure the haughtiness of certain bishops, but not of all.

12 / The passage taken from the colloquy of the Butcher and the Fishmonger brings its own remedy with it. For it continues: 'Who denies that honour is due to bishops, particularly if they do what their name

* * * * *

1108 Pio's quotation from the annotation on 2 Tim 4:13 LB VI 962D / Reeve *Galatians to the Apocalypse* 692, also quoted here by Erasmus

1109 Pio's quotation from the annotation on Matt 17:27 LB VI 93F / Reeve 74, where Erasmus quotes a problematic comment of Jerome on Matt 17:26 (*In Matthaeum* 3:439–47 CCSL 77 155) and observes: 'Indeed, Jerome seems to consider it a piece of arrogance that ecclesiastics were reluctant to pay tax to the rulers, whereas today it is considered the pinnacle of piety to fight it out for the immunity of the clergy in every respect.'

1110 Erasmus seems to be referring to his own alternative interpretation of Jerome: 'Jerome's words can also be taken in this way: Christ, who was so very great, demeaned himself on our behalf, and we who glory in the name of Christ and because of it enjoy freedom from taxation do not behave in a way corresponding to his charity towards us' (LB VI 94C / Reeve 74–5).

1111 Pio provided an abridged quotation from the annotation on 1 Pet 5:3 LB VI 1055C / Reeve *Galatians to the Apocalypse* 755, quoted by Erasmus here in full; the words *aulis uel*, omitted in 1531, are supplied in LB.

implies?[1112] But it is impious to transfer to men the honour due to God alone.'[1113] This does not apply to bishops, but to the crowd of those who fawn upon them.

13 / The thirteenth passage[1114] does not deserve to be repeated. But it is from the annotations on the Epistle to Titus, chapter 1. What offended Pio here? My reminding bishops of their duty? He says:[1115] 'You do this all too often' – but even more often Sacred Scripture provides the occasion for this. '... too fiercely' – Jerome writes fiercer stuff against virgins, monks, clerics, and bishops.[1116] Material more savage still is recorded in the decrees of the popes directed against bishops who are tyrannical, greedy, simoniacal, and so on.[1117] 'But because you do not mention anyone by name you give even greater offence to the whole order.'[1118] Jerome prescribes a different rule: a general statement is an abuse of no one.[1119] One who says 'a certain one' is

* * * * *

1112 *dicunt* 1531 *dicuntur* LB

1113 Pio's quotation of Ἰχθυοφαγία LB I 801C–D / ASD I-3 522–3:1003–4, 1006–9, 1010–13 / CWE 40 704. Eramus quotes lines 1010–12 here.

1114 Pio's quotation from the annotations on Titus 1:7 LB VI 967B–C / Reeve *Galatians to the Apocalypse* 695, which runs: 'But [Hugh of] Saint Cher expounded this in an agreeable way, "*of foul*, that is of worldly *lucre*," as if there is another sort of revenue that is churchly and decent. I suppose he was thinking about legacy-hunting and diverting goods improperly obtained to oneself, about seizing the property of others under the pretext of defending religion, about fishing for bishoprics, about milking [Pio uses the word *emulgendis* instead of Erasmus' *emungendis* 'defrauding'] gaga old women, looting dumb merchants, etc.'

1115 Paraphrasing Pio XXIII *libri* 186ᵛ–187ʳB

1116 Compare Erasmus' remark in his *antidotus* to Jerome's Ep 22: 'He seems in this letter to direct the barbs of his satire against virgins and monks and clerics and to depict their way of life rather sharply' (HO 1 61ᵛD / CWE 61 192); see also David S. Wiesen *St Jerome as a Satirist: A Study in Early Christian Latin Thought and Letters* (Ithaca 1964) 65–112.

1117 See, eg, Gratian *Decretum*, pars 2 c 1 q 1 cap 7–8 (Friedberg I cols 359–60) for denunciations of episcopal simony; pars 1 d 45 (Friedberg I cols 160–7) for denunciation of harsh episcopal discipline; pars 1 d 46 (Friedberg I cols 167–9) condemning arrogance. According to Giovanni Bertachini's famous summary of late fifteenth-century canonical thought and practice regarding bishops, *Tractatus de episcopo* (Lyon: Benedict Bonnin 1533), a bishop who sought to be feared more for his beatings (*verberibus*) than for his words (*verbis*) was to be deposed (131ᵛ).

1118 Pio XXIII *libri* 187ʳC: 'If you should say "I disparage no one, I mention no one by name," I answer that you do them far more damage, disparage them more by speaking in general terms than if you were to criticize one or the other by name. For proceeding in this way you attack the whole order.'

1119 Erasmus may have in mind Jerome's statement in Ep 125.5 CSEL 56 122:12–16: 'I know that I will be offending ever so many who regard a general discussion about vices as an attack on themselves and, while venting their anger on me, reveal their own consciences and pass a far worse judgment on themselves than on me. For I shall mention no one by name nor shall I, with the licence of old comedy, select definite persons for criticism'; see also Ep 22.32.2 CSEL 54 193:17 and Ep 40.2 CSEL 54 193:310. Erasmus' *antidotus* to

excluding the rest. 'But,' says he, 'one who says "the general run" means "the majority."'[1120] I wish I could make an exception of as many as possible! One who criticizes vices without naming names is admonishing the evil and commending the good. 'But Jerome is too reverent to say anything against clergy or bishops.'[1121] No, it is a rhetorical device; in fact, he criticizes them frankly and sharply in many passages. 'Paul does not want an accusation against an elder to be readily heard'[1122] – I am not charging anyone, but this same Paul tells us: 'Rebuke an elder in the presence of all'[1123] – nor am I mentioning anything that is not generally known. If I were enslaved by a lust for denunciation, Pio knows what I could publish about certain people, Romans in particular. And I do not preach before the people; I write for the educated, I write for future priests and bishops.[1124]

I will not pursue beyond this point his wordy disputation in which, as usual, he takes for granted things that are plainly false: that I disparage the priestly order, that I would despoil them of their tithes, that I want them to be subject to levies of princes.

PETRINE PRIMACY [FROM BOOK 14][1125]

Here he launches into a subject that is labyrinthine and endlessly convoluted, and nowhere is he more violently seized by his mania [ἐνθυσιασμὸς]

* * * * *

the notoriously satirical Ep 22 reflects Jerome's view in Ep 125 just quoted: 'But since he criticizes no one by name and censures the wicked, not the upright, those who have lived this kind of life must be blamed, not Jerome, who has painted them in their true colours' (HO 1 61ᵛD / CWE 61 192).

1120 A paraphrase of Pio XXIII libri 187ʳc

1121 Pio XXIII libri 187ʳ quotes Jerome Ep 125.17.1 CSEL 56 136:18–19; Ep 125.8.2 CSEL 56 127:11–13; Ep 14.8.1, 2 CSEL 54 55:1–7, 8–15.

1122 Pio XXIII libri 187ʳc quotes 1 Tim 5:1, 2 Tim 4:2, and, the text to which Erasmus alludes here, 1 Tim 5:19.

1123 An interpretative reworking, not a quotation, of 1 Tim 5:19–20; see Erasmus' note (LB VI 942F / Reeve Galatians to the Apocalypse 679).

1124 Alluding to a charge by Pio XXIII libri 187ʳc, worth quoting because it so clearly reveals his attitude: 'But you do not deliver a harangue before a single individual when you accuse the priests so fiercely, but you defame and accuse them before the entire world in many thousands of volumes, and this is not a case of pious remonstrance, but of tearing, rending, and stabbing. Or do you not know that there is nothing so easy as to deceive the proletariat and unlearned masses by a harangue and move them to sedition with a glib tongue, and that the subject masses love it when their princes, bishops, and magistrates are denounced?'

1125 MARGINAL NOTE: 'From book 14'; 1531 UPPER MARGINS: 'On Petrine Primacy' on pages 220–9 [= 298–308 in this volume]. Pio's book 14 (XXIII libri 194ʳ–206ʳ) is entitled 'On the Primacy of St Peter, on the Power of the Supreme Pontiff, and the Titles and Honour

than when the subject is the majesty of the pope. If he wanted to deal with this topic scientifically, he ought to have refuted the arguments of those who have set out to dispute this specific point in published books, who go so far as to make pronouncements not only against the lifestyle of the Roman pontiffs, but even against the power that they claim to possess on the basis of gospel texts. He should have explained what are the 'keys' that Christ promises to Peter, just what is the 'kingdom of heaven,'[1126] a term frequently employed for the church militant[1127] or for the preaching of the Gospel.[1128] As it is, he gives some mangled, fragmentary quotations from my works, exaggerating them by his twisted interpretation. Next he gives free rein to abuse,[1129] even though I quite frequently call the pontiff the Prince of the Church, the Supreme Vicar of Christ, endowed with a power greater than any on earth, all the while also explaining away things done in the pope's name whether without his knowledge or against his will.

If, in the course of commenting on Scripture, I offer any admonition, it does not apply to his legitimate power, but to the abuse of power, which I think ought to be blamed on those who under the pope's name advance their own business, rather than on the pope himself. Or else I annotate a passage so that the reader will examine it more carefully, for when it comes to the power of the pope and to the interpretation of these texts, the Doctors not only vary, but also contradict themselves, and sometimes even make irreconcilable statements. And so, since I have made reply to slanders like these so often, I do not have the energy to respond

* * * * *

of the Rest of the Bishops.' It opens with a series of twelve quotations from Erasmus' works, many of which are abridged and given with imprecise citations. Pio discusses these in the course of the book, supplementing them with additional quotations from Erasmus.

1126 'Keys' and 'kingdom of Heaven' are crucial terms in Matt 16:19, a central verse in any discussion of Petrine primacy. Matt 16:19 is featured in the first passage quoted by Pio at the beginning of book 14, from the *Methodus* LB V 86C / Holborn 199:4–8: 'First, in the *Method of Theological Study* you say: "What is said to Peter '[...] I will give to you the keys of the kingdom of Heaven' [...] applies to the body of the whole Christian people [*sic* for Erasmus' 'applies to the whole body of Christian people'].'' '

1127 See in the *Methodus* LB V 105C / Holborn 236:32–4.

1128 See annotations on Matt 11:12 LB VI 60D–E / Reeve 51 and Matt 22:30 LB VI 116E / Reeve 90.

1129 Eg in the typical series of rhetorical questions in Pio *XXIII libri* 195^{r}B and 204^{v}v, where Erasmus is denounced for attacking the Holy See; or the following from 195^{v}B where, speaking of Erasmus' books, Pio mentions 'your *Methodus* which would more accurately have been entitled *A Method against True Theology, A Method for Destroying All Church Doctrines*, even as you would more correctly have assigned the titles *Annotations against the New Testament* and the *Scholia against the Doctrines of Jerome*.'

again to such wordy trifles. As to the abuse that Pio, or whoever else it is, heaps upon me, I pray that the Lord will forgive him such uncontrolled invective.

Indeed, due to his mental incontinence he overleaps the pit[1130] and in savage language exaggerates what I wrote in the *Methodus*,[1131] that Christ was a Proteus,[1132] and Paul a kind of chameleon,[1133] whereas both statements were made with an especially pious intent. And it was not enough for him that I added the expression 'a kind of' to show that it was a figure of speech, or the words 'so to speak' to alleviate the opprobrium of these words. 'But you,' he says, 'manage this so cunningly, intending to appear to be composing all this to the glory of Christ.'[1134] Tell me, you unfortunate denouncer, whoever you are, do you think that I compose this for the disrepute of Christ? If this is, or ever was, my intention, may the Lord Jesus, who knows the hearts of men, today destroy me and thrust me into hell!

If I were as hostile to the Roman pontiff as this fellow pretends, I

* * * * *

1130 *Adagia* I x 93: *Ultra septa transilire* 'To overleap the pit' LB II 394F-395B / CWE 32 277–8: Erasmus says this is 'suitable for those who plan some new and incredible enterprise far beyond the capacity of ordinary men, or who depart from the plan which has been laid down, or who transgress the appointed limits and the bounds of the power entrusted to them.'

1131 Pio had quoted from the *Methodus* in XXIII *libri* 195^{v}C: 'But in the *Methodus*, to leave out other works, alas, how many difficulties you disseminate, how many snares you set, how much you detract from the authority of the Gospel, and, finally, from the deeds and words of Jesus Christ when you discuss the variants and conflicts in the deeds and words of Christ and the contradictions of the Scriptures! How many blasphemies you utter! In particular, in one passage, so great, you say, is this variety and conflict, "so much so that although nothing is more ingenuous than our Christ, by some secret dispensation he recalls a kind of Proteus because of the variety in his life and teaching."' (LB V 94B / Holborn 214:31–3). Pio goes on: 'how much you detract from the doctrine of Paul and the rest of the writings of the apostles. For you say in the same place: "Now, if it seems good, let us make a brief comparision to see how the life and teaching of the apostles correspond to the Master's model. With what craftiness Paul everywhere plays, so to speak, a kind of chameleon, and turns himself into all things," etc' (LB V 98F / Holborn 223:34–6).

1132 The allusion to Proteus is based on *Adagia* II ii 74: *Proteo mutabilior* 'More changeable than Proteus' LB II 473B–474A / CWE 33 113–14; see the analysis of Erasmus' discussion of Christ in these terms in Georges Chantraine '*Mystère*' et '*Philosophie du Christ*' selon *Erasme: Étude de la lettre à P. Volz et de la 'Ratio verae theologiae' (1518)* Bibliothèque de la faculté de philosophie et lettres de Namur 49 (Gambloux 1971) 297–301.

1133 See *Adagia* III iv 1: *Chamaeleonte mutabilior* 'More changeable than a chameleon' LB II 805E–806B. See also Ep 916:408–15.

1134 Pio XXIII *libri* 195^{v}C

would be lacking neither cleverness, nor words, nor material. But far be it from me to denounce one whom I revere as the Vicar of Christ, far be it from me to throw oil into the furnace[1135] in these inflamed and raw times, even though Pio's spite invites me to do so.

In his discussion here there are many ridiculous items which I could scarcely read without laughing. There are some sound observations as well, for example, his advice that when we undertake to discuss a writer's view, we should first of all take note of what it is the author is doing there, what his goal is, what he means, whether he is arguing or defining, whether he is combating an opponent or simply teaching, and so on.[1136] But how can people who are putting together a disputation on the basis of pathetic note-cards do this?

He adds that Jerome himself is self-contradictory in a lot of passages, not only when disputing, but also when teaching.[1137] Pio is right here. But when I said this, even more temperately than he, it seemed to him a serious blasphemy, and he spat in my face: 'You say this about Jerome, you who are not worthy to hand him a chamber pot?'[1138] I had said only that Jerome is occasionally somewhat inconsistent. But once Pio was inconvenienced a little bit by Jerome's authority, he preferred the cliché 'a reply *ad hominem*, not *ad rem*.'[1139]

He draws this conclusion: if Peter is a type of the church, then the keys

* * * * *

1135 See *Adagia* I ii 9: *Oleum camino addere* 'To add oil to a furnace' LB II 71E–F / ASD II-1 221 / CWE 31 151.

1136 A summary of Pio's observation in XXIII *libri* 199ʳG that one should attend to the rhetorical context of a statement, this preliminary to his discussion (199ʳ–200ᵛ) of a passage from Jerome's *Adversus Iovinianum* 1.26 PL 23 247A

1137 Erasmus has mistaken Pio here. What Pio argues in XXIII *libri* 199ʳG is that in the passage under consideration Jerome is using Jovinian's view against his adversary and not stating his own opinion: 'and we must acknowledge that this is what Jerome is doing in this passage, unless we are going to presume to say that he is contradicting himself in this passage, and nothing worse could be imagined. I will not mention that in a great many other passages he writes the contrary, not when disputing but when teaching. But it is obvious that in this passage he is disputing with Jovinian.'

1138 See n1064 above. Pio concludes this discussion here (XXIII *libri* 200ᵛG): 'The real meaning of Jerome's words must be deduced from these considerations. If you do not accept this, our second answer should be adequate for you, that is, that Jerome's statement did not reflect his own views, but was an *ad hominem* reply, unless you prefer, by way of defending your view, that Jerome says contradictory things.'

1139 In XXIII *libri* 199ᵛ Pio says of Jerome's argument against Jovinian: 'and this type of reply is called by the moderns [*a neotericis* 'scholastics'] a reply to the person, not to the case [*responsio ad hominem non ad rem*].'

were entrusted to any Christian.[1140] I said that they were entrusted to the church as a whole, that is, to the dove.[1141] The theologians do not deny this, nor do I, on account of it, take away from the pope his external jurisdiction.

He reproaches me for my citation of Origen's comment, and does not believe me.[1142] But as I furnish the number of the homily,[1143] why was he reluctant to examine the passage?[1144] For in it Origen plainly interprets the praise of Peter's faith and the keys given to him as belonging to devout people as individuals. But I have never supported this view. If the whole world must yield to the one pope,[1145] as Pio maintains, what need is there

* * * * *

1140 Apropos of Erasmus' remark in the *Methodus* about Matt 16:19 (see n1142 below), Pio comments in *XXIII libri* 195ʳ: 'From this it would follow that the keys of the kingdom of Heaven were no more entrusted to Peter than to any Christian of the lowest station.'

1141 The dove of *Song of Solomon* 6:7–8 was traditionally identified with the church; see, eg, Jerome Ep 65.15.3 CSEL 54 637:5–10; and Juan de Torquemada *A Disputation on the Authority of Pope and Council* trans Thomas M. Izbicki (Oxford 1988) 20.

1142 Pio in *XXIII libri* 201ʳ adds to his attack on the passage from the *Methodus* quoted above an attack on Erasmus' expression of a similar opinion in his annotation on Matt 16:18. He charges Erasmus with 'maintaining that the words whereby it [the power of the keys] was entrusted or promised apply to the entire body of the Christian people, according to the neat interpretation of Origen which you, however, do not quote. Even though, according to Jerome, Origen went wrong in many things, I can scarcely believe that his view was what your words say. But if this was not the case, he should have been interpreted in a pious way or refuted, instead of being approved of in this teaching.'

1143 The relevant section of the annotation on Matt 16:18 reads: 'Accordingly, I am amazed that there are people who twist this passage to make it refer to the Roman pontiff. Surely it does apply most particularly to him as the leading figure in the Christian faith, but it applies, not to him only, but to all Christians, as Origen neatly points out in the first homily of those we have' (LB VI 88E / Reeve 71). Erasmus' reference is to Origen's commentary on Matt 16:13–19 in book 12 9–14, the first section that survives in the old Latin translation of Origen's work (hence the reference 'the first homily [*sic*] of those we have'). There we find many statements of this kind: 'The rock is everyone who is an imitator of Christ, the one from whom they did drink who "drank from the spiritual rock that followed them [1 Cor 10:4]." And upon each rock of this sort the church of God is built. For it is in each perfected individual, who possess in themselves the assembly of all the words, works, and understandings which accomplish this sort of beatitude, that the church consists' (Origenes *Die griechischen christlichen Schriftsteller der ersten drei Jahrhunderte* (GSC) x (Leipzig 1935) 85:32–86:14).

1144 Pio seems to have had a look at Origen's commentary, for he claims in *XXIII libri* 201ᵛK that Origen acknowledges Petrine primacy, alluding to Origen's exegesis of Matt 18:18 (Origenes GCS x 269–71).

1145 This is a reply to Pio's discussion in *XXIII libri* 202ʳv of Erasmus' annotation on 2 Cor 10:8 LB V 785E / Reeve-Screech 554. Pio had earlier (194ᵛ) quoted the words: 'This passage should be noted by those who vastly exaggerate the authority of the supreme pontiff, asserting that the entire body of the Christian church should yield to one man to the degree that if he were to bring down all souls to hell, there would be no right to oppose him'; and later (202ʳL) he quoted more of the passage: 'But I wish that those

for local and ecumenical councils? And[1146] when the pope dies, what becomes of the church? Pio answers 'nothing but a headless corpse,'[1147] even though it remains joined to its living and deathless head, Christ.

He says that the hypothetical case which someone has posited, namely that the pope would lead all men to hell, would never come about.[1148] And yet one who envisions this is admitting that it could be that the pope is a hypocrite, a heretic, impious, and a tyrant. But if this were to happen, he says, 'then Christ would either remove him from the scene, or reveal remedies to the flock.'[1149] How would he remove him? With a thunderbolt? But miracles have long ceased. Through the Turk, or the princes? That is impious. Through schismatics, something we see already partially accomplished? That is even more impious. But the prevention of this rests to a large extent with the popes themselves. Thus, to admonish them is to aid them. And I do not support those who would deprive bishops of their lawful power because of immorality.[1150] Yet it was a Roman pontiff who wrote: 'If one's way of life is despised, it only remains for his teaching also to be disregarded.'[1151]

I find it droll that, for the purpose of maintaining the applicability of the terms 'dominion' and 'power' to popes,[1152] he brings in 'You call me

* * * * *

who now assign such great power to the Roman pontiff would assign the rest of the high priestly qualities as well, wisdom, purity, charity, etc'; see Erasmus' arguments against the whole world submitting to the decisions of the pope in LB VI 696E–697C / Reeve-Screech 472–3.

1146 *Et* 1531 *om* LB

1147 Pio *XXIII libri* 202ʳL

1148 Pio *XXIII libri* 202ʳM; Erasmus reported this hypothetical case in the annotation on 2 Cor 10:8 quoted in n1145 above; a likely source for this case is the *Decretum* of Gratian pars 1 d 40 cap 6 (Friedberg I 146). The great canonist and theologian Juan de Torquemada OP (1388–1468) in his *Commentarii in Decretum Gratiani* (published in 1519) held that a pope who harmed the church by acting contrary to the salvific purpose of his office, by going against divine and natural law, the sacraments, and the basic principles of Christian morality, made himself thereby subject to judgment; see Thomas M. Izbicki *Protector of the Faith: Cardinal Johannes de Turrecremata and the Defense of the Institutional Church* (Washington 1981) 87–94 especially 89.

1149 Pio *XXIII libri* 202ʳM

1150 Pio *XXIII libri* 202ʳo says those who claim great papal power take for granted the presence of those excellent qualities for which Erasmus calls, and he continues: 'But if it should be otherwise, though it is grievous, they know that nothing is withdrawn from [papal] power for this reason'; cf *Methodus* LB V 113E / Holborn 253:16–19, where Erasmus insists on obedience even to less than upright bishops so long as they are orthodox.

1151 Gregory I *Homiliae in Evangelia* 1.1 PL 76 1119A

1152 In *XXIII libri* 203ʳQ–204ʳT Pio deals with the passage he had quoted from the *Enchiridion* LB V 49A / Holborn 107:8–13 / CWE 66 101: 'For that reason I am all the more astonished that these ambitious titles of power and dominion have been transferred to supreme

Master and Lord'[1153] and 'All power in heaven and on earth has been given to me.'[1154] Why is Jerome's dictum, 'They should know that they are fathers, not masters,' quoted in the *Decretum*?[1155]

Finally he provides instruction on how the titles 'lord,' 'father,' and 'teacher' are to be taken,[1156] but I provided the same instruction as he, only much earlier.[1157] When I use the term 'dominion,' I understand 'tyrannical dominion'; when I use the term 'power,' I understand 'violent power'; when I use the term 'father,' I understand that all confidence has been placed in him; when I use the term 'teacher,' I understand haughtiness and arrogance. If Pio had read this, he would not have uselessly spilled out all these words.

It is one thing to award the title 'lord' by way of doing honour;[1158] it is quite another to seek after tyrannical domination.[1159] What, indeed, other than tyranny is a power exercised without restraint and in excess of those qualities with which the pope ought to be equipped? I am not making a statement about any particular pope. But what the whole world, and Rome too, has declared about certain ones is too well known to require reporting. I agree that one should not make rash judgments about bishops,[1160] but how I wish that they did nothing inexcusable.

* * * * *

pontiffs and bishops and that theologians are not abashed in their ignorance and ostentation to call themselves commonly "masters," although Christ forbade his disciples to allow themselves to be called either "lord" or "master," for there is one who is both Lord and Master [Matt 32:8, John 13:13].'

1153 John 13:13, quoted by Pio *XXIII libri* 203[V]

1154 Matt 28:18, quoted by Pio *XXIII libri* 203[r]Q

1155 Erasmus has confused Jerome's phrase in Ep 82.11.5: 'But let them [bishops] be content with their honour, and know that they are fathers, not masters' (CSEL 55 119:6–8), with a similar line from Ep 52.7.3: 'But let bishops also know that they are priests [*sacerdotes*], not masters' (CSEL 54 427:9–10); it is the second passage that is quoted by Gratian *Decretum* pars 1 d 95 cap 7 (Friedberg I 334).

1156 Pio discusses the use of *potestas* (*XXIII libri* 203[r]Q), *dominus* (203[V]R), *magister* and *pater* (203[V]S–204[r]T).

1157 See eg *De conscribendis epistolis* LB I 372B–C / ASD I-2 290 / CWE 25 58:19 on *dominus*; *De conscribendis epistolis* LB I 374C–E / ASD I-2 294–5 / CWE 25 61-2:2 on *dominus* and *magister*; and *De pueris instituendis* LB I 505F / ASD I-2 58 / CWE 26 328 on *pater* as opposed to *dominus*.

1158 This is Pio's explanation of the use of the title (*XXIII libri* 203[V]R).

1159 Erasmus is replying to Pio's discussion in *XXIII libri* 204[r]v of the passages he quotes from the annotation on 1 Cor 12:3 LB VI 719E / Reeve-Screech 495; Erasmus wrote: 'from this chapter and the next one can imagine the qualities of the ancient [Christian] church, before it had been loaded down with rituals, wealth, empire, troops, wars, and other things like these. Now so many wonderful functions have been reduced to virtually a single power, that is, a tyranny cloaked in the name of Christ.'

1160 Reacting, perhaps, to Pio's point in *XXIII libri* 202[r]O: 'For if, according to the blessed father Jerome whom we quoted above, it is improper to suspect wrong even of the clergy,

Is someone forthwith overturning the authority of Peter's throne when he laments the fact that there have sometimes been popes who wielded a tyranny under the pretence of religion? Erasmus has not said anywhere, O Lover of Quarrelling [φιλαίτιε], that ecclesiastical power is tyranny,[1161] but only that Roman pontiffs of this sort sometimes appear. How I wish that this happened more rarely!

Luther[1162] has made far different declarations about the Roman pontiff, and declarations in countless respects more savage. But Pio does not read these, and considers it at once more convenient and safer to quarrel with Erasmus. 'May God on high spare you, Erasmus,' says he. Rather, may God on high spare those popes who are from time to time such a cause of scandal to the church of Christ!

I am not at all angry with the City,[1163] nor did I hunt in vain for anything there.[1164] In fact, it embraced me more generously than I deserve.[1165] But he presses me with the question did I ever really see any pirate made

* * * * *

how much greater a peril is there in suspecting anything unworthy of the bishops and especially of the supreme pontiff himself?'

1161 Erasmus is echoing Pio's outcry in *XXIII libri* 204ᵛv: 'What more fierce, more bitter statement could have been made to the end not only of reducing but of destroying the authority of Peter's throne? For not even one who is extremely fierce could have said anything more bitter than that ecclesiastical power is tyranny!'

1162 Reacting to Pio *XXIII libri* 204ᵛ: 'See what great hostility you are provoking, how great a blaze you are starting up, and yet you dare to deny that you agree at all with Luther! What, indeed, has Luther ever uttered against it [the authority of Peter's throne] that is more destructive? What more rabid bite has he ever inflicted? What missile more tainted with deadly poison has he ever cast? May God on high spare you, Erasmus, etc.' Luther came to see the Roman pontiffs as anti-Christs; see Scott H. Hendrix *Luther and the Papacy: Stages in a Reformation Conflict* (Philadelphia 1981) especially 118–59.

1163 Here Erasmus takes up Pio's discussion in *XXIII libri* 205ʳ⁻ᵛy of a passage quoted from Erasmus' annotation on 1 Cor 7:39: 'As to another charge you touch upon in criticism of the supreme pontiffs: "Paul forbids that one who is a neophyte or a smiter or given to wine be made a bishop [1 Tim 3:3–6], but today the Roman pontiff accepts one who was baptized yesterday or is a notorious pirate, if he sees fit, deterred not at all by Paul's ordinance" [LB VI 696C / Reeve-Screech 472]. You have been to Rome, Erasmus, and because you were not awarded episcopal rank as you had, perhaps, hoped, you departed so angry that you have a constant, raw, and implacable hostility towards that city, so much so that you are always hurling some darts at it, not seizing opportunities to do so, but seeking them out.' Erasmus expresses similar criticisms of recent appointments to the episcopate in his annotation on 1 Tim 3:2 LB VI 934B / Reeve *Galatians to the Apocalypse* 670.

1164 Of course, Erasmus himself claimed that Pio and Aleandro schemed to prevent his own appointment as bishop; see n69 above.

1165 For Erasmus' claim that he was treated with great kindness while in Rome, see eg Ep 296:107–15.

bishop at Rome.[1166] I shall not mention what I have seen. Only let him dare to deny that, if not pirates, then certainly murderers, poisoners, simoniacs, and people subject to other vices that cannot be mentioned here were sometimes advanced to the highest ranks. Let him explain to us the basis for Jerome's anger that one from among the troops of the *exoleti* [male prostitutes] is admitted to the rank of bishop.[1167]

It is a trivial point that in this disputation Pio takes *neophyte* to mean youth,[1168] whereas it means one recently baptized, even an old man.

But what impiety is there in the passage that he quotes and quotes again from 'A Fish Diet'?[1169] Would Clement be acting impiously if while yielding something of his prerogative he would grant certain exemptions to the Bohemians, the Russians, the Ruthenians, and others for the sake of the concord of the church? That is, if he would permit the Bohemians to receive the Eucharist under both species, would forgive something from the accounting of the past-due revenues of the church, would not exact payment of annates to the last penny. I could have used these as examples. And yet Pio reckons that this was written against the authority of the Roman pontiff, and thinks that nothing more savage could be said against the pope. The misguided judgment of the man! I will say nothing about the annates. Everyone knows how they got started and under what circumstances they were introduced.[1170] Money is money – I just wish it would be put to pious

* * * * *

1166 Pio *XXIII libri* 205ʳʏ: 'When you were there, Erasmus, did you see or even hear that one baptized yesterday or a notorious pirate was advanced to episcopal rank? You surely will not claim this. Indeed, I, who have lived in that city for a great many years, cannot say that I ever saw or heard any such thing.' See Introduction n146.

1167 An inaccurate allusion to Jerome Ep 69.3.5 CSEL 54 684:11. Erasmus glosses this term: '*Exoleti* are nearly grown adolescents. But what it means is males who are shamefully obliging' (HO 3 148ʳʙ).

1168 Erasmus has been careless here: Pio *XXIII libri* 205ᵛʏ states of Timothy that he was both a neophyte and a youth (*Neophytum tamen et adolescentem*), and the same of Titus (*Titus quoque iunior erat & Neophytus*).

1169 Pio quoted Ἰχθυοφαγία LB I 793ʙ–C / ASD I-3 505:359–70 / CWE 40 687 in *XXIII libri* 195ʳ, and discusses the passage at 205ᵛz–206ʳ: '[Fishmonger] Whoever dares not recognize the Roman pontiff is outside the church. [Butcher] No objection. [Fishmonger] But he who disregards the pope's ordinances doesn't recognize him. [Butcher] And for that very reason I hope that in the future this pope (Clement by name, most clement in spirit and holiness), for the purpose of attracting all races into the fellowship of the church, may mitigate all conditions that have heretofore appeared to estrange some peoples from union with the Roman see; and that he may prefer the gain of the gospel to carrying out his own prerogative in everything. Every day I hear old complaints about annates, pardons, dispensations, and other taxes, about burdened churches; but I believe he'll so curtail all these that none but the shameless will venture to complain hereafter.'

1170 On the tolerant attitude of Rome towards the Eucharistic practices of the Bohemians and Russians, see n618 above. Pio discusses annates and their origin in *XXIII libri* 205ᵛ–

uses. Pio does well to vent his spleen about these matters against Erasmus; if he were to rage in a like way against certain other parties, he would have moved an notable Camarina.[1171]

But I cannot guess what Pio is trying to accomplish with these arguments. If he is trying to assert the authority of the pope, this has already been done long ago by others in a far more learned way than he does here. But if he is trying to refute the German sects, he ought to have presented their writings and arguments accurately and refuted them energetically. Or if he thought a brawl with Erasmus would be a source of renown, then he ought to have read the charges made by others, and my replies, and thus fought it out hand to hand. As it is, he is neither benefiting the church nor fighting fairly with me, but repeating the silliness of other denouncers like a final echo. Or if he is trying to convince people that Erasmus has unsound views about the power of the Roman pontiff, even though this is absolutely false, he would have done better to excerpt the passages in which I make lofty statements about the pope, and defend him against his enemies, than to excerpt phrases which can be twisted by a malignant interpretation to mean the opposite. For there are some, perhaps, who take Erasmus very seriously.

But who could not discern the real intent of those who battle with such jealous zeal on behalf of the outward majesty of the pope while trying to limit his inward majesty by certain convoluted arguments? In the past, he says, Christ used to do this, but it is not necessary to imitate everything that Christ taught, did, or commended, for instance his poverty, his meekness, his teaching, his refusal of the kingdom of this world. He says that the counsels apply only to professed monks,[1172] that the supreme power was entrusted by Christ, its source, specifically to the Roman pontiff[1173] (although it was to the apostles that he said 'receive the Holy Spirit,'[1174] and Paul describes himself as an apostle not by human agency nor through human

* * * * *

206ʳ; see William E. Lunt *Papal Revenues in the Middle Ages* 2 vols (New York 1934) I 93–9 (history of annates), II 315–72 (documents related to annates).

1171 See *Responsio* n82 above.

1172 Here Erasmus is repeating the distorted version of Pio's views that he had presented earlier (see 173–4 and nn 397–8 above). Pio *XXIII libri* 99ᵛN: 'But I hold that monasticism pertains to true and perfect religion since it embraces not only the common precepts of the Lord, but also the counsels and admonitions pertaining to the perfection of religion.'

1173 Pio *XXIII libri* 196ʳE–201ᵛK provides a long argument that interprets Matt 16:18 as meaning that the supreme power (a pastoral responsibility) was entrusted by Christ to the popes.

1174 John 20:22; Pio had, in fact, addressed the implications of this text in *XXIII libri* 197ʳ–198ʳ.

agency),[1175] and that, therefore, he cannot be deprived save by Christ, nor even reproached save by the greatest princes.[1176] All bishops and kings are his sheep.[1177] The boundless exaggeration of this sort of thing damages the authority of the Roman pontiff instead of strengthening it.

HUMAN REGULATIONS [FROM BOOK 15][1178]

I have replied to complaints about this all too often now, but it is very difficult to reply to snippets collected bit by bit from all over, particularly when they are garbled and the source is not cited. Yet if the attack on me is based on little fragments like these, there are even more passages which show clearly that I assign great importance to regulations, especially those introduced by the authority of the holy Fathers and approved by the general consensus of the Christian flock. If I make occasional criticisms in my writings, they are made only against certain specific regulations of abbots, bishops, or popes, either because they are wicked, or, if they are good, because they are being perverted[1179] by the quest for wealth and tyranny. Although I proclaim this and drive it home in countless places, nonetheless Pio, who is deaf to all, slings his fragmentary excerpts at me like sling-bolts, and does so neither fairly nor without the abuse to which my ears have, fortunately, long ago grown insensitive.

1 / His first quotation is from the *Methodus*;[1180] it is slanderous and shameless, running as follows: 'Regulations in the church can be viewed as being of the sort, that certainly these [*sic*] has been established integrity of the Christian religion.' What I wrote was: 'But certain things can be viewed

* * * * *

1175 Gal 1:1; Pio answered Erasmus' objection that Paul's authority came directly from God (*XXIII libri* 202[r]L,N).

1176 Pio *XXIII libri* 202[r]–[v]o provides scriptural and patristic arguments against open criticism of the faults of religious leaders; he does not mention Christ or the greatest princes but says only that 'they should be castigated by those whose prerogative it is' (206[r]z).

1177 Pio claims that bishops are under Peter's pastoral care but does not mention kings (*XXIII libri* 192[v]–193[r]E).

1178 1531 UPPER MARGINS: 'On Human Regulations' on pages 230–3 [= 308–12 in this volume]. This section responds to Pio's book 15 (*XXIII libri* 206[v]A–210[v]), entitled 'On Church Regulations and Laws, and Customary Law of Human Origin.' Pio quotes or summarizes twenty-five passages from Erasmus' works, and then takes up selected passages for discussion in the order 1, 5, 2, 17, 10, 4, 11, 15, 6, 18, 16, 7. Erasmus here deals with Pio's quotations 1–2, 4–9, 18, in the order in which Pio initially quotes them.

1179 *detorqueantur* LB *detorquentur* 1531

1180 MARGINAL NOTE: 'A garbled and irrelevant quote.' Pio's quotation in *XXIII libri* 206[r]A is quite garbled: 'Constituta in ecclesia huiusmodi videri possunt ut certa haec pure constituta sit integritas Christianae religionis,' though Erasmus makes it even worse by

as being of this sort, that the integrity of the Christian religion will endure quite nicely without them.' Is this not a handsomely done quotation? And it is not even ecclesiastical regulations that I am talking about there, but certain dogmas of the scholastics. For after a discourse on laws I begin as follows: 'Nor should one think any differently about the dogmas of the scholastics.'[1181] And he cannot blame the obscurity of my remark, for immediately following I give an example drawn from sacramental confession, which the church enjoins, at least once a year; scholastic dogmas compel us to declare that it was instituted by Christ as it is practised now.[1182] Again, I make no assertion here, but I say 'could be viewed.' Let Pio now go ahead and brandish his sword through the air. I know that he was fooled by the notecard of his servants, but this excuse is unbecoming such a man, especially when he attacks so savagely.

2 / In the second passage[1183] what am I maintaining that is different from what Pio maintains here, that is, that certain things were established, in view of the times, which had not been prescribed in the infancy of the church, for example, dietary regulations?

3 / In 'The Whole Duty of Youth' it is a boy who speaks, not Erasmus. But what awful thing does he say? 'Furthermore, I do not necessarily [*protinus*] regard as a dreadful enormity [*piaculum*] a transgression of any and every human ordinance – unless this is accompanied by a malicious contempt. No [*immo*], I hardly consider anything a serious sin unless malice – a corrupt will – is joined to it, etc.'[1184] Since I am dealing with a Latinist, what do the boy's words mean but that no human regulation whatsoever necessarily makes one guilty of a frightful crime? For this is what he means by *piaculum*, and that is the force, too, of the adverb *protinus*, which excludes precondition. Next comes *immo*, an expression used in making a *correctio*,[1185] so that one may understand that a *piaculum* is something worse

* * * * *

quoting Pio's *certa* as *certe*. Erasmus next gives a more accurate version of the passage from the *Methodus* LB v 90C / Holborn 205:24–5: 'At quaedam videri possunt huiusmodi [in *Methodus* possint eiusmodi] ut citra haec pulchre constatura sit integritas Christianae religionis.'

1181 LB v 90B / Holborn 205:15–16
1182 LB v 90C–D / Holborn 205:33–4
1183 Pio XXIII *libri* 206ᴿA quotes from the *Methodus* LB v 88A–B / Holborn 201:17–20: 'new laws have been introduced, some [*aliquae* in Pio, *aliquot* in *Methodus*] of which would appear to be in conflict with the ordinances of Christ, unless we force the Scripture into harmony with them by taking the times into account.'
1184 *Pietas puerilis* (also called *Confabulatio pia*) LB I 652B / ASD I-3 178:1742–5
1185 Erasmus describes and demonstrates this rhetorical manoeuvre in *Ecclesiastes* LB v 990C–D.

than a capital crime. But his meaning is that homicide or incest can take place without a crime being committed if malicious intent is lacking.

4 / The fourth is, again, an inaccurate quotation from the *Methodus*: 'When charity is cold, no laws, however many, are adequate; when it is ardent, there is no need of law.'[1186] I am talking about the coercion of law, and what need is there of that where charity does willingly even more than human law would dare require? Here he shouts: 'Where laws are not kept, there is no charity!'[1187] Who said that laws should not be kept by the perfect? I say that they have no need of them.

5 / In 'The Godly Feast'[1188] the one who interprets the allegory makes no assertion. He says: '"King" can be understood as the perfect man who, with his bodily passions under control, is governed solely by the power of the Holy Spirit. Moreover, to compel such a man to conform to human laws is perhaps inappropriate. Instead he should be left to his Master, by whose spirit he is led.' I know what judgment is made concerning the spiritual sort whose disposition Gerson, at least, does not want to be despised.[1189] I will leave out what Paul writes: 'Where the Spirit of the Lord is, there is liberty';[1190] and 'The spiritual man judges all things, and is judged by none';[1191] and other statements like these. And was it not the case in the past that those in the monastery who had given evidence of long-established devotion were permitted to use their own judgment in the matter of withdrawal from the community?[1192]

* * * * *

1186 Pio XXIII *libri* 206ᵣA quoted from the *Methodus* LB V 109E / Holborn 245:7–9; Erasmus quotes the same passage, replacing Pio's inaccurate *quaelibet* with *quamlibet multae*.
1187 Erasmus distorts Pio's words 'whereas charity is most ardent when the divine precepts are observed' (XXIII *libri* 207ᵥC).
1188 Pio quoted *Convivium religiosum* LB I 678B / ASD I-3 244:392–5 / CWE 39 185, repeated here by Erasmus.
1189 On the obligation of one led by the Holy Spirit to obey human laws, see Gerson's *De vita spirituali animae* 4 in *Œuvres complètes* ed Palémon Glorieux III: *L'œuvres magistrale* (Paris 1962) 157–80; for Gerson's explanation of what constitutes perfection and the law of love, see *De perfectione cordis* VIII 116–33 especially 130–3.
1190 2 Cor 3:17
1191 1 Cor 2:15
1192 Benedict allowed the spiritually more robust to advance to the more perfect life of a hermit. In the Camaldoli order only someone of mature years who has first secured the permission of his superior may become a secluded hermit. While cut off from contact with others, he remains under the rule and a superior. See Jean Leclercq *Alone with God* trans Elizabeth McCabe (New York 1961) 28, 61–2; and J. Gribomont, Ph. Rouillard, and I. Omaechevaria 'Eremitismo' in *Dizionario degli istituti di perfezione* ed Guerrino Pelliccia and Giancarlo Rocca (Rome 1976) III cols 1224–44 especially 1226–7.

6 / Again, he stitches on a patch from the *Methodus*.[1193] It was the incompetence of his employees that garbled it, but the disgrace is Pio's. 'I do not think it right that virtually all of human life is loaded down with human regulations, and that overmuch importance is assigned to them and very little at all to piety, that simple folk place their reliance on these and make no effort towards an authentic religious zeal, etc.' What impious statement did I make here? In the first part I criticize the excessive number of human regulations; in the second I censure the fact that more importance is assigned to rituals than to a zeal for authentic piety.

7 / There is an excerpt from the *Supputations*.[1194] They did not think the passage was worth citing, and I am sure that they have quoted it in garbled form. Accordingly I have nothing to say in reply.

8 / The following is quoted from the Letter to Abbot Volz:[1195] 'Do not sully that heavenly philosophy of Christ by confusing it with the decrees of men.' Why wouldn't Pio yell here 'O blasphemous words!'? Yet anyone who examines the passage will see that it is a devout statement. There I am distinguishing the categories of laws in rank order. I assign the first rank to the Gospels, which have nothing that does not savour of the heavenly Spirit. I do not want this supreme authority to be mixed up with human laws – that is, that they should have the same force – although I do assign to human laws their appropriate authority. It is a bother to report the rest of the items that have been piled up bit by bit, particularly since there is nothing there to which I will not have replied frequently.

9 / As to chapter 11 of Matthew, from which he broke off certain chunks and distorted them for purposes of slander,[1196] although I express nothing impious there, I would prefer that I had spoken more circumspectly.

* * * * *

1193 MARGINAL NOTE: 'A garbled quotation.' Pio gave an abridged and inaccurate quotation from the *Methodus* LB V 113A–B / Holborn 252:8–13; Erasmus, however, reproduces Pio's version here, correcting only the grosser inaccuracies.

1194 Pio quotes 'And in the *Supputatio*: "There are many considerations which barely allow one to argue that a regulation of the Roman See is in accord with equity according to nature."' Erasmus could not find this passage, nor can we. Erasmus does make other criticisms of papal regulations in the *Supputatio*, eg LB IX 552E–F and 641D.

1195 Pio *XXIII libri* 206ᵛ quotes from the letter to Paul Volz that prefaces the second edition of the *Enchiridion*, Ep 858:242–4 / Holborn 9:28–9.

1196 Pio *XXIII libri* 206ᵛ–207ʳA quoted a series of seven short passages from Erasmus' long annotation on Matt 11:30, 'For my yoke is easy and my burden light' (LB VI 63B–65E / Reeve 53–6), and returned to discuss some of them at 210ʳK–L.

But at the time I gauged the spirits of others according to my own, and I could not have foreseen the emergence of the sort of men whom we see in these times. How I wish that the popes and bishops had preserved the many regulations of the ancient Fathers which are preserved in the *Decretum* and in the *Decretals*, and which nowadays are functionally abrogated and not even remembered by anyone.

VOWS [FROM BOOK 16][1197]

1 / This is quoted from the book *On Christian Marriage:*[1198] 'It would be a service to all Christians to advise them to abstain from all unnecessary and indeed useless vows. If you are keen to visit Jerusalem or Compostella, you may gratify your whim, but there is no need to include a vow in your plans, etc.' Given that the world is full of superstition connected with vows like these, that the consciences of the simple are troubled because of them, and that the revenues of certain people are increasing accordingly, what harm is there if Christians are given this advice?

Pio asks me what vow is unnecessary or unprofitable.[1199] But I give an example in that passage: there is no necessity for anyone who has not vowed to do so to go to Jerusalem, since he would do better to go and see the footprints of Christ in the Gospel. It is also unprofitable if he leaves wife and children who require his presence at home. And this is, perhaps, not only unprofitable, but even destructive.

2 / No one denies the truth of what I have said about the vows of the Jews,[1200] and nonetheless many find here grounds for argument.

* * * * *

1197 1531 UPPER MARGINS: 'On Vows' on pages 234–5 [= 312–13 in this volume]. This section and the one following are a reply to Pio's book 16 in *XXIII libri* 210^{v}o–216^{v}z, entitled 'On Keeping the Vow of Chastity and Other Vows Made to God.' The book opens (210^v– 211^r) with twelve passages quoted from Erasmus' works, a selection to which Pio adds others in the course of his argument.

1198 The first quotation is from *Institutio christiani matrimonii* LB V 646D–E / CWE 69 285, quoted again here by Erasmus.

1199 This is a reply to Pio's subsequent discussion in *XXIII libri* 214^{v}T–216^{r}Y of the passage at issue here. Pio opens: 'First we must consider what you understand by an unnecessary or unprofitable vow, for these adjectives do not fit with the word "vow." What is unprofitable or unnecessary is not contained in the idea of a vow, since a vow is a promise made to God on the basis of a mature decision of the will about doing something worthy to which we are <not> constrained by necessity. For it is an empty promise that is made about a thing that is absolutely necessary.'

1200 Pio's second quotation is from *Institutio christiani matrimonii* LB V 646F / CWE 69 286: 'Jephthah's vow should not be taken as an example either. Other Jewish vows either

3/The next topic is ordinary vows.[1201] My opinion is that it is the safest course not to make them, or that those rashly made be considered void. Here, again, he will ask me what are rashly made vows. My answer is that all unnecessary vows, all unprofitable vows, are rashly made, and even more so those that are unprofitable and destructive. Vows made frivolously, while in one's cups, those made from fear and superstition, are also rashly made vows. How many young men, how many women, flock to Rome because of a vow, and how many of the young men, how many of the women, do not come back corrupted? These too are called vows, but Pio is not especially concerned about this sort of vow.

THE VOW OF CHASTITY [FROM BOOK 16][1202]

4/He adduces this fragment too:[1203] 'The cause of matrimony would be greatly advanced if the church were to decree that a vow of chastity should be invalid unless ratified by a bishop, who must be very exacting on this score, etc.' What am I arguing for here except that married people should not make a vow of chastity without the approval of the bishop, who should not give his approval readily to the making of such vows? And yet I add 'if the church were to decree.' The church surely has this power. And are not the Parisian theologians of the same opinion,[1204] whether on account of the suspect age of those making vows or on account of human frailty?

* * * * *

related to their phantom Law, or were easy to perform, or could be revoked.' Pio discusses this passage in *XXIII libri* 216ʳγ.

1201 Pio's third quotation is from *Institutio christiani matrimonii* (LB V 647A / CWE 69 286: 'I am talking about common vows, which, in my opinion, it is safest not to make at all or, if they are made in a rash moment, to consider null and void.'

1202 1531 UPPER MARGINS: 'On the Vow of Chastity' on pages 236–42 [= 313–19 in this volume]

1203 Pio's fourth quotation, again from *Institutio christiani matrimonii* LB V 647A–B / CWE 69 286, repeated by Erasmus here

1204 It is not clear if Erasmus had in mind a formal decision of the Paris theological faculty on this question or was merely inferring support among the faculty for dispensation from vows improperly taken because the faculty had not censured his questioning of such vows in the colloquy Ἰχθυοφαγία LB I 796, 798A–C / ASD I-3 512:604–18, 515–16:739–81 / CWE 40 693–4, 697–8 when censuring the colloquy (LB IX 949E–953B). The faculty found that its examination of the colloquies involved 'arduous and difficult theological questions'; and rather than condemning them outright, it urged that adolescents be prohibited from reading them (see Farge *Registre* 136–7). On Erasmus' repeated claims of support for his views among members of the Paris theological faculty, see Rummel *Catholic Critics* II 47, 51, 54.

5 / It goes on:[1205] 'Also, there need not be so sharp a distinction between a simple vow and a solemn vow. If a simple vow has been taken first, the marriage is halved, whereas if a solemn vow is taken, even afterwards, it dissolves a legally valid marriage.' Add here what preceded this: 'if the church were to decree ...' If the church has no power to do this, that is what Pio should have argued about. I think it has, but I do not assert this. 'But in fact,' he says, 'it has enacted otherwise.'[1206] I know, but the church has made many enactments, particularly about marriage, which it has changed later in consideration of the times.

6 / This follows: 'In addition, "one should consider the present conditions in most of our monasteries and nunneries, and how few of the inmates live the true monastic life."'[1207] I cannot think why Pio moved this Camarina,[1208] which is better left undisturbed.

7 / 'In addition,[1209] now I can almost hear someone say: "Parents and spouses can have no sway over someone who has been inspired by the Holy Spirit to dedicate himself entirely to God." So what happens at baptism? Is half the child dedicated to God and the other half to the devil? If, in fact, they are inspired by the Holy Spirit, no one should have the right to stand in their way [...] But it is well known that many are inspired by a spirit of folly, irresponsibility, madness, ambition, avarice, lust, and self-indulgence. Certainly those who in secret or against the will of their parents (whom God's law commands us to obey) devote themselves to a way of life from which they cannot subsequently extricate themselves seem more likely to be inspired by an evil spirit."' What is Pio condemning in these words? Does he deny that there are many who under the inspiration of the spirit of folly, etc, hurl themselves into monasteries? I suppose he is not that shameless. But he does think it was impious[1210] to say that those who

* * * * *

1205 This passage (LB V 647B / CWE 69 286) was presented by Pio continuously with the preceding, but Erasmus chose to treat it separately here. Pio discusses Erasmus' views on simple and solemn vows in XXIII libri 213^v–214^{v}s.

1206 Pio XXIII libri 213^{v}s–214^{r}t entertains the argument that the church might have the power to give dispensation from solemn vows, but asserts that in fact it does not even give dispensation from simple vows of chastity and religious life.

1207 Pio's fifth quotation, which Erasmus repeats here in Pio's abridged and slightly altered version, is from Institutio christiani matrimonii LB V 647C / CWE 69 287.

1208 See Responsio n82 above.

1209 Erasmus reproduces Pio's sixth quotation, from Institutio christiani matrimonii LB V 647D–E / CWE 69 287, with Pio's introductory 'In addition ...' and supplies the omissions.

1210 Pio XXIII libri 213^{v}R concludes: 'These words are worthy indeed to be uttered from the mouth of one who, blinded by envy against the monastic way of life, has uttered not just these, but even worse!'

enter monasteries without the knowledge or permission of their parents are under the inspiration of an evil spirit. That is not what I said, but 'it is well known' that 'many' of these are under the inspiration of an evil spirit, for the word 'many' preceded the phrases which depend upon it. An adolescent son, not yet of age, should, if he is inspired by the good Spirit, have at least this much honour for his Christian parents, that he would either become a monk with their blessing, or wait for the time of his majority, particularly if his parents require his attention. Therefore Pio is false and slanderous in concluding that [I say] those who dedicate themselves to God in a particular way are generically brought to this by a spirit of folly or insanity, or by the inspiration of a demon, rather than by that of the good Spirit.[1211]

Now as to my statement[1212] that rashly made vows should be considered void, quite apart from the fact that it is ordinary vows I am talking about, I do not exclude the authority of the church. He twists this around to apply to solemn vows with the exclusion of the authority of the church![1213]

Yet would it be a bad thing if the church were to enact also that barely grown boys who have not yet attained their majority should not make monastic profession[1214] without obtaining, or at least seeking, the consent of their parents? Béda claims[1215] that Christ owed obedience to his Mother in the business of the Gospel which his Father had assigned to him, and are we going to allow any adolescent boys or girls to bury themselves in a monastery without the knowledge and consent of their devout Christian parents?

What success will he achieve with his wordy discourse when he makes such obviously false assumptions, conjuring up shadows for himself to fight with? What reply, then, can I make to his wordy and monkish sermon, all of which rests on a rotten foundation? If he undertook to defend monastic vows, he ought to have combated Luther's arguments. If he is fighting with me, he should have read my works and refuted everything he thought

* * * * *

1211 Paraphrasing Pio's charge in *XXIII libri* 211^{V}Q
1212 Included in the third passage from *Institutio christiani matrimonii* LB V 647A / CWE 69 286, quoted in n1200 above.
1213 Pio *XXIII libri* 211^{V}Q accuses Erasmus of including vows of chastity and entry into religious life among vows rashly made; he insists (213^{V}s–214^r) that solemn vows cannot, by their very nature, be taken rashly and that even simple religious vows of chastity cannot be rashly taken and are not, thus, dispensed from by the church.
1214 Pio takes up this issue passim, but especially in *XXIII libri* 211^{r}P–213^{r}R, where he provides additional quotations from the colloquy *Virgo* μισόγαμος LB I 699–700 / ASD I-3 296:233–4, 238–40, 250–3, 294–5:188–99, 295:215–19 / CWE 39 292, 290–1.
1215 See n879 above.

was wrong, and not overwhelm the reader with clouds of smoke drawn from fragments that have been unreliably reported and distorted through malicious misinterpretation.

But in support of his contention that valid marriage should give way to monastic profession he brings forward a simile taken from the transfer of property and other unconvincing monkish fabrications.[1216] But this is the chief ruin of theology, the interpretation of divine laws on the basis of human laws. There is no loss to a field if the contract is rescinded after possession has been surrendered. But harm is done to a wife who is robbed of the husband to whom she had committed herself. He says: 'She can marry someone else.'[1217] But not without gossip. And not just any husband is to her liking, yet the one she liked is taken away.

He adds that in the marriage contract there is an unspoken exception made of the case of one's dedicating oneself wholly to God.[1218] It would have been better to mention this case openly, since ordinary people do not know this amazing doctrine. It would be more fair, perhaps, in monastic profession that a condition like this be implicit, 'unless my parents have need of my help,' or 'unless charity reveals to me something better.'

What a huge window they are opening when they say that children at however tender an age can contract marriage without the consent of their parents![1219]

He says that sons owe more to God, the 'Father of Spirits,'[1220] than to their parents.[1221] Yes, but that 'Father of Spirits' commanded absolute obedience to parents[1222] unless they keep one from Christ. But it is no impiety

* * * * *

1216 Pio XXIII libri 212^{r}R urges that, just as one might rescind a promise on the basis of subsequent events but cannot reclaim property surrendered into the possession of another, so an unconsummated marriage might be rescinded in response to a subsequent call to religious life but a consummated marriage cannot be.

1217 Pio XXIII libri 212^{v}R: 'a woman is free, or a man, if the other has entered religious life, to marry whom she/he wishes in the Lord, etc.'

1218 Pio XXIII libri 212^{r-v}R argues by analogy: 'If, then, tacit exception is made for certain circumstances [frigidity, fraud] that, if they occur, would void a marriage, how much more valid one should consider the tacit exception made in the case of a vow of chastity or monastic profession.'

1219 Not exactly Pio's point, for he does not mention age (XXIII libri 212^{v}R): 'Children are less obliged to their parents than one spouse to another ... For children have not surrendered power over their bodies and wills to their parents. And thus they may contract marriage without the consent of their parents, though it is not fitting to do so.'

1220 Alluding to Heb 12:9

1221 Paraphrasing Pio XXIII libri 212^{v}R: 'Indeed, children owe far less to their particular parents, who were originators only in reproducing the flesh according to their kind, than to the Almighty, begetter of all things, who has given both flesh and spirit ...'

1222 Exod 20:12; Deut 5:16

not to make monastic profession, and a son can achieve the same or better while living in the house of his parents, as once the sacred virgins and many monks used to do, and they were praised by praised men[1223] for the very reason that they did not abandon the fellowship of their parents.

He says: 'If his parents are suffering some necessity, a son is permitted to go out and support them.'[1224] I concur, but this is very difficult for a man with a cowl, one burdened by regulations, who owns nothing, and who is kept by his religious profession from manual labour and the ordinary means of livelihood.

Finally, it is hard to be sure that a son really wants to dedicate himself to God with his whole heart;[1225] but that it is God's will that children obey and serve their parents, this is absolutely sure.

There is not much relevance in their quotation from Jerome's letter which he wrote in a rhetorical vein when he was a youth, practically a boy.[1226] Even less relevant are the words: 'He who wishes to come after me should deny himself, take up his cross, and follow me'; and 'I have come to sunder man from man, etc'; and 'There are eunuchs who have castrated themselves, etc.'[1227] All Christians must take up the cross. This is not a matter of monastic profession but of the mortification of the flesh. And there can be eunuchs for the kingdom of God without monasticism!

I do not criticize[1228] the church's distinction of solemn vow and simple vow,[1229] but I suggest that it could be enacted that the difference between them not be so great. For the words 'solemn vow' and 'simple vow'

* * * * *

1223 Perhaps an allusion to the fragment of Naevius quoted by Cicero in *Tusculanae disputationes* 4.67.4: 'I am glad to be praised, O father, by you, by a praised man.'
1224 Paraphrasing Pio XXIII *libri* 214^{r}T.
1225 Replying, perhaps, to Pio's rhetorical question at XXIII *libri* 212^{v}R: 'Since children are free to resist their parents' commands when they urge something contrary to virtue or to their welfare, how much freer will they be to enter upon the path of a more perfect life, to embrace religious life with a whole heart, against their parents' wishes?'
1226 Pio XXIII *libri* 211^v–213^{r}R provides a pious soliloquy in the person of one who feels he is called to religious life over the claims and protests of his parents; in the course of this he quotes and paraphrases passages from Jerome's Ep 14.2, 3 CSEL 54 47–9.
1227 Paraphrases of Matt 16:24, 10:35, 19:12, from Pio XXIII *libri* 213^{r}R
1228 Replying to Pio's subsequent discussion of the passage from *Institutio christiani matrimonii* already quoted by Erasmus (see 314 above); here (XXIII *libri* 213^{v}s) he says: 'you want [a simple vow and a solemn vow] to have the same weight, and you criticize the church because it has determined that there is a great difference between them.'
1229 On solemn vows, see *Decretalium Collectiones* liber 3 t 15 (Friedberg II col 1053). For a discussion of the difference between solemn and simple vows, see Antonino Pierozzi [or Forcilioni] of Florence OP (1389–1459) *Summa theologica moralis* (Verona: Ex Typographia Seminarii 1740; ed Pietro Ballerini, 4 vols [Graz 1959]) II cols 1113C–34C especially 1113E–1114D, 1117D–1118A.

are the words of men, not of Scripture. But Pio does not solve my problems either, nor is this surprising, because they were not in the notecards.
He says: 'A solemn vow is made with mature deliberation, a simple vow
not so.'[1230] In fact, a simple vow is often made with the same deliberation
as a solemn one, as in the case of the brothers who are called the *Collationarii.*[1231] A simple vow does not generate as much scandal, because it is not
made in the presence of many witnesses. And yet it is often no less a vow
than a solemn one. And what they call a solemn vow is made in the cave
of a remote monastery, and is frequently made not only rashly, but fraud-
ulently. Pio should have considered these and many other points.[1232] But if
he wanted to publish an argument furnished by some monk, he could have
done so without slandering me.

It is not worth relating the rest, for what has it to do with the faith if
someone in the colloquies criticizes those who think that they will recover
from illness if they are clothed in the cowl of Francis?[1233] or if someone
said that he does not want to make bargains with the saints,[1234] for no one

* * * * *

1230 A bleak summary of Pio's more detailed argument in *XXIII libri* 213[v]s that solemn vows
require mature deliberation, use formulated words, involve a solemn rite with many
ceremonies, take place in a public, appointed place, require witnesses, and have as their
object serious matter, such as adopting a way of life, whereas a simple vow can be made
on impulse and indiscriminately about any number of things.

1231 The Brethern of the Common Life (also known as *Collationarii* because of the *collationes*
or spiritual meetings they held after dinner on Sundays and feast days, to which the
laity were at times invited) initially resisted taking vows. They maintained common life
without vows by making over their goods to the *provisores* of the house, and surrendering the right to reclaim them. While some brothers eventually adopted the Franciscan
or Augustinian rule with vows, others secured in 1401 episcopal approbation of their
life style that did not require vows. See M. Dortel-Claudot and A. Deblaere 'Fratelli
della Vita Comune' in *Dizionario degli istituti di perfezione* (Rome 1977) IV cols 754–62 especially 758–9; John van Engen 'Late Medieval Anticlericalism: The Case of the New
Devout' in *Anticlericalism in Late Medieval and Early Modern Europe* ed Peter A. Dykema
and Heiko A. Oberman, 2nd rev ed, Studies in Medieval and Reformation Thought 51
(Leiden 1994) 19–63 especially 45–6; and Regnerus Richardus Post *The Modern Devotion: Confrontation with Reformation and Humanism* trans Mary Foran (Leiden 1968) 272–
92, 310–13; on Erasmus' stay with the Brethren when he was a youth, see Post 395–8,
660.

1232 An unfair criticism, given Pio's description of what makes a vow solemn (*XXIII libri*
213[v]s)

1233 Pio *XXIII libri* 211[r]o had quoted from one of the colloquies of 1522, the second entitled
Male vivere: 'In the dialogue on poor health, mocking vows you say: "Some persons
have recovered their health by donning a Dominican or Franciscan cowl. Perhaps the
same thing would have happened had they put on a pimp's cloak"' (LB I 633B / ASD I-3
134:278–80 / CWE 39 14).

1234 This occurs in two of the passages from the colloquies quoted by Pio *XXIII libri* 211[r]o:
from the second dialogue entitled *Male vivere* ibidem 283–4: 'Others have been cured of

is compelled to make vows to the saints? or if someone should say that nothing surely authentic is displayed at Jerusalem,[1235] since many things of uncertain authenticity were displayed there in the age of Jerome?

VIRGINITY AND CELIBACY [FROM BOOK 17][1236]

1 / He quotes from the *Methodus*: ' "Not all receive this word." Christ said this to chosen disciples, but he did not require of them that to which he invited them, etc.'[1237]

　　Pio asks me on what basis I am sure that he did not require it. I have followed the interpretation of all the ancients. And how is Pio sure that he did require it? Because they observed it?[1238] What is his basis for this? Is it that we do not read of it? Should we here sing back to them what they usually sing to me: 'It does not follow that it did not take place just because it was not written down. Jesus did many things etc'?[1239] But granting that it has some basis, what is to prevent our observing something that is not required but recommended?

* * * * *

illness by making vows to some saint. But I make no bargains with saints'; and from *Naufragium* LB I 713E / ASD I-3 328:115–329:121 / CWE 39 256: 'What were you doing all this time? Making vows to any of the saints? Not at all. Why? Because I don't make deals with the saints. For what else is that but a bargain according to the form "I'll give you this if you do that" or "if I swim I'll go to Rome"?'

1235 Pio *XXIII libri* 211ᵣo quoted: 'And in the dialogue *Peregrinatio religionis ergo* you say about the vow of going to Jerusalem: "Anything there you consider worth seeing? To be frank with you, almost nothing. Some monuments of antiquity are pointed out, all of which I thought faked and contrived for the purpose of enticing naïve and credulous folk. What's more, I don't think it's known for certain where ancient Jerusalem was."' Pio cites this incorrectly, for this is from the colloquy known variously as *De votis temere susceptis* or *De visendo loca sacra* LB I 639B / ASD I-3 147:718–18 / CWE 39 37. Pio discusses this passage later at 216ʳ⁻ᵛz.

1236 MARGINAL NOTE: 'From book 17'; 1531 UPPER MARGINS: 'On Virginity' on pages 243–57 [= 319–33 in this volume]. Pio's book 17 (*XXIII libri* 216ᵛA–224ᵛz), entitled 'On Virginity and Celibacy,' opens with a short series of four quotations, one from the *Methodus* (actually from *Supputatio*) and three more from *Supputatio*; additional quoted passages are taken up later in the book, notably a series from *Institutio christiani matrimonii*.

1237 Pio describes his first quotation (*XXIII libri* 216ᵛA) as having come from the *Methodus*, but it is, though Erasmus does not bother to point this out, a part of Erasmus' defence of his paraphrase of Matt 19:11 from the *Supputatio* LB IX 587E.

1238 Pio *XXIII libri* 217ᵣB: 'I ask you on what basis you are so sure that he did not require this of them. For it is more likely that he did require it, since they observed it so carefully and consistently that it seems it must have been required.'

1239 Erasmus, alluding to John 20:30 and 21:25, imputes to Pio an argument that Erasmus associates with academic theology (see 332 below); Pio does not, in fact, employ that argument here, but does so elsewhere; see eg Pio *XXIII libri* 164ᵛs [misprinted 158].

2 / He quotes from the *Supputations* but does not give a citation: 'I do not quite understand what this wish of Augustine means, the one where he would prefer, if it were allowed, that all would live in chastity, unless perhaps heaven is closed to the married, etc.'[1240] There is nothing here contrary to the Catholic faith. I will give a reply as soon as I have found the passage.

3 / From the same work: 'It is likely that if Jerome had known that matrimony is one of the seven sacraments of the church, he would have been more sparing in his praise of virginity and have spoken most reverently of matrimony.'[1241] What does Pio find scandalous here? That I would suspect Jerome did not place matrimony among signs of a sacred reality? Who could be so crazy? But I do suspect that he did not place it among the sacraments of the church properly so called. Nor do I say that Augustine and Jerome either condemned matrimony or thought it better than virginity.[1242] Who could be so shameless? But Jerome, because of a certain zeal, was sometimes so carried away in his admiration for virginity that he seemed, even to devout men, to do injury to matrimony.[1243] This would be false if the same words are not found in his writings as in Tertullian, who was condemned on this score:[1244] 'What kind of a good is it that is praised by comparison to something worse? And if not to touch a woman is good,

* * * * *

1240 Erasmus quotes the passage from the *Supputatio* LB IX 587F as it is given by Pio *XXIII libri* 216ᵛ–217ʳA including the explanatory insertion 'the one ... chastity.'

1241 Again, Erasmus quotes this passage from the *Supputatio* LB IX 589C in Pio's abridged versions (*XXIII libri* 217ʳA and 218ʳC). The original is: 'And if one may make a conjecture following Béda's model, it is more likely that if Jerome had known the so very corrupt celibacy of these times, if he had seen that in the midst of so many who profess celibacy the really continent are so scarce, if he had known that matrimony is one of the seven sacraments of the church, he would have praised virginity more sparingly and have spoken more reverently of matrimony.'

1242 Reacting to what Pio says in *XXIII libri* 217ʳB apropos of the immediately preceding quotation: 'These words are absolutely silly, unless you mean that Augustine and the rest of the Fathers, all of whom agree in far preferring celibacy and virginity to marriage, in fact prefer it so much that they condemn marriage, and have fallen into the error of the Encratites, Marcion, and Manichaeus. Or, if you did not think this, you must think that their view was erroneous, and that continence is not to be preferred to intercourse.'

1243 See Erasmus' *argumentum* to Jerome's 49, where he says that people are scandalized by Jerome's *Adversus Jovinianum* 'because he seems to have been more lavish in praise of virginity than he should have been, and more severe towards marriage' (HO 3 46ʳ); and the *antidotus* to Jerome Ep 123: 'But St Jerome, because of an exceedingly ardent love of chastity, is often prejudiced against marriage' (HO 1 40ʳ).

1244 Pio *XXIII libri* 217ᵛB, 218ʳ (misprinted as 208) cites Tertullian, among other Fathers, for his respect for marriage. Tertullian praised marriage, but condemned remarriage as fornication or adultery; see the summary of his views on marriage by Paul Mattei in the introduction to SC 343.

then to touch one is evil. There is no mean between good and evil.'[1245] And there are many other things like this, but, because of the man's outstanding merits, his rare learning, and the holiness of his life, they have been overlooked, but not approved. And yet, far from exaggerating them, in my scholia I find excuse for them as best I can.[1246] I am not, then, denying that gospel virginity is worthy of the highest praise. But I suggest that due respect should be maintained for matrimony too, for no other reason than that if those who live in matrimony know that this state in life is quite dignified and spiritual, they would live a married life that is more holy and more chaste.

4 / If Augustine[1247] wrote his book *On the Threefold Good of Marriage* as a corrective to passages in Jerome's books that are abusive of marriage, as appears to be the case at first glance, does it really follow that both of them either condemn virginity or think it is better than matrimony? What need was there to teach us that each agrees on both,[1248] both that matrimony is holy, and that virginity is better?

What do I say in my preface to Abbot Volz?[1249] 'Then as concerns the vow of chastity, I would not dare to unfold how little difference there is between celibacy of the ordinary kind and chastity in wedlock.' What am I implying here, what would I not dare to make explicit? Between matrimony, as I describe it, and ordinary celibacy there is not so great a difference as certain monks commonly want there to appear to be. They think that they are angels, while they scarcely consider married people to be Christians. Did I fail to state this clearly enough? Even so, it is no less true.

* * * * *

1245 A paraphrase of Jerome's words (*Adversus Jovinianum* HO 1 7) apropos of 1 Cor 7:1–39: 'If it is good not to touch a woman, then to touch one is evil, for nothing is contrary to good except evil. But if it is evil, and it is forgiven, it is allowed to the end that nothing worse than this evil occur. Now what kind of good is this that is allowed as an option to what is worse?' (PL 23 218C–D); Erasmus had used this passage with the same paraphrase against Carvajal (*Responsio adversus febricitantis libellum* LB X 1680F), and Pio alludes to this passage shortly (*XXIII libri* 220ᵛK).

1246 Eg in his comment on Jerome's Ep 22 HO 1 62ʳA / CWE 61 192–3; and on *Adversus Iovinianum* HO 3 46ʳA

1247 As his fourth passage, Pio *XXIII libri* 217ʳA quoted from the *Supputatio*: 'It is more likely that Augustine produced the book <*On the Good of Marriage*> against Jerome, although he does not name him, <than against Jovinian>, etc'; the original (LB IX 589B) runs: 'In fact, it is more likely that Augustine produced this work against Jerome, although he does not name him.'

1248 Alluding to Pio's point in *XXIII libri* 217ᵛB apropos of the fourth passage: 'But come on, why "against Jerome," since they are in agreement, both preferring continence to marriage?'

1249 That is, the preface to the *Enchiridion*, Ep 858:634–6 / CWE 66 22

Pio tries to teach us that there is not necessarily greater dignity in things that are sacraments properly so called than in things that are not sacraments.[1250] I have never even dreamed of this,[1251] but in this respect matrimony would be the better off. What would Jovinian[1252] have left unsaid if matrimony had then been recognized as one of the seven! What follows from this, by Aristotelian reasoning? That matrimony is absolutely better than virginity? Not at all, only that it is likely that neither Jerome nor Jovinian counted matrimony among the sacraments properly so called.

So why does this statement seem false: God to some extent regards matrimony as more honourable than virginity[1253] if it was his will that the one be a sacrament and not the other? But it does not follow that it is therefore in all respects superior.

But this,[1254] however, is quite true, that at the beginning the church favoured virginity so much that it almost obscured the dignity of matrimony. This could be false were it not for the fact that Roman pontiffs have taught that intercourse cannot take place without sin, and that a man who had known his wife at night used to refrain from entering a church on the following day.[1255] I pass over the rest.

* * * * *

1250 Pio XXIII libri 217^v–218^{r}c

1251 Pio XXIII libri 218^{r}c (misprinted as 208): 'But doesn't this also follow, in your view: Matrimony is one of the seven sacraments, therefore it is more perfect than virginity?'

1252 Jovinian (d before 406), the monk whose egalitarian views of marriage and celibacy provoked Jerome's radical Adversus Jovinianum

1253 Pio XXIII libri 218^r–^{v}D presents a new set of six objectionable passages, all from the Institutio christiani matrimonii. Erasmus is here reacting to the first of these (Pio XXIII libri 218^{r}Dii): 'For you say in the Institutio christiani matrimonii, assuming what we have just rejected, namely that things which consist of or are linked to a sacrament are superior to those which lack this element: "To some extent God held matrimony in greater honour than virginity."' The passage in its original context continues immediately: 'although the latter [ie virginity] has its particular glories, by which it rises to the heights of the angels' (LB V 619E / CWE 69 225).

1254 Pio quoted (218^{v}D) from Institutio christiani matrimonii: 'The church used to hold virgins vowed to God in such honour that their exaltation almost completely overshadowed the glory of marriage' (LB V 628B / CWE 69 243).

1255 In Institutio christiani matrimonii Erasmus says that with the growth of Christianity matrimony came to be regarded more highly, but that traces of the old prejudice against it remained among the bishops, especially Roman pontiffs: 'In the time of St Gregory, for example, the custom of the Roman church laid down that anyone who had slept with his wife should refrain for a short period from entering a church and still more from partaking of the Lord's body, and was not to enter unless he had washed. St Gregory also agreed with St Augustine that intercourse between husband and wife could not be performed without sin if they had exercised their marital rights out of lust rather than to beget children' (LB V 628E–F / CWE 69 244–5). See James A. Brundage Law, Sex, and Christian Society in Medieval Europe (Chicago 1988) 157–9 on ritual impurity prohibitions to be

Moreover,[1256] when Paul advises married people to come together again lest Satan tempt them, is he not in some sense tempering an excessive zeal for chastity? Well, Pio considers this an impious statement! He asks[1257] who it was who would have denied marriage to Christians if Paul had not given marriage his approval. I am quick to answer, it was the heretics who were forbidding marriage. In the same way we would have reverted to Judaism if this same Paul had not vigorously opposed it. Let Pio come on now with his twisted corollaries and invectives, all of which have nothing to do with the case.

On the first Epistle to the Corinthians, chapter 7,[1258] to settle the disagreement between Augustine and Jerome,[1259] I suggest, apropos of the words 'It is good not to touch a woman,' that 'good' should not be taken as meaning good as opposed to sinful, which is what Jerome wants, and Augustine to an extent agrees, but as meaning 'advantageous.' Something can be disadvantageous for the time being that is not a sin. In fact, that same thing can be advantageous to one person, and disadvantageous to another. Indeed, the same thing can be advantageous to the very same man today that would be disadvantageous to him tomorrow. But I add: 'In Paul's era, since Christians were so very few in the midst of countless pagans, perhaps it was more advantageous to refrain from marriage. But now that times are changed, I would guess that it might be more beneficial even for priests to practise holy wedlock with an unstained bed in a holy and pure way, that is, if there are any priests who do not live lives of continence. But since there is everywhere such a great number of this sort, and, again, such a scarcity of those who live chastely, etc,'[1260] I make no assertion, but say 'perhaps.' I do not say that it is better, that is, more perfect, but more advantageous, that is, more beneficial in these times.

* * * * *

distinguished from sin. Erasmus acknowledges elsewhere that the church has changed its discipline in this regard; see *Declarationes ad censuras Lutetiae vulgatas* LB IX 891F.
1256 This is Erasmus' response to Pio's discussion in *XXIII libri* 218ᵛ–219ʳE–F of the last two passages quoted from *Institutio christiani matrimonii*: '[The zeal for chastity grew,] so much so that Paul was compelled to make some new rules to temper what one might almost call this immodest enthusiasm for purity'; and 'Some would have tried to dissuade all Christians from marriage, had Paul not spoken out loud and clear in its favour. But although he wishes widows to marry, at that time second marriage was held in very low esteem' (LB v 628B–C / CWE 69 243, 244). In the second passage quoted Erasmus alludes to 1 Cor 7:5.
1257 Pio *XXIII libri* 218ᵛF
1258 Replying to Pio's attack in *XXIII libri* 219ʳG on Erasmus' annotation on 1 Cor 7:1 LB VI 685C–F / Reeve-Screech 460–1.
1259 About the correct interpretation of 1 Cor 7:1
1260 LB VI 685F / Reeve-Screech 460–1

But, says he, under the Old Law they kept away from their wives for three days when they were going to take part in the rituals.[1261] What has this to do with the countless swarms of priests and monks who neither teach, nor offer mass, nor study Scripture, nor live in a sober and chaste way?

But Pio thinks that there are more chaste priests than incontinent ones.[1262] I wish God would grant that it could be Erasmus who is lying here. I will go no further into this pitfall; the prudent reader understands what I am not saying.

I do not assert that Paul had a wife;[1263] I only report what the ancients thought about this point. And it is not Clement the Martyr, the one who succeeded Peter, whom I quote; I just say 'Clement.'[1264] You would say that this fellow was a slanderer born! Moreover, Pio's explanation[1265] that the term 'woman' cannot be understood as meaning 'wife' because of the addition of the word 'sister' is worthless, since of old they called Christian women 'sisters,' just as they called all Christians 'brothers.' And what disgrace would it have been if the apostles permitted their wives, who were already living chastely, to serve the Gospel?

* * * * *

1261 Erasmus is looking ahead to Pio's citation in *XXIII libri* 221^{V}N of Exod 19:15 and 1 Sam 21:5.

1262 Erasmus has in mind Pio's claim in *XXIII libri* 223^V–224^{r}Y: 'But what you say is absolutely false, that extremely few of those who profess virginity or celibacy keep them inviolate. For those who think aright think the opposite, that, in fact, there are far more who lead a truly celibate, untainted life than there are who pollute it.'

1263 Pio *XXIII libri* 219^{r}G had quoted a passage from the *Supputatio* LB IX 588A and discussed the annotation on 1 Cor 9:5 LB VI 706C–D / Reeve-Screech 483; he later (222^{r}P) accused Erasmus of maintaining that Paul and others of the apostles were married; for Erasmus' report on patristic commentary on this point, see LB VI 706C–D. For Erasmus' own opinion that Paul could have been married, a widower, or celibate (insufficient evidence), see LB VI 687C and 687F / Reeve-Screech 462; but that he was more probably married or a widower, see LB VI 688B–C / Reeve-Screech 463 and *Declarationes ad censuras Lutetiae vulgatas* LB IX 879D–E.

1264 In the annotation on 1 Cor 9:5 (LB VI 706D / Reeve-Screech 483) Erasmus had observed: 'Surely Clement, as Eusebius reports in book 3 of the *Ecclesiastical History*, uses this passage to argue that Paul had a wife.' The Clement cited by Eusebius is Clement of Alexandria, not Clement of Rome; see *Historia ecclesiastica* 3.30.1 SC 31 140n5. Pio *XXIII libri* 219^{r}G remarks on this and charges that Erasmus uses the name Clement alone 'so that we would believe that it was the great Clement, whom Paul mentions, a martyr, and the successor to the apostolate of Peter.' Erasmus is correct in insisting that 'I just say "Clement"' in the annotation on 1 Cor 9:5, but later, in his *Declarationes ad censuras Lutetiae vulgatas*, he twice identifies this Clement as Clement the Martyr of Rome (LB IX 879D, 916B).

1265 Pio *XXIII libri* 222^{r}P

In the *Letter to the Bishop of Basel* I do not maintain that priestly celibacy should be abolished;[1266] I say only that thought should, perhaps, be given to priestly marriage as well. But if the church should permit unchaste priests to marry, what is that to Pio?

In the passage on 1 Timothy, chapter 3,[1267] there is nothing but a lament for the collapse of sobriety among monks and priests. May God grant that all return to their senses, so that I will have said all this in vain.

He reports the passage from the *Supputation* in a corrupt form, as usual. I say: 'The grace of God surely is available to all for attaining salvation. Let Béda decide whether it is available in all. Certainly it was not available to Paul when he prayed to the Lord three times.'[1268] Pio thinks that I am talking about Paul's incontinence,[1269] whereas I think this should be taken as referring to afflictions from without. Béda thinks that the gift of continence will be available to whoever asks for it.[1270] I refute this as follows: one does not always receive what he requests, for what

* * * * *

1266 Pio *XXIII libri* 219ʳH quoted *De esu carnium*: 'It might later be a greater work of piety to abrogate what it was once a work of piety to establish, in consideration of circumstances and changes in the times and the character of morals ... Maybe this ought to be the view concerning the marriage of priests' (LB IX 1200F–1201A / ASD IX-1 26:207–9, 214); and he returned to this passage in *XXIII libri* 223ᵛ. For Erasmus' advice that priests be allowed to marry, see LB IX 1201A–1202A especially 1201C–D / ASD IX-1 26–8:194–268 especially 235–46.

1267 Pio *XXIII libri* 219ʳ–ᵛH quoted three passages from Erasmus' annotation on 1 Tim 3:2 LB VI 933F–934C / Reeve *Galatians to the Apocalypse* 670.

1268 Pio quotes the passage twice (*XXIII libri* 219ᵛH, 222ᵛs), a bit differently each time; Erasmus here quotes Pio's first version, adding one corruption of his own. The original (as in *Supputatio* LB IX 586F) runs as follows: 'The grace of God surely is available to those who wish it for attaining salvation. Let Béda see if it is available to all for continence. Certainly it was not available to Paul when he prayed to the Lord three times to be freed from Satan the buffeter' ('*Gratia quidem Dei praesto est volentibus ad consequendam salutem, an praesto sit omnibus ad continentiam Beda viderit, certe Paulo non erat praesto ter dominum roganti, ut liberaretur a Satana colaphista*). Pio's first version alters *volentibus* to *omnibus*, omits *ad continentiam*, and alters *colaphista* to *colaphisante*. In his quotation of Pio's first version Erasmus adds *in* before the second *omnibus*, and drops *ut ... colaphisante*.

1269 Apropos of this passage from the *Supputatio* Pio observed in *XXIII libri* 222ᵛs: 'If by these words you mean that the grace of God was not available to Paul so that he allowed him to fall in the flesh, you would not only have uttered a complete falsehood, but blasphemed as well. But if (as I rather think) you mean that it was not available to free him from the seething of the flesh, that does not advance your case.'

1270 In his *Annotationum in Jacobum Fabrum Stapulensem libri duo, et in Desiderium Erasmum Roterdamum liber unus ...* (Cologne: Petrus Quentell 1526) 244ʳB, in the course of his criticism of Erasmus' paraphrase of Matt 19, Béda states that there is no one 'who could not, if he wished, abstain completely from carnal commerce, since the grace of God (without which, I admit, no one could do this) is available to everyone who asks for it.'

Paul sought was denied to him. See what it is to argue on the basis of snippets!

But I hardly dare bring up what follows,[1271] for it is a shocker! Writing to the feverish fellow,[1272] I say something along the lines of 'I do not know how Augustine could have the cheek to glorify continence, since he himself had two concubines, or Jerome to glorify virginity, which he, in fact, failed to keep, as he testifies in his Letter to Pammachius.'[1273] My actual words were as follows: 'How could St Jerome, who was not a virgin, have the cheek to train virgins? How could St Augustine, who had two concubines, have the nerve to teach continence to others?'[1274] Does Erasmus not get a deserved beating here? O detestable sacrilege! I defame such great men, and on a false charge! A twofold evil![1275]

I am not defaming, O amazing man, but the notecard misled you, so that you raged like a madman against your neighbour. There I am making fun of the stupid argument of Pantalabus, who had written 'Who could be the imitators of the chastity of Christ? Who finally could be the denouncers of lechery?'[1276] as he spoke of the monks, all of whom he insists are chaste

* * * * *

1271 MARGINAL NOTE: 'He criticizes words he does not understand.' Erasmus skips to Pio's attack in *XXIII libri* 224ʳz on Erasmus' writing about the sex lives of Augustine and Jerome: 'Moreover, how in error also is what you have published about Augustine and Jerome, not blushing to support the cause of the flesh by defaming such very outstanding men and renowned Fathers! For you have said that Augustine praises virginity and celibacy even though he had concubines, that Jerome praises virginity to the skies in spite of the fact that he did not himself maintain it.'

1272 Erasmus refers to his *Responsio adversus febricantis cujusdam libellum* (Basel: Froben 1529; LB X 1673–84), which was directed against the *Apologia monasticae religionis diluens nugas Erasmi* (Salamanca 1528; Paris 1529) of Carvajal, whom Erasmus calls 'feverish' (*Febricans*) and *Pantalabus* (after the buffoon of Horace *Satires* 1.8.11 and 2.1.22).

1273 It is not clear why Erasmus first gives this elaborated paraphrase and then proceeds to give a precise quotation, for Pio neither quotes nor paraphrases the passage in question. Here Erasmus alludes to Jerome's Ep 49.20.2: 'I praise virginity to the sky, not because I possess it, but because I admire it since I do not have it. It is a sincere and modest confession to praise in others what you do not yourself possess' (CSEL 54 385:11–14), a text to which Pio also alluded in his discussion.

1274 LB X 1682C

1275 Paraphrasing Pio *XXIII libri* 224ʳz: 'How much better it would have been to let these matters sleep in silence, if possible, even if they were true, than to publish them? How much more wicked to do this, since they are absolutely false? And if these words do have some element of truth, the statement was made in a very slanderous way.'

1276 Carvajal's words as quoted by Erasmus (LB X 1682B): 'He [Carvajal] wants to know who would be the lovers of purity if the monks were joined to wives. This sounds as if the monks were the only ones who love purity, or as if monks are the only ones who live chastely. There is the same stupidity in what follows: "Who would be imitators of the virginity of Christ? Who, after all, would be the denouncers of lechery?" These words

and pure, meaning that no one could be a denouncer of lust or an imitator of Christ the virgin unless they were monks. I make fun of this conclusion and show that virginity can be praised even by those who are not virgins. I cite the case of Jerome, who wrote quite a lot in praise of virginity and about the training of virgins. So too Augustine, and he was not a virgin. Therefore virginity could have been praised and lust, which he calls lechery, denounced even by those who were neither monks nor virgins. But I do wonder why the note-taker did not add what follows there immediately: 'With what assurance does Peter, who had a wife, urge purity? With what authority does Paul, who probably had a wife, praise virginity?'[1277] Now really, what defamation have we here? If I defame Jerome and Augustine, my defaming Peter and Paul is more criminal still. But this should be blamed on the note-gatherers who provided the fragment for purposes of slander.

So much for the libel of defamation – now for the charge of lying. He says that the charges I brought are absolute lies. But I made no charges, I have only reported. As to Augustine, no one can deny it. But Pio provides a plea for the defence just as if I had charged him.[1278] I also make a plea for him myself in my preface to the works of Augustine.[1279] As for Jerome, he maintains that he was a virgin and employs a convincing sleight of hand, as if Jerome, on account of modesty, said that he did not possess what, in fact, he did.[1280] Pio's fiction might be convincing were it not for the fact

* * * * *

are absolutely idiotic, unless we mean to declare that the Virgin Mother was a monk, along with the whole choir of virgins.'

1277 LB X 1682C

1278 Pio *XXIII libri* 224ʳz: 'Augustine himself admits in the *Confessions* that as a youth, when he was a pagan, he was infatuated with a woman, and he grieves over this with abundant tears. So what? He was young, he was a pagan, he had not yet received the rites of Christ, he did not consider it unsuitable nor think that he was sinning.'

1279 In the dedicatory epistle to Alfonso Fonseca in *Augustini opera* (1529) Erasmus wrote: 'As a youth he had a concubine, something that human laws allow; and when she was snatched away, not dismissed, he took another. But he kept faith with each as if married, a decency one may not readily find today in priests and abbots' (Allen Ep 2157:35–8).

1280 Pio *XXIII libri* 224ʳ–ᵛz: 'Likewise, your derogatory statement about Jerome, contrary to the common view which awards him the glory of virginity, was made in an irreverent and wanton way. You can provide no evidence of this except some few words he wrote to Pammachius, which seem to support this at first glance. But one who examines them more carefully observes that this is not what he was saying, but that he uttered these words by way of emptying himself, to avoid ostentation.' Pio explains the passage from Jerome's Ep 49 quoted above in n1273 as meaning: 'I have not extolled virginity with such great praise because I am declaring that I possess it; rather, I admire it so much the more if I lack it.'

that this same Jerome bore witness in other passages[1281] that he had fallen in his youth. In Peter's case, it is admitted that he had a wife, and it is likely that he abstained from the enjoyment of marriage after he received the Holy Spirit, but that this was on the basis of mutual consent, for if she had demanded her conjugal right, it would not have been moral for Peter to deny it. In the case of Paul, I say nothing for sure; I only say that this was likely.

But here Pio crashes on the same rock on which others have crashed, as I have shown in already published books.[1282] But Pio was not pleased to examine these, although he was pleased to slander everything. Jerome writes to Eustochium a book *On Guarding Virginity*.[1283] In it he speaks as follows: 'I not only praise virginity, I am keeping it.'[1284] 'What are you looking for,' says Pio, 'could[1285] greater or clearer evidence be provided?'[1286] O pitiable raving! In fact, Jerome used the expression 'I am keeping' by the same trope as Virgil said 'and he raises lofty alders up from the ground.'[1287] One who keeps does not necessarily possess what he keeps, for example, the shepherd of another's flock or a banker.

What thanks do they deserve who with misleading notecards and worthless arguments have shoved a man noble in lineage and letters into an arena for which he was unsuited and have shrouded him, rendered thus ignoble, in a cowl. Κλύζει θάλασσα πάντα τὰ ἀνθρώπων κακά ['The sea purges away all the sins of men'].[1288] 'These facts,' he says, 'should have been kept quiet.'[1289] If they had wanted them kept quiet, they would not have revealed

* * * * *

1281 See texts cited by J.N.D. Kelly *Jerome: His Life, Writings, and Controversies* (New York 1975) 20–1.
1282 MARGINAL NOTE: 'He crashes on a rock that had been pointed out.' See n1018 above.
1283 Pio XXIII *libri* 224ʳz insists that the passage from Jerome's Ep 49 quoted earlier (n1273 above) must be compared to other passages from Jerome, specifically to his famous Ep 22, entitled in the manuscript tradition, *Ad Eustochium de servanda virginitate*, but entitled *De custodia virginitatis* by Erasmus in his edition HO 1 62ʳA. Pio XXIII *libri* 224ᵛz cited the letter as *De custodia virginitatis* when he quoted a passage from paragraph 23.1–2 CSEL 54 175:7–12, part of which Erasmus quotes here.
1284 Jerome Ep 22.23.1 CSEL 54 175:7–8; Erasmus' text included the variant *tantum* (meaning 'only': '*non tantum efferimus*'), which was rejected in CSEL.
1285 *posset* 1531 *potest* LB
1286 Erasmus quotes Pio XXIII *libri* 224ᵛz.
1287 Virgil *Eclogue* 6.63. This tag is quoted from the so-called Song of Silenus, specifically from his reference to the fate of the sisters of Phaethon, who were turned, in this account, into alders. Erasmus' point seems to be that as Silenus does not actually himself raise up the alders but narrates their rising up, so Jerome maintains virginity, not in his own person, but for others through exhortation, teaching, etc.
1288 *Adagia* III iv 9: *Mare proluit omnia mortalium mala* LB II 807F–808A / ASD II-5 245–6
1289 Paraphrasing Pio XXIII *libri* 224ʳz, quoted in n1275 above.

them in published books. But who does not know that Augustine had concubines before baptism?

Now I do not consider it necessary to answer all the particulars of this
quite garrulous discourse, much less the abuse and lies which, of course, he
employs with excessive frequency and enthusiasm, especially since reply
has been made to virtually all of these so many times now.

He makes the false charge that I said that Paul urges the younger
widows to marry.[1290] He says that he does not urge them without qualification, but only if they cannot observe continence.[1291] But I did not say
that he urged in an unqualified way, otherwise he would have been urging all. Not even the Lord said in an unqualified way that eunuchs are
blessed, but only those who had castrated themselves for the kingdom of
God.[1292] I take the kingdom of God to mean the preaching of the Gospel.[1293]
For neither celibacy nor virginity in itself is a virtue, unless it has Christ as
its goal. I would be more ready to say that marriage is good in itself, for it
is both necessary for the propagation of human nature and has been consecrated by the words of God in paradise,[1294] and confirmed by the words of
Christ.[1295] Surely for one who cannot observe continence celibacy is not a
good thing, and therefore it is not an unqualified good.[1296] For that which
is an unqualified good is good always and everywhere; for example, not to
commit adultery is an unqualified good, to love God is an unqualified good.

* * * * *

1290 Erasmus goes back to Pio's quotation and subsequent discussion in *XXIII libri* 118ᵛD–E of
Erasmus' interpretation of 1 Cor 7:8–9 in *Institutio christiani matrimonii* LB V 628C / CWE
69 244: '[Paul] even urged younger widows to remarry, since both their vulnerable age
and their previous experience of pleasure suggested little hope of firm and permanent
continence. Some would have tried to dissuade all Christians from marriage, had not
Paul spoken out loud and clear in its favour. But although he wishes widows to marry,
at that time second marriage was held in very low esteem.' In his discussion of this
passage Pio paraphrases it (*XXIII libri* 218ᵛE): 'Paul, you say, urges young widows [*viduis
virginibus*] to marry.' For Erasmus' views on young widows remarrying, see *De vidua
christiana* LB V 758C–759C / CWE 66 243–4.

1291 Paraphrasing Pio's rhetorical question in *XXIII libri* 218ᵛE: 'Is one who says that those
who cannot be continent should marry urging them to marry without qualification?'

1292 Matt 19:12

1293 See n1128 above.

1294 Gen 1:27–8, 2:24; Eph 5:31

1295 Matt 19:4–6

1296 Echoing, perhaps, Pio's scholastic discussion in *XXIII libri* 219ᵛ1–220ᵛK (a criticism of Erasmus' annotation on 1 Cor 7:1: 'It is good for a man not to touch a woman') of whether
the 'good' to which Paul refers is *bonum absolutum* or *aliquo modo bonum, non simpliciter
bonum*, etc. Pio concludes: 'When, therefore, Paul says "It is good for a man not to touch
a woman," he means that without qualification and absolutely it is good for a man not
to touch her, and thus a good for every man, and always, not for the time being or
under the present circumstances, as you have expounded it.'

If Jerome, who called marriage an evil, is excused because in some passages he declares plainly that matrimony is worthy of honour,[1297] why am I, who nowhere called virginity evil, but praise it highly in many passages,[1298] not excused as well?

In addition, if I am viewed as hostile to virginity because I several times praise matrimony,[1299] then we may as well believe that all those who give abundant praise to virginity are hostile to matrimony. In fact, just as the ancients praised virginity or celibacy to the end that the loves of those in that state would be more holy, so I praise matrimony so that married folk may know how holy is the way of life they have professed, and that to sin in it[1300] is a most serious crime.

He is absolutely shameless in his conclusion that it is my opinion that Paul commands bishops to have wives because he warned that a bishop should duly control his wife and children.[1301] For it could be that it was before he became a bishop that he married the woman who is not taken away from him by this command now that he is a bishop. Indeed, the Greeks think that it is permitted for a bishop, and even more for a priest, to take a wife, but once only. And I did not say that Chrysostom was at

* * * * *

1297 This is a response to Pio's allusion in *XXIII libri* 220[V]K to the passage in Jerome's *Adversus Jovinianum* mentioned earlier (see n1245 above): 'for what is less good, though it is good, is, even so, sometimes called an evil by comparison with that which is more perfect. Jerome, taking this text [1 Çor 7:1] in this way when arguing against Jovinian, calls marriage an evil; yet is is incontrovertible that marriage is a good, and Jerome acknowledges this elsewhere.'

1298 Eg *Institutio christiani matrimonii* LB V 628B–C / CWE 69 243–4; *De vidua christiana* LB V 734C–735F / CWE 66 201–5; *Paraphrasis in Novum Testamentum* LB VII 879C. See August Franzen *Zolibat und Priesterehe in der Auseinandersetzung der Reformationszeit und der katholischen Reform des 16. Jahrhunderts* 3rd ed, Katholisches Leben und Kirchenreform in Zeitalter der Glaubensspaltung 29 (Munster 1969) 42–6.

1299 See Emile V. Telle *Erasme de Rotterdam et le septième sacrament* (Geneva 1954); and John B. Payne *Erasmus: His Theology of the Sacraments* (Richmond, Va 1970) 109–12.

1300 Perhaps a reference to adultery, or to a failure to honour, obey, support, etc

1301 Reacting to Pio's attack in *XXIII libri* 220[V]–221[r]L attack on two passages, one from the *Methodus* LB V 87F / Holborn 201:1–3 and one from the annotation on 1 Tim 3:2 LB VI 933F / Reeve *Galatians to the Apocalypse* 670: 'Therefore equally in error also is what you report in support of this in the *Methodus*: "Paul commands that bishops govern their wives and children well. Now subdeacons are denied the right to have a wife." With these words, as has been said, you are criticizing the practice of the church as being opposed to the precept of Paul, as if you have proven that Paul had commanded or urged bishops to have wives, and you repeat this in the *Annotationes* when you say: In this passage "Chrysostom risks Paul's appearing to require of a bishop that he have a wife," and a little later, "but now not only is chastity required of all priests, but even a marriage contracted by subdeadons is rendered invalid."'

risk because he was cause for anxiety.[1302] But if there had been no risk, he would not have rejected the view of those who thought that a bishop should be allowed to have successive wives.[1303]

Now as to his stout denial that it has ever been heard from the beginning of the Christian religion that a priest took a wife,[1304] anyone who reads all the documents of the Roman pontiffs[1305] or the *Summa* of Antoninus[1306] will find that it is false. However, here he will take evasive action.

But what does he say about the Greeks? 'They,' he says, 'did not acknowledge this law.'[1307] And so, it is not a divine law either, if it could be

* * * * *

1302 Reacting to Pio's question in *XXIII libri* 221ᴿL apropos of the annotation on 1 Tim 3:2: 'But come now, what is Chrysostom's risk and anxiety in this passage?'

1303 Eramus refers here to his annotation on 1 Tim 3:2: 'But Chrysostom interprets [this] as being about the single wife that the bishop is to have, adding that there are some who explain "the husband of one wife" as "who has been married once only." And Chrysostom risks Paul's appearing to require of a bishop that he have a wife, or that he have many at once following the example of the Jews' (LB VI 933F / Reeve *Galatians to the Apocalypse* 670).

1304 Pio's actual statement in *XXIII libri* 221ᵛM is more nuanced: 'Wherefore not even from the very beginning of the Christian religion has it ever happened that a bishop or priest joined himself in marriage after receiving ordination, neither in the churches of the East nor in those of the West, Greeks or Latins, as we have shown above, contrary to your view, since you said to me in reply that even now among the Greeks priests are joined in marriage. But this has never been done or heard of except in these lamentable times in Germany, by some abandoned men, after the emergence of the impiety of the Lutherans.'

1305 The papal documents cited in Gratian's *Decretum* trace the evolution of church discipline on clerical celibacy; see *Decretum* pars 1, d 27–32 (Friedberg I cols 98–122). By the time of Erasmus priests were not allowed to marry. Erasmus seems to be trying to respond to Pio's citation in *XXIII libri* 219ᵛI of the decretals of Innocent III (Friedberg II cols 128–9) by referring to earlier papal documents. See Erasmus' examples from canon law (LB VI 696F–697B / Reeve-Screech 472–3).

1306 Erasmus errs in making this claim. According to Antoninus, who cites the canonists, once a man has been ordained a priest, he cannot marry. Ordination has joined to it an implicit solemn vow not to marry *ex institutione ecclesiae* that invalidates any attempt at marriage. Even if someone does not wish to be obliged to live celibately or is ordained before he has attained the age of discretion, the very ordination itself, apart from the wishes of its recipient, obligates him to forgo contracting a subsequent marriage *ex statuto ecclesiae*. This applies to both the eastern and the western churches, although the Greek church does allow someone already married to be ordained and remain married. See Antonino Pierozzi *Summa theologica moralis* (cited in n1229 above) III cols 35–6. Pio (223ᵛv) echoes this position. By arguing from the Greek practice that allows someone already married to be ordained, Erasmus is not addressing Pio's point that the church does not allow someone already ordained to marry.

1307 Pio *XXIII libri* 223ᵛx anticipated an objection of Erasmus: 'But perhaps you will say the law could not oblige all priests, since it was not received by the universal church' and observed: 'I would say to this that it is true that a great many of the churches of the East regard it as invalid in respect to ordinands, but valid in respect to those who have

rejected. If it is of human origin, it can be abolished by the church. In fact, the Germans keep shouting that they never approved this regulation by unanimous consent,[1308] but that it was forced on them by their opponents.

But where now is that notorious [πολυθρύλλητον] dictum of the schools,[1309] that it does not at all follow that it did not happen because it was not written down. In point of fact, what no law forbids, even if it was not done, could have been done. But this surely has no relevance to me. Finally, the Lord does not urge absolutely that one should leave father, mother, or wife, but only if they pose an obstacle to the Gospel.[1310]

'Soldiers,' he says, 'are forbidden to take their wives with them to war.'[1311] So? Therefore it is even less permissible for clerics to take wives. Well, if it is with tropes that we are being attacked, then every Christian is a soldier of Christ to whom he was bound by oaths in baptism. Therefore no Christian is allowed to take a wife. Caesar forbade soldiers to have their wives in camp, but he did not forbid them to take wives. Lastly, what does it matter that they do not have wives if they have concubines?

He also raises as an objection the verse from Romans 12, 'that you present your bodies as a living sacrifice, etc,'[1312] as if this were not said to all Christians.

* * * * *

been ordained ... But if the churches which did not accept the law are not bound to celibacy completely and did not want a linkage between the vow of chastity and the priestly order, there is no doubt that it obliges the western churches, and all who have not rejected that law have kept it entirely by the very long-standing observance of it over many centuries.'

1308 Erasmus is responding to Pio's claim in *XXIII libri* 221ᵛM that only during these recent deplorable times in Germany is the prohibition on priests' marrying not observed. On the opposition to the imposition of clerical celibacy by the papacy in the eleventh century, see Carl Mirbt *Die Publizistik im Zeitalter Gregors VII* (Leipzig 1894) 12–15 (literary opponents of clerical celibacy) and 302–5 (complaints of the clergy of Cambrai, Noyon, and Rheims against mandatory celibacy); and also Anne Llewellyn Barstow *Married Priests and the Reforming Papacy: The Eleventh-Century Debates* Texts and Studies in Religion 12 (New York 1982), esp chapter 3, The Defense of Clerical Marriage. On the rationale behind the rejection of clerical celibacy by the Germans, see Steven Ozment 'Marriage and the Ministry in the Protestant Churches' *Concilium* 8 (1972) 39–56; and Franzen *Zolibat und Priesterehe* (cited in n1298 above) 23–41.

1309 This is Erasmus' reaction to Pio's argument from silence, that close examination of the New Testament will show 'not one little word that might be taken to mean that they [the disciples] gave the least thought to marriage' (*XXIII libri* 221ᵛO).

1310 Matt 10:35–8; Luke 14:26

1311 A paraphrase of part of Pio's argument in *XXIII libri* 222ʳP

1312 Erasmus is alluding to Pio's argument in *XXIII libri* 221ᵛN: 'but with the change in circumstances and the character of the era, it seemed good to the Holy Spirit that was inspiring the church ... to establish as law that which seemed better, that they who were to serve at the altar should abstain from marriage and that spotless sacrifices be

He says: 'The holy fathers say that those who engage in the marital act cannot offer sacrifice worthily.'[1313] Let us say that is true. They should not, therefore, engage in the marital act when they are involved in the liturgy. It is one thing to engage in the marital act, quite another to have a wife. He says: 'No one can serve two masters.'[1314] Since this statement was made about mammon and God, it follows that wealth is even more forbidden to priests than wives.

He says: 'If the divine Spirit were so to inspire the church, surely it would enact what you urge.'[1315] If the church is enabled to make changes, what sin is there in suggesting it? But if we must always await the inspiration of the Spirit, what is the point of councils? It seemed best to handle these matters in a summary and selective way to show that Pio is not here playing on his home field and how without justification is his raging[1316] so abusively against Erasmus.

MARRIAGE [FROM BOOK 18][1317]

Since I have made careful reply to many about this slander, there is no reason for me to deafen the reader by going over the same things here. Anyone who cares to should read my annotations on chapter 7 of the first

* * * * *

offered by spotless men to the most pure God.' Erasmus assumes that Pio is alluding to Romans 12:1, but Pio seems to have in mind the *hostiam puram, hostiam sanctam, hostiam immaculatam* of the canon of the mass.

1313 Pio *XXIII libri* 223ᵛx: 'But you would say, as you wrote to the bishop of Basel, that it would be beneficial that this law be relaxed for the western churches ... It seemed otherwise to Jerome, Ambrose, Augustine, Gregory, Cyprian, Tertullian, and the rest of the Fathers who affirm that those who take part in the marital act cannot offer sacrifice worthily.'

1314 Pio *XXIII libri* 223ᵛx: 'And the Lord cries out in the Gospel "no one can serve two masters [Matt 6:24; Luke 16:13]." How then will priests serve the Lord with their whole hearts if they are subject to wives and serve them?'

1315 A rather free paraphrase of Pio's words in *XXIII libri* 223ᵛx: 'But if the divine Spirit that guides the church (which distributes its gifts as it wills) and that has prompted many churches to adopt this [clerical celibacy] were to inspire it to the contrary, the church would surely follow its decree.'

1316 *Adagia* III vi 62: *In tua ipsius arena* 'In your very own arena' LB II 860F–861A / ASD II-6 374. *debacchetur* LB *debacchatur* 1531

1317 MARGINAL NOTE: 'From book 18'; 1531 UPPER MARGINS: 'On Marriage' on pages 258–60 [= 333–7 in this volume]. Pio's book 18 (*XXIII libri* 224ᵛA–227ʳF) is entitled simply 'On Marriage,' though the beginning of the book suggests greater specificity: 'let us now see also what you have written about the sacrament of matrimony.' Pio opens the book with two quotations from the scholia on Jerome and a summary of Erasmus' views on marriage and divorce in the annotations on the New Testament. Erasmus' writings on this subject have been studied in detail by Emile V. Telle *Erasme de Rotterdam et la septième sacrement* (Geneva 1954).

Epistle to the Corinthians.[1318] For either Pio has not read my remarks, or at least he does not remember them.[1319]

First he reproaches me for the inconsistency which is a disgrace to a theologian, as if it were a grace for an orator![1320] 'For what,' says he, 'is more degrading in a teacher than inconsistency in his teaching? [...] What is more silly than self-refutation?'[1321] Oh, the indignation [°Ω δείνωσιν]![1322] But where is this shameful inconsistency? In some places I declare that I place matrimony among the sacraments properly so called.[1323] That is true. Elsewhere I say that this is likely,[1324] but that Dionysius and Jerome did not express this view.[1325] Who would not be astonished at this portentous inconsistency? Which of us is being silly here?

He rejects[1326] this argument: 'If matrimony had been acknowledged as one of the sacraments properly so called at that period, its great defender

* * * * *

1318 LB VI 692D–703D / Reeve-Screech 461–81 (with indications in the margins of the chronology of additions to the annotations)

1319 Pio XXIII libri 224ᵛA mentions the annotations and gives a list of Erasmus' unorthodox views on marriage presented there; he quotes and paraphrases (225ᵛC) a passage from the annotation on 1 Cor 7:39 LB VI 692F–693A / Reeve-Screech 467.

1320 Pio XXIII libri 224ᵛB: 'Now although this practice has taken hold with the orators, that is, that they use the same texts when arguing on opposing sides in profane matters, it is surely altogether disgraceful for a philosopher or theologian when interpreting Scripture.'

1321 Direct quotation from Pio XXIII libri 224ʳB

1322 Quintilian 6.2.24 defines deinosis: 'This is the affect which is called deinosis; it adds a vehemence to things that are base, harsh, and productive of hostility.'

1323 Eg Responsio ad annotationes Lei LB IX 225F, 226B, 227C, 228D; Apologiae contra Stunicam (3) LB IX 378C. In his annotations on Eph 5:32 Erasmus states that this biblical text cannot be used to prove that marriage is a sacrament, but he does not wish to call its sacramental status into doubt or to deny that it is a sacrament (LB VI 855B–E / Reeve Galatians to the Apocalypse 615).

1324 Erasmus states that the sacramental status of matrimony was unclear in the time of Jerome and Dionysius. See his Apologiae contra Stunicam (3) LB IX 378A–B, where he states that it probably was considered a sacrament but that Jerome did not argue from this; and his Responsio ad annotationes Lei LB IX 226D, where from Dionysius' failure to mention matrimony when treating the sacraments Erasmus concludes that it probably was not then considered one of the seven sacraments.

1325 See Erasmus' scholion on Jerome's Ep 123, quoted by Pio XXIII libri 224ᵛA at the beginning of book 18: 'I am not at all sure if in the time of Jerome the church counted matrimony among the sacraments. In fact, Dionysius [the Pseudo-Areopagite], when explicitly dealing with the sacraments, makes no mention of matrimony' (HO 1 41ʳB).

1326 A response to Pio's attack in XXIII libri 225ʳB on a passage from Erasmus' antidotus to Jerome's Adversus Jovinianum, quoted at the beginning of the book (224ᵛA): 'In addition, in the fourth volume [actually in the third volume] you say "At this time matrimony is placed among the sacraments, wherefore it is amazing that neither Jovinian mentions this when he is praising it, nor Jerome when refuting Jovinian's books"' (HO 3 46ʳ).

Jovinian would not have kept silent about it.' He says: 'Jovinian saw that it was not a strong argument.'[1327] And so he presents Jovinian as a quite forbearing fellow, the sort who would not have wanted to employ any arguments but solid ones. But how does he argue that this line of reasoning is weak? Matrimony is a sacrament. Virginity is not. Therefore matrimony is superior. In fact, it is a strong argument as long as you say 'therefore superior in this respect.' But if Jovinian, even deprived of this argument, dared to make a place for matrimony on the same level as virginity, how much greater would have been his daring if aided by an irrefutable argument?

He frequently raises as objections the church's regulations and the Holy Spirit.[1328] But he should have been solving the famous questions: How has the church changed its rules, even those about matrimony?[1329] What do the lawyers mean when they maintain that the pope has the right to interpret divine Scripture,[1330] to limit and extend its applicability? Pio should have answered these and many other questions. But I define nothing, I only discuss. Pio's words 'you insist,' 'you maintain,' and 'you argue'[1331] are slanderous. In my writings nothing like this is found.

* * * * *

1327 A summary of Pio's statement in *XXIII libri* 225ʳʙ: 'I say that Jovinian did not mention this [that matrimony was a sacrament] because it did not contribute anything towards proving his position. For, as we discussed earlier, the argument "marriage is reckoned among the sacraments of the church, therefore it is more honourable than virginity" is absolutely weak and has no force. It was similarly of no use for Jerome to mention this, particularly since in Jerome's time the sacraments of the church were not restricted to the number seven.'

1328 At the beginning of book 18 Pio makes an appeal to ecclesiastical regulations passim (*ecclesiae decreta, placita ecclesiastica, summorum pontificum decreta, ecclesiasticas leges*). Pio distinguishes in *XXIII libri* 225ᵛc between the rules (*praescripta*) of the church that are of human origin and changeable, and the unalterable ones that proceed from its interpreting Scripture by divine inspiration (*afflatu divini Spiritus*), or are put forward on the basis of a subsequent inspiration (*novo illapsu sancti spiritus*).

1329 For Erasmus' survey of the church's changing rules on marriage, see his annotation on 1 Cor 7:39 LB VI 692D–703D / Reeve-Screech 467–81. For more recent surveys, see Gabriel Le Bras 'La doctrine du mariage chez les théologiens et les canonistes depuis l'an mille' in *Dictionnaire de théologie catholique* IX-2 (Paris 1927) cols 2123–2317; Joseph Freisen *Geschichte des canonischen Eherechts bis zum Verfall der Glossenlitteratur* 2nd ed (Paderborn 1983); and Jean Gaudemet 'Le lien matrimonial: Les incertitudes du haut Moyen-âge' *Revue de droit canonique* 21 (1971) 81–105.

1330 See George H. Tavard *Holy Writ or Holy Church: The Crisis of the Protestant Reformation* (New York 1959) 31–61; Remigius Bäumer 'Lehramt und Theologie in der Sicht katholischer Theologen des 16. Jahrhunderts' in *Lehramt und Theologie im 16. Jahrhundert* ed Remigius Bäumer, Katholisches Leben und Kirchenreform im Zeitalter der Glaubensspaltung 36 (Munster 1976) 34–61.

1331 That is, *vis, censes, contendis*. In the opening paragraph of book 18 Pio uses the words *censes, asseris, vis*, and *persuadere conaris*, and there are equivalent words passim in the pages which follow.

I do not want Scripture to be fitted to our morality,[1332] I want no popes to do this. But if Scripture is an unerring rule, why is its applicability extended and limited? Why does it not continue in its original force? Why do the popes' decrees on matrimony contradict one another?[1333] Of course it is better not to chafe against these matters again, but Pio ought to have settled them for good. It is he who is a man learned in every particular, who thinks it monstrous that Erasmus quotes something from secular law, as if it were not more marvellous still that Pio comments on Scripture.[1334]

But he permits only an adult, one no longer under the rule of his parents, to take a wife without his parents' knowledge or consent.[1335] But what I am complaining about there is this, that girls scarcely ten years old are, by the contrivance of bawds, married to boys of fifteen, and this is regarded as lawful matrimony.[1336] Moreover, in certain regions majority is not attained until after the twenty-first year![1337]

* * * * *

1332 In *XXIII libri* 225^{v}c Pio quotes and paraphrases the beginning of Erasmus' annotation on 1 Cor 7:39 LB VI 692E, 692F–693A / Reeve-Screech 467: 'you say that laws should be changed like medicines according to the character of the disease, for you say that some things have been received "by the authority of the church in such a way that they can be changed according to circumstances. And as it is not allowed to rescind the divine Scripture, which we have as an absolutely sure rule of life, even so it is the duty of the devout and prudent steward to fit it to public morality." You appear nowhere more foolish than in these words, for you want it to be possible for the divine Scripture, which, you admit, is a sure rule of life, to be dragged down and fitted to public morality.'

1333 Erasmus provides examples; see LB VI 696F–697B / Reeve-Screech 672–3.

1334 Erasmus alludes to Pio's gibe in *XXIII libri* 226^{v}E: 'It remains to answer your false charge about the church's enacting that matrimony can be contracted by consent alone. For the purpose of carping at this, to show that you are not unacquainted with civil law, but are "grammarian, rhetor, geometer, painter, and masseur" [Juvenal 3.76], in the book on marriage you bring in many profane laws, both of the civil lawyers and of the emperors, and a great many precedents from the Old Testament.' In *Institutio christiani matrimonii* LB V 630E–F / CWE 69 249 Erasmus discusses the jurists' maxim: *solo consensu coire matrimonium.*

1335 Summary of Pio's discussion at *XXIII libri* 226^v–227^{r}E

1336 In *Institutio christiani matrimonii* LB V 630F / CWE 69 249–50 Erasmus writes apropos of his discussion of the maxim 'marriage is made by consent alone': 'I am not saying this to condemn the decree, but to alert young men and women to the dangers of contracting marriages. It is not my place to challenge laws accepted by the church, but it is the task of the legislator to change the law to meet present circumstances.' Later he says: 'These days it is not a rare occurrence, especially in France, for a girl barely ten years old to be a wife and to become a mother in her eleventh year' (LB V 666E / CWE 69 325).

1337 According to the civil codes of Basel (1529) and Strasbourg (1530) a man must have attained twenty-four years and a woman twenty before they could marry without the consent of their parents. Nuremberg in 1534 set the ages at twenty-five and twenty-two respectively; see François Wendel *Le Mariage à Strasbourg à l'époque de la réforme 1520–1692* (Strasbourg 1928) 103–4; Steven Ozment *When Fathers Ruled: Family Life in Reformation Europe* (Cambridge, Mass 1983) 38.

Pio stipulates mature and thorough deliberation.[1338] Therefore a verbal contract of matrimony for the future or for the present made amidst cups and caresses by the intoxicated is not valid. But if intoxication, the urgings of bawds, embrace and present pleasure, or childish infatuation preclude thorough deliberation, where do the doctrines about consent and about 'words concerning present and future' belong?[1339]

CONFESSION [FROM BOOK 19][1340]

He does not seem to deal with this topic seriously. Since I have made replies to many people in a variety of ways,[1341] he should first have refuted the arguments I have presented and then proven what he asserts. No one denies that confession was public in times past.[1342] My doubts are about secret confession, when, and by whom it was instituted.[1343] And when I say 'secret,' I am talking about that which is in current use, with its particulars.

* * * * *

1338 Pio *XXIII libri* 227ʳF asserts the validity, civil law notwithstanding, of marriages made without parental consent but soundly based on thorough deliberation.

1339 Erasmus criticizes these features of marriage law in *Institutio christiani matrimonii* LB V 630F / CWE 69 250.

1340 MARGINAL NOTE: 'From book 19'; 1531 UPPER MARGINS: 'On Confession' on pages 261–3 [= 337–41 in this volume]. Pio's book 19 (*XXIII libri* 227ᵛG–230ʳN) is entitled 'On Sacramental Confession.' Pio opens with the usual series of quotations, two from the scholia on Jerome, one from the *Methodus*, two from the *Annotationes in Novum Testamentum*, and one from the colloquy *Confabulatio Pia*, and he also cites the colloquy *Militis et Cartusiani*, charging that: 'These, with the addition of what you wrote in the book on confession [*Exomologesis*], make it clear that your view is that confession in the church is based only on custom and human tradition, and enjoys no force or dignity based on divine institution. And thus you dissuade all from confessing by setting forth these propositions more effectively than if you were pleading openly' (227ᵛG).

1341 Eg *Responsio ad annotationes Lei* LB IX 255A–262E; *Declarationes ad censuras Lutetiae vulgatas* LB IX 931B–932E; and *Apologia adversus monachos* LB IX 1062C–1064B.

1342 Pio *XXIII libri* 227ᵛG quotes two passages in which Erasmus asserts the practice of public confession in the early church. The first is from a scholion on Jerome's Ep 77. Here is the passage with the words omitted from Pio's quotation indicated and supplied: 'From this passage you can observe that confession was public in time past <and concerned public sins, and that penance likewise was public>. And it appears that in the time of Jerome there had not yet been established the secret confession <of sins, which the church subsequently established, a salutary practice so long as both priests and people employ it correctly>' (HO 1 89ᵛ). The second passage is from the annotation on Acts 19:18 LB VI 507F–508A / Reeve-Screech 315. Pio (228ʳH) charges that 'you can cite no evidence that secret confession had not been established at the time of Jerome,' and asserts that 'this confession that we term ecclesiastical, which is made to a priest who is to pass judgment, had been established at the time of Jerome.'

1343 *Annotationes* LB VI 507F–508E / Reeve-Screech 315; *Responsio ad annotationes Lei* LB IX 255A–E, 259B–262E; *Apologia adversus monachos* LB IX 1062E–1063A

And it is not all theologians I condemn.[1344] I only point out that certain ones in their discussions mix up public confession with private, as Gratian does in his patchwork, the *Decretum*.[1345] And this also is not false, that sometimes the undertaking of the prescribed penance[1346] is called confession; in this way any who seek baptism are also said to confess, for whoever seeks washing is confessing that he is dirty.

One who declares that this form of confession was instituted by the Holy Spirit through the church is not far off from what the theologians have been trying to prove in a variety of ways for some time now,[1347] but they have done so neither with agreement among themselves in their proofs nor without considerable labour.

He says that confession is a part of a sacrament.[1348] But no sacrament can be instituted by men. He makes an assumption of what would have to be proven, that is, if he were dealing with someone who disagrees.[1349] Those who say that confession was not instituted by Christ are the same who say that it is not a part of a sacrament but, rather, a sacramental which is conducive to absolution. Men have established many things like this about the matter of the sacraments.

* * * * *

1344 In the second passage that Pio quotes from the same scholion on Jerome's Ep 77, Erasmus claims: 'Some not very alert theologians slip on this matter with the result that they extend what those ancients say about this sort of public, general confession, which amounted to nothing more than acknowledging through certain signs and penances imposed by the bishop that one is a sinner and unworthy of communion with good people, to this secret confession which is of a very different sort.' Erasmus' words 'slip on this' (*in hoc labuntur*) had been distorted in Pio's quotation to 'in this labyrinth' (*in hoc labyrintho*) (XXIII *libri* 227ᵛA). Pio defends the theologians a little later (228ʳH), asserting that they are 'vigilant and alert.' For Erasmus' survey of patristic and canonical opinion, see his *Responsio ad annotationes Lei* LB IX 259D–262D; for medieval scholastic opinion, see Thomas N. Tentler *Sin and Confession on the Eve of the Reformation* (Princeton 1977) 14–22.

1345 *Responsio ad annotationes Lei* LB IX 261C–E; Gratian 'Tractatus de penitencia' in *Decretum* pars 2 c 33 q 3 (Friedberg I cols 1159–1247) especially d 1 cap 1–5, 89–90 (I cols 1159, 1189–90).

1346 See Tentler *Sin and Confession* (cited in n1344 above) 4–12, 16–18.

1347 See Bernhard Poschmann 'Busse und Letzte Olung' in *Handbuch der Dogmengeschichte* IV-3 (Freiburg 1951) 83–103; Ludwig Hodl *Die Geschichte der scholastischen Literatur und der Theologie der Schlusselgewalt* I (Munster 1959); and Heinrich Karpp *Die Busse, Quellen zur Entstehung des altkirchlichen Busswessens* (Zurich 1969).

1348 Pio XXIII *libri* 228ʳH–I

1349 Pio XXIII *libri* 228ʳH–228ᵛI, 230ʳN argues that while Christ instituted the sacraments, the church can determine the sensible signifying matter of the sacrament if it has not already been constituted by Christ. The audible confession of sins to a priest is implied in the sacrament instituted by Christ since a priest cannot forgive sins that are unknown to him. Hence the church requires auricular confession as a part of the sacrament.

I am not arguing the point, but only suggesting that James is talking about reciprocal offences.[1350] I cite as the source of this interpretation Bede,[1351] who drew virtually all of his material from the commentaries of the ancients, particularly Augustine. The theologians who are most involved in the defence of confession surely do not place much reliance on this text.

I am leaving out the commonplaces he recalls about the lepers who were told 'show yourselves to the priest,'[1352] and about the keys.[1353] Not even the theologians, when they engage in serious discussion, rely very much on these foundations.[1354]

I do not know what kind of a line of argument it is when he concludes like this: 'Since penance is a sacrament, and this has been decided by the church, therefore confession was instituted by Christ, and was not thought up by the theologians,'[1355] for this does not follow, nor did I ever say it. I admit that the sacramental character of penance has been declared by the

* * * * *

1350 MARGINAL NOTE: 'Confess to one another, etc.' Pio quotes Erasmus' annotation on James 5:16 in *XXIII libri* 227^{v}G and later discusses it (229^{r-v}M). He quotes it with some omissions and inconsequential substitutions: 'Again, in the same work you argue that the words of the apostle James, namely "Confess your sins to one another," are not applicable to ecclesiastical confession, for you say "he is thinking about the daily offences of Christians among themselves; he wants them to be reconciled at once. Otherwise, if he had had in mind that which we call part of the sacrament of penance, he would not have added [...] to one another, but 'to the priests'"' (LB VI 1037D / Reeve *Galatians to the Apocalypse* 744).

1351 *Beda* 1531 *Bedda* LB. Erasmus means not Noël Béda (*Bedda*), but the Venerable Bede (*Beda*, d 735); the remainder of the annotation on James 5:16, not quoted by Pio, runs: 'Bede's exegesis does not disagree with this view. And he plainly quotes [the text] in this way in the commentary he wrote on the fifth chapter of the Epistle of John' (LB VI 1037D / Reeve *Galatians to the Apocalypse* 744); see Bede *In 1 epistolam S. Ioannis* on 1 John 5:16 PL 93 117B–C.

1352 Luke 17:14, cited by Pio *XXIII libri* 228^{r}H. For Erasmus' refutation of this argument, see *Responsio ad annotationes Lei* LB IX 257A–F.

1353 Pio *XXIII libri* 228^{r}H does not quote the first part of the classic text of Matt 16:19, but rather a passage from Jerome's Ep 14.8.1 CSEL 54 55:5–6 that alludes to this text in a discussion of priests 'who, possessing the keys to the kingdom of heaven, in a way pass judgment in anticipation of Judgment Day'; Pio himself (230^{r}N) argues from John 20:23.

1354 In his *Responsio ad annotationes Lei* LB IX 257 B–C, Erasmus points out that this argument is based on allegory and that Durandus did not use leprosy as an allegory for sin when treating this question.

1355 A parodic paraphrase of Pio's argument in *XXIII libri* 228^{r}H where he reports the opinion of the more learned theologians about auricular confession: 'they call it sacramental because it is a part of the sacrament of penance, as has been declared by the decision of the church when it defined that penance is counted among the seven sacraments ... Since, then, this is a dogma of the church and a precise decree, it is clear that it was not thought up by the theologians.'

church,[1356] but I never said that this was thought up by the theologians. I do not even say that confession was the invention of the theologians, but that it was given a narrow meaning by the monks.[1357]

He ought to have proven his point that sins not known by name and number cannot be forgiven,[1358] since in confession sins unknown to the penitent are forgiven, in baptism all sins are forgiven without[1359] confession, and a man can forgive his enemy both the injuries he knows and those he does not know. I am not saying that this is a bad argument, but that it needs underpinning.

When it comes to easing the pains of purgatory, I would wish that the pope had capacity far greater than certain people assign to him. If, as Pio says, he eases through the modality of intercession, and not by his power, they are telling lies about the bulls in which 'the angels of God are bidden, etc.'[1360]

This argument[1361] seemed to me insufficiently rhetorical: All powers in heaven and on earth were given to Christ as to a man. It is not, then, remarkable if the pope has been given the power of demolishing all the gates of hell. If God alone forgives sins, Christ does not forgive as a man. But he

* * * * *

1356 *Exomologesis* LB V 150D; *Responsio ad annotationes Lei* LB IX 255C; *Apologia adversus monachos* LB IX 1062E. For his reluctance to commit himself, see *Exomologesis* LB V 145A–146A, 147A–B; *Responsio ad annotationes Lei* LB IX 262D–E; and *Declarationes ad censuras colloquiorum* LB IX 932C–E.

1357 An ambiguous statement, perhaps referring to the evolution of confessional practices (*Apologia adversus monachos* LB IX 1064A–B), to the mendicants' faculties for absolving (*Responsio ad annotationes Lei* LB IX 258A), or to their tendency to supplant the local pastor as confessor (*Exomologesis* LB V 156C, *Responsio ad annotationes Lei* LB IX 258A).

1358 Pio insists on the precise disclosure of sin, but does not specify number and kind: 'But judgment cannot be made about things unknown. Indeed, things done in secret are hidden from the judge unless they are revealed by the mouths of sinners' (*XXIII libri* 228ʳH); 'But since this [the forgiveness of sins] takes place in sacramental confession, as is evident from the words of Christ, "Whose sins you shall forgive, etc," and they are not forgiven unless they are known, and they are known through confession ...' (230ʳN).

1359 *citra* 1531 *circa* LB

1360 Pio *XXIII libri* 229ʳL: 'What absurdity would there be, then, in saying that the same Almighty entrusts the power to snatch souls from purgatory to his only begotten Son's vicar, ... not as having power over them absolutely, but rather through the modality of intercession, that is, bringing relief, and offering to divine justice something from the treasures entrusted to him in lieu of the punishment appointed for the expiation of their sins ...?' There is no mention of papal bulls commanding angels, etc.

1361 This paragraph is part satiric distortion and part refutation of Pio's argument in *XXIII libri* 229ʳL: 'And what is surprising if he gave a man the power to accomplish this task, since he also gave to one Man the power to smash the citadel of Tartarus and the gloomy dungeon of hell, I mean to the King of glory, the Lord strong and mighty in battle [Ps 23:8, 10]. Though he was God and Man, yet this power was given to him as man ... for about him as man was it written "all power in heaven and on earth has been given

says that it has been given, not because Christ did not have it beforehand according to the hypostatic union, but because it was then revealed to the world, as when we pray to the Father for the coming of his kingdom.

Here with remarkable effrontery he says that I make fun of sacramental confession throughout the colloquy 'Military Affairs,'[1362] whereas what is made fun of there is the unprofitable confession of a stupid soldier.

It would have been better for Pio not to touch this topic at all than to ignore so many problems and preach a sermon of commonplaces. If he had provided something worth reading, he would have had me, whom he imagines to be his enemy, as his supporter.

FAITH AND WORKS [FROM BOOK 20][1363]

See how consistent he is here. There is an obvious lie at the very beginning of his treatment of this subject where he says that I deprive works of all reliability and importance,[1364] something not even Luther does.

[1] / Erasmus wrote: 'No one can become a sharer in the grace of Christ unless he has faith in him.'[1365] Where did he write this? In the *Annotations*. A real theologian's citation! How could I reply to this when I do not know in what context it was said?

Let us accept that it is the grace of Christ that justifies. Does Christ grant this to anyone who lacks faith? And yet Pio says that what I have

* * * * *

to me [Matt 28:18]." These words would not be applicable to God as God, since he has of himself power of every kind; but what is possessed is given not to the the the one who possesses, but to the one who possesses it not. What absurdity would there be . . .,' etc, as quoted in n1360.

1362 Pio *XXIII libri* 227^{V}G had refrained from quoting this colloquy: 'I am leaving out what you write in the dialogue entitled *Militaria* where you quite wantonly make fun of confession. I would have to quote the whole dialogue.' For Erasmus' defence of this colloquy (CWE 39 52–63), see *Declarationes ad censuras Lutetiae vulgatas* LB IX 931C–932A.

1363 MARGINAL NOTE: 'From book 20'; 1531 UPPER MARGINS: 'On Faith and Works' on pages 264–71 [= 341–7 in this volume]. Pio's book 20 (*XXIII libri* 230^r–239^v) is entitled 'On Faith and Works.' Pio opens with a series of seven passages quoted from Erasmus' works, three from the *Annotationes*, one from the *Supputatio*, two from the *Methodus*, and one from the *Enchiridion*. Pio closes this series of quotations with the remark: 'These are your statements. I am leaving out a great many passages in which over a long period you depreciate works to the advantage of faith. I have quoted those statements that appear more temperate and present the appearance of devotion and are, thus, more capable of deceiving less careful readers.'

1364 MARGINAL NOTE: 'Lie.' Pio's charge in the opening words of the book: 'Now that these have been taken care of, let us examine also what you have written about faith and works. You remove all reliability and importance from works.'

1365 Reacting to Pio's first quotation: 'about which [works] you say thus in the *Annotationes*: "No one can . . ."' Erasmus seems to have been unable to locate this citation; our efforts to locate its precise source have been similarly unsuccessful.

written is absolutely false and cites the prostrate Saul.[1366] Yes, but was he justified before he believed? However, I was not talking there about the few cases in which God wished to manifest his power in an extraordinary way, but about ordinary justification. Is faith not demanded before baptism is granted? And in these cases, too, it was not justifying grace that was granted without faith, was it? In fact, Paul, a man learned in the Law, had faith in the Messiah even though he did not yet know Jesus. He was in error, not in his faith, but about the person. Through his being called, however, his faith was increased so that he attained to the Gospel's grace. I can say the same about Matthew, for he was a Jew, and about the rest of the apostles called by Christ.[1367] They had, in some degree, a confused and general faith in Christ, that is, in the Messiah. When they were called, a surer faith was granted, and through it they were justified. But the gentiles who were called to the Gospel, through what were they justified? Was it not through faith? For those who did not believe perished.

The household of the centurion received the Holy Spirit before baptism.[1368] Nobody denies this. But did they receive it without faith? No. Rather, God had already purified their hearts before baptism. And all whom Christ called either had a general faith or he granted it to them, for this is what 'to call' meant, to grant a heart that is docile and obedient.[1369]

From all this he concludes that we should always be involved in zealous works, even if we are not in the state of grace, or even of faith.[1370] What has this to do with me? In fact, I defend the same view in the second book of *Hyperaspistes*,[1371] at least as being probable, for theologians differ on this

* * * * *

1366 Pio *XXIII libri* 230ᵛB: 'And at first sight what you wrote in the *Annotationes* seems to be correct and to have a pious meaning, that is, that no one can share in the grace of Christ unless he believes in him, etc. But the meaning is absolutely false. For Saul did not believe in Christ ...,' etc, paraphrasing Acts 9:3–4.

1367 Pio *XXIII libri* 230ᵛB cites the calling of Matthew (Matt 9:9) and of the rest of the apostles as having taken place 'before they acknowledged the Lord, and so they were made sharers in grace before they believed.'

1368 Pio *XXIII libri* 230ᵛB cites the example of the centurion Cornelius and his household, quoting Acts 10:44–8, as an example of grace being received first (an outpouring of the Spirit) followed by the profession of faith in baptism.

1369 Erasmus seems to be describing here the 'preparative grace' taught by Augustine, which Erasmus treats in greater detail in *Hyperaspistes* 2 (eg LB X 1457B–C, 1471E–F, 1487B–C, 1526E–F, 1528D–1529D / CWE 77 580, 610, 646, 728–9, 732–4).

1370 Pio *XXIII libri* 230ᵛB: 'By this event [the account of the centurion and his household] we are taught that we must continue in zealous works no matter in what state we are, even when we lack the grace that must make these life-giving, and what is more, even when we are without faith.'

1371 Eg LB X 1530F–1534D, 1536C–E / CWE 77 737–44, 748–9

point. Accordingly, the slander here about a 'calling grace' or a 'preparatory grace' is a sophistry,[1372] even if I think that grace is nothing other than
the rudiments of faith, that is, docility to the teachings of the faith. Whether
grace can exist without faith or faith without grace I am not going to consider now, since faith itself is a grace, that is, a gift of God.

2 / In the *Supputation*[1373] I say: 'Faith, and faith alone, cleansed the
hearts of Jews and gentiles as they came to baptism.' Even if I had said that
we are justified by faith alone after baptism, it could have a devout meaning, for 'alone' is said of that which is principal, and the term 'alone' does
not exclude those things that are associated by their very nature.

He says: 'In adults, works are required.'[1374] Who says they are not?
For they are the fruits of the faith which works through charity, and are,
therefore, God's rather than ours. For faith is a gift of God, and good works
abound in keeping with the measure of faith.

But let others decide whether a sincere faith can coexist with corrupt
works.[1375] I have never spoken about a faith like this, nor have I ever read
that it is called 'sincere,' 'alive,' and 'complete,' as he says.[1376] James calls it

* * * * *

1372 The terms Erasmus uses here, 'calling grace' (*gratia vocans*) and 'preparatory grace' (*gratia praeparans*) are an attempt at summary formulation of Pio's argument based on Rom
8:30 ('And whom he predestined, them he also called. And whom he called, them he
also justified'). Pio argues (*XXII libri* 230ᵛc–231ʳ) that Paul assigns a third place to 'justifying grace' (*gratia iustificans*) and that predestination and vocation are also grace that
is antecedent to faith and conceptually separable from it.

1373 Replying to Pio's second quotation, from the *Supputatio* LB IX 630F–631B, which Erasmus
gives here as it was given by Pio, the only changes being *etiamsi* for *si* and *cohaerent*
for *adhaerent*, *adhaerent* being the original reading. In fact, this is a combined paraphrase
and quotation from Erasmus' reply to Béda's criticism of the views of Erasmus which
Béda formulated as proposition 86: 'For faith alone purifies hearts and makes them fit
to be entrusted with the secrets of the heavenly philosophy,' which he quoted from
Erasmus' paraphrase of John 4:21 LB VII 528F / CWE 46 57. Erasmus countered with the
statement that he was talking about Jews and gentiles coming to baptism, and offers
the qualification reflected in this quotation.

1374 Erasmus replies to Pio's criticism of his remark that the statement that we are justified
by faith alone after baptism could have a pious meaning; Pio wrote (*XXIII libri* 231ʳD):
'Moreover, grace is required first of all. If it is compared to faith, it not only takes the
principal place, but itself alone justifies. And in the case of those who grow to adulthood
and have the use of reason, works are required if they are to persevere in righteousness,
for unless they keep the Lord's commandments, which are fulfilled through works, they
would soon fall away from righteousness.'

1375 A reply to Pio's attack in *XXIII libri* 231ʳve on Erasmus' defence of his use of the term
'alone': 'Faith abides along with corrupt works. For as was just said, many believe strongly
without hesitation [cf James 1:6] who, for all that, driven by their emotions, do works that
are wrong, as Paul states, etc.' Pio quotes as proof-texts 1 Cor 13:1, 2 and Romans 7:20–3.

1376 Pio *XXIII libri* 231ʳE speaks of 'a sincere faith, ... so complete and alive ...'

dead.[1377] The text he quotes from Paul, 'If I speak with the tongues of men ..., and if I have faith, etc,'[1378] has already been disposed of in a number of ways.

3 / He brings in from the *Annotations*[1379] the statement that in Matthew the Judge cites the works of charity which they themselves, however, did not know they had done.[1380] Here, not Pio, but someone denser still, declaims in a verbose but far from subtle way: 'How could it happen that the devout not know through forgetfulness that they did what they had done?' He unties the knot: 'They knew,' he says, 'that they had done them, but they did not know that they had done them for Christ.'[1381] A neat solution! How could they be the devout if they did not recall that Christ said: 'What you have done for one of these of my little ones, you have done for me'?[1382] It would be more acceptable for them to have forgotten what they had done than what Christ had taught. But who is so stupid that he does not know that there is a trope in this saying of Christ? The people on the left were boasting of their good deeds; Christ did not know these. But for those on the right, Christ recalls their good deeds, which they did not know themselves, thus placing before our eyes the hypocrites' deceptive righteousness under which they advertise themselves to men, and the modesty of the devout who depreciate their own good works and leave them to Christ's judgment.

You have heard the stupid argument, now hear the equally stupid conclusion: 'You have brought in these things to remove importance from works, since you assign all power for attaining to eternal life to faith

* * * * *

1377 James 2:17
1378 Pio *XXIII libri* 231^rE quotes 1 Cor 13:1.
1379 Pio *XXIII libri* 230^rA, 231^vF. Erasmus misread Pio, for the reference is not to the *Annotationes* LB VI 131E / Reeve 100 but to the *Supputatio* (see n1380).
1380 Replying to Pio's third quotation: 'And in the same [ie the *Supputatio*, the last cited work] in another passage on Matthew [25:37–40] concerning the Last Judgment: "The works of charity are told, but, if any had done them, they did not know that they had."' Pio's criticism of this passage follows (231^v–232^rF); he is referring to the allusion in the *Supputatio* LB IX 652A to the evangelical parable on the Last Judgment.
1381 A bleak and imprecise paraphrase of Pio's remarks in *XXIII libri* 231^vF, especially of his explanation: 'The just will not deny, then, in the Judgment, that they have done the works of charity, nor will they have forgotten them, but when they hear the Lord list them as having been done for him, in modesty and amazement at the Lord's great kindness and goodness, acknowledging their own frailty and judging themselves unworthy of the great honour and favour that will be granted them by the Lord, not elated with joy, but trembling with humility and in fervent love for the Judge, they will reply "O Lord, when did we see you hungry ..."'
1382 Matt 25:40

alone.'[1383] He repeats this shameless lie: 'Placing all confidence and worth in faith alone, and taking away all need for works ...'[1384] This would have been a big enough lie even without his adding 'all.' Whoever furnished Pio with this sort of stuff deserves to eat hay. In fact, I take nothing away from the works that proceed from faith; I only advise certain people who sell their sackcloth garments, knotted cinctures, and open sandals[1385] to other Christians and promise them what Christ promised to those who believe in him with their whole hearts.

4 / But in the *Enchiridion* I wrote that God makes little of the works of Abraham.[1386] Here he enters upon the wide-open topic, 'In Praise of Good Works,'[1387] as if it were good works I condemn! But I am talking only about Mosaic works, to which the works prescribed by men are similar. Besides, to trust in God and to love God and neighbour are works, and to come to your neighbour's aid is a work, but not a Mosaic work. Thus when Paul disputes against the Jewish reliance on Mosaic works, he makes little of the works of Abraham but he does not condemn them. He makes little of them, that is, he does not take them into account for righteousness, unless Pio is not going to forgive my use of the term 'make little of' as he overlooked Jerome's saying that matrimony is evil (that is, in comparison to virginity).[1388]

But since whoever supplied this material to Pio was stupid and arrogant, I will not try to repeat his tasteless trifles. Who said that the Lord rejects the works of the devout?[1389] But just as in Isaiah he loathes the victims

* * * * *

1383 Quoting Pio XXIII *libri* 232ʳF, giving *attribuis* for *attribuas*
1384 MARGINAL NOTE: 'Lie,' quoting Pio XXIII *libri* 237ʳG
1385 Perhaps a reference to an austere branch of the Franciscan family such as the Minims or Capuchins
1386 Pio's seventh quotation is a long passage from the *Enchiridion*: 'The Apostle makes little of the works of Abraham, which everyone will admit were extraordinary, and you have confidence in yours? God is opposed to sacrificial victims and sabbaths ... although he was the author of all these things. Will you be so bold as to compare your petty observances with the precepts of divine law? Listen to God's nausea and disgust with such things: "What is the multitude of your sacrificial victims to me?" ... When he mentions observances and sacred rites and multiplied prayers, does he not single out with his fingers, as it were, those who measure religion by the number of psalms or prayers recited?' (LB V 36C–D / Holborn 83:28–84:13 / CWE 66 80). Pio quotes this passage correctly every time it comes up; it is not clear why Erasmus garbles it here.
1387 Referring to Pio's oratory in XXIII *libri* 232ʳH–237ʳG
1388 For Paul on Mosaic works, see Rom 4:1–25; for Jerome on marriage as evil, see 330 above.
1389 Erasmus' reaction to Pio's attack on part of the passage from the *Enchiridion*: 'But as to your saying that the Lord is opposed to sacrifical victims and sabbaths, although he was the author of all these things, etc. You will not find anywhere in Scripture that the Lord disdains sacrifice devoutly offered and sabbaths duly kept. But the sacrifices of the devout have always been acceptable to the Lord' (XXIII *libri* 232ᵛ–233ʳL).

of the Jews,[1390] who, through absence of charity, made a boast of certain ritual observances and claimed a certain extraordinary righteousness for themselves on the basis of them, so God rejects the garments, cinctures, sandals, and uncomprehended psalms of those who, since they lack the Spirit, advertise themselves to the people under these disguises.

Yes, this was really the place for a declamation about the prayers and hymns of the ancients,[1391] as if anyone were attacking prayer or psalms devoutly chanted! But is it those who chant or pray devoutly who advertise their good works?

He says that Christ's teaching his disciples to say 'we are unprofitable servants' even though they had done all things has to do with inculcating modesty.[1392] I do not deny this, but did Christ teach his disciples to lie? This must be in some sense true, even if it is incongruous. They were unprofitable, because they had done nothing good of themselves. A reward was due them, because the good God chose to crown his gifts in them, as Augustine says.[1393] But our righteousness, to the extent that it is ours, is in truth only a menstrual rag.[1394]

There is a quotation from Paul: 'But if you are led by the Spirit, you are not under the Law.'[1395] Since Béda said that all men, however spiritual,

* * * * *

1390 Eg Isa 1:11–17, 58:1–7
1391 Pio *XXIII libri* 233[r–v]M parries Erasmus' attack on measuring piety by prayers with a defence of the chanting of psalms, etc, on the basis of the antiquity and ubiquity of the practice and its precedent in Scripture.
1392 This is the end of a long argument against Erasmus' statement from the *Methodus*, the fifth quotation at the beginning of book 20 (*XXIII libri* 230[r]A): 'Christ declares [Luke 17:10] that no thanks are due to servants even if they have done all their duties' (LB V 102D / Holborn 231:9–10). Pio argues (235[v]z): 'But you cannot cite even one passage in support of your view, unless, perhaps, you bring out the words "When you have done all things well, say 'We are unprofitable servants."' These words do not support you at all. For the Lord is not saying here that a reward is not due to good works, but he is teaching the servant not to be vain and boastful ... but to be humbled in his own eyes.'
1393 See Augustine Ep 194.19 CSEL 57 190:12–15.
1394 Erasmus here echoes Isaiah 64:6; Pio *XXIII libri* 236[r]C criticizes Erasmus for making this same allusion in the *Methodus* LB V 102E, 104A / Holborn 231:23–4, 234:6–7.
1395 The reply to Pio's attack in *XXIII libri* 237[r]H–239[r]M on the garbled paraphrase quoted at the beginning of book 20 (230[r]A) from the *Supputatio* LB IX 667E–668A. Pio wrote: 'And in the *Supputatio*: "Since the Apostle says 'If you are led by the Spirit, you are not under the Law,' how then is what Béda says true, that all men are subject to law? Moreover, what I said, that we are not now compelled to do good by the prescription of the Law, I do not say that about any and all of the baptized, but about those who had persevered in the grace of the Spirit once they had received it."' The text of the *Supputatio* reads: 'If what he [Paul] says here is true, "You are not under the Law" [Rom 6:14], likewise what he says elsewhere, "The Law was not appointed for the just" [1 Tim 1:9], and Galatians 5, "If you are led by the Spirit you are not under the Law" [Gal 5:18], how is what Béda says true, that all the just are subject to the Law? ... Accordingly, it is sheer slander that

are subject to human regulations, I was asking that he interpret for us Paul's words, which seem to say the opposite, and not only in this passage, but in some others, for example, 'Where the Spirit of the Lord is, there is liberty, etc,'[1396] and 'The Law was not established for the just man.'[1397]

He is talking, you say, about the Mosaic Law, that is, about circumcision, dietary restrictions, fasting, and things like these.[1398] Let us say that this is true. But some were not bound by the Law who were not led by the Spirit, I mean the gentiles. Also, this statement of the Apostle has bearing on the similar regulations of Christians. I am not saying that they should be ignored by the spiritual, but I do maintain that the spiritual are not subject to law. Those are subject to law whom law compels, who would not comply if it were not required. Those who are spiritual either comply of their own free will (and do even more than the law requires) or do what would otherwise be unnecessary for the sake of avoiding scandal. For some rituals were devised on account of the weak, but the perfect man is not obligated to these, either because he does not need them or because charity summons him to even better things. But he does observe them, to the degree that he can, to avoid scandalizing the weak. Now let this fellow go ahead and discourse about various laws and obligations[1399] to his heart's content, and draw conclusions of which I have never even dreamed.

THE RIGHT TO MAKE WAR [FROM BOOK 21][1400]

So as not to repeat the replies I have made so often to complaints on this topic, I simply admit that I have written some rather distasteful things for the purpose of frightening Christians away from the insanity of war, for I observed that the largest part of the evils of the Christian community take

* * * * *

Béda interprets me as alleging that they who have professed the law of Christ are not compelled by the commandments of the Law to do good. For I am not talking about all the baptized, but about those who had persevered in the grace of the Spirit once they had received it.'
1396 2 Cor 3:17
1397 1 Tim 1:9
1398 A paraphrase of Pio's argument in *XXIII libri* 237[r]–[v]H
1399 Pio *XXIII libri* 237[v]I–239[v]N discusses the law of nature, the law of the Gospel, and human laws in all their variety, to support Béda's position that all are subject to law.
1400 MARGINAL NOTE: 'From book 21'; 1531 UPPER MARGINS: 'On the Right to Make War' on pages 272–5 [= 347–51 in this volume]. Pio's book 21 (*XXIII libri* 239[v]O–244[v]Z) is entitled 'On War and the Right to It.' It opens: 'Now let us get to the few things which remain to be examined, as we review what you have written about war and your assertion that it is absolutely forbidden to Christians.' Pio then presents twelve quotations or summaries: two from the *Methodus*, two from the *Institutio principis christiani*, three from the *Annotationes*, one from the *Enchiridion*, and four from the scholia on Jerome.

their origin from the wars which we have seen for all too many years. It was necessary, however, to cry out, not only against those deeds which are invariably criminal, but also against the enterprises which can hardly ever be undertaken without many crimes. But when I discuss seriously the question of whether war is permissible for Christians under any circumstance, I declare that it should be allowed either if terrible necessity requires it or if a notable benefit, a commendable one, suggests it.

But here I am charged with inconsistency.[1401] I will allow myself to be so charged, so long as they admit that I have this in common with Jerome and Augustine. In the course of discouraging clerics from the love of wealth Jerome clearly states that clerics cannot possess anything but the Lord.[1402] And in the course of urging them to Christian meekness he concludes that the Jews were allowed to make war, but that Christians are not.[1403] If he were being pressed on a point of doctrine, he would have declared otherwise. We often encounter hyperbole like this in Scripture. If they grant the same indulgence to me as to the other sacred Doctors, my views on war cannot be unclear to them, since both in *The Christian Prince*[1404] and in the book *On Undertaking War against the Turks*[1405] I give an account of the basis for war, and I would hardly do otherwise if I condemned war absolutely.

* * * * *

1401 After the series of quotations Pio (*XXIII libri* 240ʳo) expresses his amazement that some actually deny that Erasmus teaches that war is forbidden, but says that they should be forgiven this 'since you are inconsistent in some places and speak in such a way that you seem to hold this view [that war is not forbidden]. When they note these passages, and fail to compare them to your other statements, they make their claim to assert your freedom from so obvious an error, not seeing that while they try to pull you back from the whirlpool of Charybdis, they are thrusting you into the maw of Scylla. In fact, inconsistency is far more disgraceful in an author than ignorance of some doctrine, for ignorance proceeds from a lack of knowledge that is often pardonable, but inconsistency only occurs due to weakness and frivolity of mind, and it is absolutely disgraceful.'

1402 A reference to Jerome Ep 52.5.2 CSEL 54 421:18–19

1403 Jerome urges clerics to meekness in Ep 52.5.4, but makes no mention of war in that connection. Earlier, in urging against the pursuit of wealth, he says: 'Do not regard the work of a cleric as a type of old-time soldiering, that is, do not seek the profits of the world in the soldiery of Christ' (Ep 52.5.3 CSEL 54 422:8–9). Erasmus is probably confusing Ep 52 with Ep 123.12.3–4, where the views of war in the Old and New Testaments are contrasted. Pio quoted from Erasmus' *antidotus* to Jerome's *De monogamia ad Gerontiam viduam* HO 1 41ʳB at the beginning of book 21 (240ʳo): 'In the same letter it appears that he held the opinion that it was permitted to make war in the Mosaic law, but the same is not permitted in the Law of the Gospel, since Christ ordered Peter to put away his sword.'

1404 *Institutio principis christiani* LB IV 561A–612A / ASD IV-1 136–219 / CWE 27 206–88

1405 *De bello Turcico* LB V 345C–368B / ASD V-3 31–82

I say that all of Christian philosophy (that is, the gospels and the writings of the apostles) discourages war.[1406] Is this surprising, since they are always urging us to harmony with one another and to love even of enemies? But if all Christians were such as Christ wanted them to be, there would be no war among them, not even a quarrel.

He says: 'There is no prohibition of war anywhere in the gospels.'[1407] Christ says: 'If anyone strikes you on the right cheek, etc';[1408] and Paul says: 'Do not avenge yourselves, beloved, but give place to anger.'[1409] 'These,' he says, 'are counsels.'[1410] A very large loophole! But how is it an error to want all Christians to observe all the counsels?

'But Christ,' he says, 'rebuked the one who struck.' Yes, but he did not hit him back, just as Paul did not hit the high priest.[1411]

But Michael made war with the dragon.[1412] We too should wage war against Satan and sin.

* * * * *

1406 Pio's third quotation is the following passage from *Institutio principis christiani*: 'It can be argued that papal laws do not condemn all wars. Augustine too approves it somewhere. Again, St Bernard praises some soldiers. True enough, but Christ himself, and Peter, and Paul always teach the opposite. Why does their authority carry less weight than that of Augustine or Bernard? Augustine does not disapprove of war in one or two passages, but the whole philosophy of Christ argues against war. Nowhere do the Apostles approve it, etc' (LB IV 608E / ASD IV-1 215:511–17 / CWE 27 283–4). Erasmus noted that Jerome also followed Paul's teaching which is not observed today; see his scholia on Jerome's Ep 120 HO 4 70^{r}A.

1407 Not a verbatim quotation from Pio, but probably prompted by his remark that soldiering is not forbidden by Scripture: 'In addition, Scripture does not even in a single passage express an abhorrence of soldiering. The gospel nowhere does this. It is not forbidden in the Acts of the Apostles, nor by the Apostle Paul' (XXIII *libri* 241^{r}Q); or by Pio's observation (242^{r}s) that 'you cannot cite one single passage that asserts this' when commenting on Erasmus' claim that 'the whole of the philosophy of Christ teaches against war, etc.'

1408 Matt 5:39

1409 Romans 12:19

1410 Pio XXIII *libri* 243^{r}v, apropos of Matt 5:39: 'I answer that this was not a precept of the Lord, but a counsel, one to be practised not always and in every sort of activity, but when time or circumstances require.'

1411 A reply to Pio's argument in XXIII *libri* 243^{r}v: 'And the Lord himself made this clear by his example: when he was struck on the cheek before the high priest, he did not turn the other cheek, but restrained the one who was striking him by the nobility of his protest [see John 18:23]. And Paul, after this example, when he was struck before the proconsul Festus rebuked Ananias the high priest with equally stern words [Acts 23:2–3].'

1412 Pio XXIII *libri* 243^{v}Y, after describing Michael's combat with Lucifer and his followers (quoting Rev 12:7), argues: 'By this too is made the point that making war is not forbidden to Christians, since this is reported by John the Evangelist in the Apocalypse, and that being a soldier is not a disgrace, since the angelic spirits were soldiers and carried out their duties.'

Christian princes have made war. Yes, but not as Christians; they followed the same set of rules as they would if they had been pagans.

Everywhere St Jerome declares explicitly that Christians may not wage war. Pio interprets him as speaking about either unjust wars or wars undertaken frivolously.[1413] Will he not interpret my words in the same way?

I say that war should not be undertaken rashly even against the Turks.[1414] By 'undertaken rashly' I mean when one challenges those who are more powerful, or against treaties.

He says: 'It is beneficial to plunder them of property and livelihood and to drive them to the Catholic faith by force of arms.'[1415] If this has seen good success so far, let us continue with it!

But, in the Letter to Volz I denounce war against the Turks absolutely.[1416] A quick reply: This is a lie. I say nothing there but that it is more prudent to invite the Turks to our religion with letters and books before attacking them with arms,[1417] and that if it turns out that they must be attacked, this should be done after our morals are set in order.

He asks: 'How can we repel violence without violence?'[1418] I answer: 'How did the apostles do it?'

* * * * *

1413 In *XXIII libri* 241^v–242^r s, in the course of discussing (the twelfth quotation, the one from Erasmus' *antidotus* to Jerome Ep 123 (quoted above, see n1403), Pio argues that Jerome had in mind wars that were undertaken unjustly or frivolously (241^v s).

1414 Pio's seventh quotation consists of several passages from *Institutio principis christiani* LB IV 610C–D / ASD IV-1 219:601–3, 606–7, 608–9, 616–19 / CWE 27 287, and in *XXIII libri* 242^r T he criticizes in particular line 616: 'However, I do not think, either, that war against the Turks should be hastily undertaken.' For two studies of Erasmus' views on war with the Turks, see A.G. Weiler 'Einleitung' in ASD V-3 especially 20–7; and Weiler 'The Turkish Argument and Christian Piety in Desiderius Erasmus "Consultatio de Bello Turcis inferendo"' in *Erasmus of Rotterdam: The Man and the Scholar* ed J. Sperna Weiland and W.T.M. Frijhoff (Leiden 1988) 30–9.

1415 A parodic paraphrase of Pio's remarks in *XXIII libri* 243^v x, where he argues that in the case of those who oppose the glory of Christ and the spread of the Gospel 'it is permitted and it is beneficial, according to place and time, to attack them and subdue them to Christ's easy yoke.'

1416 MARGINAL NOTE: 'Lie'; Erasmus reacts here to Pio's prejudicial summary of the sections having to do with war against the Turks in Erasmus' preface to the second edition of the *Enchiridion* (Ep 858:78–165 / Holborn 5:14–7:32). Pio had given this summary at the beginning of book 21 (240^r o) and repeats it on 242^r T: 'in the letter to Volz you discourage all from undertaking wars against the Turks.'

1417 Erasmus summarizes the statement he made in the preface to the *Enchiridion* (Ep 858:139–53 / Holborn 7:3–20); but he does not say here that if this strategy fails, warfare may follow. The previous section (lines 78–129) only prays that the war being prepared against the Turks may turn out well and urges that, instead of resorting to arms, Christians win over the Turks by good example.

1418 A direct quotation from Pio *XXIII libri* 240^v P

He says that bishops also are allowed to make war.[1419] They may be allowed to, but I think it is disgraceful. And I was convinced of this by Ambrose, who wrote that the arms of a bishop are prayers and tears![1420] 'But,' he says, 'some of them have secular jurisdictions.' But the same ones have their civil governors. And not even secular princes are at war all the time! This will be enough for the Father, my reader, who seems to be a young man giving himself some practice in theological argument under the name of Pio.

OATHS [FROM BOOK 22][1421]

Here we have very little disagreement. We cannot deny the words of the Lord and of James.[1422] I was looking for a way out of the problem,[1423] the same thing he attempts here.

* * * * *

1419 Erasmus' alludes to Pio's comment (*XXIII libri* 244^{V}z) on the annotation on 2 Cor 10:4 LB VI 784E / Reeve-Screech 553, quoted at the beginning of book 21: 'But as to your statement in the *Annotationes* in criticism of bishops who have troops, swords, lances, and cannons, etc, this was dealt with earlier, when it was made clear that they are allowed to have these things and that it is appropriate and beneficial, especially when they direct the government.'

1420 Erasmus probaby has in mind Ambrose's remark in *Contra Auxentium* 2 (Ep 75a:18–20): 'also against the Gothic soldiers tears are my arms, for such are the defences of a bishop' (CSEL 82 10:83); or his repeated descriptions of prayer as a shield, eg in Ep 14 (63):1082–3: 'Prayers are a good shield by which abuse is warded off, invective repelled' (CSEL 82 10:291).

1421 MARGINAL NOTE: 'From book 22,' with no running title in 1531 on 275–6 (= 351–2 in this volume) because of the brevity of this section; Pio's book 22 (*XXIII libri* 244^{V}A–248^{r}K) is entitled 'On Swearing.' It opens: 'But to this view you attach another no less erroneous, asserting that this [swearing] is absolutely forbidden for Christians in the way that war is.' Pio begins with only two quotations, both from the *Supputatio* LB IX 575D, and a summary plus quotation from the annotation on 1 Cor 15:31 LB VI 737F–738C / Reeve-Screech 513.

1422 Referring, probably, to Matt 5:33–7, quoted by Pio *XXIII libri* 245^r; and James 5:12

1423 Pio *XXIII libri* 244^{V}A: 'For you say in the *Supputatio*: "If swearing was acceptable to Christ, he spoke in vain the words 'I, however, say to you not to swear at all [Matt 5:34].' If it is beneficial for Christians to swear, why did he so teach? Why did the Lord make an addition to the Mosaic law? It both prohibits perjury and forbids us to take the name of God in vain."' Pio has garbled Erasmus' statement to some extent. What Erasmus wrote in the *Supputatio* was: 'If swearing among Christians was acceptable to Christ, he spoke in vain the words "I, however, tell you not to swear at all, etc" (LB IX 575D); and 'Let Béda teach us at least this: Since Christ says he did not come to destroy the Law but to fulfil it, if it is fitting and beneficial for Christians to swear, as he [ie Béda] has taught, why did the Lord make an addition to the Mosaic law? It both prohibits perjury and forbids us to take the name of God in vain, that is, as I maintain, in a slight matter, although now people routinely swear an oath on account of two drachmas' (LB IX 576B–C).

This seems to be an unconvincing fabrication: Moses allowed the Jews to swear even for a trifling reason, so long as they swore by God.[1424] But in Exodus he commands: 'You shall not take the name of your God in vain,'[1425] that is, rashly and unnecessarily. Accordingly, he seems to forbid, not only perjury, but even readiness to swear an oath.

This too is weak: the Lord did not forbid the swearing of oaths absolutely, only oaths on heaven or on the life of another; but he permitted oaths on God.[1426] If this were true, then the everyday public practice of swearing on the gospel book or the cross should be abolished. For in Matthew 23, when rebuking the Pharisees' foolish teaching on oaths, he says: 'He who swears on the temple is swearing by it and by him who dwells in it. And he who swears by heaven is swearing by the throne of God and by him who sits upon it.'[1427] But if it was licit to swear by God himself, and it makes no difference whether you swear by God or on heaven or the temple, then he seems to have forbidden, not some forms of swearing, but the swearing of oaths itself.[1428]

LYING [FROM BOOK 23][1429]

The real lie is that I advocate lying. But occasionally we employ the term *lying* in an extended sense to mean *feigning*,[1430] quite within the customary use of the Latin language, whereby we speak of 'the lying mirror,' and Virgil said 'she lied the face of Lycisca,'[1431] meaning 'imitated with intent to deceive.' So one who presents a happy face, although he is sad of soul,

* * * * *

1424 This is Erasmus' incorrect inference of the meaning of the passage quoted by Pio (*XXIII libri* 245^{v}D) from Jerome's comments on Matt 5:34 CCSL 77 32.

1425 Exodus 20:7

1426 Again an incorrect summary of what Pio says (*XXIII libri* 245^v–246^{r}D) in summarizing Augustine *De sermone domini in monte* 1.17:55 CCSL 25 59–60

1427 Matt 23:21–2

1428 Erasmus deals with Pio's arguments up to *XXIII libri* 246^{r}D, but ignores altogether Pio's nuanced and well-argued defence of limited swearing on 246^{r}E–248^{r}K.

1429 MARGINAL NOTE: 'From book 23'; 1531 UPPER MARGINS: 'On Lying' on pages 276–83 [= 352–8 in this volume]. Pio's book 23 (*XXIII libri* 248^{r}L–251^{r}V) is entitled 'On Lying.' It begins: 'It remains to examine what you have written about lying. For your writings attest that you must be counted in the number of those who believe that lying is not a sin when circumstances require it.' Pio provides only two quotations, from the scholia on Jerome and the *Methodus*, and remarks: 'With these words you make it clear that you feel that lying is not only permitted according to circumstances and times, but that it is even beneficial.'

1430 Pio *XXIII libri* 248^{r}M–249^{r}N provides a sketch of non-sinful uses of various sorts of deceit.

1431 Erasmus has garbled his reference; the text referred to is Juvenal *Satires* 6.122, which reads *titulum mentita Lyciscae*, not *faciem mentita Lyciscae* as Erasmus quotes it.

is in a way making a lie of joy; just as face and eyes are said to speak, so too they can correctly be said to lie. Or, someone who pledges that he will deliver what he promises says 'I shall not fail/lie.'[1432] And 'Every man is a liar'[1433] is said, not because there is no one who does not lie, but because God alone can deliver everything which he promises.

In the same way, 'truth' is taken as meaning 'absolutely'; Terence says 'where there is truly no life.'[1434] Augustine and Jerome use 'lie' in this way in their argument about the feigning of Peter and Paul.[1435]

Moreover, I never said or wrote that lying in its precise meaning, when a falsehood is uttered with intent to deceive,[1436] can be altogether without sin. Therefore this person, whoever it is, as he embarked upon a sharp disputation against lying, would have behaved more correctly if he had not defamed his neighbour with a lie. And if he is as pious as he wants to appear, he could have left out the tasteless witticism that I was taught by the orators and think that one may lie as often as it is advantageous.[1437] But an orator who lies in aid of a defendant is more deserving of excuse than this theologian who, not in only a single tribunal, but before the theatre of the whole world, alleges such a dreadful lie against one whom he pretends is his friend.

And Plato does not teach that one should lie,[1438] which, strictly speaking, is the use of language to deceive, but that one should make use of sham

* * * * *

1432 *Non mentiar* means both 'I shall not fail' and 'I shall not lie.'

1433 Rom 3:4; see also *Lingua* LB IV 702C / ASD IV-1 302:310 / CWE 29 332

1434 Terence *Heauton Timoroumenos* 154

1435 See Erasmus' annotation on Gal 2:11–14 in LB VI 807D–808B / Reeve *Galatians to the Apocalypse*; and *The Correspondence, 394–419, between Jerome and Augustine of Hippo* ed and trans Carolinne White (Lewiston, NY 1990).

1436 Compare Pio's observation in *XXIII libri* 248ᵛN: 'Therefore, though every lie is uttered to deceive, not everyone who deceives is a liar, but only the one who asserts something false with the intention of deceiving.'

1437 Pio *XXIII libri* 249ᵛP: 'But it is the habit of orators and those who plead cases to lie whenever it is useful in advancing the cause which they are supporting. It is, perhaps, therefore not surprising that you defend lying since you were trained by them.'

1438 Erasmus' response to Pio's criticism in *XXIII libri* 250ᵛ–251ʳv of his scholion on Jerome's Ep 84.3.7, quoted by Pio at the beginning of book 23 (248ʳL): 'For you say in the scholia of the second volume of Jerome: "[Plato] in the books of the *Republic* wants the guardians of the state to be skilled in the art of lying and to fool the people as often as [their] welfare requires it, but forbids the rest to lie ... For the dull masses are not influenced by the truth ...'; the rest of the passage, which is found, in fact, in the third volume of the 1516 edition of Jerome (HO 3 88ᵛ) is quoted by Erasmus below. Erasmus cited Plato's position as one of his justifications for urging princes, under the pretext of praising them for virtues they in fact lack, to adopt a better course of action; see the dedicatory letter to Jean Desmarez in *Panegyricus* LB IV 551B–C / Ep 180:49–52, 71–2.

and pretence. For he is talking about the pretended use of lots,[1439] craftily feigned so that the people will not become angry with the magistrate, and about similar shams[1440] for tricking the dull masses for their own benefit, as parents deceive children, or physicians the ill.[1441]

But is Erasmus in error if in his scholia he occasionally points out the teaching of Plato?[1442] Supposing that Plato did mean a lie which deceives purposely with words,[1443] does one necessarily give his approval to what is reported to explain a passage?

But I go on: 'for the dull masses are not influenced by the truth.' Maybe he wants to oppose 'truth' to lies. But 'truth' is an expression for what is absolute and uncovered. And so the dull masses are in need of temporary promises, figures, allegories, parables, and some oversimplified doctrines, so that they can advance bit by bit to things more lofty. In this way Paul among the weak did not judge that he knew anything but Jesus Christ, and him crucified.[1444]

The scholion continues: 'Otherwise, what need was there for so many rituals in the Mosaic code? Jerome writes that these were imparted so that the people would not lapse into idolatry.' Here they reason as follows: 'If the people are not influenced by the truth, and there was need of rituals, then the rituals were deceptive.'[1445] My answer: if they were beneficial to the people earlier, why should the truth of the Gospel be set in opposition to the figures of the Law? For this truth does not make them deceptions, but adumbrations. But when shadows are compared to real bodies, they are, in a sense, untrue.

* * * * *

1439 *sortibus* 1531 *sordibus* LB; Erasmus is alluding to *Republic* 460A.
1440 Plato discusses lying as an instrument of government in the *Republic* 382C–D, 389B–C, 414B–415D.
1441 Erasmus makes use of the same analogy in the letter to Desmarez LB IV 551B–C / Ep 180:74–6.
1442 Pio *XXIII libri* 250ᵛv remarks that many, especially his disciple Aristotle, have criticized Plato's radical provisions in the *Republic*, and observes: 'Indeed, it is not the teachings of Plato that constitute our law, but those of Jesus Christ.'
1443 Pio *XXIII libri* 250ᵛv defends Plato: 'But surely not even Plato ever meant that we should lie strictly speaking, even though he concluded that it is appropriate to deceive people for their own benefit by means other than lying as when we perhaps set fables before them or present plays.'
1444 See 1 Cor 2:2.
1445 An imprecise paraphrase of Pio's argument in *XXIII libri* 251ʳv, which concludes: 'But what is the point of your suggesting that rituals were provided in the Mosaic law to fool the people, unless it would be your meaning that Almighty God gave the law to fool his own special people?'

The passage in the scholia continues: 'Moreover, what Augustine writes about lying, and what the theologians of our time teach about it, are in part such that, even if they are mostly true, they are quite repugnant to common sense.' Do my words here seem to approve of lying? Not at all. But as to the hypothetical cases they pose, for example, 'Even if one could preserve the bodies and souls of the whole human race by a single lie, however harmless, an evil however slight cannot be committed for the sake of a good however great,[1446] and so it must not be done,' and ones like it, I do not say that these are false, but I do say that they are repugnant to common sense. The Critic does not disagree with me here,[1447] and Augustine thinks that in a like case charity will not be afraid of a slight bit of sin.[1448]

But I wrote in the *Methodus*: 'Although we read "You will destroy all who lie," we also read that there were good people who were liars, like Abraham and Judith, etc.'[1449] This is how these people who hate lying, however harmless, quote me. I did not write *mendaciosos* [liars] but *mendacio usos* [employed lies].[1450] They will say that this is a slip on the part of the typesetters, but they slip all too often in quotations of my words. It is more likely that this is the fault of those on Pio's staff who had charge of the defamatory notecards. Their negligence is the responsibility of their master.

As it is, since he admits that Judith, Rahab, and certain others who are praised in Scripture made use of lying,[1451] how is it that I am the advocate

* * * * *

1446 This seems to echo a scholastic *casus conscientiae*; it is not found in Pio's discussion in *XXIII libri* 250[r]–[v]s of whether one can lie to attain a greater good or avoid a greater evil. Although earlier Erasmus cites Antonino's *Summa* approvingly (at 331 above), he does not here follow Antonino's opinion that, while dissimulation is allowable, lying is always a sin and is contrary to charity; see Antonino Pierozzi *Summa theologica moralis* (cited in n1229 above) II 1044B, 1053A–1056B.

1447 Pio *XXIII libri* 250[r]–[v]s argues that there can be no necessity to lie, and hence to sin, but he allows: 'But it also seems to suit divine goodness that if one had lied simply to save the life of another for God's sake, God would attend more to the work of mercy that was done than to the lie that had been uttered, and that the former would be more pleasing than the latter offensive, as was made clear in the cases of the midwives and of Rahab.'

1448 Both Pio and Erasmus seem to have in mind Augustine's *Contra mendacium* 15.32–16 CSEL 41 512–16.

1449 Erasmus repeats the text as quoted by Pio at the beginning of book 23 (*XXIII libri* 148[r]L); Pio discusses the issues raised by this passage on 249[v]–250[r]R.

1450 For the text of the *Methodus*, 'tamen quosdam probatos viros mendacio usos legimus' (LB V 131F / Holborn 293:7–8), Pio substituted 'quosdam tamen probos viros mendaciosos legimus.'

1451 Erasmus is referring to Pio's statement in *XXIII libri* 250[r]R: 'But as to what was written about the midwives and Rahab, they are not commended or rewarded because they

of lying and he the one who hates it? They will say: 'These things support your argument that lying is permissible.'[1452] No, they support my position that not every lie is a capital crime. He clears Abraham of lying because it was a matter of Hebrew practice that a closely related woman was called 'sister.'[1453] But here Pio himself is maintaining that speech must be fitted to the understanding of the one to whom we are speaking.[1454] For example, one who swears to a pagan by Jupiter the Stone is bound by his oath.[1455] But Abraham purposely deceived the Egyptian, knowing that he would understand 'sister' to mean a real sister. And so he has not adequately cleared Abraham of lying.[1456]

From these little fragments my friend the brawler concludes that I think that it is not only permissible to lie, according to subject and circumstance, and without sin, but even beneficial to lie.[1457] The heresy of the Priscillianists[1458] was the same as that of the Origenists. In fact, I reported what Plato thought. I never said myself that lying properly so called was free of any sin; but I do raise the question whether a harmless and trivial lie is not negligible when great danger threatens or the greatest benefit presents itself. But not even on this point do I settle anything finally.

* * * * *

lied, but on account of the work of mercy that they performed, and for this reason God forgave their lie. For one sins more lightly who lies to accomplish good than one who does so to accomplish evil.'

1452 Not in Pio save by implication; Erasmus seems here to be anticipating an objection.

1453 Pio *XXIII libri* 250ᵣR, citing Jerome

1454 Pio does not say this explicitly. Erasmus may be extrapolating this from Pio's general insistence that speech cannot be used to deceive, or from his statement (250ᵣR) that figurative expressions are clearly not lies, 'since they quite truly signify that for which they were uttered; this is recognized by those who understand them, although others who lack this capacity take them to mean something different.'

1455 See *Adagia* II vi 33: *Jovem lapidem iurare* 'To swear by Jupiter the stone' LB II 593B–D / CWE 33 306–7, where Erasmus makes this same point, basing his argument on a text of Augustine.

1456 Erasmus does not do justice to Pio's sophistic exculpation of Abraham (*XXIII libri* 250ᵣR): 'But father Abraham did not say that Sarah was not his wife, for that would have been a lie, but he said "Say that you are my sister," which was true on their father's side [Gen 20:12].'

1457 Erasmus has shifted his focus back to the beginning of book 23; these words are a virtual quotation of Pio's opening charges (*XXIII libri* 248ᵣM), which follow the quotations from Erasmus that were presented at the beginning of the book.

1458 A reference to Pio's next words after the opening charges of book 22: 'This was the most detestable heresy of the Pricillianists, as Augustine attests' (*XXIII libri* 248ᵣM); this was only one of many heterodox tenets of the Pricillianists (whom Erasmus discussed in *De concordia* LB V 478C). Pio has in mind Augustine *Contra mendacium* 2.2 CSEL 41 471.

But at this point he enters into the broad topic of how very counter-productive lying is,[1459] and he makes me out to be a defender of lying.[1460] He says: 'Who is so stupid as to say that allegories are lies?'[1461] Who was it who said this, stupid? I have indicated that the types and shadows of the Old Testament were given to the dense Jewish people so that they could advance gradually to the truth of the Gospel.[1462] Truth, as I have shown, is not always the opposite of lying.

He presents me as replying that there were very many highly esteemed persons in the Scripture who were liars.[1463] Who ever dreamed of this? Are *mendacio uti* [make use of a lie] and *mendacem esse* [be a liar] so very much the same?[1464] But it was his notecard that read *mendaciosum* that deceived him![1465] Yes, this is a fine exculpation for one who brings so many charges of heresy and does it so insolently.

I accept the account he gives of speech, which is the proper instrument provided by nature for manifesting the dispositions of the soul.[1466] He should have added 'or for manifesting them more effectively.' Now if someone deceives by writing, will it not be a lie? He will say: 'Writing has the same force as speech.' But what if the person is a mute? What if someone deceives by saying what is true, as the character Davus,[1467] is he not abusing an instrument provided by nature?

Pio should have made a distinction in the case of *fallere* [to deceive],[1468] for not everyone who deceives means to harm. A father deceives his son

* * * * *

1459 Pio XXIII libri 248ᵛM argues that lying subverts social life and commerce.
1460 Pio XXIII libri 249ᵛo and P describes Erasmus as defending lying: 'Yet against such express evidence you dare to defend lying ... Wherefore perhaps it is less than surprising that you, trained by these [orators and lawyers] defend lying.'
1461 Erasmus is paraphrasing Pio's remark in XXIII libri 250ʳR: 'For no one is so dim that he thinks that metaphorical expressions are lies.'
1462 Agreeing with a similar statement by Pio in XXIII libri 251ʳv
1463 Pio XXIII libri 249ᵛQ, where he uses the expression '... quos tamen legimus fuisse mendaces'
1464 Erasmus alludes to the correction he made earlier (355 n1450 above), changing mendaciosos in Pio's quotation from Methodus to mendacio usos.
1465 That is, Erasmus' words mendacio usos had been corrupted by the notetaker to mendaciosos, which Pio, in turn, had revised to mendaces.
1466 A paraphrase of Pio's account in XXIII libri 248ᵛM: 'words are, as the Prince of Philosophers has so wisely taught, the signs and indications of the soul's perceptions and thoughts, and of its dispositions.'
1467 The devious slave in Terence's Andria, notorious for his use of a selective presentation of the truth in order to deceive
1468 Erasmus ignores Pio's distinction in XXIII libri 248ᵛN: 'lying (mendacium) is not condemned because it deceives (fallat) the one to whom one is speaking, since to be deceived (falli) is not always harmful, but is beneficial in many cases. Therefore, though

so that he will learn his letters; he promises him a sword although he has planned better things for him. Thus did God nurse along the dense people with promises of things earthly for the purpose of leading them along finally to a knowledge of things spiritual.

The one who is writing this charges me with inconsistency. He is, of course, wholly consistent, for just as he began with lies, so too does he lie stoutly through the whole course of the work, and bring it to an end with a slanderous lie.

PERORATION[1469]

In spite of the fact that the whole work teems with countless incorrect quotations, ones that are incorrectly understood, shamelessly twisted, obviously misleading, with so much insolent abuse, and with so many tasteless witticisms, he denies that he has written anything from a zeal for contradiction or in a less than kindly spirit, etc.[1470]

Next he wants to convince me that, if any rather intemperate statements should be found here, this is unintentional, an accident.[1471] But how can something that occurs constantly seem accidental?

And I am not offended because he was excessive in quoting from my works,[1472] but because so many of his quotations were falsified, so many were incomplete, so many were without citations and could not be verified, and, finally, so many were not even understood. It is for these that he should beg pardon.

* * * * *

all lying is uttered to deceive (*ad fallendum*), not everyone who deceives is a liar, but only the one who makes a false statement with intent to deceive.'

1469 1531 UPPER MARGINS: 'Peroration' on pages 284–5 [= 358–60 in this volume]. Erasmus replies to Pio's 'Conclusion of the Work and End of the Argument, in place of book 24' (*XXIII libri* 251ʳx–252ʳz). Pio begins: 'But what do I achieve by delaying still? For the subject and the constraints of time demand that my toil finally be concluded, since we have completed the journey we planned and can, from out at sea, catch sight of the longed-for shore.' At the end Pio dates the work 'Farewell. The day before the Lord's Resurrection, in the year 1530.'

1470 Paraphrase of Pio's disclaimer on *XXIII libri* 251ʳx

1471 Pio *XXIII libri* 251ᵛ

1472 Pio *XXIII libri* 251ᵛ–252ʳz: 'In addition, if I have been, perhaps, excessive in quoting your statements and views, first bringing together those appropriate to each topic and then answering them when reviewed separately, and have wearied the reader in the process, I think that forgiveness is due. For I thought it an advantage to gather and present all your arguments and passages that asserted the same view before I replied to any one of them, because your developed view and complete meaning could better be extracted from all of them than from a single statement taken separately, since statements clarify and interpret one another.'

He is right to give thanks to God by whose aid he completed so marvellous a work.[1473] But he shows ingratitude in not thanking the Spaniard
who polished up his style,[1474] those in charge of note taking, and the theologians and monks who supplied the subject matter.

But I am delighted at his prayer that God may grant me 'always to
think and teach what is best and to say what is most beneficial, and hoped
for happiness along with these.'[1475] I, in turn, pray for Pio that he will find
the Lord a more fair and merciful judge that Pio has shown himself to be
in my case.

Now the mechanics [βάναυσοι] even summon me to sing a palinode,[1476] if it please the gods! But what is it they want me to recant? What
Pio charges me with: Wanting the order of theologians eliminated? De-
spoiling the priests of their property? Rescinding the authority of Scripture? Condemning the church's rituals and all its regulations? Defending the teachings of the Arians? Reviving the teachings of the Priscillianists and the Epicureans, and other things even more absurd? I have
never even conceived of his versions of these things, and so far he has
proven not a one of all the charges he has made so often, so strongly,
so savagely. It would be a novel kind of palinode if I were to accommo-
date people who are raving mad with the disease of slander by becoming my own false accuser. No, it is they who are singing a palinode, for

* * * * *

1473 A malevolent allusion to Pio's thanksgiving at the beginning and end of this conclusion: (251ᴿx) 'rendering boundless thanks to Almighty God who enabled us to complete the course we had begun'; (252ᴿz) 'I have written what seemed to me in agreement with the truth and conducive to piety. If I have achieved this in most instances,
I must thank Almighty God and the Lord and Saviour Jesus Christ who enabled me
to utter them; but as to statements that were less than correct, because of my ignorance, I reject them altogether. For I submit everything to the pronouncements of Holy
Mother Church and to the sounder judgment of all, especially of the professors of sacred theology.'
1474 See *Responsio* n74 above and *Apologia* n15 above.
1475 Pio *XXIII libri* 252ᴿz
1476 Pio (*XXIII libri* 251ᵛy) had assured Erasmus that he wrote 'to motivate you to review your
works critically ... and to remove from your writings those elements which provide
a multiplex occasion for endangering many and sound so terrible that they cannot be
defended by any orthodox interpretation'; but this passage seems more likely to be a
reference to Bade and others (real and imagined) who were associated in the publication
of Pio's work, which Bade had entitled *Alberto Pio's ... three and twenty books against
passages in various works of Desiderius Erasmus of Rotterdam that he thinks he should review
and correct*. Erasmus' reference to a palinode recalls Bade's remark in the letter he had
prefixed to *XXIII libri* (sig āiᵛ) in which he referred to Pio's citations of numerous suspect
and evil-sounding passages excerpted from Erasmus' works which harm the Catholic
faith and the dignity of the church and which Pio hoped he could persuade Erasmus to
retract.

they are repeating for us the behaviour of the devil, publicly and shame-lessly hurling slanders like this at one who is innocent, and either seek-ing renown or promising themselves vengeance and victory from their uncivil behaviour, all as if Christ were dead and had no care for his church.

THE END

AGAINST THE INDEX TO ALBERTO PIO, VERY BRIEF NOTES

In Elenchum Alberti Pii
brevissima scholia

64 R· Nihil impie dixit Moria. Quid aūt
recipit Pius,cui propositum erat in Eraſ
mo calumniari omnia.

F.

65 Fallere non semper est peccatum.

65 R· Quur ergo offendit ꝙ aio Chriſtū
in quibuſdā fefelliſſe diſcipulos ad tēp?

66 Funeralia & quæ in funeribus Chri
ſtianorum fiunt omnia ridet Eraſmus.

66 R· Legat dialogum qui uolet,& com
periet hæc omnia eſſe falſa,Sacerdotum
quæſtū illic noto, qui & ſepulchri locū.
& ſtrepitū campanarum & preces ſuas
uendunt. G.

67 Gratia dei præſto fuit Paulo apoſtolo
ad ſeruandū eū à lapſu in peccatum car
nis, licet illi nō abſtulerit ſtimulū carnis.

67 R· Rurſus ſomniat Paulum fuiſſe ten
tatum de libidine turpi,quaſi caro,in ſa
cris litteris nihil declaret niſi coitum.

H.

68 Hominū uitam,quam non niſi fabulā
cenſet,taxat uniuerſam Eraſmus.

68 R· Moria fabulam appellat nō ſermo
nem anilem, ſed actionē argumenti, ex
uarijs perſonis conſtantē,hoc eſt comœ
diam,

diam,quæ rectè ſpeculū dicitur humanæ
uitæ.Prodigium uero, ſi Moria dicit ui
tam humanam eſſe fabulam , quū toties
eccleſiaſtes clamet, Vitam humanam &
quicquid in ea geritur eſſe uanitatē ua
nitatum . Bona autē eſſet fabula, ſi ſuas
quiſꝗ partes decenter ageret.

Hieronymum falſò ſugillat Eraſmus. 69

R· Imò illic petulantiſſime debacchat 69
Pius,cū ſua matula.Quaſi theologi,nūc
nō multa reijciāt ab Hieronymo tradita.

Hieronymo irreuerenter, & falſò uir 70
ginitatem detrahit.

R· Qui falſò,quum ipſe fateatur?Nec 70
id uno in loco. I.

Ieiunium & ciborum delectū ab Eraſ 72
mo irriſum,commendat Pius.

R· Ieiunia Chriſtiana nunquam uerbo 72
taxat Eraſmus.De delectu ciborū diſpu
tat,nō tamen irridet, niſi accedat ſuper
ſtitio.Pius putat omne ieiunium conuer
ti cum abſtinentia carnium.

Imagines ſanctorū uenerari licet, con 75
uenit,decet,& expedit.

R· Quanta ſolicitudo de ueneratione 75
imaginū,quum uix explicari poſſit,quō
n 3 ueneratio

Erasmus *Brevissima scholia* (Basel: Froben 1532)
signatures n2^{v}–n3^{r}
By permission of the Folger Shakespeare Library, Washington, DC

AGAINST THE INDEX TO ALBERTO PIO, VERY BRIEF NOTES BY ERASMUS OF ROTTERDAM

I cannot keep from grieving with all my heart whenever I ponder the extent of malice to which the characters of Christians have degenerated. For now one wants it considered an act of devotion, please God, to denounce one's neighbour before the uneducated masses with out-and-out slanders, lies, and buffoonish abuse, while the whole criminal business is carried out under the guise of that best of enterprises, religion. Not the least part of this poison is imparted to titles and indexes, because most people read only them. What devil inspired this contrivance in the minds of those who want to be considered the teachers of holiness? I have written a brief answer to the slanders of Alberto Pio on account of his renown, this in spite of the fact that they deserve no reply. But an index was sent to me afterwards.[1] I do not know whose work it is, nor does it matter, but whoever the author was,[2] he behaved disingenuously. I have made notes on a few items to make this more apparent.

Notwithstanding that a large part of the book is an attack on Luther, the first part of the title on the first page displays the name of Erasmus.[3] Here you have the first indication of a Christian mind at work.

* * * * *

1 See *Apologia* n3 above.
2 See Introduction cviii–cix.
3 This complaint is a restatement of one Erasmus had made earlier in the *Responsio* 3–4 above.

A⁴

1   Erasmus is wrong in writing that the works of Abraham and the sanc-
     tification of the Sabbath were despised, the former by the Apostle, the
     latter by the Lord.⁵ [232G]
     REPLY: Instead of 'Sabbath' he used the appealing phrase 'sanctifica-
     tion of the Sabbath' to make the matter more hateful, and instead of
     'considered less important,' the more moderate expression, he used
     the more hostile 'despised.' We correctly consider something less im-
     portant by comparison to what is better, but we do not, for all that,
     despise it.

2   The Apostle nowhere criticizes the works of Abraham, as Erasmus
     maintains again, but even more openly.⁶ [233N]
     REPLY: Error begets error. As if I said that the works of Abraham
     were criticized by God! But according to Paul God does not account
     them for righteousness; this is, in a way, to consider them less impor-
     tant, to regard them as of less value.

3   Erasmus ungratefully criticizes Aldo Manuzio because he revised his
     Grammar repeatedly.⁷ [74D]
     REPLY: I do not criticize him, but Folly reports, by way of a jest, what
     Aldus told me himself, namely that he revised his *Grammar* nine times,
     and that was no disparagement of Aldus. So far as ingratitude goes,
     Aldus owed me a lot more that I owed him. A harmless jest of Folly
     is distorted for the purpose of slander, but he ignores the fact that I
     praise Aldus, pulling out all the stops, as the saying goes,⁸ in countless
     passages, particularly in my treatment of the proverb *Festina lente.*⁹

* * * * *

4 In each of the scholia Erasmus quotes the entry from the Index to Pio's *XXIII libri* that
   he wishes to refute and then gives his response. At the end of the quoted Index entries
   we have inserted in brackets the folio number and section letter as given in the Index
   to Pio's *XXIII libri*. We have also provided footnotes that refer to the relevant passages
   in Erasmus' *Responsio* and *Apologia* (citing the page numbers of this volume) where the
   reader will find notes that cite or even quote the passages in *XXIII libri* to which the
   Index entries refer.

5 See 345 nn1386, 1389.

6 See 345 n1386; Erasmus' opening rejoinder here is puzzling, for Pio does not claim that
   God criticized the works of Abraham. Perhaps Erasmus is lampooning the contrived
   word order of item 1 above.

7 See 161–2 n314.

8 See *Adagia* I v 96: *Apertis tibiis* 'Using all the holes (With all the stops out)' LB II 219A–B
   / ASD II-1 565–6 / CWE 31 466–7.

9 *Adagia* II i 1 LB II 397C–407D, in particular 402C–403B / CWE 33 3–17, 9–10

4   Ambition is not always associated with sin, and it was not so in the
     Virgin Mary.[10] [168EF]
     REPLY: As if it were I who attributed a sinful ambition to the most
     holy Virgin! I report what others have attributed, but, even so, I do
     not agree with them.

5   Erasmus ridicules the anathemas and excommunications issued by
     bishops.[11] [82z]
     REPLY: Reader, please note here not only the malice but also the cun-
     ning of the fellow. Folly jokes about the *ex voto* offerings [*anathemata*],
     usually candles, which are put up in churches according to Italian
     practice. He, losing the way altogether, criticizes making fun of ex-
     communications by bishops.

6   Erasmus complains wickedly about the exacting of the so-called an-
     nates, and about many burdens being imposed on the church of Christ
     by the pope, whereas the actual situation is quite different.[12] [205E]
     REPLY: It is not I who complain, for these things are no burden to me.
     Rather, I say that the complaints of others are heard, and this situation
     is long-standing and no secret. Let others decide whether the annates
     are just or not.

7   The apostles and evangelists uttered or wrote nothing influenced by
     a lapse of memory.[13] [179I]
     REPLY: I have never written or thought this, and yet Augustine writes
     that this could have taken place, and almost thinks that it did.

8   The apostle Paul distorted nothing in the Scriptures, and employed no
     cunning, and made no statement contrary to what he really thought.[14]
     [179K]
     REPLY: But Chrysostom and Jerome and even Ambrose frequently re-
     mark on this. Nor is it remarkable if the Apostle did this in the process of
     adapting himself to the catching of men, since Christ did the same thing.
     Furthermore, this is not a question of deceit but of accommodation.

9   Nor did he employ cunning against the Athenians when he said that
     he had seen an altar inscribed 'To the Unknown God' [Acts 17:23].[15]
     [180L]

     * * * * *

10 See 247 n841.
11 See 178 n428.
12 See 306 n1169.
13 See 273–4 n992.
14 See 272 n981.
15 See 272 n981.

REPLY: Let him quarrel about this point with Jerome and not with me. I followed the great Doctor of the church.

10 Pio replies to the arguments made by Erasmus against the veneration of images.[16] [381KL]
REPLY: Erasmus nowhere rejects this, but he does ridicule the superstitious cult of images.

11 Pio refutes Erasmus' arguments whereby he maintains that God is to be invoked, and not the saints.[17] [157O]
REPLY: Erasmus does not maintain this anywhere, though he thinks that it is more perfect and more in agreement with the Scriptures to ask all things of Christ.

12 Arguments that beneficial lying is licit, and their refutation[18] [249QR]
REPLY: I have never praised beneficial lying as a virtue among Christians. Moreover, I understand lying, in an extended sense, to embrace even pretending and dissimulation that are accomplished by means of actions or by words that are not understood.

13 Bishops are permitted to own aggressive armaments like swords, missiles, cannon, etc, so long as they use them properly.[19] [244E]
REPLY: Why not? They are also permitted to be killed in action, as happened to some bishops of Liège and Utrecht![20]

14 Erasmus has incorrectly stated that the Arian heresy should be called a faction and schism rather than a heresy, and he is incorrect in saying that its defenders were equal in number to the Catholics and superior to them in eloquence and learning.[21] [181P]
REPLY: Both are true. It is Chrysostom who admits the first and Jerome who writes the second. I term it a faction because the affair was conducted by the favour and weaponry of monarchs rather than on the basis of the Scriptures. Indeed, I still think that neither side was adequately understood by the other.

15 Erasmus makes fun of the professors of liberal arts in the *Folly* by calling them *logodaedali*.[22] [73C]

* * * * *

16 The Index refers to *XXIII libri* 138^r–139^{r}IKL; Erasmus mentions these arguments on 234–5.
17 See 242.
18 See 352 n1430.
19 See 171 n379, 172 n389, 351 n1419.
20 Thibaut van Bar of Liège, who died in battle in 1312, and Willem van Mechelin of Utrecht, who died while fighting in 1301; see Pius Bonifacius Gams *Series episcoporum Ecclesiae Catholicae* (Regensburg 1873–86) 249, 255.
21 See 276 n1008.
22 The Index cites Pio's quotation (*XXIII libri* 73^{v}C) of *Moria* LB IV 433C / ASD IV-3 110:717 / CWE 27 106, where the term *logodaedali* 'verbal wizards' occurs.

REPLY: Come now! What is abusive about *logodaedali*, which means artificers of language. Is it any wonder that Folly calls them *logodaedali* in jest when Virgil [*Aeneid* 7.282] calls the goddess Circe *daedala*.[23]

16   Erasmus' allegation that Augustine had two concubines is false, and his charge that he had one when he was still a pagan is impudent, for once he became a Christian and a bishop he lived chastely.[24] [224Z]
REPLY: Here the index-composer [ἐλεγχογράφος] is simply raving mad, along with Pio. Augustine himself in the sixteenth chapter of book 6 of the *Confessions* reports that before his baptism he had two concubines; the first of these, from whom he had received a son, bound herself by a vow of continence. Who does not know that he lived quite chastely after baptism? But he cleaved to this woman alone! Certainly, for he was faithful to his concubine as if to a wife, as he himself tells; but, as I have said, when she left he took himself the second concubine. Is what Augustine states about himself false? Oh, but it must be emphasized that he did this when he was a pagan! As if there is anyone who did not know this. Yes, but the matter should be kept quiet for the sake of his good name! How is one to keep quiet what he himself wanted everyone to know by publishing his book.

17   Erasmus makes fun of courtiers.[25] [80s]
REPLY: Not Erasmus, but Folly! What is left but for all courtiers to stone Erasmus?!?

B

18   Eternal blessedness is due to no one as merited because of his good works, but it has been promised by God out of his generosity as a reward for good works.[26] [236AB]
REPLY: But it is Béda[27] who is energetic in denying this absolutely.

19   He proves that Christians can justly make war on the Turks, against Erasmus' position.[28] [242TV]

* * * * *

23 See *Adagia* II 3 62: *Daedali opera* 'The works of Daedalus' LB II 509A–510A / CWE 33 168–9, where Erasmus mentions the expression *logodaedali* (see its pedigree in TLL sv) and refers to the Virgilian passage cited here.
24 See 326 n1271.
25 See 168 n359.
26 See 341 n1364.
27 See Béda *Annotationum ... in Desiderium Erasmum Roterodamum liber unus ...* (Cologne: Petrus Quentell 1526) 180ʳC, 207ʳv–208ᵛz, 215ᵛT, 265ʳD, 269ᵛP, 275ᵛN.
28 See 350 n1415.

REPLY: Not against Erasmus, who nowhere maintains this, so long as the war is undertaken and waged for just causes and sound reasons, for I call a war undertaken otherwise a war undertaken rashly. 'Who,' says Pio, 'starts war rashly?' Who has looked for water in the sea?[29] A more likely question would have been 'Who starts war properly?'

C

20   Erasmus considers marriage superior to celibacy, and thinks that celibacy should be abolished and that priests should be allowed to marry.[30] [216A, 219H]
REPLY: Erasmus nowhere maintains this absolutely, but he does wonder in some places whether it might be more tolerable to allow priests and monks to take wives than to have so many unchaste celibates. Nor do I think that celibacy should be abolished, but I set forth for the consideration of the leaders of the church whether it is more tolerable to allow them to marry or to suffer such a vast Lerna of filth.[31]

21   The law that celibacy must be observed by those in major orders is not to be abolished, nor is it an evil rule, even though Erasmus maintains both of these.[32] [222S, 223V]
REPLY: This too is a twofold lie. In fact, I admit that celibacy was the correct temporary measure; I leave the question of its abolition to the judgment of the church.

22   Pio does not much approve of the use in church of *cantus perfractus*, what they call *discantus*.[33] [37Z]
REPLY: But the fellows of the Sorbonne[34] are very much in favour of it. See the agreement of the censors on this point.

23   Erasmus' assertion, 'Wherever there is charity, there is no need for law,' is false.[35] [207C]
REPLY: But this has seemed to be absolutely true to the most holy and learned men. Not even Thomas Aquinas, the *doctor scholasticus*,

* * * * *

29 See *Adagia* I ix 75: *In mari aquam quaeris* 'You seek water in the sea' LB II 360B–C / ASD II-2 392 / CWE 33 221.
30 See 320 n1240, 323 n1260.
31 The swamp of Lerna, home of the many-headed Hydra of Hercules' labours; see *Adagia* I iii 27: *Lerna malorum* 'A Lerna of troubles' LB II 122D–F / ASD II-1 338–40 / CWE 31 258.
32 See 325 n1266.
33 See 30 n144, 219 n669.
34 See *Declarationes ad censuras Lutetiae vulgatas* LB IX 899E–900C, where the censures of the Sorbonne are quoted.
35 See 310 n1186.

denies this, so far as the coercive force, not the directive force, of law is concerned, and he maintains in different words the same thing[36] as the ancients whom I have followed.

24   Erasmus is speaking derisively when he says that Christ always approved of fools and that the founders of religious orders were fools, and the like.[37] [83D]

REPLY: It is Folly, not Erasmus, nor does she speak so impiously as this slanderer. Paul says: 'God has chosen the foolish things of the world'[38] and 'The foolishness of God is wiser than men.'[39] Distinguish the types of foolishness, and Folly's words are devout.

25   His report of why the foolish pleased Christ so very much is a mockery of Christ, as if he too were a fool.[40] [83B]

REPLY: Mockery of Christ is the work of people like you, you shameless slanderer! Moreover, Christ was a complete fool in the eyes of the world.

26   Christ was not allowing his disciples a temporary error when they assumed that Christ himself was ignorant of when the Day of Judgment would come in the same way that the rest of men were. But this is what Erasmus has stated.[41] [180M]

REPLY: What Erasmus has stated is quite true, and not here only. Erasmus, however, has never imagined that Christ did not know the day in the way that the rest of men did not know. No, Pio added this element on his own to create an opportunity for slander.

27   Erasmus is incorrect in saying that Christ, according to some hidden plan, reminds one of a kind of Proteus because of the inconsistent character of his life and teaching, whereas nothing could be more straightforward than he.[42] [195C]

REPLY: Paul too was straightforward in this way, and yet he became all things to all men, following, of course, the model of Christ. This inconsistency does not contradict singlemindedness.

* * * * *

36 Thomas Aquinas *Contra gentiles* 3.128 in *Opera omnia* Leonine ed 14 (Rome 1926) 393: 'The first, then, are a law unto themselves (Rom 2:14), possessing charity, which in place of a law inclines them and causes them freely to act. So it was not necessary that external law be established on their account, but on account of those who were not of themselves inclined towards the good.'
37 See 179 n434.
38 1 Cor 1:27
39 1 Cor 1:25
40 See 179.
41 See 274 n996.
42 See 300 n1131.

28 Because all things belong to Christ, the Turks, his foes, cannot justly hold property.[43] [243x]

REPLY: Well, then, do we plunder everything straightaway? Even as he lay dying he gave unlucky advice.

29 Erasmus disparages auricular, secret confession to a priest.[44] [227G]

REPLY: This is a lie. In fact he teaches how it is done properly in both ways.

30 Erasmus errs when he says that auricular confession did not exist in the time of St Jerome.[45] [227H]

REPLY: It was not the sort that exists now. Erasmus nowhere states that it did not exist at all.

31 What words Erasmus uses to depreciate the rules of the church.[46] [206A]

32 There are many types of ecclesiastic rules and laws. And it is, therefore, unsuitable for Erasmus to censure them all with the same degree of criticism.[47] [207B]

REPLY: Both [31, 32] are absolutely false. For he neither scorns the rituals received by the church, nor condemns all without distinction, but in some places expresses his clear approval. What he criticizes is reliance on them and superstition.

33 If, in fact, those in charge of the church are more concerned with the pursuit of gain than with the glory of Christ, it is an abuse, and Pio himself does not approve of it either.[48] [209G]

REPLY: But it is only such as these whom Erasmus, the object of Pio's carping, holds up for criticism.

34 The new regulations are not in conflict with the old ones.[49] [208E]

REPLY: I admit that this is true about some, but I say it is not true of most of them.

35 No one would approve the whole life of Christians being burdened by regulations of human origin, though Erasmus suggests that this is the case.[50] [210N]

* * * * *

43 See Pio XXIII libri 243^{v}x and 350 above.
44 See 337 n1340.
45 See 337 n1342.
46 See Pio XXIII libri 206^{r}A and 308 n1178 above.
47 See Pio XXIII libri 207^{v}B.
48 See 308; the Index cites Pio's remarks on XXIII libri 209^{r}F.
49 The Index cites XXIII libri 208^{r}C.
50 See 311.

REPLY: In fact, all who obtain revenues or tyrannical dominion from this applaud, and those who really love the house of God complain.

36 Erasmus ridicules the approved customs, rites, and rituals of the church.[51] [82Y]

REPLY: An out-and-out lie, since in this area what Erasmus holds up for criticism is the trafficking in indulgences, something of which not even the theologians approve.

37 It is incorrect for Erasmus to raise the difficulty of perpetual continence as an objection against the rule of celibacy.[52] [222S]

REPLY: As if every celibate were chaste, or as if many people could maintain something so easy![53]

38 Pio admits the difficulty of maintaining perpetual continence.[54] [223T]

REPLY: But Béda denies it; Josse [Clichtove], however, calls it an arduous undertaking.[55]

D

39 He denies absolutely that he harmed Erasmus' reputation among the purple-clad fathers in Rome.[56] [3C]

* * * * *

51 See 175 n406.

52 See 323.

53 As Erasmus noted in his colloquy *Concio sive Merdardus* 'The Sermon' of September 1531 (LB I 855F–856A / ASD I-3 663:360–2 / CWE 40 950): 'How often has it been protested to them that a "celibate" is one who has no wife even if he keeps six hundred concubines. And yet to them celibacy is no different from continence and chastity'; see his comments at 206 and 323.

54 The Index cites Pio's remarks in *XXIII libri* 223^{r}T.

55 The censures of the Sorbonne state: 'Potest namque quilibet, qui ad continentiam sua sponte se devinxit, cum Dei adjutorio se contenere'; see *Declarationes ad censuras Lutetiae vulgatas* LB IX 902D–E. Erasmus felt that the censures reflected Béda's views; eg Allen Epp 1902:260–4, 287–90; 1909:80–8, 135–8, 185–7, 242–3. For Josse Clichtove's views that priestly celibacy can be observed and that the church should continue to require it, see his *Propugnaculum ecclesiae adversus Lutheranos* (Cologne: Hieronis Alopecii 1526) book 2 'Continentiam sacerdotum contra eius improbatores defendens' 195–405 especially 269–81 (the pope cannot grant a dispensation to a priest from his obligation of celibacy because it is *de jure divino*); 298–313 (given that many clerics do not observe celibacy, they should be punished and taught how to discipline their sexual drive, eg avoid the company of women, guard one's senses and thoughts, mortify one's body, pray, etc); 341–7 (if one practises mortifications, continence will not be difficult). Erasmus is thus less than straightforward in citing Clichtove's views as being similar to his own.

56 The Index refers to Pio's *Responsio* as printed in *XXIII libri* 3^{v}C; the text occurs in the same location (3^{v}C) in Bade's earlier printing (Paris 1529).

REPLY: But I discovered this from the letters of many people. But he denies everything.

40 He ridicules dialecticians and dialectics.[57] [74G]

REPLY: What an outrage! Sentence Folly to hard labour in the quarries.

41 Erasmus reviles the liberal arts.[58] [73C]

REPLY: No, it is Folly; let her be stoned as a poisoner.

42 Ecclesiastical dispensation in the case of non-solemn vows is duly given for the reasons stated.[59] [214T]

REPLY: But the pope does give dispensations in the case of solemn vows, and some of the canonists and theologians (among them Cardinal Cajetan)[60] say that this is correct. And it is the judgment of many that the pope is acting piously if he restores to their original status any boys or youths who were inducted into an unsuitable way of life by force, fear, or subterfuge.

43 Ecclesiastics are allowed to possess wealth.[61] [81x]

REPLY: Why not? Let them have wealth beyond that of Croesus[62] for the adornment of the church. Otherwise not at all.

44 Whether divorce can be granted for any cause other than fornication, in such a way that the divorced woman may marry another, as Erasmus maintains.[63] [225C]

REPLY: It is untrue. Erasmus nowhere maintains this; he only discusses it, referring judgment to the church.

45 It is not inappropriate that the pope, the bishops, and the other authorities of the churches are called 'lord.'[64] [203R]

* * * * *

57 See 165.
58 The implication of Pio's criticism (60 n287), to which Erasmus replies in part at 161.
59 See Pio XXIII libri 214ʳT.
60 See Cajetan's De dispensatione matrimonii in occidentali ecclesia (1505) in his Opuscula omnia (Lyon: Apud haeredes Jacobi Iunctae 1567) 1.27.122 cols a:81–b:16. When another Dominican tried to defend the proposition that the pope can dispense a priest, bishop, or archbishop from the obligation of celibacy so that he may legitimately marry, the Faculty of Theology of the University of Paris ruled on 2 July 1524 that the friar should be punished for his boldness, that his position could bring no little harm to the church at this time, and that no one should hold that the pope can dispense a priest so that he may contract a true marriage. On 1 August 1528 it made a similar ruling regarding dispensing a deacon. See Farge Registre 34 number 29D, 196–7 number 240E.
61 See 171–4.
62 See Adagia I vi 74: Croeso, Crasso ditior 'As rich as Croesus or Crassus' LB II 251D–E / ASD II-2 100 / CWE 32 50–1.
63 Erasmus did not respond specifically to Pio's discussion of divorce (XXIII libri 225ʳ⁻ᵛc); his response to this section of Pio's argument is at 336–7.
64 See 303–4.

REPLY: Certainly, call them 'kings of kings'! But Jerome says: 'Let them bear in mind that they are bishops, not lords.'

46 Erasmus disapproves of gifts to churches.[65] [82A]
REPLY: The Index is lying; rather, Folly holds only the superstitious up for criticism.

47 Even though he had no prior acquaintance with him, Pio wishes Dorp well because of his vigorous admonition to Erasmus.[66] [56L]
REPLY: Even as now the flocks of monks offer prayers[67] for the well-being of Pio's soul because of his vigorous raving against Erasmus. But Dorp returned to grace, while Pio died in the midst of his slanders.

E

48 Erasmus ridicules even the approved rituals of the church.[68] [82Y]
REPLY: It would have been enough to put down this lie just once, as also the following one.

49 Erasmus would be more just in complaining that alms are misappropriated for the luxurious living of bishops, for their keeping hounds and falcons, and maintaining retinues, than that they are misappropriated for the adornment and building of churches.[69] [132X]
REPLY: Erasmus nowhere expresses disapproval of the building of churches, where they are needed, and of their ornamentation with suitably restrained decoration. The Index is engaging in slander.

* * * * *

65 See 224.
66 The Index refers to Pio's marginal note at *XXIII libri* 56ʳl, where Dorp's name is mentioned in Erasmus' *Responsio* (59 above): 'I had not before this heard the name of Dorp, but I wish him well, for he did this with good reason.'
67 The 'monks' who principally prayed for Pio's soul were the Franciscan friars whose habit became his shroud and whose church in Paris his final resting place. Their affection for him was probably based more on his benefactions to them (he hosted at his own expense in Carpi their general chapter of 1521) and family relations (his brother Teodoro was a Franciscan bishop) than on his literary attacks on Erasmus. See Guaitoli 'Sulla vita' 232, 307; Eubel III 248. In his own city of Carpi Pio had the Franciscan church of San Nicolò di Bari and the monastery attached to it renovated and beautified at his own expense. So too their smaller church of the Beata Vergine della Rosa; see Adriana Galli 'Chiese e conventi come elementi di organizzazione urbana fino al XVII secolo' in *Materiali per la storia urbana di Carpi* ed Alfonso Garuti et al, Quaderni dell' Assessorato ai Servizi Culturali del Comune di Carpi (Carpi 1977) 71–88 especially 72–3.
68 See 175 n406.
69 The Index refers to Pio's remarks at *XXIII libri* 132ʳx, a passage not discussed by Erasmus.

50   Erasmus is unfair in criticizing the bishops because they have abun-
     dant sense [sic; the Index reads sensu, an obvious typographical error
     for censu 'wealth'].[70] [81x]
     REPLY: I wish they might all have abundant sense instead of wealth,
     and that the Index slipped only this once.

51   Erasmus diminishes the power of bishops and other authorities in the
     church.[71] [186A]
     REPLY: He does not diminish it; rather, he laments the corruption of
     some, and leaves the good ones alone.

52   Bishops, even if they are evil and corrupt, should not be impudently
     and thoughtlessly reviled by a private person like Erasmus.[72] [186B]
     REPLY: Jerome's frequent criticisms of them are neither thoughtless
     nor impudent, and he was a private person just like Erasmus. Pio,
     twice made a private person[73] himself, does the same thing in this
     work. One who talks about a class is criticizing vices, not individuals.

53   If Erasmus' excuse is that he lacks sufficient resources to attack Luther,
     he rejects it, since Luther's teachings are so impious and stupid.[74] [8L]
     REPLY: It would be quite easy to refute Luther if he accepted scholastic
     teachings, and the decretals, regulations, and constitutiones camerae.[75]
     But in so easy an enterprise Pio failed, and he is a man learned beyond
     human capacity! Even so, none of the Lutherans thought it worthwhile
     to reply to him.

54   If the pope approved of Erasmus' writings, he did not do so because
     of their content, but because of their elegant style.[76] [60D]
     REPLY: Shameless! In fact, the pope stated the reason for his approval.
     He says nothing there about elegance of language, but about the ben-
     efit to Christendom.

55   Many theologians who do not know Erasmus have observed that he
     is acting in collusion with Luther.[77] [50S]
     REPLY: They did not observe this; rather, they either dreamed it up or
     believed the people who keep on falsely declaring it. The books pub-
     lished on each side give evidence of the extent of our collusion, as does
     the remarkable good will with which Luther's friends persecute me.

* * * * *

70 See 171.
71 See 290–8.
72 See 170 n375.
73 See Introduction xxviii and Apologia 117 n50.
74 See Responsio 91–4.
75 See 184 n460.
76 See 58 n278, 143 n201.
77 See 36, 84 n417, 99 n493.

56   Erasmus is wrong when he says that, had he been urged, he would
      have done his duty; in fact he did not do so, although he was urged
      repeatedly, even by the pope.[78] [63C]
      REPLY: This too is an out-and-out lie.

57   Erasmus has written nothing against Luther specifically.[79] [64M]
      REPLY: A neat loophole for escaping from an obvious lie. Pio did
      not write anything against him specifically either, and yet he slew the
      Giant.

58   Erasmus' statements against the veneration of the images of the saints,
      and a criticism of them[80] [end of 133, 134B]
      REPLY: He criticizes, not the veneration, but the superstition.

59   Erasmus' derogatory remarks about new developments in theology
      and about the theology now in use[81] [172L]
      REPLY: No devout person disparages scholastic theology. Rather the
      personalities of certain people and their meddlesome questions are
      held up for criticism. One hundred years ago John Gerson complained[82]
      that theology had been reduced to sophistry. What would he say if he
      were alive now?

60   Erasmus' statements about reducing the power of the pope[83] [194A]
      REPLY: Not the legitimate power bestowed by Christ, but the tyran-
      nical power usurped by certain people

61   The reason, as it seemed to Pio, for Erasmus' great hatred for the
      Roman pontiff[84] [205Y]
      REPLY: You wretch, who hates the pontiff? All who sincerely love the
      Bride of Christ hate the behaviour of certain parties.

62   Erasmus' statements on the nullification of vows[85] [210O, 211P]
      REPLY: In fact he says that ordinary vows should not be made, or that
      they should be nullified if they were made rashly.

63   Contradictory statements by Erasmus[86] [222 after S]
      REPLY: They are not contradictory; rather, Pio has misunderstood. For

        * * * * *

78 See 34 n158, 130 n121, 150 n247.
79 See 128 n117.
80 See 233 nn758–61.
81 See 255 n892.
82 See Gerson's *Contra curiositatem studentium* of 8–9 Sept 1402 in *Œuvres complètes* ed P.
    Glorieux (Paris 1960–73) III 224–49; and Christoph Burger *Aedificatio, Fructus, Utilitas:
    Johannes Gerson als Professor der Theologie und Kanzler der Universität Paris* Beiträge zur
    historischen Theologie 70 (Tübingen 1986) 120–5.
83 See 298–9 nn1125–6 etc.
84 See 305 n1163.
85 See 313–15 nn1204–14.
86 See eg 348 n1401.

the 'goad of the flesh' [2 Cor 12:7] was not the temptation of lust, but an affliction by which Paul was buffeted from every side.

64 Pio does not accept Erasmus' explanation for the impious statements in *Folly*.[87] [84F]

REPLY: Folly said nothing impious. But, since it was his plan to slander everything in Erasmus, what does Pio accept?

F

65 Deception is not always a sin.[88] [248N]

REPLY: Why then is he scandalized that I say that Christ temporarily deceived his disciples in certain matters?

66 Erasmus makes fun of funeral rites and all the things that take place in the funerals of Christians.[89] [132Y]

REPLY: Let anyone who wishes to read my dialogue, and he will find that all these things are false. There I criticize the moneymaking of the priests who are selling the burial site and the tolling of the bells and their prayers.

G

67 The grace of God was there for Paul the Apostle to save him from falling into a sin of the flesh, though it did not remove the 'goad of the flesh' from him [see 2 Cor 12:8–9].[90] [222 after s]

REPLY: Again he imagines that Paul was tempted by a base lust, as if the word *flesh* in Scripture means nothing but sexual intercourse.

H

68 Erasmus derides all of human life, which he thinks is nothing but a *fabula*.[91] [81T]

REPLY: By *fabula* Folly means not an old wives' tale [1 Tim 4:7],

* * * * *

87 The reference given by the Index for this item, 84F, cannot be correct. Item 74F is more likely, for this falls within book 3, which is devoted to the *Moria*; but the defence of jurists taken up there does not correspond to the term *impie* used in this entry.

88 See 352 n1429.

89 See 229 n739.

90 See 325 nn1268–90.

91 Pio XXIII *libri* 81ʳᵀ quotes *Moria*: 'Now, what else is the whole of life of man but a sort of play?' (LB IV 428C / ASD IV-3 104:599 / CWE 27 103).

but the working out of a plot by a variety of characters,[92] that is, a comedy, which is rightly called a mirror of human life. But when Folly calls all human life a *fabula* it is monstrous, despite the fact that Ecclesiastes so often cries out that human life and its activities are the vanity of vanities [Eccles 1:2 and passim]! But it would be a good play [*fabula*] if everyone played his part in a becoming way.

69 Erasmus mistakenly scoffs at Jerome.[93] [183T]

REPLY: In fact it is Pio who is raving there in the most wanton way with his chamber pot! As if the theologians do not now reject many things taught by Jerome!

70 He is irreverent and false in denying Jerome virginity.[94] [222 after z]

REPLY: How false, when he himself admits it? and not in just one place!

I

[71 Missing]

72 Pio praises fasting and dietary observances, which Erasmus ridicules.[95] [84A through whole section]

REPLY: Erasmus never speaks a word in criticism of Christian fasting. He discusses dietary observances, but does not ridicule them unless there is an admixture of superstition. Pio thinks that fasting of every kind amounts to abstinence from meat.

73 It is permissible, fitting, proper, and wholesome to venerate the images of the saints.[96] [136G]

REPLY: What anxious concern over the veneration of images, although how veneration applies to an inanimate object can scarcely be explained!

74 Erasmus is wrong to argue that, because the Jews were not allowed to make and venerate images, Christians are likewise not allowed to do so.[97] [140C]

REPLY: Erasmus nowhere draws this conclusion. And if he did reach this conclusion, it would not be absurd.

* * * * *

92 Similar to Erasmus' definition at 57 n271
93 See 288 n1064.
94 See 377 nn1273–5.
95 The Index cites *XXIII libri* 84[r-v], the beginning of book 4; see 182 n454.
96 See eg 234 n764.
97 The Index cites Pio's argument in *XXIII libri* 140[v]0–150[r] incorrectly as 'cxxxx.c.'

75 Erasmus is incorrect in maintaining that no oath is allowed to Christians, as if forbidden by Christ.[98] [244A]

REPLY: Erasmus nowhere maintains that it is absolutely not permitted, but he does deny that there is room for swearing among the perfect. Everyone agrees that Christ forbade all oath taking. But the knot is untied variously by different people.

L

76 Pio shows from the words of the Apostle about the law of circumcision how it is true that we are subject to the Law, something that Erasmus denies.[99] [237HI]

REPLY: He is subject who does what the law prescribes out of fear of punishment, but real Christians are motivated by the spirit of freedom, not by fear.

77 It is not likely that the apostles carried books around with them.[100] [193P]

REPLY: But this is an established fact in the case of Paul,[101] and what danger would follow from our believing the same about the rest? They did not carry Aristotle and Averroes around with them,[102] as now do some Franciscans who advertise themselves as 'observant.'

78 A knowledge of many languages is not required for a theologian, though it is often useful.[103] [64v]

REPLY: But, of course, the philosophy of Aristotle is necessary! And yet the fact that a knowledge of languages is necessary is proven if only by the countless mistakes made by theologians which have no other cause than an ignorance of languages.

* * * * *

98 See 351 n1423.
99 See 346–7 nn1395, 1399.
100 Pio XXIII libri 193ᵛP asserts that the apostles did not carry around books because they had Christ and the Holy Spirit to instruct them. This is an element of Pio's discussion of clerical property that Erasmus ignores in his reply (see 293).
101 See 2 Tim 4:13; see Erasmus' annotation on this passage quoted at 296 n1108.
102 See Erasmus' discussion of these philosophers at 261–2. The majority of Franciscan theologians studied, taught, and wrote on the teachings of John Duns Scotus; see John Moorman A History of the Franciscan Order from Its Origins to the Year 1517 (Oxford 1968) 534–9. In the faculty of arts at Paris Averroes furnished the traditional commentary on the texts of Aristotle that were the basis of the university curriculum; see Augustin Renaudet Préréforme et Humanisme à Paris pendant les premières guerres d'Italie (1494–1517) 2nd ed rev (Paris 1953; repr Geneva 1981) 59–60, 92. Carrying around their books may indicate, not an inordinate love of non-Christian learning, but a desire for a solid philosophical basis on which to do Christian theology in the traditional scholastic style.
103 See 96 nn478–9.

M

79 Erasmus diminishes the titles of Mary.[104] [162s]
REPLY: They make no diminution who do not assign those that are unsuitable.

80 Mary is not venerated extravagantly or excessively by the devout, nor is it unsuitable to sing her praises towards evening.[105] [166AB]
REPLY: Not by the devout, but by the superstitious, and it is they whom Erasmus criticizes.

81 He clears Mary of the charges which Erasmus makes on the basis of Chrysostom and Augustine.[106] [168E, 169F]
REPLY: Erasmus does not make charges against Mary, but he reports the statements of others, although, for all that, he does not agree with them.

82 Christ's reply to Mary at the wedding feast had no harshness.[107] [170G]
REPLY: Erasmus did not say this anywhere either; rather, he says that the reply was less than pleasant. Ambrose calls it a reproach, Chrysostom a warning not to try anything similar in the future.

83 Erasmus maintains that matrimony had not been received among the sacraments of the church at the the the time of Jerome, and that it can be dissolved, and not only on account of fornication, in such a way that after divorce either partner can be allowed to marry.[108] [224A]
REPLY: The first is likely, the second is false. Erasmus nowhere maintains this, but brings it up for discussion.

84 Erasmus argues that marriage should be preferred to virginity.[109] [218D]
REPLY: He argues this, but in a fictitious declamation, one coloured by its particular circumstances. But Erasmus has never even dreamed that marriage is absolutely preferable to a virginity vowed of one's free will out of a love of piety.

85 The ancient Christian Fathers would not have forbidden marriage even if Paul had not commended it [1 Cor 7:9, 25-8, 38], since the Lord himself adequately commended and expressed approval of it [Matt 19:4-12].[110] [218F]

* * * * *

104 See 244 n825.
105 See 250 nn861-4.
106 See 247 nn841, 843.
107 See 253.
108 See 334 nn1325-6.
109 See 322 n1253, 335 n1327.
110 See 320 n1242, 323 n1256.

REPLY: He is a prophet too! But Paul attests [1 Tim 4:1–3] that such people would appear, and he did not lie.

86 Erasmus slanders the church when he says that it holds that marriage can by contracted by consent alone.[111] [226E]

REPLY: He is not slandering, but discussing, the matter, and it is not any sort of consent that is at issue, but legal consent.

87 Erasmus argues that a trivial or beneficial lie is not a sin, as his writings show.[112] [249L]

REPLY: Erasmus does not maintain even this anywhere. To deceive those who deserve it by actions or words is not a lie properly understood.

88 Erasmus makes fun of commerce and business.[113] [79 0]

REPLY: Folly, not Erasmus. Let it be she who is stoned!

89 Since there is military service in heaven, and we read that Michael fought a great battle there with the Dragon, and the Lord Almighty himself is called 'The Lord of Hosts,' it is clear that military service is not unbecoming for Christian princes nor have military titles improperly been assigned to the angels.[114] [243Y]

REPLY: I wish that all, the sacred and secular alike, would fight against that Dragon.

90 Erasmus intemperately and impiously attacks and censures monks and their way of life.[115] [93A]

REPLY: Where does he attack good monks? And what impiety is it to attack the corruption of the bad ones so that they can be straightened out?

91 Erasmus impiously ridicules and censures the activities and austerities of the monks as if they were like the Essenes.[116] [98M]

REPLY: The Essenes are praised by Josephus.[117]

92 Erasmus is critical of the fact that monks are encountered on the public roads, in transport, and mounted on horses and mules.[118] [103V]

REPLY: This is offensive to many other good men.

* * * * *

111 The Index refers to Pio's discussion of this point at XXXIII libri 226ᵛE–227ʳ, a matter Erasmus mentions only tangentially at 336 n1334.
112 See 352 n1429.
113 See 166 n346.
114 See 349 n1412.
115 The Index refers to the collection of passages at the beginning of book 5; see 192 n507.
116 See 203 nn578–9.
117 Pio XXIII libri 94ʳA mentions Josephus' account of the Essenes, a text to which Erasmus alludes passim. See Josephus The Jewish Wars 2.120–61 Loeb (1927) 368–85.
118 The Index refers to XXIII libri 102ᵛv–103ʳ, where Pio criticizes the view of Erasmus to this effect without citing a source; he may have in mind Moria LB IV 471A–B / ASD IV-3 158:1526, 160:1535 / CWE 27 130–1.

93 Erasmus unfairly criticizes the way of life, the practices, and the common property of the cenobites.[119] [100P]

94 Pio concludes that Erasmus condemns monks and monasticism.[120] [103Y]

REPLY: Both [93, 94] are completely false, as is that which immediately follows:

[95] Erasmus reviles monastic life and those who have professed a life in religion.[121] [79P]

96 Erasmus unfairly maintains that there is a due limit to [the number of] monasteries in his book on the *Education of a Prince*.[122] [101S]

REPLY: Even in the finest of things a limit is best. Whatever is overdone is heading straight for corruption.

O

97 Everyone can dismiss his wife because of fornication, but no one is compelled to, and thus the married state is not as harsh as Erasmus considers it.[123] [226D]

REPLY: But it seemed that way to the apostles in the Gospel [Matt 19:10].

98 Erasmus considers anyone's good works as the unproductive works of servants [Luke 17:10].[124] [234S]

REPLY: This is an out-and-out lie. I maintain that we ought to belittle our good works so that God can value them highly.

99 Good works deserve recompense, but not eternal salvation, etc, just as the good works of slaves are not absolutely deserving of freedom.[125] [235A]

REPLY: Noël Béda maintains the opposite.[126] Note the agreement of the censors.[127]

100 Erasmus is unfairly critical of the ornamentation of churches, because of baptismal fonts, candelabra, images of the saints, chanting, organs, etc.[128] [128 OP]

* * * * *

119 See 202 nn571, 573.
120 See 199 n554, 209 n603.
121 See 167 n348.
122 See 205 n588, 206 n593.
123 The Index cites 226^{r}D–226^v, an argument not taken up by Erasmus in his reply.
124 See 346 n1392.
125 See *XXIII libri* 236^{r}B; Erasmus did not reply directly to this point.
126 See entry 18 n27 above.
127 Censures quoted in *Declarationes ad censuras Lutetiae vulgatas*, eg LB IX 858A, 885F, 886F–887B, 888B–D, etc
128 See *Apologia* 223–4.

REPLY: No, this is an impious lie of the Index. Certain characters in the colloquies criticize the excessive expenditures of a particular monastery in which there were a gilded baptismal font, a silver altar worked with gold, golden candlesticks, and countless other items. And I do not condemn chant, or organs if they are correctly used.

P

101 The forms of address of pope and bishops; they are not undeserving of the title 'lord.'[129] [194S]
REPLY: As I say before, it seemed otherwise to Jerome, who advises them to bear in mind that they are bishops, not lords.

102 It is the 'fishy' days that bother Erasmus.[130] [89R]
REPLY: No wonder, they are bad for my health. And it is not Erasmus who said this, but some Epicurean in the colloquies who is refuted in the same place.[131]

103 Erasmus laughs as he catalogues the corrupt lives of princes and bishops.[132] [81v]
REPLY: Not Erasmus, but Folly. And it is urbane to correct what is corrupt with laughter.

Q

104 Erasmus is unfairly critical of theological arguments.[133] [78K]
REPLY: Only the silly or needless ones

R

105 Erasmus' Folly criticizes all kings and princes and courtiers.[134] [80C (actually s)]

* * * * *

129 See eg 303 n1152.
130 With this entry the Index refers to XXIII libri 89R, meaning 89ʳK, where Pio takes up what Erasmus wrote in De esu carnium ASD IX-1 46:824–6. Pio begins: 'Inasmuch as Augustine wrote so explicitly that fasting on these days was, in his era, an ancient practice, why are you so foolish as to say that Friday was gradually rendered fishy for us ...?'
131 Erasmus missed the reference to his De esu carnium here. It is not clear to which colloquy he refers, but he may mean the Butcher in the colloquy Ἰχθυοφαγία LB I 787E–810A / ASD I-3 495–536 / CWE 40 675–762.
132 See 170 n375.
133 See 166 n345.
134 See 168 n359.

REPLY: Oh the hussy! Let her tongue be cut off. But it is the Index that is lying here.

106 Erasmus makes fun of the monastic way of life, and calls it artificial.[135] [99N]

REPLY: Here the Index is daft along with Pio, for it is not I who call it so. Rather, I am criticizing Béda, who used this expression,[136] one whereby contemporary theologians indicate rules of life added by human agency.

107 The See and power of the Roman pontiff are impiously termed by Erasmus a tyranny cloaked in the name of Christ.[137] [204V]

REPLY: Power without knowledge and devotion is tyranny. If there were no popes like this, then he has something to complain of. It is the abuses that are criticized, not the power, which should not be absolute. Nowhere is the See blamed, but many others have complained about many of its occupants.

S

108 Not all *sacerdotes* are bishops, nor were they ever.[138] [187D]

REPLY: And that is not what I said, though Jerome maintains this when he calls them presbyters. Later on the presbyters were distinguished from the bishop, who was then called *sacerdos*.

109 Erasmus makes fun of the fact that different saints are invoked in different places both for the same and for different purposes.[139] [155M]

REPLY: Folly ridicules superstition in these matters.

110 Erasmus greatly diminishes the authority of the Holy Scriptures and of their writers.[140] [177AB]

REPLY: Here the Index is lying a lot.

111 It was not right for Erasmus to rely on his own understanding against the opinion of Augustine and all theologians on the topic of lying.[141] [249P]

REPLY: My statements were not untrue, but they departed considerably from the usual understanding.

* * * * *

135 See 199 nn551–3, 200 n563.
136 See 199 nn552–3.
137 See 304 n1159.
138 See 131, 290–1 n1083.
139 See 178–9 nn424 and 431–2, 239 n797.
140 The Index cites the beginning of Pio's book 11; see 265 nn942, 944.
141 See 352 n1429, 355 n1448.

112 Pio does not approve of the practice of burying things profane in community churches.[142] [189 after F]

REPLY: Then he condemns the public practice of the church, something he is always charging me with.

T

113 Erasmus criticizes the adornment of churches and funeral rites.[143] [119A *sic*]

REPLY: Yes, but excessive or superstitious ones.

114 It is Erasmus' impious view that the theologians should be destroyed by Christ.[144] [176v]

REPLY: I ask: Are there now or were there ever no monks and theologians who behaved like tyrants? If there are – as, in fact, there are, alas, all too many – then how does one sin who prays that Christ will free his people from the tyranny of such men? Nor is it necessary that they be done away with by Christ; it is quite enough if they are straightened out.

V

115 Nothing excessive is expended in the veneration of the saints by their devotees.[145] [150E]

REPLY: I wish that he were stating the truth here, but reality clearly denies this.

116 Erasmus assigns virginity and celibacy second place to marriage.[146] [216A]

REPLY: It would have been sufficient to mention this lie just once.

117 Erasmus' understanding of the unity of the Father, Son, and Spirit is incorrect.[147] [182QR]

REPLY: No, it is quite correct, in the matters of nature and will.

* * * * *

142 See 294 n1098; we take 'things profane' (*prophanorum*) to mean corpses; thus the Index refers to 189^{V}F, where Pio objects to the pollution of churches by burials within them.
143 The Index refers to the opening of Pio's book 7 (120^{V}A); see 223 n691.
144 See 263 n937.
145 The Index refers to the formal argument by Pio on 150^{V}E–151^V, which Erasmus ignored in his reply.
146 The Index cites the opening of Pio's book 17 (216^{V}A–217^r); see 320 nn1241–2 and 1244, 335 n1327.
147 See 275 n1000, 277 nn1012, 1014.

118 Erasmus despises and ridicules vows.[148] [210o, 211P]

REPLY: This is false, unless you understand it to be about vows that were made thoughtlessly.

119 One bound to the religious life by a solemn vow cannot be dispensed by the church in those matters which are the substance and principal features of the vow.[149] [214 after T]

REPLY: Cardinal Cajetan thinks the opposite,[150] and the pope practises it.

120 He ridicules making vows to the saints.[151] [82B]

REPLY: Folly does, if superstition is involved.

121 Erasmus impiously makes fun of the vow of pilgrimage to Jerusalem, and of the holy things that are preserved there.[152] [216z]

REPLY: Someone dissuades someone from taking that vow; and rushing off to Jerusalem should not be praiseworthy for all.

122 The distinction between a solemn vow and a simple vow, denied by Erasmus[153] [213s]

REPLY: Scotus rejects[154] the distinction, which Thomas and others posit. The distinction that Scotus suggests is of no importance; let something else be sought.

THE END

* * * * *

148 The Index refers to the opening of Pio's book 16 (210^{v}A–211^{r}); see 312 n1197.
149 The Index refers to Pio's discussion on 214^{v}T[bis]–215^{r}, which Erasmus ignored in his reply; see 315 n1213.
150 See his *De dispensatione matrimonii* (see entry 42 n60 above), where he agrees with Thomas Aquinas that the church can grant dispensation from a solemn religious vow, and cites the example of Ramiro II of Aragon (d 1147), who had been dragged from his monastery in Narbonne to succeed his brother Alfonso I in 1134 and obtained from Innocent II a dispensation from his vow of celibacy in order to provide an heir. In 1137, two years after the birth of his daughter Petronilla, Ramiro retired into a monastery. See Roger Bigelow Merriman *The Rise of the Spanish Empire* (New York 1918) I 277.
151 See 318 n1234.
152 See 312, 319 n1235.
153 See 314 n1205.
154 See Cajetan's summary of the debate in his *De dispensatione matrimonii* (cited in entry 42 n60 above) 122:5–34.

LAST WILL AND TESTAMENT OF ALBERTO PIO

Bronze sepulchral statue of Alberto Pio, 1535,
possibly by Gian Battista di Jacopo, known as 'il Rosso Fiorentino' (1495–1540),
or by Paolo Ponzio di Firenze and Andrea Coria
The figure of Pio, clad in the armour of a warrior, reclines on a couch and reads
from an opened book. The epitaph on the statue reads: *Quem prudentia
Clarissimum reddit / Doctrina fecit immortalem* 'Whom prudence
renders most renowned, / learning makes immortal.'
By permission of the Musée du Louvre, Paris

LAST WILL AND TESTAMENT OF ALBERTO PIO[*]
Paris, 21 July 1530

We make known to all and each who will inspect whether by seeing or hearing the present letters that today, the twenty-first of the month of July in the year from the nativity of the Lord 1530, in the third indiction, and the seventh year of the pontificate of the highest father in Christ and of our lord, the lord Clement vii, pope by divine providence, in the presence of my public notary and of the below listed witnesses, I, Alberto Pio of Savoy, count of Carpi, of the royal order of Saint-Michel, being of sound mind but infirm of body, not wishing to depart from human affairs without a will, having invoked the name of Christ, I wish this to be my testament and last will.

First of all, I commend my soul to the infinite goodness and mercy of almighty God who created it and of our most compassionate Lord Jesus Christ who redeemed it by his most holy blood, and I humbly beg him that he grant to me to close the last day of my life in this perfect and authentic faith, and I declare my desire to live as much as I can and die a faithful and orthodox Christian and in the bosom of holy mother church. And I wish and request to have administered to me the sacred sacraments of that church at an opportune time and to be buried according to the custom of Christians. And I call upon and even cry out especially to Mary, the most glorious ever-virgin Mother of God, to be my special advocate for the salvation of my soul before God, Jesus Christ her son, almighty God, to the most holy apostles Peter and Paul, John the Evangelist, James and Andrew, together with the rest of the apostles, and to the most holy archangels Gabriel and Michael, to the spouse of the glorious Virgin, Joseph, to the martyrs Stephen and Lawrence, to the most holy confessors Francis and Martin, and to all the

* * * * *

[*] Pio's Last Will and Testament is preserved in the Ambrosian Library in Milan. The library collocation for the will is Biblioteca Veneranda Ambrosiana, Archivio Falcò Pio di Savoia, 1 sezione, prima parte (Famiglie e persone), scatola 277, entry 6, folios 1^r–4^v: Testamento de lo illustrissimo signore Alberto Pio.

male and female saints of the heavenly court; I supplicate them all that they pray for me to almighty God and defend me from the snares and attacks of the devil and of all evil spirits, all of whom I renounce just as I also did when bathed by sacred baptism. My body, however, I wish to be buried among the Friars Minor at Paris, in the decent and seemly place that will seem best to the father guardian and friars to grant to me, and according to the disposition of the executors of this my will. And I wish when death draws near that the care of me be granted to the friars of the said order, who are to be with me day and night and to pray for me and console me spiritually, who are to clothe me, before my soul breathes forth completely, in the habit of the Friars Minor placed over my bare flesh, my night shirt having been removed, and gird me with a cord. And in that habit I wish to be buried according to the custom of the friars, whom I wish to carry the bier of my corpse and to bury my body in a sepulchre with their hands. The funeral rites, however, and procession are to be respectable, as the executors decide. And I want twelve of the friars of the said convent to be newly clothed at my expense, that is, with an outer mantle (*cappa superior*); let the first of these be the father vicar, Master Martin[1] my confessor, and the Italian friars studying in the same convent, let the rest be from among the poor youths to be chosen by their master and by my father confessor, and let the divine office be kept according to the custom followed in Paris regarding funerals, with this addition, that there be celebrated the seventh, thirtieth, and anniversary offices[2] as is done in Italy, and I wish that in the place where my corpse is buried there be set up some stone as a memorial of my name.

I wish and appoint, however, as my heirs my daughters, Caterina[3] and Margherita,[4] and my brother Lord Leonello, according to the shares and distributed estate as will be stated below.

* * * * *

1 Friar Martin Rogerii, later identified as *doctor in sacra pagina*

2 Pio is here requesting the traditional series of masses to be celebrated for the repose of his soul, on the seventh and thirtieth days after his burial and on the anniversary of his death. He does not seem to be requesting the trentals or Gregorian masses, that is thirty masses spread over as many days or concentrated in fewer days; see Philippe Ariès *The Hour of Our Death* trans Helen Weaver (New York 1981) 158, 174–5, 178; and Samuel K. Cohn JR *Death and Property in Siena, 1205–1800* (Baltimore 1988) 65.

3 Caterina Pio, the elder daughter of Alberto Pio and Cecilia di Franciotto Orsini Pio, was born c 1519 and married Bonifazio Gaetani di Roma, duke of Sermonetta; see Litta 'Pio di Carpi' table III; and Sabattini 79.

4 The younger daughter of Alberto and Cecilia di Franciotto Orsini was born c 1527 and married in 1544 Giangerolamo Acquaviva di Napoli, duke of Atri. Their son Rodolfo (1550–83) became a Jesuit missionary to India where he was martyred; he was later canonized as a saint. See Litta 'Pio di Carpi' table III, and 'Acquaviva di Napoli' table V; and Sabattini 79.

It is my wish that Lord Leonello[5] be the heir of the whole dominion and county of Carpi with all the revenues, jurisdictions, fortresses, and villages subject to that town and county of Carpi itself, among which are to be understood all the fortress of Montana and half of the town of Soleria, which is owed to me although it is now occupied by others.[6] Likewise I wish Lord Leonello to have possession of [the town of] Rovereto with the house and all the rights acquired by me from the transaction that was made and the inheritance coming to me together with Lords Giovan Marsiglio, Niccolò, and Bernardino Pio,[7] as appears in the documents that established this. Likewise, I wish to belong to Lord Leonello the greenery or gardens which are called della Gabarda,[8] and I wish him to be content with these, because if he complains I wish him to be deprived of half of that dominion of Carpi purchased by me with my own money; and I also wish to belong to the same Lord Leonello that house located in the citadel of Carpi towards the east which was called the house of Madonna Benedetta.[9]

My daughters, however, I wish to be heirs of whatever other goods I have, both immovable and movable. Of the immovable I name these as the more important: namely, the fortress of Novi with the citadel and all possessions of whatever kind and whatever revenues, along with the curia of Santo Stefano and everything pertaining to it.[10] Also the possessions

* * * * *

5 Son of Lionello and Caterina Pico Pio, brother and heir of Alberto, father of Rodolfo, who died in 1535; see Litta 'Pio di Carpi' table III.

6 When Giberto III Pio in 1499 exchanged his half of Carpi for the fiefdom of Sassuolo and other territories in Ferrara, he and his heir Alessandro retained possession of Soliera, a town southeast of Carpi; see Garuti 16, 31.

7 These men are the sons of Alberto's cousin Galasso II Pio; in 1500 they formally renounced, on the granting of allodial lands and the faculty to reside permanently in Carpi, their rights to these possessions, but then had recourse to the pope and the emperor to retain certain rights to them. The house in Roverto was called Casa delle Vacche, a palace that served as a country residence and administrative centre. See Alfonso Prandi 'Il Patrimonio fondiario dei Pio nello "stato" di Carpi' *Medioevo e umanesimo* 47 (1981) 469–502, especially 483–4 n35, 493; and Garuti 15. For a picture of Pio's palace in Roverto, which survives to this day, see Dante Colli, Alfonso Garuti, and Romano Pelloni *Carpi: l'anima della città* (Modena 1982) 142, 150.

8 La Gabarda was the name of a garden located next to the eastern wall of Carpi; see Prandi 484.

9 Apparently a reference to the home of Benedetta Pio, the daughter of Galeotto del Carretto (the marchese of Finale), widow of Marco II Pio (co-ruler of Carpi 1465–94) and mother of Giberto Pio (co-ruler 1494–9, d 1500). The family of Giberto, including Benedetta, left Carpi in 1500 to relocate in Bologna and in Sassuolo, one of the territories Giberto had received in return for ceding his portion of Carpi to Ercole d'Este. See Litta 'Pio di Carpi' table IV; Guaitoli 'Sulla vita' 152–6; and Sammarini 380.

10 The Castello of Novi and the *pieve* of San Stefano were fiefdoms of the bishop of Reggio; see Garuti 30–1.

of Fossoli, Budrione, and Migliarina, and all the lands of Groppo and of the bridge of La Preda, and the other places which are called the forests, namely Grosso and Calesella and Pinzono and Molendino di Novi.[11] Also those goods which are in Mantua and the curia or possession that is called La Bardellona. Likewise the county of Meldola[12] and the goods located in Rome and in other places near there, and also the account at [the Bank of] San Giorgio in Genoa. On which daughters I similarly impose the penalty of losing the possession of Fossoli should they complain or in any way lodge a suit against the above-mentioned Lord Leonello. To which daughters of mine I also leave my own home in Carpi (or, as they call it, the Palace) entire, together with the large tower and gardens, which home I wish to be finished by the said daughters according to the plan I established, which home I wish that they be able to alienate.

This, however, my estate, I wish to be so divided, and because there may be a place for substitution, about which below; namely, in case the above-mentioned Lord Leonello should die, I wish his son Rodolfo[13] to succeed him regarding all the goods of my estate, who [Rodolfo] dying without masculine and natural sons, let [Leonello's] third son[14] succeed, and thus successively as far as there are males. Should it be the case, however, that there is no male, I wish all my estate to devolve on my daughters, or, if they do not survive, on their sons, on this condition, however, that the one

* * * * *

11 Alberto Pio received Rovereto, Migliarina, Budrione, Fossoli, and Novi as part of his inheritance; see Garuti 30, 32.

12 Although Alberto wished his daughters to inherit Meldola, his brother Leonello, who had settled there since 1528 and served since 1529 as president of Romagna and governor of Bertinoro, was invested with not only Meldola but also Sarsina, Bertinoro, and other fortress towns (*borghi*) by Clement VII on 31 March 1531. See Comune di Roma *Registri di bandi editti notificazioni e provvedimenti diversi relativi alla città di Roma ed allo stato pontificio* vol I: *Anni 1234–1605* (Rome 1920) 11 entry 57; Garuti 154; and Paolo Mastri *Leonello Pio da Carpi in Meldola (1531–1571)* (Bologna 1941).

13 Rodolfo Pio (1500–64), the eldest son of Leonello Pio and Maria, the daughter of Bernardino Martinengo. In 1516 Rodolfo was a knight of St John of Jerusalem and Rhodes, later secret chamberlain of Clement VII, and in 1528 bishop of Faenza. He had a distinguished diplomatic career (he was the extraordinary nuncio of Clement VII to the French court from 26 July to 28 November 1530 and from May to July 1533, and served as Paul III's legate from 9 January 1535 to 1537), and in 1535 was named cardinal by Paul III. See Litta 'Pio de Carpi' table III; Fraikin xxxii; and Pier Giovanni Baroni *La nunziatura in Francia di Rodolfo Pio (1535–1537)* Memorie storiche e documenti sulla città e sull' principato di Carpi (Bologna 1962) xxiii–xxxviii.

14 Leonello's third son (the first by his new wife, Ippolita Comneno, widow of Zanobi dei Medici and heiress of the county of Verucchio in Romagna) was Alberto who succeeded to his father's feudal estates, married Ippolita, daughter of Giulio Cesare Rossi, count of Cajazzo, and died in 1580; see Litta 'Pio di Carpi' table III.

succeeds the other, if the other dies without children, and, in turn, if they die without sons, that their estate should devolve on the legitimate sons of the above-mentioned Lord Leonello; the care of which daughters of mine and the administration of their goods I wish to be in the keeping of their mother and of the Lord Teodoro,[15] bishop of Monopoli, my brother, and of Angelo Sagacini, the master of my house. And I wish that in the division of the estate my first-born daughter Caterina be preferred over my daughter Margherita in such a way that the greater half belong to that Caterina, and even the sixth part of the remaining half be given to her, and this by reason of primogeniture and because she should be married more quickly and is already almost old enough for a husband. But the other daughter is still a young child, not to be given in marriage to a man even for many years hence. On account of which reason I wish that the revenues and fruits of her portion be reserved for her until the age of marriage and that they increase her dowries.

Moreover, I wish that the Lady Cecilia, my wife, be able to carry on and that everything necessary for her sustenance in keeping with her respectable and appropriate status be provided her from my whole estate if she should live on as a widow. If, however, she should move on to a second marriage, I want returned to her her dowries which I received, and also that part of the additional dowry which was appointed for her according to her portion which I received. To this wife of mine I bequeath all her garments and jewellery which she had in Lombardy, for she lost all the better items in Rome.

Likewise I bequeath to Lord Teodoro, the above-mentioned bishop, the income of the possessions La Brada, La Savana, and La Furgera, which were pledged to Lord Andrea de Cropis,[16] and the gardens called Caval, upon

* * * * *

15 Teodoro was the natural son of Lionello Pio and Polissena, daughter of Johannes Richem-bach. Polissena was already married to Michele da Berino of Cremona. Teodoro entered the Franciscans, was elected in 1504 guardian of their monastery of San Nicolò in Carpi, was appointed on 9 April 1513 bishop of Monopoli, was governor of Carpi from 1518 to 1522, and died in 1546. See Litta 'Pio di Carpi' table III; John R.H. Moorman *Medieval Franciscan Houses* Franciscan Institute Publications, History Series 4 (St Bonaventure, NY 1983) 111; Eubel III 248; Gabriella Zarri 'La proprietà ecclesiastica a Carpi fra quattrocento e cinquecento' *Medioevo e umanesimo* 47 (1981) 503–59, especially 535 n86; and Guaitoli 'Sulla vita' 225.

16 Apparently Andrea Crotti of Cremona, who was Alberto Pio's secretary and trusted negotiator and who became governor of Carpi in 1507 and remained so until his death on 8 October 1518. Teodoro's succession of Crotti as governor of Carpi suggests that the revenues from the property may have been attached to the office. See Guaitoli 'Sulla vita' 179, 225; and Sabattini 33.

whose [Teodoro's] death, the said goods should come to my daughters. To him I also bequeath my silver cross which I have in my chapel and this in my memory and as a token of love.

Should, however, my heirs die without sons, I wish my estate to be divided, half to one of the sons of Lord Gianfrancesco,[17] count of Mirandola and Concordia, that is to his second son, or in the event of his death, to his third son; the other half to one of the sons of Lord Giovanni Francesco Gonzaga,[18] lord of Luzzara, or of Lord Luigi,[19] his brother, lord of Castelgoffredo and Castiglione, my womb brothers. And on this condition, however, I make my bequest, as much regarding those of Gonzaga as those of Mirandola, that if they would be my heirs, they should adopt the name of my family with its arms and insignia and put aside those of their own family so that they should thereafter no longer be called Picos of Mirandola nor those of Gonzaga but Pios of Carpi and neither otherwise nor in any other way. So that if any one of them should refuse to do this, I do not wish him to share in my estate. And since the first-born of the said families may do this perhaps with greater difficulty or with reluctance, for this reason I chose for this the second-born instead, provided they were not enrolled in the service of the church but were to be married.

Should, however, these last heirs of mine die without offspring, I leave as my heir the church, on these conditions, that a request be made to the holy apostolic See and to our highest lord the pope that he set up in the land

* * * * *

17 Gianfrancesco Pico was the son of Galeotto I, ruler of Mirandola and brother of Caterina Pico Pio Gonzaga, hence Alberto's first cousin. He succeeded his father in 1499, but was expelled from Mirandola by his brothers in 1502. Alberto Pio gave him military assistance in his two attempts of 1503 and 1505 to retake the city. Julius II briefly restored Gianfrancesco in 1511, but he did not regain control of Mirandola until 1514. He was married to Giovanna Carafa, the daughter of Giantommaso Carafa, the count of Maddaloni in Naples. They had three sons: Paolo (d 1567), Giantommaso (d 1567), and Alberto (d 1533). On 16 October 1533 his nephew Galeotto II Pico murdered Gianfrancesco and his son Alberto. See Charles Schmitt *Gianfrancesco Pico della Mirandola (1469–1533) and His Critique of Aristotle* International Archives of the History of Ideas 23 (The Hague 1967) 11–30; and Litta 'Pico della Mirandola' table IV.
18 Gianfrancesco Gonzaga, lord of Luzzara, was the son of Caterina Pico Pio Gonzaga and Rodolfo Gonzaga. He was married to Laura Pallavicino, the daughter of Galeazzo Pallavicino, the lord of Borgo di S. Donnino, by whom he had four sons: Massimilliano (d 1578), Guglielmo (d young), Galeazzo (d young), and Rodolfo (d after 1553). See Litta 'Gonzaga di Mantova' table XVI.
19 Luigi Alessandro Gonzaga, lord of Castelgoffredo, Castiglione, and Solferino, was the son of Caterina Pico Pio Gonzaga and Rodolfo Gonzaga. By his wife Caterina, the daughter of Count Giangiacomo Anguissola of Piacenza, he was father of three sons: Alfonso (d 1592), Ferdinando (d 1586) whose son was the Jesuit St Luigi Gonzaga (1568–99), and Orazio (d 1589). See Litta 'Gonzaga di Mantova' tables XVI and XVII.

of Carpi a bishopric whose incumbent is to be bishop and count of Carpi and who is to be required to reside in his see and not elsewhere. If, however, he wishes to be absent, I wish that he have given to him each year from my goods only five hundred ducats, all the rest indeed I wish to be given to an abbatial monastery of the order of St Benedict under the discipline of the Rule to be built in Carpi under the name of St John the Evangelist, to which religious I commend my soul and those of all my ancestors, and I ask them to be willing to be upright stewards of the goods that I leave to them, distributing them partly for divine worship, partly for alms, and for their upkeep and necessary things.

But what am I doing! I designate heirs and divide an estate which I do not at all possess, for all my goods and fortune are held by others, by whom I was evicted because of my loyalty and service to the most Christian King, and this is obvious from the actual situation and from the privations and spoilations done to me by the emperor, no just reason for this however existing, but only this pretext: that I was committed to the service of the most Christian King to whose faithfulness, justice, and mercy I commit and humbly commend my poor wife and poor little daughters who are not yet adults, all of whom I leave empty-handed and deprived of all property, in the greatest poverty and destitution, and my whole house destroyed and lost, something that afflicts me in an extraordinary way. I am consoled however only by the hope I have placed in the goodness and liberality of the king and of his most serene Lady Mother, to whom I once more commend with many tears the most miserable state of my affairs. Surely his Majesty will be able at least to order that what is owed to me by his Majesty be from justice granted to them, having seen the letters patent of the promises of his Majesty when he called me to his service. And that also which is owed to me for that time when I was in his service, and that which is owed to me for my own monies given in loan and expended otherwise at his order, and also for the interest and exchange fees of those monies that were distributed in Rome for the Neapolitan expedition which the lord duke of Albany,[20] in the name of his Majesty, led and made, which exchanges and

* * * * *

20 John Stuart (1481–1536), son of Alexander, the brother of King James III of Scotland. John, who was raised in France where his father lived in exile, served with distinction in the French army. In 1525 he was supposed to attack Naples and thus attract the imperial forces to its defence, while Francis I invaded Milan. John's army (200 lancers, 600 light horsemen, and 4,000 infantry) moved slowly down the Italian peninsula to the vicinity of Rome where it was to join the forces of Renzo da Ceri. But the opposition of the imperial Colonna forces amassed there and news of the capture of Francis at Pavia led John to retreat back to France. John, who was himself childless, remained the

interest it was necessary for me to pay to Lord Ansaldo Grimaldi and to other merchants as is established from their financial records.

Heirs having been designated, it is fitting that I should set up those things that I wish to be observed and carried out by delegates and by those heirs of mine. The first of which is that there be concord among them and they be content with that portion which I leave to them, that if they act otherwise, if they complain and lodge law suits against each other, I wish that they incur the above-mentioned penalty, namely that if Lord Leonello complains against my daughters, let him lose half of the estate that I leave to him, which I bought with my own monies in the county and dominion of Carpi, and that it devolve to those daughters of mine together with rights and possessions of Rovereto. And, in turn, if they protest against that Lord Leonello, they lose as was said and forfeit the possession of Fossoli, Migliarina, and Budrione which would devolve to him. And I ask and beg both parties that for the love of me they come to agreement with one another in all things and that they treat one another with love and let that Lord Leonello regard my daughters as his own daughters, and that they honour and regard him as their own father and that they regard his sons as brothers and have special regard for the above-mentioned Rodolfo.

Bequeathed debts and obligations, however, I wish to be brought to a conclusion and completely paid in equal shares by my heirs, that is, that Lord Leonello be bound to a half and my daughters to the other half according to the portion of my estate that is to be divided among them; namely, Caterina more than Margherita, since her portion of the estate is larger than Margherita's. In first place, however, I obligate these heirs to be held to pay all my debts, and if it is discovered that I received or retained something knowingly or unknowingly from someone, let them return it and satisfy everything.

In addition, concerning the obligation of building the monastery of the canons regular of St Augustine, to which my father bound me by his will, an obligation that I was unable to fulfil because of various disturbances of fortune although I very often wished to begin,[21] I want them to fulfil this

* * * * *

presumptive heir to the Scottish throne until James v produced an heir, Mary, born in 1542, after John's death. See Gordon Hamilton *Scotland: James v to James vii* (Edinburgh 1965), 18–63; and Guicciardini *Storia d'Italia* iv, 240–1, 247–9, 271–5.

21 Alberto's efforts to found this collegiate church, begun in 1512 with the bull of Julius ii that raised the ancient *pieve* of Santa Maria to a collegiate church, were related to his efforts to remove various benefices in his territory from the control of the bishops of Modena and Reggio and place them under the control of the collegiate church whose patron he would be and whose juridic status would be *nullius dioesesis*, a step

obligation immediately, in the beginning of the first year in which (God willing) they will have recovered the Carpian county and state. I wish that the monastery be set up in the ancient canonry situated in the citadel of the land of Carpi, which they should purchase from the chapter of the major church, for thus I had determined to do, and in that place are already constructed dwellings almost sufficient for the habitation of those canons, whom I wish to have for their church that temple begun but not finished within the confines of my house under the name of our Saviour, Lord Jesus Christ, for the finishing of which I wish my daughters to be bound since they also are the heirs of that house. To which temple, or monastery, I bequeath half of all the ornaments and sacred vestments of my chapel. The other half I leave to the major church, to which I leave a thousand ducats for the building, the town of Carpi nonetheless having been first recovered and not otherwise, and then a hundred each year until the construction is finished.

Also I bequeath and wish that a college [be established] for instructing twelve boys in grammar and the good arts, for whom sustenance and an honest salary for a teacher are to be provided. To which college I bequeath a share in the income of the foundation of the house of the public meat market, which I constructed, and of the inn or public hospices joined to it, with all their revenues and incomes. I wish, moreover, that there be selected poor boys of good natural abilities and suited for learning letters, whom I wish to be chosen by those mentioned below: first by my heirs, then by the chapter of the major church, by the guardian of the monastery of San Nicolò, and by the superior of the canons regular and by the prior of San Agostino, if the convent will have been reformed. For I wish those fathers to be the correctors and conservators of the said college. The administration of its goods and building I wish to be entrusted to the members of the third order of St Francis whose oratory is San Giuseppe.[22] To the said *reformatores*, moreover, I grant the faculty also of admitting besides the said boys two or three externs and wandering paupers whom they judge will make progress in letters and good arts, with which choice whether of

* * * * *

towards raising the status of Carpi from a town to a city with its own bishopric; see Antonio Bellini 'Alberto Pio III, restauratore della Chiesa di Carpi' in *Alberto Pio III, Signore di Carpi (1475–1975)* Deputazione di storia patria per le antiche provincie Modenesi, Biblioteca, new series 36 (Modena 1977) 48–56; and Zarri 'Proprietà ecclesiastica' (cited in n15 above) 503–59, especially 505–8, 515–17, 521; for Pio's own donations to the collegiate church, see Zarri 517 n39.

22 The oratory of San Giuseppe was built in 1504 by the Friars Minor of the third order; Galli 'Presenza religiosa' 75, 88.

Carpians or of externs I burden their conscience. All of this, however, is to happen once the town and county of Carpi is recovered and not otherwise.

Likewise, I wish and bequeath that they finish that which remains to be done on the church of San Nicolò,[23] which I built: that is, the windows, door, floor and vault, plaster and decoration. And that in the hollow frame of the high altar let there be painted an image of the Saviour, our Lord Jesus Christ seated on the throne of majesty, and to the side on the right panel an image of the glorious Virgin adoring him on bended knee and behind her in the same posture the image of San Nicolò, the patron of the church; on the left I wish there to be similar adoring images of St Francis and of St Anthony of Padua, and I wish at the foot of St Francis there be painted my figure upon which he extends an arm either on the head or on the shoulders as if presenting me and commending me to the Lord Jesus. And because my body could not be buried in the tomb of my ancestors, especially of my father interred in the same church, lest [my] memory quickly perish among the friars and people who should pray for me, I wish a *xenopaphium* to be constructed in some place in that church and in that let the bones of my only son, Francesco, deceased before me, who rests in the parish church, be transferred and redeposited, and [I wish] that in that church there be said for me the offices of the dead which are customarily done for the soul of my father.

I likewise leave to the church of San Francesco di Carpi[24] two hundred ducats for making a vault in the said church, as I had determined to be done. Likewise that the marble tomb of my grandfather[25] be set up in the place elsewhere designated for me, that is, in the middle of the wall of the chapel of San Bonaventura, or of Santa Caterina joined to the chapel of San Bernardino.

Likewise I leave to the church of San Agostino[26] in which is buried my paternal grandmother[27] one hundred ducats to be spent on the repair of the monastery, provided it will be reformed, and in any case, that a marble tomb be located there with the bones of my said grandmother in the chapel

* * * * *

23 On the church of San Nicolò di Bari and its Observant Franciscan monastery, founded in 1449/51, which was 'a well-known Scotist academy,' see Galli 'Presenza religiosa' 71–88, especially 71, 73–4, 83; and Moorman *Medieval Franciscan Houses* (cited in n15 above) 111.

24 Galli 'Presenza religiosa' 71–2, 83–4

25 Alberto II Pio di Carpi who died 1463/4; see Litta 'Pio di Carpi' table II.

26 Galli 'Presenza religiosa' 73, 84–5

27 Camilla Contrari of Ferrara who married Alberto II Pio di Carpi; see Litta 'Pio di Carpi' table II.

Madonna and Child with Saints and Donors,
attributed to Vincenzo Catena (c 1470–1531)
The kneeling figures of the donors in the painting, which was formerly
in the cathedral in Carpi, are considered to be Alberto Pio and his wife
Cecilia Orsini Pio. St John the Baptist presents the male donor to the
enthroned Madonna and the Christ Child, who places his hand in blessing
on the donor's head, while two saints – a friar, possibly St Anthony of Padua,
and a young lay woman, possibly St Catherine – look on.
By permission of the Galleria Estense, Modena

of Santa Maria Magdalena in which they now lie, and that that chapel be adorned by a decent picture.

Likewise I leave to the church of the friar Servites of the Blessed Virgin Mary[28] one hundred ducats for the fabric.

Likewise to the monastery of Santa Chiara[29] another one hundred ducats to be spent on the fabric.

Likewise another hundred to the monastery of San Sebastiano[30] to be spent similarly on the fabric.

To the remaining churches or oratories and confraternities existing in the town of Carpi, however, twelve ducats to each, to be spent in some ornamentation relevant to the divine worship, as will seem proper to my heirs.

Likewise to Angelo Sagacini, master of my house, if he remains in the service and household of my daughters, I leave one thousand ducats in goods or in money, and in whatever case five hundred ducats. I do not wish to obligate my heirs to satisfy everything unless the state of Carpi is recovered, but [obligate them] to half in whatever case, that he may have forever an appropriate and respectable sustenance in the house and two ducats per month as long as he lives.

Likewise I leave to Laelio Garuffo[31] one hundred ducats, and respectable upkeep in the house, together with two ducats per month as long as he lives.

So much again I leave to Bernardo Tirello from Piacenza who has served me for a long time.

Likewise I leave to my cook Andrea, called 'a little sugar,' twenty five ducats and perpetual sustenance in the house together with one ducat per month as long as he lives, also that in the house he may do nothing but thoroughly rest.

* * * * *

28 Galli 'Presenza religiosa' 74, 86

29 Largely through the financial support of Alberto, the convent of Santa Chiara, founded by his cousins Camilla and Violante Pio with a licence granted in 1490, was built in 1500 and opened in 1510; see Galli 'Presenza religiosa' 73, 85; Moorman, *Medieval Franciscan Houses* (cited in n15 above) 567; and Zarri 'Proprietà ecclesiastica' (cited in n15 above) 510–12, 519, especially 512 n23 for bibliography. Alberto's niece was a nun there; see n38 below.

30 This monastery was built by the third-order Franciscans starting in 1457/8, but in 1503/4 was taken over by the Servite sisters; see Galli 'Presenza religiosa' 73, 87.

31 On Garuffo from Romagna, see *Actes de François Ier* VII 459 n25707; and Jean Gribomont 'Gilles de Viterbe, le moine Elie, et l'influence de la littérature maronite sur la Rome érudite de 1515' *Oriens Christianus* 54 (1970) 125–9, especially 127 where Garuffo is identified as having copied a Maronite liturgical text apparently for Alberto Pio on 23 August 1517.

Likewise I leave to the wife and son of the former Boccalini, my familiar, sustenance and clothing as long as she lives.

Likewise I bequeath that, if the town of Carpi is recovered, to my familiar the Florentine Baldessarri,[32] who recently took a wife, be given a respectable house in Carpi; that, if it is not recovered, let there be assigned to him in that part towards the stable of my home in the city of Rome in the corner bordering it towards the east enough space and land so that he can there build a suitable home for his own habitation.[33]

Likewise I bequeath that my house never be closed to the heirs of Lord Sigismondo dei Santi,[34] but that they be raised and nurtured in that house, if they need it and especially if they will not have recovered their goods.

Likewise I bequeath that master Bernardo Loschi of Parma,[35] painter, may have sustenance and clothing from my heirs for the whole time of his life.

* * * * *

32 Perhaps the Balthazar Piat who is listed by Fraikin (xxxi) as 'gentleman of Alberto di Carpi' and extraordinary nuncio of Clement VII to the French court, who arrived there on 11 February 1528. By 1535 Rodolfo Pio employed as his secretary in Rome a messer Baldassaro, a friend whom he claims to have known for thirty years, that is, since he was five; see Baroni (cited in n13) 13, 113 n a.

33 According to Garuti 62, Alberto Pio's house in Rome was a palace built into the ancient theatre of Pompey bordering on the Campo de' Fiori. This assertion does not agree with the 1527 census records, according to which Pio lived in Regio de Ponte, apparently next to the church of Santa Maria in Posterola; the area in and around Pompey's theatre was apparently inhabited by Cardinals de Valle, Piccolomini, Giacobazzi, and Cesarini. See *Descriptio urbis: The Roman Census of 1527* ed Egmont Lee (Rome 1985), 67 (no 3295), 101 (nos 6544–7).

34 In 1502 Dottore Sigismondo Santi came from Ferrara, where he taught philosophy at the university, to Carpi, where he settled in a house given to him by Pio in 1508 and by 1512 had become one of Pio's secretaries. In 1510 he was one of Pio's three trusted negotiators with Alfonso d'Este to regain half of Carpi. His dedication, fidelity, and wisdom in difficult negotiations led d'Este to hate him so much that he confiscated his possessions in Ferrara. Pio compensated him for his losses and assigned him a generous stipend and annual pension and eventually the post of chancellor. In the early morning of 29 November 1525 (others give the month as July), in Brescian territory (near Lago d'Iseo between Verona and Trent), while he was on his way as Pio's courier to secure a new French alliance in the wake of defeat at Pavia, Santi was murdered for his money by the guide he had hired. He left behind in poverty a family with many children. See Guaitoli 'Sulla vita' 179, 224, 269; Sabattini 33; Fraikin xxix (lists him as extraordinary nuncio of Clement VII to the French court from 10 July to August 1525); and Julia Haig Gaisser *Pierio Valeriano on the 'Ill Fortune of Learned Men': A Renaissance Humanist and His World* (Ann Arbor 1999) 209–11, 323.

35 Although he was born in Parma, the son of the painter Jacopo, Bernardino Loschi made his career in Carpi where he was court painter and married Margherita (sister of the printer Dolcibelli alias del Manzo). For examples of Alberto's payments to Bernardino for artistic work done in Carpi, see Contini [8], [18–19].

Likewise I leave to my sister Lady Angela Gabriella,[36] a sister of the monastery of Corpus Christi[37] or of San Paolo of Mantua, the hours or office of the Blessed Virgin written by hand on parchment, painted and decorated, covered in purple silk, and one crystal rosary from the two which I have and this as a token of love. And in addition let there be given to her each and every month a gold piece for her needs.

One matching rosary with an enameled figure or *pax* I leave to my niece, Sister Caterina Angelica, nun of Santa Chiara, daughter of the above-mentioned Lord Leonello.[38]

Likewise I leave to the reverend Lord Giammatteo Giberti,[39] bishop of Verona, in my memory and as a token of love that silver globe on which is depicted the representation of the whole world or, as it is called, a *mappamondi* and in which there is a clock.

Likewise I leave to my above-mentioned nephew Rodolfo Pio my vineyard with the house I have on the outskirts of the city of Rome at the foot of Monte Mario and all my books and codices of whatever kind and also all the statues and paintings and monuments of antiquity that were mine and belonged to me in whatever place they may have been, and I urge him, once the town of Carpi has been recovered, to set up a library with the above-mentioned books for the adornment and benefit of the above-mentioned town, just as I had planned to do, handing over to the library of San Nicolò of Carpi a share from the books of which he has duplicates.

Again, it is my wish that here in Paris in the parish church of Saint-Pol my funeral rites be celebrated as if I were to be buried in that church.

* * * * *

36 Sister Angelica Gabriella (c 1493–1544) was the half-sister of Alberto, daughter of Rodolfo Gonzaga and Caterina Pico Pio Gonzaga, whom they named Giulia and who at the age of ten entered the monastery of San Paolo. She died on 25 November 1544 with a reputation for sanctity. See Litta 'Gonzaga di Mantova' table XVI.

37 On the Clarissa monastery of Corpus Christi or St Paul, which was founded in 1414, see Moorman *Medieval Franciscan Houses* (cited in n15 above) 612; and *Regesta Leonis x* ed Joseph Hergenroether (Freiburg im Breisgau 1884–91), no 12806 (1514 papal permission for a mother to visit her daughter and dine with her in the monastery four times per year).

38 Leonello had two unmarried daughters, Costanza and Laura, according to Litta; see Litta 'Pio di Carpi' table III. On the monastery of Santa Chiara in Carpi to which Alberto left a hundred ducats, see n29 above.

39 Gian Matteo Giberti (1495–1543), a curial prelate and trusted adviser of Clement VII, was Alberto's personal friend in Rome (a fellow patron of humanists) and political ally in fashioning the pro-French League of Cognac in 1526. When the League resulted in the disastrous Sack of Rome in 1527, he left the curia in 1528 for his diocese of Verona. Erasmus had earlier solicited his assistance in silencing his critics at Louvain. See CEBR II 94–6.

Likewise I wish that at the time of my funeral there be given five golden *scudi* for each of their monasteries to the sisters of the monasteries of Ave Maria[40] and Saint-Marcel-lez-Paris[41] of Paris and of Longchamp[42] outside the walls of Paris that they may pray and beseech God for the salvation of my soul and chant an office of the dead.

Moreover, I want the following to be the executors and entrusted commissioners of this my will and I entreat them to accept this responsibility, namely the most illustrious Lord Montmorency,[43] the grand master of France, and the reverend lord bishops, Lord Antonio Pucci[44] bishop of Pistoja and Lord Gian Matteo Giberti bishop of Verona and Lord Lorenzo Toscani[45] bishop of Lodève, and the most illustrious lord Count Gianfrancesco of Mirandola, to these I commend my soul, my heirs, and the members of my household. And I ask that they most humbly commend them at the exalted feet of our lord pope;[46] being away from him, I kiss those exalted feet with whole heart and every affection and I beg that he would deign to turn the eyes of his clemency towards me as his most devoted servant dying in such a miserable state, leaving all those dear to me in the greatest need and extreme unhappiness.

In addition, I leave to the church of Saint-François in Paris[47] for the decoration and ornamentation of the high altar that silver image of Jesus

* * * * *

40 The Ave Maria convent, located near Alberto's parish church of Saint-Pol in Paris, earlier founded for Beguines, had become a Clarissa house by 1485; see Moorman *Medieval Franciscan Houses* (cited in n15 above) 642; and Laurent Henri Cottineau and Grégoire Poras *Répertoire topo-bibliographique des abbayes et prieurés* 3 vols (Maçon 1935–70) II col 2194.

41 The Clarissa convent of Saint-Marcel-lez-Paris was founded in Paris c 1270; see Moorman *Medieval Franciscan Houses* 641–2.

42 On the royal Clarissa abbey of Longchamp, located in Boulogne-sur-Seine, Neuilly, Saint-Denis, Seine, in the diocese of Paris, see Moorman *Medieval Franciscan Houses* 640–1; and Cottineau and Poras *Répertoire* I, col 1644.

43 On Anne de Montmorency (1493–1567), see Introduction nn 16 and 64.

44 On Antonio Pucci (1484–1544), see CEBR III, 122–3; and Friedrich Lauchert *Die italienischen literaischen Gegner Luthers* (Freiburg im Breisgau 1912), 464–6. He was the extraordinary nuncio of Clement VII to the French court in March 1525, February 1528, and September or October 1528; see Fraikin xxix, xxxi.

45 Toscani, originally a cleric of Milan, succeeded Giberti as bishop of Lodève from 1528 until his death in 1537; see Eubel III, 227. He was the extraordinary nuncio of Clement VII to the French court in the spring of 1527; see Fraikin xxx.

46 Clement VII (1523–34) was a personal friend of Pio; see CEBR I 308–11.

47 At the time of Pio's death there was no church in Paris under the patronage of St Francis of Assisi (the church of Saint-François-d'Assisi in the 7th arrondissement was built in 1622). The principal monastery of the Franciscans in Paris, whose church was dedicated to St Mary Magdalen, was known as the Grand Couvent de l'Observance de Saint-François. Pio was buried there as he had wished, and presumably it was to

the Saviour rising from the dead, with two angels, similarly of silver, that attend this image.

And I commend to my heirs all those citizens of Carpi, and also the other subjects of that state, who on account of me and my cause which they followed have, perhaps, steadfastly suffered many misfortunes. For I wish that the above-mentioned be favoured and protected by my heirs, and that assistance be provided them in their need, according to the discretion of my heirs.

And if for the greater strengthening of this my testament, there be required some dispensation or confirmation by the highest lord our pope or by the emperor, I wish that [illegible: the rights be retained?] by my heirs.

This document was enacted in Paris under the year, indiction, month, day, and pontificate mentioned above in the house of the dwelling of the most illustrious lord, Lord Alberto Pio, count of Carpi, in the parish of Saint-Pol, in the presence there of the reverend father in Christ and lord, Lord Lorenzo Toscani bishop of Lodève, and the masters in sacra pagina doctors Friar Martin Rogerii, vicar of the convent of the Friars Minor in Paris, the confessor of the same lord testator, and Friar René Perusselle, a religious of the convent of the Friars Minor of Angers, and the noble lord Ippolito de Petra Santa of Milan, Jean Charrey of the diocese of Rodez, Jean Macorelle, cleric of the diocese of Verdun, and Antoine Boutonnet of Mont Pessulane of the diocese of Maguelone, witnesses called and requested especially for this purpose, and by me the public notary signed below.

And I, Jean Moucaud, priest of the diocese of Autun, licensed in civil law, a public notary by apostolic authority, sworn to the above-said statements of the testament and last will of the most illustrious lord, Lord Alberto Pio, count of Carpi, to the setting up and substitution and to each and all other things said above, when they were thus being discussed, described, and accomplished, I was present along with the above-named witnesses, and I noted those things that I saw, knew, and heard thus to be done, and on that account have I accordingly made, undersigned, and brought into this public form this public instrument written with my own hand, and I requested and asked, signed in trust, confirmation, and testimony of the afore-set things by my customary mark and name joined together with the hand-written signature of the above-mentioned most illustrious lord Alberto Pio. Thus signed [special mark] Jean Moucaud.

So it is, Alberto of Carpi.

* * * * *

the monastery's church that he donated the silver image of the risen Christ. See Sivry and Champagnac II, cols 187, 215.

WORKS FREQUENTLY CITED

SHORT-TITLE FORMS
FOR ERASMUS' WORKS

INDEX OF SCRIPTURAL REFERENCES

GENERAL INDEX

WORKS FREQUENTLY CITED

This list provides bibliographical information for the publications referred to in short-title form in the introduction and notes. For Erasmus' writings see the short-title list on 416–19.

AASS
: *Acta sanctorum* ed Jean Bolland and Gottfried Henschen 3rd ed (Paris 1863–70) 60 vols

Actes de François Ier
: *Collection des ordonnances des rois de France. Catalogue des actes de François Ier* ed Academie de sciences morales et politiques (Paris 1887–1908) 10 vols

Aldo Manuzio Editore
: *Aldo Manuzio Editore: dediche, prefazioni, note ai testi* intro Carlo Dionisotti, ed and trans Giovanni Orlandi, Documenti sulle arti del libro 2 (Milan 1975) 2 vols

Allen
: *Opus epistolarum Des. Erasmi Roterodami* ed P.S. Allen, H.M. Allen, and H.W. Garrod (Oxford 1906–47) 11 vols with Index vol by Barbara Flower and Elisabeth Rosenbaum (Oxford 1958)

Amerbachkorrespondenz
: *Die Amerbachkorrespondenz* ed Alfred Hartmann and B.R. Jenny (Basel 1942–   )

Andrés
: Melquiades Andrés Martín *La teología española en el siglo XVI* Biblioteca de Autores Cristianos Maior 14 (Madrid 1976) 2 vols

Apologia
: Erasmus *Apologia adversus rhapsodias Alberti Pii*

ASD
: *Opera omnia Desiderii Erasmi Roterodami* (Amsterdam / Oxford 1969–   )

ASV
: Archivio Segreto Vaticano. Three collections are cited in this volume: Fondo Pio; Reg Vat (Registri Vaticani); and ss (Segretariato di Stato).

Avesani
: *Società, politica e cultura a Carpi ai tempi di Alberto III Pio: Atti del convegno internazionale (Carpi, 19–21 maggio 1978)* ed Rino Avesani, Giuseppe Billanovich, Mirella Ferrari, and Giovanni Pozzi (2 vols) *Medioevo e umanesimo* 46–7 (1981)

Balan
: Peter Balan *Monumenta Reformationis Lutheranae ex Tabulariis Secretioribus S. Sedis 1521–1525* (Regensburg 1884)

Bauch

Gustav Bauch 'Ritter Georg Sauermann, der erste adelige Vorfahr der Grafen Saurma-Jeltsch' *Zeitschrift des Vereins für Geschichte und Altertum Schlesiens* 19 (1885) 146–81

Bäumer 'Lutherprozess'

Remigius Bäumer 'Der Lutherprozess' in *Lutherprozess und Lutherbann: Vorgeschichte, Ergebnis, Nachwirkung* ed Remigius Bäumer, Katholisches Leben und Kirchenreform im Zeitalter der Glaubensspaltung 32 (Münster 1972) 18–48

BAV

Biblioteca Apostolica Vaticana

Beaumont-Maillet

Laure Beaumont-Maillet *Le grand couvent des Cordeliers de Paris: Étude historique et archéologique du XIIIe siècle à nos jours* Bibliothèque de l'Ecole des hautes études, ive section, Sciences historiques et philosophiques 325 (Paris 1975)

Bentley *Humanists and Holy Writ*

Jerry H. Bentley *Humanists and Holy Writ: New Testament Scholarship in the Renaissance* (Princeton 1983)

BHR

Bibliothèque d'humanisme et renaissance (Geneva 1941–   )

Biblia sacra vulgata

Biblia sacra iuxta vulgatam versionem ed Bonifatius Fischer, Johannes Gribomont, H.F.D. Sparks, and W. Thiele; 2nd rev ed Robertus Weber (Stuttgart 1975) 2 vols

Biondi

Albano Biondi 'Alberto Pio nella pubblicistica del suo tempo' *Medioevo e umanesimo* 46 (1981) 95–132

Catalogus

Catalogus omnium Erasmi Lucubrationum, also known as the letter to Johann Botzheim, Basel, 30 January 1523, Allen I 1–46 / CWE Ep 1341a

Causa Lutheri

Dokumente zur Causa Lutheri (1517–1521) ed Peter Fabisch and Erwin Iserloh Corpus Catholicorum 41–2 (Münster 1988, 1991) 2 vols

CC

Corpus catholicorum ed Joseph Greving et al (Münster 1919–   )

CCSL

Corpus Christianorum, series Latina (Turnhout 1953–   )

CEBR

Contemporaries of Erasmus: A Biographical Register of the Renaissance and Reformation ed Peter G. Bietenholz and Thomas B. Deutscher (Toronto 1985–7) 3 vols

Ceresa-Gastaldo   Jerome *De viris illustribus / Gli uomini illustri* ed and trans
                    Aldo Ceresa-Gastaldo, Biblioteca patristica 12 (Florence
                    1988)

COD   *Conciliorum oecumenicorum decreta* ed Giuseppe Alberigo et
                    al 3rd ed (Bologna 1973)

Contini *Alberto III Pio*   Carlo Contini *Alberto III Pio e il suo tempo: iconografia con note
                    storiche nel quinto centenario della nascita, Carpi, 1475–1975*
                    (Cassa di Risparmio di Carpi 1975)

Cortesi   Paolo Cortesi *De cardinalatu* (In Castro Cortesi: Symeon
                    Nicolai Nardi 1510)

CSEL   *Corpus scriptorum ecclesiasticorum Latinorum* (Vienna
                    1866–   )

CWE   *Collected Works of Erasmus* (Toronto 1974–   )

D'Amico *Renaissance   John F. D'Amico *Renaissance Humanism in Papal Rome:
Humanism*   Humanists and Churchmen on the Eve of the Reformation*
                    (Baltimore 1983)

d'Ascia *Erasmo e   Luca d'Ascia *Erasmo e l'umanesimo romano* Biblioteca della
l'umanesimo*   Rivista di storia e letteratura religiosa. Studi 2 (Florence
                    1991)

DBDI   *Dizionario biografico degli Italiani* (Rome 1960–   )

de la Brosse   Olivier de la Brosse *Le pape et le concile: La comparaison de
                    leur pouvoirs à la veille de la Réforme* Unam Sanctam 58 (Paris
                    1965)

DTC   *Dictionnaire de théologie catholique* ed A. Vacant, E. Mangenot,
                    and E. Amann (Paris 1903–50) 15 vols

Duffy   Eamon Duffy *The Stripping of the Altars: Traditional Religion
                    in England c.1400–c.1580* (New Haven, Conn 1992)

Duplessy *Collectio   Charles Duplessy d'Argentré *Collectio judiciorum de novis
judiciorum*   erroribus qui ab initio duodecim seculi post Incarnationem Verbi
                    usque ad annum 1713 in Ecclesia proscripti sunt et notati ...
                    I: In quo exquisita monumenta ab anno 1100 usque ad annum
                    1542 continentur* (Paris 1724)

Erasmi opuscula   *Erasmi opuscula: A Supplement to the opera omnia* ed Wallace
                    K. Ferguson (The Hague 1933)

Eubel	Konrad Eubel, Wilhelm van Gulik, and Ludwig Schmitz-Kallenberg *Hierarchia catholica medii et recentioris aevi ...* III: *Saeculum XVI ab anno 1503 complectens* 2nd ed (Münster 1923)
Farge *Biographical Register*	James K. Farge *Biographical Register of Paris Doctors of Theology 1500–1536* Subsidia Mediaevalia 10 (Toronto 1980)
Farge *Registre*	James K. Farge ed *Registre des procès-verbaux de la Faculté de Théologie de l'Université de Paris de janvier 1524 à novembre 1533* (Paris 1990)
Fiorina	Ugo Fiorina *Inventario dell'Archivio Falcò Pio di Savoia* Fontes Ambrosiani 64 (Vicenza 1980)
Florido *Apologia*	Francesco Florido Sabino *In M. Actii [Titi Maccii] Plauti aliorumque Latinae Linguae Scripto cum calumniatores Apologia, nunc primum ab autore aucta atque recognita ...* (Basel 1540)
Florido *Libri tres*	Francesco Florido Sabino *C. Iulii Caesaris Praestantia libri tres ... Lectionum succisiuarum libri III, iam quoque primum et nati et in lucem editi* (Basel 1540)
Forner	Fabio Forner 'Genesi ed elaborazione della *Responsio* di Alberto Pio a Erasmo' *Giornale storico della letteratura italiana* 177 (2000) 200–24
Fraikin	J. Fraikin *Nonciatures de France: Nonciatures de Clément VII* I: *Depuis la bataille de Pavie jusq'au rappel d'Acciaiuolo (25 février 1525–juin 1527)* (Paris 1906)
Friedberg	*Corpus iuris canonici: Editio Lipsiensis secunda* ed Emil Ludwig Richter and Emil Friedberg (Leipzig 1879) 2 vols. I: *Decretum Magistri Gratiani*; II: *Decretalium collectiones*
Galli 'Presenza religiosa'	Adriana Galli 'Presenza religiosa: Chiese e conventi come elementi di organizzazione urbana fino al XVII secolo' in *Materiali per la storia urbana di Carpi* ed Alfonso Garuti et al (Carpi 1977) 70–6, 82–8 (section D of the catalogue)
Garuti	Dante Colli, Alfonso Garuti, and Pietro Parmiggiani *Le torri perdute: Rocche e castelli dei Pio* (Modena 1986)
Gilmore 'Italian Reactions'	Myron P. Gilmore 'Italian Reactions to Erasmian Humanism' in *Itinerarium Italicum: The Profile of the Italian Renaissance in the Mirror of Its European Transformations. Dedicated to Paul Oskar Kristeller on the Occasion of His 70th Birthday*

ed Heiko A. Oberman and Thomas A. Brady JR Studies in Medieval and Reformation Thought 14 (Leiden 1975) 61–115

Grendler, 'How to Get a Degree' Paul F. Grendler 'How to Get a Degree in Fifteen Days: Erasmus' Doctorate of Theology from the University of Turin' Erasmus of Rotterdam Society Yearbook 18 (1998) 40–69

Grendler, Universities Paul F. Grendler The Universities of the Italian Renaissance (Baltimore 2002)

Guaitoli 'Sulla vita' Paolo Guaitoli 'Memorie sulla vita d'Alberto III° Pio' in Memorie storiche e documenti sulla città e sull' antico principato di Carpi: studi e indagini I (Carpi 1877)

Guicciardini Storia d'Italia Francesco Guicciardini Storia d'Italia ed Costantino Panigada, Scrittori d'Italia 120–4 (Bari 1929; repr 1967) 5 vols

Gundersheimer Werner L. Gundersheimer 'Erasmus, Humanism, and the Christian Cabala' Journal of the Warburg and Courtauld Institutes 26 (1963) 38–52

Hallman Barbara M. Hallman Italian Cardinals, Reform, and the Church as Property (Berkeley 1985)

Heesakkers 'Erasmus' Suspicions' Christian Lambert Heesakkers 'Erasmus' Suspicions of Aleander as the Instigator of Alberto Pio' in Acta Conventus Neo-Latini Torontonensis: Proceedings of the Seventh International Congress of Neo-Latin Studies, Toronto, 8 August to 13 August 1988 ed Alexander Dalzell, Charles Fantazzi, and Richard J. Schoeck, Medieval and Renaissance Texts and Studies 86 (Binghamton, NY 1991) 371–84

Hefele-Leclercq Karl Joseph Hefele Histoire des conciles d'après les documents originaux trans and rev Henri Leclercq (Paris 1907–38) 10 vols. Vols VII and VIII by Joseph Hergenroether; vol IX by P. Richard; vol X by A. Michel

HO Omnium operum Divi Eusebii Hieronymi Stridonensis ed Desiderius Erasmus et al (Basel: Froben 1516) 9 vols in 5

Hofmann Forschungen Walther von Hofmann Forschungen zur Geschichte der kurialen Behörden vom Schisma bis zur Reformation Bibliothek des königlich-preussischen historischen Instituts in Rom 12, 13 (Rome 1914) 2 vols

Holborn Desiderius Erasmus Roterodamus Ausgewählte Werke ed

Hajo Holborn and Annemarie Holborn (Munich 1933; repr 1964)

Horst   Ulrich Horst *Zwischen Konziliarismus und Reformation: Studien zur Ekklesiologie im Dominikanerorden* Institutum Historicum FF. Praedicatorum Romae ad S. Sabinae, Dissertationes Historicae 22 (Rome 1985)

Klaiber   Wilbirgis Klaiber ed *Katholische Kontroverstheologen und Reformer des 16. Jahrhunderts: Ein Werkverzeichnis* Reformationsgeschichtliche Studien und Texte 116 (Münster 1978)

Klawiter   Randolph J. Klawiter ed and trans *The Polemics of Erasmus of Rotterdam and Ulrich von Hutten* (Notre Dame, Ind 1977)

Kristeller   Paul Oskar Kristeller *Iter Italicum: A Finding List of Uncatalogued or Incompletely Catalogued Humanistic Manuscripts of the Renaissance in Italian and Other Libraries* (Leiden 1977–97) 6 vols and Index vol

LB   Erasmus *Opera omnia* ed Jean Leclerc (Leiden 1703–6; repr Hildesheim 1961–2) 10 vols

Liddell and Scott   *A Greek-English Lexicon* compiled Henry George Liddell and Robert Scott, corrected Henry Driseler; new rev ed Henry Stuart Jones and Roderick McKenzie with supplement 1968 (Oxford 1968)

Litta   Pompeo Litta et al *Famiglie celebri italiane* 1st series (Milan / Torino 1819–99) 11 vols; 2nd series (Naples 1902–23) 2 vols

Loeb   Loeb Classical Library (London and Cambridge, Mass 1912–)

Losada   Angel Losada Garcia *Juan Ginés de Sepúlveda a traves de su 'epistolario' y nuevos documentos* (Madrid 1949)

Lowry   Martin Lowry *The World of Aldus Manutius: Business and Scholarship in Renaissance Venice* (Ithaca, NY 1979)

Mangan *Life*   John Joseph Mangan *Life, Character, and Influence of Desiderius Erasmus of Rotterdam* (New York 1927)

McSorley   Harry J. McSorley *Luther: Right or Wrong? An Ecumenical-Theological Study of Luther's Major Work, 'The Bondage of the Will'* (New York / Minneapolis 1969)

Minnich 'Debate on
Images'

Nelson H. Minnich 'The Debate between Desiderius
Erasmus of Rotterdam and Alberto Pio of Carpi on the Use
of Sacred Images' *Annuarium Historiae Conciliorum* 20 (1988)
379–413; repr in Minnich *The Catholic Reformation: Council,
Churchmen, Controversies* (London 1993) entry 11

Minnich '*Protestatio*'

Nelson H. Minnich 'The "*Protestatio*" of Alberto Pio (1513)'
in *Società, Politica e Cultura a Carpi ai Tempi di Alberto* III
*Pio: Atti del convegno internazionale (Carpi 19–21 maggio 1978)
Medioevo e umanesimo* 46 (1981) 261–89; repr in Minnich
*The Fifth Lateran Council (1512–17): Studies on Its Membership,
Diplomacy, and Proposals for Reform* (London 1993) entry 3

Minnich 'Underlying
Factors'

Nelson H. Minnich 'Some Underlying Factors in the
Erasmus-Pio Debate' *Erasmus of Rotterdam Society Yearbook*
13 (1993) 1–43

Moncallero

G.L. Moncallero *Epistolario di Bernardo Dovizi da Bibbiena*
2 vols, Biblioteca dell' Archivum Romanicum 1st series:
Storia-Letteratura-Paleografia vols 44 and 81 (Florence
1955–65)

OER

The Oxford Encyclopedia of the Reformation ed Hans J. Hille-
brand (New York 1996) 4 vols

Paquier *Jérôme
Aléandre*

Jules Paquier *L'Humanisme et la Réforme: Jérôme Aléandre de
sa naissance à la fin de son séjour à Brindes (1480–1529)* (Paris
1900)

Paulus

Nikolaus Paulus *Geschichte des Ablasses im Mittelalter vom
Ursprunge bis zur Mitte des 14. Jahrhunderts* (Paderborn
1922–3) 3 vols

PG

Patrologiae cursus completus ... series Graeca ed Jacques-
Paul Migne (Paris 1857–66) 167 vols. *Registers*: 2 vols (Paris
1928–36)

Pio *Libro*

Alberto Pio *Libro del muy Illustre y doctissimo Señor Alberto
Pio Conde de Carpi: que trata de muchas costumbres y estatutos
de la Iglesia, y de nuestra religion Christiana, mostrando su
autoridad y antiguedad: contra las [malditas] blasphemias de
Lutero, y algunos dichos de Erasmo Rotherodamo* (Alcalá de
Henares: Miguel de Eguya 1536)

Pio *Response* –
Chantilly MS

Albert Pio *La response de magnifique et noble homme Albert
Pius, Compte de Cappe, sur lepistre a luy envoyee par maistre
Dydier Herasme ...* Chantilly, Musée Condé, MS 187, 203
folios

Pio *Response* – Chantilly Plon	Alberto Pio *La Response de magnifique et noble homme Albert Pius, compte de Cappe, sur lepistre a luy envoyee par maistre Dydier Herasme* ... dedicatory letter to Guillaume Montmorency, printed in Gustave Maron *Chantilly. Les Cabinet des livres. Manuscrits* (Paris: Librairie Plon 1900) I 168–9
Pio *Response* – Paris MS	Alberto Pio *La response de magnifique et noble homme Albert Pius, conte de Carpe, sur lepistre a luy envoiee par maistre Didier Herasme* ... Paris, Bibliothèque nationale de Paris, Fonds français, Ancien fonds, MS 462, 221 folios
Pio *Responsio paraenetica*	*Alberti Pii Carporum Comitis illustrissimi ad Erasmi Roterodami expostulationem responsio accurata et paraenetica, Martini Lutheri et asseclarum eius haeresim vesanam magnis argumentis, et iustis rationibus confutans* (Paris: Josse Bade 1529)
Pio XXIII *libri*	*Alberti Pii Carporum Comitis illustrissimi et viri longe doctissimi, praeter praefationem et operis conclusionem, tres et viginti libri in locos lucubrationum variarum D. Erasmi Roterodami, quos censet ab eo recognoscendos et retractandos* ed Joost Bade van Assche and Francesco Florido de Sabina (Paris: Josse Bade 1531)
Pio *Will*	Alberto Pio 'Testamento de lo illustrissimo signore Alberto Pio' Paris, 21 July 1530: Milan, Biblioteca Veneranda Ambrosiana, Archivio Falcò Pio di Savoia, I sezione, prima parte (Famiglie e persone), scatola 277, entry 6, folios 1^r–4^v. Translated 387–404 above
PL	*Patrologiae cursus completus ... series Latina* ed Jacques-Paul Migne (Paris 1841–64) 221 vols; *Supplementa* (Paris 1958–70) 5 vols
Reeve	Anne Reeve ed *Erasmus' Annotations on the New Testament: The Gospels. Facsimile of the Final Latin Text with All Earlier Variants* intro Michael A. Screech (London 1986)
Reeve *Galatians to the Apocalypse*	Anne Reeve ed *Erasmus' Annotations on the New Testament: Galatians to the Apocalypse. Facsimile of the Final Latin Text with All Earlier Variants* Studies in the History of Christian Thought 52 (Leiden 1993)
Reeve-Screech	Anne Reeve and Michael A. Screech eds *Erasmus' Annotations on the New Testament: Acts – Romans – I and II Corinthians. Facsimile of the Final Latin Text with All Earlier Variants* Studies in the History of Christian Thought 42 (Leiden 1990)

Renouard *Imprimeurs*   Philippe Renouard et al *Imprimeurs et libraires parisiens du
                         XVIe siècle: Ouvrage publie d'après les manuscrits de Philippe
                         Renouard* (Paris 1964–91) 8 vols

Responsio   Erasmus *Responsio ad epistolam Alberti Pii*

Roberts   Michael Roberts *Biblical Epic and Rhetorical Paraphrase in Late
                         Antiquity* (Liverpool 1985)

Rummel *Catholic   Erika Rummel *Erasmus and His Catholic Critics* Bibliotheca
    Critics*   Humanistica et Reformatorica 45 (Nieuwkoop 1989) 2 vols

Rummel *Erasmus'   Erika Rummel *Erasmus' Annotations on the New Testament:
    Annotations*   From Philologist to Theologian* (Toronto 1986)

Rummel *Erasmus as   Erika Rummel *Erasmus as a Translator of the Classics* (Toronto
    a Translator*   1985)

Rummel *Humanist-   Erika Rummel *The Humanist-Scholastic Debate in the Renais-
    Scholastic Debate*   sance & Reformation* (Cambridge, Mass 1995)

Sabattini   Alberto Sabattini *Alberto III Pio: Politica, diplomazia e guerra
                         del conte di Carpi. Corrispondenza con la corte di Mantova,
                         1506–1511* (Carpi 1994)

Sammarini   Achille Sammarini 'Lettre inedite dei signori Pio di Carpi
                         ai principi Gonzaga di Mantova dall' anno 1366 al 1518' in
                         *Memorie storiche e documenti sulla città e sull' antico principato
                         di Carpi* (Carpi 1877) I, 333–95

Sanuto   Marino Sanuto *I diarii di Marino Sanuto 1496–1533* ed
                         Rinaldo Fulin, Federico Stefani, Nicolò Barozzi, Guglielmo
                         Berchet, and Marco Allegri 58 vols (Venice 1879–1903)

SC   *Sources chrétiennes* (Paris 1941–   )

Scholia   Erasmus *Brevissima scholia*

Seidel Menchi   Silvana Seidel Menchi *Erasmo in Italia 1520–1580* (Torino
                         1987)

Semper   Hans Semper 'Alberto Pio III. als Herrscher und Staats-
                         mann. Seine äusseren Schicksale' in *Carpi: Ein Fürstensitz
                         der Renaissance* ed H. Semper, F.O. Schulze, and W. Barth
                         (Dresden 1882) 3–18

Sepúlveda *Antapologia* *Io. Genesii Sepulvedae Cordubensis artium et theologiae magistri
                         Antapologia pro Alberto Pio Comite Carpensi in Erasmum
                         Roterodamum* (Rome: Antonio Blado 1532)

Sivry and Champagnac	Louis de Sivry and M. Champagnac *Dictionnaire géographique, historique, descriptif, archéologique des Pèlerinages ancien et modernes, des lieu de devotion les plus célèbres de l'Univers* 2 vols (Paris 1850–1)
Smith *Erasmus*	Preserved Smith *Erasmus: A Study of His Life, Ideals and Place in History* (New York 1923)
Solana	Juan Ginés de Sepúlveda *Obras completas* VII: *Io. Genesii Sepulvedae Cordubensis, artium & theologiae magistri, Antapologia pro Alberto Pio, Comite Carpensi, in Erasmum Roterodamum/ Antiapología del cordobés Juan Ginés de Sepúlveda, Maestro en artes y teología, en defensa de Alberto Pío, Conde de Carpi, frente a Erasmo de Rotterdam* ed and trans Julián Solana Pujalte (Salamanca 2003), vii–cxi, 112–217
TLL	*Thesaurus linguae latinae* (Leipzig 1900–)
Tuetey	*Registres des déliberations du Bureau de la Ville de Paris 1499– 1628* ed F. Bonnardot et al, Histoire générale de Paris (Paris 1883–1958) 19 vols. Vol II: *1527–1539* ed Alexandre Tuetey (Paris 1886)
vander Haeghen	Fernando vander Haeghen *Bibliotheca Erasmiana: Répertoire des œuvres d'Erasme, 1st and 2nd series* (Ghent 1893; repr Nieuwkoop 1961)
Vasoli *Pio*	Cesare Vasoli *Alberto III Pio da Carpi* (Carpi 1978)
WA	*D. Martin Luthers Werke, Kritische Gesamtausgabe* (Weimar 1883–)
WA *Briefwechsel*	*D. Martin Luthers Werke, Briefwechsel* (Weimar 1930–78) 15 vols
Zúñiga *Blasphemiae et impietates*	Diego López Zúñiga *Erasmi Roterodami Blasphemiae et Impietates per Iacobum Lopidem Stunicam nunc primum propalatae ac proprio volumine alias redargutae* (Rome: Antonio Blado 1522)

Titles following colons are longer versions of the same, or are alternative titles. Items entirely enclosed in square brackets are of doubtful authorship. For abbreviations, see Works Frequently Cited 406–15.

Acta: Acta Academiae Lovaniensis contra Lutherum *Opuscula* / CWE 71

Adagia: Adagiorum chiliades 1508, etc (Adagiorum collectanea for the primitive form, when required) LB II / ASD II-1, 4, 5, 6 / CWE 30–6

Admonitio adversus mendacium: Admonitio adversus mendacium et obtrectationem LB X

Annotationes in Novum Testamentum LB VI / CWE 51–60

Antibarbari LB X / ASD I-1 / CWE 23

Apologia ad Caranzam: Apologia ad Sanctium Caranzam, or Apologia de tribus locis, or Responsio ad annotationem Stunicae ... a Sanctio Caranza defensam LB IX

Apologia ad Fabrum: Apologia ad Iacobum Fabrum Stapulensem LB IX / ASD IX-3 / CWE 83

Apologia adversus monachos: Apologia adversus monachos quosdam Hispanos LB IX

Apologia adversus Petrum Sutorem: Apologia adversus debacchationes Petri Sutoris LB IX

Apologia adversus rhapsodias Alberti Pii: Apologia ad viginti et quattuor libros A. Pii LB IX / CWE 84

Apologia contra Latomi dialogum: Apologia contra Iacobi Latomi dialogum de tribus linguis LB IX / CWE 71

Apologia de 'In principio erat sermo' LB IX

Apologia de laude matrimonii: Apologia pro declamatione de laude matrimonii LB IX / CWE 71

Apologia de loco 'Omnes quidem': Apologia de loco 'Omnes quidem resurgemus' LB IX

Apologiae contra Stunicam: Apologiae contra Lopidem Stunicam LB IX: (1) Apologia respondens ad ea quae Iacobus Lopis Stunica taxaverat in prima duntaxat Novi Testamenti aeditione ASD IX-2; (2) Apologia adversus libellum Stunicae cui titulum fecit Blasphemiae et impietates Erasmi; (3) Apologia ad prodromon Stunicae; (4) Apologia ad Stunicae conclusiones; (5) Epistola apologetica adversus Stunicam [= Ep 2172]

Apologia qua respondet invectivis Lei: Apologia qua respondet duabus invectivis Eduardi Lei *Opuscula*

Apophthegmata LB IV

Appendix de scriptis Clithovei LB IX / CWE 83

Appendix respondens ad Sutorem LB IX

Argumenta: Argumenta in omnes epistolas apostolicas nova (with Paraphrases)

Axiomata pro causa Lutheri: Axiomata pro causa Martini Lutheri *Opuscula* / CWE 71

Brevissima scholia: In Elenchum Alberti Pii brevissima scholia per eundem Erasmum Roterodamum CWE 84

Carmina LB I, IV, V, VIII / ASD I-7 / CWE 85–6
Catalogus lucubrationum LB I / CWE 9 (Ep 1341A)
Ciceronianus: Dialogus Ciceronianus LB I / ASD I-2 / CWE 28
Colloquia LB I / ASD I-3 / CWE 39–40
Compendium vitae Allen I / CWE 4
Concionalis interpretatio (in Psalmi)
Conflictus: Conflictus Thaliae et Barbariei LB I
[Consilium: Consilium cuiusdam ex animo cupientis esse consultum] Opuscula /
 CWE 71

De bello Turcico: Consultatio de bello Turcico (in Psalmi)
De civilitate: De civilitate morum puerilium LB I / CWE 25
Declamatio de morte LB IV
Declamatiuncula LB IV
Declarationes ad censuras Lutetiae vulgatas: Declarationes ad censuras Lutetiae
 vulgatas sub nomine facultatis theologiae Parisiensis LB IX
De concordia: De sarcienda ecclesiae concordia, or De amabili ecclesiae concordia
 (in Psalmi)
De conscribendis epistolis LB I / ASD I-2 / CWE 25
De constructione: De constructione octo partium orationis, or Syntaxis LB I /
 ASD I-4
De contemptu mundi: Epistola de contemptu mundi LB V / ASD V-1 / CWE 66
De copia: De duplici copia verborum ac rerum LB I / ASD I-6 / CWE 24
De esu carnium: Epistola apologetica ad Christophorum episcopum Basiliensem de
 interdicto esu carnium LB IX / ASD IX-1
De immensa Dei misericordia: Concio de immensa Dei misericordia LB V / CWE 70
De libero arbitrio: De libero arbitrio diatribe LB IX / CWE 76
De praeparatione: De praeparatione ad mortem LB V / ASD V-1 / CWE 70
De pueris instituendis: De pueris statim ac liberaliter instituendis LB I / ASD I-2 /
 CWE 26
De puero Iesu: Concio de puero Iesu LB V / CWE 29
De puritate tabernaculi: De puritate tabernaculi sive ecclesiae christianae (in
 Psalmi)
De ratione studii LB I / ASD I-2 / CWE 24
De recta pronuntiatione: De recta latini graecique sermonis pronuntiatione LB I /
 ASD I-4 / CWE 26
De taedio Iesu: Disputatiuncula de taedio, pavore, tristicia Iesu LB V / CWE 70
Detectio praestigiarum: Detectio praestigiarum cuiusdam libelli germanice scripti
 LB X / ASD IX-1
De vidua christiana LB V / CWE 66
De virtute amplectenda: Oratio de virtute amplectenda LB V / CWE 29
[Dialogus bilinguium ac trilinguium: Chonradi Nastadiensis dialogus bilinguium
 ac trilinguium] Opuscula / CWE 7
Dilutio: Dilutio eorum quae Iodocus Clithoveus scripsit adversus declamationem
 suasoriam matrimonii / Dilutio eorum quae Iodocus Clithoveus scripsit ed Emile V.
 Telle (Paris 1968) / CWE 83
Divinationes ad notata Bedae LB IX

Ecclesiastes: Ecclesiastes sive de ratione concionandi LB V / ASD V-4, 5
Elenchus in N. Bedae censuras LB IX
Enchiridion: Enchiridion militis christiani LB V / CWE 66
Encomium matrimonii (in De conscribendis epistolis)
Encomium medicinae: Declamatio in laudem artis medicae LB I / ASD I-4 / CWE 29
Epistola ad Dorpium LB IX / CWE 3 / CWE 71
Epistola ad fratres Inferioris Germaniae: Responsio ad fratres Germaniae Inferioris
    ad epistolam apologeticam incerto autore proditam LB X / ASD IX-1
Epistola ad graculos: Epistola ad quosdam imprudentissimos graculos LB X
Epistola apologetica de Termino LB X
Epistola consolatoria: Epistola consolatoria virginibus sacris, or Epistola consolato-
    ria in adversis LB V / CWE 69
Epistola contra pseudevangelicos: Epistola contra quosdam qui se falso iactant
    evangelicos LB X / ASD IX-1
Euripidis Hecuba LB I / ASD I-1
Euripidis Iphigenia in Aulide LB I / ASD I-1
Exomologesis: Exomologesis sive modus confitendi LB V
Explanatio symboli: Explanatio symboli apostolorum sive catechismus LB V /
    ASD V-1 / CWE 70
Ex Plutarcho versa LB IV / ASD IV-2

Formula: Conficiendarum epistolarum formula (see De conscribendis epistolis)

Hyperaspistes LB X / CWE 76–7

In Nucem Ovidii commentarius LB I / ASD I-1 / CWE 29
In Prudentium: Commentarius in duos hymnos Prudentii LB V / CWE 29
Institutio christiani matrimonii LB V / CWE 69
Institutio principis christiani LB IV / ASD IV-1 / CWE 27

[Julius exclusus: Dialogus Julius exclusus e coelis] Opuscula / CWE 27

Lingua LB IV / ASD IV-1A / CWE 29
Liturgia Virginis Matris: Virginis Matris apud Lauretum cultae liturgia LB V /
    ASD V-1 / CWE 69
Luciani dialogi LB I / ASD I-1

Manifesta mendacia CWE 71
Methodus (see Ratio)
Modus orandi Deum LB V / ASD V-1 / CWE 70
Moria: Moriae encomium LB IV / ASD IV-3 / CWE 27

Novum Testamentum: Novum Testamentum 1519 and later (Novum instrumentum
    for the first edition, 1516, when required) LB VI

Obsecratio ad Virginem Mariam: Obsecratio sive oratio ad Virginem Mariam in
    rebus adversis, or Obsecratio ad Virginem Matrem Mariam in rebus adversis
    LB V / CWE 69

Oratio de pace: Oratio de pace et discordia LB VIII
Oratio funebris: Oratio funebris in funere Bertae de Heyen LB VIII / CWE 29

Paean Virgini Matri: Paean Virgini Matri dicendus LB V / CWE 69
Panegyricus: Panegyricus ad Philippum Austriae ducem LB IV / ASD IV-1 / CWE 27
Parabolae: Parabolae sive similia LB I / ASD I-5 / CWE 23
Paraclesis LB V, VI
Paraphrasis in Elegantias Vallae: Paraphrasis in Elegantias Laurentii Vallae LB I /
    ASD I-4
Paraphrasis in Matthaeum, etc (in Paraphrasis in Novum Testamentum)
Paraphrasis in Novum Testamentum LB VII / CWE 42–50
Peregrinatio apostolorum: Peregrinatio apostolorum Petri et Pauli LB VI, VII
Precatio ad Virginis filium Iesum LB V / CWE 69
Precatio dominica LB V / CWE 69
Precationes: Precationes aliquot novae LB V / CWE 69
Precatio pro pace ecclesiae: Precatio ad Dominum Iesum pro pace ecclesiae LB IV,
    V / CWE 69
Psalmi: Psalmi, or Enarrationes sive commentarii in psalmos LB V / ASD V-2, 3 /
    CWE 63–5
Purgatio adversus epistolam Lutheri: Purgatio adversus epistolam non sobriam
    Lutheri LB X / ASD IX-1

Querela pacis LB IV / ASD IV-2 / CWE 27

Ratio: Ratio seu Methodus compendio perveniendi ad veram theologiam (Methodus
    for the shorter version originally published in the Novum instrumentum of 1516)
    LB V, VI
Responsio ad annotationes Lei: Liber quo respondet annotationibus Lei LB IX
Responsio ad collationes: Responsio ad collationes cuiusdam iuvenis geronto-
    didascali LB IX
Responsio ad disputationem de divortio: Responsio ad disputationem cuiusdam
    Phimostomi de divortio LB IX / CWE 83
Responsio ad epistolam Alberti Pii: Responsio ad epistolam paraeneticam Alberti
    Pii, or Responsio ad exhortationem Pii LB IX / CWE 84
Responsio ad notulas Bedaicas LB X
Responsio ad Petri Cursii defensionem: Epistola de apologia Cursii LB X / Allen
    Ep 3032
Responsio adversus febricitantis libellum: Apologia monasticae religionis LB X

Spongia: Spongia adversus aspergines Hutteni LB X / ASD IX-1
Supputatio: Supputatio calumniarum Natalis Bedae LB IX

Tyrannicida: Tyrannicida, declamatio Lucianicae respondens LB I / ASD I-1 / CWE 29

Virginis et martyris comparatio LB V / CWE 69
Vita Hieronymi: Vita divi Hieronymi Stridonensis *Opuscula* / CWE 61

Index of Scriptural References

General Index

Achates (Aeneas' companion) 121 and
n69
Adams, H.M. cxxxvi
Adams, Robert P. 174 n401
Adler, Ada 107 n1
Adler, Sara 107n1
Adrian VI, pope: appoints Pio his
governor and intercedes for him
xxviii; prohibits Zúñiga's publication
against Erasmus 15 n69; Sauermann's
oration dedicated to 15 n69; urges
Erasmus to write against Luther 34
n158, 87 n430, 129–30 and n121 n123,
150 n247; praises Erasmus' writings
and supports his scriptural work 77
n380; Erasmus dedicates edition of
Arnobius to 141 n196
Aeschines (fourth-century BC Athenian
orator-politician) 123 and n81
Albert of Brandenburg (Albrecht Ho-
henzollern, Markgraf von Brandenburg), cardinal-archbishop of Mainz
5 n16, 55 n263, 85 n421
Albert of Mainz. *See* Albert of Brandenburg
Albrecht IV, duke of Bavaria cxxx
Aldus (Aldo Manuzio) (Venetian
printer): tutor of Pio, and dedicates
works to xvii–xviii and n7; refers to
Pio as philosopher-prince xxv n14;
praises Erasmus xl; Erasmus denies
that he was taught by xcviii n131, 14,
86; Erasmus comments on 161–4, 364
Aleandro, Girolamo (humanist, prelate): refers to Pio as a philosopher-

prince xxv n14; compares Pio to
Plutarch xxxi and n17; shares a room
with Erasmus xxxv n23; secretary
of Giulio dei Medici xxxvi; suspects
Erasmus of Lutheran sympathies
xlii; letters of xlii n31; as papal nuncio meets with Erasmus xliii n32;
confronts Erasmus for spreading rumours that he is of Jewish origin xliii
n32; expresses disapproval of those
who openly attack Erasmus xliii n32;
referred to by Erasmus as Bullbearer
xliii n33, 11, 14, 15, 27 n126, 38, 42–3,
87, 92, 305 n1164; suspected by Erasmus of assisting Pio lxiii; succeeded
as papal librarian by Steuco cxix;
identified by Erasmus as one of his
Roman critics 114 n32
Alexander of Hales OFM (medieval theologian) 78, 109
Allen, P.S. xv n1, xxix n16, xxxvii n26,
xliii n32, lviii n62, lxv n68, lxvi n69,
lxvii n70, lxxii n77, lxxiii n79, cxx
nn181–2, cxxv n192, cxxviii n199, 30,
31 n146, 34 n158, 56 n267, 78 n389
Almain, Jacques (Paris theologian) 41
n200
Alonso, Joaquin Maria 243 n813
Ambrose, bishop of Milan (fourth-
century Father of the church): helps
to confound the Pelagians cxvii; does
not use the Vulgate version of the
Bible 48 n224; quotes the Bible in
different ways 146; on fasting 187
n480; on virgins taking public vows of

upon the gods at the beginning of their orations 251 n866; the church Fathers use humanistic theology in their struggles against the pagans 262; Paul quotes pagan poets out of context 273 n985; he speaks of the demons of the pagan gods 285; as a young pagan Augustine had a concubine 327 n1278, 367; a Christian swearing by a pagan god before a pagan is bound by his oath 356 n1455;

– modern paganism: Erasmus claims the Roman Academy is a neopagan sodality of scholars led by Aleandro and Pio xliii, 62 n297; he opposes bringing back old paganism xlvii, 35 n169; criticizes the pagan elements in Roman Ciceronianism 14 n58, 62 n298; criticizes neopagan sermons in Rome 35 n169; claims that Roman humanists are cryptopagans 35 n169, 62 n297; scholastic theology is appropriate for refuting paganism 262; Christian rulers act as pagans in making warfare 350

Pahl, I. 76 n378

Pallai, Biagio (member of the Roman Academy, editor of *Coryciana*) xxxvii and n28, lxxv n81

papacy, pope, Roman pontiff: treated in book 14 of Pio's *XXIII libri* 298 n1125; Erasmus' response 298–308

– in antiquity the pope is called *summus sacerdos*, a title Erasmus also uses 40 n195, 84, 98 n488, 299; as successor of Peter 42 n203; his titles have varied over the centuries 44; referred to as bishop of Rome 98

– Cyprian's controversies with 30 n138; his supreme power not acknowledged in antiquity 38, 42; Ockham undermines papal authority 40 and n197; many over the centuries have disputed his power 40, 299; *Liber pontificalis* teaches papal pre-eminence 46 and n220; whether the power of a pope is above that of a church council 49 n233, 303; Pope Gregory XI successfully pressured by council to abdicate 49 n233; whether the pope's supreme power is of divine origin 130–1 n130 n133, 307 and n1173; the pope's position at the top of the spiritual hierarchy 133 n150, 173; his authority should be compatible with the salvation of souls 176; papal primacy 298–308; as receiver of the keys 302 n1143; the proper limits of his power 302–3 and n1145 n1148, 304–5

– papal ordinances 181–2, 312; they carry great weight lxxvi, cv; the pope is guided by the opinions of orthodox Fathers and saints lxxxi; contradictions in papal statements over the centuries ciii, 336; papal authority should be used to correct abuses 34; Erasmus wants the pope to value his office appropriately 42; papal decretals criticize clerical neglect of duty, tyranny, simony, and excessive concern for wealth 48, 54, 297; pope has power to enjoin clerical celibacy and to dispense from it 67, 332 n1308, 371 n55, 372 n60; supreme charity demanded of the pope 130; the salvation of souls should take precedence over papal financial gain in granting indulgences 176, 177, 230; some of the pope's ordinances are open to just criticism 184 and n461, 308; papal grants of exemptions and privileges to mendicant orders 193; pope as steward of the treasury of the church 231; can canonize saints 255; great virtue should accompany his great power 303 n1145; imposes annates taxes on church appointments 306, 365; should mitigate ruling on reception of Eucharist under only one species 306 n1169; failure to recognize papal ordinances can put one outside the church 306 n1169; popes have taught that intercourse cannot take place without sin 322 and n1255;

baptism 137 n168; Pio wrongly accuses Erasmus of condemning the taking of vows 180; whether the vows of monks in the early church could be dissolved 193 and n513; in the early church they did not take vows of obedience, poverty, or common life 193 and n514; holy virgins made solemn vows only of chastity, and some fell from them 197 and n543, 198; some consider religious vows as more important than baptism 202, 208; Pio's definition of a vow 312 n1199; many superstitions connected with specific vows 312; one should abstain from all unnecessary and useless vows 312, 313, 385; vows of the Jews 312–13 and n1200; vows made rashly should be considered void, and it is safest not to vow 313 and n1201, 315, 375; vows of chastity made by married persons should be invalid unless ratified by the bishop 313; the position that the church has the power to render invalid both solemn and simple vows made from human frailty or by those under age supported by Erasmus and the Paris theologians 313–14 and n1204, 372, 385 and n1503; Pio claims the church does not dispense from religious vows 314 n1206, 315 n1213, 385; Erasmus does not reject monastic vows 315; those below the age of majority should have their parents' consent before taking vows 315; monastic vows should be considered to have implicit or tacit exceptions 316 and n1218; the Brethren of the Common Life do not take vows 318 and n1231; no one is compelled to make vows to saints 318–19 and n1234; Antonino holds that ordination to the priesthood has attached to it an implicit solemn vow not to marry 331 n1306; Erasmus claims this does not apply to priests of the eastern churches 332 n1307; Augus-

tine's first concubine took a vow of continence 367; Erasmus holds that virginity freely vowed out of love of piety is preferable to marriage 379; Folly ridicules vows made to saints out of superstition 385

Wackernagel, Rudolf 69 n343
war, warfare: treated in book 21 of *XXIII libri* lxxv, 174 n401, 347 n1400; Erasmus' response 347–51; Erasmus' views contained principally in *De bello Turcico, Institutio principis christiani*, and the letter to Volz 174 n401, 348, 350 n1416
– Pio involved in the wars of the League of Cambrai and the League of Cognac xxvi–xxviii; Pio engaged in civil war over the control of Carpi xxix; the Peasants' War in Germany li; Guillaume de Montmorency's interest in lxi; princes disturb the world with wars lxxv n80; reputation of the French aristocracy for enjoying war cxv n170; the devil's style of warfare cxvii; Henry VIII as a warrior king 115 n37; Pio wants the pope and bishops to be warriors 173; Erasmus claims the modern way of waging war is pagan and criminal, more criminal than Jewish war 174, 183, 348; Christ and war mutually exclusive 174; Christ and Paul did not strike back, and Peter told to put up his sword 174, 349; Erasmus holds it inappropriate for bishops to make war, which Pio denies 174 and n403, 351; whether clerics may make war in defence of church property 174 n401 n403, 207; Erasmus never declares warfare is without qualification illicit for Christians 174, 348; Folly never condemns warfare absolutely 175; the Jews allowed to make war 181 n447, 183; would be better if princes spent their money on churches instead of on wars 224; unlike the ancient church the modern church is

This book

was designed by

VAL COOKE

based on the series design by

ALLAN FLEMING

and was printed by

University

of Toronto

Press